Third International Conference on Chemical Engineering Sciences and Applications (3rd ICChESA 2017)

IOP Conference Series: Materials Science and Engineering
Volume 334

Banda Aceh, Indonesia
20 - 21 September 2017

Editors:

Sri Aprilia
Yunardi
Cut Meurah Rosnelly
Wahyu Rinaldi

ISBN: 978-1-5108-6114-5
ISSN: 1757-8981

Printed from e-media with permission by:

Curran Associates, Inc.
57 Morehouse Lane
Red Hook, NY 12571

Some format issues inherent in the e-media version may also appear in this print version.

Printed by Curran Associates, Inc. (2018)

For permission requests, please contact the Institute of Physics
at the address below.

Institute of Physics
Dirac House, Temple Back
Bristol BS1 6BE UK

Phone: 44 1 17 929 7481
Fax: 44 1 17 920 0979

techtracking@iop.org

Additional copies of this publication are available from:

Curran Associates, Inc.
57 Morehouse Lane
Red Hook, NY 12571 USA
Phone: 845-758-0400
Fax: 845-758-2633
Email: curran@proceedings.com
Web: www.proceedings.com

Table of contents

Volume 334

The 3rd International Conference on Chemical Engineering Sciences and Applications 2017 (3rd ICChESA 2017)

20–21 September 2017, Banda Aceh, Indonesia

Accepted papers received: 9 March 2018
Published online: 6 April 2018

Preface

Papers

Energy Resources

Control System and Computational Fluid Dynamics

Nanotechnology

Chemical Engineering Fundamental

IOP Conf. Series: Materials Science and Engineering **334** (2018) 011001 doi:10.1088/1757-899X/334/1/011001

THE THIRD INTERNATIONAL CONFERENCE OF CHEMICAL ENGINEERING SCIENCES AND APPLICATIONS

(The 3rd ICChESA 2017)

September 20-21, 2017, Banda Aceh - INDONESIA

Accepted, peer reviewed papers from the 3rd ICChESA 2017

Edited by

Yunardi, Sri Aprilia, Cut Meurah Rosnelly, and Wahyu Rinaldi
Chemical Engineering Department
Syiah Kuala University
Banda Aceh - INDONESIA

IOP Conf. Series: Materials Science and Engineering **334** (2018) 011001 doi:10.1088/1757-899X/334/1/011001

Preface

The Third International Conference on Chemical Engineering Sciences and Applications 2017 (The 3rd ICChESA 2017) taking place in Banda Aceh, the Province of Aceh, Indonesia during 20-21 September 2017 was organized by the Chemical Engineering Department, Syiah Kuala University, Indonesia together with the Faculty of Chemical Engineering, Universiti Teknologi MARA (UiTM), Shah Alam, Malaysia in cooperation with the other two institutions: Chemical Engineering Department, Malikussaleh University and Lhokseumawe State Polytechnics of which both are from North Aceh, Indonesia. The conference addressed the important theme in supporting the sustainable development: "Strengthening Chemical Engineering Education, Research, and Innovation for Sustainable Development". The objective of the 3rd ICChESA 2017 was to gather scientists, researchers, scholars, educators, professionals, and students in different fields of chemical engineering from ASEAN and beyond to share knowledge, exchange ideas, and explore possibilities of building cooperation networking.

The Proceedings consists of accepted papers and presented by researchers from different countries, among others, Indonesia, Malaysia, Thailand, Pakistan, Iran and Japan. The papers are categorized under seven topics in the area of chemical engineering, namely: Energy Resources, Control System and Computational Fluid Dynamics, Environmental Engineering, Nan otechnology, Chemical Engineering Fundamental, Food and Biochemical Engineering, and Chemical Reaction Engineering, Polymer and Composite Materials.

It is fair to conclude that the conference was a great success! So many people and organizations have contributed in so many ways to turn this event into a smoothly running meeting with many very interesting presentations and a very good atmosphere for discussion and networking. I would like to take this opportunity to express my deep gratitude to all parties who have made this event a success, particularly to Syiah Kuala University and partner universities, keynote speakers, sponsors, reviewers, participants, members of organizing committee, and volunteers of Chemical Engineering students of Syiah Kuala University.

We sincerely hope the 3rd ICChESA 2017 Conference would provide a good opportunity to share experiences and ideas as well as to establish new academic collaborations among participants leading to exchange program and joint research and publication for the benefit of all.

Yunardi
Chairman of the 3rd ICChESA 2017
Chemical Engineering Department
Syiah Kuala University
Banda Aceh – INDONESIA

IOP Conf. Series: Materials Science and Engineering **334** (2018) 011001 doi:10.1088/1757-899X/334/1/011001

Organizing Committee

Advisory Chairman

1. Prof. Dr. Samsul Rizal, M.Eng (Rector of Syiah Kuala University)
2. Dr. Ir. Mirza Irwansyah, MBA. MLA (Dean of Faculty of Engineering, Syiah Kuala University)
3. Dr. Ir. Darmadi, MT (Chairman of Chemical Engineering Department, Syiah Kuala University)

Chairman
Dr. Yunardi

Co-Chairman
Dr. Fachrul Razi, MT

Secretary and Registration Division

1. Dr. Ir. Sri Aprilia, MT

Treasurer

1. Dr. Ir. Cut Meurah Rosnelly, MT
2. Ir. Pocut Nurul Alam, MT

Publicity and Promotion Division

1. Wahyu Rinaldi, ST, M. Sc
2. T. Mukhriza, ST, M.Sc

Sponsorship Division

1. Dr. Hesti Meilina, ST., MT
2. Dr. Ir. Muhammad Zaki, M.Sc

Logistic and Exhibition Division

1. Dr. Ir. Asri Gani, M.Eng
2. Dr. Ir. Edi Munawar, M.Eng

Event and Program Division

1. Suraiya, ST., MT

International Scientific Committee

1. Prof. Dr. Shaobin Wang, (Department of Chemical Engineering, Curtin University, Australia)
2. Prof. Dr. José M. Encinar (Department of Chemical Engineering and Physical Chemistry, Extremadura University, Spain)
3. Prof. Dr. Sharifah Aishah Syed A Kadir (Faculty of Chemical Engineering, Universiti Teknologi MARA,UiTM, Shah Alam, Malaysia)
4. Prof.Dr.Ir. Herri Susanto (Chemical Engineering Department, Bandung Institute of Technology, Indonesia)
5. Prof. Dr. Ku Halim Ku Hamid (Faculty of Chemical Engineering, Universiti Teknologi MARA,UiTM, Shah Alam, Malaysia)
6. Prof. Dr. Arshad Hussain (Chemical Engineering Dept., National University of Sciences and Technology, Islamabad, Pakistan)
7. Prof. D. Eng. Ir. Arief Budiman, M.S (Chemical Engineering Department, Gadjah Mada University, Indonesia)
8. Prof.K. Dossumov (Center of Physical-chemical Methods of Investigations and Analysis of Al-Farabi Kazakh National University, Almaty, Kazakhstan)
9. Prof. Ir. Dr. Mohd. Azlan Bin Hussain (Dept. Chemical Engineering, Universiti Malaya, Kuala Lumpur, Malaysia)
10. Prof. Dr.-Ing. Misri A. Gozan, M.Tech (Chemical Engineering Dept., University of Indonesia, Indonesia)
11. Assoc. Prof. Dr. Johann Fellner (Vienna University of Technology, Austria)
12. Assoc. Prof.Dr Abdul Hadi (Faculty of Chemical Engineering, Universiti Teknologi MARA,UiTM, Shah Alam, Malaysia)
13. Prof.Tulay Ozcan (Department of Food Engineering, Uludag University, Bursa, Turkey)
14. Assoc. Prof.Dr. Kamariah Noor Ismail (Faculty of Chemical Engineering, Universiti Teknologi MARA,UiTM, Shah Alam, Malaysia)
15. Prof. Dr. Ir.H. M. Djoni Bustan, M.Eng (Chemical Engineering Department, Sriwijaya University, Indonesia)
16. Assoc. Prof. Yasumasa Tojo, PhD. (Laboratory of Solid Waste Disposal Engineering, Hokkaido University, Japan)
17. Assoc.Prof Dr Noor Fitrah Abu Bakar (Faculty of Chemical Engineering, Universiti Teknologi MARA,UiTM, Shah Alam, Malaysia)
18. Asst. Prof. Dr. Tanvir Sowgath, Department of Chemical Engineering, Bangladesh University of Engineering and Technology, Dhaka, Bangladesh)
19. Dr. Didin Agustian (School of Environment, Resources and Development Asian Institute of Technology, Bangkok, Thailand)
20. Dr. Faizal Abnisa (Department of Chemical Engineering, University of Malaya, Kuala Lumpur, Malaysia)
21. Dr. Suhas Patil (Rajendra Mane Polytechnic, Ambav, Maharashtra, India)
22. Dr Atikah Kadri (Faculty of Chemical Engineering, Selangor, Universiti Teknologi MARA,UiTM, Shah Alam, Malaysia)
23. Dr. Sofia B. Yarusova (Institute of Chemistry, Far-Eastern Branch of Russian Academy of Sciences, Vladivostok, Russia)
24. Dr. Hadi Fauzi (Mechanical Engineering Dept., University of Malaya, Kuala Lumpur, Malaysia)
25. Dr. Hanggara Sudrajat (Chulalongkorn University, Bangkok, Thailand)

26. Dr. Angelo Matos (Laboratory of Food Biotechnology, Federal University of Santa Catarina, Brazil)
27. Dr. Irvan Dahlan (School of Chemical Engineering Universiti Sains Malaysia, Pulau Pinang, Malaysia).
28. Dr. Prapat Pongkiatkul (Department of Environmental Engineering, King Mongkut's University of Technology Thonburi, Thailand)
29. Dr. Azil Bahari (Faculty of Chemical Engineering, Universiti Teknologi MARA,UiTM, Shah Alam, Malaysia)
30. Dr. Abdullah T. Al-fawwaz (Microbial Biotechnology,Faculty of Science, Al al-Bayt University, Jordan)
31. Dr. Syam Syafiie (Chemical and Environmental Engineering, Universiti Putra Malaysia, Serdang, Malaysia)
32. Dr. Wichitra Singhirunnusorn (Multidisciplinary Research Center for Environmental Sustainability, MRCES, Mahasarakham University, Thailand)
33. Dr Siti Shawalliah Idris (Faculty of Chemical Engineering, Universiti Teknologi MARA,UiTM, Shah Alam, Malaysia)
34. Dr. Chutimon Satirapipathkul (Chemical Engineering Department, Chulalongkorn University, Bangkok Thailand)
35. Dr Siti Wahidah Puasa (Faculty of Chemical Engineering, Universiti Teknologi MARA,UiTM, Shah Alam, Malaysia)
36. Dr.Duminda Liyanage (SPS-AIT Global Partnership Office, Pathum Thani, Thailand)

National Scientific Committee

1. Prof. Ir. Suryo Purwono, MA.Sc (Chemical Engineering Dept., Gadjah Mada University , Yogyakarta, Indonesia)
2. Prof. Dr. Heri Hermansyah, S.T., M.Eng (Chemical Engineering Dept., University of Indonesia, Jakarta, Indonesia)
3. Prof. Dr. Tati Suryati Syamsudin (School of Life Sciences and Technology, Bandung Institute of Technology, Bandung, Indonesia)
4. Prof. Dr. Husni Husin, MT (Chemical Engineering Dept., Syiah Kuala University, Banda Aceh, Indonesia)
5. Prof. Dr. Udin Hasanudin (Department of Agricultural Products Technology, Lampung University, Bandar Lampung, Indonesia)
6. Prof. Dr. Rosdanelli Hasibuan, MT (Chemical Engineering Dept., North Sumatera University, Medan, Indonesia)
7. Setiyo Gunawan, ST. PhD (Chemical Engineering Department, Sepuluh November Institute of Technlogy, Surabaya, Indonesia)
8. Dr. Syaifullah Muhammad, M.Eng (Chemical Engineering Dept., Syiah Kuala University, Banda Aceh, Indonesia)
9. Dr Meika Syahbana Rusli (Director of PT. Bogor Life Science and Technlogy, Bogor, Indonesia)
10. Dr. Nasrul A Rahman (Chemical Engineering Dept., Syiah Kuala University, Banda Aceh, Indonesia)
11. Dr. Edy Saputra (Chemical Engineering Dept., Riau University, Pekan Baru, Indonesia)
12. Dr. Elda Melwita, S.T, Phd (Chemical Engineering Dept., Sriwijaya University, Palembang, Indonesia)

IOP Conf. Series: Materials Science and Engineering **334** (2018) 011001 doi:10.1088/1757-899X/334/1/011001

13. Dr. Azhari Muhammad Syam, MSc (Chemical Engineering Dept., Malikussaleh University, Lhokseumawe)
14. Dr. Nita Aryanti, MT (Chemical Engineering Dept., Diponegoro University, Semarang, Indonesia)
15. Dr. Dani M. Supardan, MT (Chemical Engineering Dept., Syiah Kuala University, Banda Aceh, Indonesia)
16. Dr. Elida Purba, MSc (Chemical Engineering Dept., Lampung University, Bandar Lampung, Indonesia)
17. Dr. Izarul Machdar, M.Eng (Chemical Engineering Dept., Syiah Kuala University, Banda Aceh, Indonesia)
18. Dr. Aswati Mindaryani (Chemical Engineering Dept., Gadjah Mada University, Yogyakarta, Indonesia)
19. Dr. Teuku Rihayat, MSc (Chemical Engineering Dept., State Polytechnics of Lhokseumawe, Lhokseumawe, Indonesia)
20. Dr. Setyawati Yani, MT (Chemical Engineering Dept., Indonesian Muslim University, Makassar, Indonesia)
21. Dr. Sri Mulyati, MT (Chemical Engineering Dept., Syiah Kuala University, Banda Aceh, Indonesia)
22. Dr. Novizar Nazir (Agriculture Engineering, Andalas University, Padang, Indonesia)
23. Dr. M. Faisal, ST., M.Eng (Chemical Engineering Dept., Syiah Kuala University, Banda Aceh, Indonesia)
24. Dr. Zuchra Helwani, MSc (Chemical Engineering Dept., Riau University, Pekan Baru, Indonesia)
25. Dr. Marwan (Chemical Engineering Dept., Syiah Kuala University, Banda Aceh, Indonesia)
26. Chairul Irawan, ST., MT., Ph.D (Chemical Engineering Dept., Lambung Mangkurat University, Banjarmasin, Indonesia)
27. Dr. Rachmad Imbang., MT (Mechanical Engineering Dept., Bandung Polytechnics, Bandung, Indonesia)
28. Dr. Irwan Nurdin, MT (Chemical Engineering Dept., State Polytechnics of Lhokseumawe, Lhokseumawe, Indonesia)
29. Dr Dewi A Iryani (Chemical Engineering Dept., Lampung University, Indonesia)
30. Dr Dewi Yunita (Post Harvest Technology Department, Syiah Kuala University, Banda Aceh, , Indonesia)
31. Dr Agus Arip Munawar (Post Harvest Technology Department, Syiah Kuala University, Banda Aceh, , Indonesia)
32. Dr Yuli Amalia Husnil (Chemical Engineering Dept., Indonesia Institute of Technology, Serpong, Indonesia)
33. Dr Sidik Marsudi (Chemical Engineering Dept., Indonesia Institute of Technology, Serpong, Indonesia)
34. Dr Aniek Sri Handayani (Chemical Engineering Dept., Indonesia Institute of Technology, Serpong, Indonesia)
35. Dr Yanna Syamsuddin (Chemical Engineering Dept., Syiah Kuala University, Banda Aceh, Indonesia)

IOP Conf. Series: Materials Science and Engineering **334** (2018) 011001 doi:10.1088/1757-899X/334/1/011001

Keynote Speakers

1. Professor Said Irandoust, Accelerated Innoveation AB Sweden; " Innovations and Universities: How to Reduce the Gap Between Knowledge and Innovations"

2. Professor T. M. Indra Mahlia, Mechanical Engineering Department, Universiti Tenaga Nasional, Malaysia; "Potential Fuel Savings and Emissions Reduction by Implementing Fuel Economy Standards and Labels for Motor Vehicles In Indonesia"

Peer review statement

All papers published in this volume of *IOP Conference Series: Materials Science and Engineering* have been peer reviewed through processes administered by the proceedings Editors. Reviews were conducted by expert referees to the professional and scientific standards expected of a proceedings journal published by IOP Publishing.

Pyrolysis of plastic waste for liquid fuel production as prospective energy resource

S D A Sharuddin[1,4], F Abnisa[1,2], W M A W Daud[1] and M KAroua[3]

[1] Department of Chemical Engineering, Faculty of Engineering, University of Malaya, 50603 Kuala Lumpur, Malaysia
[2] Department of Chemical Engineering, Faculty of Engineering, University of Malikussaleh, Lhokseumawe 24351, Indonesia
[3] School of Science and Technology, Sunway University, No.5 Jalan Universiti, 47500 Selangor Darul Ehsan, Malaysia

E-mail: dayanareen@gmail.com

Abstract. The worldwide plastic generation expanded over years because of the variety applications of plastics in numerous sectors that caused the accumulation of plastic waste in the landfill. The growing of plastics demand definitely affected the petroleum resources availability as non-renewable fossil fuel since plastics were the petroleum-based material. A few options that have been considered for plastic waste management were recycling and energy recovery technique. Nevertheless, several obstacles of recycling technique such as the needs of sorting process that was labour intensive and water pollution that lessened the process sustainability. As a result, the plastic waste conversion into energy was developed through innovation advancement and extensive research. Since plastics were part of petroleum, the oil produced through the pyrolysis process was said to have high calorific value that could be used as an alternative fuel. This paper reviewed the thermal and catalytic degradation of plastics through pyrolysis process and the key factors that affected the final end product, for instance, oil, gaseous and char. Additionally, the liquid fuel properties and a discussion on several perspectives regarding the optimization of the liquid oil yield for every plastic were also included in this paper.

1. Introduction

Plastic plays an important role in improving our lifestyles in numerous sectors such as healthcare, construction, packaging, electronic, automotive and many more. The rise of the world population has caused the demand of commodity plastics to further increase. According to statistic, the global production of plastic has reached about 299 million tons in 2013 and has increased by 4% over 2012. Since most of the plastics were thrown out after single use, the amount of plastic waste accumulated in the environment each year was at alarming level. In Europe, 25 million tons of plastic ended up in waste stream during the year of 2012. Based on the statistic, about 38% of the plastic waste still went to the landfill, 26% were

[4] To whom any correspondence should be addressed.

recycled while 36% were utilized for energy recovery [1]. This signifies that the percentage of plastic waste dumped in the landfill was high and it occupied a large space. Besides, the degradation of plastics may take up billions of years, thus the continuous disposal of plastic in the landfill would definitely impact the environment negatively.

In order to overcome the challenges faced for recycling such as the needs of sorting that is labour intensive, a much more reliable method is established. Nowadays, converting the waste into valuable energy resource has been a brilliant way to fully utilize the waste in order to meet the increased energy demand. Plastic wastes can be turned into valuable energy since they are derived from petrochemical source which having significant calorific value. The conversion can be made possible through several thermal treatments technologies such as gasification, pyrolysis, plasma process and incineration [2]. Among all these methods, pyrolysis is the most desirable process since the initial volume of the waste is significantly reduced, more energy can be recovered from the plastic waste by producing varieties of products, requires lower decomposition temperature and low capital cost [2, 3].

Pyrolysis is the process of thermally degrading long chain polymer molecules into smaller, less complex molecules through heat. The process requires intense heat with shorter duration and in absence of oxygen. The three major products that are produced during pyrolysis are oil, gas and char which are valuable for industries especially production and refineries. Pyrolysis was chosen by many researchers since the process able to produce high amount of liquid oil up to 80 wt% at moderate temperature around 500^0C [4]. In addition, pyrolysis is also very flexible since the process parameters can be manipulated to optimize the product yield based on preferences. The liquid oil produced can be used in multiple applications such as furnaces, boilers, turbines and diesel engines without the needs of upgrading or treatment [5]. The oil produced from biomass pyrolysis has received positive notice as a more environmentally friendly fuel because it contributes to reducing the amount of CO_2 in the atmosphere [6]. Unlike recycling, the process handling is also much easier and flexible than the common recycling method since it does not need an intense sorting process, thus less labour intensive.

Several research papers have been published regarding the potential of different types of plastics in pyrolysis processes for liquid production. It has to be noted that the product yield and quality mainly relies upon the set up parameters. Therefore, this paper reviewed the potential of thermal degradation of plastic with the important parameters in pyrolysis process that need to be considered to obtain an optimal liquid oil production and improve the oil quality. Selective catalysts and their influences to the product yield were briefly discussed. Additionally, some relevant discussion regarding the optimization of liquid oil yield and the potential fuel quality were also presented in this paper.

2. Process parameters influence on product yield

Parameters play major role in optimizing the product yield and composition in pyrolysis. There are several parameters that influence liquid oil production in pyrolysis such as temperature, type of reactors, pressure, residence time, type and rate of fluidizing gas, and catalyst selection. In thermal degradation of plastics, temperature is one of the most significant operating parameters in pyrolysis since it controls the cracking reaction of the polymer chain. Different plastics have different degradation temperature depending on the chemical structure. For common plastics such as PET, HDPE, LDPE, PP and PS, the thermal degradation temperature started at 350^0C except for PVC which the degradation started at lower temperature of 220^0C. Additionally, the operating temperature required relies strongly on the product preference. If gaseous or char product was preferred, higher temperature more than 500^0C was suggested. If liquid was preferred instead, lower temperature in the range of 300-500^0C was recommended and this condition is applicable for all plastics [7, 8].

Besides temperature, right reactor selection for the process also crucial to increase efficiency of the reaction towards achieving the final desired product. Most plastic pyrolysis in the lab scale were performed in batch, semi-batch or continuous-flow reactors such as fluidized bed, fixed-bed reactor and conical spouted bed reactor (CSBR). Each reactor may have its own advantages and disadvantages depending on the application. Batch or semi-batch reactors were likely used in thermal pyrolysis since the parameters could be easily controlled. Nevertheless, these reactors might not be suitable for catalytic pyrolysis because of the potential of coke formation on the catalyst outer surface when they were mixed together with the plastics that would reduce the efficiency of the catalyst. Fluidized bed reactor is concluded to be the best reactor to perform catalytic plastic pyrolysis since the catalyst can be reused many times without the need of discharging, considering catalyst is a very expensive substance in the industry. On the other hand, CSBR also provides good mixing with the ability to handle large particle size distribution and low bed segregation than the bubbling fluidized bed as claimed by Olazar and Lopez [9]. However, a variety of technical challenges during operation of this reactor have been encountered such as its complicated design with many pumps, catalyst feeding, catalyst entrainment and product (solid and liquid) collection that make it less favourable [10].

Furthermore, pressure and residence time also governed the performance of pyrolysis process. Pressure and residence time are both temperature dependence factors that may have potential influence on product distribution of the plastic pyrolysis at lower temperature. Higher pressure increased the gaseous product yield and affected the molecular weight distribution for both liquid and gaseous products but only apparent at high temperatures. Based on the literature, most researchers conducted their plastic pyrolysis studies at atmospheric pressure and focused more on the temperature factor. The residence time was not brought up to attention while carrying out the experiment since the effect would become less apparent at higher temperatures. Moreover, in terms of economic viewpoint, additional units such as compressor and pressure transmitter need to be added into the overall system, thus increase the operation cost if the factor of pressure is considered. However, it should be noted that these two factors should be put under consideration based on the product distribution preference especially when running at temperature below 450^0C.

Apart from that, type and rate of fluidizing gas used during pyrolysis also influenced the pyrolysis product. Fluidizing gas is an inert gas (also known as carrier gas) which only engaged in transportation of vaporized products without taking part in the pyrolysis. There are many type of fluidizing gas that can be used for the plastic pyrolysis such as nitrogen, helium, argon, ethylene, propylene and hydrogen. Each type of fluidizing gas has different reactivity based on its molecular weight as reported by Abbas-Abadi et al. [11]. Nevertheless, of all those gases, nitrogen was commonly used by most researchers as fluidizing gas in plastic pyrolysis since it was easier and safer to handle than the high reactivity gas like hydrogen and propylene due to their flammability hazard. In terms of fluidizing gas flow rate, Lin and Yen [12] found that the rate of degradation dropped instantly at the lowest fluidizing flow rate of 300 ml/min. The contact time for primary product is high at lower flow rate, causing the formation of coke precursor (BTX) to increase with the secondary product obtained even though the overall degradation rate is slower [13]. At the highest flow rate of 900 ml/min, the gasoline and hydrocarbon gases fraction were maximized. This clearly shows that the flow rate of fluidizing gas also affected the final product distribution.

As for the catalytic degradation, catalyst is used in the thermal process to speed up chemical reaction and improves the hydrocarbon distribution in order to obtain pyrolysis liquid that had similar properties to the conventional fuel such as gasoline and diesel. The three types of catalysts that are widely used in plastic pyrolysis are zeolites, fluid cracking catalyst (FCC) and silica-alumina catalysts. The usage of zeolite catalyst in pyrolysis of real municipal plastic waste may also help to reduce the impurities in the oil produced and this was proven in the study conducted by Miskolczi et al. [14]. Besides zeolite, the usage of spent FCC catalyst could also improve the plastic conversion in pyrolysis. According to Kyong et al. [15], the amount of solid residue left reduced drastically from 4.5 to 0.9 wt% and this indicates that the conversion was maximized. Other than these two catalysts, some researchers also chose to use silica-alumina catalyst that is an amorphous acid type. The acid concentration of silica alumina catalyst is determined by the mole ratio of SiO_2/Al_2O_3. Different strength of acidity in catalyst has great influence in the final end product of plastic pyrolysis. Sakata et al. [16] explored the effect of catalysts acidity (SA-1, SA-2, ZSM-5) on the product distribution of HDPE pyrolysis and found that SA-2 catalyst with lower acidity produced higher amount of liquid oil (74.3 wt%), followed by SA-1 (67.8 wt%) and ZSM-5 (49.8 wt%). ZSM-5 possessed strong acid sites, thus produced more gaseous products than the other two acid catalysts but very less liquid yield. Therefore, it is worth noting that that the usage of zeolite catalyst in plastic pyrolysis only maximized the production of volatile hydrocarbon while FCC and silica-alumina maximized the liquid oil production. However, for higher efficiency and longer cycle time usage, HZSM was recommended since the deactivation rate of the catalyst was extremely low and thus, more efficient for regeneration.

3. Liquid fuel properties from pyrolysis of plastics

Table 1 summarized the fuel properties of the liquid oil produced in pyrolysis process. The experimental calorific value of HDPE, PP and LDPE are all above 40 MJ/kg and were considered high for energy utilization. The calorific value of PS was commonly lower than the polyolefin plastic due to the existence of the aromatic ring in the chemical structure which had lesser combustion energy than the aliphatic hydrocarbon [17]. Overall, PET and PVC had the lowest calorific value below 30 MJ/kg due to the presence of benzoic acid in PET and chlorine compound in PVC that deteriorated the fuel quality. Benzoic acid also consisted of aromatic ring that explained the low calorific value in PET. Overall, Table 1 clearly depicted that the physical properties of plastics pyrolysis oil were very close to the properties of commercial gasoline and diesel. Therefore, plastic pyrolysis oil has very high potential to be used as new energy resources.

Table 1. Fuel properties of plastic pyrolysis oil.

Physical Properties	Type of Plastics (Experimental typical value)						Commercial Standard Value (ASTM 1979)	
	PET [18],[19]	HDPE [20]	PVC [18],[21]	LDPE [22]	PP [20]	PS [23],[24]	Gasoline [20]	Diesel [20]
Calorific value (MJ/kg)	28.2	40.5	21.1	39.5	40.8	43.0	42.5	43.0
API gravity @ 60^0F	n.a	27.48	38.98	47.75	33.03	n.a	55	38
Viscosity (mm^2/s)	n.a	5.08[a]	6.36[b]	5.56[c]	4.09[a]	1.4[d]	1.17	1.9-4.1
Density @ 15^0C (g/cm^3)	0.90	0.89	0.84	0.78	0.86	0.85	0.780	0.807
Ash (wt%)	n.a	0.00	n.a	0.02	0.00	0.006	-	0.01
Octane number MON (min)	n.a	85.3	n.a	n.a	87.6	n.a	81-85	-
Octane number RON (min)	n.a	95.3	n.a	n.a	97.8	90-98	91-95	-
Pour point (^{0}C)	n.a	-5	n.a	n.a	-9	-67	-	6
Flash point (^{0}C)	n.a	48	40	41	30	26.1	42	52
Aniline point (^{0}C)	n.a	45	n.a	n.a	40	n.a	71	77.5
Diesel index	n.a	31.05	n.a	n.a	34.35	n.a	-	40

*n.a., not available in the literature
[a] Viscosity at 40^0C
[b] Viscosity at 30^0C
[c] Viscosity at 25^0C
[d] Viscosity at 50^0C

4. Discussion on plastic pyrolysis scenarios

Table 2 summarized the optimum temperature required to optimize liquid oil yield in thermal and catalytic pyrolysis at different conditions. Other affected parameters include the type of reactors, pressure, heating rate and pyrolysis duration for each type of plastics. All experiments carried out were using nitrogen gas as the fluidizing medium. Based on table 2, PET and PVC are two plastics that produced very low yield of liquid oil in comparison with other plastic types, which made these plastics infrequently explored by researchers. It also should be noted that not all plastic types are recommended for pyrolysis. PVC was not preferred in pyrolysis since it produced the major product of harmful hydrochloric acid and very low yield of liquid oil. Additionally, the pyrolysis oil also contained chlorinated compound that would degrade the oil quality and also toxic to the environment.

As summarized in table 2, it can be concluded that the most effective temperature to optimize the liquid oil yield in plastic pyrolysis would be in the range of 500-550^{0}C for thermal pyrolysis. However, with the usage of catalyst in the pyrolysis, the optimum temperature could be lowered down to 450^{0}C and higher liquid yield was obtained. In most plastics, the usage of catalyst in the process might improve the liquid oil yield, but PS was exceptional. This is because PS degraded very easily without the needs of any catalysts to speed up the reaction and yet 97 wt% of oil was produced [17]. Therefore, PS was the best plastic for pyrolysis since it produced the highest amount of liquid oil production among all the plastics. As for the polyolefin plastic type, LDPE produced the highest liquid oil yield (93.1 wt%), followed by HDPE (84.7 wt%) and PP (82.12 wt%) in thermal pyrolysis. However, with addition of catalyst such as FCC and at the right operating temperature, the liquid yield could be further maximized to above 90 wt%. Besides, the energy potential of pyrolysis oil can also be calculated based on total amount of plastic waste available in Malaysia. For instance, the amount of PS waste available in Malaysia is 280,000 tons/year [25]. Pyrolysis oil obtained from PS manage to contribute 11.68 PJ/year to Malaysia's energy supply based on the HHV of 43 MJ/kg as shown in Table 1 and the liquid oil production of 97wt% as reported by Onwudili, Insura [17], indicated in table 2.

Table 2. Summary of studies on plastic pyrolysis.

Ref.	Type of plastic	Reactor	Process parameters				Yield (wt%)			Additional information
			Temp (^{0}C)	Pressure	Heating rate (^{0}C/min)	Duration (min)	Oil	Gas	Solid	
[18]	PET	Fixed bed	500	-	10	-	23.1	76.9	0	
[4]	PET	-	500	1 atm	6	-	38.9	52.1	8.9	
[20]	HDPE	Horizontal steel	350	-	20	30	80.9	17.2	1.9	
[26]	HDPE	Semi-batch	400	1 atm	7	-	82	16	2	Stirring rate 200 RPM, FCC catalyst 10 wt%
[27]	HDPE	Batch	450	-	-	60	74.5	5.8	19.7	
[28]	HDPE	Semi-batch	450	1 atm	25	-	91.2	4.1	4.7	Stirring rate 50 RPM, FCC catalyst 20 wt%
[29]	HDPE	Fluidized bed	500	-	-	60	85	10	5	Silica alumina catalyst
[30]	HDPE	Batch	550	-	5	-	84.7	16.3	0	
[31]	HDPE	Fluidized bed	650	-	-	20-25	68.5	31.5	0	
[18]	PVC	Fixed bed	500	-	10	-	12.3	87.7	0	
[32]	PVC	Vacuum batch	520	2 kPa	10	-	12.8	0.34	28.1	Also yield HCl=58.2 wt%
[17]	LDPE	Pressurized batch	425	0.8-4.3 Mpa	10	60	89.5	10	0.5	
[33]	LDPE	Batch	430	-	3	-	75.6	8.2	7.5	Also yield wax= 8.7 wt%
[4]	LDPE	-	500	1 atm	6	-	80.4	19.4	0.2	
[34]	LDPE	Fixed bed	500	-	10	20	95	5	0	
[30]	LDPE	Batch	550	-	5	-	93.1	14.6	0	
[35]	LDPE	Fluidized bed	600	1 atm	-	-	51.0	24.2	0	Also yield wax=24.8 wt%
[20]	PP	Horizontal steel	300	-	20	30	69.8	28.8	1.3	
[36]	PP	Batch	380	1 atm	3	-	80.1	6.6	13.3	

[26]	PP	Semi- batch	400	1 atm	7	-	85	13	2	Stirring rate 200 RPM, used FCC catalyst 10 wt%
[11]	PP	Semi-batch	450	1 atm	25	-	92.3	4.1	3.6	Stirring rate 50 RPM, used FCC catalyst 10 wt%
[4]	PP	-	500	1 atm	6	-	82.1	17.76	0.12	
[37]	PP	Batch	740	-	-	-	48.8	49.6	1.6	
[26]	PS	Semi-batch	400	1 atm	7	-	90	6	4	Stirring rate 200 RPM, used FCC catalyst, cat/poly=10 w/w
[17]	PS	Pressurized batch	425	0.31-1.6 MPa	10	60	97	2.50	0.5	
[38]	PS	Batch	500	-	-	150	96.7	3.27	0	Used Zn catalyst, Cat/poly=5 w/w
[37]	PS	Batch	581	-	-	-	89.5	9.9	0.6	64.9 wt% of liquid comprised of styrene

*All experiments used nitrogen gas as fluidizing medium.

5. Conclusion

Pyrolysis process was chosen by most researchers among other thermal treatment technologies mainly because of its potential to convert the most energy from plastic waste to valuable liquid oil, gaseous and char. The flexibility that it provides in terms of product preference could be achieved by adjusting the parameters accordingly. The pyrolysis could be done in both thermal and catalytic process. However, the catalytic process provided lower operating temperature with greater yield of liquid oil for most plastics with the right catalyst selection. The sustainability of the process is unquestionable since the amount of plastic wastes available in every country is reaching millions of tons. With the pyrolysis method, the waste management becomes more efficient, less capacity of landfill needed, less pollution and also cost effective. Moreover, with the existence of pyrolysis method to decompose plastic into valuable energy fuel, the dependence on fossil fuel as the non-renewable energy can be reduced and this solves the rise in energy demand.

Acknowledgments

The authors would like to thank the University of Malaya for fully funding the work described in this publication through the PPP project number PG066-2015B. A special thanks is also extended to the University Technology Mara (UiTM) and Ministry of Higher Education (MOHE) for providing financial support under the Young Lecturer's Scheme (TPM).

References

[1] Association of Plastic Manufacturers Europe. An analysis of European plastics production, demand and waste data. European Association of Plastics Recycling and Recovery Organisations, Belgium, 2015. pp. 1-32.

[2] K. Moustakas, M. Loizidou. Solid waste management through the application of thermal methods. in: S. Kumar, (Ed.). Waste Management. InTech, Greece, 2010. pp. 89-124.

[3] N. Patni, P. Shah, S. Agarwal, P. Singhal. Alternate Strategies for Conversion of Waste Plastic to Fuels. *ISRN Renewable Energy*. **2013** (2013) 1-7.

[4] S.M. Fakhrhoseini, M. Dastanian. Predicting pyrolysis products of PE, PP, and PET using NRTL activity coefficient model. *Hindawi Publishing Corporation*. **2013** (2013) 1-5.

[5] A.V. Bridgwater. Review of fast pyrolysis of biomass and product upgrading. *Biomass Bioenergy*. **38** (2012) 68-94.

[6] F. Abnisa, W.M.A. Wan Daud, A. Arami-Niya, B.S. Ali, J.N. Sahu. Recovery of liquid fuel from the aqueous phase of pyrolysis oil using catalytic conversion. *Energy & Fuels*, *28* (2014), 3074-3085.

[7] S.D. Anuar Sharuddin, F. Abnisa, W.M.A. Wan Daud, M.K. Aroua. A review on pyrolysis of plastic wastes. *Energy Convers Manage.* **115** (2016) 308-26.

[8] F. Abnisa, S.D. Anuar Sharuddin, W.M.A. Wan Daud. Optimizing the use of biomass waste through co-pyrolysis. INFORM - International News on Fats, Oils, and Related Materials. American oil chemists' society Press2017. pp. 16-9.

[9] M. Olazar, G. Lopez, M. Amutio, G. Elordi, R. Aguado, J. Bilbao. Influence of FCC catalyst steaming on HDPE pyrolysis product distribution. *J Anal Appl Pyrolysis.* **85** (2009) 359-65.

[10] H.S. Fogler. Elements of chemical reaction engineering. 4th ed. Pearson Education Inc., New Jersey, 2010.

[11] M.S. Abbas-Abadi, M.N. Haghighi, H. Yeganeh, A.G. McDonald. Evaluation of pyrolysis process parameters on polypropylene degradation products. *J Anal Appl Pyrolysis.* **109** (2014) 272-7.

[12] Y.H. Lin, H.Y. Yen. Fluidised bed pyrolysis of polypropylene over cracking catalysts for producing hydrocarbons. *Polym Degrad Stab.* **89** (2005) 101-8.

[13] Y.H. Lin, M.H. Yang. Catalytic pyrolysis of polyolefin waste into valuable hydrocarbons over reused catalyst from refinery FCC units. *Applied Catalysis A: General.* **328** (2007) 132-9.

[14] N. Miskolczi, A. Angyal, L. Bartha, I. Valkai. Fuels by pyrolysis of waste plastics from agricultural and packaging sectors in a pilot scale reactor. *Fuel Process Technol.* **90** (2009) 1032-40.

[15] H.L. Kyong, G.J. Sang, H.K. Kwang, S.N. Nam, H.S. Dae, J. Park, et al. Thermal and catalytic degradation of waste high density polyethylene (HDPE) using spent FCC catalyst. *Korean J Chem Eng.* **20** (2003) 693-7.

[16] Y. Sakata, M.A. Uddin, A. Muto, Y. Kanada, K. Koizumi, K. Murata. Catalytic degradation of polyethylene into fuel oil over mesoporous silica (KFS-16) catalyst. *J Anal Appl Pyrolysis.* **43** (1997) 15-25.

[17] J.A. Onwudili, N. Insura, P.T. Williams. Composition of products from the pyrolysis of polyethylene and polystyrene in a closed batch reactor: Effects of temperature and residence time. *J Anal Appl Pyrolysis.* **86** (2009) 293-303.

[18] O. Cepeliogullar, A. E.Putun. Utilization of two different types of plastic wastes from daily and industrial life. in: C. Ozdemir, S. Sahinkaya, E. Kalipci, M.K. Oden, (Eds.), ICOEST Cappadocia 2013. ICOEST Cappadocia, Turkey, 2013. pp. 1-13.

[19] M. Sarker, A. Kabir, M.M. Rashid, M. Molla, A.S.M. Din Mohammad. Waste Polyethylene Terephthalate (PETE-1) Conversion into Liquid Fuel. *Journal of Fundamentals of Renewable Energy and Applications.* **1** (2011) 1-5.

[20] I. Ahmad, M.I. Khan, H. Khan, M. Ishaq, R. Tariq, K. Gul, et al. Pyrolysis Study of Polypropylene and Polyethylene Into Premium Oil Products. *Int J Green Energy.* **12** (2014) 663-71.

[21] E. Manickaraja, S. Tamilkolundu. Catalytic degradation of waste PVC into liquid fuel using BaCO3 as catalyst and its blending properties with diesel fuel. *Discovery.* **23** (2014) 74-8.

[22] S. B.Desai, C. K.Galage. Production and analysis of pyrolysis oil from waste plastic in Kolhapur city. *International Journal of Engineering Research and General Science.* **3** (2015) 590-5.

[23] F. Pinto, P. Costa, I. Gulyurtlu, I. Cabrita. Pyrolysis of plastic wastes. 1. Effect of plastic waste composition on product yield. *J Anal Appl Pyrolysis.* **51** (1998) 39-55.

[24] M. Blazso. Composition of liquid fuels derived from the pyrolysis of plastics. in: J. Scheirs, W. Kaminsky, (Eds.), *Feedstock recycling and pyrolysis of waste plastics: Converting waste plastics into diesel and other fuels.* John Wiley & Sons, Hungary, 2006. pp. 315-42.

IOP Conf. Series: Materials Science and Engineering **334** (2018) 012001 doi:10.1088/1757-899X/334/1/012001

[25] F. Abnisa, W.M.A.W. Daud, J.N. Sahu. Pyrolysis of mixtures of palm shell and polystyrene: An optional method to produce a high-grade of pyrolysis oil. *Environ Prog Sustain Energy.* (2013) DOI: 10.1002/ep.11850.
[26] H.L. Kyong, S.N. Nam, H.S. Dae, Y. Seo. Comparison of plastic types for catalytic degradation of waste plastics into liquid product with spent FCC catalyst. *Polym Degrad Stab.* **78** (2002) 539-44.
[27] N. Miskolczi, L. Bartha, G. Deák, B. Jóver, D. Kalló. Thermal and thermo-catalytic degradation of high-density polyethylene waste. *J Anal Appl Pyrolysis.* **72** (2004) 235-42.
[28] M.S. Abbas-Abadi, M.N. Haghighi, H. Yeganeh. Evaluation of pyrolysis product of virgin high density polyethylene degradation using different process parameters in a stirred reactor. *Fuel Process Technol.* **109** (2013) 90-5.
[29] G. Luo, T. Suto, S. Yasu, K. Kato. Catalytic degradation of high density polyethylene and polypropylene into liquid fuel in a powder-particle fluidized bed. *Polym Degrad Stab.* **70** (2000) 97-102.
[30] A. Marcilla, M.I. Beltrán, R. Navarro. Thermal and catalytic pyrolysis of polyethylene over HZSM5 and HUSY zeolites in a batch reactor under dynamic conditions. *Applied Catalysis B: Environmental.* **86** (2009) 78-86.
[31] F.J. Mastral, E. Esperanza, P. Garcia, M. Juste. Pyrolysis of high-density polyethylene in a fluidized bed reactor. Influence of the temperature and residence time. *J Anal Appl Pyrolysis.* **63** (2001) 1-15.
[32] R. Miranda, Y. Jin, C. Roy, C. Vasile. Vacuum pyrolysis of PVC kinetic study. *Polym Degrad Stab.* **64** (1998) 127-44.
[33] M.A. Uddin, K. Koizumi, K. Murata, Y. Sakata. Thermal and catalytic degradation of structurally different types of polyethylene into fuel oil. *Polym Degrad Stab.* **56** (1996) 37-44.
[34] R. Bagri, P. T. Williams. Catalytic pyrolysis of polyethylene. *J Anal Appl Pyrolysis.* **63** (2001) 29-41.
[35] P. T. Williams, E. A.Williams. Fluidised bed pyrolysis of low density polyethylene to produce petrochemical feedstock. *J Anal Appl Pyrolysis.* **51** (1998) 107-26.
[36] Y. Sakata, M.A. Uddin, A. Muto. Degradation of polyethylene and polypropylene into fuel oil by using solid acid and non-solid acid catalysts. *J Anal Appl Pyrolysis.* **51** (1999) 135-55.
[37] A. Demirbas. Pyrolysis of municipal plastic wastes for recovery of gasoline-range hydrocarbons. *J Anal Appl Pyrolysis.* **72** (2004) 97-102.
[38] Adnan, J. Shah, M.R. Jan. Thermo-catalytic pyrolysis of polystyrene in the presence of zinc bulk catalysts. *Journal of the Taiwan Institute of Chemical Engineers.* **45** (2014) 2494-500.

Geochemistry of hot springs in the *Ie Seu'um* hydrothermal areas at Aceh Besar district, Indonesia

R Idroes[1,4], M Yusuf[1], M Alatas[1], Subhan[1], A Lala[1], Saiful[1], R Suhendra[1], G M Idroes[2] and Marwan[3]

[1] Department of Chemistry, Faculty of Mathematics and Natural Sciences, Syiah Kuala University, Kopelma Darussalam, Banda Aceh 23111, Indonesia
[2] Department of Chemical Engineering, Faculty of Engineering, Syiah Kuala University, Kopelma Darussalam, Banda Aceh 23111, Indonesia
[3] Department of Geophysics Engineering, Faculty of Engineering, Syiah Kuala University, Kopelma Darussalam, Banda Aceh 23111, Indonesia

E-mail: rinaldi.idroes@unsyiah.ac.id

Abstract. Indonesia geothermal resources are the largest in the world, about 40 percent of the total geothermal resources worldwide with a potential energy of 28,617 MW. Geothermal energy is one of the renewable energy in the world that can be developed sustainably. This kind of energy is not only environmentally friendly but also highly prospective compared to fossil energy. One of the potential geothermal energy in Indonesia is *Seulawah Agam* geothermal field with some manifestation areas. The fluid type of *Ie Seu'um* manifestation was chloride (Cl^-) obtained from the ternary diagram Cl^--SO_4^{2-}-HCO_3^-, using UV-Vis spectrophotometry, argentometry and acidimetry method. The reservoir range temperature was 188,7 ± 9,3 °C calculated using geothermometer Na-K-Ca, Na-K Fournier and Na-K Giggenbach by applying Atomic Absorption Spectroscopy method. This data processing was carried out using liquid chemistry plotting spreadsheet version 3 powell geoscience Ltd.3 September 2012 by Powell & Cumming. The potential in the geothermal manifestation of *Ie Seu'um* was estimated about 50-100 MW (medium enthalpy).

1. Introduction

Geothermal resources in Indonesia have been explored since 1970 with approximately over 200 geothermal systems throughout Indonesia utilizing surface manifestations. Most geothermal systems in Indonesia manifested the high-temperature fluid which exists along the volcanic line area [1].

There are not many countries in the world that have potential geothermal energy reserves, except a few countries which passed by the ring of fire. This Ring of Fire also known as the circum-Pacific belt, it is a significant area in the cavity of the Pacific Ocean where a great number of earthquakes and volcanic eruptions occur. Indonesia as a country that passed by the ring of fire is currently identified to have about 117 active volcanoes as the potential sources of geothermal energy [2].

[4] To whom any correspondence should be addressed.

IOP Conf. Series: Materials Science and Engineering **334** (2018) 012002 doi:10.1088/1757-899X/334/1/012002

The potential resources of Indonesian's geothermal predicted was about 40% of the total share of the world's geothermal reserves or circa 28,617 MW. Whereas, only about 4.5% is being exploited as electrical energy in the country. The power plant comes from a minimum of 10 locations: Darajat (260 MWe), Dieng (60 MWe), Kamojang (200 MWe), Gunung Salak (377 MWe), Sibayak (11 MWe), Lahendong (87 MWe), Wayang Windu (227 MWe), Ulu Belu – South Sumatra (110 MWe), Ulumbu – Flores (5 MWe) and Mataloko (2.5 MWe) [3]. Therefore, it is necessary to explore and exploit other geothermal sources especially in the Aceh province.

Nearly half of Indonesia's geothermal potential is on the island of Sumatera. One of the areas in Sumatera that has a potential geothermal energy is the Seulawah Agam mountain with the discovery of many manifestations around it, such as *Ie seu'um, Ie Jue, Ie Busuk, Alue Ie Suum, Alue Utuen Pineung* and much more. *Ie Seu'um* was the explored manifestation in this study that located circa 20 km from the top of a mountain and about 35 km from the city of Banda Aceh. Until now, the area of *Ie Seu'um* manifestation or hydrothermal area has been utilized by the district government and society as a hot spring and one of tourism objects close to Krueng Raya harbor, Mesjid Raya sub-district, Aceh Besar regency, Aceh province.

Geochemistry studies of geothermal manifestations have been conducted in many countries including the Obock coastal hot springs in the Republic of Djibouti [4], The Sabatini Volcanic District and the Tolfa Mountains, Italy [5], geothermal fields in Salihli [6], Mapamyum Western Tibet, China [7], Tang-Bijar oilfield springs, Zagros region, Iran [8], Southern Saint Lucia, Lesser Antilles island arc [9], Domuyo volcanic complex, Argentina [10], Southern Sula graben, Honduras [11] and Eastern Himalayas, China [12]. Related to Seulawah Agam geothermal field of Ie-Seu'um manifestation area, Aceh Besar district, Indonesia, up to now has never been published particularly concerning the determination of the fluid type and reservoir temperature. This research began with estimating bottom surface reservoir temperatures based on geochemical composition. Several geothermometers were used in this research include Na-K-Ca, Na/K Fournier (1973) and Na/K Truesdell (1973), while geothermal type determination using ternary diagram Cl^{-}-SO_4^{2-}-HCO_3^{-}.

2. Materials and Methods

2.1. Water Sampling

Hot spring samples were collected on October 4, 2016, from *Ie Seu'um* hot spring at *Ie Seu'um* Village, Mesjid Raya Sub-district, Aceh Besar Regency, Aceh Province by purposive sampling method. Hot spring sampling technique is done by observing the springs debit coming out from hot spring resources. The *Ie Seu'um* manifestation area is categorized in the outflow geothermal zone of Mount Seulawah Agam. The hot spring samples were found from two geothermal sampling site, which is the sampling point 1 (S1) at 06 ° 13'77.8 "E and 07 ° 82'38.2" N and at sampling point 2 (S2) at 06 ° 13 ' 73.9 "E and 07 ° 82'38.7" N coordinates (Figure 1).

2.2. Materials and Instruments

The instruments used among others are Portable Temperature (Fisher Scientific Traceable), pH meter (Schott Instrument) and Conductometer (Schott Instrument), Atomic Absorption Spectroscopy (AAS) instruments (Shimadzu 6200), UV-Vis (DR / 2010), GPS, Spectrophotometer (Garmin 62S), 1000 ml polyethylene bottle, glass mouthpiece, universal pH indicator, glove, dropper, plastic bucket, 50 mL burette, 250 mL Erlenmeyer flask, 100 ml and 1000 mL flask, 100 mL measuring cup, 25 mL measuring pipette, spatula, electric heaters, analytical scales, watch glasses, spray bottles and cool boxes.

IOP Conf. Series: Materials Science and Engineering **334** (2018) 012002 doi:10.1088/1757-899X/334/1/012002

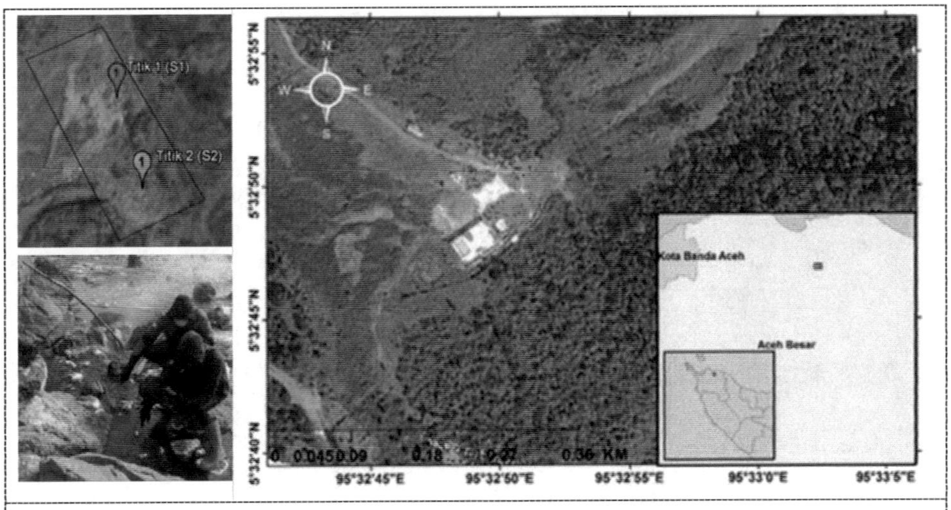

Figure 1. Satellite imagery of google earth of hot spring sampling point location at *Ie Seu'um* hydrothermal areas.

The materials used were chloride-free and sulphate-free distilled water, HNO_3 solution, 0.045 N NaCl solution, 0.45 µm filter paper, chromate potassium indicator, phenolphthalein indicator, methyl orange indicator, sulfaver®, 0.051 N $AgNO_3$ solution, HCl 0.1 N, standard solution of calcium (1000 mg/L), magnesium (1000 mg/L), potassium (1000 mg/L) and sodium (1000 mg/L).

2.3 *Measurements and Data Processing*

Analysis methods for sodium, magnesium, calcium and potassium cations was carried out using AAS instrument. Analysis of sulphate, chloride and bicarbonate anions was performed by UV-Vis spectrophotometer, argentometry (Mohr) and acidimetry methods respectively. The data processing was done using liquid chemistry plotting spreadsheet version 3 powell geoscience Ltd. 3 September 2012 by Powell & Cumming. The fluid type was determined using the triangular ternary diagram Cl^--SO_4^{2-}-HCO_3^-. The chemical equilibrium of *Ie Seu'um* manifestation reservoir was specified with a ternary diagram Na-K-Mg while geothermometer Na-K-Ca estimated the temperature reservoir manifestation, geothermometer Na/K (Fournier 1979) and geothermometer Na/K Giggenbach (1988), described by equation (1), (2), and (3).

$$T = \frac{1390}{1.750 + \log\left[\frac{Na}{K}\right]} - 273.15 \text{ (Na - K Giggenbach 1988) [14]} \quad (1)$$

$$T = \frac{1217}{\log\left[\frac{Na}{K}\right] + 1{,}483} - 273.15 \text{ (Na - K Fournier 1979) [15]} \quad (2)$$

$$T = \frac{1647}{\log\left[\frac{Na}{K}\right] + \beta(\log[Ca/Na]^{0.5} + 2.06) + 2.47} - 273.15 \text{ (Na - K - Ca) [14]} \quad (3)$$

3. Results and Discussion

3.1. *Characteristics of Ie Seu'um Surface Manifestation*

The measurement of the physicochemical parameter (temperature, pH and conductivity) of *Ie Seu'um* field was conducted with a repetition of five times by insitu analysis. Measurements were made at the manifestation that has the highest debit and temperature. The average results of measurement of the manifestation can be seen in table 1.

Table 1. Locations and manifestations surface characteristic of *Ie Seu'um* carried out on October 4, 2016 at 72-meter elevation.

No	Location	Coordinate		pH	Temperature (^{0}C)	Conductivity (mV)
		E	N			
1	*Ie Seu'um* (S1)	0°61'37.78	0°78'23.82	6,61 ± 0,005	84,20 ± 0,06	27,65 ± 0,46
2	*Ie Seu'um* (S2)	0°61'37.39	0°78'23.87	6,65 ± 0,003	82,55 ± 0,01	26,43 ± 0,13

The condition around *Ie Seu'um* manifestation showed the clear water and deposits on the soil and rocks were reddish brown.

3.2. The Manifestation Fluid Type of Ie Seu'um

The fluid type determination of *Ie Seu'um* hot springs was done using the triangular ternary diagram equation based on the concentration of sulphate, bicarbonate and chloride anions. The Results of these anions were then plotted into the equation triangular ternary diagram Cl^--SO_4^{2-}-HCO_3^- (Figure 2). The highest anion concentration was chloride with the value of 2656,99 ± 0,577 mg/L that is shown in table 2.

Table 2. The concentration of the sulphate, bicarbonate and chloride anions of hot spring samples at *Ie Seu'um.*

No	Anion	Concentration (mg/L)
1	Chloride	2656,99 ± 0,577
2	Sulphate	240,67 ± 0,236
3	Bicarbonate	122,81 ± 0,575

Based on the plotting results in the diagram ternary Cl^--SO_4^{2-}-HCO_3^- it was known that the type of hot fluid indicated was chloride. The fluid from this hot springs area indicated that the water came from the deep reservoir without rocks and other fluids mixing. So, this type reflected the condition of the deep reservoir and suitable for calculating the reservoir temperature using geothermometer equation.

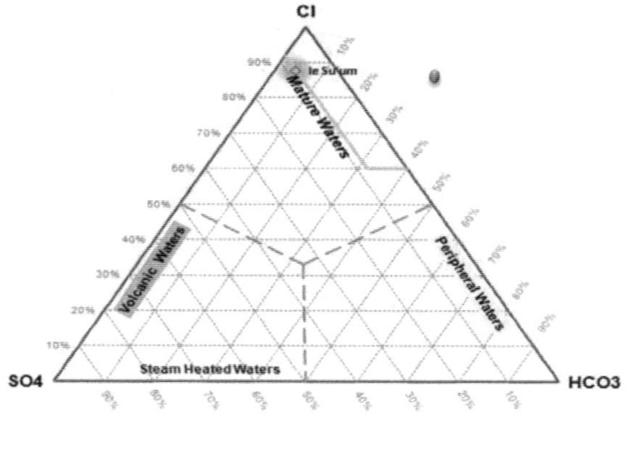

Figure 2. Ternary Diagram Cl^--SO_4^{2-}-HCO_3^- at *Ie Seu'um* manifestation Plotting point for hot spring *Ie Seu'um* sample.

3.3. Ternary Diagram Na–K–Mg

The Na-K-Mg ternary diagram is an equation used to estimate water equilibrium in the lithology. Table 3 showed the laboratory measurements for potassium, calcium, magnesium and sodium ions respectively. The Results of these cations were plotted based on the triangular ternary diagram Na–K–Mg (Figure 3). The highest cation concentration was sodium with the value of 2630.03 ± 0,15mg/L that shown in table 3.

Table 3. The concentration of the potassium, calcium, magnesium and sodium cations of hot spring samples at *Ie Seu'um* manifestation.

No	Cation	Concentration (mg/L)
1	Potassium	167,91 ± 0,13
2	Calcium	178,61 ± 0,08
3	Magnesium	7,08 ± 0,01
4	Sodium	2630,03 ± 0,15

The content of magnesium ion concentration was minimal, it can indicate that the hot spring *Ie Seu'um* less influenced by surface water (meteoric water). The content of each ion then used to plot the ternary triangular diagram Na-K-Mg.

Based on the plotting results of the ternary diagram Na-K-Mg, the sample *Ie Seu'um* hot springs were in the area of partial equilibration. Equilibration partial area is an area in which the concentration of metal ions is in the chemical equilibrium, it means the dissolved cations in the geothermal water directly comes from the deepwater reservoir (magmatic water).

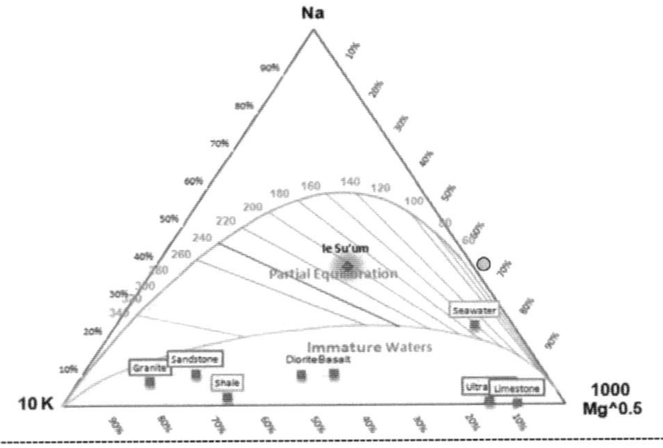

Figure 3. Ternary Diagram Na-K-Mg *Ie Seu'um* manifestation Ploting point for hot spring *Ie Seu'um* sample.

3.4. Reservoir Temperature of Ie Seu'um Geothermal Manifestation

Geothermometer is an equation that is widely used in predicting the reservoir temperature in the geothermal system. Up to now, the geothermometry has been proposed by many authors with different geothermal fields. Table 4 exhibited the plotting results that used several equations for fluid geothermometer at *Ie Seu'um* manifestations.

Table 4. Plotting result by using the geotermometer Na-K-Ca, Na/K Fournier (1979) and Na/K Giggenbach (1988) equations for *Ie Seu'um* geothermal manifestations.

No	Geothermometer	Temperature Estimation (^{0}C)
1	Na-K-Ca	186
2	Na/K Fournier (1979)	181
3	Na/K Giggenbach (1988)	199

Based on the plotting results on geothermometer Na-K-Ca, the temperature of *Ie Seu'um* manifestation reservoir approximately 186 °C. Na-K-Ca geothermometer is suitable to be applied in water that has high calcium concentration. For *Ie Seu'um* hot spring manifestation, geothermometer Na-K-Ca can be utilized since the result in the temperature range of 120 - 200°C. Based on the equation geothermometer Na/K Fournier (1979) then obtained the estimated reservoir temperature circa 181 °C. From its usefulness, this geothermometer equation also can be applied to estimate the temperature of the reservoir at *Ie Seu'um* manifestations due to exceeding the minimum threshold of 180 °*C*. By the plotting of geothermometer equation, Na/K Giggenbach (1988) was the estimated reservoir temperature of about 199 °C. This geothermometer Equation Na/K is commonly used to estimate the temperature of the geothermal reservoir system that generated the highest temperature value. According to the results comparison obtained from the three equations of table 4, the average value of geothermal field geothermometer *Ie Seu'um* is 188.7 ± 9.3 (°C).

The reservoir temperature can be classified into three geothermal systems, namely high enthalpy (temperature > 225 °C), medium enthalpy (temperature 125 - 225 °C) and low enthalpy (temperature < 125 °C) Hochstein (1990). Based on the results, the reservoir temperature of *Ie Seu'um* hot spring manifestation is in the *medium enthalpy* area and can be predicted according to reference table [13] then have the potential of geothermal energy range between 50-100 MW.

4. Conclusion

The fluid type of *Ie Seu'um* manifestation is chloride (Cl^-) which was obtained from the ternary diagram Cl^--SO_4^{2-}-HCO_3^-, using spectrophotometry, argentometry and acidimetry method. The reservoir range temperature was 188,7 ± 9,3 °C calculated using geothermometer Na-K-Ca, Na-K Fournier (1979) and Na-K Giggenbach (1988) by Atomic Absorption Spectroscopy method. Finally, the potential in the geothermal manifestation of *Ie Seu'um* was estimated approximately 50-100 MW (medium enthalpy).

References

[1] M. P. Hochstein and S. Sudarman, "History of geothermal exploration in Indonesia from 1970 to 2000," *Geothermics*, vol. 37, no. 3, pp. 220–266, 2008.
[2] Nasruddin *et al.*, "Potential of geothermal energy for electricity generation in Indonesia: A review," *Renew. Sustain. Energy Rev.*, vol. 53, pp. 733–740, 2016.
[3] R. Bertani, "Geothermal Power Generation in the World 2010 – 2014 Update Report," *Proc. World Geotherm. Congr. 19-25 April 2015*, no. 19–25 April 2015, pp. 1–19, 2015.
[4] M. O. Awaleh *et al.*, "The geothermal resources of the Republic of Djibouti - I: Hydrogeochemistry of the Obock coastal hot springs," *J. Geochemical Explor.*, vol. 152, pp. 54–66, 2015.
[5] D. Cinti *et al.*, "Fluid geochemistry and geothermometry in the western sector of the Sabatini Volcanic District and the Tolfa Mountains (Central Italy)," *Chem. Geol.*, vol. 284, no. 1–2, pp. 160–181, 2011.
[6] T. Özen, A. Bülbül, and G. Tarcan, "Reservoir and hydrogeochemical characterizations of geothermal fields in Salihli, Turkey," *J. Asian Earth Sci.*, vol. 60, pp. 1–17, 2012.
[7] P. Wang, X. Chen, L. Shen, K. Wu, M. Huang, and Q. Xiao, "Geochemical features of the geothermal fluids from the Mapamyum non-volcanic geothermal system (Western Tibet,

China)," *J. Volcanol. Geotherm. Res.*, vol. 320, pp. 29–39, 2016.
[8] Y. Rafighdoust, Y. Eckstein, R. M. Harami, M. H. M. Gharaie, E. M. Griffith, and A. Mahboubi, "Isotopic analysis, hydrogeochemistry and geothermometry of Tang-Bijar oilfield springs, Zagros region, Iran," *Geothermics*, vol. 55, pp. 24–30, 2015.
[9] E. P. Joseph, N. Fournier, J. M. Lindsay, R. Robertson, and D. M. Beckles, "Chemical and isotopic characteristics of geothermal fluids from Sulphur Springs, Saint Lucia," *J. Volcanol. Geotherm. Res.*, vol. 254, pp. 23–36, 2013.
[10] O. Benavente *et al.*, "Chemical and isotopic features of cold and thermal fluids discharged in the Southern Volcanic Zone between 32.5°S and 36°S: Insights into the physical and chemical processes controlling fluid geochemistry in geothermal systems of Central Chile," *Chem. Geol.*, vol. 420, pp. 97–113, 2016.
[11] B. Capaccioni, F. Tassi, A. Renzulli, O. Vaselli, M. Menichetti, and S. Inguaggiato, "Geochemistry of thermal fluids in NW Honduras: New perspectives for exploitation of geothermal areas in the southern Sula graben," *J. Volcanol. Geotherm. Res.*, vol. 280, pp. 40–52, 2014.
[12] Q. Guo, M. Liu, J. Li, X. Zhang, W. Guo, and Y. Wang, "Fluid geochemical constraints on the heat source and reservoir temperature of the Banglazhang hydrothermal system, Yunnan-Tibet Geothermal Province, China," *J. Geochemical Explor.*, vol. 172, pp. 109–119, 2017.
[13] Kasbani, "Tipe Sistem Panas Bumi di Indonesia dan Estimasi Potensi Energinya," *Kelompok Progr. Penelian Panas Bumi, PMG –Badan Geol.*, pp. 64–73, 2009.
[14] H. Baioumy, M. Nawawi, K. Wagner, and M. H. Arifin, "Geochemistry and geothermometry of non-volcanic hot springs in West Malaysia," *J. Volcanol. Geotherm. Res.*, vol. 290, pp. 12–22, 2015.
[15] T. Powell and W. Cumming, "Spreadsheets for Geothermal Water and Gas Geochemistry," *Thirty-Fifth Work. Geotherm. Reserv. Eng.*, vol. February, no. 1–3, p. 10, 2010.

Oil palm biomass utilization as an energy source and its possibility use for polygeneration scenarios in Langsa City, Aceh Province, Indonesia

M R Hani[1], M Mahidin[1,2,4], H Husin[1,2], H Hamdani[1,3] and K Khairil[1,3]

[1] Doctoral Program in Engineering, Syiah Kuala University, Banda Aceh 23111, Indonesia
[2] Department of Chemical Engineering, Syiah Kuala University, Banda Aceh 23111, Indonesia
[3] Department of Mechanical Engineering, Syiah Kuala University, Banda Aceh 23111, Indonesia

E-mail: mahidin@unsyiah.ac.id

Abstract. This article deals with the discussion on the recent status of oil palm biomass utilization as an energy source and its possibility use for polygeneration system. The discussion focused only on the energy viewpoint. At this point, many projects and research have been developed in order to utilize the oil palm biomass to meet the energy demand of industries and communities, especially in the largest producing countries: Indonesia and Malaysia; and a few in Thailand, Africa, Latin America and Europe. Through the simulation work in the case study, it is evident that the government of Langsa City can fulfill the fresh water to their community and electricity to Langsa Harbor only by using EFB and PKS from one POM with the generated power of 12 MW, while the desalination plant consumes about 7 MW of electricity. If all potency of biomass from all POMs in Aceh Timur and Aceh Tamiang, without the combination of other primary energy sources is used, Langsa City might earn surplus of energy. The use of the oil palm biomass for polygeneration scenarios is possible and feasible from the technical point of view.

1. Introduction

Indonesian oil palm industry (planting and milling) produces about 4.4 million tons of solid waste (biomass) per year (or 32,654 MWe), and Aceh Province produces about 116.5 thousand tons per year. This kind of biomass consists of empty fruit bunches (EFB), mesocarp fiber (MF), palm kernel shells (PKS), oil palm fronds (OPF), oil palm trunks (OPT), oil palm leaves (OPL) and oil palm stone (OPS). This biomass can be used as a source of energy to produce power [1-5] and heat [5-13]. Heat (thermal) and power are the only direct energy products resulted from biomass utilization.

Furthermore, most of the products from the oil palm biomass are categorized as fuel such as bio-oil [14-38], biofuel [39-41], syngas [42-47], hydrogen [48-55], biogas [56-57], biochar [58-63, briquette and pellet (in some cases biomass added as co-fuel) [64-74], etc.

---

[4] To whom any correspondence should be addressed.

On the other hand, even though Indonesia has a lot of renewable energy sources (one of them is oil palm biomass), this country still faces the energy crisis problem until now. From 32,654 MWe of resources, it was only 1,717.1 MWe at a stage of installed capacity for power generation (1,626 MW off grid and 91.1 MW on grid), as was reported by Indonesia Energy Outlook 2016. In Aceh, the biomass-based energy projects can only be found generally in palm oil mills (POMs), and the use of this energy is still limited for their own needs, both for heating (steam) and generating power (electricity). Only a little was addressed to the independent power plant (IPP). Installed capacity for heat and power plants in the POMs is more or less 65 MW, whereas the capacity of IPP to the grid is about 10 MW. The main current energy issues in Indonesia appear due to the fact that fossil energy reserves depletion and limited access to energy, especially in remote and border areas. Another issue is that the investment for the renewable energy development is still high; however, the return rate and profitability are relatively long and uncertain.

This article deals with the discussion on the study of oil palm biomass uses in the last ten years; including some commercial scale implementations and/or pilot projects. The focus of this study is only limited in the scope of energy purpose. At the end of this article, the case study on the electricity generation fueled by the oil palm biomass in Langsa City and the possibility of its use on polygeneration technology implementation are presented.

2. Oil palm biomass utilization

Pattern of the oil palm biomass utilization is almost the same in all producing countries. PKS and MF are mostly served as boiler fuel to produce steam and electricity, while EFB is burned in the incinerator, and some of them are sent to the plantations for mulching. PKS is sometimes used as a conditioner for pathways within plantations [75] and exported to the target countries such as Japan. OPF and OPL (also OPT after replanting) are usually left in the field [76].

2.1. As a fuel in heat and power plants

Almost all POMs have their own heat and power plants with the capacity around 2.0 to 2.5 MW fueled by MF and PKS. The plants are operated as the utility units in POMs and some energy of the POMs are shifted to electricity for the staff dormitories or housing. In addition, as the fuel in power plants, the biomass is also used to generate the steam, which is consumed in processing fresh fruit bunches (FFBs). Nevertheless, only a little amount was addressed to the IPP. Table 1 lists some of the commercial projects on heat and power plants in Indonesia and Malaysia.

Table 1. Oil palm biomass-based heat and power plants in Indonesia and Malaysia.

Fuel	Capacity	Location	Owner
EFB, MF and PKS	14 MWe (cogeneration system)	Sabah, Malaysia	TSH Bio Energy Sdn Bhd
MF and PKS	2 MWe (cogeneration system)	Kedah, Malaysia	Kumpulan Guthrie Bhd
EFB	6 MWe	Perak, Malaysia	Bumibiopower Sdn Bhd
EFB	7 MWe	Sabah, Malaysia	Potensi Gaya Sdn Bhd
EFB	8 MWe	Sabah, Malaysia	Alaff Ekspresi Sdn Bhd
EFB	12 MWe	Johor, Malaysia	Naluri Ventures Sdn Bhd
EFB	11.5 MWe	Sabah, Malaysia	Seguntor Bioenergy Sdn Bhd
EFB	11.5 MWe	Sabah, Malaysia	Kina Biopower Sdn Bhd
PKS	10 MWe	Kalimantan Barat, Indonesia	PT Rezeki Perkasa Sejahtera Lestari
EFB, MF, PKS (and wood chip)	6 MWe	Bangka Barat, Indonesia	Kencana Agri Ltd
EFB, MF, PKS (and wood chip)	6 MWe	Belitung, Indonesia	Kencana Agri Ltd

Fuel	Capacity	Location	Owner
PKS	2 MWe	Kalimantan Tengah, Indonesia	PT Atman
EFB and PKS	2 x 10 MWe	Dumai, Indonesia	Wilmar Group
EFB and PKS	2 x 10 MWe	Kuala Tanjung, Indonesia	Wilmar Group
EFB and PKS	6 MWe	North Sumatra, Indonesia	PT Medan Sugar Industry
PKS	10 MT/hour of sat. steam at 71.4 kg/cm^2	North Sumatra, Indonesia	PT Multimas Nabati Asahan

Meanwhile, the laboratory and pilot scale studies have been conducted widely and intensively in these areas. A few numbers of the studies on this topic are tabulated in the following table 2.

Table 2. Study on oil palm biomass utilization for heat and power generations.

Fuel	Study	Result	Reference
EFB (and POME)	Simulation on integrated small-scale power generation of EFB gasification and POME digestion; gasification temperature: 800, 900 & 1000 °C; flow rate of wet EFB 10 tones/hour.	Total gasification yields: 68.24%, 80.05% & 91.7% at 800, 900 & 1000 °C, respectively; generated powers: 3.05 - 7.46 MW at 800 – 1000 °C; power generation efficiencies: 11.2, 16.3 & 24.6%.	[2]
EFB and MF	Assessment study based on the survey data analysis from 21 POMs in North Sumatera with capacity of 30 to 60 tones FFB/hour.	It was found that generated power in POM with capacity of 30 tons/hour ranges from 20 to 25 MW.	[4]
EFB, MF, PKS (and biogas)	Simulation study using ECLIPSE software in order to investigate and optimize power and heat generation (case study); the model was validated using the practical data of the CHP plant.	EFB and PKS fuel can generates 2.2 MW power; biogas can replace PKS to produce about 1.3 MW power; the highest power provided by fuel combination of EFB, PKS and biogas of 2.4 MW.	[5]
OPS	Burning of OPS in the pilot-scale fluidized bed combustor; bed material: sand (average size 850 μm); initial bed temperature was 950 °C; fuel rate was 1.2 kg/h.	Bed and bed's surface temperatures decrease when primary airflow rate increases; stable combustion was recorded at 950 °C.	[6]
EFB, MF and PKS	Oxidation was done in Thermo-gravimetric Analysis (TGA) at temperature ranges 25 – 1100 °C.	EFB and PKS have additional peak besides drying, devolatilization and char oxidation during combustion.	[12]
PKS	Combustion of PKS was done in a conical fluidized-bed combustor (FBC) at excess air of 20%, 40%, 60%, and 80%; fuel feed rate is 45 kg/h; superficial air velocity at the air distributor exit was varying from 5.2 to7.7 m/s.	Recommended excess air is 60% to get combustion efficiency up to 98.9%; alumina sand, dolomite, and limestone beds can be used to burn PKS safely in the FBC; Alumina bed lead for K_2O, SiO_2 and CaO formation.	[13]
EFB	Theoretical study for steam power plant by using wet EFB (60% moisture content).	Generated electricity is 418 kWh per tons wet EFB with the overall efficiency 20%.	[77]

2.2. As feedstock in fuel production

In this part, discussions will be put on the gasification, pyrolysis, liquefaction, carbonization, torrefaction, etc. A lot of studies have been established in order to convert the oil palm biomass to be the most useful and valuable fuels in the last ten years as cited in the introduction. Some of them are presented in Table 3 as examples. There is no information about commercial or pilot plants scale on this field so far.

Table 3. Study on oil palm biomass for fuel source.

Feedstock	Study	Result	Reference
PKS	Torrefaction in batch reactor; temperature: 200 – 350 °C; retention time: 10 – 60 min; nitrogen flow rate: 100 – 1000 ml/min.	Maximum torrefied solid (semi-char) mass yielded is 73% found at temperature of 300 °C, retention time of 20 min and nitrogen flow rate of 300 ml/min.	[23]
PKS	Microwave assisted pyrolysis in nitrogen gas that applied the central composite design.	Yield of bio-oil is 32.34% at 45.0 5% carbon loading and 4.9 l/min nitrogen flow rate.	[31]
MF	Torrefaction using nitrogen and air at 300 °C.	Torrefied solid (semi-char) mass yields of 51.7% and 30.2% given by nitrogen and air carrier gases, respectively.	[32]
EFB	Pyrolysis in rotating cone reactor	Gas yield is about 51.4 wt% (mf).	[33]
PKS	Hydrothermal liquefaction using an 8.8 ml Inconel batch reactor placed in a reactor furnace; resident time of 1 h; temperatures: 330, 360 and 390 °C; pressures: 25, 30 and 35 MPa; biomass-to-water ratios: 0.20, 0.35 and 0.50.	Optimum conditions are obtained at 390 °C, 25 MPa and biomass-to-water ratio of 0.20 with bio-oil yield 15.55%.	[35]
MF dan OPF	Pyrolysis was done in a slow-heating (heating rate: 10 °C/min) fixed-bed reactor using N_2 gas at flow rate 100 - 300 ml/min; experimental temperature: 400 – 600 °C; reaction time: 15 min.	Highest bio-oil yield of 50 and 47 wt%, respectively for MF (at 550 °C) and OPF (at 600 °C) appear at N_2 gas flow rate of 200 ml/min; highest biochars for both biomass detect at 400 °C and the same N_2 gas flow rate of 40 wt% and 31 wt%, respectively.	[36]
MF	Thermo-liquefaction at range of temperature: 300 – 500 °C; heating rate: 10 – 30 °C/min; cracking time: 10 – 30 min; solvent: supercritical ethanol.	The greatest yield of bio-oil is 84% recorded at temperature of 500 °C, heating rate of 10 °C/min and cracking time of 10 min; gas yield is 11% and residual solid is 5%.	[37]
EFB	Metal oxide - catalytic super-critical hydrothermal liquefaction in Inconel batch reactor; catalysts content: 1 wt%; temperature: 360 – 450 °C; pressure: 25 MPa; reaction time: 15 – 960 min.	The highest bio-oil yield was observed at temperature of 390 °C, pressure of 25 MPa and reaction time of 60 min, in which relative yield (ratio of yield with catalyst to yield without catalyst) is about 1.4.	[38]
EFB	Supercritical water gasification to produce hydrogen at 380 °C and 240 bar for 8 min.	Reducing the carbon was occurred when loading of biomass increased from 0.05 to 0.5 gram.	[54]
OPF	Non-catalytic and catalytic supercritical water gasification (SCWG) at temperature of 380 °C and pressure about 22.1 Mpa for	Catalytic gasification provides higher yield of H_2 than non-catalytic one.	[55]

Feedstock	Study	Result	Reference
	hydrogen production.		

Figure 1 describes a pathway of the oil palm biomass conversion processes that comprise three main processes: physical, thermo-chemical and biological processes. It can be seen that the only possible way to directly generate the heat and power is through the thermo-chemical process (direct combustion or co-combustion). This is a very simple way to generate heat and power. Additionally, all the commercial heat and power plants listed in Table 1 are run through this way. In differ to the heat and power; all the fuels might be produced by all other processing ways. However, effort on the fuel production is mostly still in the investigation stage, only a few is already in commercial level such as densification and gasification.

3. Possibility use of oil palm biomass for polygeneration scenarios in Aceh Province

The term of polygeneration is usually adopted to identify an energy supply system which simultaneously delivers more than one form of energy to the final user, for example: electricity, heat and cold, and in many cases it also uses one or a combination of primary energy sources such as several fossil fuels (oil, natural gas, coal, etc.) and primary renewable energy sources (solar energy, biomass, biogas, bio-oil, biofuel, syngas, geothermal, wind energy, fuel cell, etc.).

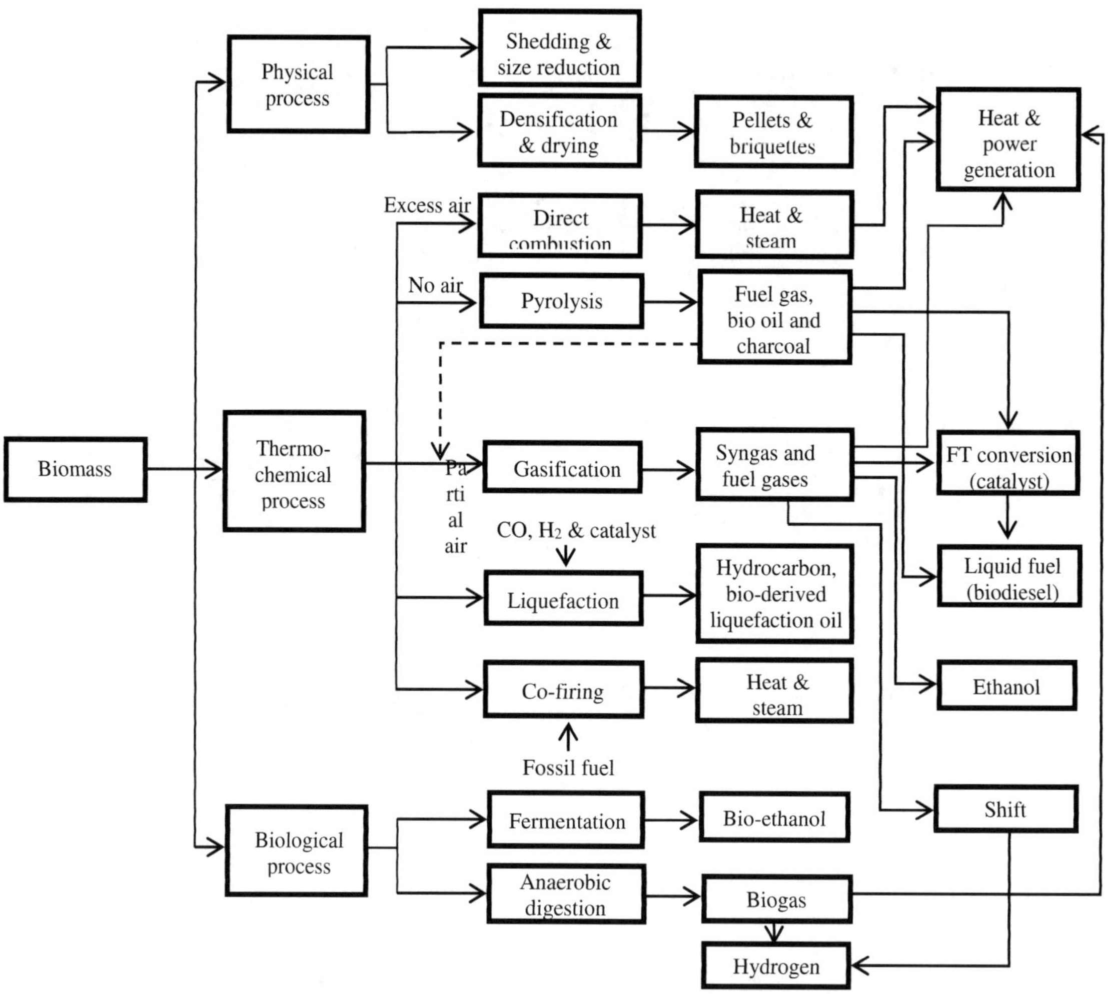

Figure 1. Pathway of oil palm biomass utilization [Source: Ref. 78].

Table 4. Potency of the oil palm biomass in Aceh Tamiang.

Type of oil palm biomass	POM Tj. Seumantoh (capacity 45 ton FFB/hour)	POM Pulo Tiga (capacity 30 ton FFB/hour)	Total (ton/year)
EFB	41,623.00	31,455.00	73,078.00
MF	24,450.00	18,476.00	42,927.00
PKS	11,763.00	8,889.00	20,652.00

Since direct combustion is simple and commonly used to convert biomass to heat and power, this method is applied in this case study. Langsa City has been chosen for the case study area. The oil palm biomass is available in Aceh Timur and Aceh Tamiang districts, a neighborhood of Langsa City. The potency of EFB, MF and PKS in two POMs of PTPN 1 (a state company) operated in Aceh Tamiang is listed in Table 4 and the proximate and ultimate data of the biomass is shown in Table 5. It is clear that the proximate, ultimate and calorific value data for all selected biomass exhibit the similarity. By this reason, in this preliminary study, two of them (EFB and PKS) are chosen and burned in the simulation works to generate the electricity (for the desalination plant and Langsa Harbor) (Figures 2 and 3). So far, a two-feedstock model only gives the reasonable results on the simulation.

Because the primary energy source is only biomass and the energy delivery is only electricity, the polygeneration scenario might not be achieved yet in the case study. The polygeneration scenario (as briefly discussed in part 3.2) will be further examined in the future investigation in detail. Observation in that part is limited to cost of energy (COE) as initial description only.

Table 5. Proximate and ultimate analysis data of EFB, MF and PKS used the study.

Parameter (dry sample)	Type of biomass			Unit	Basis	Reference
	EFB	MF	PKS			
Proximate						
Moisture in air (M)	9.38	9.35	9.76	%	adb	ASTM D.3173
Ash (A)	5.38	3.87	1.19	%	adb	ASTM D.3174
Volatile Matter (VM)	68.47	71.47	69.95	%	adb	ISO D.562
Fixed Carbon (FC)	16.77	15.31	19.10	%	adb	ASTM D.3172
Calorific Value	4,469	4,278	4,515	Cal/gr	adb	ASTM D.5865
Ultimate						
Carbon (C)	46.50	44.97	45.74	%	adb	ASTM D.3178
Hydrogen (H)	7.13	6.99	5.54	%	adb	ASTM D. 3179
Nitrogen (N)	0.89	0.45	0.25	%	adb	ASTM D. 3179
Total Sulphur (S)	0.21	0.14	0.09	%	adb	ASTM D. 3177
Oxygen (O)	39.89	43.58	47.19	%	adb	ASTM D. 3176

3.1. Case study

The area of Langsa City is about 262.4 square kilometers with the population of 165.890 people. Water consumption for people in this city is supplied by the district government company, PDAM Tirta Keumuneng. The company has their own water treatment plant (WTP) to treat the raw water from Alur Gampu River with debit about 160 liter/second. Unfortunately, the company has the capability to serve only approximately 30% of the water needs in the city. It is predicted that about 23 thousand households in Langsa are not connected to the government pipeline system. Most of them consume water directly from well and river. A future challenge that the Langsa City Government will face is how to solve the problems on limitation of water resources. One of the possible raw water resources is seawater. In this article, the simulation was done on the seawater desalination using membrane technology (figure 4).

Figure 2. Langsa City and Kuala Langsa location.

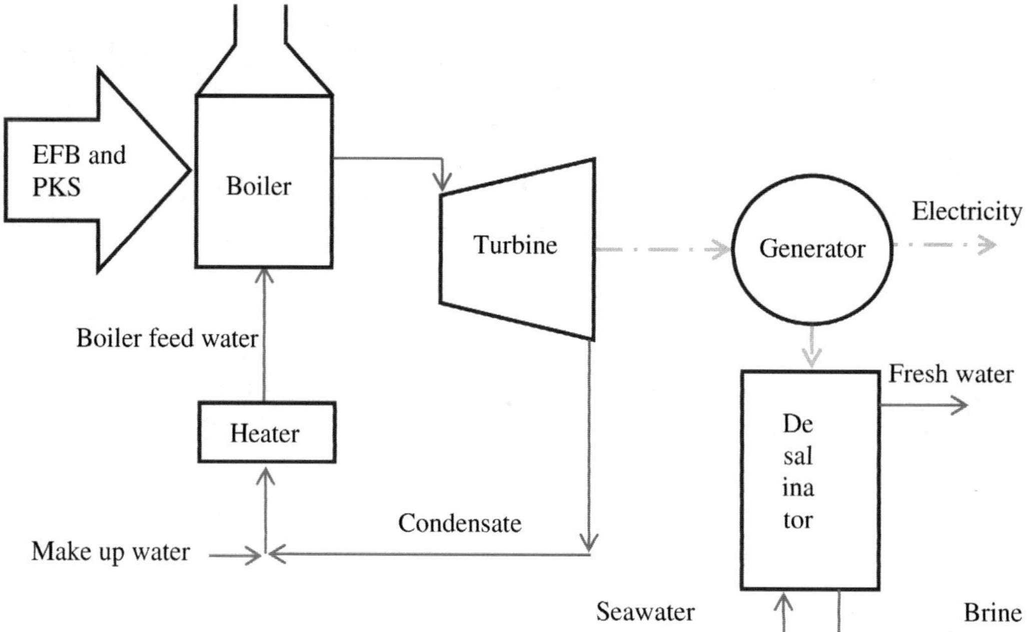

Figure 3. Oil palm biomass utilization for electricity generation and seawater desalination.

Fan et al. [79] said that water consumption for the city household is plus minus 116 m^3 per household per year. Thereafter, to cover the water for 23 thousand households, PDAM Tirta Keumuneng has to produce about 2.7 million m^3 water per year or 7,300 m^3 water per day. In the simulation, the reverse osmosis (RO) system (one type of the membrane technology) is applied to meet the Republic of Indonesia Government Regulation No. 82 year 2001 (read as PPRI No. 82/2001) on the water quality management and pollution control. Figure 4 presents the RO system developed by Oh et al. [80]; the system consists of high-pressure (HP) pump, booster pump, RO unit and energy recovery device. Seawater is pumped to reach a pressure of > 50 bar and is passed through RO unit to produce a fresh water (permeate) and brine. Due to the fact that the rejected brine has a relatively high pressure and temperature, an energy recovery device is attached to the system to recover the energy from the braine to lower the power consumption. The characteristics of seawater for the desalination feed are tabulated in Table 6.

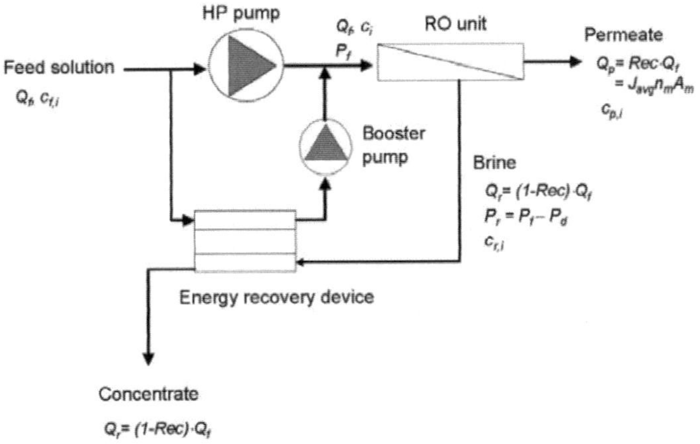

Figure 4. Schematic diagram of RO desalination process.

Table 6. Mineral content in seawater for desalination feed.

Constituent	Content (mg/l)	Constituent	Content (mg/l)
Ammonium ($NH4^{+}$ + NH_3)	-	Nitrate (NO_3)	0.50
Potassium (K)	390.00	Chloride (Cl)	19,700.00
Sodium (Na)	10,900.00	Fluoride (F)	1.40
Magnesium (Mg)	1,310.00	Sulphate (SO_4)	2,740.00
Calcium (Ca)	410.00	Silica (SiO_2)	0.05
Strontium (Sr)	13.00	Boron (B)	3.44
Barium (Ba)	0.05	TDS	35,661.50
Carbonate (CO_3)	24.88	pH	8.10
Bicarbonate (HCO_3)	152.00		

Source: Government Fishery Agency, Langsa City (2016)

At the first step of the simulation task, the performance of the desalination unit and its energy need is predicted by using ROSA 9.0 software. The second step is the optimization of the energy generating system based on the oil palm biomass firing both for the energy in desalination process and electricity in Langsa Harbor using System Advisor Model (SAM) software. The desalination system is analyzed to meet the fresh water product with RO permeate silt density index (SDI) < 1 and type of membrane module of SW30XHR-440i. The specifications of feed are shown in Table 7.

The results of this simulation are presented in the below part of Table 7. It is obvious that the permeate flow reaches a point of ≅300 m^3/h or 7,300 m^3 per day as expected by PDAM Tirta Keumuneng. In addition, the permeate TDS is 40.02 mg/l, lower than the limit issued in PPRI No. 82 (50 mg/l for water class I and II). Furthermore, it is found that the energy consumption for the desalination plant is 6,960.83 kW or ≅7 MW, with the specific energy consumption of 23.2 kWh/m^3 fresh water.

The final step (the second step) is the simulation for power plant in the SAM simulator. The simulation starts by feedstock specification, including mass flow rate; moisture, carbon, hydrogen and nitrogen contents both for feedstock 1 (EFB) and feedstock 2 (PKS). Calculation of higher heating value (HHV) proceeds based on the input data. Finally, combustion process is conducted in the Fluidized Bed Boiler (FBB) and continued with the optimization as well.

The results show that the FBB can generate a superheated steam of 600 psig (40.82 bar) and 750 °F

(398.9 °C) with the boiler efficiency of 84.2%. The output power is 19,644 kW (19.6 MW). Because the desalination plant consumes the energy of 7.0 MW, the remaining of 12.6 MW is then transmitted to Langsa Harbor in Kuala Langsa. It is such an interesting fact that, the number of energy is already enough to run the desalination plant and to supply 12.6 MW to the harbor by considering only two of biomass sources.

Table 7. Simulation result of RO desalination process.

Parameter	Value/type	Unit
Feed specification		
Flow rate	2000	m³/h
Flow factor	0.85	-
TDS	35,661.50	mg/l
Type of membrane module	SW30XHR-440i	-
Result		
Feed pressure	100.22	bar
Total active area	8,502.21	M²
Number of module	208	#PV
Permeate flow	299.98	m³/h
Recovery	15	%
Permeated TDS	40.02	mg/l
Power	6,960.83	kW
Specific energy	23.20	kWh/m³

3.2. Polygeneration scenarios

In order to achieve the Indonesian government goal in energy mix by 2025 share of renewable energy (bio-oil, biofuel, geothermal, biomass, wind energy, solar cell, hydropower, etc.) is expected 17%. Currently, the utilization of the energy sources (both renewable and non-renewable) as a fuel is only achieved through the single processing or co-processing system, such as combustion or co-combustion, gasification or co-gasification, etc. Generally, these processing systems have low efficiency. To improve the disadvantages of those technologies, energy hybrid is one of the interesting topics to be paid attention from the research institutions and energy industries. Polygeneration (hybrid of energy sources as a part of technology) is known as an energy-efficient technology, which plays a key role in the sustainable energy system (SES). The additional target from the polygeneration technology is eliminating/reducing CO_2 emissions [81-82]. Therefore, in the future work, the primary energy source will be combined with wind energy and solar cell. Moreover, the source of biomass itself will also be extended to four kinds (EFB, MF, PKS and OPF). The following Figures 5 and 6 describe the polygeneration scenarios for the next investigation on the biomass hybrid with solar cell and biomass hybrid with wind energy and solar cell, respectively.

Sahoo et al. [83] have developed an innovative polygeneration process in hybrid solar and biomass system for combined power, cooling and desalination. They did the thermodynamics evaluation (energy and exergy) and optimization by a simulation. Thereafter, Edwin and Sekhar [84] have also simulated a combination of biomass, biogas and solar energies to run the milk chilling units in remote villages. Finally, hybrid of biomass and wind energy to produce power, heat, cold and synthetic natural gas have been attempted in Aspen Plus, a commercial simulator, by Rudra et al. [85]. Different from the past studies, our simulation on the polygeneration scenarios is pointed in the COE, as one of parameter consideration for the future examination.

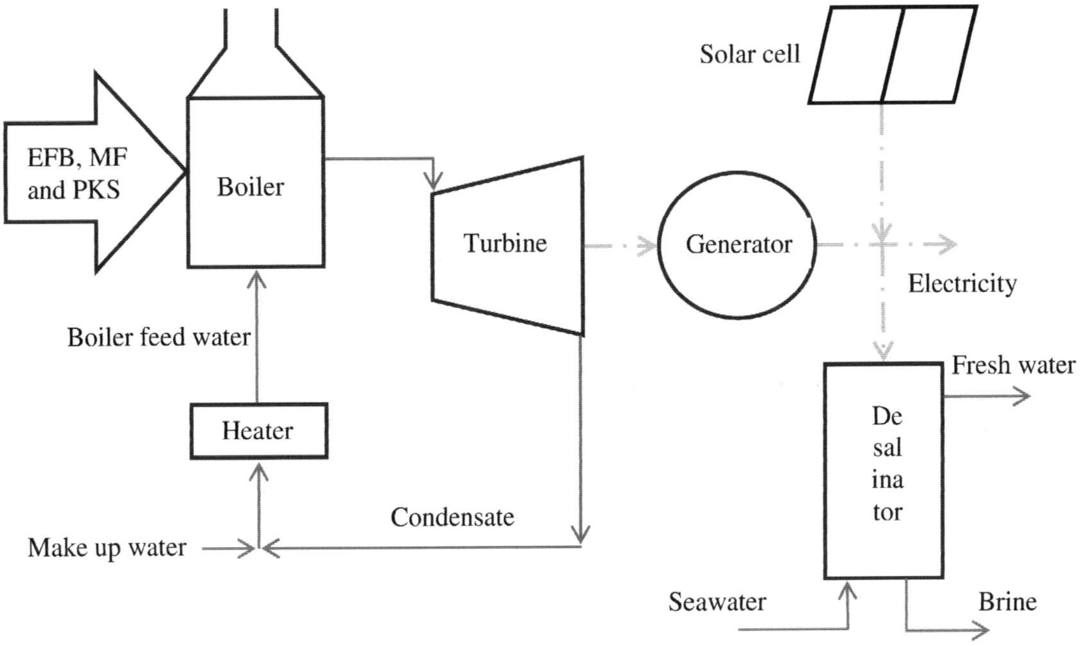

Figure 5. Scenario-1: a combination of primary energy sources, biomass hybrid with solar cell.

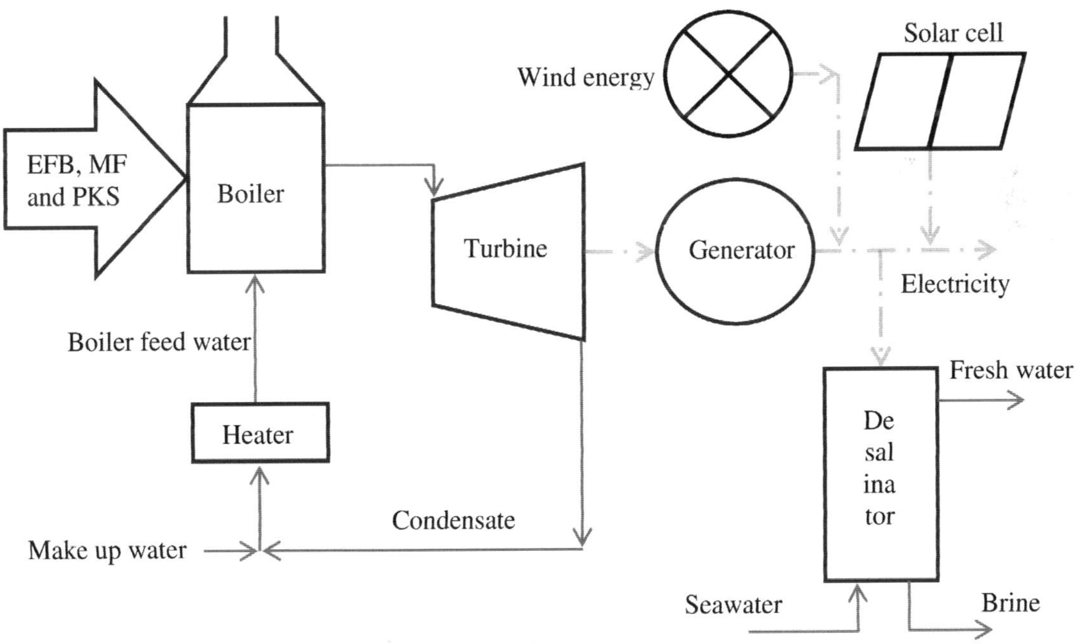

Figure 6. Scenario-2: a combination of primary energy sources, biomass hybrid with wind energy and solar cell.

For COE estimation process, the work starts with the prediction of daily electricity consumption in the Langsa Harbor. It is obtained that total power consumption is 3,172 kW daily, in which the daily electricity used is 1,600 kWh. Afterwards, it was recorded by the Surface Meteorology and Solar

Energy-NASA that the monthly average solar radiation ranges from 3.82 kWh/m^2/day and 5.22 kWh/m^2/day, and the annual average solar radiation is 4.55 kWh/m^2 /day. Meanwhile, the annual average wind speed at 50 m above the earth surface of Langsa is about 2.95 m/s. The lowest wind speed and the highest are 2.13 m/s and 3.83 m/s, respectively.

The configuration of the hybrid system in HOMER software (the type of software applied in the study) consists of PV arrays, wind turbines, biomass gasifier, battery banks and inverters. The economic data inputs in HOMER include capital, maintenance and replacement costs of components. The project life fixes for 25 years and the annual real interest rate is 6%.

As a result, HOMER shows as the top ranked system configurations according to total net present cost (NPC) and HOMER defines the enhanced cost of energy (COE) as the average cost per kWh of the useful electrical energy produced by the system. Two best scenarios resulted from the simulations are showed in Table 8.

Table 8. Simulation result on hybrid system.

	PV (kW)	Wind (kW)	Generator Biomass (kW)	Battery H3000	Converter (kW)	Total NPC ($)	COE ($/kW)
Scenario-1	240	-	200	119	99	3,818,947	1,320
Scenario-2	150	250	200	115	122	3,684,119	1,086

1. Hybrid PV-biomass power system composed of 240 kW PV panels, 200 kW biomass generator, 119 battery banks and 99 kW converter.
2. Hybrid PV-wind-biomass power system composed of 150 kW PV panels, a 250 kW wind turbine, 200 kW biomass generator, 115 battery banks and 122 kW converter.

References

[1] Bazmi A A, Zahedi G and Hashim H 2011 *Renew. Sustain. Energy Rev.* **15** 574–83
[2] Aziz M, Kurniawan T, Oda T and Kashiwagi T 2017 *Appl. Thermal Eng.* **114** 1378–86
[3] Rahman N A, Atan M F, Low C M, Baini R and Mat Nfc Salleh S F 2013 *Eur. Int. J. Sci. Technol.* **2** 111–18
[4] Nasution M A, Herawan T and Rivani M 2014 *Energy Proc.* **47** 166–72
[5] Wu Q, Qiang T C, Zeng G, Zhang H, Huang Y and Wang Y 2017 *Int. J. Hydrogen Energy* ***XXX*** 1–7 (article in press)
[6] Razuan R, Chen Q, Finney K N, Russell N V, Sharifi V N and Swithenbank J 2011 *Fuel Process. Technol.* **92** 2219–25
[7] Uemura Y, Saadon S, Osman N, Mansor N and Tanoue K I 2015 *Fuel* **144** 171–179
[8] Uemura Y, Omar W, Othman N A, Yusup S and Tsutsui T 2013 *Fuel* **103** 156–60
[9] Uemura Y, Omar W N, Tsutsui T and Yusup S B 2011 *Fuel* **90** 2585–91
[10] Aziz M, Prawisudha P, Prabowo B and Budiman BA 2015 *Appl. Energy* **139** 188–95
[11] Guangul F M, Sulaiman S A and Ramli A 2012 *Biores. Technol.* **126** 224–32
[12] Idris SS, Rahman NA and Ismail K 2012 *Biores. Technol.* **123** 581–91
[13] Ninduangdee P and Kuprianov VI 2015 *Biores. Technol.* **182** 272–81
[14] Mazaheri H, Lee KT, Bhatia S and Mohamed A R 2010 *Biores. Technol.* **101** 7641–47
[15] Mazaheri H, Lee K T, Bhatia S and Mohamed A R 2010 *Biores. Technol.* **101** 745–51
[16] Mazaheri H, Lee KT, Bhatia S and Mohamed A R 2010 *Biores. Technol.* **101** 9335–41
[17] Abnisa F, Daud WMAW, Husin WNW and Sahu JN 2011 *Biomass Bioenergy* **35** 1863–72
[18] Abnisa F, Arami-Niya A, Daud W W, Sahu J and Noor I 2013 *Energy Convers Manage.* **76** 1073–82
[19] Salema A A and Ani F N 2011 *Biores. Technol.* **102** 3388–95

[20] Salema A A and Ani F N 2012 *J. Anal. Appl. Pyrolysis* **96** 162–72
[21] Aziz M A, Sabil K M, Uemura Y and Ismail L 2012 *J. Appl. Sci.* **12** 1130–35
[22] Asadullah M, Ab Rasid N S, Kadir Sas A and Azdarpour A 2013 *Biomass Bioenerg.* **59** 316–24
[23] Asadullah M, Adi A M, Suhada N, Malek N H, Saringat M I and Azdarpour A 2014 *Energy Convers. Manage.* **88** 1086–93
[24] Ja'afar M M, Rahman A A and Shamsuddin A 2013 *IOP Conf. Ser. Earth Environ. Sci.* 16
[25] Sabil K M, Aziz M A, Lal B and Uemura Y 2013a *Biomass Bioenerg.* **56** 351–60
[26] Sabil K M, Aziz M A, Lal B and Uemura Y 2013b *Appl. Energy* **111** 821–26
[27] Sing C Y and Aris M S 2013 *Asian J. Sci. Res.* **6** 537–45
[28] Kim S W, Koo B S and Lee D H 2014 *Biores. Technol.* **167** 425–32
[29] Lim, Xiaoyien, Sanna A and Andrésen J M 2014 *Fuel* **119** 259–65
[30] Saadon S, Uemura Y and Mansor N 2014 *Proc. Chem.* **9** 194–201
[31] Mushtaq, Faisal, Abdullah T A T, Mat R and Ani F N 2015 *Biores. Technol.* **190** 442–50
[32] Lu K-M, Lee W-J, Chen W-H, Liu S-H and Lin T-C 2012 *Biores. Technol.* **123** 98–105
[33] Venderbosch Brh, Gansekoele E, Florijn J F and Assink D 2016 *PyNe Newsl Issue* 19 n.d. 2–4
[34] Sulaiman M H, Uemura Y and Azizan M T 2016 *Proc. Eng.* **148** 573–79
[35] Chan Y H, Quitain A T, Yusup S, Uemura Y, Sasaki M and Kida T 2017 *J. The Energy Inst.* 1-12
[36] Kabir G, Mohd Din A T and Hameed B H 2017 *Biores. Technol.* **241** 563–72
[37] Oliveira, A L P C, Almeida P S, Campos M C V, Franceschi E, Dariva C and Borges G R 2017 *Biores. Technol.* **230** 1–7
[38] Yim S C, Quitain A T, Yusup S, Sasaki M, Uemura Y and Kida T 2017 *The J. Supercritical Fluids* **120** 384–94
[39] Abdullah N, Gerhauser H and Sulaiman F 2010 *Fuel* **89** 2166–2169
[40] Chin K L, Hng P S, Go W Z, Wong W Z, Lim T W and Maminski M 2013a. *Ind. Crops Prod.* **49** 768–74
[41] Matali S, Rahman N A, Idris S S, Yaacob N and Alias A B 2016 *Proc. Eng.* **148** 671–78
[42] Lahijani P and Zainal Z A 2011 *Biores. Technol.* **102**(2) 2068–76
[43] Nipattummakul N, Ahmed II N, Kerdsuwan S and Gupta A K 2012 *Appl. Energy* **92** 778–82
[44] Atnaw S M, Sulaiman SA and Yusup S 2013 *Energy* **61** 491–501
[45] Ogi T, Nakanishi M, Fukuda Y and Matsumoto K 2013 *Fuel* **104** 28–35
[46] Guangul F M, Sulaiman S A and Ramli A 2014 *Renew. Energy* **72** 271–83
[47] Moghadam R A, Yusup S, Uemura Y, Chin B L F, Lam H L and Al Shoaibi A 2014 *Energy* **75** 40–4
[48] Akhtar J, Kuang SK and Amin N S 2010 *Renew. Energy* **35** 1220–7
[49] Fan S P, Zakaria S, Chia C H, Jamaluddin F, Nabihah S and Liew T K 2011 *Biores. Technol* **102** 3521–6
[50] Mohammed M A A, Salmiaton A, Wan Azlina W A K G, Amran M S M, Fakhru'l-R A and Taufiq-Yap YH 2011 *Renew. Sustain. Energy Rev.* **15**(2) 1258–70
[51] Ismail K, Yarmo M A, Taufiq Yap Y H and Ahmad A 2012 *Int. J. Hydrog. Energy* **3**(7) 3639–44
[52] Taufiq-Yap Y H, Sivasangar S and Salmiaton A 2012 *Energy* **47**(1) 158–65
[53] Khan Z, Yusup S, Ahmad M M and Chin B L F 2014 *Energy Convers. Manage.* **87** 1224–30
[54] Sivasangar, Zainal, Salmiaton and Taufiq-Yap Y 2015 *Fuel* **143** 563–9
[55] Mastuli M S, Kamarulzaman N, Kasim M F and Sivasangar S 2017 *Int. J. Hydrogen Energy* **42**(16) 11215–28
[56] Chin M J, Poh P E, Tey B T, Chan E S and Chin K L 2013b *Renew. Sustain. Energy Rev.* **26**717–26
[57] Chaikitkaew S, Kongjan P and Sompong O-T 2015 *Energy Proc.* **79** 838–44
[58] Sulaiman F and Abdullah N 2011 *Energy* **36** 2352–59

[59] Sulaiman F, Abdullah N, Gerhauser H and Shariff A 2011 *Biomass Bioenerg.* **35** 3775–86
[60] Harsono SS, Grundman P, Lau L H, Hansen A, Mam S and Meyer A A 2013 *Resour. Conserv. Recycl.* **77** 108–15
[61] Sulaiman F, Abdullah N and Rahman A A 2013 *Am. Inst. Phys.* **248** 248–53
[62] Idris J, Shirai Y, Andou Y, Ali A A M, Othman M R, Ibrahim I and Hassan M A 2015 *J. Cleaner Prod.* **89** 257–61
[63] Idris J, Shirai Y, Andou Y, Ali A A M, Othman M R, Ibrahim I and Hassan M A 2015 *J. Cleaner Prod.* **104** 475–79
[64] Chaiyaomporn K and Chavalparit O 2010 *Environ. Asia* **3** 103–10
[65] Nasrin AB, Ma Ah, Ng An, May C Y, Mohd B W, Li J and Sew M 2010 *MPOB Inf. Ser.* **456** 3–6
[66] Nasrin A B, Choo Y M, Lim W S, Joseph L, Michael S and Rohaya M H 2011a. *J. Eng. Appl. Sci.* **6** 446–51
[67] Nasrin A B, Ravi N, Lim W S, Choo Y M and Fadzil AM 2011b. *J. Eng. Appl. Sci.* **6** 433–439
[68] Jaafar A and Ahmad M 2011 *World Acad Sci. Eng. Technol.* **60** 554–57
[69] Forero N C A, Jochum J and Forero V F E 2012 *Biotechnol.* **8** 67–76
[70] Kavalek M, Havrland B and Pecen J 2012 *Eng. Rural Dev.* **24** 468–74
[71] Lai Z Y, Chua H B and Goh S M 2013 *J. Mater. Sci.* **48** 1448–56
[72] Na B-I, Kim Y-H, Lim W-S, Lee S-M, Lee H-W and Lee J-W 2013 *Biomass Bioener.* **52** 159–65
[73] Salomon M, Gómez M F, Erlich C and Martin A 2013 *Biomass Convers. Biorefin.* **3** 213–29
[74] Mahidin, Gani A, Muslim A, Husin H, Hani M R, Syukur M, Hamdani, Khairil and Rizal S 2016 *J. Eng. Technol. Sci.* **48** (4) 465-81
[75] Er A C, Nor A R M and Rostam K 2011 *Am. J. Appl. Sci.* **8** 436–40
[76] Umar M S, Jennings P and Urmee T 2013 *Renew. Energy* **60** 107–15
[77] Wijono A 2014 *Simposium Nasional RAPI XIII* ISSN 1412-9612
[78] Samiran N A, Jaafar M N M, NgSu J H and Chong S L C T 2016 *Renew. Sustain. Energy Rev.* **62** 1047–62
[79] Fan L, Lingtong Gai T, Tong Y and Li R 2017 *J. Cleaner Prod.* **166** 124-33
[80] Oh H J, Hwang T M and Lee S 2009 *Desalination* **238** 128–39
[81] Rong A and Lahdelma R 2016 *Renew. Sustain. Energy Rev.* **53** 363–72
[82] Salkuyeh Y K, Elkamel A, Thé J and Fowler M 2016 *Energy* **113** 861–74
[83] Sahoo U, Kumar R, Pant P C and Chaudhary R 2017 *Applied Thermal Eng.* **120** 560-67
[84] Edwin M and Sekhar S J 2015 *Energy* **91** 842–51
[85] Rudra S, Rosendahl L and Blarke M B 2015 *Energy Convers. Manage.* **106** 1276–85

Natural rubber, a potential alternative source for the synthesis of renewable fuels via Hydrous Pyrolysis

N Ahmad[1,2], S A S Dayana [1], F Abnisa[1,3,4] and W A W D Mohd[1]

[1] Department of Chemical Engineering, Faculty of Engineering, University of Malaya, Kuala Lumpur, Malaysia.
[2] Department of Chemical Engineering, COMSATS Institute of Information Technology, Lahore, Pakistan.
[3] Department of Chemical Engineering, Faculty of Engineering, University of Malikussaleh, Lhokseumawe 24351, Indonesia.

E-mail: faisal.abnisa@gmail.com

Abstract. Natural rubber is a humid agricultural harvest, which mostly contains hydrocarbon cis-1, 4-Poly isoprene. Through depolymerisation technology, the natural rubber can be changed into liquid product, and then it can be subsequently utilized as a fuel or chemical feedstock. This article aims to provide an outlook on the natural rubber and its sources, which are available globally. Numerous depolymerisation processes, which include pyrolysis, gasification, chemical degradation, catalytic cracking and hydrogenation, were introduced in this paper, while the focus of discussion was emphasized on the hydrous pyrolysis process. Many studies have shown that the use of hydrous pyrolysis able to improve the depolymerisation process, e.g. the raw material can be feed without drying, the process can be carried out at lower temperature, only the water is used as the reaction medium, and it is easy to separate the water from oil product. The effect of operating parameters such as temperature, water to rubber mass ratio, reaction time and type of gases on the product yield and composition were reviewed in this paper. In addition, this paper also highlighted the eco-friendly and economic viability of the hydrous pyrolysis process.

1. Introduction

Shortening of fossil fuels and their effects of greenhouse gas emission has become a very famous and leading problem from the last few years, which subsequently convinced the scientists to develop alternative fuels to substitute these fossil fuels. It is reported that the fossil fuels like gas, coal, and petroleum are available up to next 30, 96, and 29 years respectively [1]. The global consumption of natural gas is around 9,413.69 million m^3/day and that of crude oil is estimated around 77.83 million barrel/day [2]. Because of this high consumption and depletion, the prices of fossil fuel are predicted to rise excessively in the next decades. In order to replace the fossil fuels, efforts are being made to invent alternate fuels which are effectual, environmental friendly and economically feasible. The environmental pollution has also become a main issue in the past decades globally due to the excessive use of fossil fuels. The direct combustion of fossil fuels has significantly increased the environmental

[4] To whom any correspondence should be addressed.

IOP Conf. Series: Materials Science and Engineering **334** (2018) 012004 doi:10.1088/1757-899X/334/1/012004

complications due to the release of injurious contaminants such as SOx, NOx, CO_2 and hydrocarbons which consequently caused global warming, acid rain, and ozone depletion. The utilization of fossil fuels is contributing to 62% of the global CO_2 emissions [3]. In this regard, the United States of America and China are the major emitters (approximately 44% of world's emission) of the greenhouse gas emissions from conventional fuels. US Energy Information Administration (EIA) stated that, the U.S. had presently produced about 19% of global CO2 emissions [3]. It is expected that by the year 2030, the CO_2 emission levels will rise to 40 billion Mg/year [4], which will be a very shocking situation. Barbose et al. proposed that to avoid and to combat these environmental issues, the consumption of fossil fuels needs to be minimized and some new renewable fuels would be deployed for their environmental and social benefits [5, 6].

Many alternative energy resources are available worldwide for the substitution of fossil fuels. The user employs fuels to generate mechanical work, power generation, and heat energy, for successive use in daily life. A perfect ideal fuel exhibits good properties such as high hydrocarbon contents, high heating value, low price, less moisture contents, combustion should be controllable, products of combustion shouldn't be harmful, easy and less transportation and less storage cost. Therefore, an appropriate alternative fuel needs to fulfil the criteria of perfect or an ideal fuel. Several other renewable sources of energy like biomass, bio-diesel, solar energy, hydro-power, and wind-power are available as the substitute of the fossil fuels [3].

Natural rubber (NR) is a biomass generally extracted from Hevea rubber tree, guayule plant, russian dandelion and rubber rabbit brush [7]. NR is also a unique available source that has a potential to meet upcoming energy requirements. NR is a vigorous, premeditated and exceptional feedstock used in huge quantities globally. It is essential and vital agricultural crop in tropical Asia and is found plentifully in south East Asian tropical countries and several central African countries. Hevea rubber tree cultivates finely in a warm humid climate [8]. NR is a polymeric compound of Isoprene, i.e. Poly-isoprene $(C_5H_8)n$ with small quantity of impurities such as other organic compounds and moisture [7]. In the year of 1997, the worldwide manufacturing of NR was calculated about 5000 thousand metric tons, which is rising progressively on account of its high demand and utilization [3]. Because of its large availability, the NR has attracted the researchers to utilize it as a valuable raw material for the synthesis of valuable chemical commodities such as bio-fuels, fertilizers, and other chemical derivatives by depolymerisation process.

Depolymerisation is a process that decomposes the polymeric organic material into small molecules or into its corresponding monomers. Thermal decomposition (Liquefaction, Pyrolysis, hydrogenation, gasification), Photo-induced decomposition, Chemical decomposition (solvolysis, hydrolysis, ozonolysis, oxidation), and biological decomposition are very renowned depolymerisation processes. Thermal depolymerisation process (TDP), commonly known as hydrous pyrolysis is a suitable technique to convert NR to liquid fuels [9]. During hydrous pyrolysis, carbonaceous organic materials like plastics, rubbers or biomasses etc. are heated in the presence of water to produce valuable chemical commodities. This process has an advantage over the rest of depolymerisation techniques since it uses water that is found in abundance, low cost, and eco-friendly. Furthermore, the separation of product liquid oil from water is easy and inexpensive [10]. This conversion process has been employed on industrial scale to convert the agricultural organic waste into valued products like fuels, fertilizers, and other chemicals [9].

In the present article, the decomposition of NR via hydrous pyrolysis was reviewed. The influence of operating conditions such as temperature, pressure as well as water to rubber mass ratio, on the product quantity and quality were explained. In addition, some essential information associated to NR statistics and its alternative sources were also presented.

2. Natural rubber and its sources

2.1. Natural rubber

Typically, the rubber tree (Heveabrasiliensis tree), rubber rabbitbrush, sunflower, Guayule plant, Fig tree, Goldenrod, and Russian dandelion etc., are the sources to obtain NR. It is a hydrocarbon and also recognized as cis-1,4 Polyisoprene represented by chemical formula (C_5H_8) n, the molecular weight of 100% cis -1,4 Polyisoprene which ranging from 1 to 2.5e10^6 [7, 8].

Rubber is obtained from the liquefied latex through coagulation technique which involves the addition of formic acid. The coagulum is a soft solid slab that is pressed by series of rollers to expel an excessive amount of water and to increase the surface area. After that, smoke ovens are employed to dry the rubber sheets.

Table 1. The available sources of natural rubber [7].

Rubber Source	Rubber MW (KDa)	Production (Tones/Year)	Rubber Contents (%)	Yield (kg ha^{-1} year^{-1})	Comments
Rubber tree H. Brasiliensis	1,310	9,000,000 (2005)	30 to 50 in latex. 2% of tree dry weight	500 to 3000	Usually ripeness period of rubber tree is 6 years. It can live over 100 years, but it is cut mostly after 30 years because tapping reduces its productivity of latex, rebirth of latex takes a few days depending upon the condition of the tree.
Guayule shrub P. argentatum Gray	1,280	10,000 (1910)	3 to 12 of plant	300 to 1000	2-5year production time and re-generation time is 12 to 18 months.
Russian dandelion T. kok-saghyz	2,180	3000 (1943)	0 to 15 of root	150 to 500	Its plantation is usually done in early spring. Its takes approximately 85 to 95 days to get mature.
Rubber RabbitbrushC. nauseous	585	n.a	less than 7 of plant	n.a	Rubber rabbit brush reaches ripeness in 2 to 4 years. It has a life period of 5 to 20 years. It harvests seed at the age of 2 years or more than that.
Goldenrod S. virgaurea minuta	160-240	n.a	5 to 12 of root dry weight	110 to 155	Low-quality rubber producer. Demonstration Project in 1931.
Sunflower Helianthus sp.	279, 69	Research stage	0.1 to 1 of plant	n.a	It contains low molecular weight rubber.
Fig tree Ficuscarica	190	n.a	4 in latex	n.a	It's R and D is related to biochemistry.
Lettuce Lactucaserriola	1,380	Research stage	1.6 to 2.2 in latex	n.a	It's R and D is related to genetic engineering and characterization.

Table 2. Various reported investigations on hydrous pyrolysis of rubber.

Ref	Materials	Process parameters					Product analysis		Comments
		Type of reactor	Temp (°C)	Press (MPa)	Water as an agent	Reaction time	Quantitative analysis	Qualitative analysis	
[16]	• Rubber tire • Natural rubber	All experiments were conducted in a 180 mL Hastelloy autoclave.	380°C	27.6 MPa	Water at the super-critical condition	0.5–3 h	Oil was obtained as the main pyrolysis product. Gases were not collected. Solid residues were also obtained and identified as carbon black (about 30 wt.%).	• The NMR analysis of the products showed the presence of aromatic, olefinic, and aliphatic groups. • Natural rubber produces homogeneous organic liquid. The carbon content of natural rubber is low (<1 wt.%); consequently, Consequently, no solid is produced. • Tire rubber product fraction includes the aqueous phase; if SCF water is used, organic fraction is present as a free layer or absorbed on carbon black (around 70 wt.%).	In this work, super-critical H_2O and CO_2 were used to depolymerize tire and natural rubber in a controllable manner. Molecular weight analysis of the degraded material indicates that reaction time can be used to control the degree of breakdown; materials in the molecular weight range of 10^{-3}–10^{-4} were obtained. In addition, both material composition and super-critical fluid affect the rate at which the material is depolymerized. Functional analysis of the processed material shows carbonyl and aromatic groups.
[10]	Rubber tire	Hydrous pyrolysis was conducted using a batch reactor. The reactor was heated using a molten salt bath. The temperature was controlled using a thermocouple.	200 – 430°C	0–28 MPa	• Water at a sub-critical and super-critical condition • H_2O/rubber tire mass ratio is 0/3 to 12/3.	20–120 min	• Maximum liquid product yield was 52.7 wt.% at 370°C. • At 390°C, the lowest yield of char was achieved (about 38 wt.%).	Under the optimum conditions, the calorific value of oil, energy recovery efficiency, carbon content, and hydrogen content were reported to be 44.09–45.09 MJ/kg, 62.49%, 86.66 wt.%, and 10.99 wt%, respectively.	Hydrous pyrolysis of rubber tire was performed to produce liquid fuels. The effect of different parameters, such as temperature (200°C to 430°C), corresponding pressure (0 to 28 MPa), water to rubber tire ratio (0/3 to 12/3), time (20 to 120 min), and environment (air, CO_2, CO, H_2, and N_2) was studied. Temperature significantly influenced the yield and characteristics of the end product.
[23]	Rubber tire	Rubber tires and water were placed into a batch reactor and heated with the inlet opening. After water boiling and vapour was used, 30 min was needed to exclude air in the reactor. The reactor was sealed and heated electrically. After residence time, the reaction products were drained when the temperature and pressure reached the requirement.	350 – 550°C	13.9–25 MPa	Water at sub-critical and super-critical conditions	30 min	• Oil, gas, and solid residues were obtained as the main products. • Oil was not collected because of some limitations. • Gas products consist of CO_2, CO, H_2, CH_4, and other low-molecular-weight hydrocarbons. The amounts of CH_4 and H_2 are comparatively higher than the other components of the gas.	• Carbon contents were high in solid residues. The average value of carbon contents was estimated to be higher than 80 wt.%. • The hydrogen content was estimated to be 0.58–4.76 wt.%. • At 500°C and 22.2 MPa, the percentages of CH_4, H_2, and CO_2 were approximately 38%, 38%, and 16%.	Pyrolysis efficiency of used rubber tires and the effect of temperature and pressure on sub-critical/super-critical water were studied. Products were analysed by GS-Mass, FTIR, and elementary analyser.
[17]	Rubber tire	Thermal depolymerization was conducted using a U-shaped stainless steel batch reactor consisting of electrically heated fluidized sand bath, cold water bath, and thermocouple. The working volume of the reactor was 35.2 mL. About 1 g of the rubber sample was used with 0.1 g of catalyst and 6.5 mL of water. The experiment was conducted under different atmospheres, such as air and helium.	400 – 450°C	22.7–31.1 MPa	• Water at super-critical conditions • 1 g of rubber:6.5 mL of water	15–30 min	The maximum oil yield and conversion of rubber tire were reported to be 44 wt.% and 58%, respectively at 400°C in helium atmosphere.	–	Thermal degradation of rubber tire in a batch reactor using super-critical water was performed to investigate the effect of process variables on product yield and conversion. The temperature and gas atmosphere were the most significant factors affecting oil yield and conversion.
[14]	Rubber tire	An autoclave with a capacity of 0.7 L was used to conduct hydrous pyrolysis of rubber tire. About 1 g of the sample was loaded in an autoclave.	330 – 450°C	5.2–23 MPa	Water at a sub-critical and super-critical conditions	5.5 h	The yield of the oil product and solid residues was estimated to be 47.6 and 42.9 wt.%, respectively, at 380°C and 22 MPa.	• Oil product consisting of about 85 wt.% and 10 wt.% of carbon and hydrogen, respectively. • Solid residues consisted of 91.6 and 1.48 wt.% carbon and hydrogen, respectively.	Autoclave as a batch reactor was used to perform the thermal degradation of rubber tire (containing 52.2 wt.% natural rubber) by using super-critical water and n-pentane. Although the magnitude of pressure of super-critical water is higher than the super-critical pentane (critical pressure is 3.37 MPa), water at super-critical condition was as effective as super-critical pentane.
[15]	Rubber tire	Hydrous pyrolysis experiments were conducted using 316 stainless steel vessels as a batch reactor. The reactor was capable of handling 4137 MPa pressure. The volume of the reactor is 400 ± 20μL.	150 – 400°C	–	**Exp Set 1:** • Double distilled water • 30 mg of rubber tire and 480 mg of water **Exp Set 2:** • Double distilled water and oxalic acid • 30 mg of rubber tire and 480 mg of water with 4 mg of oxalic acid	48 h	The product yields obtained in the presence and absence of oxalic acid were reported to be 47.53 wt.%. and 41.95 wt.%, respectively, at 350 °C.	Alkanes, methyl n-alkanoates, plasticizers, n-alkanoic acids, phenols, and polycyclic aromatic hydrocarbons were found to be the major products of hydrous pyrolysis.	This research mainly aimed to study the effect of temperature and different experimental conditions in hydrous pyrolysis on the chemical composition of products from scrap tire.

2.2. Natural rubber production statistics and its sources
In 2015, the worldwide production of NR was approximately twice from the year 2000 to 2014. In 2000, it was estimated 6,811 thousand metric tons and in 2014, it was reported to be 12,070 thousand metric tons [11]. In 2000, the global NR consumption was calculated approximately 7,108 thousand metric tons globally [12], and it was increased to 12,159 thousand metric tons in 2014 [11].
Thailand, Indonesia, and Malaysia are the main manufacturers of NR. In 2000, the NR production in Malaysia was reported to be 927,608 tons. Whereas, in the year 2014 the production of NR was estimated about 668,613 tons [11]. In the year 2014, the utilization of NR in Chain, India, USA, Thailand and Malaysia was valued to be 4760, 1015, 932, 541 and 447 tons respectively [13]. Food and Agriculture organization reported that, the production of NR in Indonesia was improved from 1,792.35 to 3,108 metric tons in 2003 and 2013 respectively [12]. The accessible sources of NR are mentioned in Table 1.

3. Hydrous pyrolysis

Numerous methods, like chemical decomposition, pyrolysis, catalytic cracking, hydrogenation, gasification, and hydrous pyrolysis are available to convert the NR into fuels and beneficial products. Among them, hydrous pyrolysis is the most feasible technique for the conversion of organic material (rubbers, biomasses, plastics etc.) into liquid fuels. It is conducted in the presence of water at elevated temperature and pressure conditions which outcomes in the degradation of long-chain polymeric compounds into smaller chain petroleum based products (monomers) [9]. This process is performed in water at elevated temperature (250 to 400°C) and elevated pressure (4 to 22MPa). This process can also be carried under self-generated pressure. The main advantage of this process is that, it can utilize raw material with high water contents, thus no need for drying. Generally, the products of hydrous pyrolysis are liquid-oil, solid residual, and gases. However, the yield of liquid oil achieved is high as compared to other by-products. Furthermore, the oil produced by hydrous pyrolysis shows similar characteristic as naturally occurring crude oil [10]. Although the hydrous pyrolysis process gives high value of liquid yield, the quality of liquid still depends upon the type of raw material used e.g., the oil obtained from scrap tire has good heating value but it also contains high sulfur.

3.1. Utilization of natural rubber in hydrous pyrolysis
There is not much research available on the conversion of 100 percent pure NR into fuel. However, some literature is available on the conversion of tyre rubber (where tyre consists of 40-52.2 wt. % of rubber [14]) using hydrous pyrolysis. Several investigations have been made on the hydrous pyrolysis of tyre rubber to evaluate the effects of different operating parameters for the synthesis of valuable liquid products [10, 14-17].

There are many parameters that affect in hydrous pyrolysis process such as size of particle, residence time, water to rubber mass ratio, heating rate, and reaction atmosphere. In addition, the quality of raw material also has a significant influence on the final product, however, the temperature is found to be the most effective parameter that has an influence on the product quantity and quality. Studies on the hydrous pyrolysis of rubber are shown in Table 2.

3.1.1. Use of water and its effect. During hydrous pyrolysis of rubber tyre, water is utilized as subcritical and supercritical conditions for the decomposition of variety of materials [10]. When temperature of the water is more than its atmospheric boiling point (100°C) but lower than its critical temperature (374°C) in that case the water is in sub condition. However, when temperature as well as pressure of water is equivalent to or more than 374°C and 22.1MPa respectively, it is called as supercritical condition[18]. Supercritical water is extra reactive than subcritical water as at supercritical condition the water transform its structure and most of the hydrogen-bonding associations are fragmented, thus changing the polar forces of attractions into dipole-dipole attractions, which consequently reduces the dielectric constant of water [19, 20]. Besides water, the degradation of rubber tyre can be carried out using other various compounds. However, water is the most suitable solvent for the decomposition, which is not only eco-friendly but also cost effective and easily accessible.

Several studies have been done to use water at subcritical and supercritical conditions for the depolymerisation of rubber tyre [14, 16, 21-23]. Rubber is decomposed almost completely in super-critical water as compared to the subcritical water and the formation of char is prohibited [23]. The amount of water has an influence on the liquid product yield. Zhang et al. observed that the yield of liquid products rose as the water to rubber tyre mass ratio increased during hydrous pyrolysis of rubber tyre. The liquid products yield was improved from 45.9 to 52.9 wt.% when water/rubber tyre mass ratio was increased from 1:1 to 4:1 at 390°C. Furthermore, the large quantity of water increases the nucleophiles concentration, which subsequently elevates the pressure inside the reactor. This is poor condition for gas formation [10]. Similar studies are shown in Table 2.

3.1.2. Influence of temperature. Several investigations have been made to study the influence of temperature on hydrous pyrolysis products. Zhang et al. performed the hydrous pyrolysis of rubber tyre under a temperature range of 200 to 430°C. It was observed that the temperature had high influence on the product quantity. The liquid yield increased from 13.5 wt.% to 52.7 wt.% due to rising in temperature from 200 to 370°C. A further rise in temperature to 400°C did not show any significant increase in the liquid product yield. However, the decrease in liquid yield product was observed at a temperature above 400°C. This decline in yield was recognised due to the decomposition of oil into light fractions and volatiles [10].

Park and Gloyna conducted the depolymerisation of rubber tyre in an autoclave using supercritical water. The maximum oil yield was achieved at 400°C. However, the oil yield reduced within temperature range of 400°C to 450°C due to the transformation of oil into light compounds and volatiles by thermal depolymerisation. At a temperature of 450°C, the conversion of rubber was 5 to 6% greater than that obtained at 400°C. The highest conversion and oil yield of rubber tyre was reported to be 58% and respectively 44 wt.% at 400°C [17]. Same investigations that have been made on rubber are shown in Table 2.

3.1.3. Influence of gas atmosphere. Several gas atmosphere or environment can be adopted to conduct hydrous pyrolysis of rubber such as He_2, N_2, CO, CO_2, and H_2. However, it was reported that gas atmosphere had very little effect on final product yields [10]. Zhang et al. performed the hydrous pyrolysis under various atmosphere gases such as air, CO_2, CO, H_2 and N_2. The authors observed that gas environment didn't effect or enhanced the liquid yield. The liquid products obtained with CO and H_2 atmosphere showed the high proportion of light fractions as compared to CO_2 or N_2 [10]. Also, Park and Gloyna carried out the hydrous pyrolysis of rubber tyre under air and He_2 atmosphere. It was found that the gas atmosphere was the second best significant operating variable after temperature that affected the liquid yield. However, atmosphere gas showed very little influence on the oil yield. The usage of He_2 as an alternative of air was more adequate for the recovery of oil. The highest oil yield obtained was 44 wt.% at 400°C under He_2 environment (non-reactive atmosphere). The usage of air as an atmosphere gas affects in the oxidation reaction that enhances the volume of discharge gas [17].

3.1.4. Influence of residence time. Various studies also have been made on the effect of residence time on product yield and composition in hydrous pyrolysis. Zhang et al. investigated the influence of residence time from 20 to 120 minutes at the temperature of 390°C. It was observed that at 390°C the residence time of 20 minute was adequate for the decomposition of rubber tyre and the liquid product yield obtained was 40 wt.%. Further increase in reaction time to 60 minute with slightly increase in temperature enhanced the liquid product yield to maximum of 53 wt.% [10]. On the other hand, Chen et al. showed that the residence time could be used to control the extent of depolymerisation, which was subsequently useful to control the molecular mass for the synthesis of useful chemicals. The decrease in molecular mass was observed throughout the time interval of 30 to180 minutes [16]. Similarly, Funazukuri et al. also described, the decrease in molecular mass was observed as the function of time at 380°C and 22 MPa [14]. Zhang et al. found that high residence time led to low N and S contents in the liquid product [10].

3.2. Characteristics of products
The various investigations have been made on the characteristic of products obtained by pyrolysis and hydrous pyrolysis of rubber. Zhou et al. stated that at optimum conditions, HHV of oil achieved was 44.09-45.09 MJ/kg [10]. The HHV achieved was very similar to petroleum obtained from crude oil like diesel (44.8MJ/kg). Zhou et al. reported, the ultimate analysis of oil product obtained at optimum conditions contained C and H contents of 85.96wt% and 11wt% correspondingly [10]. Also, Funazukuri et al. performed the elemental analysis of oil products consisting about 85wt.% and 10wt.% of carbon and hydrogen contents at 380°C and 22MPa [14]. For solid products, Funazukuri et al. reported the elemental analysis of solid residue consisting about 90.9 wt.% and 1.49 wt.% of C and H contents at 380°C and 22MPa [14]. Also, Yi et al. described that the C contents were high in solid residual. The mean value of C contents were estimated greater than 80wt.% [23]. Yi et al. studied that most of the product gas were comprised of CH_4, H_2, CO, CO_2 and hydrocarbons with lower molecular weight. The concentration of CH_4 and H_2 was high in the product gas. The yield of CH_4 increased between the temperature range of 400 to 550°C. At 500°C & 22.2MPa, CH_4, H_4 and CO_2 were approximately 38, 38 and 16%. NOx and SOx were not produced in gas products [23]. Chen et al. conducted the NMR study of the oil products, which showed the presence of aromatics, olefin and aliphatic groups. The reported composition of HC, aromatic rings, and olefins in products were 80.9 wt.%, 18.8 wt.%, and 0.39 wt.% correspondingly at 380°C and 27.6MPa [16]. Similar investigations made on the characteristic of products obtained by the hydrous pyrolysis can be seen in Table 2.

3.3. Environment and economical viability
Hydrous pyrolysis is a feasible and eco-friendly technique for the conversion of waste into useful liquid fuels. Liquid oil products, gas products and solid residue are three major products that obtained in hydrous pyrolysis. In hydrous pyrolysis of rubber tyre, most of the S and N contents are taken by solid residue thus considerably reducing the SOx and NOx emission in gas products [16, 17, 23]. However, further developments are essential for the decrease of N and S contents in the solid residual. Similarly, for liquid products little developments are required for the reduction of N and S contents [10]. Hydrous pyrolysis process of rubber tyre gives high quality and quantity of oil products and this method can play a vital role in the development of energy industry. It has been reported that the hydrous pyrolysis of rubber tyre is a cheap method to transform rubber tyre into fuel [15]. Normally, the pyrolysis of rubber tyre is carried out at elevated temperatures above 400°C, which consequently increases the operational cost of pyrolysis. Therefore, the investigations have been made to optimize this process using H_2O below 400°C. Rushdi et al. have studied the economic feasibility of the hydrous pyrolysis of rubber tyre and it was shown that the yearly profit of $1.19–1.77M could be made by the hydrous pyrolysis of 30MT rubber tyre/day. This process can be made extra profitable by utilizing solid residual and product gases through process upgrading [15].

4. Conclusions
NR is one of the perfect alternative sources to substitute fossil fuels, which can potentially resolve the issues of energy crisis. The review article showed that there were several methods available for the decomposition of NR such as pyrolysis, chemical decomposition, gasification, catalytic cracking, hydrogenation and hydrous pyrolysis. Among them, hydrous pyrolysis of NR was found to be a low cost process because it gave high liquid yield at lower temperature conditions. From literature survey, it was seen that most of the research work have been done on rubber tyre. The highest liquid yield of about 44 wt.% was achieved under optimal conditions, which were shown to be at temperature of 400°C, a H_2O to rubber mass ratio of 2:1, and a residence time of 60 minutes. Additionally, this

technique could utilize a raw material with high H_2O contents without the need for drying. Furthermore, the usage of H_2O as a reactive medium has attained much attention since it makes easy on the separation of H_2O from oil products and much cost-effective as compared to other organic solvents. For environmental concern, it is important to note that during the hydrous pyrolysis of rubber tyre the NOx and SOx were not obtained in the gas product. In addition, the usage of pure NR as a raw material is eco-friendly as pure NR doesn't contain S contents. A very little literature is available on the hydrous pyrolysis of NR so, a lot of works are needed to be performed in the future on the hydrous pyrolysis of NR to evaluate the effect of parameters on product yield and composition. Moreover, the NR is more environmentally friendly as compared to rubber tyre.

Acknowledgments
Authors would like to acknowledge the financial support provided by the University of Malaya through PPP grants (PG074-2016A).

References

[1] Shafiee S and Topal E 2009 When will fossil fuel reserves be diminished? *Energy Policy* **37** 181-9
[2] 2013 The World Factbook 2013-14. (Washington, DC: Central Intelligence Agency)
[3] Ahmad N, Abnisa F and Daud W M A W 2016 Potential use of natural rubber to produce liquid fuels using hydrous pyrolysis - A review *RSC Advances* **6** 68906-21
[4] Fang Z ed 2013 *Biofuels - Economy, Environment and Sustainability*: InTech
[5] Barbose G, Bird L, Heeter J, Flores-Espino F and Wiser R 2015 Costs and benefits of renewables portfolio standards in the United States *Renewable and Sustainable Energy Reviews* **52** 523-33
[6] Abnisa F, Wan Daud W M A, Arami-Niya A, Ali B S and Sahu J N 2014 Recovery of Liquid Fuel from the Aqueous Phase of Pyrolysis Oil Using Catalytic Conversion *Energy & Fuels* **28** 3074-85
[7] Mooibroek H and Cornish K 2000 Alternative sources of natural rubber *Appl Microbiol Biotechnol* **53** 355-65
[8] Rogers T H and Cornish K 2005 *Van Nostrand's Scientific Encyclopedia*: John Wiley & Sons, Inc.
[9] Basu P 2010 *Biomass gasification and pyrolysis : practical design and theory*: Elsevier
[10] Zhang L, Zhou B, Duan P, Wang F and Xu Y 2016 Hydrothermal conversion of scrap tire to liquid fuel *Chemical Engineering Journal* **285** 157-63
[11] 2015 Natural Rubber Statistics. Malaysian Rubber Board)
[12] FAO. 1992 *Commodity Review and Outlook: 1991/92*: Food and Agriculture Organization of the United Nations
[13] 2015 Top consumers of natural rubber worldwide in 2013 and 2014 (in 1,000 metric tons). Statista Inc.
[14] Funazukuri T, Takanashi T and Wakao N 1987 Supercritical extraction of used automotive tire with water *Journal of Chemical Engineering of Japan* **20** 23-7
[15] Rushdi A I, BaZeyad A Y, Al-Awadi A S, Al-Mutlaq K F and Simoneit B R T 2013 Chemical characteristics of oil-like products from hydrous pyrolysis of scrap tires at temperatures from 150 to 400 °C *Fuel* **107** 578-84
[16] Chen D T, Perman C A, Riechert M E and Hoven J 1995 Depolymerization of tire and natural rubber using supercritical fluids *Journal of Hazardous Materials* **44** 53-60
[17] Park S and Gloyna E F 1997 Statistical study of the liquefaction of used rubber tyre in supercritical water *Fuel* **76** 999-1003
[18] Brunner G 2014 *Hydrothermal and Supercritical Water Processes* vol 5: Elsevier Science

[19] Fang Z and Xu C C 2014 *Near-critical and Supercritical Water and Their Applications for Biorefineries*: Springer Netherlands

[20] García J, Pérez O, Fdez-Polanco F and Cocero M Supercritical water as reaction media. Physical properties at supercritical conditions an overview

[21] Leung D Y C and Wang C L 1998 Kinetic study of scrap tyre pyrolysis and combustion *Journal of Analytical and Applied Pyrolysis* **45** 153-69

[22] Park Y, Hool J N, Curtis C W and Roberts C B 2001 Depolymerization of Styrene–Butadiene Copolymer in Near-Critical and Supercritical Water *Industrial & Engineering Chemistry Research* **40** 756-67

[23] Yi B-k, Ma C-y, Chen G-f and Chen S-y 2009 Study on Pyrolysis of Used Tyre in Subcritical and Supercritical Water. In: *Energy and Environment Technology, 2009. ICEET '09. International Conference on,* pp 779-82

IOP Conf. Series: Materials Science and Engineering **334** (2018) 012005 doi:10.1088/1757-899X/334/1/012005

A Preliminary study of deoxygenation of *Calophyllum inophyllum L.* oil for green diesel production

J Prasetyo[1,3], Adiarso[1], S D S Murti[1], S P Senda[1], S M RFDH[2], Y E Prada[2] and E Oktariani[2]

[1] Center of Energy Resources Development Technology and Chemical Industry, BPPT
[2] Polytechnic STMI Jakarta

E-mail: joni.prasetyo@bppt.go.id

Abstract. Biofuel is a solution to reduce the dependence of fossil fuels. Pure Plant Oil (PPO) of *Calophyllum inophyllum L.* is a potential raw material for green diesel through the processes of deoxygenation, hydrotreating, and isomerization. Deoxygenation of the PPO with NiMo / Al_2O_3 catalyst was carried out in a stirred autoclave reactor at a temperature of 300 - 400°C for 3 hours, and the water/PPO ratio was 1:2 and 1:4. The result showed that deoxygenation would work more effective at high temperature as indicated by higher CO and CO_2 resulting from carboxylation and carbonylation. In addition, raising the reaction temperature from 300 to 400°C succeeded in increasing the diesel fraction of C16 - C20 by 33.01% and decreased the fraction of C21-C25 by 2.41%. Increasing water/ppo ratio did not give any significant improvement on green diesel products.

1. Introduction

Energy crises especially fuel occurs in the world in last recent decades and has significant impact on the hiking fuel price, including in Indonesia. Indonesia which is not only in the list of crude fossil fuel exporting countries but also in the list of importing countries. Therefore, the price of the crude fuel currently affects further to domestic market.

Alternative fuel such as bio-fuel that treated from renewable resources is expected to reduce dependence on fossil fuel. Pure Plant Oil (PPO) that purified of crude plant oil is one of renewable bio-fuels [1]. PPO of *Calophyllum inophyllum L.*'s (PCO) derivative such as green diesel can replace fossil diesel. The green diesel can be achieved through a series processes of PCO through deoxygenating, hydrotreating and isomerization [2].

Indonesia as a tropical country is a heaven for plants for growing in. Almost all plants can grow in Indonesia. Biomass resources are scattered throughout the mainland of Indonesia but the utilization is not maximized yet. One potential plant that produces oil is *Calophyllum inophyllum L. Calophyllum inophyllum L.*oil is suitable as resource for green diesel because *Calophyllum inophyllum L.* oil is categorized as non-edible oil. Oil content in *Calophyllum inophyllum L.* seed is quite high at level 50%-73% [3]. In addition, *Calophyllum inophyllum L.* can grow in marginal areas. The extent of marginal area that can be utilized in Indonesia currently reaches 255,350 ha. The area is spread wide

[3] To whom any correspondence should be addressed.

in West Sumatra, in Riau, in Jambi, in South Sumatra, in Lampung, in Java, in West Kalimantan, in Central Kalimantan, in Sulawesi, in Maluku and in Nusa Tenggara Timur [4].

In this study of deoxygenation of PCO as the first step of upgrading PCO process was done to produce green diesel. Green diesel is known as the second generation of bio-diesel that has very similar characteristic with diesel. Therefore, green diesel can replace fossil-diesel rather than partially blended like biodiesel (Fatty Acid Methyl Esther)[2]. Deoxygenation is an effective process in order to economization H_2 need to produce green diesel that usually through two processes HydroDeOxygenation (HDO) and isomerization of PCO. As is in HDO, deoxygenation is separation process of HDO with hydrotreating. Two reactions that expected to occur in deoxygenation as well as in HDO are decarboxylation and decarbonylation. This work studied the effect of temperatures and water-oil ratios in conversion of PCO to green diesel.

2. Materials and Methods

2.1 Materials

2.1.1 Crude of Calophyllum inophyllum L. oil, was gotten from Cilacap regency. This crude oil was purified so that the FFA content was less than 1%, namely PCO. The characteristic of Crude of *Calophyllum inophyllum L.* oil is shown on table 1.

Tabel 1. Characteristic of *Calophyllum inophyllum L***Oil**[5].

Characterization	Annotation
Color	Dark green
Viscosity	High viscosity
Iodinenumber(mg iod/g minyak)	100 – 115
Density at 20°C (g/cm3)	0,920 - 0,940
Refraction Index	1,4750 - 1,4820
PeroxideNumber(meq/kg)	<20,0
Lipid fraction	98 – 99,5 %
Fatty Acid (%)	
• Palmitic Acid (C16 : 0)	15 – 17 %
• Palmitoleic Acid (C16 : 1)	0,5 – 1 %
• Stearic Acid (C18 : 0)	8 – 16 %
• Oleic Acid (C18 : 1)	30 – 50 %
• LinoleicAcid(C18 : 2)	25 – 40 %
• ArachidAcid(C20 : 0)	0,5 – 1 %
• GadoleicAcid(C20 : 1)	0,5 – 1 %

Unfortunately, the PCO still had high viscosity as its crude. Analyzing of PCO for total C, H, N, and S were 59.72, 8.65, 8.27% and 60 ppm, repectively. By this content, calorie content was 39,329 J/g.

2.1.2 A commercial catalyst, NiMo/γAl₂O₃, was used in this deoxygenation. In general, this NiMo/γAl_2O_3 catalyst has been analyzed and the composition is shown at tabel 2.

IOP Conf. Series: Materials Science and Engineering **334** (2018) 012005 doi:10.1088/1757-899X/334/1/012005

Tabel 2. Specification of commercial NiMo/γAl_2O_3 catalyst.

Compound	% weight
NiO	4,0 - 6,0
MoO_3	2,0 - 24,0
γAl_2O_3	*Balance*

2.2 Methods

2.2.1 Catalyst Preparation. Catalyst preparation was carried out by putting the catalyst on the stirrer that covered by wire mesh in order to keep the catalyst from scattered and eroded. However, the catalyst was still able to have interaction with solution water-PCO. Next is the catalyst reduction to activate the NiMo/γAl_2O_3 catalyst with H_2 (T-01) in condition of 30 bar for 3 hours.

2.2.2 Deoxygenation. After catalyst reduction, reactor's cover is slightly opened to minimize contact with oxygen of air when the mixing of water-PCO was poured into reactor (R-01). The batch reactor was a stainless steel cylindrical shape with a volume of 500 ml. The stirrer was equipped to make the mixture homogeny. Figure 1. shows the batch reactor system. Water-oil ratio was adjusted according experimental design, at ratios 1:2 and 1:4. Deoxygenation was carried out by regulating temperatures at 300°C, 325°C, 350°C, 375°Cand 400°C. Water would be evaporated and reach superheated steam. Prior to each experiment, to minimize oxygen content in the reactor, the rest air inside the reactor was purged using H_2 for 3 times before deoxygenation. The deoxygenation as prescribed in experimental planning was conducted within a 3-hour.

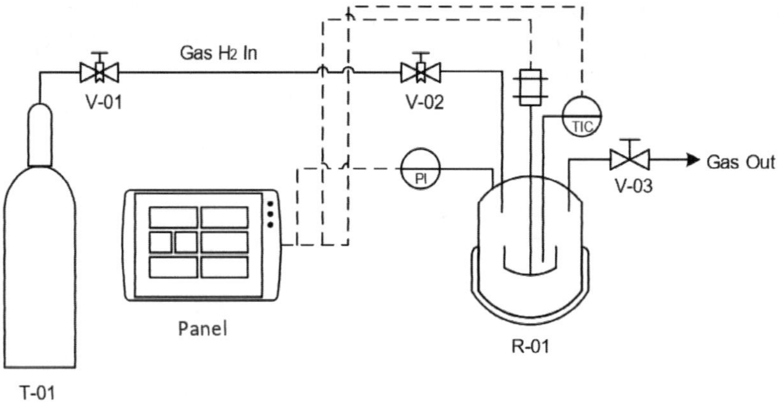

Figure 1. Schematic of Deoxygenation System.

2.2.3 Separation Process. During the deoxygenation process, agitation in the reactor would cause some of the catalyst to be crushed. Therefore, the catalyst particle in the water-PCO solution should be filtered gradually. The last filtering was using filter-vacuum system to remove the rest impurities.

2.2.4 Analysis. Viscosity was measured using the Lauda Viscotemp 18s Viscometer. 10 mL sample was inserted into the capillary tube using a pipette and suck the sample over the second limit (red ribbon mark). The time it took PCO dropped between the first marker and the second marker in the capillary was recorded. Each PCO was measured using different capillaries constants: 0.1199, 0.1198 or 0.1197 for various temperatures.

Gas chromatograph thermal conductivity detector (GC-TCD 8A) Shimadzu 2014 was used to analyze H_2, CO, CO_2 and CH_4. Injection temperature, cooling temperature and final temperature were set to 100°C, 50°C and 50°C, respectively. Gas in sample bag was inserted by pushing the sample bag smoothly for 30 seconds.

Gas Choromatograph Flame Ionization Detector (GC-FID)was used to analyze upgraded PCO with GC-FID SCION 456 appliance. The column was BR-1 with 100% dimethyl polysiloxane as stationary phase. The gases supply of He as mobile phase and H_2 and compressed air was used as make up to support the ignition. Vial that incorporated 0.05 gr sample wasmixed with 1 mL hexane.

Flash Point, the point of flame of the upgraded PCO was measured using flash point AMETEK, Gabrer instruments mini flash FLP touch. Anisole was the standard at flash point 42-44°C and the setting of standard D6450 ASTM 930 was applied.The flash point apparatus has specification range of characteristic from of 50°C to 200°C. Sample was analyzed by putting1 ml sample in a cup and inserting into the apparatus.

3. RESULT AND DISCUSSION

Deoxygenation PCO with superheated steam was be started with the hydrolysis of triglyceride. The hydrolysis would produce FFA [6]. Based on the product analyzing, the discussions will be explained by the gas analysis formed during the deoxygenation and the analysis of changes in structure or composition in PCO.

3.1 Observation at gas phase

Three moles superheated steam and one mole triglyceride would produce three moles of FFA and one mole of glycerol as described at equation 1[7]:

$$C_3H_5(COOR)_3 + 3H_2O \leftrightarrows C_3H_5(OH)_3 + 3RCOOH \quad (1)$$

The presence of NiMo/γAl_2O_3 catalyst would cause further the reaction of decarboxylation and decarbonylation. Once FFA was formed from triglycerides, the FFA would then break down catalytically to straight-chain alkenes and gases CO and CO_2, decarboxylation and carbonylation. At the first pathway was decarboxylation. The product formed was paraffinic hydrocarbon induced by carboxymethyl clumps and releasing carbon dioxide [8]:

$$RCOOH \longrightarrow R + CO_2 \quad (2)$$

The second pathway was decarbonylation. The product formed was olefinic hydrocarbon induced also by the removal of the carboxyl group and releasing carbon monoxide and H_2O [8].

$$RCOOH \longrightarrow R + CO + H_2O \quad (3)$$

Therefore, this work observed the deoxygenation of PCO in batch process over NiMo/γAl_2O_3 through gases formation. The presence of unsaturated FFA like oleic acid and linoleic acid, they would also experience the same process, converted also to saturated diesel range hydrocarbons through hydrolysis, decarboxylation and carbonylation.

In the raising water amount as described by water-PCO ratio, for example from ratio = 1:4 and 1:2, the gases formation would shift to tend CO rather than CO_2, especially at high temperature as shown at figure 2. However, the magnitude of this process shift was not so ssignificant. It can be seen in the amount of CO_2 that remains much higher than CO. Therefore, it should be more paraffinic hydrocarbon rather than olefinic hydrocarbon. The decrease of CO_2 caused by addition of water in decarboxylation was between 13.75% and 23.66%. The sharpest lowering was 7.36% at condition

375°C, from 31.09% to 23.73%. In addition, lowering CO_2 formed at 400°C was from 37.50% to 30.27% [9,10,11,12].

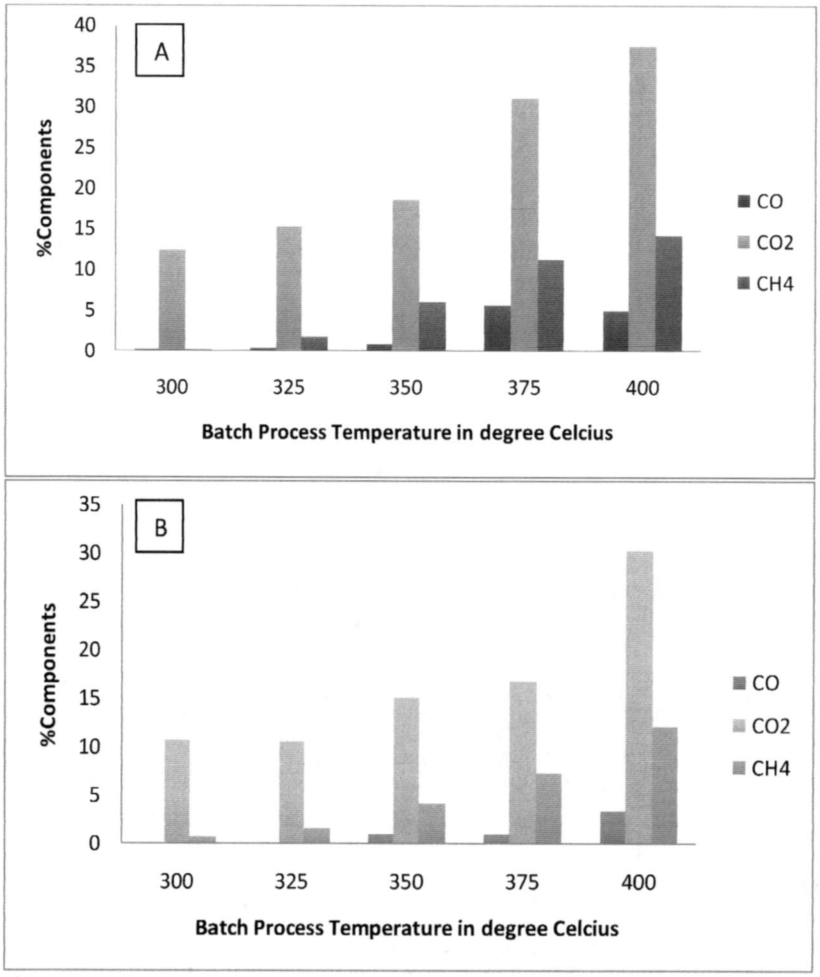

Figure 2. Deoxygenation of PCO with Superheated Steam (A) Water-PCO Ratio = 1:4, (B) Water-PCO Ratio = 1:2.

The composition of effluent gas was not only carbon dioxide and carbon monoxide but also methane [11,12]. Methanation occurred because of the presence of CO and CO_2 that react with steam, shift water reaction. Shift water gas reaction is an equilibrium reaction where H_2 would be produced with the existence of those components as described by equation 4.

$$CO + H_2O \rightleftarrows CO_2 + H_2 \quad (4)$$

The subsequent process due to the formation of CO, CO_2, and H_2 was methanation as described at equation 5 and 6 [13,14,15]. CO and CO_2 would produce CH_4 and H_2O through hydrogenation.

$$CO + 3H_2 \rightleftarrows CH_4 + H_2O \quad (5)$$

$$CO_2 + 4H_2 \rightleftarrows CH_4 + 2H_2O \quad (6)$$

3.2 Observation at liquid phase

Discusion for the upgraded PCO is based on GC-FID analysis. As shown at figure 3, the column cannot detect triglyceride but only detect hydrocarbon. Therefore, there is no peak at original PCO analysis (black line). Upgraded PCO, red line, shows that paraffinic hydrocarbon and olefinic hydrocarbon was detected.

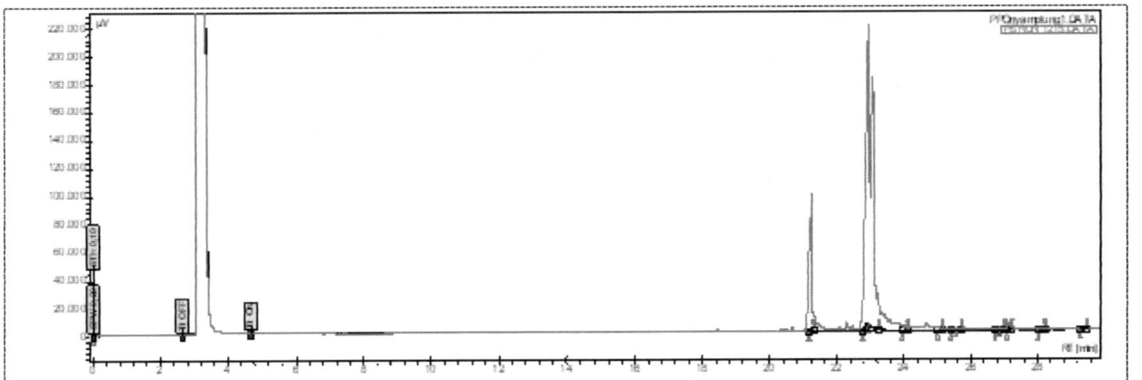

Figure 3. GC-FID analysis of PCO (A) Black Line for Untreated PCO (B) Red Line for Upgraded PCO by deoxygenation using superheated steam.

The presence of paraffinic hydrocarbon and olefinic hydrocarbonwere is the evidence of the carbonylation and carboxylation occurrence in this deoxygenation. Estimation of the paraffinic hydrocarbon and olefinic hydrocarbon formation were calculated based on the peak area of GC FID analysis.

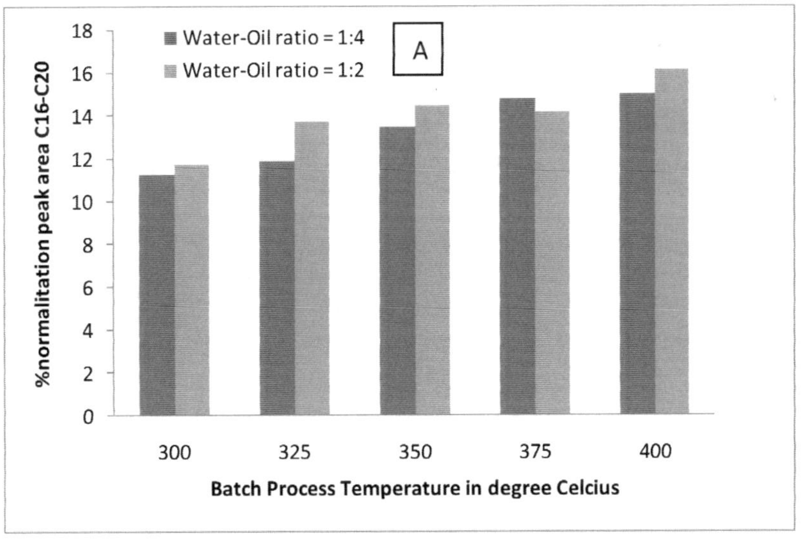

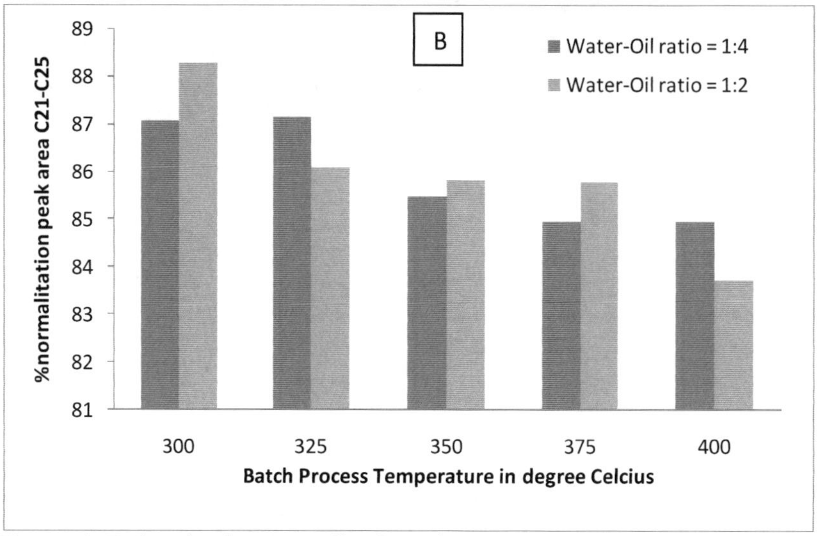

Figure 4. Estimation by Normalization of Conversion to Paraffinic Hydrocarbon and Olefinic Hydrocarbon in Upgraded PCO (A) C16-C20 (B) C21-C25.

As shown at figure 4, raising temperature would improve the convertion to longer hydrocarbon chain. However, water quantity was not significantly influence the hydrocarbon conversion, both C16-C20 yield and C21-C25 yield. C16-C20 yield increased only 3.74%, from 11.33% to15.07% at operating conditions 300°C and 400°C, respectively. Meanwhile, C21-C25 yield decreased only 2.1%, from 87.08% to 84.98% at the same conditions.

However, deoxynated PCO analyzation was not improved significantly. The deoxynated PCO for total C, H, N, and S were 59.67, 8.71, 8.27, 7.91% and 50 ppm, repectively. Calorie content was 39,400 J/g. This calorie content was almost the same as untreated PCO. It might be some water trapped in upgraded PCO. Moreover, the properties of upgraded PCO have changed from its original PCO. Based on the regulation of fuel in Indonesia stated on Decree of Director General of Oil and Gas No. 978.K/10/DJM.S/2013: "Standard and Quality (specification) of fuel: Type Diesel 48 marketed in the country", table 3 shows the comparison of the product deoxygenation with reference standards issued by the government.

Table 3. Properties of Upgraded PCO.

Characteristic	PCO	Upgraded PCO	Decree Oil and Gas No. 978.K/10/DJM.S/2013	
			Minimum	Maximum
Density, kg/m^3	912	892	815	860
Visosity, 40°C in mm^2/S	38.24	21.52	2	4.5
Sulphur content, ppm	<50	<50	-	50
Flash point, °C	>200	>200	52	-

4. Conclusion

Deoxygenation using superheated steam is complex and interconnected reactions. Starting from the hydrolysistri of glyceride to FFA formation, the catalytic reaction of carboxylation and carbonylation, shift water gas reaction and methanation. The properties of deoxygenation product and HDO product was similar. The effluent gases of both process was also almost the same. Therefore, deoxygenation offers efficiency process in H_2 usage.

Raising water-PCO ratio gave increasing C16-C20 yield 33.01% and decreasing C21-C25 yield 2.41% at temperature 400°C. The calorie content reached up to 39.400 Joule/gr because of Oxygen reduction in upgraded PCO. However, this work did not improve the calorie content in PCO yet. Further suggested process to achieve green diesel including improving calorie value is treated with hydrotreating and isomerization.

References

[1] R. Prihandana dan R. Hendroko 2008 *Penebar Swadaya*, Jakarta.
[2] E. Santillan-Jimenez and M. Crocker 2012 *Chemical Technology and Biotechnology* **87(8)** 1041-50.
[3] A.C. Dweek and T. Meadows 2002 *International J. Cos. Sci.* **24** 1-8.
[4] Balai penelitian dan pengembangan hutan 2008 *Forestry Department*.
[5] V.J. Debaut, Y.B. Jean and S.A. Greentech 2005 *Greentech* St. France.
[6] R. Brockmann, G. Demmering, U. Kreutzer, M. Lindemann, J. Plachenka, U. Steinberner 2005 *Fatty acids, Ullmann's encyclopedia of industrial chemistry* Weinheim Germany: Wiley-VCH Verlag GmbH & Co. 1–34.
[7] P.H.L. Moquin and F. Temelli 2008 *Fluids* **45** 94–101.
[8] M. Snåre, I. Kubickova, P. Mäki-Arvela, K. Eränen andD.Y. Murzin 2006 *Ind. Eng Chem Res* **45** 5708–15.
[9] M. Snåre, I. Kubickova, P. Mäki-Arvela, D. Chichova, K. Eränen andD.Y. Murzin 2008 *Fuel* **87** 933–45.
[10] S. Lestari, I. Simakova, A. Tokarev, P. Mäki-Arvela, K. Eränen and D.Y. Murzin 2008 *Catal Lett* **122** 247–51.
[11] P. Mäki-Arvela, I. Kubickova, M. Snåre, K. Eränen and D.Y. Murzin 2007 *Energy Fuels* **21** 30–41.
[12] W.C. Wang, et.al. 2012 *Fuel* **95** 622-29.
[13] A. Srifa, et al. 2014 *Bioresource Technology* **158** 81-90.
[14] W. Kiatkittipong, et al. 2013 *Fuel Processing Technology* **116** 16-26.
[15] A. Guzman, et al. 2010 *Catalysis Today* **156(1-2)** 38-43.

Biodiesel production methods of rubber seed oil: a review

M Ulfah[1,2], **Mulyazmi**[1], **Burmawi**[1], **E Praputri**[1], **E Sundari**[1] **and Firdaus**[1]

[1] Faculty of Tecnology Industrial, Bung Hatta University, Jl. Gajah Mada No. 19 Olo Nanggalo Padang, Sumatera Barat, Indonesia

E-mail: ulfah_nad@yahoo.com

Abstract. The utilization of rubber seed as raw material of biodiesel production is seen highly potential in Indonesia. The availability of rubber seeds in Indonesia is estimated about 5 million tons per annum, which can yield rubber seed oil about 2 million tons per year. Due to the demand of edible oils as a food source is tremendous and the edible oil feedstock costs are far expensive to be used as fuel, production of biodiesel from non-edible oils such as rubber seed is an effective way to overcome all the associated problems with edible oils. Various methods for producing biodiesel from rubber seed oil have been reported. This paper introduces an optimum condition of biodiesel production methods from rubber seed oil. This article was written to be a reference in the selection of methods and the further development of biodiesel production from rubber seed oil. Biodiesel production methods for rubber seed oils has been developed by means of homogeneous catalysts, heterogeneous catalysts, supercritical method, ultrasound, in-situ and enzymatic processes. Production of biodiesel from rubber seed oil using clinker loaded sodium methoxide as catalyst is very interesting to be studied and developed further.

1. Introduction

Recently, production costs between biodiesel and diesel indicate that diesel fuel from fossil sources is lower than biodiesel. An effective way to reduce the cost of biodiesel is to use raw materials that have high free fatty acids (FFAs) or non-edible oils [1] [2]. The price of non edible oil is much lower than edible oil whose demand continues to increase, so the use of non edible oil such as rubber seed oil as a feedstock of biodiesel production is very promising.

Biodiesel from rubber seed oil had tested as a fuel in diesel engines. Rubber seed biodiesel can be used as a partial substitute for diesel fuel. The blending of the biodiesel with diesel fuel can be used to fuel diesel engines providing comparable performance, reduced emissions, and wear reduction of engine components and neutral effects of lubricating oil. Engine diesel modifications are not required [2], [3], [4], [5]. The utilization of rubber seed as raw material of biodiesel production is highly potential in Indonesia. The availability of rubber seed in Indonesia is estimated about 5 million tons per year [6]. The yield of brown colored oil extracted from the rubber seed kernel is about 40-50% [7].

Therefore, Indonesia can yield rubber seed oil about 2 million tons per year. It can be utilized as the main feedstock for biodiesel production. The usage of rubber seed oil as an alternative fuel has been investigated extensively due to the objective of ensuring energy security and reducing the environmental impact of diesel emissions. This article introduces the biodiesel production methods of

[2] To whom any correspondence should be addressed.

rubber seed oil. Aspects written on the method include the ratio of methanol to oil, the amount of catalyst, the type of catalyst, reaction time and reaction temperature. The purpose of this article is to find out the methods that have been developed in the production of biodiesel from rubber seed oil and to choose the method for further development so that the method can be applied on a commercial scale. The structure of the paper as follows: section 2 describes the fatty acid composition of rubber seed oil, section 3 deals with alternative transesterification methods for biodiesel from RSO and, section 4 for the conclusion.

2. Fatty acid composition of rubber seed oil (RSO)

Normally, the rubber seed oil is acquired by disposing of seed. The colour of the oil depends on the pre-extraction of the history of the kernels. It ranges from white to pale yellow for low FFA content of approximately 5% to dark for high FFA content of 10-40% [7]. The composition of fatty acid in the rubber seed oil described in table 1.

Table 1. Fatty acid composition of rubber seed oil (RSO).

Palmitic acid C16:0	Stearic acid C18:0	Oleic acid C18:1	Linoleic acid C18:2	Linoleic acid C18:3	Ref.
10.2	8.7	24.6	39.6	16.3	[8]
0.23	35.85	12.7	52.85	-	[9]
10.102	10.704	25.605	37.054	15.099	[10]
13.11	12.66	39.45	33.12	-	[11]
19.3	3.9	23.7	37.9	15.1	[12]
9.1	5.6	24.0	46.2	14.2	[13]

It is obvious in the table that the rubber seed oil consists of mainly fatty acid with 18 carbon atoms. Meanwhile, petroleum diesel composes of mainly hydrocarbon with carbon chain length of 8 to 10 atoms. Cetane number, which refers to indicator to fuel diesel engine quality, increases along with the increases in carbon number. Branched chains and double bonds improve low temperature, flow properties. So, biodiesel produced by rubber seed oil will have a higher cetane number as its fatty acid contains of 18 carbon atoms [9].

3. Alternative transesterification methods for biodiesel from RSO

There are many methods that have been used recently to produce biodiesel from various non-edible feedstock. These methods are pyrolysis, micro-emulsification, dilution, and transesterification [1]. The most common way to produce biodiesel is transesterification. Transesterification is the chemical reaction between one mol triglycerides and three mol alcohol to produce methyl ester and glycerine in the presence or no of catalyst. Alcohols used in transesterification include methanol, ethanol, propanol, butanol, and amyl alcohol. Methanol and ethanol are used most frequently in both laboratory scale and the biodiesel industry scale. Methanol becomes the first choice for transesterification reaction as its low cost [2].

A literature survey reveals that some transesterification methods available to produce biodiesel from rubber seed oil such as: conventional transesterification (homogeneous catalyst), heterogeneous catalyst, supercritical method (SCM), in-situ process, ultrasound assisted process, and enzymatic process.

3.1. Conventional transesterification methods (homogeneous catalyst)

Rubber seed oil contains high amount of Free Fatty Acid (FFA), in the range 5-40 % [7]. An effective way to produce biodiesel from raw materials of non-edible oil is through two stages [2]. The first step is esterification to reduce the FFA content of the oil. The second step is converting the triglyceride portion of the oils to mono-alkyl ester and glycerol by transesterification [2].

The FFA emerges serious problem on transesterification reaction catalyzed by base catalyst. FFA is able to react with the base catalyst to produce soap. It certainly leads to a loss of catalyst, while the soap produced deactivates the catalyst. Therefore, the FFA content has to be removed or converted to alkyl esters via acid esterification. Conventionally, sulfuric acid catalyst is used in acid esterification and alkali metal hydroxides and alkoxides used as catalyst for transesterification reaction [14]. Condition operation of biodiesel production from rubber seed oil by conventional methods as shown in table 2.

Table 2. Operating condition of biodiesel production from RSO by homogeneous catalyst.

Optimum operating condition	Initial FFA, %	Conversion efficiency, %	Ref
Esterification: 0.5 % (v/w) H_2SO_4 catalyst, 6:1 molar ratio of methanol to RSO, 45 ± 5°C and 30 min.	17	99[a]	[15]
Transesterification: 0.5 % (w/w) NaOH catalyst, 6:1 molar ratio of methanol to RSO , 45 ± 5°C and 30 min.		99[b]	
Esterification: 1 % (v/v) H_2SO_4 catalyst, 0,75 v/v methanol-RSO ratio, 63 ± 2 °C and 60 min.	24	96[a]	[16]
Transesterification: 0.5% w/v KOH catalyst, methanol–RSO ratio of 0.30 v/v, 55±2°C and 40 min .		98-99[b]	
Esterification: 10 % (w/w) H_2SO_4 catalyst, 15:1 molar ratio of methanol to RSO, 45°C and 90 min.	42	98[a]	[17]
Transesterification: 1 % (w/w) KOH catalyst, 6:1 molar ratio of methanol to RSO, 55°C and 67.5 min.		81.8[b] 96.8[c]	
Esterification: 2.5 % (w/w) H_2SO_4 catalyst, 6:1 molar ratio of methanol to RSO , 60°C and 30 min	20	85[a]	[18]
Transesterification: 1.5 % w/w KOH catalyst, 4.5:1 molar ratio of methanol to RSO , 60°C and 30 min.		90[b]	
Esterification: Para-toluene sulphonic acid catalyst.	16.58	-	[19]
Transesterification: 1 % (w/w) KOH catalyst, 9:1 molar ratio of methanol to RSO, 65°C and 100 min.		92.8[b]	

[a]FFA conversion
[b]FAME yield
[c]FAME conversion

In an aspect of the yield of FAME, the homogeneous catalyst method is superior to the other five transesterification methods [15], [16] but using sulfuric acid catalysts provides high biodiesel production costs [14].

3.2. Heteregenous catalyst

The weakness of homogeneous acid catalysts is difficult to recycle and they also impact to serious corrosion and environmental problems. Meanwhile, homogeneous base-catalysts are reactive on free fatty acids to form unwanted soap as by-products that need expensive separation cost. Thus, the biodiesel production using solid catalysts is preferred. Moreover, the main advantages of heterogeneous reaction come from compatibility with the environment and reduction of manufacturing cost by elimination of the expensive process steps such as separation and purification of biodiesel [20-

22]. Different heterogeneous catalytic transesterification for producing biodiesel from rubber seed oil are summarized in table 3.

Table 3. Optimum parameter condition and conversion efficiency via heterogeneous catalytic.

Optimum operating condition	Initial FFA, %	Conversion efficiency, %	Ref
5 wt.% of clinker loaded with methanol as catalyst; methanol to RSO molar ratio of 4:1, 65°C and 4 hours.	17.75	96.9[a] 75[b]	[20]
0.6 wt% of clinker loaded with 0.5 wt% of sodium methoxide as catalyst; methanol to RSO molar ratio of 7:1, 60°C and 4 hours	17.75	100[a] 88[b]	[21]
14.5 wt.% SO_3H-MCM-41 catalyst; methanol to RSO molar ratio of 16:1, 129.6°C and 48 hours	-	83[b]	[22]
9 wt.% sodium metasilicate catalyst; methanol to RSO molar ratio of 9:1, 65°C and 40 minute	5.2	98[b]	[13]
9 wt.% CaO-AR grade catalyst; methanol to RSO molar ratio of 15:1, 65°C and 180 minute		97[b]	
100 wt.% CaO-based coral fragment catalyst; methanol to RSO molar ratio of 15:1, 65°C and 210 minute		98[b]	
3 wt.% poly (sodium acrylate) supporting NaOH (NaOH/NaPAA) catalyst, methanol to RSO molar ratio of 6:1, 60°C and 120 minute	0.6	96[b]	[23]

[a]FFA conversion
[b]*FAME yield*

Biodiesel from rubber seed oil has been able to be produced through one step transesterification even though the rubber seed oil contains high free fatty acids. Transesterification of rubber seed oil using clinker loaded with sodium methoxide as catalyst doesn't take place under extreme conditions; total mole ratio of methanol to oil, the amount of catalyst and reaction temperature is low. The reaction time of transesterification of rubber seed oil using clinker loaded methoxide catalyst is long, but total biodiesel production time is relatively shorter than via homogeneous catalyst process [21]. The ability reuse of clinker loaded methoxide catalysts in the rubber seeds transesterification reaction has not been reported by Muhammad S A W [21] and yield FAME from the reaction results only reached 88%, so the development of the solid catalyst needs to be studied further.

3.3. Supercritical methanol method (SCM)

The content of water in rubber seed oil is about 3.71% [21]. The water existed in conventional base catalyst transesterification reaction of biodiesel causes the formation of free fatty acids. The FFA formed, then reacts with the base catalyst to form saponification materials in the final product that is normally avoided or limited. After the transesterification reaction takes place, separation between saponification materials emulsify the biodiesel product and other materials turn to be very difficult. Meanwhile the oil contains amount water, when the non-catalyzed supercritical methanol treatment is applied, the yield of biodiesel has no less significant [21-23]. By this process, the transesterification of feedstock oil with methanol is done under temperature and pressure, which exceeding the critical properties of methanol. Table 4 summarizes the works have been done in the area of biodiesel production from rubber seed-oils by using non-catalyzed supercritical methanol method.

Table 4. Operating condition by SCM methods.

Optimum operating condition	Initial FFA, %	FAME yield, %	Ref
methanol to RSO molar ratio of 42:1, 350°C, 43 MPa, 9 min	17	90.86	[21]
methanol to RSO molar ratio of 40:1, 260°C, 160 bar, 10 min	12	87.54	[22]
methanol to RSO molar ratio of 40:1, 250°C, 100 bar, 5 min	16.58	87.1	[16]
methanol to RSO molar ratio of 42:1, 280°C, 85 bar, 20 min	21.35	92.7	[7]
1:40 (w/v) seeds-to-methanol ratio, 280°C, 12 Mpa, 30 min	-	97.9	[23]

The consequence of transesterification reactions that use supercritical conditions is an expensive investment in process equipment.

3.4. In situ-transesterification

Another alternative method of producing ester transport fuels is in-situ transesterification, in which the oil–bearing material contacts with acidified or alkalized alcohol directly. A catalyst is needed to activate the reaction. Acid or alkali catalyst also helps to hit the cell wall of the seed and improve the ability of the solvent to access the oil. The operating conditions of biodiesel production from rubber seed oil using in-situ method are presented in Table 5.

Table 5. Operating condition by in situ transesterification methods.

Optimum operating condition	FAME yield, %	Ref
0,25 % (w/v) H_2SO_4 catalyst, methanol to rubber seed weight ratio of 3:1, 60°C, 1 atm, 120 min	91.5	[29]
3 % (w/w) KOH catalyst, methanol to oil weight ratio of 6:1, 60°C, 1 atm, 120 min	96	[30]
0,75 % (w/v) KOH catalyst, methanol to rubber seed weight ratio of 3:1, 60°C, 1 atm, 120 min	52.86	[31]

The problem arising from this in-situ method is relatively similar to that of a homogeneous catalyst process; difficult to recycle and impact of serious corrosion and environmental problems.

3.5. Ultrasound assisted process

The interactions between alcohols and vegetable oils are weak because of their difficult mixed properties and different reactant densities lead to low reaction rates and increased reaction time. One of the most effective and interesting procedures to overcome this problem is by using ultrasonication (ultrasound) that produces sound waves. The ultrasound phenomenon has its own physical and chemical effects. The chemical effect comes from radicals such as H^+ and OH^- produced during a transient implosive collapse of bubbles (in a liquid irradiated with ultrasound), accelerate chemical reaction in the bulk medium. Meanwhile, the physical effect comes from emulsification, in which the microturbulence generated due to radial motion of bubbles leads to intimate mixing (homogenize the mixture) of the immiscible reactants. Considering all these principles, increases intensively the interfacial region between the oil and alcohol, resulting in faster reaction kinetics and higher conversion of oil and product yield [24]. Reaction parameters and results of some studies that are focused on the use ultrasonic irradiation on biodiesel production from rubber seed oil are summarized in table 6.

Table 6. Performance of ultrasonic assisted transesterification for biodiesel production from RSO.

Optimum operating condition	Initial FFA, %	Conversion efficiency, %	Ref
Esterification: 1 % (v/v) H_2SO_4 catalyst, ratio of rubber seed to methanol was 1:2 (wt/v), 60°C, 1 atm, 30 min and 42 kHz frequency. **Transesterification:** 0.1% (w/v) KOH catalyst, rubber seed to methanol was 1:1.72 (wt/v), 60°C, 1 atm, 30 min and 42 kHz frequency.	16.48	97[a] 92.5[b]	[25]
Esterification: 1 % w/w H_2SO_4 catalyst, methanol to RSO molar ratio of 9:1, 50±2°C, 20 min and amplitudo 30% **Transesterification:** 0.3-0.7 % (w/w) KOH/EFB (empty fruit bunch) catalyst, methanol to RSO molar ratio of 3:1, 50±2 °C, 20 min and amplitudo 30%.	42	99.52[a] -	[26]
Esterification: 1 % (w/w) HCl catalyst, methanol to RSO molar ratio of 20:1, 45°C, 1 atm, 30 min and amplitudo 40%. **Transesterification:** 0.5 % (w/w) NaOH catalyst, methanol to RSO molar ratio of 6:1, 45°C, 1 atm, 15 min and amplitudo 40%.	14.42	79.14[a] 78,84[b]	[27]
Esterification: 2 % w/w HCl catalyst, methanol to RSO molar ratio of 15:1, 95°C and 60 min. **Transesterification:** 0.8 % (w/w) KOH catalyst, methanol to RSO molar ratio of 6:1, 65°C, 15 min and frequency of 33±3 kHz.	23.21	97.84[a] 95[b]	[20]

[a]FFA conversion
[b]FAME yield

Table 6 shows that the type of catalyst used in the esterification and transesterification of the ultrasonic method is a homogeneous catalyst so that the cost of producing biodiesel from this method is not much different from the homogeneous method. The ultrasonic method can be considered for transesterification of rubber seed oil using clinker loaded methoxide to shorten reaction time.

3.6. Enzymatic transesterification

Many investigations have been conducted for the synthesis of biodiesel mediated by lipases. The enzymatic approach for the modification of oils and fats takes advantage of the specificity of some lipases, leading to high purity products, therefore reducing post processing operations and costs. Due to the mild conditions under which the enzymatic reaction occurs, oil oxidation becomes a negligible problem and less energy is required. Moreover, wastes are minimized and by-products are more easily purified [32].

Thermomysis Lanugonosus Lipase was found to be the most suitable for the transesterification of rubber seed oil with a biodiesel conversion of 92.83% at a molar ratio of 4 and 5%(w/v) enzyme concentration in solvent free reaction medium [33]. Biodiesel conversion with rubber seed oil as a feedstock of more than 90% was attained for chemical-catalysed transesterification, whereas the conversion rate was 85% for enzyme-catalysed method [34].

4. Conclusion

This review aims to develop further biodiesel production method from the rubber seed oil, which is potential as a raw material. The literature search results show that the biodiesel production methods of the rubber seed oil include homogeneous catalyst processes, heterogeneous catalysts, supercritical methods, in-situ processes, ultrasound processes and enzymatic processes. The method of producing biodiesel from the rubber seed oil, which is considered to be efficient is to use a clinker loaded methoxide solid catalyst. Using this method, the transesterification of rubber seed oil containing 17.75% free fatty acids can be carried out in one step, the mole ratio of methanol to oil is not high, the reaction temperature is low and the amount of catalyst use is small. The correlation between the characteristic of clinker loaded methoxide catalyst with free fatty acid conversion up to 100% and the low FAME yield (88 %) of the method compared to the homogenous catalyst method (99 %) has not been studied.

References

[1] Atabani A E, Silitonga A S, Ong H C, Mahlia T M I, Masjuki H H, Badruddin I A and, Fayaz H 2013 *Renewable and Sustainable Energy Reviews* **18** 211-245.
[2] Ramadhas A S, Jayaraj S, and Muraleedharan C 2005 *Renewable Energy* **30** 795-803.
[3] Babu B D, Raju R V S and Reddy K T 2014 Engine *International Journal of Engineering Research* **3** 58-64.
[4] Harikrishnan S N and Sabarish R 2014 *Middle-East Journal of Scientific Research* **20** 709-714.
[5] Narayan S A and Shobana S 2013 *International Journal of Science and Modern Engineering* (IJISME) **1** Issue-6.
[6] Ulfah M and Subagjo 2016 *Reaktor* **16 (3)** 114-120.
[7] PandeyA 2008 *Handbook of Plant-Based Biofuels* CRC Press Boca Raton FL.
[8] Ramadhas A S, Jayaraj S, and Muraleedharan C 2005 *Fuel* **84** 335-340.
[9] Krishnakumar U, Sivasubramanian V,and Selvaraju N 2013 *International Journal of Engineering Research and Application* (IJERA) **3** 2206-2209.
[10] Hui H T, Viet T T, and Phung L T K 2014 *International Journal of Renewable Energy and Enviromental Engineering* ISSN 2348-0157 **2** 306-314
[11] Setywardhana D A, Distantina S, Utami M D, dan Dewi N 2009 *Prosiding Simposium Nasional RAPIV III.*
[12] Bello E I, Danayan I A, and Otu F 2014 *Research Journal in Enginnering and Applied Science* **3(2)** 114-117.
[13] Roschat W, Siritanon T, Yoosuk B, Sudyoadsuk T, and Promarak V 2017 *Renewable Energy* **101** 937-944.
[14] Lotero E, Liu Y, Lopez D E, Suwannakarn K, Bruce D A, and Goodwin J G Jr 2005 *Ind. Eng. Chem. Res* **44** 5353-5363.
[15] Ramadhas A S, Jayaraj S, and Muraleedharan C 2005 *Fuel* **84** 335-340.
[16] Satyanarayana M and Muraleedharan C2011 *Energy Sources, Part A: Recovery, Utilization, and Environmental Effects* **33** 925-937
[17] Ahmad J, Yusup S, Bokhari A and Kamil R N M 2014 *Energy Conversion and Management* **78** 266–275.
[18] Thaiyasuit P, Pianthong K, and Worapun I 2012 *J. Oleo Sci* **61(2)** 81-88.
[19] Ezeanayanso C S, Agbaji E B, Ajibola V I, and Okonkwo E M 2011 *Journal of Petroleum Technology and Alternative Fuels* **2(10)** 190-199.
[20] Gimbun J S, Ali S, Kawal C C S C, Shah L A, Hidayah N, Ghazali M, Cheng C K and, Nurdin S 2013 *Procedia Engineering* **53** 13-19.
[21] Muhamad S A W, Gimbun J, Cheng C K, and Nurdin S 2013 *Journal of Engineering and Technology* **4(1)** 123-132.
[22] Samart C, Karnjanakom S, Chaiya C, Reubroycharoen P, Sawangkeaw R, Charoenpanich M M, 2015 *Arabian Journal of Chemistry.*
[23] Yang R, Su M, Zhan J, Jin F, Zha C, Li M and, Hao X 2011 *Bioresource Technology* **102** 2665-2671.

Optimization of binder addition and particle size for densification of coffee husks briquettes using response surface methodology

Raudah[1,2] and Zulkifli[1]
[1] Department of Chemical Engineering, Lhokseumawe State Polytechnic, Buketrata, 24301, Indonesia;

E-mail: raudah_pnl@yahoo.co.id

Abstract. The present research focuses on establishing the optimum conditions in converting coffee husk into a densified biomass fuel using starch as a binding agent. A Response Surface Methodology (RSM) approach using Box-Behnken experimental design with three levels (-1, 0, and +1) was employed to obtain the optimum level for each parameter. The briquettes wereproduced by compressing the mixture of coffee husk-starch in a piston and die assembly with the pressure of 2000 psi. Furthermore, starch percentage, pyrolysis time, and particle size were the input parameters for the algorithm. Bomb calorimeter was used to determine the heating value (HHV) of the solid fuel. The result of the study indicated that a combination of 34.71 mesh particle size, 110.93 min pyrolysis time, and 8% starch concentration werethe optimum variables.The HHV and density of the fuel were up to 5644.66 calgr^{-1} and 0.7069 grcm^{-3},respectively. The study showed that further research should be conducted to improve the briquette density therefore the coffee husk could be convert into commercialsolid fuel to replace the dependent on fossil fuel.

1. Introduction

Coffee husk is abundantly available in coffee plantation, unfortunately no treatment and used of it. Coffee husk have a large contribution to the waste material in Middle Aceh, Indonesia, as it is well known for coffee-producing region in the Aceh Province. The production of coffee in Aceh generated significant quantities of coffee husk biomass as agro-residues.However, this material has a good potential to be analternativefuel raw materials. It contains high carbon which can be convert to solid energy source. Therefore, in one hand this might solve the environmental issues related to huge amount coffee husks waste produced each year, in another hand, it creates renewable energy.

Agro residues are usually burned in open area, which pollutes the environment. Moreover, the ash contains a largeportion of unburned carbon due to incomplete combustion.Using agricultural residues directly as energy source is not recommender due to some disadvantages: low bulk density, difficult to handle and the transport and storage are costly[1]. Therefore, biomass densification, which apply mechanical pressure, enhance the density of solid material. This process overcome these limitations [2].Using coffee husk as raw material for bio briquettes has been performed by [1]and [3]. According to Suarez et al., [1], it is possible to convert coffee husks into briquettes using the piston with preheated technology and it can be used for combustion purposes in bakery furnaces.

[2] To whom any correspondence should be addressed.

According to Suarez et al., (2003) [1], and study of [3]which found that the briquetting preassure can increase the durability of coffee husk briquettes.Additionally, the optimum response of coffee husk biobriquettes obtained in briquetting pressure, moisture content and sugarcane molasses levelswere14.91 MPa, 8% and 45% respectively. In these conditions, the density of the briquettes obtained upto 718.09 kg / m^3, 80.77% durability and 14.98%.stability

Coffee husks biobriquettes can be produced directly from raw materials without pyrolysis process. However, this process has some disadvantages, high volatile matter in the material which produce smoke in the briquettes. Pyrolysis is one of the solution to increase the briquettes quality.Pyrolyslis eliminates unwanted material, such as moisture, volatile matter, and increase the briquette HHV.

Adapa, et al, [4] stated that the briquetting pressure significantly affected the density, durability and specific energy of bio briquettes. Dense briquettes showed high energy per unit volume of material. Hence, durability is the resistance of the briquettes to hold strain and force collisions that may occur during handling and transport the products [5]. Regarding to[6], the optimum conditions to produce solid fuels from rice straw biomass were 3002.8 psi of briquetting pressure, 34. 90 mesh particle size, and 7.775 % maximum raw material water content.

Previous studies about biomass pyrolysis focused on the parameter of optimization of biomass pyrolysis using response surface methodology [6, 8, 9]. Briquetting pressure on this research was relatively low (2000 psi) so that can be easily applied by society. Sustainability-income research will certainly provide a very promising prospect for the community in the countryside.

This research aimed to study the impact of particle size of coffee husk charcoal and binder concentrations on the briquettes quality. Response surface methodology was applied in order to optimize the interaction between the parameters. This result would be on behalf of promoting clean energy applications in the future.

2. Materials and Methods

The coffee husk biomass was collected from coffee hullers in the rural areas of Gayo Highlands, Aceh, Indonesia. The biomass was dried in the open air for a week, then pyrolysed in the furnace into three pyrolysis time categories, 50,100, and 150 minutes. The charcoal then grounded in a milling machine and screened to 20, 40, and 60 mesh particle size. The concentration of 8, 10, and 12 % (w/w) of starch binder, were applied in the densification process.

Gross calorific value was measured using adiabatic bomb calorimeter according to ASTM standard D- 5685. The independent variables were pyrolysis time X_1, particle size X_2, and starch binder concentration X_3 with a 2000 psi briquetting pressure as a constant variable. The dependent variables were density and heating value of briquettes. The densification experiments were conducted using bench type manually operated, laboratory hydraulic press with 20 ton (Hydraulic press Shop CMC ISO9002) capacity and a densification die. The densification die was constructed from stainless steel cylinder of 30 mm in internal diameter and 250 mm in length, equipped with stamp of 30 mm in external diameter.

A Bob-Behnken technique in RSM with three levels (low, medium and high) being coded as -1, 0, and +1 and a total of 17 runs were carried out in repetitious to optimize the level of chosen variables, pyrolysis time (min), particle size (mesh) and starch concentration (%). The test carried out according to [6], the range and the level used in the experiments are select and listed in Table 1.

Table 1. Experimental range and levels of independent variables.

Independent variable	**Coded level and range**		
	-1	**0**	**+1**
Pyrolysis time, X_1(min)	50	100	150
Particle size, X_2 (mesh)	20	40	60
Starch concentration, X_3 (%)	8	10	12

The density of the briquettes was measured by using the ratio of weight and volume determined from the briquette geometric shape. The research carried out by conducting the parameters from the Box-Behnken Design Experiment where consisted of 17 treatments randomly (Table 2). Tabel 2 presented the actual and prediction of heating value and density from the experiment. A total of 17 experiments had to be conducted to determine 10 coefficients of second order polynomial reationship, which modeled according to Equation (1).

$$Y_k = \beta_o + \sum_{i=1}^{3} \beta_i X_i + \sum_{i=1}^{3} \beta_{ii} X_i^2 + \sum\sum_{i<j} \beta_{ij} X_i X_j + \varepsilon_j \qquad (1)$$

In the experimental design model, Y is the predicted response and pyrolysis time, X_1 (min), particle size, X_2 (mesh), and starch, X_3 (%) were taken as independent variables. Constants β_o, β_i and β_{ij} are linier term, quadratic term and cross product term coefficients, respectively.

Table 2. Bob-Behnken Design matrix along with experimental and predicted results.

No.	Dependent variable			Independent Variable			
	Pyrolysis time, X_1(min)	Particle size, X_2 (mesh)	Starch conc, X_3 (%)	Density (gcm^{-3}) Exp.	Density (gcm^{-3}) Predicted	Heating value ($calg^{-1}$) Exp	Heating value ($calg^{-1}$) Predicted
1	100	60	12	0.8204	0.7870	5433.27	5485.16
2	50	40	8	0.6565	0.6463	5238.37	5246.88
3	100	20	8	0.6262	0.6286	5711.04	5659.15
4	150	40	12	0.7288	0.7510	5503.97	5495.46
5	150	60	10	0.8011	0.8092	5282.79	5239.41
6	150	40	8	0.7680	0.7799	5548.15	5496,35
7	100	40	10	0.7402	0.7225	5665.90	5583.83
8	100	40	10	0.7063	0.7225	5539.55	5583.83
9	100	60	8	0.8208	0.7976	5323.16	5418.34
10	150	20	10	0.7392	0.7217	5552.21	5655.90
11	50	60	10	0.7452	0.7755	5337.01	5233.32
12	100	20	12	0.6846	0.6768	5634.61	5539.43
13	50	40	12	0.7129	0.7129	5143.07	5194.87
14	100	40	10	0.7016	0.7225	5523.31	5583.83
15	50	20	10	0.5790	0.5837	5068.55	5111.93
16	100	40	10	0.7286	0.7225	5634.61	5583.83
17	100	40	10	0.7233	0.7225	5555.79	5583.83

The individual desirability functions (DF) is usually performed as the optimum criteria for the purpose of optimizing multiple response variable. The Desirability Functions (DF) approach has been successfully applied for optimizing process variables for densification of rice straw as a rural alternative solid fuel [6]. Such an operating approach would be adopted in this research to get maximum or minimum value of the variable response for technical and economical considerations.

3. Results and Discussion

3.1 Calorific analysis of coffee husk biobriquettes

Direct combustion of biomass is not preferable due to the negative aspects coming from the intrinsic properties of biomass such as low density, low calorific value in a unit volume, and high moisture [10].

Therefore, it is important to develop strategies where biomass is converted to secondary fuels, which have better characteristics in comparison to the parent material[11]. For this purpose, biomass is firstly carbonize to eliminate the moisture and volatile matter contents and then the volatile matter-free solid char, which is called as "smokeless fuel", is briquetted to form firm bio-fuel briquettes.

Table 3. Calorific analysis of coffee husk.

Parameter	Heating value ($calg^{-1}$)
• Coffee husk before carbonization (original sample)	3971.052
• Coffee husks after carbonization	5713.672

Table 3 showed the calorific analysis of the coffee husk before and after carbonization. The calorific value of the coffee husk briquettes was 5713,672 $calg^{-1}$. It was higher than the original sample 3971,052 $calg^{-1}$. According to these result it is clear that the coffee husks have a typical biomass structure with high heating value after carbonization process.

The briquetting of a smokeless fuel that obtained from the carbonization of coffee husk at 350°C gave very good results. It contained high calorific value and almost no volatile matter content. The coffee husk briquettes is a solid fuel geen energy.

3.2 Regression Model and ANOVA for heating value and density of biobriquettes

The results of the each experiment performed was given in Table 2. Empirical relationship between the response and the independent variables have been expressed by the quadratic model Equation (2) and Equation (3).

$$Y_{1k} = 4358{,}199 + 24{,}48398\,X_1 + 8{,}75582\,X_2 - 40{,}8725\,X_3 - 0{,}088164\,X_1^2 - 0{,}1332\,X_2^2 - 1{,}25775\,X_3^2 - 0{,}13447\,X_1.X_2 + 0{,}1278\,X_1.X_3 + 1{,}16588\,X_2.X_3 \quad (2)$$

$$Y_{2k} = -\,0{,}040288 + 4{,}29175 \times 10^{-3}\,X_1 + 9{,}77312 \times 10^{-3}\,X_2 + 0{,}0433\,X_3 - 2{,}6075 \times 10^{-5}\,X_1.X_2 - 2{,}39 \times 10^{-4} X_1.X_3 - 3{,}675 \times 10^{-4}\,X_2.X_3 \quad (3)$$

The response plots and a mathematical model were generated for the process. The results were determined in the correlation to coefficient of determination (R2) in the model.

Table 4. The Model fit Summary for (a) Heating Value and (b) Density

(a) Heating Value

Model Summary Statistics

Source	Std. Dev.	R-Squared	Adjusted R-Squared	Predicted R-Squared	PRESS	
Linear	170.01	0.3431	0.1915	-0.2036	6.884E+005	
2FI	171.48	0.4859	0.1774	-0.9284	1.103E+006	
Quadratic	100.00	0.8776	0.7203	-0.5613	8.930E+005	Suggested
Cubic	62.70	0.9725	0.8900		+	Aliased

(b) Density

Model Summary Statistics

Source	Std. Dev.	R-Squared	Adjusted R-Squared	Predicted R-Squared	PRESS	
Linear	0.029	0.8360	0.7981	0.6722	0.021	
2FI	0.022	0.9261	0.8817	0.6848	0.021	Suggested
Quadratic	0.022	0.9460	0.8765	0.3615	0.042	
Cubic	0.016	0.9844	0.9374		+	Aliased

The model fit summary was in Table 4. According to heating value, the quadratic model suggested model with adjusted R^2 of 0. 8776 and predicted R^2 of -0.5613. This shows that adding the quadratic (squared) terms to the mean, block, linier and the two factor interaction terms in the model is significant. Additionally, the model fit summary in Table 4 for density shows that two Factor Interaction (2FI) is selected with adjusted R^2 of 0.8817 and predicted R^2 of 0.6848. The correlation coefficients might have resulted from the insignificant terms in Table 5. It is most likely due to three different variables selected in wide ranges with a limited number of experiments as well as the nonlinier influence of the investigated parameter on the response process.

For each source of terms as depicted in Table 5, the probability ("PROB> F") was examined to see if it falls below 0.05. So far, the quadratic model and Two Factor Interaction model looks the best. These terms are significant, but adding the cubic other terms will not significantly improve the fit since it is aliased.

The SS for lack and fit is the difference between the residual SS for the simple linier regression model and the residual SS. The word "lack of fit" refers to the fact that the simple linier regression model may not adequately fit the data. If the SS for lack of fit is small there is evidence of simple linier regression model is more appropriate to explain the relationship of the parameter on coffee husk biobriqettes.

Analysis of variance (ANOVA) used to evaluate the effect linearity, quadratic, or the interaction of independent variables (factors) on the dependent variable (response). Analysis of variance for heating value and density can be seen in Table 4 and Table 5. As indicated by Table 4 and Table 5, all the linearity and quadratic model terms found to be significantly affect on heating value and density of coffee husk briquettes. In addition, interaction of pyrolysis time with particle size also have significant effect ($p>0.05$) on the heating value and density.

Table 5. Analysis of variance (ANOVA) to describe the heating value briquettes using starch as binding agent.

Variation	Sum of Squares	Mean Square	F, Value	Prob> F	Judgement
Model	5.020E+5	55773.61	5.58	0.0169	Significant
A	1.513E+5	1.513E+5	15.13	0.0060	Significant
B	3539.05	43539,05	4.35	0.0754	
C	1399.20	1399.20	0.14	0.7194	
A^2	2.046E+5	2.046E+6	20.45	0.0027	Significant
B^2	11953.12	11953.12	1.20	0.3105	
C^2	106.57	106.57	0.011	0.9207	
AB	72328.72	72328.72	7.23	0.0311	Significant
AC	653.31	653.31	0.065	0.8056	
BC	8699.29	8699.29	0.87	0.3820	
Residual	70002.59	10000.37			
Lack of Fit	54278.87	18092.96	4.6	0.0872	**Not significant**
Pure Error	15723.72	3930.93			
Cor Total	5,720E+005				

Quadratic models
$R^2 = 0.8776$; adj. $R^2 = 0.7203$; Std. Dev. = 100; C.V. = 1.83

Table 6. Analysis of variance (ANOVA) to describe the density briquettes using starch as binding agent.

Variation	Sum of Squares	Mean Square	F, Value	Prob> F	Judgement
Model	0.060	0.010	20.88	<0.0001	Significant
A	0.015	0.015	30.63	0.0002	Significant
B	0.039	0.039	80.97	<0.0001	Significant
C	7.069E–4	7.069E–4	1.47	0.2535	
AB	2.720E–3	2.720E–3	5.65	0.0388	Significant
AC	2.285E–3	2.285E–3	4.74	0.0544	
BC	8.644E–4	8.644E–4	1.80	0.2100	
Residual	4.815E–3	4.815E–3			
Lack of Fit	3.796E–3	6.327E–4	2.48	0.1989	**Not significant**
Pure Error	1.019E–3	2.5484			
Cor Total	0.065				

Model Two-Factor Interaction (2FI)
$R^2 = 0.9261$; adj. $R^2 = 0.8817$; Std. Dev. = 0.022; C.V. = 3.04

Values of "Prob > F" less than 0.0500 indicated the model terms are significant. In the heating value case A, A^2 and AB are significant model terms. Whereas, in the density case A, B and AB are significant model terms. When the value was greater than 0.100 , it indicated the model terms are not significant.

The important diagnostic tool shown in Figure 1 is the normal probability plot of the studied residuals. The data points from the plot is approximately linier, which shows that the quadratic model developed for heating values and the two factor interaction developed for density are a good representation on the process.

(a)

(b)

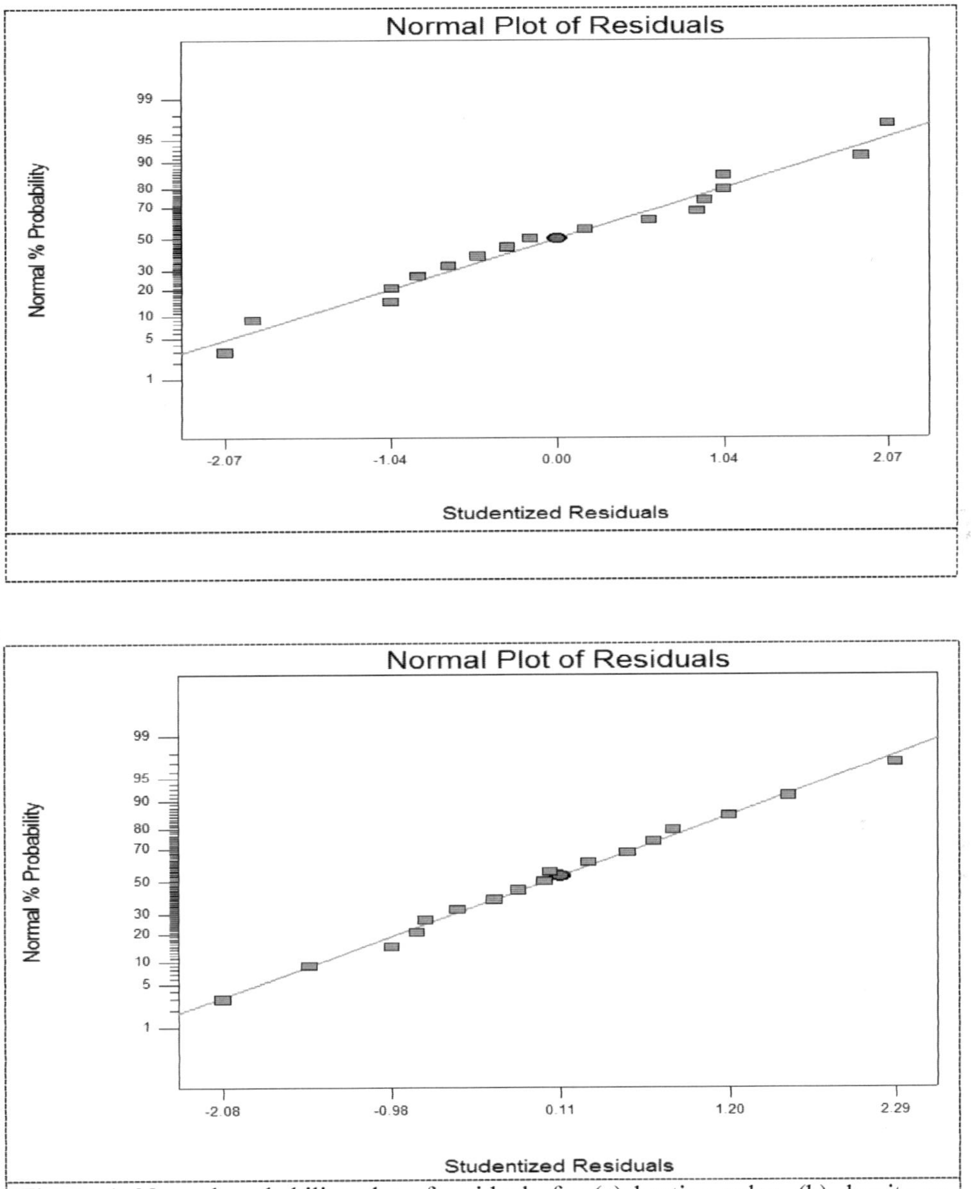

Figure 1. Normal probability plot of residuals for (a) heating value, (b) density or coffee husk biobriquettes.

3.3 Estimation of quantitative Effects of the Factors

The three dimensional response surface and contour plot obtained from second polynomial equations are shown in Figures 2-3. The values of the variable obtained as the response is optimized using the RSM technique. The best response range obtained by analyzing the response surface plots. The combined effect of particle size of coffee husk biobriquettes and pyrolysis time and binding agent 10 % on heating value is shown in Figure 2. The heating value increased with increasing the particle size of coffee husk biobriquettes. In addition, the heating value increases with the longer pyrolysis time of

IOP Conf. Series: Materials Science and Engineering **334** (2018) 012007 doi:10.1088/1757-899X/334/1/012007

raw coffee husk from 50 min to 150 min. It is noted that the maximum heating value at all the particle size achieved at 110 minute pyrolysis time of raw coffee husk.

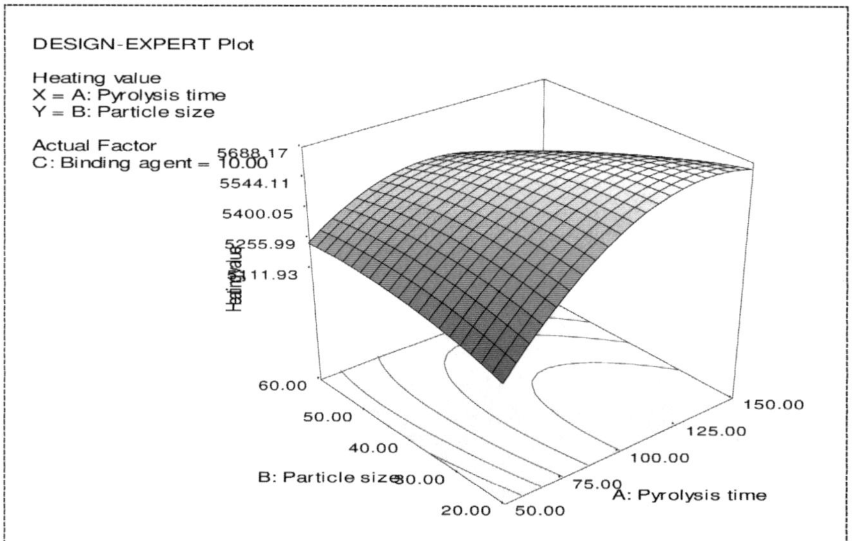

Figure 2. Response plots between pyrolysis time and particle size for heating value of coffee husks briquettes.

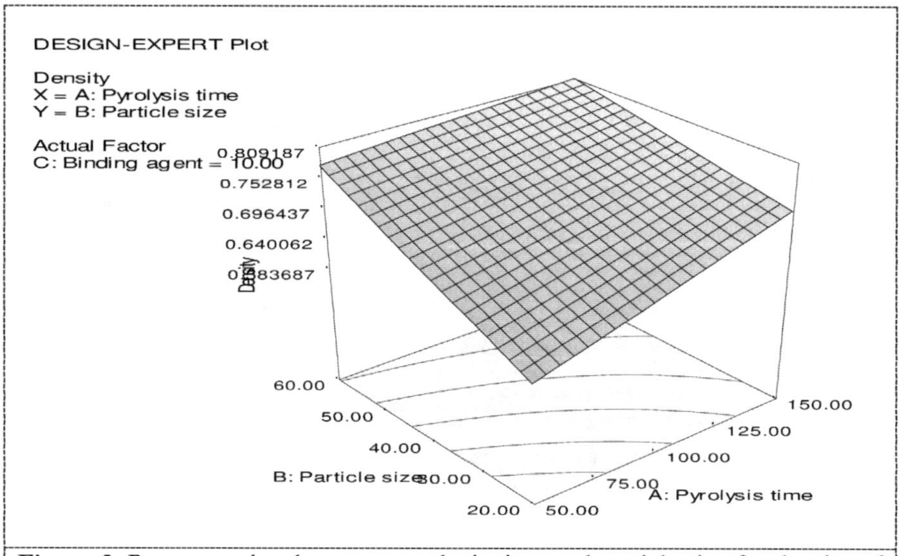

Figure 3. Response plots between pyrolysis time and particle size for density of coffee husks briquettes.

Figure 3 shows three dimensional response srfaces for the interaction effect of pyrolysis time and particle size on density at using 10% starch concentration. Increasing the pyrolysis time from 50 min to 150 min produced more dense biobriquettes. Consequently, the density increases with the increasing particle size from 20 mesh to 60 mesh.

3.4 Optimization of Variable Response

The optimization process aimed to get the technical parameters of the conditions that can provide optimum response with consider to economical operating cost. The quality of solid fuels was determined by its calorific value (at least 21000 J / g, SNI. No. 1/6235/2000). In order to reduce the cost of operation, the pyrolysis time and particle size should be keep to a minimum, because the longer pyrolysis process taking place and the smaller particle size, the greater energy and cost would be required. Therefore, to get the optimum response at low cost, the set-up goal pyrolysis time should be set at the "minimum" with a lower limit of 50 min and an upper limit of 150 min. For more details, refer to the table 7.

Table 7. Optimization of dependent variable of coffee husk briquettes.

	Parameters			Response	
No	Pyrolysis time (min)	Particle size (mesh)	Starch content (%)	Heating Value (Calgr^{-1})	Density (grcm^{-3})
1	82.93	57.82	8.00	5415.06	0.7735
2	83.20	57.59	8.00	5417.83	0.7747
3	84.46	55.66	8.00	5437.76	0.7648
4	88.58	54.85	8.00	5456.08	0.7650
5	110.93	34.71	8.00	5644.66	0.7069

In applying the desirability function (DF) method, the Design Expert software produced 5 solutions. Each has a DF = 1, as shown in Table 7. As the heating value of the fifth solution is the highest among the available solutions, as well as particle size, this yields the lowest density of the product. Although the fifth solution produce the longest pyrolysis time, it gives the lowest operational cost. It is therefore appropriate to select the fifth acceptable solution as the optimum. The optimum conditions to produce solid fuels from coffee husk biomass were obtained at 110.93 min pyrolysis time particle size of 34.71 mesh and starch concentration of 8%. The heating value obtained up to 5644.66 calg^{-1} and density about 0.77354 gcm^{-3}.

4. Conclusions

A desirability function approach has been used to optimize the process variables of pyrolysis time, particle size, and starch content on the multiple-response variables of density and heating value of coffee husk briquettes produced through a mechanical densification. The optimum conditions to produce solid fuels from coffee husks biomass were obtained at a pyrolysis time of 110.93 min, particle size of 34.71 mesh and starch content of 8%. The heating value and density of briquette were accordingly 5644.66 calg^{-1} and 0.77354 gcm^{-3}.

Acknowledgements

The authors wish to express their gratitude to Directorate General of Higher Education (DGHE), Ministry of National Research and Education (MNRE) of Republic of Indonesia for providing research grant under Hibah Bersaing Project 2015-2016.

References

[1] Suarez, JA(a)., and Pedro Anibal Beaton. (2003). Coffee Husk Briquettes: A New Renewable Energy Source. Energy Sources, 25:961–967, Copyright © Taylor & Francis Inc. ISSN: 0090-8312 print/1521-0510 online DOI: 10.1080/00908310390232415.

[2] Munoz-Hernandez, G., M. Gonzalez-Valadez & J. Dominguez-Dominguez. 2004. An Easy Way to Determine the Working Parameters of the Mechanical Densification Process. Agricultural Engineering International: the CIGR Journal of Scientific Research and Development.

Manuscript FP 03 013. Vol. VI.

[3] Okello, Collins., Levi Lukoda Kasisira & Mackay Okure. (2010). Optimising Densification Condition Of Coffee Husks Briquettes Using Response Surface Methodology. Second International Conference On Advances In Engineering And Technology.

[4] Adapa, Phani K., Lope G. Tabil & Greg J. Schoenau. (2013). Factors Affecting The Quality of Biomass Pellet for Biofuel And Energy Analysis of Pelleting Process. International Journal Agricultural And Biological Engineering. Vol. 6 No. 2.

[5] Larsson, S. H, Thyrel M, Geladi P, Lestander, T. A. (2008). High Quality Biofuel Pellet Production from Pre-compacted Low Density Raw Materials. Bioresource Technology.

[6] Yunardi., Zulkifli dan Masrianto. (2011). Response Surface Methodology Approach to Optimizing Process Variables For the Densification of Rice Straw as a Rural Alternative Solid Fuels. Journal of Applied Science 11 (7): 1192-1198, ISSN 1812-5654/DOI:10.3923/jas.2011.1192.1198.

[7] Suarez, JA(b)., and Pedro Anibal Beaton. (2003). Physical Properties Of Cuban Coffee Husk For Use As An Energy Source. Energy Sources, 25:953–959, Copyright © Taylor & Francis Inc. ISSN: 0090-8312 print/1521-0510 online DOI: 10.1080/00908310390232406.

[8] Donghui, L., Tabil, L. G., Decheng, W., Guanghui, W., Zhiqin, W., (2014). Optimization of Binder Addition and Compression Load for Pelletization of Wheat Straw Using Response Surface Methodology. Int J Agric & Biol Eng Vol 7 No 6: 67-78.

[9] Svatek, M and krizan, P., (2015). Mutual Interaction of Selected Parameters of Oak sawdust Densification Process by The Density Response Surface. Acta technica corvininsis-Bulletin of Engineering. ISSN: 2067-3809.

[10] Haykiri-Acma, H and Yaman, S. (2010). Production of Smokeless Bio-briquettes from Hazelnut Shell. Proceedings of the World Congress on Engineering and Computer Science Vol II. WCECS San Francisco, USA.

[11] A.Kurklu, S.Bilgin, "Biomass briquetting machines and applications", M.Sc. Thesis, Akdeniz University, Antalya, Turkey.

Torrefaction of Durian peel and bagasse for bio-briquette as an alternative solid fuel

S Haryati[1,3], Rahmatullah[2] and R W Putri[2]

[1] Chemical Engineering Department, Agroprocess Laboratory, Engineering Faculty, Sriwijaya University, Indralaya ,st Km.32 Ogan Ilir (OI) 30622, Palembang, Indonesia
[2] Chemical Engineering Department, Engineering Faculty, Sriwijaya University Indralaya-Prabumulih,st Km.32 Ogan Ilir (OI) 30622, Palembang, Indonesia

E-mail: haryati_djoni@yahoo.co.id

Abstract. Biomass waste of durian (*Durio zibethinus*) peel and bagasse could be used as solid fuel by a toreffaction process. Durian peel and bagasse were washed and crushed into small sizes then dryed in order to remove water content. The treated biomass was burned at varied temperature of 200 – 350 °C and a residence time of 30 min prior to producing torrified charcoal as intermediate product. Torrified charcoal was ground into a powder blended with tapioca glue followed by casting into a cylinder to form a bio-briqquette. The bio-briquette was characterized by determining its calorific value via bomb carolimeter analysis. The key parameter of bio-briquette are calorific value and combustion rate. The result that as the burning temperature was increased the calorific value of bio-briquettes also increased. The maximum calorific value was achieved at 350°C whereas the maximum calorific value of durian (6,157 cal/gr) is higher than bagasse (6,109 cal/gr). The minimum combustion rate was attained in durian peel torrefaction at 350 °C with the rate 0.0398 g/s. The result showed that bio-briquette of durian peel and bagasse have calorific values equivalent to that of subbituminus coal in the range of 4,900 - 6,800 cal/gr.

1. Introduction

Energy consumption in Indonesia still dominates with fossil fuel like natural oil and gas or coal as solid fuel. Whilst the fossil fuel is non-renewable energy and the supply decreases all over the time. This condition needs researchers to find out the alternative energy to ensure energy consumption. Biomass waste has the high prosperity to be raw material for alternative energy. Two of them are Durian Peel and Bagasse.

Durian (*Durio zibethinus*) is the local fruit of Indonesian and it grows well in Indonesia especially in South Sumatera. Durian peel is a waste that can rise environment problems but it is high prospect as bio-briquette raw material for improving its added value.

Bagasse is the by-product (waste) in sugar industry which 35% - 40% of sugarcane weight generate bagasse. Biomass waste from bagasse (with 37% cellulose content) and durian peel (with 60% cellulose content) potentially proceed as solid fuel (bio-briquette) with high calorific value and environment benign without emmision of combustion via torrefaction process [1], [2].

[3] To whom any correspondence should be addressed.

Torrefaction is a thermochemical treatment of biomass at 200 to 300 °C. It is carried out under atmospheric conditions and in the absence of oxygen. In addition, the process is characterised [3]. After passing through torrefaction, the moisture contained in the biomass is substantially reduced and the components of low weight organic volatile are liberated. As a result, hydrophobic bio-briquette with higher content of fixed carbon are produced [4]. Torrefied biomass has advantageous properties (such as low moisture, lower emission of smoke and increased heating values) that improve fuel quality [5]. Torrefaction also makes biomass hydrophobic, that is, it practically does not reabsorb moisture during storage [6].

Several researches have been also established for producing briquette from various biomass. Nugraha [7] informed that baggase charcoal briquettes production via carbonization process with adhesive material Lapindo Mud generate calorific value of 3,564 cal/g and combustion rate of 0.142 g/min. Production of briquette from empty fruit bunches torefaction has been reported by Irawan et al [8] with temperature and holding variation. The result showed the higher temperature had higher calorific value 4,469 cal/g. Ridhuan and Suranto [9] compared carbonization and pyrolysis process in producing biobriquette from durian peel. It showed that carbonization has lower calorific value (3,418 cal/g) than pyrolisis (5,609 cal/g).

In this present work, biomass waste of durian peel and bagasse was torrefied in the purpose of comparing calorific value and combustion rate of durian peel and bagasse bio-briquette. Furthermore, this work aims to find out the influence of temperature and residence time in calorific value and combustion rate.

2. Methodology/Experimental

2.1. Samples

Durian peel samples which grow in Medan were collected directly from Kuto Fruit Market, Palembang, Indonesia. Baggase samples were collected from sugar industry PT. Pemuka Sakti Manis Indah (PT.PSMI), Lampung, Indonesia. All samples were cut into 5 cm then dried in sunny side for a week until attain constant weight.

2.2. Torrefaction Process

250 g dried material enters the reactor then burnt on 8.5 A with varied temperature of 200- 350 °C and residence time of 30 min. After the recidence time, the product is in a form of charcoal. Figure 1 shows the experiment equipment scheme for torrefaction. This equipment includes reactor that completed with heating element in a whole reactor wall. This reactor is connected with thermocouple in order to control temperature during torrefaction process.

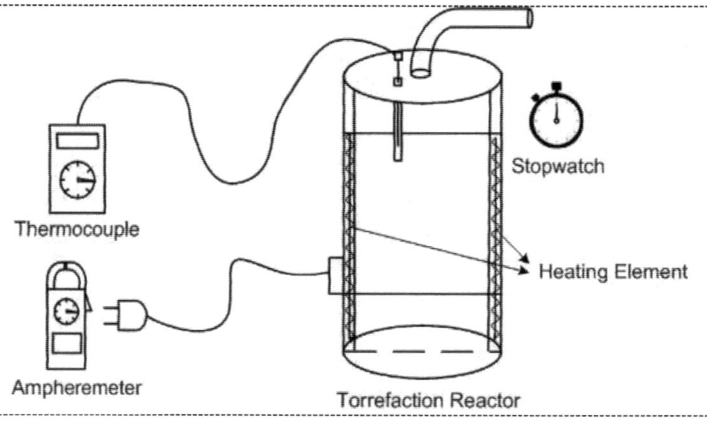

Figure 1. Experimental equipment scheme.

2.3. Biobriquette Casting
The torified charcoal of biomass durian peel and bagasse were ground into powder. The charcoal powder samples blended with tapioca glue which is tapioca flour was mixed with boiled water in ratio of 1 : 10 and followed by casting in pressed cylinder tube to produce bio-briquette with diameter 2.5 in and height 10 cm. The cylinder bio-briquette was dried in an oven for 1 h to ensure the bio-briquette has low water content for improving storage time.

The bio-briquette products were characterized by determining the calorific value with bomb calorimeter analysis. The sample briquette was burnt in atmospheric condition for establishing combustion rate.

2.4. Calorific Value Analysis
All the biobriquete samples were analyzed using a bomb calorimeter (Parr) in Jasa Mutu Mineral Indonesia Laboratory, Inderalaya, South Sumatra, Indonesia. The bomb calorimeter was equipped with an integrating ignition unit (Parr 2901) and oxygen bomb (Parr 1108).

Bomb calorimeters require 0.5 g of sample matter weighed in a crucible. The sample was placed crucible inside the stainless steel container (the bomb vessel) and filled bomb vessel with 30 bar (435psi or 30 atm) of pure oxygen. Once the bomb vessel temperature has stabilized in the bomb well, the sample was then ignited. During the burning of the fuel sample, the crucible could momentarily rise to above 1000 °C with the pressure spiking to 3 x the initial pressure. Once the determination is complete, typically within 4 min, the calorimeter calculated the calorific value of the fuel sample. At this point in time, the bomb vessel was removed from the bomb well to be cooled. The heating value on calorimeter can be calculated as follows [7]:

$$HHV = \frac{(c \; x \; \Delta T) - (Acid) - (Fulse)}{Mass \; of \; sample} \tag{1}$$

where HHV is Highest Heating Value (cal/g), Acid is ash residue (cal/g), Fulse is length of burnt wire (1 cm = 1 cal/g), ΔT is temperature differences (°C), c is constanta (2,401.46 cal/g).

2.5. Combustion Rate Calculation
Briquette combustion rate was determined by equation (2). Briquette sample was put on the analytic scale to know the weight. Briquette sample of known weight was placed on bucket completed with the burner ignited. The mass balance was monitored to record measurements of the weight loss throughout the combustion process using stopwatch, until the briquettes were completely burnt.
The weight loss at specific time was calculated from this expression [7]:

$$M = \frac{M_o}{\Delta t} \tag{2}$$

Where M is Briquette combustion rate (g/s), M_o is mass of briquette sample (g), Δt is time of combustion (s).

3. Result and Discussion

3.1. Calorific Value Analysis with different biomass waste
Figure 2 shows the calorific value of biobriquette with two different biomass waste (durian peel and baggase). Based on Figure 2, it can be seen that calorific value of biobriquette is significantly influenced by temperature differences and raw material. The calorific value of durian peel samples with more cellulose content is higher than baggase that of smaller one. It could be explained that the greater cellulose content, the greater fixed carbon and the greater calorific value. The maximum calorific value was attained at 350°C, as the burning temperature increased, the calorific value of bio-briquettes also increased. This phenomenon occurs due to accomplished decomposition of celullose to carbon content at 350°C. Generally, the thermal decomposition of hemicellulose occurs at

temperatures ranging from 150 to 350° C; cellulose is decomposed for the temperatures in the range of 275-350°C, and lignin is featured by gradual decomposition for the temperatures between 250 and 500°C [10], [11]. The high fixed carbon carried out high calorific value as indicator of bio-briquette quality.

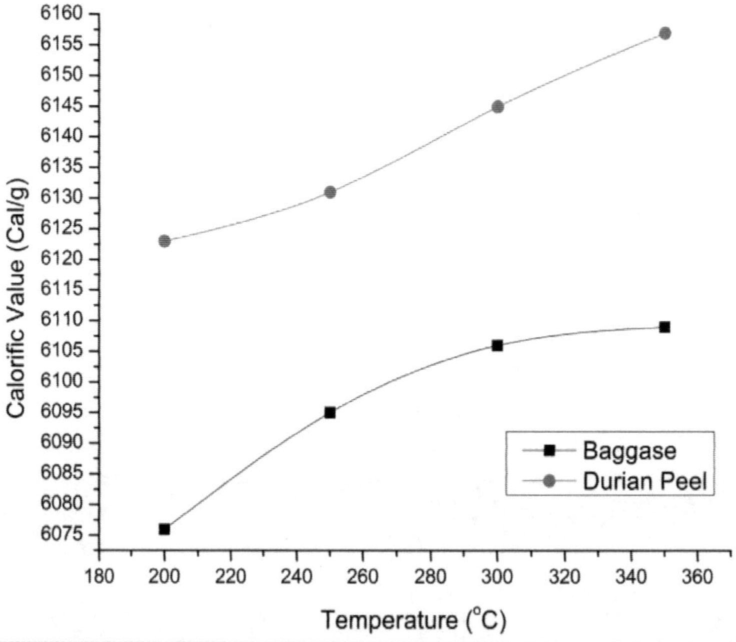

Figure 2. The influance of temperature variations with calorific value of baggase vs durian peel.

Compare to previous research [7], [8], [9], this present study indicates that torrefaction process in ranges temperature of 200 - 350 °C is better than carbonization process (high temperature, with oxygen presence). This is because of the vacuum condition in torrefaction process lead high calorific value of 6,157 cal/g than others only 3,000 - 5,000 cal/g.

3.2. Combustion Rate Analysis

Figure 3 indicated combustion rate of durian peel bio-briquette and bagasse bio-briquette. This figure shows the rising temperature which conduct the increased combustion rate. It is caused by the increasing of fixed carbon and the decreasing of bio-briquette volatile mater. The combustion rate of baggase bio-briquette is higher than durian peel bio-briquette with the value of 0.0509 g/s and 0.0398 g/s, respectively. Low combustion rate leads the briquette burning in longer time consistent with Nugraha's statement [7].

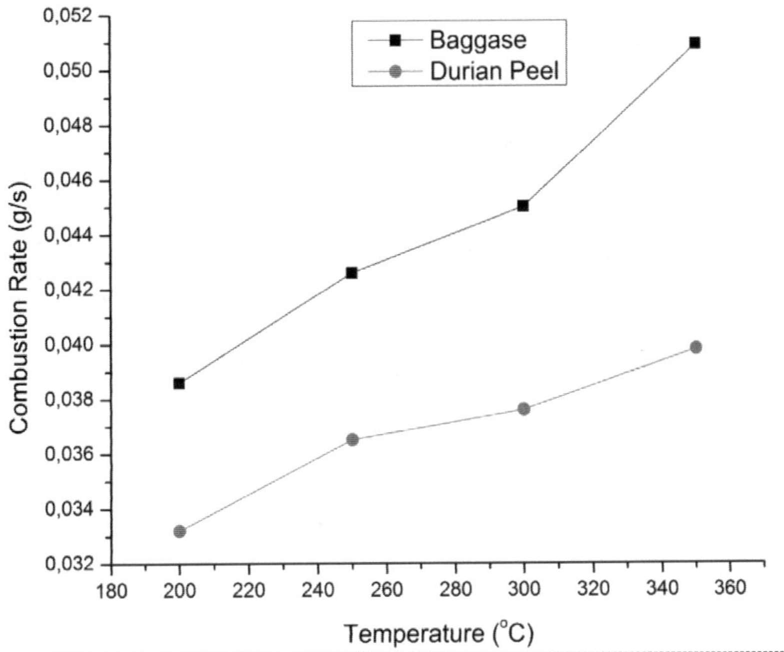

Figure 3. The influence of temperature with combustion rate of bagasse vs durian peel.

4. Conclusion

The comparison of durian peel and bagasse bio-briquette indicates that durian peel is better than bagasse as torefaction material. The calorific value and combustion rate is highly influenced by a variety of temperature. The calorific value increased with rising of temperature that achieved the maximum value to 6,157 cal/g at 350 °C over durian peel bio-briquette and 6,109 cal/g over bagasse. This result shows both biomass waste potentially as an alternative energy to substitute coal (fossil fuel) due to the calorific value of durian peel and bagasse bio-briquette equals to subbituminus coal in the range 4,900-6,800 cal/gr according to International standard.

Acknowledgement

The researchers would like to extend their deepest gratitude to the Ministry of Research, Technology, and Higher Education, Indonesia for this research grant.

References

[1] Wijayanti., 2009, Activated Charcoal of Baggase as Absorbent on Refining Used Cooking Oil (In Indonesian), Mathematic and Science Faculty, IPB : Bogor.

[2] Wahidin, N., Nurfa, A., and Martana, T., 2013, Durian Peel Biobriquette Characteristic as An Alternatif Renewable Energy (In Indonesian), Journal of Agroindustrial Technology, vol 23 (1).

[3] Basu, Prabir. 2013. Biomass Gasification, Pyrolysis, and Torrefaction. Practical Design and Theory (Second Edition). Elsevier Inc, UK.

[4] Felfli, F.F., Luengo, C.A., Suárez, J.A., and Beatón, P.A., 2005, Wood briquette torrefaction, Energy for Sustainable Development , 9(3).

[5] Pach, M., Zanzi, R., and Björnbom, E., 2002, Torrefied biomass as substitute for wood andcharcoal, 6th Asian Pacific Int. Symp. on Combustion and EnergyUtilization, Proc. Int. Conf., Kuala Lumpur,Malaysia, pp. 285-290.

[6] Li, H., Liu, X., Legros, R., Bi, X.T., Lim, C.J., and Sokhansanj, S., 2012, Torrefaction of sawdust in a fluidized bed reactor, Bioresource Technology,103, pp. 453-8.

[7] Nugraha, J.R., 2013, The Themal Characteristic Of Bagasse Charcoal Briquettes of Adhesive Materials Lapindo Mud (In Indonesian), Departement Mechanical Egineering of Engineering Faculty of Jember University.

[8] Irawan, A., Tubagus, R., and Nurmalisa, 2015, Empty Fruit Bunches Torrefaction for Hemicellulose Content and Water Absorption Test, Reaktor, Vol 15 (3), pp.190-195.

[9] Ridhuan, K and Suranto J. 2016. Comparison of Pyrolysis Combustion and Carbonization from Durian Peel to Calorofic Value (In Indonesian). Turbo, Vol 5 (1), pp.50-56.

[10] Antal, M.J., and William, S.L. Mok., 1990. Review of methods for improving the yield of charcoal from biomass, Energy & Fuel, 4(3), pp. 221-226.

[11] Mansaray K.G and Ghaly A.E. 1998. Thermal Degradation of Rice Husks in Nitrogen Atmosphere. Bioresoource Technology , 65, pp.13-20.

Material flow analysis for resource management towards resilient palm oil production

H Kamahara[1], M Faisal[2,6], U Hasanudin[3], K Fujie[4], and H Daimon[5]

[1] *Toyohashi University of Technology*, Toyohashi, Japan
[2] *Syiah Kuala University*, Banda Aceh, Indonesia
[3] *University of Lampung*, Bandar Lampung, Indonesia
[4] *Yokohama National University*, Yokohama, Japan
[5] *Toyohashi University of Technology*, Toyohashi, Japan

E-mail: mfaisal@unsyiah.ac.id

Abstract. Biomass waste generated from palm oil mill can be considered not only as the feedstock of renewable energy but also as the nutrient-rich resources to produce organic fertilizer. This study explored the appropriate resource management towards resilient palm oil production by applying material flow analysis. This study was conducted based on two palm oil mills in Lampung, Indonesia. The results showed that the empty fruit bunch (EFB) has the largest potential in terms of amount and energy among the biomass waste. The results also showed that the palm oil mills themselves had already self-managed their energy consumption thatwas obtained from palm kernel shell and palm press fiber. Finally, this study recommended the several utilization options of EFB for improvement of soil sustainability to contribute towards resilient palm oil production.

1. Introduction

Agro-industries such as palm oil mill, sugar cane mill and cassava mill are recognized as the leading industries in Indonesia, because the main products such as palm oil, sugar and tapioca have been widely used as beneficial dairy products. On the other hand, generally, only 25% or less amount of the raw materials are converted into the beneficial products while the rest of them is discharged as waste. The effective use of these wastes can contribute not only for sustainable production of renewable energy and renewable resources but also to prevent general waste issues such as waste pollution and illegal disposal.

Palm oil mill is well known as a surplus energy industry by utilization of these biomass waste. The potential of waste biomass utilization in palm oil mill has been shown by several studies [1, 2, 3, 4, 5, 6]. These studies indicated that the surplus energy can be obtained by utilizing biomass waste such as palm kernel shell (PKS), palm pressed fiber (PPF) and empty fruit bunch (EFB), and also the biogas produced from palm oil mill effluent (POME).

However, they paid less attention on the soil sustainability of oil palm plantation. For example, soil pH, organic C and total N in Johor, Malaysian oil palm plantation have seen decline from 1967 to 2000 [7]. Losing nutrients in the soil can be the cause of the decline of palm oil productivity. For this reason, maintaining the soil organic matter and nutrients in soils are important for the sustainability of

[6] To whom any correspondence should be addressed.

palm oil production. The alternative resource management for resilient palm oil production is required to contribute to archive renewable energy production and soil sustainability. Therefore, the purpose of this study is to explore the appropriate resource management for sustainable palm oil production by evaluating the energy as well as nutrient balances of palm oil mill and plantation. Understanding these balances are possible to lead to resilient palm oil production.

2. Experimental procedure and methodology

2.1. Material flow analysis and indicators

Material flow analysis (MFA) is defined as a method to analyse the flow and stock of materials in a system systematically [8]. MFA describes the amount of material in each process, namely material generation, intermediate and final form. MFA is a useful tool to determine resource management, waste management and environmental management because it is possible to determine input and output in a system based on the law of conservation of mass [8]. This study applied MFA to resource management for sustainable palm oil production. First, this study identified material flow of biomass through palm oil production. Secondly, this study evaluated energy and nutrient potential of biomass wastes. Finally, this study discussed about the appropriate resource management by reviewing several relating researches.

2.2. Palm oil mills and plantation

This research was conducted based on two palm oil mills with production capacity of 25 t and 40 t of fresh fruit bunches (FFB) per hour, respectively. The latter palm oil mill has palm kernel milling process. These mills are managed by Indonesian state own company that are located in Lampung Province, Indonesia. Composition of biomass residue was obtained from reference elsewhere [9].

3. Results and discussions

3.1. Material flow in palm oil mills

Productivity of palm oil varies from company to company due to different factors such as soil, climate, and processing conditions. Companies that have better management would result into higher productivity. For instance, certain company produced an average of 23 t-FFB/hectare/year with an average of 24% of oil extraction rate (OER). Some companies produced less, for instance only an average of 13 t-FFB with 18% of OER. Based on this information, it was estimated that about 80% of FFB are considered as biomass waste.

Figure 1 and Figure 2 showed the material and energy flow in the observed palm oil industries. It was shown that the OER value was about 21.8% from FFB. Table 1 showed biomass waste generation rate in palm oil mills, Lampung (this study) and Aceh [6], Indonesia and Thailand [3]. Biomass waste generation rates were not much different in each palm oil mills. Based on the calculation, palm oil process produced solid waste in the forms of EFB, fiber and shell, which were about 40-45% of the total FFB. From the observation, palm oil mills in Lampung have already used the biomass waste (PPF and PKS) to generate electricity and steam for palm oil processing. Diesel oil was used to generate electricity only in the start-up stage. After the start-up was achieved, steam which is generated from PPF and PKS would take place as the source of energy for palm oil processing. 85% of total PPF and 55% of total PKS were directly used as fuel in the boiler system of the factory to generate electricity and steam for palm oil processing. It was predicted that 15.6 t/day of PPF and 18 t/day of PKS would be remained in the cases of production capacity 40 t-FFB/hour or about 800 t-FFB/day. In common practice, EFB is used as a mulching for soil conditioner and nutrients recycling in oil palm plantation, nevertheless such practice might cause increasing of soil toxicity. Energy consumption of palm oil mill came only from PKS and PPF except diesel oil during start-up. Furthermore, biogas production from POME was also expected as a renewable energy production. POME can produces a tremendous amount of biogas by anaerobic digestion method for self utilization in the palm oil mill. However for small and medium scale factory, the investment cost for biogas utilization is considered as relatively high.

3.2. Palm oil biomass potential for soil quality improvement

Palm oil mill produced several biomass waste containing energy and nutrients. At the same time, oil palm plantation itself applies several chemical fertilizers. Hence the biomass waste can not only used to displace the chemical fertilizers, but also to maintain the soil fertility. Maintaining soil health of the plantation is one of the important aspects for the sustainability of crop production. Therefore, this paper considered both energy and nutrients potential of biomass waste from the palm oil mill.

Table 2 showed the energy and nutrients potential of biomass waste from palm oil mill. FFB yield of this plantation was 17.6 t/ha/year. Based on this yield, EFB has the highest potential of energy. Generally, nitrogen fertilizer consumes more energy than other fertilizers in the production process. The oil palm plantation applied about 280 kg/ha/year of urea relevant with 131 kg/ha/year of nitrogen [4]. On the other hand, total nitrogen potential of biomass waste generated from palm oil was 57 kg-N/ha/year (see Table 2). Therefore, the estimation of this study can roughly fulfill 44% of chemical fertilizers use in oil palm plantation.

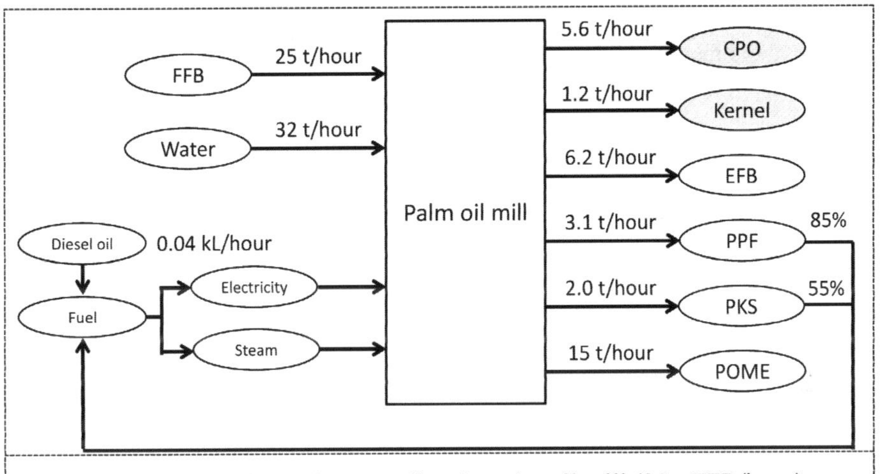

Figure 1. Material and energy flow in palm oil mill (25 t-FFB/hour).

Table 2 showed the moisture content and C/N ratio as well. Generally, C/N ratio of 30 and moisture content of 50-60% are considered as the suitable conditions for compost production [10]. Producing compost from EFB has been succeeded by co-composting with partially tresated POME [11] and anaerobic treated POME [12]. The products of co-composting of EFB and anaerobic treated POME have a nutrient value of 2.31% N, 1.36% P, and 1.04% K [12]. It was assumed that 1 t of EFB compost was equivalent to the chemical fertilizer, i.e., 47.8 kg of N fertilizer, 25.6 kg P fertilizer, and 46.5 kg K fertilizer [13]. Additionally, co-fermentation with POME on anaerobic digestion is also possible to produce biogas as renewable energy, as well as the production of organic fertilizer [14]. This integrated waste treatment technology is also expected to reduce greenhouse gas emissions of biomass waste treatment in palm oil mill [15]. Further study is expected to clarify the dynamic effects of this technology to the life cycle of palm oil production.

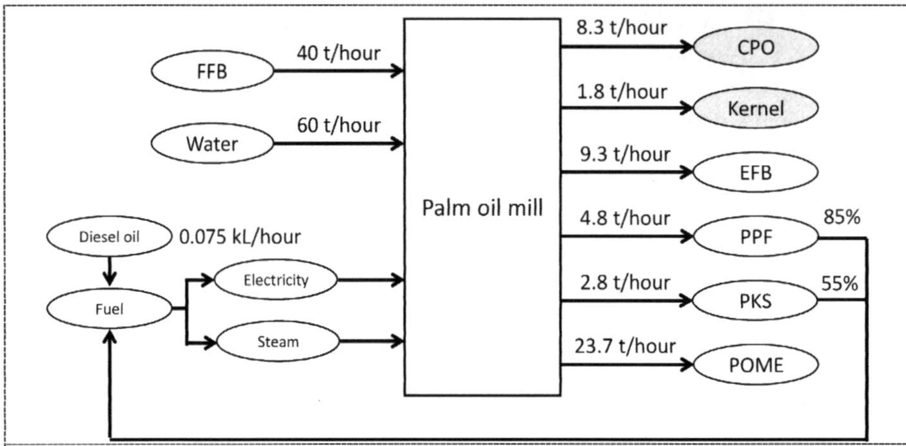

Figure 2. Material and energy flow in palm oil mill (40 t-FFB/hour).

Table 1. Biomass waste generation rate (%-FFB) of palm oil mills.

	Lampung, Indonesia (25 t-FFB/hour) This study	Lampung, Indonesia (40 t-FFB/hour) This study	Aceh, Indonesia, [6]	Thailand, [3]
EFB	24.8	23.3	21.5	24.0
PPF	12.4	12.0	12.3	14.0
PKS	8.0	7.0	6.1	6.1
Total	45.2	42.3	39.9	44.1

Table 2. Energy and nutrients potential of biomass waste from palm oil mill.

Parameters		PPF	PKS	EFB
Amount	t/ha/year	2.5	0.9	4.4
Nitrogen potential	kg-N/ha/year	28	9.5	20
Carbon potential	kg-C/ha/year	1,164	440	2,003
Energy potential	GJ-LHV/ha/year	49	17	75
Moisture content	%	37	21	57
C/N ratio	-	42	46	101
Lower heating value (LHV)	MJ/kg	20	18	17
Carbon content per 1 MJ of LHV	g-C/MJ	24	25	27
Nitrogen content per 1 M of LHV	mg-N/MJ	571	546	264

4. Conclusion

This study evaluated the energy and nutrient balance of palm oil production by applying material flow analysis for two palm oil mills in Lampung, Indonesia. The results highlighted the importance of biomass residue utilization to contribute on the soil sustainability towards resilient palm oil production. The potential of nitrogen fertilizer was roughly estimated as 44% of total nitrogen chemical fertilizer. These palm oil mills were already self-managed their energy consumption by burning PKS and PPF. Therefore, the utilization of EFB is a crucial task with high priority. The co-composting of EFB with POME would be an atrractive alternative technique to produce biogas and compost from EFB. Further practice was recommended to enhance the resilience of palm oil production.

References

[1] Mahlia T M I, Abdulmuin M Z, Alamsyah T M I and Mukhlishien D 2001 *Energy Conv. and Manage.* **42**(18) 2109-2118
[2] Yusoff S 2006 *J. of Cleaner Product.* **14**(1) 87-93
[3] Chavalparit O, Rulkens W H, Mol, A P J and Khaodhair S 2006 *Environ. Develop. and Sustain.* **8**(2) 271-287
[4] Kamahara H, Hasanudin U, Widiyanto A, Tachibana R, Atsuta Y, Goto N, Daimon H and Fujie K 2010 *Biom. and Bioene.* **34**(12) 1818-1824
[5] Ng F Y, Yew F K, Basiron Y and Sundram K 2011 *J. of Oil Palm and the Environ.* **2** 1-7
[6] Faisal M and Mahidin B 2013 *Int. J. on Adv. Science, Eng. and Inform. Tech.* **3**(3) 32-38
[7] Ng P H C, Gan H H and Goh K J 2011 *J. of Oil Palm and the Environ.* **2** 93-104
[8] Brunner P H and Rechberger H 2003 *Practical Handbook of Material Flow Analysis: Advanced Methods in Resource and Waste Management Series 1* (CRC Press LLC, Florida)
[9] Umemura Y, Omar W N, Tsutsui T and Yusup S B 2011 *Fuel* **90**(8) 2585-2591
[10] Chen L, de Haro M M, Moore A and Falen C 2011 *The Composting Process* (Available online: http://www.cals.uidaho.edu/edcomm/pdf/CIS/CIS1179.pdf)
[11] Baharuddin A S, Wakisaka M, Shirai Y, Abd-Aziz S, Abdul R and Hassan M A 2009 *Int. J. of Agri. Res.* **4** 69-78
[12] Baharuddin A S, Hock L S, Yusof M Z, Rahman N A A, Shah U K, Hassan M A, Wakisaka M, Sakai K and Shirai Y 2010 *African J. of Biotech.* **9**(16) 2427-2436
[13] Chiew Y L, Shimada Y 2013 *Biom. and Bioene.* **51**(1) 109-124
[14] Hasanudin U, Sugiharto R, Haryanto A and Fujie K 2015 *Water Science and Tech.* **72**(7) 1089-1095
[15] Amelia J R, Suprihatin S, Indrasti N S, Hasanudin U and Fujie K 2017 *IOP Conf. Series: Earth and Environ. Science* **65** 1-8

Biohythane system using two steps of POME fermentation process for supplying electrical energi : economic evaluation

P Zuldian[1,2], Z D Hastuti[1], S D S Murti[1] and Adiarso[1]

[1] Center of Technology Energy Resources and Chemical Industry,
Agency for Assessment and Application of Technology.
625 Building, Technology Energy Cluster, PUSPIPTEK, South Tangerang

E-mail: prima.zuldian@bppt.go.id

Abstract. Indonesia as the largest producer of palm oil in the world has the prospective to generate additional benefits such as electricity by utilizing Palm Oil Mill Effluent (POME). The high Chemical Oxygen Demand (COD) content of 35,000 ppm POME is a great potential for conversion to hydrogen and methane through a fermentation process. In this study, two stages of fermentation using a microbial consortium have been performed in the 1 m^3 BioHythane reactor system to produce biohydrogen and biomethane. After two-stage fermentation process for 24 hours in this system, the microbial consortium succeeds in producing biohydrogen and biomethane of 32 and 60 vol. %, respectively. This gas product after the purification process could be converted to electricity to be 0.02 and 0.75 kWe, respectively. Furthermore, as result of economic calculation analysis, this biohythane system showed up the value of Capital Expenditures (CAPEX) of US $ 26,39540 and Operating Expenses (OPEX) of US $ 14,712 per year, and resulted total generated electricity cost of US $ 2.478 / kWh.

1. Introduction

Palm plant has been an interesting commodity to be developed continuously due to its high potency even in Indonesia which has the biggest production in the world. During its development, the use of palm plant is mostly prioritized as an raw material for household necessity, although in fact it has high potency to produce renewable energy product based on waste which are not still optimal yet to be developed that consists of kernel, Fresh Fruit Bunch FFB, branch, and POME. POME provides potency as high supplier emission in plantation sector. It is caused by still high CO_2, CH_4 and other gases (in CO_2 equivalent unit) which still be large to be released to surrounding without additional treatment process.

In West Java at Bogor regency has been established Cikasungka palm oil plant underauspice of PTPN VIII company which produce Crude Palm Oil (CPO) having capacity 30 ton FFB/hour which has potency to produce POME amount 300 liter/ton FFB with COD content around 35.000 ppm. The steps for processing POME were analyzed at BPPT by conducting two steps fermentation of POME using consortium bacteria to produce biohydrogen continued by biomethane production. Naturally, the process of POME degradation were started from hydrolysis, acidification, acetat acid formation until methanation process (figure 1). Biohydrogen formation occurs in acetate acid formation step from

[2] To whom any correspondence should be addressed.

which the content of biohydrogen could achieve 32 vol. % of first-fermentation product. In this step there would be conversion of COD process by 25% of total COD remaining recidual-fermentation process which still has high COD composition by 75% of total COD in POME. The specification of recidual waste still could not to be disposed directly to surrounding because the required standardization was not complete yet. Therefore, the second fermentation step for producing methane were carried out to degrade remained COD which could produce CH_4 by 60 vol. % of second-fermentation product.

Biohydrogen become prospective commodity which could be used as source of new energy source (fuel), raw material, reductor and chemical. Meanwhile, hydrogen is still produced dominantly from natural-gas-reforming process in chemical and petrochemical industries. Furthermore, there is another technology growing up hydrolysis technology, due to the high operational cost makes this technology could not to be developed well.

Related to availability of natural gas as principal hydrogen supplier, Indonesian dependence against natural gas would be unstable due to the decrease of natural gas sources. Even Indonesia was forecasted to be net natural gas importer in 2024 based on Indonesia Energy Outlook 2016 analysis [1]. Another effect that would be problem was the purchasing price of natural gas by industry sector which tends to increase, thus only big industries having big funding for providing natural gas which would keep existing.

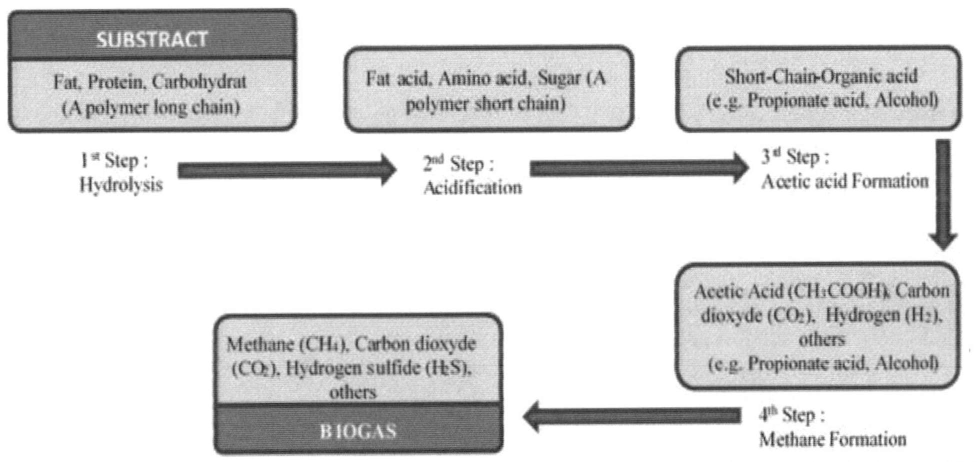

Figure 1. Anaerobic decomposition processes of POME.
(Source : adapted from nachwaschender-rohstoffe.de)

The fact became a reason why the use of renewable energy for natural gas substitution like biohydrogen and biomethane should be realized soon. It could be proper solution in current energy crisis. In the other hand, dissemination of palm plant areas and palm industries from Sabang until Merauke shows up that a big-potencial-POME reserve have been available. Recorded in 2015 that total palm plant area was 11.300.370 hectare with total CPO production achieved 31.284.306 tonne [2].

Currently, utilization of hydrogen is dominantly used as feedstock and fuel cell power plant system which mostly used as portable electricity supply in Base Transceiver Station (BTS) and installations which are placed or not connected with electricity grid yet. Meanwhile, the utilization of hydrogen has been started and continuously developed as future transportation fuel having low emission content too.

Regarding to utilize Hydrogen from POME, BPPT has been developing biohythane prototype system with capacity 1 m^3/day. The operation of this system has been started by conducting initial laboratory scale research began from 5 ml, 50 ml, 2.5 liter, 40 liter and 1 m^3 capacities. Each different capacity consist of POME was added by microbes contained inside sludge which was obtained from an anaerobic pond at PKS Cikasungka as well. These obtained microbes were bred, screened and inoculated in to get consortium microbes which support hydrogen formation process. Within 24 hours, consortium microbes would convert COD become H_2 and CO_2 as dominant compounds which were detected during anaerobic fermentation condition.

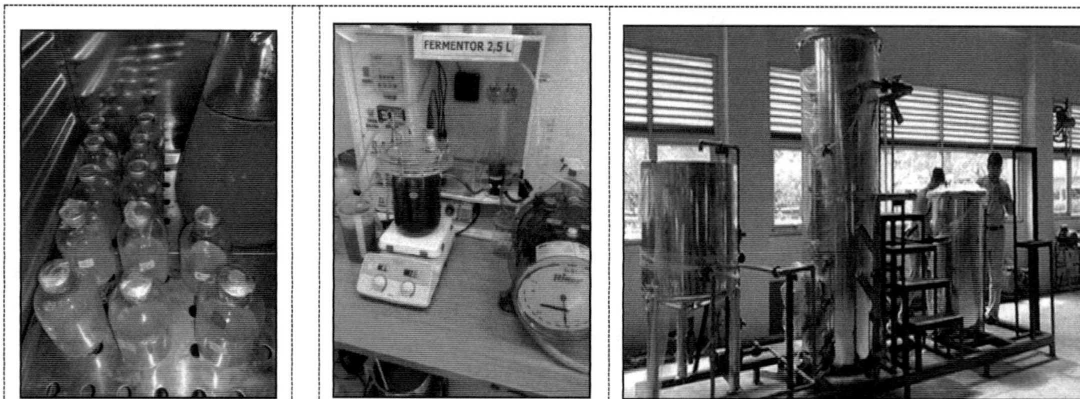

(a) (b) (c)

Figure 2. a). 50 ml Biohydrogen fermentation process, b). 2.5 liter Biohydrogen fermentation process, c). One m^3 prototype of Biohythane system (Source : private document).

Output POME from first fermentor was chaneled into second RTD more than 24 hours, the composition of gas from fermentation had changed where methane had been formed become dominant product in line with the increase of RTD. The peak composition of methane could achieve 60 vol. % and CO2 40 vol. %.

Based on the results of research above. The highest composition of Biohythane and Biomethane amount 32 vol. % and 60 vol. % respectively would be purified to achieve 99% wt and used to estimate potencial energy especially electricity which could be generated from prototype 1 m^3 Biohythane system. Furthermore, economic calculation and analysis were conducted for estimating the value of Capital Expenditures (CAPEX), Operational Expenditures (OPEX) and electricity generation cost resulted from Biohythane system.

2. Experimental procedure and methodology

Methodology of analysis and calculation was conducted by using the data of research result. It was supported by additional data and assumptions eferred to engineering references related with electricity and economic calculation methods. Detail methodologies were explained together with explanation of result calculations in sub points at below.

2.1. Analysis Electricity Potency

Calculation for electricity analysis did not only use some internal data but also use external-source data derived from some relevant research and engineering references that have been published for substitute data against unknown information data in this economic calculations. Calculation of electricity capacity involved electrical calculations of biohydrogen and biomethane which were channeled toward fuel cell engine and biogas/biomethane engine to generate electricity.

A. There were some data resulted from research and calculation on prototype 1 m^3 Biohythane, namely :=

1. HHV of Biohythane (Bio-H_2 99% wt) = 10.1 MJ/m^3
2. HHV of Biomethane (Bio-CH_4 99% wt) = 33.6 MJ/kg

B. There were some data from PKS Cikasungka, namely:
1. Production capacity of PKS Cikasungka = 30 ton FFB/hour.
2. POME waste production = 300 liter POME/ton FFB.
3. Total COD = 35,000 ppm

C. There were some assumptions used for electricity potency analysis, namely :
1. POME was supplied 0.6 m^3/day or 60% of total prototype capacity.
2. H_2 density = 0.08698 kg/m^3
3. CH_4 density = 0.656 kg/m^3
4. COD consumption for biohydrogen synthetis amount 25% of total COD [3].
5. COD consumption for biomethane syhnthetis amount 75% of total COD [3].
6. Ratio of H_2/COD = 0.0892 nm^3/day
7. Ratio of CH_4/COD = 0.35 nm^3/day [3]
8. H_2 target for fuel cell engine = 99%
9. CH_4 target for biogas/biomethane engine = 99%
10. Gas engines efficiency = 34.6% [3]

2.2. Analysis of Economic Calculation

Economic calculation and analysis were classified into three calculation parts on 1 m^3 biohythane prototype system which will be planned to be placed at PKS Cikasungka. These parts were Capital Expenditures (CAPEX), Operational Expenditures (OPEX) and electricity generation cost calculation. There were real data used in calculation which supported by some external data and relevant assumptions. Base year calculation used the year of 2015 because the availability of external data and assumptions were the most complete.The following real and supporting data were described in detail as follows:

D. There were some real data were obtained from direct survey which closely related to Biohythane prototype information, namely:
1. The land cost was assumed free of charge.
2. Electricity consumption was supplied from external plant distributed by Indonesian Electricity State Enterprise (PLN) where the cost of purchasing electricity from PLN was
3. Water consumption was assumed free of charge.
4. Employee salary info was the real data obtained from a small scale power plant info.

E. There were some assumptions which used in economic calculation including as follows :
1. One m^3 plant was operated for 365 days having life time 16 years.
2. The capital was owned 100%, not a loan.
3. Exchange rate was IDR 13,000/US$.
4. POME was obtained free in charge and transported from POME pond to Biohythane prototype by using truck with length-distance around 200 meters.
5. All equipments were not imported.
6. Calculation of working capital cost before prototype could continously operate was assumed 1 month.
7. Annual tax income was 10% of total obtained profit which referred to government regulation.
8. Electricity generation in prototype used capacity factor by 90% of total generated electrical capacity
9. Depreciation cost was calculated by straight line method [4].
10. Electricity cost and Indonesian electricity consumer indices were used to project future costs especially on data which were available before year 2015 [5].
11. Return on Investment Rate (ROR) was stated 13%.

3. Results and Discussion

The result of calculation in this part was obtained using information from calculation methodology above with some equations which will be further explained in this part. The result of CAPEX and OPEX calculations, and economical analysis would be analyzed in detail as source information to explain potency utillization of Biohydrogen and Biomethane according to economic aspect.

3.1. Electricity Potency Estimation

The calculation of electricity capacity potential was referred to detail POME info. It consists of some equations which closely related to some info in part 2.1 namely as follows :

1. Daily wastewater flow

   = Daily waste water volume *(No.C1)* [3] (1)

2. COD Loading (kg.COD/day)

   $$= \text{COD POME } (No.B3) \text{ x Daily wastewater flow } (eq.1) \text{ x } \frac{\text{kg}}{1.000.000 \text{ mg}} \text{ x } \frac{1000 \text{ l}}{\text{m}^3} \quad [3] \quad (2)$$

3. Bio-H_2 or Bio-CH_4 Production (nm^3 H_2 or CH_4/day)

   = COD loading *(eq.2)* x COD eff *(No.C8 or C9)* x H_2 or CH_4/COD *(No.C6 or C7)* [3] (3)

4. Generated Electricity Capacity (kWe)

   $$= \frac{\text{H2 or CH4 production } (eq.3) \text{ x HHV} \frac{H_2}{COD} \text{ or } \frac{CH_4}{COD} (No.A1 \text{ or } A2) \text{x Eff.Gas Engine } (No.C10)}{24 \text{ hr x } 60 \text{ minutes x } 60 \text{ seconds}} \quad [3] \quad (4)$$

According to equation 1 to 4 above, the calculation of electricity power potency on 1 m^3 biohythane system was conducted by different calculation from POME fermentation for biohydrogen and POME fermentation for biomethane. The result of calculations were given in table 1.

Table 1. Calculation of potencial-electricity generation from prototype 1 m^3 biohythane system.

No	Remark	Biohydrogen	Biomethane
1	Daily wastewater flow (m^3 POME/day)	0.6	
2	COD loading (kg.COD/day)	5.25	15.75
3	H_2 production or CH_4 (Nm^3/day)	0.464	5.457
4	Electricity Generation Capacity (kWe)	0.019 kWe	0.734 kWe
	Total electricity capacity (kWe)	0.753 kWe	

According to calculation, one m^3 biohythane system had potency to generate 0.735 kWe electricity capacity. This capacity showed up potencial-capacity value was big enough to generate electricty although it was only generated from 1 m^3 plant capacity.

3.2. The Estimation of CAPEX

The estimation of CAPEX was closely determined by two-important components namely Fixed Capital Investment cost which having cost by 98% of CAPEX and Working Capital cost by 2% of CAPEX. By inputting data both real and assumption data based on engineering references, then the result of CAPEX calculation could be summarized into table 2.

IOP Conf. Series: Materials Science and Engineering **334** (2018) 012010 doi:10.1088/1757-899X/334/1/012010

Purchased equipment cost (No.A.I.1) in fixed-capital investment was real data which considered safety cost by assuming it by 5% of total purchase equipment too. Electrical component system (No.A.I.2) was obtained from real data for electricity distribution having length-distance 200 meters connected with PLN network electrical distribution. Contingency component was calculated by assuming it at amount of 15% of total fixed-capital investment (No.A).

In Working Capital calculation, transportation POME cost (No.B.4) was obtained from fuel vehicle consumption for length-distance 200 meters/day by truck. The intensity of solar consumption for truck was assumed 0.33 liter/km and diesel/solar price was stated amount IDR 6,950/liter based on diesel/solar price in 2015.

Monthly regular-employee salaries (No.B.5) consist of financing expense for three employees. These employees were assumed as permanent employee who have been involved before starting prototype operation. General administration cost was important-enough component which should be available in each project. It was calculated by assuming it by 15% of total monthly regular-employee salaries (No.B.5). Start-up cost component (No.B.7), it has close relationship with monthly regular-employee salaries (No.B.5) where involved employees as well. The difference was mechanism of salary payment where start-up cost was paid daily (for 14 days), not monthly.

According to the estimation results, total CAPEX was US$ 26,395. It consists of US$ 25,833 and US$ 563 for Fixed Capital Investment and Working Capital Costs, respectively

Table 2. CAPEX estimation of Biohythane system.

No	Component	Cost ($)
I	**Direct Cost**	
1	Purchase Equipment (PE), including instalation, instrumentation & control, piping and safety percentage costs (5% of PE)*	19,304
	1.1. Biohythane production system*	11,923
	1.2. Biohythane purification system*	3,462
	1.3. Biohythane electricity generation system*	3,000
2	Electrical system cost [6],**	2,654
	Total Direct Cost *(Total No.1 and 2)*	**21,958**
II	**Indirect Cost**	
1	Contingency [6]	3,875
	Total Indirect Cost *(Total No.3)*	**3,875**
A.	**Fixed Capital Investment Cost *(Total No. I.1, I.2 and II.3)***	**25,833**
1	Transportation POME cost (around 200 metes)*	1
2	Monthly regular-employee salaries **	462
3	General administration cost	69
4	Start-up cost (including interim laborer salaries)**	31
B.	**Working Capital Cost (WC) *(Total No. 4,5,6)***	**563**
	Total CAPEX *(Total A + B)*	**26,395**

* = Data was closely related with process of prototype 1 m^3 biohythane.
** = Data for these components were external data obtained from direct survey

3.3. The Estimation of OPEX

On the calculation of OPEX, it was calculated by summing Variable Production Cost, Fixed Cost and Plant Overhead Cost. The result of calculations as shown in table 3.

In principal, POME transportation cost calculation (table 3) used the same method as transportation cost in the working capital cost of CAPEX calculation (table 2) for annual calculation. Annual regular-employee salary (table 2) in this calculation used the same method as monthly regular-employee salary (table 2) in CAPEX calculation .

The calculation of utilities cost (table 3) were highlighted on the consumption of electricity consist of heater, generation system and purification system pumps as shown in table 4 which were calculated by summing total electricity consumption including additional safety cost by 5% of total utilities cost. The electricity was supplied from PLN by using electricity tariff in 2015 amount US$ 1,116/kWh for consumer with grade I-3.

Table 3. The estimation of OPEX on Biohythane system.

No	Component	Cost (US$/year)
1	POME transportation cost*	13
2	Annual regular-employee salaries**	7,385
3	Utilities (e.g. electricity)**	4,951
4	Maintenance and Repair cost [6]	1,056
5	Operational supply (e.g. lubricant, chemicals) [6]	158
A.	**Total Variable Production Cost** ***(Total No. 1,2,3,4,5)***	**13,560**
6	Depreciation [6],[10]	1,149
B.	**Total Fixed Charge** ***(Total No.6)***	**1,149**
	TOTAL OPEX ***(Total A + B)***	**14,712**

* = Data was closely related with process of prototype 1 m^3 biohythane.
** = Data for these components were external data obtained from direct survey.

Meanwhile, maintenance and repair cost (No.A.4) were calculated with assumption by 4% of Fixed Capital Investment (table 2). Operational supply cost was calculated by using assumption as well. Operational supply consists of some purchased chemicals and lubricants which were needed during operational process. Especially in Biohythane system, some chemicals consist of alkaline solutions, alcohol and aquadest. In this calculation, operational supply was calculated by 15% of maintenance and repair cost.

Table 4. Electricity consumption on 1 m^3 Biohythane prototype's equipments.

No	Equipment	Unit	Power consumption (KW)
1	Heater	1	5
2	Generation system pumps	2	0.42
3	Purification system pumps	2	0.42
	TOTAL		**5.84**

The depreciation cost as the one component which was considered in fixed charge calculation (table 3) of biohythane system equipment. It was calculated by referring to depreciation tariff based on Finance ministry regulation of Indonesia KMK.138/KMK.03/2002 as shown in table 5 at below.

Table 5. Depreciation tariff based on Finance ministry regulation of Indonesia.

Components	(year)	Tariff	Detail components for each list
Non Building			
1st Category	4	25%	Office machines, spareparts, tools
2nd Category	8	12.5%	Car, vehichles
3rd Category	16	6.25%	Industry machines

4th Category	20	5%	Heavy machines for construcution
Building			
Permanent	20	5%	Building and supporting building
Non Permanent	10	10%	-

On the other hand, deprecation cost on 20 kV electricity distribution equipments (table 2) was calculated too according to real data from PLN. Depreciation cost was calculated using equation 5 and 6.

$$\text{Salvage value} = \left(\frac{\text{Real component life}}{\text{Component Life in regulation}}\right) x \text{ Tariff x Initial component cost} \quad (5)$$

$$\text{Depreciation} = \frac{\text{Initial component cost} - \textit{Salvage value}}{\text{Real component Life}} \quad (6)$$

According to overall calculation, total resulted OPEX was US$ 14,712 /year consist of total Variable Production Cost and total Fixed Charge Cost amount US$ 13,563/year and US$ 1,149/year respectively.

3.4. The estimation of Economic Analysis
Economic analysis was conducted to determine Electricity Generation Cost (EGC) calculation. In other hand, it involved profit, payback of period time and return of investment cost (POT & IRR) even Break Even Point (BEP) calculations as well. EGC value was calculated using equation 7.

$$\text{EGC} = \frac{\text{Total OPEX}}{\text{Capacity factor x 0.753 kWe x 8,760 hr/year}} \quad (7)$$

Rate on Return (ROR) became an absolute parameter whose its value referred to an engineering reference which presented good calculation about economic prospect of biogas power plant. Although it has different type with Biohythane, ROR value from biogas power plant could be a parameter which has close relativeness with other economic parameters.

By using biogas references, ROR value used was 13% [8]. The ROR was used as principal parameter which determined the value of other parameters using iteration method. As long as iteration, the value of ROR keep the value of ROR always keep 13%. The value of all economic parameters after iteration were shown in table 6.
Based on calculation, the value of EGC achieved US$ 2.478/kWh. Meanwhile, others parameters could be determined as well including annual cash flow, Pay Out Time value (called POT) was 7.55 years, IRR was 11% and BEP was 23% or around 3.71 years. POT, IRR and BEP were calculated by equation 8,9 and 10.

Annual Cash Flow (ACF) was calculated from difference of total Annual Electricity Sale Price (ESP) which had been multiplied with annual total generated electricity to total OPEX (see Table 3) and income tax (10% of ACF-(total annual ESP x annual total generated electricity (kWh/year))). Annual generated electricity amount 5,938 kWh/year was calculated by multiplying between assumed capacity factor (90%) and total electricity capacity 0.753 kWe (see Table 1). The value of ESP obtained was US$ 3.118/kWh which was calculated from iteration of ROR = 13% as well. In order to achieve profitability, the value of ESP always have to set be higher than EGC.

$$\text{POT} = \frac{\text{FCI } (\textit{see No. A in Table } 2) + \text{Interest of } \textit{CAPEX}}{\text{ACF } (\textit{No.3 in Table } 6)} \quad (8)$$

Remark : Interest of CAPEX was zero due to do not have a loan

$$\text{CAPEX } (\textit{see Table } 2) = \text{ACF } (\textit{No.3 in Table } 6)\left[\frac{1}{(1+i)^1} + \frac{1}{(1+i)^2} + \dots\dots + \frac{1}{(1+i)^n}\right]$$

(9)

Where i = IRR, n = life time of prototype (16 years)

$$\text{BEP} = \left[\frac{\text{Total Fixed Charge } (\textit{see No.B in table 3})}{\text{Total Electricity Sale Cost} - \text{Total Variable Production Cost } (\textit{see No.A in Table 3})}\right] \times 100\% \quad (10)$$

The whole calculations showed that this power plant already has good prospect to be built although the value of EGC was high enough. Moreover, if it was compared with some fossil and renewable energy power plants in commmercial scale which has lower EGC value. However it still could be understood because prototype capacity was small. Thus, it might be possible to result lower cost when the prototype capacity was increased.

Table 6. The economic calculation of biohythane system.

By iteration with ROR = 13%			
No	Parameter	Unit	Result
1	Electricity Generation Cost (EGC)	US$/kWh	2.478
2	Electricity Sale Price (ESP)	US$/kWh	3.118
3	Annual Cash Flow (ACF)	US$/year	3,422
4	Payout Time (POT) [7]	years	7.55
5	Internal Rate of Return (IRR) [8]	%	13%
6	Break Even Point (BEP)	% or years	23% or 3.71 years

4. Conclusion and Recommendation

The prototype of 1 m^3 biohythane system could generate electricity 0.735 kWe resulted from two steps anaerobic fermentation process of POME with COD content 35,000 ppm. It was transported from PKS Cikasungka which have been fermented achieving H_2 and CH_4 amount 32 vol. % and 60 vol. % from first and second fermentations, respectively which have been purified by biohytane purification system. According to economic calculation, the prototype has CAPEX and OPEX values amount US$ 26,395 and US$ 14,712/year, respectively. On the other hand, it resulted EGC at amount of US$ 2.478/kWh.

The increase of generated electricity capacity by increasing prototype capacity could be an effort to decrease EGC value. Calculation and analysis using higher capacity become recommendation to achieve EGC even ESP value which could compete with other EGC an ESP values from other power plant types. Regarding to effort to increase H_2 and CH_4 compounds from two steps fermentation process, then the use of different source POME pond which would probably produce higher COD. The COD used in calculation was 35,000 ppm. This value was categoried small because in theory that general COD from POME could achieve composition more than 50,000 ppm. Therefore, by increasing COD then it would affect the increase of Biohydrogen and Biomethane compounds which could increase generated electricity from these gases as well.

Acknowledgement

The authors do thank and fully give high appreciation and honor to Agency for Assessment and Application of Technology (BPPT) and Ministry of Higher Education and Culture of Indonesia which have given financial support during this research conducted

References

[1] Sugiyono, Agus, Anindhita, Laode M.A Wahid, Adiarso 2016, *Indonesia Energy Outlook 2016, Energy Development in Supporting Green Industry*, Jakarta, Indonesia
[2] Directorate of Estate Crops 2015, *Indonesia Estate Crops Statistic 2014-2016*, Palm tree:

Jakarta, Indonesia

[3] Rahayu, A.S., Karsiwulan, D., Yuwono, H., Trisnawati, I., Mulyasari, S., Rahardjo, S., Hokermin, S., and Paramita, V 2015, *Buku Panduan Konversi POME menjadi Biogas: Pengembangan Proyek di Indonesia (A convertion POME becomes biogas's guidebook)* , USAID & WINROCK International, Jakarta, Indonesia

[4] The 1945 Constitution of the Republic of Indonesia 2010, *A constitution No.36/2008 about Income Tax*, Jakarta, Indonesia

[5] Indonesian Ministry of Energy and Mineral Resource (MEMR) 2015, *Handbook of Energy and Economics Statistic Indonesia 2015*. Jakarta, Indonesia

[6] Peters, M.S., Timmerhaus, K.D. dan West, R.E. 2003, *Plant Design and Economic for Chemical Engineers, 5th Ed: New York, ch. 6-9,12.H. B. Maynard and W. K. Hodson. Maynard's Industrial Engineering Handbook, 4th Ed. 1992, pp.1293.* New York, United States

[7] Garrett, Ed 1989, *Chemical Engineering Economics pp. 354.*New York, USA

[8] Sinnot, R. dan Towler, G. 2009, *Chemical Engineering Design, 5th Ed, p.359*, Burlington

IOP Conf. Series: Materials Science and Engineering **334** (2018) 012011 doi:10.1088/1757-899X/334/1/012011

Synthesis of K_2O/Zeolite catalysts by KOH impregnation for biodiesel production from waste frying oil

N Fitriana[1], H Husin[1,2,3] D Yanti[1], K Pontas[1,2], P N Alam[2], M Ridho[2] and Iskandar[2]

[1] Graduate Program in Chemical Engineering, Syiah Kuala University, Darussalam, 23111, Banda Aceh, Indonesia
[2] Department of Chemical Engineering, Syiah Kuala University, Darussalam, 23111, Banda Aceh, Indonesia

Email: husni_husin@che.unsyiah.ac.id

Abstract. K_2O/Zeolite compounds were successfully synthesized using KOH as starting material and natural zeolite as support. The catalysts were calcined at 500°C for 3 h and then characterized by X-Ray Diffractometer (XRD) and Scanning Electron Microscopy (SEM). The SEM images reveal that the zeolite and K_2O/zeolite particles are irregular in shape (100 to 400 nm). The independent variables were impregnated amounts of KOH (15 - 25%), catalyst to oil ratios of 1.0 - 6.0 wt.%, and reaction time of 2 h. The highest biodiesel yield of 95% was produced from the reaction with 2.1 wt.% catalyst of 25% KOH impregnated. The properties of produced biodiesel complied with SNI. The catalytic stability test showed that the 25% KOH impregnated catalyst was stable.

1. Introduction

Petroleum fuels sources that use as energy sources are currently depleted, while the needs for fuels increase rapidly with increasing population. Reduced petroleum feedstock causes increasing the price of petroleum fuel [1-5]. Biodiesel is one of the types of renewable energy of biofuels that can replace the fuel oil of diesel without requiring engine modifications and produce cleaner emissions. In general, the properties of biodiesel are very similar to diesel, the difference is that biodiesel has a higher cetane number than diesel and high flash point it can be degraded easily (biodegradable), does not contain sulfur (or very low, if exist), and aromatic compounds so that the resulting combustion emissions are more environmentally friendly than diesel fuel [6-9].

The homogenous catalyst is more widely used in biodiesel production processes, such as KOH and NaOH. The use of these homogeneous catalysts raises another problem in the resulting product. The resulting biodiesel still containing the catalyst so it supposed to separated and also can cause side reactions like saponification reactions that affected the process of biodiesel production. Heterogeneous catalysts have the advantage of separating products from the catalyst more easily and simply, which can be separated by filtration or precipitation [10-12].

In this work, the potassium was impregnated on natural zeolite materials (KOH/Zeolite) for producing biodiesel from waste vegetable oil. Waste vegetable oil is one of the raw materials for making biodiesel, because the material still containing triglycerides, aside of free fatty acids. The

[3] To whom any correspondence should be addressed.

availability is quite abundant, causing the oil to be a waste that has a potential to pollute the environment such as rising levels of COD (Chemical Oxygen Demand) and BOD (Biology Oxygen Demand) in the waters, also cause the smell of decay due to biological degradation [13]. During producing biodiesel process catalysts heterogeneous have lower yields, so recent studies have focused more on developing potassium-solid catalysts support into zeolite to obtain optimal yield. Several type of solid based catalysts have been reported in recent years, such as metal oxides, CaO, MgO, K_2CO_3/MgO, KNO_3/Al_2O_3, KOH/Al_2O_3, and NaY zeolite solid based catalyst shown a great activity of catalytic in biodiesel production [14].

This paper discusses the influence of active phase content in KOH/zeolite catalysts and the catalyst performance on the biodiesel production yield. The prepared catalyst was characterized by XRD and SEM methods. The effect of catalyst preparation conditions and various reaction variables on biodiesel yield were investigated. The catalyst reusability is also investigated for several consecutive runs as it represents a key advantage for industrial biodiesel production.

2. Materials and methods

2.1. Preparation of K_2O/Zeolite
Prior to use, the crushed natural zeolite was placed into the oven to reduce the water content. Then various KOH solutions (15%, 20%, and 25%) were impregnated into the natural zeolite mixed with aquadest and stirred at room temperature. The purpose of this method was to fill the pores using a metal salt solution of a certain concentration. After impregnation, the catalyst was dried in an oven at 110 °C for 24 h. The purpose of drying was to reduce the water content and crystallize the salt on the surface of the pores. Then the solids were calcined for 3 h at 500 °C which is aimed to enhance the contact surface area and to obtain K_2O on the catalyst with zeolite as support. The KOH as a source of potassium oxide which added to the zeolite structure during calcination will be converted to K_2O.

2.2. Catalyst characterization
The calcined catalyst was characterized by X-ray diffractometer (XRD) to identify the crystal phase of the sample. In the XRD method, the samples were analysed using Bruker diffractometer with Cu-Kα radiation (λ = 0.154 nm) operating at 40 kV and 100 mA. The diffractograms were recorded in the 2θ ranges of 10 – 60° at a scanning speed of 5° min^{-1}. The shape and size of the catalyst was recorded by Scanning Electron Microscopy (SEM).

2.3. Waste frying oil characterization
Kinematic viscosity of frying oil is determined using the Oswald viscometer. The density of used cooking oil is determined using a pycnometer. The acid number is determined by the method of acid base titration analysis.

2.4. Esterification reaction
The 50 g of waste vegetable oil is put into a two-neck flask. Then, methanol was added with a mole ratio oil-to-methanol of 1:8 and 1% of 0.1 N HCl as a catalyst. The two-neck flask was then assembled on the tool set. The reaction was carried out at 60 °C for 2 h then after the esterification process completed, the solution was poured into the separating funnel and precipitated for 24 h. Finally, oils were dried in the oven to remove water.

2.5. Transesterification reaction
The reaction took place in a two-neck flask in where 1- 6 % of the catalyst were placed in the reactor and mixed with 50 g esterified waste vegetable oil and methanol with the ratio of oil-to-methanol of 1: 8 [15]. After that, the reactor was closed and then heated to a temperature of 60 °C and stirred using a magnetic stirrer with varying time for 2, 4, and 6 h. During the process, the operating temperature was kept constant. After the reaction, stirring and heating were stopped. Then the result was inserted into

IOP Conf. Series: Materials Science and Engineering **334** (2018) 012011 doi:10.1088/1757-899X/334/1/012011

the separation funnel and allowed for 12-24 h to achieve a perfect separation between two phases. The separation/purification process of the methyl ester layer needed to be resumed by drying it in an oven at 100 °C for 3 h to evaporate methanol remaining in methyl ester product. The yield of biodiesel was calculated by using the equation given in the literature.

$$\%\,\text{Yield} = \frac{Weight\ of\ biodiesel\ produced}{Weight\ of\ oil\ used}\ x\ 100\% \qquad (1)$$

A mixture of K_2O/Zeolite catalyst with methanol was filtrated in a vacuum condition. A rotary evaporator was used to separate the excess methanol from the filtrate. The residue was then centrifuged for 20 minutes to separate the biodiesel from glycerol. The used catalyst was filtered, heating at 500 °C and then fed to the next run.

3. Results and discussion

3.1. Characterization of waste frying oil

The characteristic data of waste frying oil are demonstrated in Table 1.

Table 1. Properties of waste frying oil.

Parameter	Value
Kinematic viscosity, at 40 °C (cSt)	17.24
Density ($g.cm^{-3}$)	0.9166
Acid number (mg $KOH.g^{-1}$)	2.92

The result of the analysis of waste frying oil showed acid number reach 2.92 mg $KOH.g^{-1}$, kinematic viscosity of 17.24 at 40 °C (cSt) and density 0.9166 $g.cm^{-3}$. The high acid number in the frying oil requires the esterification process to reduce the acid number and the transesterification process to decrease the viscosity so that it can produce the final product in the form of biodiesel that meets the quality of SNI. 04-7182-2006. The acid number of waste frying oil decrease of 0.83 mg $KOH.g^{-1}$ after esterification process.

3.2. Characterization of natural zeolite and K_2O/zeolite catalyst by XRD

The XRD pattern of synthesized catalyst is shown in Figure 1. The natural zeolite exhibits a very

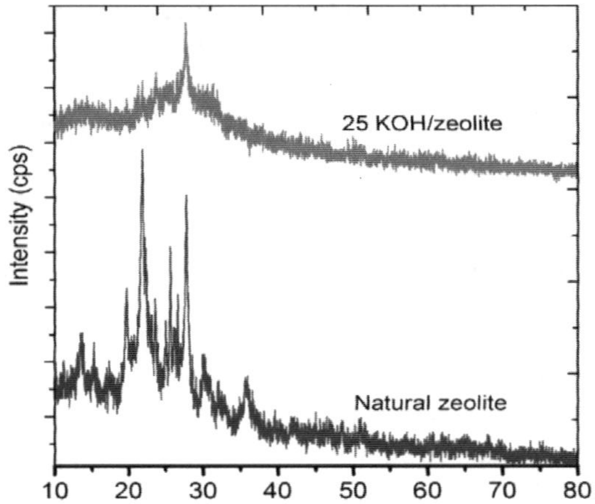

Figure 1. XRD patterns of natural zeolite and 25KOH/zeolite.

intense diffraction peak in the XRD pattern indicating a high crystallinity of the support.The diffraction peaks of the 20 wt.% KOH/zeolite sample are less intense than those of the support indicating that the zeolite crystallinity decreased upon KOH impregnation. These results are in accordance with previous study [16]. Moreover, no KOH or K_2O diffraction peaks are observed on the XRD patterns of the prepared catalysts. This might be due to the presence of highly dispersed and/or amorphous K_2O particles [17-19].

3.3. Morphology of catalyst by SEM

In order to elucidate the morphology and particle sizes of prepared sample, the SEM measurement of ash catalysts was also performed. The SEM morphologies of zeolite and KOH/zeolite are shown in Figure 2 & 3. The zeolite and KOH/zeolite catalyst were previously calcined in a muffle furnace at 500 °C for 4 h. Both zeolite and KOH/zeolite particles seem to agglomerate. The KOH/zeolite prepared samples show a difference in the morphology, as illustrated in Figure 3. KOH particles highly dispersed in the structure of zeolite support are nearly undistinguished with K_2O particles. The SEM images reveal that the zeolite and KOH/zeolite particles are irregular in shape (100 to 400 nm), as shown in Figure 2 & 3. The catalyst in nanosizes is able to increase the catalyst surface area significantly.

Figure 2. SEM Images of natural zeolite.

Figure 3. SEM images of K_2O/Zeolite.

3.4. Effect of mass KOH on zeolite and catalyst to oil ratio

Figure 4 is a plot of the influence of the amount of impregnated KOH and catalyst to oil ratio on the yield of biodiesel. All transesterification reaction conditions were carried out with a methanol to oil ratio of 8:1, temperatures at 65 °C, reaction time of 2 h, and various catalysts to oil ratio (1.0 - 6.0, wt.%). The transesterification reactions were carried out with the zeolite support, 15KOH/zeolite, 20KOH/ zeolite and 25KOH/zeolite catalysts. As can be seen in Figure 4, the amount of KOH loaded on zeolite affected the conversion of waste frying oil into biodiesel. The zeolite support used alone is inactive in the transesterification reaction. However, by increasing the loading amount of KOH from 15 wt.% to 25 wt.%, a significant increase in biodiesel yield was observed. For instance, by varying the loading amount of KOH from 15 wt.% to 25 wt.%, the biodiesel yield increased from 54,02% to 75% by using a catalyst to oil ratio of 1.0 wt.%. This increase in biodiesel yield is attributed to the higher K_2O active phase content in the catalyst [16]. The possible reason is that enough KOH with good dispersion on the surface and in pores of zeolite is needed for the catalysis of transesterification reaction [20].

The effect of catalyst amount on biodiesel yield is demonstrated in Figure 4. As seen in Figure 4, the biodiesel yield on 35% KOH/zeolite is raised from 75.0% to the maximum yield of 95% with catalyst amount increasing from 1.0 to 2.5 wt.%. However, biodiesel yield dropped to 94% as the catalyst amount further increased to 4.0 wt.%. The reasons for yield decrease is that when the solid catalyst was excessive, the slurry (mixture of solid catalyst and reactants) was too viscous, therefore leading to a problem of mass transfer [21].

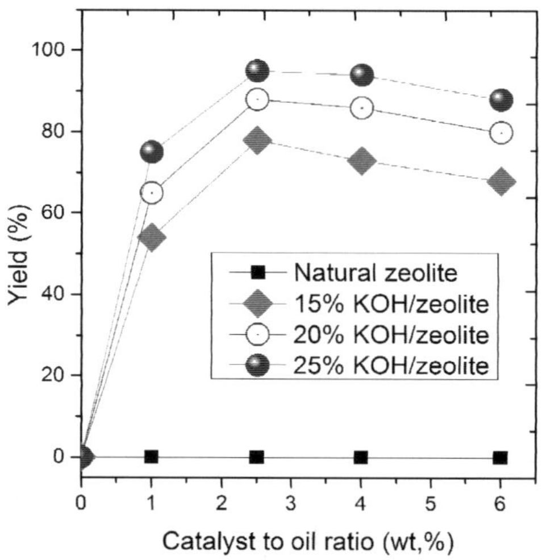

Figure 4. Effect of KOH loaded on zeolite and catalyst amount on biodiesel yield. Reaction condition: methanol: oil ratio = 8:1, T =65 °C, P = 1 atm, t = 2 jam.

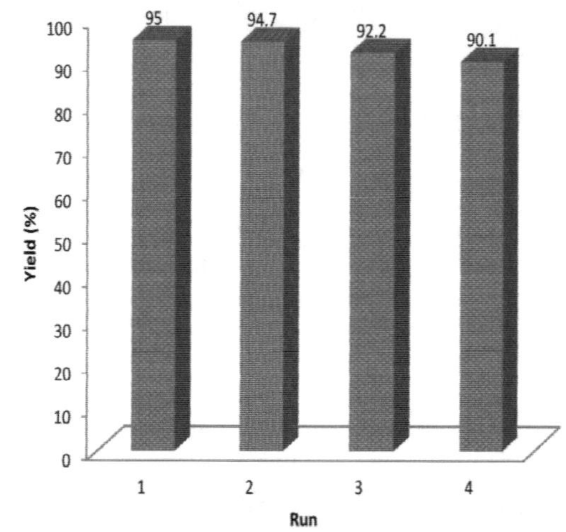

Figure 5. Reuse of the catalyst: Reactionn condition: Catalyst 2.5 wt.%, over the 25KOH/zeolite catalyst, methanol: oil ratio = 8:1, T =65 °C, P = 1 atm.

3.5. Reusability Catalyst

In this case, reusability of KOH with loading of 25%, 3% catalyst ratio and a reaction time of 2 h was investigated by obtaining and reusing the catalyst in four consecutive runs. Figure 5 shows the biodiesel yield with four successive runs. After four experiments, a catalyst with a 25KOH/zeolite shows good stability in maintaining its catalytic performance, where the yield of biodiesel decreased by only 2.8%. The decrease of catalytic activity after the four recycling was caused by the loss of catalyst during the process of filtration, activation, and the morphology change of used catalyst. The other reason is heating process would developed crystals agglomeration which decreasing catalyst surface area and subsequently suppressed nanocatalyst reactivity and lead to decline biodiesel yield [19].

3.6. Characteristic of biodiesel

To study the physicochemical properties, the transesterification reaction was carried out in the presence of the catalyst under calcined at 500 °C. The characteristics of the biodiesel sample were shown in Table 2.

Table 2. Characteristic of biodiesel of this work, SNI standard.

Parameter	This work	SNI
Kinematic viscosity, at 40 °C (cSt)	2,6 – 3,8	2,30 – 6,00
Density ($g.cm^{-3}$)	0,886 – 0,873	0,85 – 0,89
Acid number ($mg\ KOH.g^{-1}$)	0,21 – 0,42	max 0,80

As listed in Table 2, the fuel properties of the prepared biodiesel have been found to comply with the SNI standards. Refer to Table 2, its can be seen that the kinematic viscosity of biodiesel is ranging from 2.6 - 3.8cSt. The density of biodiesel is ranging from 0.866 - 0.873 g.cm^{-3} and the acid number is ranging from 0.21 - 0.42 mg KOH.g^{-1}. These values are also in compliance with SNI standards.

4. Conclusions

Preparation and the use of zeolite impregnated with KOH as heterogeneous catalysts for the transesterification of waste frying oil with methanol has been evaluated. The higher the number of KOH impregnations and the more catalysts used, the higher the yield of biodiesel. The optimum condition was obtained at 25% KOH impregnation indicated a very high catalytic activity (95.0% biodiesel yield) with the optimum reaction conditions determined: the molar ratio of methanol to oil by 8:1, the amount of catalyst by 2.5%. Reusability of a 25% KOH load catalyst demonstrates good stability in maintaining its catalytic performance, in which the heterogeneous catalyst is able to remain stable for 4 consecutive runs. Characteristics of the biodiesel, such as kinematic viscosity, density and acid number are sufficient in accordance with the requirements of SNI.

Acknowledgment

We acknowledge the financial support through the Research Grants of Professor of Syiah Kuala University, Republic of Indonesia (No: 1442/UN11/SP/PNBP/2017).

References

[1] Zhang H, Li H, Pan H, Liu X, Yang K, Huang S, et al. 2017, Efficient production of biodiesel with promising fuel properties from Koelreuteria integrifoliola oil using a magnetically recyclable acidic ionic liquid. Energy Conversion and Management.**138:**45-53.

[2] Husin H, Adisalamun, Sy Y, Asnawi TM, Hasfita F. 2017, Pt nanoparticle on $La_{0.02}Na_{0.98}TaO_3$ catalyst for hydrogen evolution from glycerol aqueous solution. AIP Conference Proceedings: AIP Publishing; p. 030073.

[3] Husin H, Mahidin M, Yunardi Y, Hafita F. 2015, Visible Light Driven Photocatalytic Hydrogen Evolution by Lanthanum and Carbon-co-Doped $NaTaO_3$ Photocatalyst. Key Engineering Materials. **659**:231-6.

[4] Husin H, Mahidin M, Zuhra Z, Hafita F. 2014, H_2 evolution on Lanthanum and Carbon co-doped $NaTaO_3$ Photocatalyst. Bulletin of Chemical Reaction Engineering & Catalysis. **9**:81-6.

[5] Husin H, Syamsuddin Y. 2010, Pembuatan Katalis $Cu/ZnO/Al_2O_3$ untuk Proses Steam Reforming Metanol menjadi Hidrogen sebagai Bahan Bakar Alternatif. Jurnal Rekayasa Kimia & Lingkungan. **7**:98-104.

[6] Azizi F, Kahoul A, Azizi A. 2009, Effect of La doping on the electrochemical activity of double perovskite oxide Sr_2FeMoO_6 in alkaline medium. Journal of Alloys and Compounds. **484**:555-60.

[7] Husin H, Hasfita F, Rinaldi W. 2013, Waste shells of cockle (Clinocardium nuttalli) as solid catalysts for transesterification of calophyllum inophyllum L. oil to biodiesel production. Proceedings of The Annual International Conference, Syiah Kuala University-Life Sciences & Engineering Chapter. 14-8.

[8] Manique MC, Lacerda LV, Alves AK, Bergmann CP. 2017, Biodiesel production using coal fly ash-derived sodalite as a heterogeneous catalyst. Fuel. **190**:268-73.

[9] Piker A, Tabah B, Perkas N, Gedanken A. 2016, A green and low-cost room temperature biodiesel production method from waste oil using egg shells as catalyst. Fuel. **182**:34-41.

[10] Syamsuddin Y, Husin H. 2010, Pembuatan Katalis Padat ZrO_2/Al_2O_3 untuk Produksi Biodiesel dari Minyak Jarak. Jurnal Rekayasa Kimia & Lingkungan. 7.

[11] Zuhra Z, Husin H, Hasfita F, Rinaldi W. 2015, Preparasi Katalis Abu Kulit Kerang Untuk Transesterifikasi Minyak Nyamplung Menjadi Biodiesel. Agritech. **35**:69-77.

[12] Mahlinda M, Husin H, Riza M. 2017, Transesterifikasi In Situ Biji Pandan Laut (Pandanus Tectorius) Menjadi Biodiesel Menggunakan Gelombang Ultrasonik. Jurnal Teknologi Industri Pertanian. **26**:294-300.

13] Maneerung T, Kawi S, Dai Y, Wang C.H. 2016, Sustainable biodiesel production via transesterification of waste cooking oil by using CaO catalysts prepared from chicken manure. Energy Conversion and Management. **123**:487-97.

[14] Noiroj K, Intarapong P, Luengnaruemitchai A, Jai-In S. 2011, A comparative study of KOH/Al_2O_3 catalysts for biodiesel production via transesterification from palm oil. Renewable Energy, **34**:1145-50.

[15] Husin H, Marwan, Mahidin. 2011, Studi penggunaan katalis abu sabut kelapa, abu tandan sawit, dan K_2CO_3 untuk Konversi Minyak Jarak Menjadi Biodiesel. Jurnal Reaktor. **13**:254-61.

[16] Saba T, Estephane J, El Khoury B, El Khoury M, Khazma M, El Zakhem H, 2016, Biodiesel production from refined sunflower vegetable oil over KOH/ZSM5 catalysts. Renewable Energy. **90:**301-6.

[17] Intarapong,P, Iangthanarat,S, Phanthong,P, Luengnaruemitchai,A, Jai-In, S.,2013, Activity and basic properties of KOH/mordenite for transesterification of palm oil, J. Energy Chem. 2 690-700.

[18] Husin H, Zuhra, Hasfita F. 2007, Oksidasi parsial metana menjadi methanol dan formaldehida menggunakan katalis $CuMoO_3/SiO_2$: Pengaruh rasio Cu : Mo, temperatur reaksi dan waktu tinggal Jurnal Rekayasa Kimia & Lingkungan. **6**:21-7.

[19] Bet-Moushoul E, Farhadi K, Mansourpanah Y, Nikbakht A, Molaei R, Forough M. 2016, Application of CaO-based/Au Nanoparticles as Heterogeneous Nanocatalysts in Biodiesel Production. Fuel. **164**:119-27.

[20] Li Xf, Zuo Y, Zhang Y, Fu Y, Guo Q-x. 2013, In situ preparation of K_2CO_3 supported Kraft lignin activated carbon as solid base catalyst for biodiesel production. Fuel. **113**:435-42.

[21] Alonso DM, Granados ML, Mariscal R, Douhal A. 2009, Surface chemical promotion of Ca oxide catalysts in biodiesel production reaction by the addition of monoglycerides, diglycerides and glycerol. J Catal. **276**:229-36.

Economic analysis of biomass gasification for generating electricity in rural areas in Indonesia

H Susanto[1,3], T Suria[1] and S H Pranolo[2]

[1] Study Program of Chemical Engineering, Institut Teknologi Bandung, Bandung, Indonesia, email: herri@che.itb.ac.id
[2] Study Program of Chemical Engineering, Faculty of Engineering, Universitas Sebelas Maret, Surakarta, Indonesia

Abstract: The gaseous fuel from biomass gasification might reduce the consumption of diesel fuel by 70%. The investment cost of the whole unit with a capacity of 45 kWe was about IDR 220 million in 2008 comprised of 24% for gasification unit, 54% for diesel engine and electric generator, 22% for transportation of the whole unit from Bandung to the site in South Borneo. The gasification unit was made in local workshop in Bandung, while the diesel-generator was purchased also in a local market. To anticipate the development of biomass based electricity in remote areas, an economic analysis has been made for implementations in 2019. A specific investment cost of 600 USD/kW has been estimated taking account to the escalation and capacity factors. Using a discounted factor of 11% and biomass cost in the range of 0.03-0.07 USD/kg, the production cost of electricity would be in the range of 0.09-0.16 USD/kWh. This production cost was lower than that of diesel engine fueled with full oil commonly implemented in many remote areas in Indonesia at this moment. This production cost was also lower than the *Feed in Tariff* in some regions established by Indonesian government in 2017.

1. Introduction

Biomass has gained much attention as an energy resource in Indonesian following the national energy policy, in which biomass accounts for 5% in the national energy mix [1]. Meanwhile there are large amount of various biomass as renewable energy source in Indonesia. Biomass conversion to electricity particularly for rural area gets much attention at this moment. A common technique for the utilization of biomass to electrical energy is steam power plant. Many palm oil mills and sugar cane factories produce power and heat using this technique with a capacity in the range of 1-3 MW [2].

A gasification of biomass may have several advantages over a steam power plant for applications in a rural or remote area, where the capacity is relatively small. This paper deals with economic evaluation and feasibility study on the application of an appropriate technology of biomass gasification to electricity for the near future. The feasibility study was prepared for a project started for operation in 2019. An optimum capacity of this rural electricity was about 500 kWe (kW-electricity). The economic evaluation was based on literature data and our previous experiences on implementation of rural electricity generation units with capacities in the range of 10-100 kWe.

[3] To whom any correspondence should be addressed.

IOP Conf. Series: Materials Science and Engineering **334** (2018) 012012 doi:10.1088/1757-899X/334/1/012012

2. Biomass to Electricity System

Gasification process deals with many chemical reactions, and an overall reaction may be represented as follows [3]:

$$C_xH_yO_x + a\,O_2 + b\,H_2O \leftrightarrow c\,CO + d\,H_2 + e\,CH_4 + f\,CO_2$$

The coefficients of reaction (a, b, c, d, e and f) in the above overall reaction depends on temperature and pressure of the process. The desired combustible gaseous components are H_2, and CO, as well as CH_4. The molar ratio of H_2/CO may be dictated by the following.

$CO + H_2O \leftrightarrow CO_2 + H_2$; *homogeneous water shift reaction*

Methane, CH_4 is mainly generated in the pyrolysis, which is a step of gasification process, or probably produced through the following methanation reaction in the presence of a suitable catalyst:

$C + 2H_2 \leftrightarrow CH_4$; *methanation reaction*

The gasification process is usually operated as auto-thermal process, in which heat for the following endothermic reduction reactions are supplied by oxidation of some parts of biomass.

$C + CO_2 \leftrightarrow 2CO$; *Bouduard reaction*

$C + H_2O \leftrightarrow CO + H_{2=}$; *heterogeneous water shift reaction.*

The exothermic oxidation reactions may be represented with the following reactions:

$C + O_2 \leftrightarrow 2\,CO$

$H + 0.25\,O_2 \leftrightarrow 0.5H_2O$

If air is used as the oxidizing agent, the gaseous product contains significant amount of N_2 gas thus reducing its heating value (see Table 1). This gas is usually known as *producer gas* and it is used as gaseous fuel. If steam or/and O_2 are used, the gaseous product contains high concentrations of H_2 and CO, with a very little amount of N_2, which is called as *synthesis gas* with a heating value up to 10.5 MJ/Nm^3 (medium calorific gaseous fuel).

Down draft fixed bed gasifiers have been used in all of our field units with capacities up to 100 kWe. Basicaly, this type of gasifier is expected to produce a combustible gaseous fuel with a low tar content. Unfortunately for some reasons a small concentration of tar may still present in the gas e.g. due to improper preparation of biomass or improper operation and maintenance.

Table 1. Typical composition of producer gas (our experiences).

No.	Gas composition (mol fraction)	types of biomass		
		Coconut shell	Rubber wood	Rice husk
1.	CO	25%	18%	20%
2.	H_2	12%	16%	11%
3.	CH_4	1%	2%	2%
4.	CO_2	10%	10%	11%
5.	N_2	52%	54%	56%
6.	Gas *Low Heating Value*, kJ/Nm^3	4,900	4,600	4,350

A typical set up of a gasification and diesel engine-generator is presented in Figure 1. This unit produced net electricity of about 45 kW distributed to about 60 houses of oil palm farmers. Main components of this unit are:

a. down draft fixed bed gasifier fueled with corn cobs with a capacity of about 80 kg/h
b. screw-feeder to bring biomass to the top of the gasifier
c. water scrubber for gas cooling and tar removal
d. filter box filled with locally available filter medium such as coconut husk, saw dust, etc
e. start-up blower for sucking gas during the first 15-30 minutes operation
f. diesel genset (diesel engine and electric generator) with a rated capacity about 60 kWe

During start-up, low quality gas was drawn from the gasification unit using a start-up blower. The gas was blown through a burner and flared. In this short period, electricity had already been generated using a diesel engine and electric generator set. The electricity was used for running the start-up blower and other electricity-needs in the power house. As soon as the gas quality better (heating value of at least 3500 kJ/Nm^3 as indicated in flare), the diesel engine could be operated *dual fuel,* and the start-up blower was then stopped. At the end of operation, the diesel engine and generator set was also operated *full oil* again for the the last 15 minutes, while the gasification process was stopped already. The purpose of this operation was purging tar from the internal parts of engine.

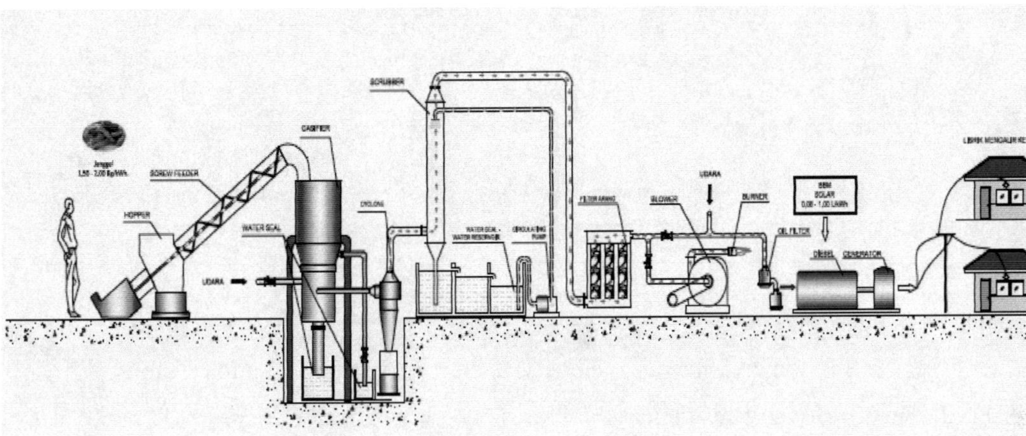

feeding system gasifier gas cleaning and cooling blower engine-generator

Figure 1. Gasification of corn cobs for rural electricity in Pelaihari, South Kalimantan.

The oil replacement during *dual fuel* operation of the diesel engine was about 70%. In dual fuel operation, fluctuations of gas quality may be compensated automatically with consumptions of diesel fuel. This integrated biomass gasification and diesel engine-generator set consumed about 1.5-2.5 kg dry biomass per 1.0 kWh depending especially on the biomass quality.

Gas engine and diesel engine are basically internal combustion engines. A gas engine is a spark-ignition and requires gaseous fuel with a high methane number. While a diesel engine is compression-ignition and requires fuel with a high cetane number. This moment, gas engines are getting popularity for medium and low calorific gases including producer gas. The price of a gas engine is still higher and the number of manufacturer is limited. On the other hand, diesel engines have already been popular for rural electricity all over the country. Conversion diesel engine into gas engine by some modification with a minimum cost has also been studied [4].

3. Investment

This feasibility study was prepared in 2016 for a project started in operation in 2019 and hence currency values in 2019 are used in the following sections. Capacity of this proposed unit was 500 kWe, which was considered as an optimum based on the availability of biomass at the site. Fixed capital cost was estimated based on our experiences on the 45 kWe gasification unit in South Kalimantan, in 2008 (Figure 1). The total investment for this unit was about IDR 220 million in 2008, and the unit comprised of:

a. 24% for gasification unit
b. 54% for engine-generator and electrical system
c. 22% for transportation from Java to an oil palm plantation close to city of Pelaihari, in South Borneo, including the installation cost at the site.

The investment cost was extrapolated for scaling up using the sixth power law and also corrected for inflation using CEPCI (*Chemical Plant Cost Index*). The investment for 500 kWe biomass gasification coupled with gas or diesel engine was estimated to be USD 250,000 (IDR 3,375 million), or a specific investment cost of about 600 USD/kWe. This investment was much lower than those available in several reports as presented in Figure 2 [5, 6, 7].

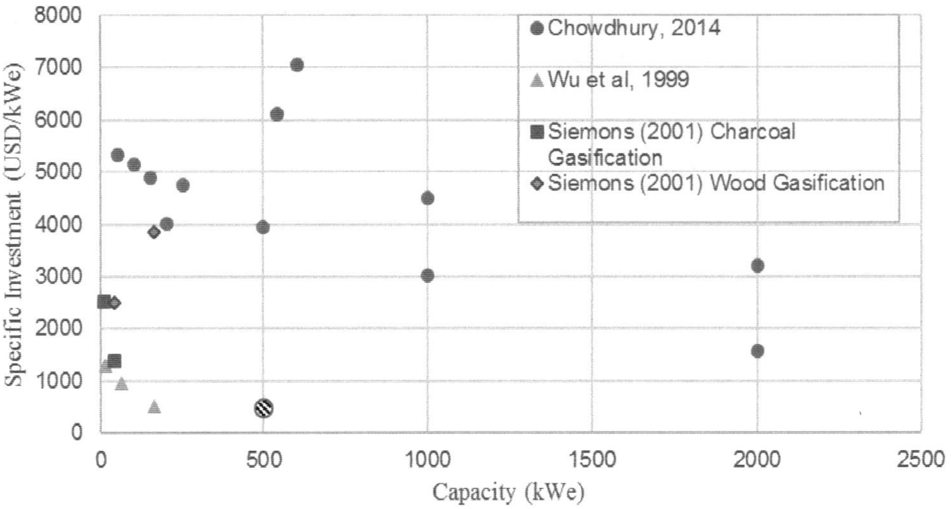

Figure 2. Specific investment for biomass to electricity via gasification [5, 6, 7].

The specific investment cost for biomass to electricity via gasification were in the range of USD 2100-6000/kW. Although there is a relatively low specific investment of USD 1000/kWe in USA was also found [8], which was more or less the same as our experience and as reported in China [8]. A similar value of investment was also found for a fluidized bed gasifier with capacity around 3 MW [9].

The following three configurations have been considered in this economic evaluation (Tabel 2).

a. Case A. A gas engine is used as a primer mover to drive for the electric generator. This system can be operated fully using the producer gas.
b. Case B. A diesel engine is used as a driver for the electric generator. This diesel engine is operated in the *duel fuel*. An oil replacement of 70% was assumed.
c. Case C. A diesel engine is operated fully with diesel liquid fuel as commonly implemented rural or remote areas.

Table 2. Investment cost for 500 kWe gasification system (values in USD).

	gasification unit type of engine	yes gas	yes diesel	no diesel	
No.	Components	Case A	Case B	Case C	References
1.	Gasification unit: gasifier, biomass feeder, cyclone, gas cleaning system using water scrubber, cooling water system, etc	47,919	47,919	0	authors' experiences
2.	Gas engine or diesel engine and electric generator, simple power house, and electrical box and wiring in power house	123,696	123,696	123,696	Wu, 1999
3.	Waste water treatment and solid waste handling (from gasification unit)	27,350	27,350	0	Wu, 1999
4.	Miscellaneous costs, such as: transportation of equipments, workshop, etc	49,444	49,444	49,444	authors' experiences
5.	Total Fixed Capital Cost	248,409	248,409	173,140	
6.	Working Capital	32,963	42,296	65,852	estimated
7.	Total Investment	281,372	290,705	238,992	calculated
8.	Specific Fixed Capital Investment (USD/kWe)	497	497	346	
9.	Specific Total Capital Investment (USD/kWe)	563	581	478	

The price of a gas engine was assumed the same as a diesel engine in this study. Fixed bed gasifier and cleaning systems account about 20% of the fixed capital investment. A waste water system also accounted for a significant portion of the total investment (about 12%). This water scrubber should not be utilized anymore in the near future for the environment reason.

4. Production Cost of Electricity

Economic evaluation was carried out using the LCOE (*Levelized Cost of Electricity*), i.e. the price per unit of the net sold electricity for which the NPV (*Net Present Value*) of an investment is zero [10]. Thus finding LCOE is equivalent to finding the electricity price for a given IRR (*internal rate of return*) of a specified discount factor. The LCOE is less detail than Discounted Cash Flow having the real profitability for the developer [7]. The LCOE is a kind of breakeven price of electricity. If the electricity price is above the LCOE, the project gives good return, because the revenue from electricity is greater than the cost.

The LCOE is calculated using the following equations:

$$\mathrm{LCOE} = \frac{\mathrm{FCI} \cdot \mathrm{CRF} + \mathrm{OC}}{\mathrm{E}} = \frac{\mathrm{A}}{\mathrm{E}} + \frac{\mathrm{OC}}{\mathrm{E}} \qquad (1)$$

$$\mathrm{CRF} = \left(\frac{r}{1-(1+r)^{-n}} \right) \qquad (2)$$

FCI = Fixed Investment Cost
CRF = Capital Recovery Factor
OC = Operating cost: fixed and variable cost
A = Annualized capital cost
n = life time of equipment
E = net sold electricity
r = interest rate, %/year
n = life time of project, years.

The simplified LCOE can also be used for modeling or policy development. The parameters and basic assumptions for calculation of LCOE are presented in Table 3.

The calculated production cost of electricity (LCOE) for 500 kWe electricity from biomass via gasification is presented in Table 4. Two modes of operation, i.e. 8 or 24 hour/day were evaluated in this study. This first mode of operation has been commonly implemented in rural area, where people use electricity only in the evening and in maybe also in the early morning. The second mode of operation is implemented in more developed areas, where electricity is utilized for various activities such as local offices, clinics, shops and market. Of course, the later mode of operation may be more economically attractive (see Table 4).

Fuel cost was found to be the main component of LCOE. The fuel cost might account for 44%-58% of the LCOE depending on the biomass cost of 0.03-0.06 USD/kg (a mid value was used in Table 4). In Case A (gasification unit coupled with gas engine), the biomass cost was more than a half of LCOE. While in Case C, the cost of diesel fuel was the most affecting parameter in the cost of electricity generation. The effect of the biomass cost on the LCOE in Case A and Case B were examined further (see Figure 3).

Oil palm plantation area would be very attractive for the implementation of the biomass to electricity via gasification. Five 100 kWe demonstration units were installed in the Province of Riau (Sumatra mainland) in 2011 – 2013. The oil palm biomass could be:

(a) kernel fibre: already collected in palm oil mill, but utilized totally as fuel in mill power plant
(b) palm kernel shell (PKS): already collected in palm oil mill, but utilized part as fuel in mill power plant
(c) empty fruit bunches: already collected in palm oil mill, but very poor characteristic for gasification and combustion
(d) oil palm fronds: left in the plantation, must be collected and pre-treated
(e) oil palm trunk: biomass generated during replanting every 20 years.

Comparison of LCOE against the *Fit in Tariff* for Sumatra where abundant biomass is available from oil palm plantation are presented in Figure 4.

In the cases of the daily operation mode of 8 hours, the biomass cost should not be more than 0.054 and 0.072 USD/kg for Case A and Case B respectively (Figure 3.a). While for the daily operation of 24 hours, the biomass cost might be as high as 0.08 USD/kg (Figure 3.b).

Table 3. Basic assumptions for economic evaluation (USD 1 = Rp 13,500).

No.	Parameters	Case A	Case B	Case C
1.	Electricity generation, kW			
	a. Designed capacity	500	500	500
	b. Sold (net production)	450	450	450
	c. In house consumption	50	50	50
2.	Costs of fuel			
	a. Biomass (IDR/kg)	400-800	400-800	400-800
	b. Diesel fuel (subsidized, IDR/L)		7,000	7,000
	c. Diesel fuel (not subsidized, IDR/L)	-	10,500	10,500
3.	Operation time			
	a. Daily (hours/day)	8 24	8 24	8 24
	b. Yearly (days/year)	360 330	360 330	360 330
4.	Specific fuel consumption			
	a. Biomass consumption (kg/kWh)	1.30	1.30	1.30
	b. Diesel fuel consumption (L/kWh)	-	-	0.29
	c. Dual fuel operation:			
	i). Diesel fuel consumption (L/kWh)	-	0,09	-
	ii). Biomass consumption (kg/kWh)	0.90	0,90	-
5.	Utilities: water cost (IDR/kW produced)	15	15	15
6.	Number of operators and staffs			
	a. operators (persons)	2 6	2 6	2 6
	b. non-operator staff (persons)	1 3	1 3	1 3
	c. supervisors (persons)	1 1	1 1	1 1
7.	Salaries (1000 IDR/ man-month)			
	a. operators	2,000	2,000	2,000
	b. non-operator staffs	2,000	2,000	2,000
	c. supervisors	5,000	5,000	5,000
8.	Fixed capital investment:			
	a. case of low investment (USD/kWe)	497	497	-
	b. case of high investment (USD/kWe)	523	523	-
9.	Maintenance cost (% of fixed capital cost)	6.5%	6.5%	6.5%
10.	Interest rate	11%	11%	11%
11.	Capital Recovery Factor (USD/kWe)	0.117	0.117	0.117
12.	Economic Plant Life, years	20	20	20
13.	Working capital (% of yearly operating cost)	25%	25%	-

Based on our experiences, the operation of gasification process required discipline operators and needed intensive maintenance, such as removal of ash from the gasifier, and removal of tar from the gas cleaning system. Automatic operation was possible, but it would increase the investment. Lub-oil of the engine had to be replaced about twice or three times more frequently than the engine normally. These technical aspects were decisive factors affecting the on-stream hours and days, and maintenance cost. So they eventually affected the economic feasibility.

5. Scheme of investment

Since the implementation of the biomass gasification technology requires an additional investment, and its implementation needs demonstration and promotion, the following three scheme has been proposed to extent Case A and Case B.

a. Scheme-1: the business entity makes investment on both gasification unit and engine-generator. These are the base case as discussed above.
b. Scheme-2: the business entity makes investment on engine-generator only, while the gasification unit is granted by the government or institution.
c. Scheme-3; both gasification and engine-generator set were grants. So the business entity manage only the operating cost.

The proposed investment schemes represent one type of subsidy from the government. These scheme made differences in the annualized cost. Cost of electricity generation are presented in Table 5.

Table 4. Production cost of electricity at 500 kW.

No.	Operating time (hours/day)	Case A 8	Case A 24	Case B 8	Case B 24	Case C 8	Case C 24
1.a	Biomass (USD 0.045/kg)	0.063	0.063	0.043	0.043	-	-
1.b	Diesel Fuel (0.4 USD/L, Indonesia 2016)	-	-	0.051	0.051	0.163	0.163
2.	Utilities, water treatment, etc (USD/kWh)	0.001	0.001	0.001	0.001	0.001	0.001
3.	Salary of operators and staffs (USD/kWh)	0.008	0.006	0.008	0.006	0.008	0.006
4.	Maintenance (USD/kWh)	0.014	0.005	0.014	0.005	0.009	0.003
5.	Annualized capital cost (USD/kWh)	0.024	0.009	0.024	0.009	0.017	0.006
6.a	LCOE (USD/kWh)	0.110	0.084	0.141	0.115	0.199	0.180
6.b	LCOE (IDR/kWh)	1685	1,134	1904	1553	2430	2430
Feed in Tariff, category of a low voltage grid		USD/kWh (IDR/kWh)					
1.	Riau, Sumatra	0.0862 (1161)					
2.	Jambi, Sumatra	0.0668 (901)					
3.	Bangka Island	0.1161 (1567)					

Source: Indonesian Government Regulation, Ministry of Energy and Mineral Resource 12/2017

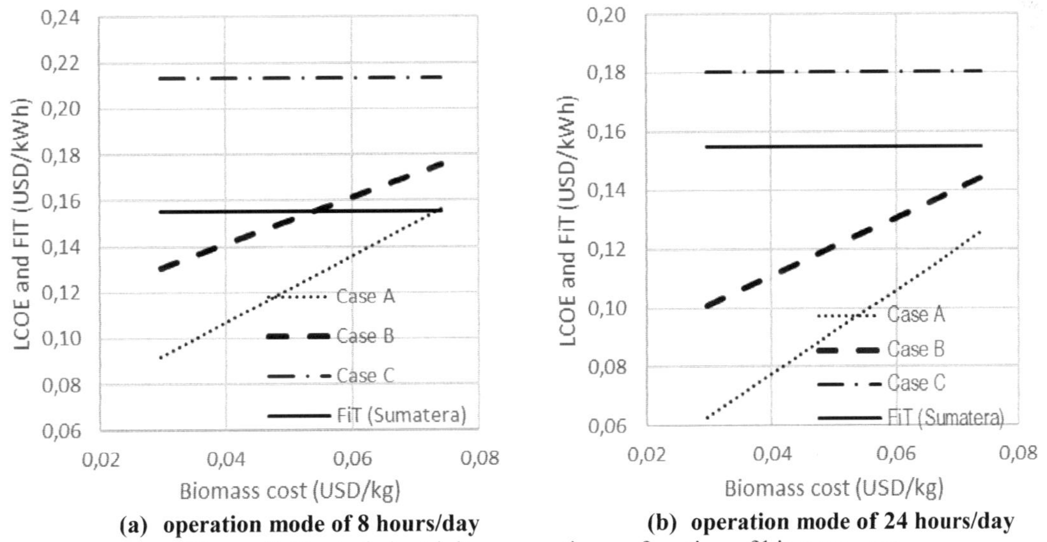

(a) operation mode of 8 hours/day **(b) operation mode of 24 hours/day**

Figure 3. Cost of electricity generation as function of biomass cost.

Table 5. Cost of electricity generation at three different scheme of investment.

		Case A			Case B			
No.	Parameters	A1	A2	A3	B1	B2	B3	C
1.	Scheme of investment:							
	a. Gasification Unit	inv.	grant	grant	inv.	grant	grant	-
	b. Engine and Generator	inv.	inv.	grant	inv.	inv.	grant	inv.
2.	LCOE in USD/kWh							
	a. operation mode: 8 h/day	0.110	0.103	0.086	0.141	0.134	0.117	0.199
	b. operation mode: 24 h/day	0.084	0.078	0.075	0.115	0.109	0.106	0.180
3.	LCOE in IDR/kWh							
	a. operation mode: 8 h/day	1,485	1,385	1,155	1,909	1,809	1,579	2,688
	b. operation mode: 24 h/day	1,132	1,052	1,012	1,556	1,476	1,436	2,434

Case B3 was implemented by the Ministry of Energy and Mineral, thorugh Directorate of New and Renewable Energy, in 2011 – 2013. Five units of 100 kWe fueled with oil palm fronds were installed in Riau Province, in the main land of Sumatra. These units were designed, manufactured and constructed at the site by a domestic company in Jakarta. Training for operators and supervision for two months were also given.

A unit of 30 kWe fueled with rice husk was installed in Sumba Island in 2014, by the same domestic company as those in Riau. Another unit with a capacity of 1 MW was installed recently, in Sumba, 2016. The later was intended to be fueled with *Caliandra* wood from a pilot project of energy plantation. The electricity system comprised of a fixed bed gasification unit and five gas engine generator sets. In order to ensure tar removal, a refrigerated gas cooling was used. Tar removed the gas stream is dumped in a pond. This electricity generation plant was bought completely from a foreign manufacturer, probably with an expectation to have a reliable operation.

Case B2 is very interesting to be considered, since diesel engine and generator sets have already been popular in many remote places. So the government or funding institutions may contribute only for the investment on gasification unit and addition facilities, including and a little modification of the diesel engine.

6. Conclusions

Economic evaluation for gasification of biomass has been studied based on our field experience and information available in open literatures. An appropriate capacity for this rural electricity was in 100-500 kWe. The investment unit was approximated to be about 600 USD/kWe installed. The biomass cost affected electricity cost significantly. The biomass gasification unit was preferable coupled with a diesel engine to drive an electric generator, rather than with gas engine.

Following to implementing a biomass to electricity, local capacity building should also be developed for instance in creating economic activities for the optimum utilization of electricity. Local officers have to understand the philosophy of new and renewable energy resource in the near future. and understanding the principle of gasification process, and maintenance of engine running with gaseous fuel. The gasification unit may be manufactured in metal workshops in order to reduce the transportion cost, so the design of a gasification unit must be prepared and ready to be used.

Budget for operator training and a certain period of technical monitoring should also be provided in implementation of biomass to electricity. Participation of local agencies was very important in the implementation of the biomass gasification for rural electricity. The local agencies may play important roles, such as establishing management system, environment evaluation, and legal administration. Local academic or technical persons should also be invited for developing technical aspects and and development of economical activities.

IOP Conf. Series: Materials Science and Engineering **334** (2018) 012012 doi:10.1088/1757-899X/334/1/012012

Acknowledgements
This feasibility study on biomass to electricity was conducted under cooperation between Chemical Engineering ITB and State Company of Gas (PGN) in 2015. The authors like to thanks to the Directorate General of New and Renewable Energy, and *PT Perkebunan Nusantara XIII* for the challenging opportunities in the implementation of biomass gasification for rural electricity.

References

[1] Government Regulation (Peraturan Pemerintah) No. 79/ 2014, on Energy Mix in Indonesia
[2] Md Jaye, I F, Sadhukhan J and Murphy R J, 2016, *International Journal of Smart Grid and Clean Energy*, **vol. 5**, no. 2
[3] Basu P, *Biomass Gasification and Pyrolysis: Practical Design and Theory*, 2nd Ed, 2013, *Elsevier,* Elsevier
[4] Perez, N S, Machin, EB, Pedroso, DT, Roberts, J J, Antunes, J S and Silveria, J L, 2015, *Applied Thermal Engineering*, **90**
[5] Chowdhury, Nahid-ur-rahman, 2014, Advances and Trends in Woody Biomass Gasification, Thesis, Tecnico Lisboa, 2014, https://fenix.tecnico.ulisboa.pt/downloadFile/1689244997255169/Thesis.pdf, accessed in January, 2017
[6] Siemons, R V, 2001, *Biomass and bioenergy*, **20**
[7] Wu, C Z, Huang, H, Zheng, S P and Yin X L, 2002, *Bioresource Technology*, **83**
[8] International Renewable Energy Agency (IRENA), *Renewable Power Generation Cost in 2014*, http://www.irena.org/documentdownloads/publications/irena_re_power_costs_2014_report.pdf, accessed in July, 2016
[9] Guangzhou Institute of Energy Conversion (GIEC), 2013, http://understandchinamachinery.org/wp-content/uploads/2013/10/CRESP-Consultation-on-biomass_power_generation.pdf, accessed in July, 2016
[10] Department of Energy and Mineral Engineering, htpps://www.e-education.psu.edu/eme801/node.560, accessed in January, 2017

Hydrogen production from water-glucose solution over $NiO/La\text{-}NaTaO_3$ photocatalyst

R Mardian[1], H Husin[1,2,3], K Pontas[1,2], M Zaki[1,2], T M Asnawi[2] and Ahmadi[2]

[1]Graduate Program in Chemical Engineering, Syiah Kuala University, Darussalam, 23111, Banda Aceh, Indonesia
[2]Department of Chemical Engineering, Syiah Kuala University, Darussalam, 23111, Banda Aceh, Indonesia

Email: husni_husin@che.unsyiah.ac.id

Abstract. This paper reports the evaluation of $La\text{-}NaTaO_3$ photocatalyst performance in producing hydrogen from water-glucose solution. The main goal of the studies is to investigate the influence of glucose as a sacrificial reagent on the photocatalytic efficiency in water splitting reactions under ultraviolet (UV) irradiation. Photocatalyst has been fabricated via sol-gel method and being confirmed using x-ray diffraction (XRD) and scanning electron microscopy (SEM). Nickel loaded $La\text{-}NaTaO_3$ photocatalyst are prepared by impregnation method. It was observed that the prepared photocatalysts displayed particle sizes in the 30-250 nm range with orthorhombic structure. Their photocatalytic activity for hydrogen production via water splitting was conducted in a Pyrex glass reactor under UV light irradiation. The aqueous solution contained glucose employed as a renewable organic scavenger. A significant improvement in hydrogen production was observed in glucose-water mixtures and NiO loaded photocatalyst. The prepared $La\text{-}NaTaO_3$ showed that the highest activity for hydrogen generation of 35.1 mmol $h^{-1}.g^{-1}$ was obtained at 0.10 $mol.L^{-1}$ glucose and 0.3 wt.% NiO. This suggests the important role played by the glucose as electron donor and loading nickel on $La\text{-}NaTaO_3$ as a cocatalyst increasing electron storage and suppressing electron-hole recombination.

1. Introduction

Human civilization builds by an energy system which complexly integrated in which facilitated by an advancing technology to conduct an easier activity. It is important for us to have a breakthrough in an intensively researches to find, optimize, and use renewable resources. Now a days, an alternative energy resources is a big vision in most countries, especially in developing countries is hydrogen production from water [1]. Currently, over 50% of the world's hydrogen supply is from catalytic steam reforming (SR) of hydrocarbons [2-4]. Great efforts have been made in developing economical, environmentally friendly, and competitive processes for hydrogen production from renewable sources [5, 6].

Hydrogen has been considered as the chemical fuel of the future to replace fossil fuels as an energy resource in many fields. Through an intensive research, photocatalytic technology has been a

[3] To whom any correspondence should be addressed.

prepared technology for producing hydrogen on a commercial scale which can be used as an alternative for energy source replacing fossil fuel. Photocatalytic process in hydrogen production using semiconductor photocatalyst has been intensely studied and that photocatalytic water splitting using oxide semiconductor as one of the promising schemes in producing hydrogen from water [7]. Semi-conductor photocatalyst consists of an oxide or a sulfide. From various photocatalyst materials, $NaTaO_3$ oxide has been reported having excellence activity in producing hydrogen using UV lights. The addition of doping or strange substance into $NaTaO_3$ crystal structure can narrow band gab, increase crystallinity and smaller particle sizes, thus increasing the production level of hydrogen. Furthermore, the structure of the crystal photocatalyst also affecting photocatalysts band gap [8-11]. However, the efficiency of photocatalytic hydrogen evolution is in general very low due to the recombination of the photoinduced electron and hole, and the back reaction of H_2 and O_2. For the photocatalyst, to perform efficiently, the electron-hole recombination must be avoided.

To inhibit the above disadvantageous processes, an electron donor can be added into the reaction system as a hole scavenger to achieve high efficiency for hydrogen evolution. Many organic pollutants in wastewater are good electron donors. Methanol and glucose are known as a potential donor in decreasing the level of recombination for having Gibbs energy lower than water splitting (ΔG^o = 228.64 $kJ.mol^{-1}$). The success for glucose as a sacrificial reagent for rapidly increasing the rate for hole photo induction has been reported [12]. NiO_x has been found a more effective cocatalyst and it is necessary to provide an active site in photocatalytic of hydrogen production [13].

The aim of the present work is to study the influence of glucose as sacrificial reagents on photocatalytic efficiency in water splitting reactions under ultraviolet (UV) irradiation. In this paper, the photocatalytic hydrogen production over NiO/La-$NaTaO_3$ photocatalyst using glucose as electron donor is investigated. The H_2 production has been stable and the photo induction recombination between electron-hole has been avoided.

2. Experimental

2.1. Preparation of La-$NaTaO_3$

All reagents were of analytical grade and were used without further purification. Photocatalyst La-$NaTaO_3$ was synthesis by a sol-gel method [14]. A typical synthesis was carried out as follow: 1.5 g of tantalum chloride ($TaCl_5$ 99,9%) was dissolved and stirred in 150 ml ethanol (C_2H_5OH 99,9%), the stirring continuous for several hours until the solution became transparent (denoted solution A). 0,8 gram of sodium hydroxide (NaOH 99,9%) was dissolved in 50 ml distilled water and stirred for 15 minutes (denoted: solution B). Then, both solutions were mixed using magnetic stirrer (denoted: solution C). To this solution, 0,0362 grams lanthanum ($La(NO_3)_3.6H_2O$ (99,9%) was dissolved in 15 ml distilled water and stirred for 15 minutes then added to solution C by using pipette. Afterward, 4 g of citric acid was dissolved in 50 ml distilled water, poured into burette and introduced to the mixture by drop wise to obtain sol liquid at room temperature. The pH level was adjusted at 4 by adding NH_4OH to ionize the citric acid. The mixture was then heated and maintain at 80°C until white gel formed. The gel then was dried in the oven at 110°C and calcined at 450°C for 4 h. Lastly it was calcined at 800°C for 8 h while flowing air in the reactor to increase crystallinity[15].

2.2. Preparation of NiO

NiO co-catalyst was prepared by dissolving $Ni(NO_3)_2.6H_2O$ in water and deposited in La-$NaTaO_3$ via impregnation method [16]. Nickel between 0.1 – 1.0 % wt. from $Ni(NO_3)_2.6H_2O$ was dissolved in distilled water to mixed with 1 gram of La/$NaTaO_3$. The mixture stirred for 6 h at room temperature. Nickel-loaded $NaTaO_3$ was heated in oven at 110°C overnight, and then calcined at 300°C for 1 h to form NiO oxide.

2.3. Catalyst Characterization

There are a few characterization methods for photocatalyst such as X-Ray *diffraction* (XRD) and scanning electron microscopy (SEM) observation to study composition and morphology which contain important information. XRD measurement was conducted by using Shimadzu XRD 600 X-ray diffractometer (30 kV, 30 mA). The morphology of the prepared sample was recorded by Scanning electron microscope (SEM) using a Philips XL-30 (15 kV, 70 Pa). A prepared sample was filled through gold sputtering in the surface of La-$NaTaO_3$. The picture was gained through scanning electron microscope gained at 15 kV with 10000 times enlargement.

2.4. Evaluation of photocatalytic reaction

Photocatalytic reaction was carried out in Pyrex glass reactor under UV light irradiation for H_2 evolution. The reaction was performed in 500 ml reactor by taking of 350 ml distilled water containing of glucose between 0.010 - 0.125 mol L^{-1}. Before irradiation, 0.30 g of catalyst was suspended by using a magnetic stirrer and Ar gas was bubbled through the reaction mixture to remove oxygen. The irradiation source was using 400-Watt Mercury Lamp ($\lambda \geq 250$ nm). The photocatalytic activity was determined by measuring the amount of hydrogen production for 2 h irradiation. The reaction was conducted at 30°C. The amount of hydrogen measured using gas chromatography (Shimadzu GC-8A) using molecular sieve 5A column, thermal conductivity detector, and nitrogen as a carrier gas.

3. Results and discussion

3.1. Morphology

The La-$NaTaO_3$ photocatalyst in this research was prepared by sol-gel method and calcined at 800°C. SEM characterization was performed to the sample was meant to gain information about the morphology, shape, and size of the particle. The SEM characterization of $NaTaO_3$ and La-$NaTaO_3$ are shown in Figure 1 & 2.

Figure 1. SEM Images of $NaTaO_3$.

Figure 2. SEM Images of La-$NaTaO_3$.

From Figure 1, $NaTaO_3$ without lanthanum has irregular shape with agglomeration. The particle size is formed with the range of 300 – 700 nm. Figure 2 shown SEM images of La$NaTaO_3$ sample. From the SEM images, an orderly-crystallized particle was formed with orthorhombic structure. The surface seems clean, with clear angle indicating the synthesized material has a high degree of crystallinity. The particle sizes of the sample were ranging between 30 – 250 nm. The result confirmed that with the addition of lanthanum, the crystallinity become well and the size become smaller. It is due to the ability of lanthanum to preserve the stability of substance avoiding the agglomeration of synthesized particles when introduce to high temperature. The higher degree of crystallinity contributed to the photogenerated electron and hole separating efficiency and H_2 evolution enhancement. It is in accordance to the report from Husin et al. that $LaNaTaO_3$ calcined at 800°C has orthorhombic perovskite structure with high crystallinity [15]. Calcination at high temperature is a factor in structure formation into orthorhombic crystal. Nickel particles are not seen through SEM recorded due to its small amount loading.

3.2. Crystal Structure

Figure 3 shown XRD patterns of $NaTaO_3$ standard and La-$NaTaO_3$ sample. From Figure 3, its can be seen, both sample have the same profile. It is confirmed by 3 main characteristic peaks in the sample matching the 3 peaks of the $NaTaO_3$ standard powder data file. Those characteristic peaks are identified at 2θ:22.7°, 33.6°, and 52.5° with the intensity of 33977.59 cps, 29636.75 cps, and 9414.01 cps, respectively. The crystallinity of the samples is greatly improved, as indicated by stronger and narrower XRD peaks in Figure 3 [7].

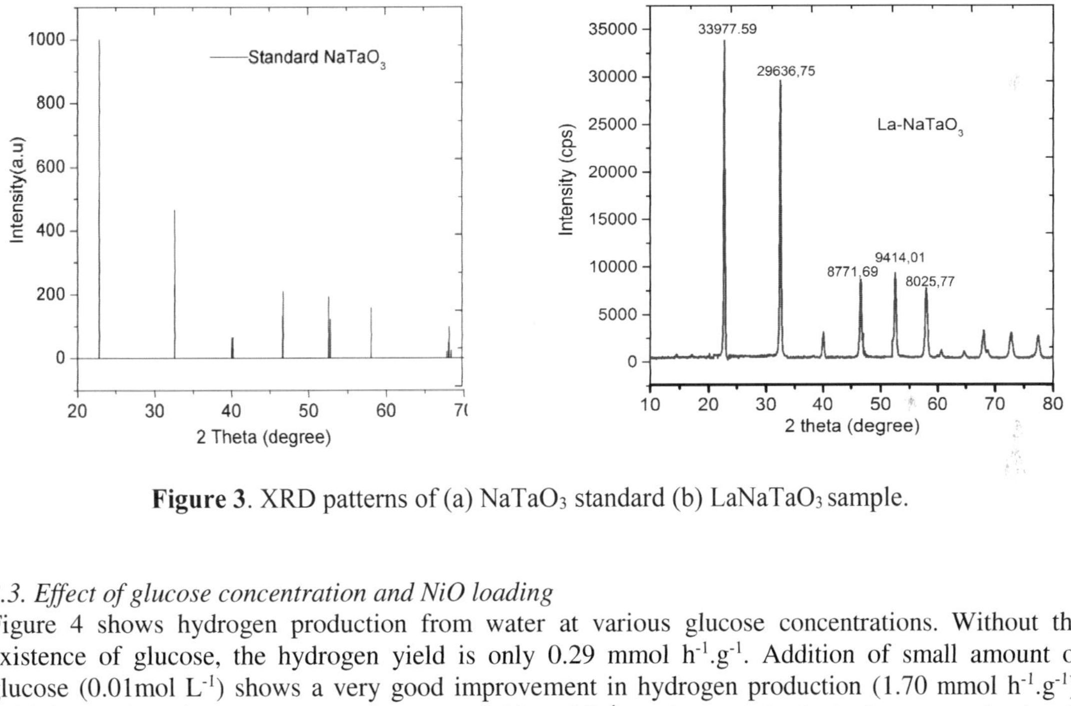

Figure 3. XRD patterns of (a) $NaTaO_3$ standard (b) $LaNaTaO_3$ sample.

3.3. Effect of glucose concentration and NiO loading

Figure 4 shows hydrogen production from water at various glucose concentrations. Without the existence of glucose, the hydrogen yield is only 0.29 mmol $h^{-1}.g^{-1}$. Addition of small amount of glucose (0.01mol L^{-1}) shows a very good improvement in hydrogen production (1.70 mmol $h^{-1}.g^{-1}$). With increasing glucose concentration up to 0.10 mol L^{-1}, an increase in the hydrogen production is observed (5.70 mmol $h^{-1}.g^{-1}$), i.e. 19.7 times higher than the non-glucose solution. It can be seen that the relation between hydrogen production rates in the presence of glucose as a sacrificial reagent are proportionally increased. When the photocatalytic reaction done with the accompany of electron donors such as glucose, photogenerated hole in the valence band irreversibly oxidized glucose instead of H_2O, thus facilitating water reduction by potential electron conduction. Further increase of glucose concentration, the hydrogen production activity is seen decreasing and this may be due to saturation of catalyst active sites.

Photocatalytic hydrogen production from glucose aqueous solution is carried out over 0.3 wt% NiO loaded La-$NaTaO_3$ photocatalysts, as can be seen in Figure 5. Hydrogen production is 0.29 on bare La-$NaTaO_3$ catalyst. The co-catalyst of NiO on La-$NaTaO_3$ from glucose aqueous solution resulted in a substantial improvement in the hydrogen evolution. In nickel loaded on La-$NaTaO_3$, the samples showed much higher photocatalytic activity from glucose water mixture compared with pure water, as illustrated in Figure 5.

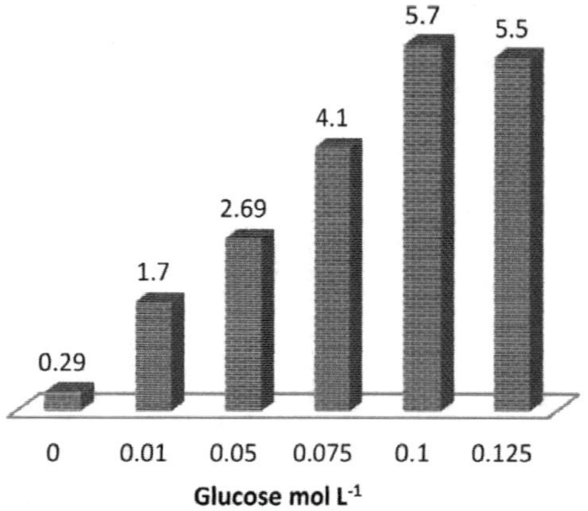

Figure 4. Effect of glucose concentration on hydrogen production without nickel loaded; Reaction condition: Glucose + water mixture of 350 ml, catalyst 0.3 g. UV light source.

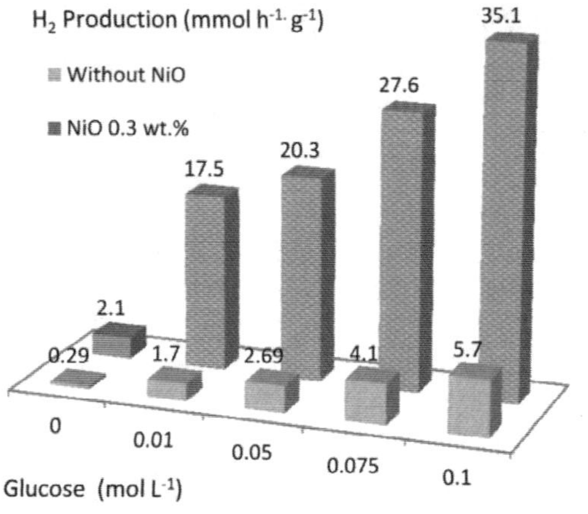

Figure 5. Effect of glucose concentration at NiO 0.3 wt.%, on hydrogen production; Reaction condition: Glucose + water mixture of 350 ml, catalyst 0.3 g. UV light source.

Figure 6 shows the hydrogen production over various NiO loaded La-$NaTaO_3$ photocatalysts in 0.05 mmol L^{-1} glucose aqueous solution. The higher photocatalytic activity of nickel loaded samples with increasing NiO loading up to 0.3 wt% could be due to the improved dispersion of NiO particles over La-$NaTaO_3$ photocatalyst. This may be seen as that at 0.3 wt% nickel loading on La-$NaTaO_3$, the light harvesting as well as the lower electron hole recombination are satisfied.

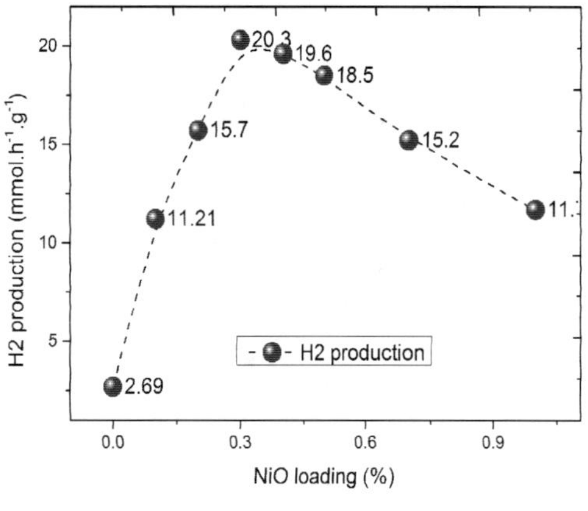

Figure 6. Effect of various NiO loading on hydrogen production; Reaction condition: Glucose + water mixture of 350 ml, 0.05 mol.L^{-1} of glucose, 0.3 g catalyst in the solution, UV light source.

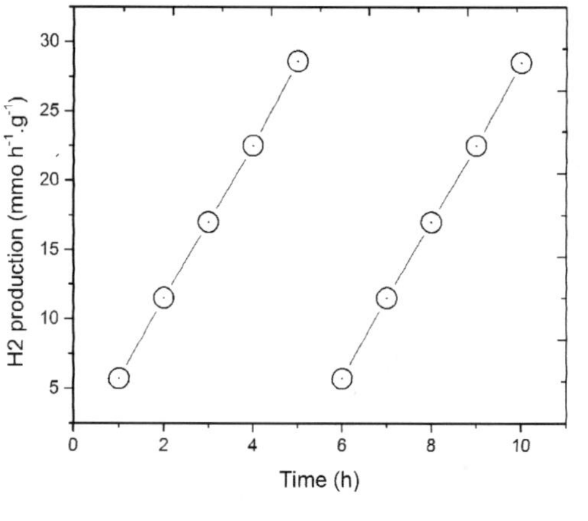

Figure 7. Hydrogen production activity: Reaction condition: reaction glucose + water mixture of 350 ml, 0.10 mol.L^{-1} of glucose, 0.3 g in the solution, UV light source.

Co-catalyst loading is very effecting to increase photocatalytic activity at certain condition. If the amount is in the right portion then the production can be maximize [17]. The amount of co-catalyst in the mixture are small, the active site for the process to take place are inadequate resulting low production. On the other hand, when the NiO concentration increased above 0.3 wt%, the photocatalytic activity decreased gradually. The phenomenon suggest that loading co-catalyst continuously will cover photocatalyst surface and decrease light absorbance capacity by La-$NaTaO_3$ semiconductor [18]. This fact explains nickel loading above 0,3% has negative result in the H_2 production[19, 20]. Nickel has probably cover the photocatalyst surface thus reducing photocatalyst effectiveness in absorbing light affecting the performance of the photocatalyst itself [21].

3.4. Photocatalytic Performance Stability
Figure 7 shown cumulative H_2 productions over La-$NaTaO_3$ catalyst during 5 h. Two cycles of reaction were carried out. It can be found that hydrogen evolves steadily during the reaction; no decline of activity has been detected during the span of time about 10 h. The H_2 generation rate of La-$NaTaO_3$ is about 5.7 mmol $h^{-1}.g^{-1}$, which is comparable with some photocatalysts reported before [22]. It can be concluded that the catalyst is stable enough in its performance.

4. Conclusions
NiO/La$NaTaO_3$ crystall photocatalyst has been sintesized through the sol-gel method. Using glucose as electron donor, photocatalytic hydrogen generation over NiO/La-$NaTaO_3$ was investigated. Lanthanum has been proven to change the morphology, particle size, and crystallinity of the substance into a much suitable for the photocatalytic process. The amount of hydrogen production was found to be depending on the glucose concentration and NiO content. The optimum amount of glucose for NiO/La-$NaTaO_3$ was found to be about 0.10 (mol L^{-1}). Deposition with NiO particles was found to be greatly enhancing the photocatalytic activity of La-$NaTaO_3$.The highest activity for hydrogen evolution of 35.1 mmol $h^{-1}.g^{-1}$ was obtained at 0.10 mol.L^{-1} of glucose and 0.3 wt.% of NiO. The NiO particles were oxidized and self-regulated to achieve the optimum valence for maximum activity during the course of reaction.

Acknowledgment
This work was supported by grant (No: 723/UN11/S/LK-PNBP/2015) from the Syiah Kuala University Darussalam Banda Aceh, Republic of Indonesia.

References

[1] Hu S, Jia L, Chi B, Pu J, Jian L. 2014,Visible light driven (Fe, Cr)-codoped $La_2Ti_2O_7$ photocatalyst for efficient photocatalytic hydrogen production. *Journal of Power Sources*. **266**:304-12.

[2] Gomaa MA, Abed RMM. 2017, Potential of fecal waste for the production of biomethane, bioethanol and biodiesel. *Journal of Biotechnology*.**253**:14-22.

[3] Husin H, Su WN, Hwang BJ. 2012, Hydrogen production over $La_{0.02}Na_{0.98}TaO_3$ photocatalysts from pure water and an aqueous methanol solution, *Jurnal Ilmiah Sains dan Teknologi*.**11**.21-26.

[4] Husin H, Syamsuddin Y. 2010, Pembuatan Katalis Cu/ZnO/Al_2O_3 untuk Proses Steam Reforming Metanol menjadi Hidrogen sebagai Bahan Bakar Alternatif. *Jurnal Rekayasa Kimia & Lingkungan*. **7**: 98-104.

[5] Husin H, Chen HM, Su WN, Pan CJ, Chuang WT, Sheu HS, 2011, Green fabrication of La-doped $NaTaO_3$ via H_2O_2 assisted sol-gel route for photocatalytic hydrogen production. *Appl Catal, B*. **102**:343-51.

[6] Jeong K.E, Kim H.D, Kim T.W, Kim J.W, Chae H.J, Jeong S.Y, 2014, Hydrogen production by aqueous phase reforming of polyols over nano- and micro-sized mesoporous carbon supported platinum catalysts. *Catalysis Today*. **232**:151-7.

[7] Husin H. 2012, Produksi Hidrogen Secara Fotokalitik dari Air Murni Pada Katalis $NaTaO_3$. Jurnal Rekayasa Kimia & Lingkungan. **9**:51-6.
[8] Husin H, Mahidin M, Yunardi Y, Hafita F. 2015, Visible Light Driven Photocatalytic Hydrogen Evolution by Lanthanum and Carbon-co-Doped $NaTaO_3$ Photocatalyst. *Key Engineering Materials.***659:** 231-6.
[9] Husin H, Mahidin M, Zuhra Z, Hafita F. 2014, H_2 evolution on Lanthanum and Carbon co-doped $NaTaO_3$ Photocatalyst. *Bulletin of Chemical Reaction Engineering & Catalysis.* **9**:81-6.
[10] Konta R, Ishii T, Kato H, Kudo A. 2004, Photocatalytic Activities of Noble Metal Ion Doped $SrTiO_3$ under Visible Light Irradiation. *The Journal of Physical Chemistry B.* **108**:8992-5.
[11] Li W, Wu D, Yu Y, Zhang P, Yuan J, Cao Y, 2014, Investigation on a novel ZnO/TiO_2 photocatalyst with enhanced visible photocatalytic activity. *Physica E: Low-dimensional Systems and Nanostructures.* **58**:118-23.
[12] Li Y, He F, Peng S, Gao D, Lu G, Li S. 2011, Effects of electrolyte NaCl on photocatalytic hydrogen evolution in the presence of electron donors over Pt/TiO_2. *Journal of Molecular Catalysis A: Chemical.* **341**:71-6.
[13] Husin H, Su WN, Pan CJ, Liu JY, Rick J, Yang SC, 2013, Pd/NiO core/shell nanoparticles on $La_{0.02}Na_{0.98}TaO_3$ catalyst for hydrogen evolution from water and aqueous methanol solution. *International Journal of Hydrogen Energy.* **38**:13529-40.
[14] Husin H. 2011, Preparasi dan karakterisasi fotokatalis $NaTaO_3$ didoping ion lantanum ($La/NaTaO_3$) dengan metode sol-gel. *Jurnal Rekayasa Kimia & Lingkungan.* **1:** 60-5.
[15] Husin H, Pontas K, Sy Y, Syawaliah S, Saisa S. 2014. Synthesis of Nanocrystalline of Lanthanum Doped $NaTaO_3$ and Photocatalytic Activity for Hydrogen Production. Journal of Engineering and Technological Sciences. **46**:318-27.
[16] Husin H, Marwan M. 2017, Studi Penggunaan Katalis Tembaga Molybdenum Oksida Berpenyangga Silica ($CuMoO_3/SiO_2$) Untuk Oksidasi Metana Menjadi Methanol dan Formaldehida. *Reaktor.* **8**:37-42.
[17] Maeda K, Abe R, Domen K. 2011, Role and Function of Ruthenium Species as Promoters with TaON-Based Photocatalysts for Oxygen Evolution in Two-Step Water Splitting under Visible Light. *The Journal of Physical Chemistry C.* **115**:3057-64.
[18] Husin H. 2012, Produksi Hidrogen Secara Fotokalitik dari Air Murni Pada Katalis $NaTaO_3$. *Jurnal Rekayasa Kimia dan Lingkungan*, **9**:53-8.
[19] Chen W.T, Chan A, Sun Waterhouse D, Moriga T, Idriss H, Waterhouse GIN. 2015, Ni/TiO_2: A promising low-cost photocatalytic system for solar H_2 production from ethanol–water mixtures. *Journal of Catalysis.* **326**:43-53.
[20] Liu B, Yang H, Wei A, Zhao H, Ning L, Zhang C, 2015, Superior photocatalytic activities of NiO octahedrons with loaded AgCl particles and charge separation between polar NiO surfaces. *Applied Catalysis B: Environmental.* **172**:165-73.
[21] Husin H, Su WN, Chen HM, Pan CJ, Chang SH, Rick J, 2011. Photocatalytic hydrogen production on nickel-loaded $La_xNa_{1-x}TaO_3$ prepared by hydrogen peroxide-water based process. *Green Chemistry.* **13**:1745-54.
[22] Yin M, Jia F, Qiao F, Zheng P, Zhang W, Fan Y. 2017. Facile wet-chemical synthesis and efficientphotocatalytic hydrogen production of amorphous MoS_3 sensitized by Erythrosin B. *Materials Characterization.* **128**:148-55.

IOP Conf. Series: Materials Science and Engineering **334** (2018) 012014 doi:10.1088/1757-899X/334/1/012014

The influence of temperature on the formation of liquid fuel from Polypropylene plastic wastes

M. Martynis[1,2], Mulyazmi[1], E. Praputri[1], R. Witri[1] and N. Putri[1]
[1] Department of Chemical Engineering, Faculty of Industrial Technology, Bung Hatta University, Padang West Sumatera

E-mail: martynismunas@yahoo.co.id

Abstract. The current trend of municipal waste management in urban areas is caused by rapid changes in social, economic, political and cultural life. As a non-biodegradable polymers that have become essential materials, plastic wastes have created a very serious environmental challenge because of the huge quantities and their disposal problems. Recycling of plastics is seen as one method for reducing environmental and resource depletion. The most attractive technique of plastics recycling is pyrolysis involving the degradation of the polymeric materials by heating in the absence of oxygen. This study investigated the characteristics of pyrolysis liquid fuel (PLF) produced from polypropylene plastic wastes with temperature variations. Pyrolisis was carried out on 200 grams of polypropylene waste plastics at the operating temperature of 200°C, 250°C, 300 °C and 350 °C for 45 minutes. The liquid products were found to have carbon chain length in the range of C8-C9, similar with gasoline. The maximum density, volume and calorific value of the oil obtained were 0.8 g/cm^3, 61 ml and 1307 cal/gr, respectively.

1. Introduction

During the last few decades, the production and consumption of plastics are increasing because of modern lifestyle and an urgent need for comfort. As a consequence, responsible disposal of plastic waste has created serious social and environmental problems. According to the Environment and Forestry Ministry, in 2016 Indonesia produced 64 million tons of waste with 14% being of plastic wastes [1]. Plastics degrade gradually since the molecular bonds containing carbon, hydrogen and other elements such as nitrogen, chlorine and others that make plastic very durable. The disposal of plastic by landfilling and incineration will lead to environmental pollution. Disposing of the waste to landfill is becoming undesirable due to ineffectiveness and can cause the reduction of soil nutrients. Furthermore, incineration stimulates the growing emission of harmful greenhouse gases, e.g. NO_x, SO_x, CO_x .Therefore, recycling is seen as one method for reducing environmental and resource depletion.

Thermoplastics are general-purpose plastics used in our daily lives and usually end up as municipal solid waste with 96% of this compound being consists of polyethylene (PE), polypropylene (PP), polystyrene (PS) and polyvinylchloride (PVC) [2]. Recently, recycling of thermoplastics already occurs on wide scale. Recycling has many types of techniques including pyrolysis, gasification, hydrolysis and etc. The most attractive technique to waste minimization that has been gaining interest

---

[2] To whom any correspondence should be addressed.

recently is pyrolysis. Pyrolysis involves the degradation of the polymeric materials by heating in the absence of oxygen [3]. The pyrolysis process begins at a temperature of about 230 °C, when thermally unstable components and volatile matters in the waste will break and evaporate along with other components [4]. The three major products that are produced during pyrolysis are condensable liquids (tar), char and non-condensable gases (52.2%, 25.2% and 22.6%, respectively) [5]. Pyrolysis was chosen by many researchers since the process able to produce high amount of liquid oil up to 80 wt% at moderate temperature around 500°C [6] that can be used as chemical feedstock or fuel. The pyrolysis process has been described as one of the environmentally acceptable treatment of waste plastics and at the same time preserving the petroleum resources [7,8].

Thermal cracking or pyrolysis operating temperature ranges between 400 and 700 °C [9]. The macromolecular structures of polymer are broken down into low molecular weight hydrocarbon oils and gases. The pyrolysis reaction consists of four progressive steps: initiation, propagation, decomposition, and termination [10].

2. Materials and methods

Plastic cup wastes (polyethylene) were collected from Bung Hatta University campus area, cleaned and dried. The dried plastic wastes were cut into small square shaped pieces (about 2 cm side). The pyrolysis setup consists of batch reactor made of stainless steel as shown in Figure 1. 200 g of plastic cup sample was loaded into the reactor in each pyrolysis reaction. The condensable liquid products were collected through the condenser and the volume was measured. After pyrolysis, the char left was weighed and the gaseous product was calculated. The pyrolysis reactions were carried out at temperature variations of 200, 250, 300 , and 350°C for 45 minutes.

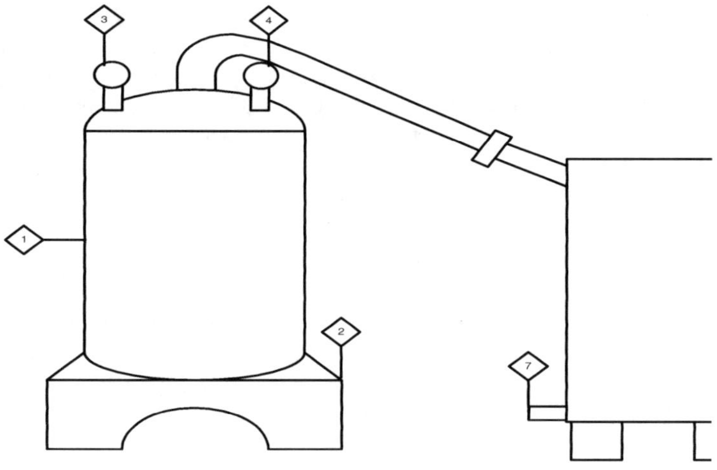

Figure 1. Pyrolysis setup.

The calorific value of the obtained liquid fuel was performed using Bomb Calorimeter. The density of fuel produced was measured using pycnometer and calculated by using Equation 1.

$$\rho s = \frac{mp+s - mp}{mp+w - mp} \times \rho w \quad (1)$$

The components of the liquid product were analyzed by Gas Chromatography-Mass Spectrometry analysis (GC-MS).

3. Results and discussions

3.1 Effect of temperature on liquid fuel volume

The effect of temperature on liquid fuel volume for the pyrolysis of plastic cup wastes is shown in Figure 2. High temperature supports the easy cleavage of bonds and thus speeds resulted in more liquid fuel produced. This shows that temperature has a significant effect on liquid fuel volume. The volume of liquid fuel produced is 23, 35, 44, and 61 ml, respectively.

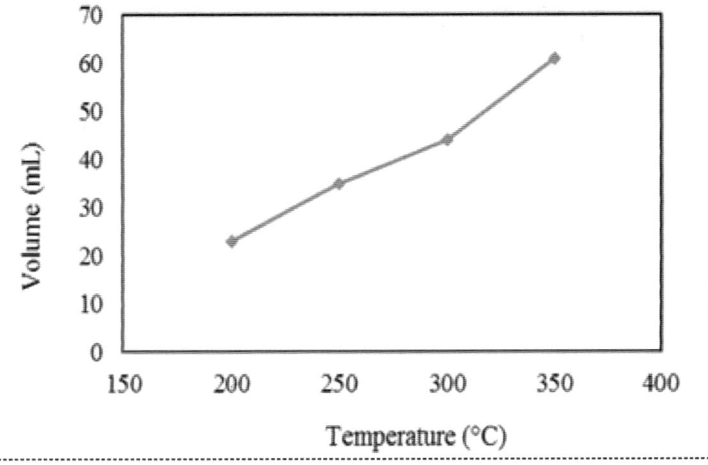

Figure 2. Effect of temperature on liquid fuel volume.

3.2 Effect of temperature on product yields

The pyrolysis of plastic cup waste yielded three different products, those are, oil (liquid fuel), char and gas. The distributions of these fractions are different at various temperatures and are shown in the Figure 3.

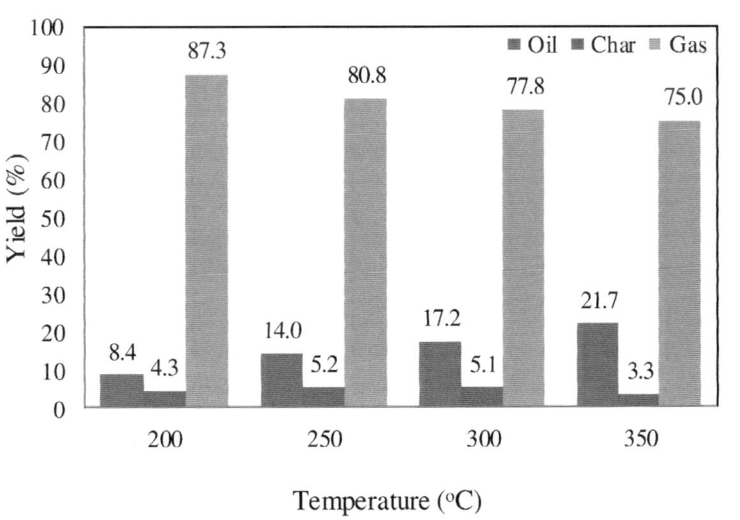

Figure 3. Effect of temperature on product yields.

The oil and gas constituted major product as compared to the char fractions. The recovery of condensable fraction (liquid fuel) increased with gradual increase of temperature. A maximum of 21.7% of the liquid fuel has been obtained at temperature of 350°C. The maximum yield of char 5.2% was obtained at the temperature of 250°C. The yield of char was increased with increasing pyrolysis temperature up to 250°C and then decreases. The highest yield of gas 87.3% was obtained at a pyrolysis temperature of 200°C. With the increasing of temperature, the gas yield is decreased. From low to moderate temperature, the liquid fuel yield increases resulting in a decrease of char yield. As a result, the pyrolysis process showed an inverse ratio between solid yields (char) and liquid products (oil). It is due to the fact that the higher the pyrolysis temperature, the components in the plastic cup wastes would be decomposed sharply. As more components were decomposed, the char production is lessened and the oil produced is rising.

3.3 Effect of temperature on calorific value

For each pyrolysis temperatures, the liquid fuel calorific value was tested using a bomb calorimeter. The calorific values for oil produced is shown in Figure 4. The calorific value of oil produced declined up to a temperature of 250°C, after that, it would increase significantly. The highest calorific value of 1307 cal/gr was achieved using pyrolysis reaction temperature of 350°C.

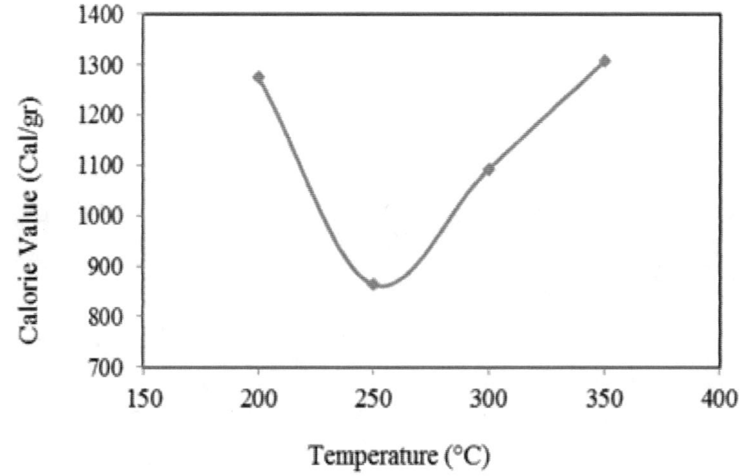

Figure 4. Effect of temperature on calorific value.

3.4 Effect of temperature on liquid fuel density

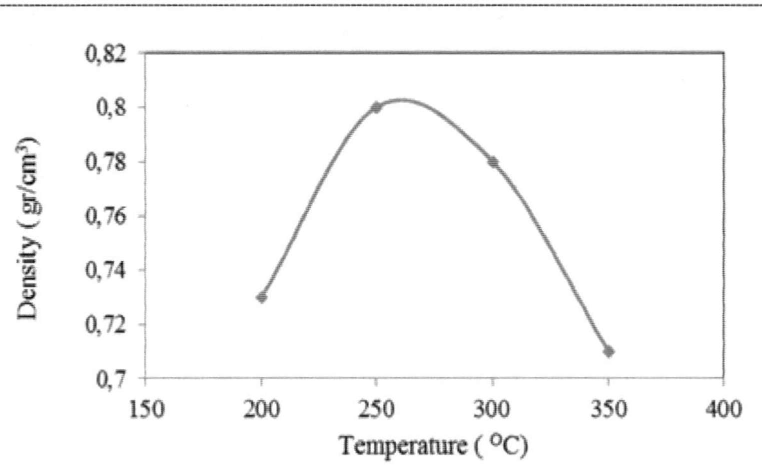

Figure 5. Effect of temperature on liquid fuel density.

In liquid fuel yield, as illustrated in Figure 5, the density inclined up to temperature of 250°C, and then it would decrease. However, the density of liquid fuel produced was in the range of 0.71-0.8 gr/cm^3, which is close to the density of gasoline. According to Director General of Oil and Gas Decree No.3674K/26/DJM/2006, the standard gasoline has a density of 0.715-0.770 gr/cm^3 (15°C) [11].

3.5 GC-MS of liquid fuel produced

Assessment of chemical composition features of oil the oil sample obtained from plastic cup wastes can be illustrated by qualitative examination of their GC traces. The hydrocarbon oil can have dramatically varied compositions in C_5 to C_{40} carbon range. Light oils are typically products in the C_3 to C_{12} carbon range. They include aviation gas, naphta, and automotive gasoline. Mid-range products are typically in a relative broad carbon range (C_6 to C_{26}) and include kerosene, jet fuel, and diesel products. Classic heavy oils include fuel No. 6 and lube type oil [12].

The GC-MS analysis were carried out to verify the exact composition of the liquid fuel obtained at 350°C (Figure 7) and summarized in the Table 1.

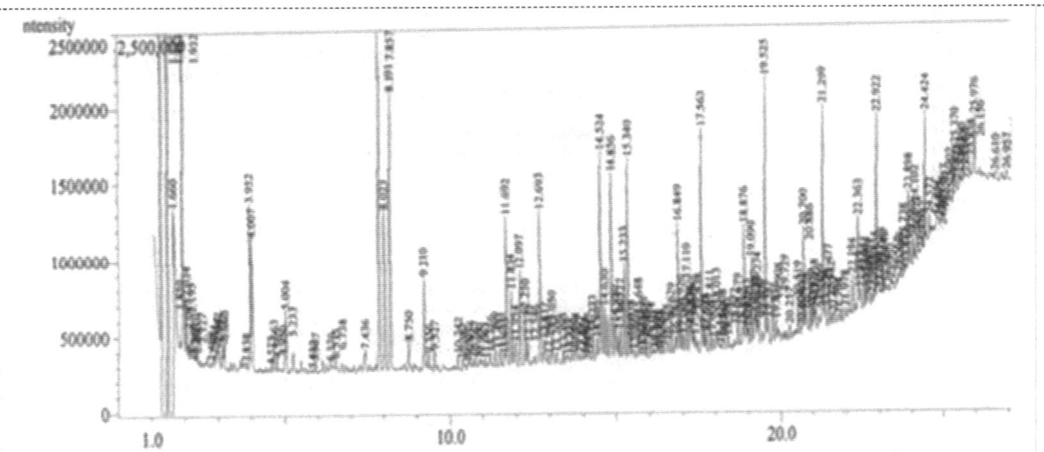

Figure 6. GC-MS chromatograms of liquid fuel obtained at 350°C.

Using GC-MS, 6 components were detected from the liquid fuel obtained at 350°C. The highest peak areas of total ion chromatogram (TIC) of the compounds displayed by Cyclohexene, 1,3,5-Trimethyl- (CAS) 1,3,5- trimethylcyclohexane at retention time of 2.034. The components present in plastic cup waste pyrolytic oil are mostly derivatives of alkenes, aromatic and aliphatic hydrocarbons with carbon number of C_8-C_{13}.

Table 1. GC-MS composition of liquid fuel obtained at 350^{0}C.

No	Retention Time	% Height	Compound	Formulas	Relative Molecules
1	2.034	0.61	Cyclohexene, 1,3,5- trimethyl- (CAS) 1,3,5- trimethylcyclohexane	C_9H_{18}	126.24
2	3.122	0.17	Undecane, 2,8- dimethyl- (CAS)	$C_{11}H_{24}$	156.18
3	4.663	0.15	1-nonene, 4,6,8- trimethyl- (CAS) 4,6,8 trimethylnonene-1	C_9H_{22}	154
4	8.750	0.19	1- tridecanol (CAS) n-tridecanol	$C_{13}H_{28}O$	200.36
5	10.614	0.08	Octadecane,3- ethyl- 5,2 ethylbutyl (CAS)	C_8H_{16}	112.10
6	10.910	0.06	Dodecane, 2,6, 11- trymethyl- (CAS) 2,6, 11 trymethyldodecane	$C_{10}H_{22}$	142.29

4. Conclusion

Thermal pyrolysis of plastic cup wastes were performed in a batch reactor at a temperature variation of 200, 250, 300 and 350°C for 45 minutes. The maximum liquid fuel yield was 21.7% at temperature of 350°C. The pyrolysis process showed an inverse ratio between solid yields (char) and liquid products (oil). The functional groups present in the pyrolytic oil are mostly derivatives of alkenes, aromatic and aliphatic hydrocarbons. It was found that the pyrolytic oil contains compounds having carbon chain length in the range of C_8-C_{13}. The density of liquid fuel produced was in the range of 0.71-0.8 gr/cm^3, which is close to the density of gasoline. A simple batch pyrolysis method can convert plastic cup waste to liquid hydrocarbons with yields which varies with temperature.

Acknowledgement

The authors would like to thank the Department of Chemical Engineering Bung Hatta University for supporting this project.

References

[1] Sari R A 2016 Indonesia Produce 64 Million Tons of Waste Annually: Ministry (Indonesia: Tempo).
[2] Mikata N, Takeuchi T, Hashimoto S and Nishiyama H 1996 *Nippon Steel Tech. Rep* 70.
[3] Kamal D M and Zainuri F 2014 *J Ener.Pow. Eng* 9.
[4] Perdana A R and Ali M 2012 Pengolahan Sampah Plastik Menjadi Minyak Menggunakan Proses Pirolisis. Universitas Pembangunan Nasional (Jawa Timur: Veteran).
[5] Rachmawati Q and Herumurti W 2015 *J Teknik ITS* 4 1 ISSN 2337-3539 (2301-9271 Printed).
[6] Fakhrhoseini S M and Dastanian M 2013 *J Chem.* 1-5.
[7] Miandad R, Barakat M A, Aburiazaiza A S and Nizami A S 2017 *Int. Biodeter. Biodegr* 119 239-252.
[8] Osueke C O and Ofondu I O 2011 *Int. J Adv. Eng. Sci. Tech* 4 1 021-024.
[9] Adrados A, de Marco I, Caballero B M, Lopez A, Laresgoiti M F and Torres A 2012 *Waste Manage.* 32 5 826-832.
[10] Miandad R, Nizami A S, Rehan M, Barakat M A, Khan M I, Mustafa A, Ismail I M I and Murphy J D 2016 *Waste Manage.* 58 250-259.
[11] SK Dirjen Minyak & Gas Bumi, E. S. D. M.: 2008.
[12] Wang Z, Fingas M and Sigouin L 2002 Environ Forensics 3 251-262.

IOP Conf. Series: Materials Science and Engineering **334** (2018) 012015 doi:10.1088/1757-899X/334/1/012015

High density Polyethylene plastic waste treatment with microwave heating pyrolysis method using coconut-shell activated carbon to produce alternative fuels

S R Juliastuti[1,2], **M I Hisbullah**[1] **and M Abdillah**[1]
[1] Department of Chemical Engineering, Institut Teknologi Sepuluh Nopember, Kompleks ITS, Keputih Sukolilo, Surabaya, Indonesia, 60111

E-mail: julizainul@gmail.com

Abstract. Pyrolysis is a technology that could crack polimer such as plastic waste into alternative fuels. This research uses microwave heating methode, which more efficient than conventional heating methode. The plastic waste used is 200 grams of HDPE, with feed to catalyst weight ratio are 1:1 , 0.6:1 , 0.4:1. Pyrolysis was run at temperatures of 250, 300, 350, & 400 °C for 15, 30 and 45 min. From the experimental result, the best variable of pyrolysis process with microwave method is at 45 minutes, at 400°C, and 1:1 feed to catalyst weight ratio. Result shows that yield of liquid and gas product is 99.22%; yield of residue is 0.78%; value of liquid product's composition (cycloparaffin and n-paraffin) is 54.09% and concentration of methane gas is 10.2%.

1. Introduction

Pyrolysis is one of the effective ways to process palstic waste, it can crack polymer such as HDPE plastic by just using heat energy [1]. Generally, there is two methode of pyrolysis that is conventional heating and microwave heating methode. For conventional heating method, heating process is carried out in the container by conduction while in microwave heating method heating process is carried out directly on the material. By heating directly on the material pyrolysis process occur less time and energy. So, it can be concluded that microwave heating method more effective and efficient than conventional heating method [2]. In the pyrolysis process occurs polymer scissions. For polyethylene pyrolysis occurs *random scission* that a lot kind of scissions happen but dominated by *beta scission* which is produce olefins group of hydrocarbons. Heat will continue to cut the bound of carbon groups so that will form simpler range of hydrocarbon groups [1]. For microwave heating method pyrolysis need a catalyst as a thermal assist and a cracking catalyst. For thermal assist function, catalyst must contain organic compound so it can absorb microwave and convert it into heat and for cracking catalyst function it must contain Al and Si compound so it can reduce activation energy of cracking reaction [3]. However, most of catalyst used for pyrolysis have a high price. In the other hand CSAC (Coconut Shell Activated Carbon) that known as a catalyst which made from a waste material with a low price have a potential as catalyst for pyrolysis. Because CSAC have an enough Al and Si contained and basically CSAC is an organic compound [4].

---

[2] To whom any correspondence should be addressed.

IOP Conf. Series: Materials Science and Engineering **334** (2018) 012015 doi:10.1088/1757-899X/334/1/012015

2. Materials And Methods

2.1 High Density Polyethylene (HDPE)

HDPE plastic used in this research was a transparent HDPE plastic bottle which usually used for chemicals and drugs container. The bottle chopped manually into small section of 1x1 cm^2. HDPE was a thermoplastic material which formed from carbon and hydrogen compound [5]. Density of HDPE was about 0,964 g/cm^3 [6].

2.2 Coconut Shell Activated Carbon (CSAC)

Catalyst that used in pyrolysis process usually a zeolite that contained a lot of Si and Al compound but it's expensive [7]. Coconut shell waste had a potential to be used as pyrolysis catalyst because it contains a lot of C, Al, and Si [8]. Research used a local commercial CSAC which then reactivated by burned in a furnace for two hours at 700°C. CSAC that used had 334.771 m^2/g of surface area and contained 5.08 % of Si, 1.49 % of Al and 68% of C. Material that used as a catalyst must have a surface area about 300-1000 m^2/g [9]. Could be conclude that CSAC used in this research were qualified as a pyrolysis catalyst

2.3 Microwave pyrolysis instrument construction

This research used instrumentation such as reactor, microwave, connectors, thermocouple, condenser and vacuum pump. The microwave used was a domestic microwave which was perforated with two holes on the top so the connectors could pass trough of it. The reactor used is 500 ml two-neck flask which then assembled with glass connector, condenser, liquid product container that shaped like Erlenmeyer that connected to vacuum pump and then constructed as the construction in the figure 1.

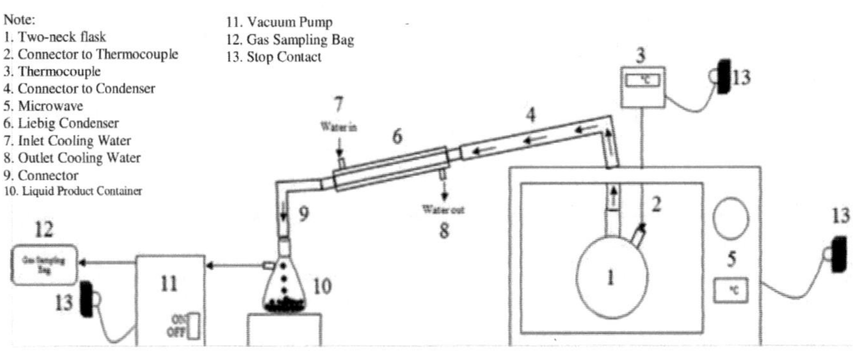

Figure 1. Pyrolysis instrumentation construction .

2.4 Pyrolysis Procedure

After all instrument and materials were set-up, vacuum pump and cooling water circulation turned on. Then the microwave turned on and keep on raising the set of microwave power until reach the temperature variable that was 250°C, 300°C, 350°C, and 400°C. at the moment that system reach the temperature that maintained the temperature by set down or up the microwave power specified time variable which was 15 minutes, 30 minutes, and 45 minutes.

After the pyrolysis process was completed the liquid product and the residue collected, then weighted it to get the yield of the process. To get the compositions of the product took a sample of liquid and gas product to be analyzed later.

2.5 Product Composition Analysis
To analyzed liquid composition of liquid product used GC-MS (***Gas Chromatography – Mass Spectrometry HP 6890 GC Method: Liquid 16 M***) analysis. Liquid product would be analyzed thoroughly and quantitatively. The gas product would only be analyzed methane content quantitatively used GC (**Gas Chromatography Agilent Hewliet Packard 6890 Series**). Because methane was the simplest compound for the result of hydrocarbon cracking process and methane was one of the alternative fuel that abundant [10].

3. Result And Discussion

3.1 Principle
Result's quality of pyrolysis could be seen from the product yield and the composition. Liquid product contains a lot of components that would be divided into 5 groups of hydrocarbons which were n-paraffin, cycloparafin, olefin, alkynes, and aromatics. For alternative fuels liquid must contained a lot of cycloparafin and n-paraffin and prohibited to contain too much olefin, alkyne and aromatic groups. So, more quantity of n-paraffin and cycloparafin showed a better quality of liquid product produced. For gas product, amount of methane would represent amount of other desired gas product such as ethane, propane, butane, etc. Because methane was the simplest result for pyrolysis and the most difficult result to form. So, more amount of methane compositions showed a better quality of gas product.

3.2 Effect of Operation Time on Results
In the research to determine the effect of the operation time, the temperature and number of catalysts were determined at 300°C and 1:0,6 feed to catalyst weight ratio. Operation time would affect both yield and the composition of product. After the result were obtained the best time would be used as a fixed variable in the next run of the research for determined effect of others variable.

3.2.1 Effect of Operation Time on Yield of the Product, The result of the research for determined effect of operation time could be seen on figure 2. From figure 2 could be observed that at 15 minute of operation time liquid product produced just 8.16 gram/gram feed and continue increasing with the increase of operation time used, and reach 46.9 gram/gram feed at 45 minute of operation time. Residue kept decreasing with increasing of operation time from 50.51 gram/gram feed at 15 minutes to 9.36 gram/gram feed. It happened because of more time operation that used more energy that given to plastic to crack into simpler compound. [1]. But, a different thing happened to gas product that given a fluctuate pattern. At 15 minute of operation time yield gas product reach 41.69 gram/gram feed and increased at 30 minute of operation time into 65.67 gram/ gram feed then decreased at 45 minute of operation time into 43.74 gram/ gram feed. Indicated during a short operation time pyrolysis tend to produce a lot of gas that easy to be produced such as CO_2, CO, etc. With the increase of operation time pyrolysis get more energy to start crack the plastic to heavy fraction which was a liquid product.

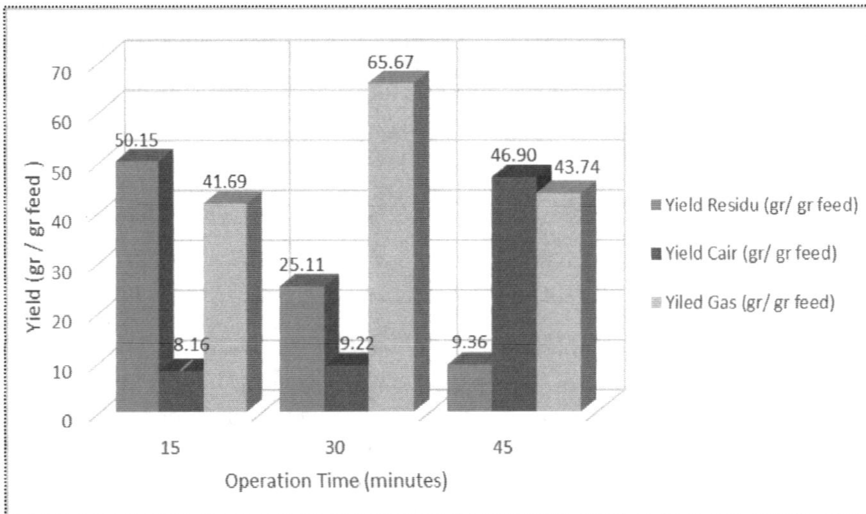

Figure 2. Effect of operation time on yield of the product.

3.2.2 Effect of Operation Time on Composition of the Product, From figure 3, could be seen that the composition of liquid product at 15 minute of time operation only legible of 13.7% of n-paraffin. It happened because the low quality of product which cause a poor analysis of GC-MS. Quality of product was indicated from the physical appearance which was higher the quality, the color getting closer to yellow and being more dilute. Product from 15 minutes time operation shows dark green color and very viscous liquid. The liquid product from 30 minutes of operation time had a lot better quality and thus affect the analysis of GC-MS. Total amount of n-paraffin and cycloparaffin was 40.84% at 30 minutes and continued increase into 41.38 % at 45 minute of operation time. So, could be conclude that the longer the operation time better the result obtained.

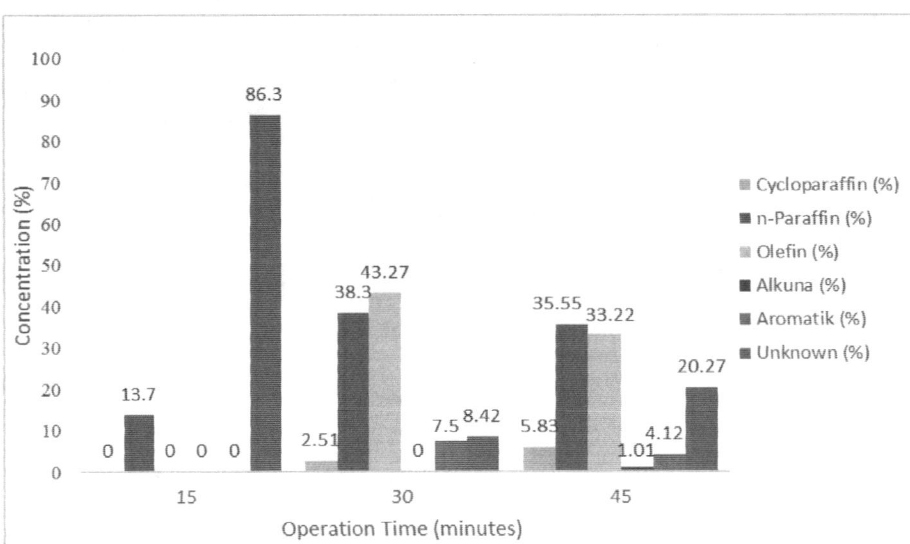

Figure 3. Effect of operation time on composition of the liquid product.

For the result of the composition of the gas could be seen in figure 4 which showed that the methane content continues to increase with the increase of the operating time used. The composition of methane increases from 2.4 % at 15 minutes of operation time into 7.1 % at 45 minutes of operation time.

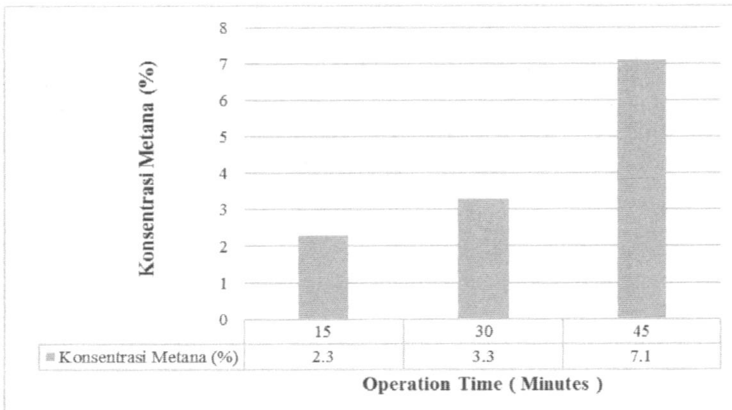

Figure 4. Effect of operation time on methane concentration.

3.3 Effect of Temperature on Results

In the research to determine the effect of the temperature, the operation time and number of catalysts were determined at 45 minutes and 1:0,6 feed to catalyst ratio temperature would affect both yield and the composition of product. After the result are obtained the best temperature variable would be used as a fixed variable in the next run of the research for determined effect of others variable.

3.3.1 Effect of Temperature on Yield of the Product, Figure 5 shows the yield change in each temperature variable. Yield of liquid product continue increasing from 9.43 gram/ gram feed at 250°C into 52.24 gram/ gram feed at 350°C but decreasing to 47.59 gram/ gram feed at 400°C while the yield of gas initially decreased and start to increase at 300°C. It can happen because higher temperature that used, more energy given to pyrolysis process and cause more liquid product produced, but at a certain point the energy will excess and thus make the product that initially a heavy fraction of hydrocarbon (liquid product) being re-crack into a lighter fraction of hydrocarbon (gas product) [1].

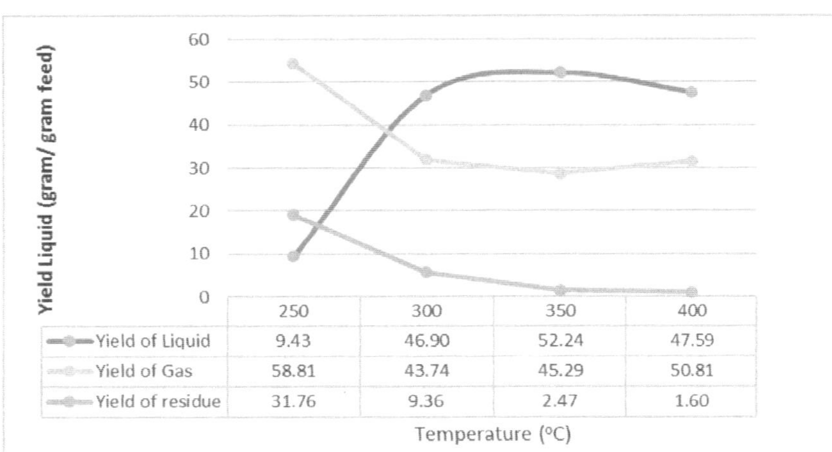

Figure 5. Effect of Temperature on yield of the products.

3.3.2 Effect of Temperature on Composition of the Product, From figure 6 could be seen that the total of cyclopaaffin and n-paraffin was 50.29 at 250°C but if it was reviewed that yield on 250°C was so little compared to yield of another variable. From a temperature of 300°C total amount of cycloparaffin and n-paraffin continue increase to a temperature of 400°C. It shows that the higher temperature of the pyrolysis the better composition of liquid product.

For the result of gas product composition could be seen at **figure** 7. It shows that the methane composition increase form 5.2% at 250°C into 7.4% at 400°C. It was due to the higher the temperature the higher the energy given and will cause more of light fraction that produced. So can be conclude that higher temperature of the pyrolysis the better composition of gas product and the best temperature is 400°C

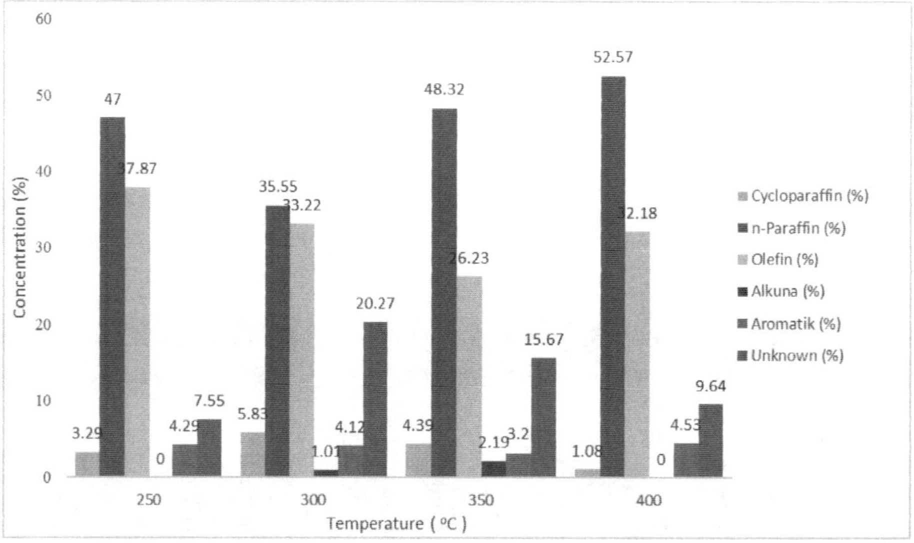

Figure 6. Effect of temperature on composition of the liquid product.

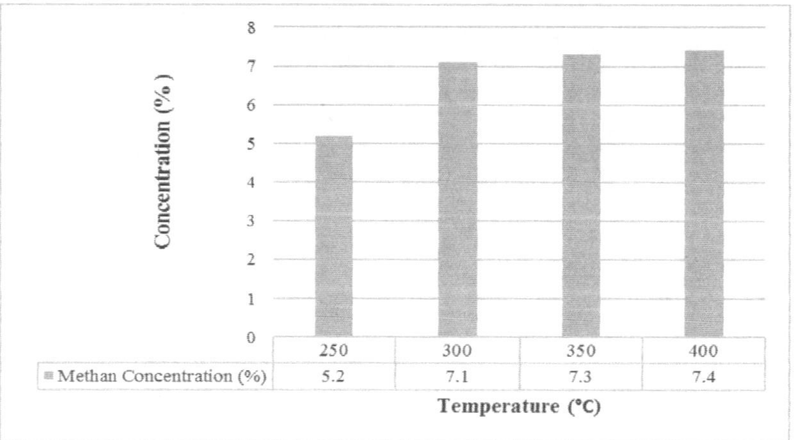

Figure 7. Effect of temperature on methane concentration.

3.4 Effect of Feed to Catalyst Weight Ratio on the Results In the research to determine the effect of the catalyst ratio, the operation time and temperature were determined at 45 minutes and 400°C. Temperature will affect both yield and the composition of product.

3.4.1 Effect of Feed to Catalyst Weight Ratio on Yield of the Product, Figure 8 shows the yield change in each feed to catalyst weight ratio variable. The yield of the liquid product continues to decrease as the number of catalyst ratios increases while the yield of gas product continues to increase as the number of catalyst ratios increase. It happens because with increases of the amount of the catalyst will increase efficiency of heating process and help of pyrolysis cracking process to produced lighter fraction product which is more inclined to from gas and make a better composition of liquid product [3]. The best result was found in the 1:1 feed to catalyst weight ratio which is has 36.58 gram/gram feed yield of liquid product, 62.64 gram/ gram feed yield of as product and 0.78 gram/ gram feed yield of residue.

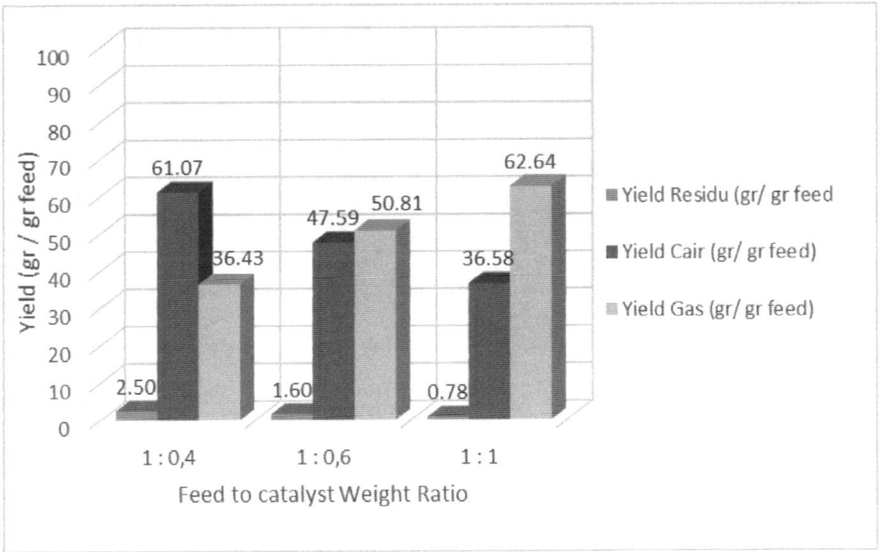

Figure 8. Effect of feed to catalyst weight ratio on yield of the product.

3.4.2 Effect of Feed to Catalyst Weight Ratio on Composition of the Product, From figure 9. can be seen that the quality of the liquid product continues to be better by the higher amount of catalyst that used. Best result is at 1:1 feed to catalyst weight ratio: catalyst with total amount of cycloparafin and n-paraffin is 54.09%.

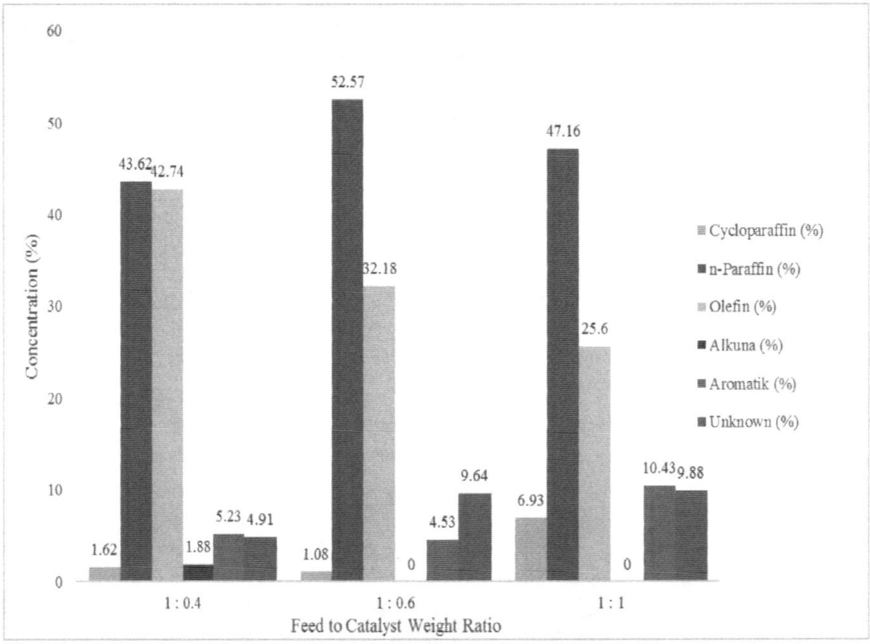

Figure 9. Effect of feed to catalyst weight ratio on yield of the product.

For the effect of the ratio catalyst to concentration of methane of the gas product could be seen at figure 10. The concentration of methane continues to increase by the increases of the ratio of catalyst with the best result at 1:1 feed to catalyst weight ratio which was 10.2 % concentration of methane.

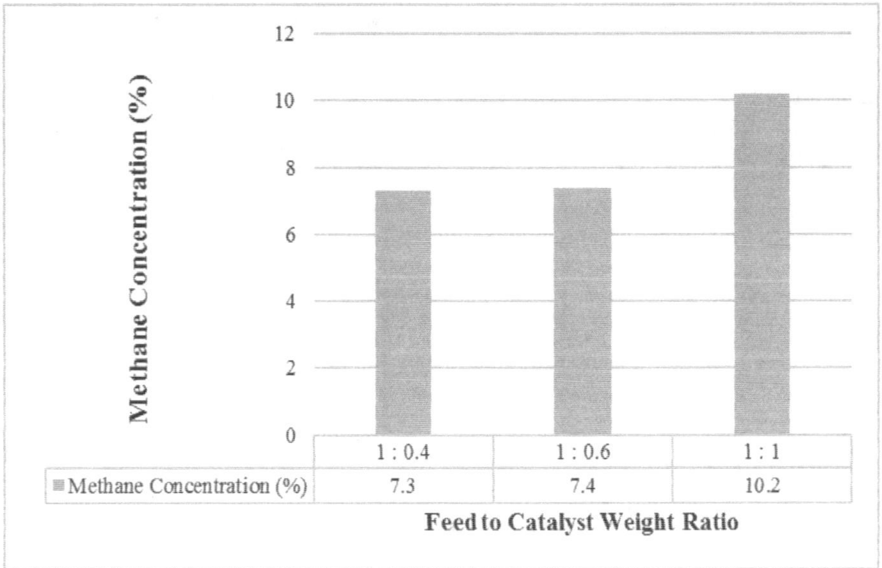

Figure 10. Effect of feed to catalyst weight ratio on methane concentration.

4. Conclusion

In this research can be conclude that HDPE plastic waste can be treat with microwave heating method pyrolysis to produce alternative fuels. Best result obtained at 400°C temperatures of operations with 1:1 feed to catalyst weight ratio for 45 minutes of operation time with results; 36.58 gram/gram feed yield of liquid product, 62.64 gram/gram feed yield of gas product, 0,78 gram/gram feed yield of residue, 54.09% desired liquid product composition and 10.2% methane concentration of gas product.

Acknowledgments

We would thank to Wastewater Treatment Laboratory, Chemical Engineering Department, Institut Teknologi Sepuluh Nopember, for the facilities provided to accomplish the research.

References

[1] Wampler, T. P. (2007). Applied Pyrolysis Handbook Second Edition. Boca Raton: Taylor & Francis Group, LLC.
[2] Undri, A., Rosi, L., Frediani, M., & Frediani, P. (2013). Microwave pyrolysis of polymeric materials: Waste tires treatment and characterization of the value-added products
[3] Almeida, D., & Marques, M. d. (2015, August 31). Thermal and Catalytic Pyrolysis of Plastic.
[4] Madakson, P. B., Yawas, D. S., & Apasi, A. (2012, March). Characterization of Coconut Shell Ash for Potential Utilization in Metal Matrix Composites for Automotive Applications. International Journal of Engineering Science and Technology (IJEST), 4.
[5] Gabriel, L. H. (n.d.). History and Physical Chemistry of HDPE
[6] Salih, Sihama E., Hamood, Abdulkhaliq F., & Abdaslalam, Alyaa H. (2013). Comparison of the Characteristics of LDPE: PP and HDPE: PP Polymer Blends
[7] Leong, S. K., Lam, S. S., Ani, F. N., Ng, J.-H., & Chong, C. T. (2016). Production of Pyrolyzed Oil from Crude Glycerol Using a Microwave Heating Technique. International Journal of Technology, 323-326.
[8] Iqbaldin, M. M., Khudzir, I., Azlan, M. M., Zaidi, A., Surani, B., & Zubri, Z. (2013). Properties of Coconut Shell Activated Carbon. Journal of Tropical Forest Science.
[9] Leofanti, G., Padovan, M., Tozzola, G., Venturelli, B. (1998). Surface area and pore texture of catalysts
[10] Rutuja Bhoje, G. R. (2013). Chemical Looping Combustion of Methane: A Technology Development View. Journal of Energy, 2013.

Characterization of bio char derived from tapioca skin

F I Hasnan[1,2], K N Iamail[1], M Musa[1], J Jaapar[1], H Alwi[1] and K K K Hamid[1]
[1] Fakulti Kejuruteraan Kimia, Universiti Teknologi MARA, 40450 Shah Alam, Selangor

E-mail: mohibah@salam.uitm.edu.my

Abstract. Pyrolysis of tapioca skin was conducted to produce bio chars in the range between 500°C–800°C. Surface modification treatment were performed on bio chars by using chemicals within 24 hours at 30°C and hot water within 1 hour to enhance the bio char's adsorption properties according to surface area, pore volume, pore size, crystallinity structure and functional groups. The samples were characterized by using BET, XRD, FTIR and Methylene Blue adsorption. Based on BET result, it showed the surface area increased as the pyrolysis temperature increased followed by pore volume and pore size for S_0. The optimum temperature for S_{NaOH}, S_{HW} and S_{MeOH} was at 600°C, 700°C and 800°C with the surface area of 75.9874, 274.5066 and 351.5531 m^2/g respectively compared to S_0 while S_{P3HO4} has the worst result since it felt on macroporous structure. The percentage of MB adsorption was followed the size of bio chars surface area. Based on FTIR result, at temperature 500°C to 700°C, the bio chars still have functional groups while at 800°C, many functional groups were diminished due to high temperature struck on them. XRD result showed all the bio chars were amorphous. In conclusion, the best surface modification treatment was by Methanol followed by hot water and Sodium Hydroxide at temperature of 700°C and 800°C while Ortho-Phosphoric acid was the worst one and was not suitable for bio char's surface modification for adsorption purpose.

1. Introduction

In the era of globalization now, world-wide carbon dioxide CO_2 emissions from energy uses are increasing, and estimated at 2020, the world will produce 33.8 billion metric tons of CO_2 [1]. Regarding this issue, it can be seen the world strive for high and infinite development from developed countries which uses fossil fuel to generate energy for their development. Therefore, this action causes the CO_2 gas release to surrounding become increase from year to year. The increases of CO_2 emission have contributed to the research in alternative energy, which is from biomass [2]. Biomass energy is abundant, cheap and clean since it trapped solar energy through photosynthesis process to produce sugar and oxygen.

The agriculture waste also known as biomass, which can be converting into value-added products. Waste of agricultural product such as apricot kernel shell, hornbeam sawdust, rice straw, rice bran and rice husk has been widely used in research in order to produce bio char. According to [3], organic biomass which undergoes pyrolysis process has high carbon content is called bio char. As stated by [4], bio char also known as activated carbon. Bio char was widely used as a tool for waste management, sequestration and mitigation of climate change, treatment of waste water, building

[2] To whom any correspondence should be addressed.

sector, cosmetics industries, metallurgy, food industry, energy production, and last but not least as a support for catalyst development. This can be show bio char has various purposes in human life.

On the other hand, a developed country has improper solid and agricultural waste management including in Malaysia. Since many Small and Medium Industries (SMI) present [5], many agriculture wastes such as tapioca skins were produced which contribute to the increasing of waste and affect the usage area of disposal site.

Other than that, water pollution was one of the problems faced by many countries and need more attention to treated in a safe way with lowest cost. Therefore, tapioca skin can be converting into bio chars due to its ability in treating wastewater. The properties of bio-char depend on types of biomass used and the production condition like temperature, pressure and moisture content. The method of producing bio char by using pyrolysis may remove the carbon dioxide from the air, increase the crops growth and keep the carbon on the soil from returning to the atmosphere.

In a view by [7], bio char is the most effective and widely used method for removal of organic compound such as chlorine, pesticides, volatile compounds, certain metal and many more since it has high content of activated carbon. In wastewater treatment scope, bio char can help in treating water by adsorption process that is by carbon filtering. This can proved that active bio-char filters are most effective in removing of odour, sediments, volatile organic compounds (VOCs), chlorine and taste. According to [8] due to the presence of anions like hydroxyl and carboxyl groups, bio-char acts as cation exchanger. In addition to that, classical graphite structure of carbon in bio-char enables the carbon to connect with neighbouring atoms or atoms from foreign molecules, which increases adsorption capacity. This has been proven by [6], potato peel waste has the ability in removing cobalt ions from synthetic wastewaters (composed of various Co(II) concentrations in distilled water which adjusted previously at different pH values) without any co-existing ions.

Bio char can be used as a material to treat wastewater due to its large surface area and high porosity if it produced by pyrolysis at certain condition. Large surface area of bio-char helps in increasing water holding capacity by combining the impurities with its active sites, cation exchange capacity (CEC) and microbial activity. Therefore, tapioca skin waste produced from SMI can be converted into bio chars in treating wastewater besides can decrease the emission of CO_2 in the air. Many other studies have done to produce bio char for treatment of wastewater. However, the search for process to produce bio-char with better yield for absorption properties still face a major challenge in this field.

The main objectives of this study is to prepare bio char from tapioca skin by using pyrolysis at difference temperature (500°C, 600°C, 700°C and 800°C) and to enhance bio char properties for adsorption purpose by using surface modification treatment that is hot water treatment and chemical treatment that is alkali, acid and solvent.

2. Methodology

2.1. Sample Preparation

Dried tapioca skins were cut into small pieces and were compressed into 4 different crucibles to be pyrolyzed in Carbolite Furnace at 500°C, 600°C, 700°C and 800°C within 30 minutes respectively. The samples were collected and were kept in sealed plastic after it was cooled to room temperature.

2.2. Surface modification Treatment

Three samples of 1.0 g of bio chars at temperature 500°C were weighted and were put into 100 ml solution of NaOH.H2O, H3PO4 and Methanol respectively. All the three samples were stirred within 24 h at 30°C. Next, the samples were filtered and were dried at 80°C for 24 h. The steps were repeated for bio chars at 600°C, 700°C and 800°C. Similarly, for hot water treatment, 1.0 g of bio char was weighted and was put into 100 ml of hot water at temperature 90°C for 30 minutes. Next, the sample was filtered and dried in oven at 80°C for 24 h. All the samples were kept in desiccator for further used. S0 was stand for untreated bio chars while SMeOH, SNaOH, SHW and SH3PO4 were stand for bio chars treated with Methanol, Sodium Hydroxide, hot water and Ortho-Phosphoric acid respectively. All the methods were adopted from [9].

2.3. Characterization

The samples were characterized by using BET, Methylene blue adsorption, XRD and FTIR. BET characterization was performed at 250°C digest temperature within 30 minutes. Based on MB adsorption, 5 different concentrations of MB at 2,4,6,8 and 10 ppm were prepared to create calibration curve. Then, 0.2 g of bio char samples were put into 20 ml of 10 ppm MB solution and were left for 1 h for absorption process. Lastly, the filtrates were filtered before analyzed by using UV-Vis at 645 nm. Other than that, for XRD characterization process, the initial and end angle used was 100 and 700 respectively with 100/min speed. The Voltage used was 20 Watt and the amplitude was 20 mA.

3. Result And Discussion

3.1. BET Result Regarding Surface Area, Pore Volume and Pore Size

Based on Figure 1, surface area of bio char samples was increase as the pyrolysis temperature increased. This can be shown surface area for S0 at 500°C, 600°C, 700°C and 800°C was 1.7007, 6.0385, 134.4512 and 206.3966 m^2/g respectively. The highest S0 surface area was at 800°C. Other than that, for SMeOH, the surface area at 500°C to 800°C was 3.2426, 10.3776, 237.2951 and 351.5531 m^2/g respectively, compared to S_0, S_{MeOH} at 800°C has the highest surface area that was 351.5531 m^2/g. This can be concluded Methanol solvent has the ability in widen the pores wall from micro-pores to meso-pores. According to [10], the presence of Methanol or other alcohols contributed to mesoporous materials with fibrous, high surface area, large pore volume and open web-like structures. In a view [11], the presence of hydrophilic and the strongly polarity compound with the polarity index of 6.6 of sugar in Methanol solvent has proved its ability to increase the pore size and surface area. Based on Graph 1, SMeOH showed the largest surface area compared to others bio chars.

Similarly, surface area of SHW at 500°C, 600°C, 700°C and 800°C was 2.4024, 2.9233, 274.0566 and 188.2421 m^2/g respectively. It showed at 700°C, the surface area produced was the highest compared to S0 at 700°C. This can be shown hot water increased the surface area from temperature of 500°C to 700°C compared to S0, while at 800°C, the surface area decreased from 206.3966 to 188.2421 m^2/g. This proven that hot water has the ability to remove any minerals deposited on the pore's wall. In accordance to [12], bio chars were heterogeneous materials contain high carbon and minerals. Related to the statement, [12] proved hot water vapor penetrated bio char layers during water quenching process. This seen that hot water has an activating effect and expulse minerals out from bio chars in his analysis. In conclusion, hot water has the ability in increasing the surface area, pore volume and pore size of bio chars pyrolyzed at 500°C to 700°C. Briefly, the optimum temperature for S_{HW} was at 700°C since it has the highest surface area compared to bio chars at 800°C.

Other than that, for S_{NaOH}, the surface area at 500°C, 700°C and 800°C was decreased a little compared to the original ones except at temperature 600°C. This clearly seen in Graph 1, the surface area was 2.3235, 75.9874, 88.3366 and 121.7390 m^2/g respectively. From the observation and comparison between the S_{NaOH} against S0, it can be concluded NaOH was not suitable to modify bio chars compared to the Methanol. In conclusion, the optimum S_{NaOH} was at temperature 600°C compared to the original bio chars.

Lastly, surface areas of S_{H3PO4} were 1.5183, 1.9945, 7.1621 and 1.2931 m^2/g for temperature of 500°C to 800°C respectively. Ortho-phosphoric acids were used in treating bio chars since it was a triprotic acid. Triprotic acid was an acid that has highest proton and readily to donate it to reagents, with the three protons. Based on the surface treatment, compared to the S_0, S_{H3PO4} has the lowest surface area and the worst result. The acid was too strong until diminish the surface wall. Hence, it indicated S_{H3PO4} at temperature 500°C-800°C best as solid fuel as the porosity produced was not in the range for adsorption purposes.

In conclusion, the higher the pyrolysis temperature, the higher the surface area of the bio chars. Besides that, it's can be concluded S_{MeOH} has the highest surface area followed by S_{HW}, S_0, S_{NaOH} and S_{H3PO4}.

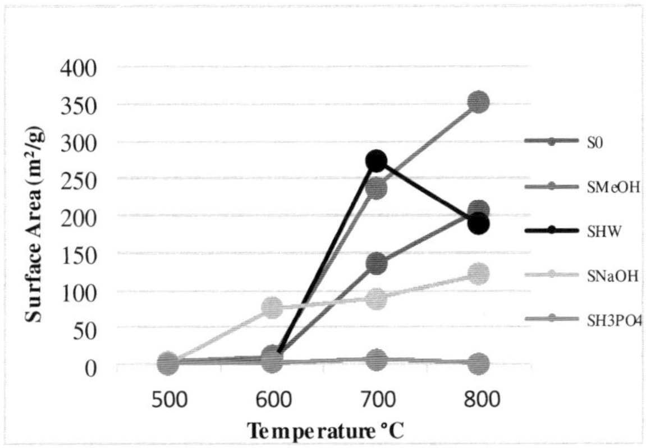

Figure 1. Bio chars versus Surface Area.

Referring to Figure 2, the pore volume also increased according to the pyrolysis temperature that was 0.000572, 0.002743, 0.043207 and 0.062053 cm^3/g respectively for S_0 at temperature 500°C to 800°C. Other than that, pore volume for S_{MeOH} also increased according to the pyrolysis temperature that was 0.001135, 0.003444, 0.071476 and 0.175246 cm^3/g at 500°C to 800°C respectively. Next, the pore volume for S_{HW} also increased according to the pyrolysis temperature that was 0.000566, 0.000890, 0.082317 cm^3/g for temperature of 500°C to 700°C compared to the S_0 but decrease at 800°C with 0.055122 cm^3/g. This is due to bio char has reached its resistance temperature at 700°C.

Meanwhile, for the pore volume of S_{NaOH}, the result also showed decreasing in value that was 0.000618, 0.024615, 0.027207 and 0.033994 cm3/g respectively. Last but not least, for S_{H3PO4}, no reading showed since BET only cover the micro-porous and meso-porous structure only. In conclusion based on pore volume size, this can be concluded pore volume was related to the surface area.

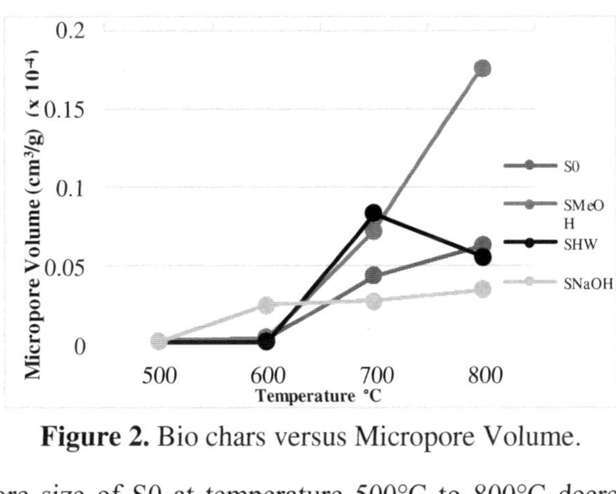

Figure 2. Bio chars versus Micropore Volume.

Based on table 1, pore size of S0 at temperature 500°C to 800°C decreased. At temperature of 500°C and 600°C the pore size was 105.1759 and 35.8640 Å respectively which felt on mesoporous line while for temperature of 700°C and 800°C lied in microporous category with the pore size of 18.4649 and 18.6010 Å respectively. Other than that, for SMeOH, at 500°C, 600°C and 700°C the pore was in mesopores category which has pore size of 104.6779, 31.6062 and 20.2156 Å. Meanwhile at temperature of 800°C lied in micropores type with the pore size of 19.9396 Å. Hence, the small pore diameters showed that narrow and deep pores of bio char was formed.

Other than that, pore type for SHW at temperature of 500°C to 600°C were in mesopores category with the pore size of 118.4134 and 76.7675 Å. Meanwhile, at temperature of 700°C and 800°C, SHW were in micropores category with the pore size of 18.7016 and 19.7218 Å. In addition to that, SNaOH, lied on mesopores category that was 150.4447, 24.6137, 23.2339 and 22.4776 Å for temperature of 500 to 800°C. Based on observation and comparison between SNaOH against the S0, it can be concluded NaOH was not really suitable in modifying bio chars compared to Methanol and hot water.

Lastly, pore size for SP3HO4 were lied on macropores line. This can be concluded, SP3HO4 was not suitable to be used in adsorption purposes in treating waste water but can be used as fuel solid that was charcoal.

Table 1. Pore Size of Bio Chars.

Sample	Pore Size (4V/A by BET) (Å)
S_0: 500	105.1759
600	35.8640
700	18.4649
800	18.6010
S_{MeOH}: 500	104.6779
600	31.6062
700	20.1258
800	19.9396
S_{HW}: 500	118.4134
600	76.7665
700	18.7016
800	19.1278
S_{NaOH}: 500	150.4447
600	24.6137
700	23.2339
800	22.4776
S_{P3HO4}: 500	28.7092
600	12.7470
700	2.7643
800	18.6344

3.2. Methylene Blue Analysis

Based on figure 3, the percentage of MB adsorption for S_0 was 87.43, 87.48, 95.97 and 99.45% for temperature 500°C to 800°C respectively. Other than that, for S_{MeOH}, the percentage of MB adsorption was 86.86%, 91.66%, 98.27% and 99.50%. Compared to S_0, S_{MeOH} has the higher absorption at temperature from 600°C to 800°C and the maximum absorption was at temperature 800°C with 99.50%. Therefore, based on the percentage of the MB absorption by S_{MeOH}, this data related and was supported the surface area data obtained from BET result.

The percentage of MB absorption by S_{HW} was 87.71%, 91.82%, 95.11% and 95.59%. Other studies indicated bio chars treated with hot water possess higher adsorption capacities. However, opposite results were observed in this experiment. At temperature of 500°C to 800°C the percentage of MB absorption showed a little decrement compared to the S_0 even though the surface area of the treated one was larger. This can be concluded S_{HW} followed the rule in which the highest surface area that can be yield by S_{HW} was 700°C, since it was the maximum temperature resistance.

Other than that, for percentage of MB absorption by S_{NaOH}, the value was 84.14%, 92.26%, 94.93% and 97.53%. The percentage of the MB absorption was tele with the surface area as shown in Figure 1 1 that was 2.3235, 75.9874, 88.3366 and 121.3790 m^2/g at temperature of 500°C to 800°C. In addition to that, the percentage of MB absorption by S_{P3HO4} was 57.11%, 50.79%, 58.42% and 64.12%. The percentage of the MB absorption was also tele with the surface area as shown in Figure 1 that was 1.5183, 1.9945, 7.1621 and 1.2931 m^2/g.

In conclusion, the bio chars surface area gave a very important impact on the percentage of MB absorption. The result showed every temperature gave different surface area for adsorption purposes. The larger the surface area of bio chars yield the larger percentage of MB absorption. From the analyzed data in Figure 3, the best bio char performance in MB absorption was S_{MeOH} followed by S_{NaOH} and S_{HW}. This can be concluded the optimum temperature for S_{NaOH} was at temperature 600°C, while for S_{HW} and S_{MeOH} was at temperature 700°C and 800°C. The worse bio char in MB absorption was bio char treated by Ortho-Phosphoric acid since it produced smallest surface area resulted too strong acid which destroyed the samples wall. Therefore, the optimum temperature of bio chars in MB absorption was 700°C and 800°C.

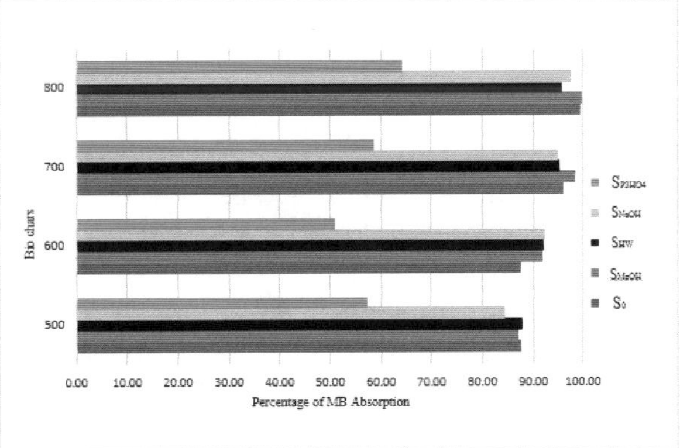

Figure 3. Chart of Bio-Chars at Different Temperature Versus Percentage of MB Absorption.

3.3. XRD analysis

Based on the results at Figure 4, it showed all bio chars has the amorphous structure since at 500°C, all the crystalline structures were diminished due too high temperature penetrate on bio chars surface.

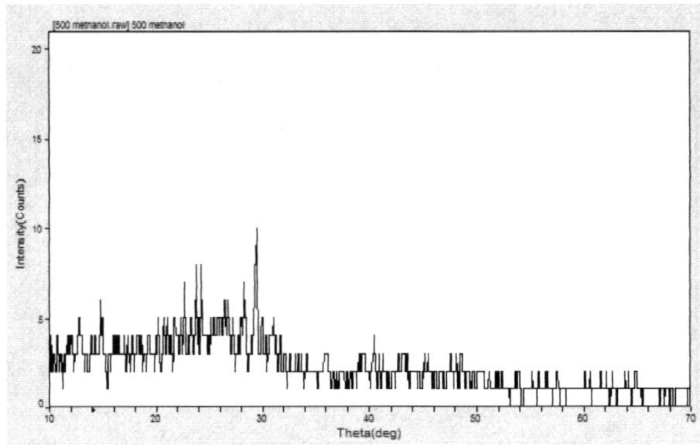

Figure 4. Bio Chars at 500°C.

3.4. FTIR analysis

FTIR was conducted to determine the functional group presented on bio chars surface.

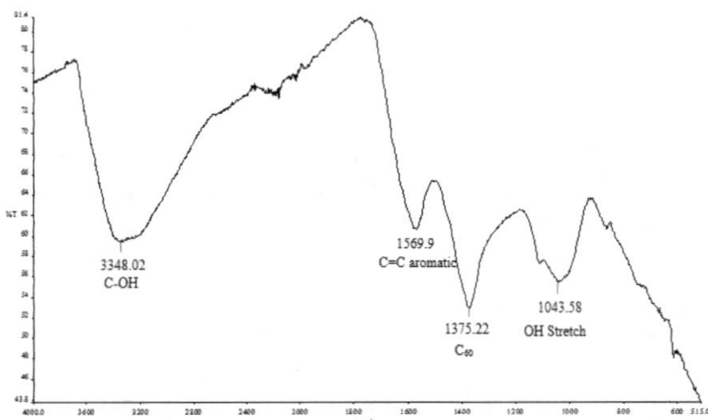

Figure 5. Bio chars at 500°C.

Based on Figure 5, the band at 3348 cm^{-1} represented the stretching vibrations of the OH groups, which could be attributed to the adsorbed water on the bio char. It showed the intensity and shaped was strong. Other than that, the band at 1569.92 cm^{-1} represented the stretching variations C=C aromatic. The double bonds appeared as medium to strong absorptions. Based on the C=C bond, it indicated the presence of adjacent carbon in the bio chars. At band 1043.58 cm^{-1}, it represented the stretching variations of the C-OH. It showed strong intensity of hydroxyl group.

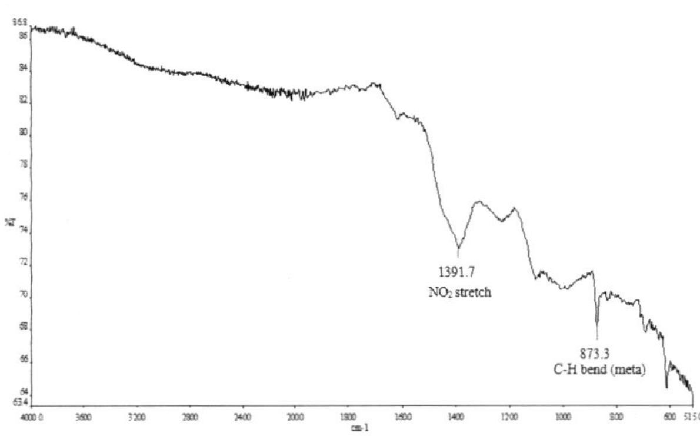

Figure 6. Bio Chars at 600°C.

Regarding Figure 6, the band at 1391.76 cm^{-1} represented the stretching vibrations of NO_2 (aliphatic). It belonged to nitro group. It showed the intensity was strong. Other than that, at band 873.39 cm^{-1} represented C-H bend (meta) aromatics.

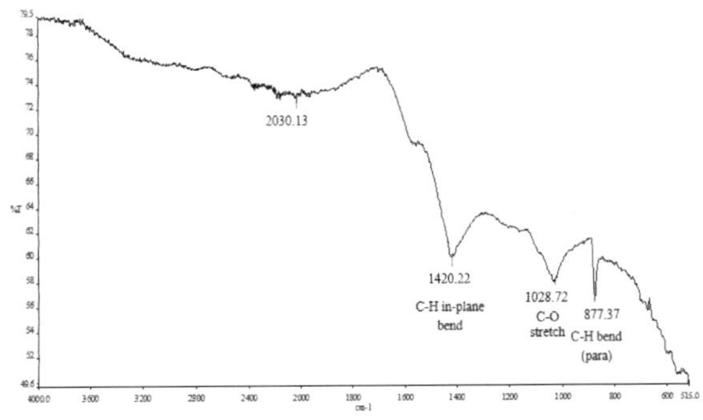

Figure. 7 Bio Chars at 700°C.

Based on figure 7, the band at 1420.22 cm^{-1} represented the vibrations of C-H in-plane bend. At band 1028.72 cm^{-1} represented C-O stretch. This can be concluded alcohol presented on the bio char after treated with Methanol. At band 877.72 cm^{-1} represented the vibrations of C-H bend (para). This can be concluded many aromatics present in the bio chars.

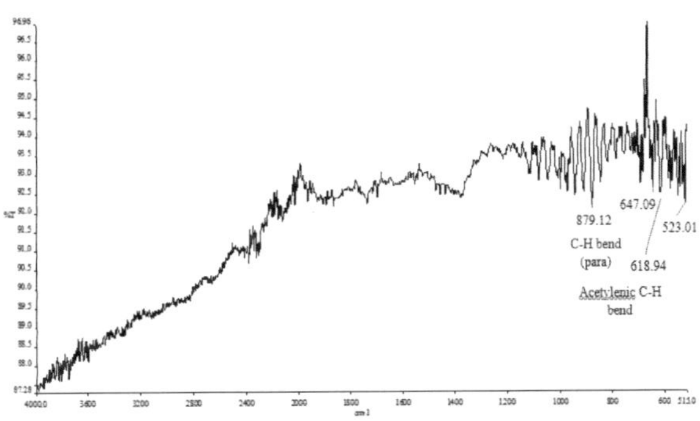

Figure 8 bio chars at 800°C.

Referring to figure 8, the band at 879.12 cm^{-1} represented the vibrations of C-H bend (para). It showed it belonged to aromatic group. At band 647.09 cm^{-1} and 618.94 cm^{-1} represented acetylenic C-H bend. This showed alkynes group present in the bio chars. At the end of the data, it showed the graph was interrupted due to high temperature during pyrolysis until diminish all the functional groups.

4. Conclusion

Optimum temperature for S_0 was at 800°C, while for S_{MeOH}, S_{HW} and S_{NaOH} was at 800°C, 700 and 600°C respectively. The best surface modification treatment for adsorption purpose was by Methanol solvent, followed by hot water and Sodium Hydroxide. The largest surface area and pore volume size was achieved by S_{MeOH} and S_0 at 800°C that was 351.5531 m^2/g and 206.3966 m^2/g respectively. Meanwhile, surface area for S_{HW} at 700°C and S_{NaOH} at 600°C was 274.5066 m^2/g and 75.9874 m^2/g compared to S_0. The sample's pore volume also related to the surface area and majority of samples were in microporous and mesoporous line which indicated the bio chars suitable for adsorption purposes.

Based on MB adsorption, percentage of MB adsorption by S_{MeOH} was the highest followed by S_0, S_{NaOH} and S_{HW} at 800°C that was 99.50, 99.45, 97.53 and 95.59 %.

Lastly, S_{P3HO4} has the smallest surface area and pore volume compared to S_0. The pore size also decreased and lied on macroporous line. Therefore, Ortho-Phosphoric acid was not suitable in modifying bio chars, since it was too strong until break the micropores and mesopores wall. Hence, it indicated the modified bio chars with Ortho-Phosphoric acid at temperature 500°C-800°C best as solid fuel as the porosity produced was not in the range for adsorption purposes.

XRD result showed bio chars have amorphous structure since all the crystallinity has been diminished due to high temperature started at 500°C. FTIR result showed some functional group presents on bio chars surface at 500°C to 700°C. However, it became diminished at temperature 800°C due too strong temperature resistance. In a nutshell, all the bio chars have the ability in adsorption purposes except for S_{P3HO4}.

Acknowledgment
Thank you to Fakulti Kejiriteraan Kimia, Universiti Teknologi MARA to facilitate the facility in accomplishment of these research activity

Reference

[1] Narzari, R., Bordoloi N., Chutia R.S., Borkotoki B., Gogoi N., Bora A., Kataki R., (2015). Biochar: An Overview On Its Production, Properties and Potential Benefit. *Biology, Biotechnology and Sustainable Development,* 13-31.

[2] Moralı, Uğur, Şensöz, Sevgi (2015). Pyrolysis of hornbeam shell (Carpinus betulus L.) in a fixed bed reactor: Characterization of bio-oil and bio-char. *Fuel, 150*, 672-678 doi: http://dx.doi.org/10.1016/j.fuel.2015.02.095

[3] Sobhy M. Y, Abd El Hakim M. Daifullah, Sohair A. El-Reefy. (2015). Pore Structure Characterization of Chemically Modified Biochar Derived from Rice Straw, *Environmental Engineering and Management Journal.* 73-480

[4] Shafeeyan, M. S., Daud, W. M. A. W., Houshmand, A., & Shamiri, A. (2010). A review on surface modification of activated carbon for carbon dioxide adsorption. *Journal of Analytical and Applied Pyrolysis, 89*(2), 143-151. Doi http://dx.doi.org/10.1016/j.jaap.2010.07.006

[5] Joseph S.H. Lim, Dominic C.Y. Foo, Denny K.S. Ng, A. Ramlan, Raymond R. Tan (2014). Graphical tools for production planning in small medium industries (SMIs) based on pinch analysis. *Journal of Manufacturing Systems,*33(4),639-646. Doi https://doi.org/10.1016/j.jmsy.2014.06.001

[6] Kyzas, George Z., Deliyanni, Eleni A. Matis, Kostas A. (2016) Activated carbons produced by pyrolysis of waste potato peels: Cobalt ions removal by adsorption *Colloids and Surfaces A: Physicochemical and Engineering Aspects* p. 74-83

[7] T. A. Ghezzehei, D. V. Sarkhot, and A. A. Berhe (2014) Biochar Can Be Used to Capture Essential Nutrients from Dairy Wastewater and Improve Soil Physico-Chemical Properties. *Solid Earth*, p. 953–962

[8] Lee, J.Y., Lee, K.K. (2004) A short note on investigation and remediation of contaminated groundwater and soil in Korea. *Engineering Geology 14* p. 123-130.

[9] Gong, H., Chen, Z., Fan, Y., Zhang, M., Wu, W., & Wang, W. (2015). Surface Modification of Activated Carbon for Siloxane Adsorption. *Renewable Energy,* p. 144-150.

[10] Sinkó K. (2010). Influence of Chemical Conditions on the Nanoporous Structure of Silicate Aerogels. *Materials*, p. 704-740.

[11] Rodolfo A. V., Carlos F.P.M., Vera L. P. (2016). Characterization of Chemical Compounds with Antioxidant and Cytotoxic Activities in Bougainvillea x buttiana Holttum and Standl, (var. Rose) Extracts. *Antioxidants,* p. 1-11.

[12] Cornelissen G., Pandit N. R., Taylor P., Pandit B. H., Sparrevik M., Schmidt H. P. (2016). Emissions and Char Quality of Flame-Curtain "Kon Tiki" Kilns for Farmer Scale. *Charcoal/Biochar Production.*

Thermal characteristic investigation of eutectic composite fatty acid as heat storage material for solar heating and cooling application

R Thaib[1], H Fauzi[2,5], H C Ong[2], S Rizal[1], T M I Mahlia[3] and M Riza[4]

[1] Department of Mechanical Engineering, Syiah Kuala University, Banda Aceh 23111, Indonesia.
[2] Department of Mechanical Engineering, University of Malaya, Kuala Lumpur 50603, Malaysia.
[3] Department of Mechanical Engineering, Universiti Tenaga Nasional, Kajang 43000, Selangor, Malaysia
[4] Department of Chemical Engineering, Syiah Kuala University, Banda Aceh 23111, Indonesia.

E-mail: hadidoank@gmail.com

Abstract. A composite phase change material (CPCM) of myristic acid/palmitic acid/sodium myristate (MA/PA/SM) and of myristic acid/palmitic acid/sodium laurate (MA/PA/SL) were impregnated with purified *damar* gum as called *Shorea Javanica* (SJ) to improve the thermal conductivity of CPCM. The thermal properties, thermal conductivity, and thermal stability of both CPCM have investigated by using a Differential Scanning Calorimetry (DSC) thermal analysis, hot disc thermal conductivity analyzer, and Simultaneous Thermal Analyzer (STA), simultaneously. However, a chemical compatibility between both fatty acid eutectic mixtures and SJ in composite mixtures measured by Fourier Transform Infra-Red (FT-IR) spectrophotometer. The results were obtained that the thermal conductivity of MA/PA/SM/SJ and MA/PA/SL/SJ eutectic composite phase change material (CPCM) were improved by addition 3 wt.% and 2 wt.% of *Shorea javanica* (SJ), respectively, without occur a significant change on thermal properties of CPCM. Moreover, the absorbance spectrum of FT-IR shows the good compatibility of SJ with both MA/PA/SM and MA/PA/SL eutectic mixtures, the composite PCM also present good thermal performance and good thermal stability. Therefore, it can be noted that the purified Shorea Javanica proposed, the as high conductive material in this study was able to improve the thermal conductivity of eutectic PCM without any significant reduction on its thermo-physical and chemical properties and can be recommended as novelty composite phase change material for thermal energy storage application.

1. Introduction

Thermal energy storage (TES) plays an important role in generating the solar energy to use in a wide range application such as heating and cooling, solar power, and industrial waste heat recovery [1-4]. The latent heat thermal energy storage (LHTES) material known as phase change materials (PCMs)

[5] To whom any correspondence should be addressed.

offer more advantage over sensible heat thermal energy storage (SHTES) materials, particularly the small temperature different between melting and solidifying point, small volume, and low weight per unit of storage capacity. PCMs absorb heat as storage energy during the heating process and release it during cooling [5]. However, phase change material (PCM) that used in thermal energy storage (TES) application have classified into two categories which are inorganic and organic phase change materials (PCMs). Although the inorganic PCMs group have great heat storage capacity with a wider range of phase transition temperature. But they have numbers of disadvantages such as subcooling, corrosion, phase separation, phase segregation, and lack thermal stability, compared to organic PCMs which only have a problem with low heat transfer rate due to low thermal conductivity [6-8].Organic phase change materials (PCMs) such as paraffin and fatty acids have been used in a wide application of thermal energy storage due to their high thermal storage capacity, a wide range of phase transition temperature, low subcooling, and have a good reliability [7]. However, the common problem of these organic PCMs is their low heat transfer rate due to low thermal conductivity, which leads to large temperature gradients. Therefore, enhancement the thermal conductivity of organic phase change materials (PCMs) became one of the main issues in a wide application of thermal energy storage (TES) [9].

Numbers of different methods have been proposed for enhancing the thermal conductivity of PCMs. Encapsulation was one of few methods proposed in order to improve the thermal conductivity of PCMs. On the other hand, this method was handled the leakage problem properly and bring a high cost of production [8]. The thermal conductivity of PCM can be improved as well by adding high conductive porous material using simple mixing method, solution casting method or impregnation method [10-13].

However, in this work, we propose a modification by combination two existed methods of simple mixing and impregnation in order to prepare the composite fatty acids eutectic mixture as composite phase change material (CPCM) by adding Shorea javanica, which is obtained from purified damar gum. The study was performed to analyze the improvement of thermal conductivity of prepared eutectic composite fatty acid with different composition of Shorea Javanica and to evaluate its thermal performance and chemical compatibility of prepared CPCM.

2. Materials and methods

2.1. Materials

Two kind of fatty acid eutectic mixture, myristic acid/palmitic acid with 5 wt.% sodium myristate (MA/PA/5 wt.%SM) and myristic acid/palmitic acid with 10 wt.% sodium laurate (MA/PA/10 wt.% SL) were propose as based phase change material (PCM) as conducted by Fauzi, Metselaar in previous study [14]. A natural damar gum or *Shorea javanica* which obtained from conventional plantation farm in Indonesia was proposed as high conductive porous material. However, additional purification processes were required in order to remove the impurity content from *Shorea Javanica* (SJ).

The purification process was conducted by using an organic solvent method which was introduced previously by Setianingsih, 1992 [15]. Toluene (C_6H_5-CH_3) with a molecular weight (MW) 92.14 g.mol^{-1} (Fisher Scientific) and *Shorea Javanica* (SJ) in composition ratio 1:8 wt.% stirred at 2000 rpm for 20 minutes at ambient temperature in order to obtain a completed dissolves solution of *Shorea Javanica* (SJ). Moreover, 1 wt.% of activated charcoal (AC) manufactured by Acros Organic was added for decolorizing of SJ solutions which stirred for 15 minutes at 45 °C. The sedimentation and filtration process was conducted to separate the impurities from the solutions. Subsequently, the pure solution of Shorea Javanica (SJ) was placed in a rotary evaporator to vaporize the solvent toluene, and thus the button product of evaporation process was dried with oven with temperature 80 °C for 8 hours in order to obtain pure *Shorea Javanica* powder. A dried powder of *Shorea Javanica* needs to be grinded in a rotary ball mill for further separated by sieve shaker to obtain the same size particle of *Shorea Javanica* in 100 μm [16].

2.2. *Preparation and analysis methods*

A modification mixing-impregnation method was used to prepare a novel eutectic composite phase change material (CPCM) of MA/PA/SM/SJ and MA/PA/SL/SJ. The eutectic mixtures of MA/PA/5wt.%SM and MA/PA/10wt.%SL separately were placed into a jacketed flask reactor along with different mass composition of SJ in 1, 2, 3, 4, and 5 wt.% under vacuum pressure and hot heat transfer fluid (HTF) circulation was maintained the temperature stable at 70 °C. However, a stirrer mixing was applied to homogenous distribution of SJ in MA/PA/SM and MA/PA/SL eutectic mixtures. These preparation processes performed for 2 hours and ended by drying process at ambient temperature.

Each composition of prepared eutectic composite MA/PA/SM and MA/PA/SL with different mass fraction SJ simultaneously were evaluated their thermal conductivity, thermal properties, and thermal stability using thermal conductivity analyzer (Hot Disc, TPS 2500 S), thermal Differential Scanning Calorimetric (DSC) analyzer (Metler Toledo, DSC[1] Stare system), and Simultaneously Thermal Analyzer (STA 6000, Perkin Elmer), respectively. Furthermore, the Fourier transform infrared spectroscopy (FT-IR, Bruker Tensor 27) used to identify the chemical compatibility between both fatty acid eutectic mixtures and *Shorea Javanica*. The analysis was performed using ATR sample compartment with MIR spectra in the wavenumber range of 4000 – 400 cm^{-1}.

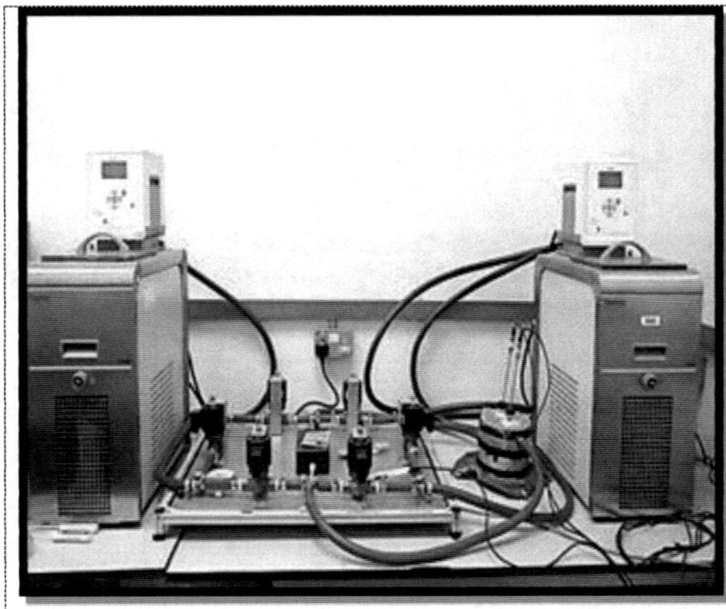

Figure 1. Thermal performance test setup.

Moreover, the thermal performance of MA/PA/SM/SJ and MA/PA/SL/SJ eutectic composite were evaluated with customized test setup device as shown in Fig. 1. The setup was consisting of 2 fluid circulation bath for both hot and cold medium circulated into the chamber. The 2 cylindrical glass tubes were attached inside the chamber and direct contact with heat transfer fluid (HTF). Subsequently, 6 gr of two couple samples of MA/PA/SM; MA/PA/SL eutectic mixtures and MA/PA/SM/SJ; MA/PA/SL/SJ eutectic composites were placed separately in those glass tube in different turn. The alterations temperature of these samples against endothermic and exothermic time was recorder by thermocouples (J-type, Omega) connected to data acquisition (DAQ) system. The setting temperature of hot and cold heat transfer fluid (HTF) set above of melting temperature and below of solidification temperature of each prepared samples which are 65 °C and 30 °C, respectively.

3. Results and discussion

3.1. Thermal characteristic of Eutectic composites

In the previous study, Fauzi et al. was reported that the myristic acid/palmitic acid (MA/PA) eutectic mixture have a better improvement on their thermal properties by adding 5 wt.% sodium myristate (SM) and 10 wt.% sodium laurate (SL), respectively, in term of reducing the phase transition temperature and increasing the latent heat of fusion [14]. The studies were obtained the melting temperature of MA/PA reduced 11.5% and 10.12% by adding 5 wt.% SM and 10 wt.% SL, while the latent heat of fusion improved about 15.53 % and 12.24%, respectively. Nevertheless, these additions of acid-based surfactants do not show a significant improvement in thermal conductivity of eutectic PCM. Therefore, *Shorea Javanica* was proposed in this study as a high conductive material to enhance the thermal conductivity of prepared PCM with consideration no much effect on degrading the thermal properties.

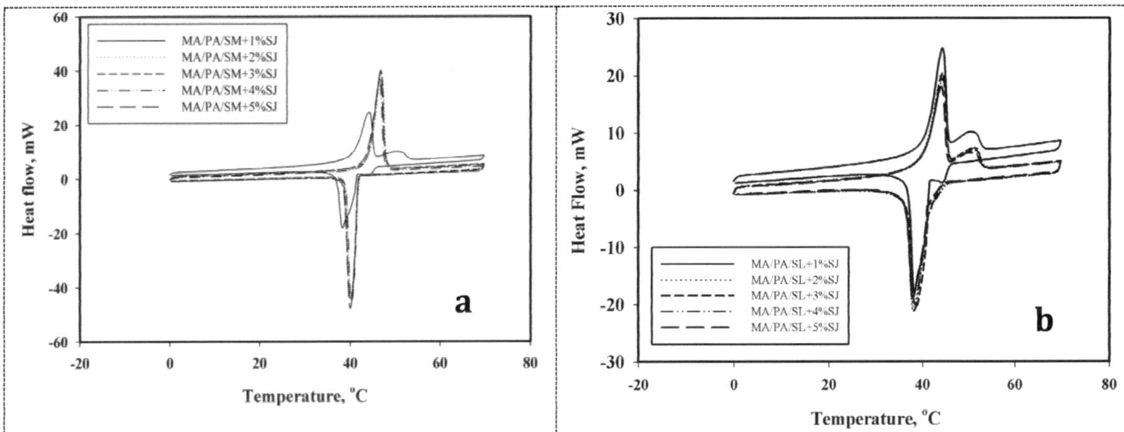

Figure 2. DSC curves of eutectic mixture composites with different mass fraction SJ: a). MA/PA/SM/SJ; b). MA/PA/SL/SJ.

The DSC curve in Fig. 2.a and Fig. 2.b presents the thermal properties of CPCM in a different composition of SJ. The melting temperature (T_m) and solidification temperature (T_s) of MA/PA/SM/SJ and MA/PA/SL/SJ interpreted from onset point, while the latent heat of fusion during melting ($\Delta H_{f,m}$) and solidification phase (ΔH_{fs}) were obtained from interpolating of charts peak area. The result as shown in Table 1 indicated that the phase transition temperature of MA/PA/SM/SJ and MA/PA/SL/SJ eutectic composite rises up against increasing percentage composition of SJ, and in the same time, its latent heat of fusion tend to decrease with increasing SJ composition.

The thermal conductivity of MA/PA/SM/SJ and MA/PA/SL/SJ eutectic composite in a different composition of SJ were listed in Table 2 which shows that the thermal conductivity of those composite mixture has increased simultaneously with increasing the percentage composition of SJ. These results noted that the highest amount of thermal conductivity was obtained at MA/PA/SM and MA/PA/SL with addition 5 wt. % SJ which are 0.492 and 0.393 $Wm^{-1}K^{-1}$, respectively. Nevertheless, the thermal properties of MA/PA/SM and MA/PA/SL with 5 wt.% SJ present an unexpected decrease on latent heat of fusion which has a significant drop to 167.38 and 153.91 Jg^{-1},compared to the latent heat of fusion of initial MA/PA/SM and MA/PA/SL eutectic mixture as 179.12 and 174.47 Jg^{-1}, respectively. However, an improvement of thermal conductivity without a significant drop of latent heat of fusion instead shows by MA/PA/SM+3%SJ and MA/PA/SL+2%SJ eutectic composite mixtures. Moreover, the thermal conductivity of MA/PA/SM/3%SJ and MA/PA/SL/2%SJ eutectic composite shows a great improvement by increased 101.65% and 62.55%, respectively, compared to thermal conductivity value of their initial mixture [14]. Meanwhile, the latent heat of fusion only presents a slight decreased which is 0.93% and 7.50%.

Tabel 1. Thermal properties of MA/PA/SM and MA/PA/SL/SJ eutectic mixture with Shorea Javanica.

Composite phase change materials (CPCM)	T_m (°C)	$\Delta H_{f,m}$ (J/g)	T_s (°C)	$\Delta H_{f,s}$ (J/g)
MA/PA/SM + 1% SJ	40.54	179.92	41.46	183.55
MA/PA/SM + 2% SJ	43.20	176.39	41.95	179.95
MA/PA/SM + 3% SJ	43.96	177.45	41.73	180.85
MA/PA/SM + 4% SJ	43.89	169.35	41.75	172.29
MA/PA/SM + 5% SJ	43.75	167.38	41.67	175.63
MA/PA/SL + 1% SJ	40.64	169.26	41.56	169.70
MA/PA/SL + 2% SJ	40.20	161.27	41.36	159.26
MA/PA/SL + 3% SJ	40.48	158.69	41.50	156.68
MA/PA/SL + 4% SJ	40.26	157.45	41.40	155.68
MA/PA/SL + 5% SJ	40.08	153.91	41.30	153.14

Numbers of studies have been using a different kind of porous materials to improve the thermal conductivity of PCMs. Karaipekli, Sarı [17] proposed a different mass fraction of expanded graphite (EG) and carbon fiber (CF) to the enhanced thermal conductivity of stearic acid (SA). The result of these study reported that thermal conductivity of SA was increased with increasing the mass fraction of EG and CF and indicated an insignificant decrease in its latent heat of fusion at the same time. Sarı and Karaipekli [18], studied the effect of addition expanded graphite (EG) into paraffin in improvement the thermal conductivity of paraffin. The results were indicating that the thermal conductivity of paraffin/EG composite mixture was increased with increasing the mass fraction of EG. The decreasing of latent heat of fusion also occurred by increasing the composition of EG in paraffin in this study. In the extended work, these authors evaluated the thermal characteristic and thermal conductivity improvement of capric-myristic acid/expanded perlite and some fatty acid compounds with expanded graphite. They also obtained that these porous material were able to improve the thermal conductivity of PCMs and also reduced its latent heat of fusion insignificantly [19, 20].

Tabel 2. Thermal conductivity MA/PA/SM/SJ and MA/PA/SL/SJ composite PCM with different composition SJ.

Composite phase change materials (CPCMs)	Thermal conductivity, $Wm^{-1}K^{-1}$
MA/PA/SM + 1% SJ	0.463
MA/PA/SM + 2% SJ	0.475
MA/PA/SM + 3% SJ	0.488
MA/PA/SM + 4% SJ	0.489
MA/PA/SM + 5% SJ	0.492
MA/PA/SL + 1% SJ	0.373
MA/PA/SL + 2% SJ	0.382
MA/PA/SL + 3% SJ	0.387
MA/PA/SL + 4% SJ	0.387
MA/PA/SL + 5% SJ	0.393

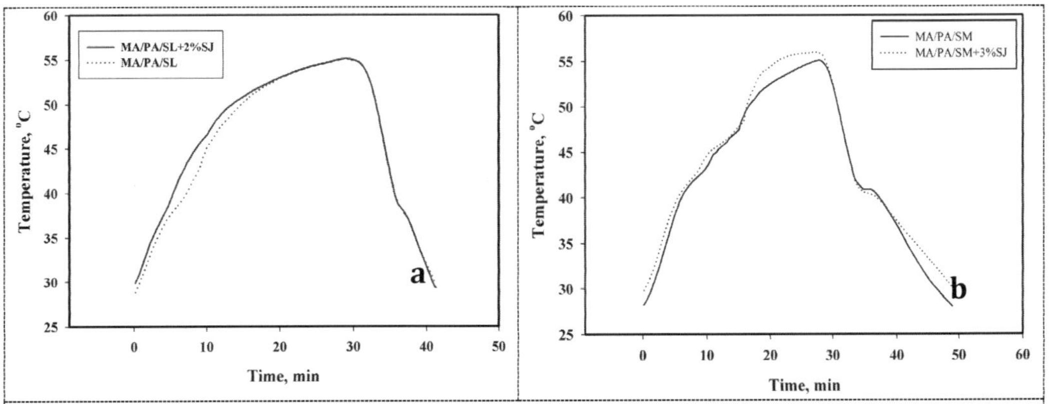

Figure 3. Curves of thermal performance heat storage and heat release: a). MA/PA/SM/SJ; b). MA/PA/SL/SJ.

Moreover, the improvement of thermal conductivity of MA/PA/SM+3%SJ and MA/PA/SL+2%SJ composite mixtures were proved by the improvement of their heat transfer rate as seen in Fig. 3.a and Fig. 3.b. The figures show the heat storage and heat release duration time needed for MA/PA/SM+3%SJ and MA/PA/SL+2%SJ eutectic composites comparison with their initial non-composite eutectic mixtures MA/PA/5wt.%SM and MA/PA/10wt.%SL to change the phase from solid to liquid and vice versa. Both curves show the phase transition time of MA/PA/SM+3%SJ and MA/PA/SL+2%SJ eutectic composites, and noticed that the duration time for both composite PCMs to reach the melting and solidification points were 8.4 and 6.7 minutes; 8 and 6 minutes, respectively. This phase transition time shows shortest than their initial mixtures of MA/PA/5wt.% SM and MA/PA/10wt.%SL which is 9.7 and 10.6 minute; 11.3 and 7.16 minute, respectively. However, the same identic result also reported by Zang and Fang [10] in other work proposed paraffin and paraffin/expandable graphite (P/EG) and reported that the heat storage time of paraffin was longer 27.4% than P/EG and heat release time paraffin was longer 56.4% than P/EG. It is noted that the heat transfer rate of CPCM was higher than PCMs. In addition, the heat transfer process during the heat storage was controlled by natural convection, whereas in the heat release, the heat transfer controlled by thermal conduction. Hence, the increase of conductivity coefficient of eutectic CPCM had brought a significant effect on enhancement the heat transfer rate during release the heat than heat storage process. Therefore, the CPCM need shorter time once heat release process than storage process [10].

3.2. Compatibility SJ with MA/PA/SM

The FT-IR spectra as seen in Fig. 4.a and Fig. 4.b shows the absorbance peak for each functional group of chemical structures of MA/PA/SM, MA/PA/SM+3%SJ, and MA/PA/SL, MA/PA/SL+2%SJ compared to pure SJ spectrum, respectively. The spectra indicated have the same frequency in every peak range of wavenumber between MA/PA/SM eutectic mixture and MA/PA/SM+3%SJ eutectic composite; MA/PA/SL eutectic mixture and MA/PA/SL+2%SJ eutectic composite, respectively. It is mean that the addition of 3 wt. % and 2% of *Shorea Javanica* (SJ) in MA/PA/SM and MA/PA/SL eutectic mixtures did not occur any chemical reaction in the composite mixtures. Therefore, it can be noted that the changing of thermal properties of MA/PA/SM/SJ and MA/PA/SL/SJ eutectic composites was not caused by the chemical interaction between SJ and initial eutectic PCM. Subsequently, it was due to physical properties of *Shorea Javanica* (SJ) which has a high melting point and low latent heat of fusion, and thus contributed to changing the thermal properties of eutectic CPCM.

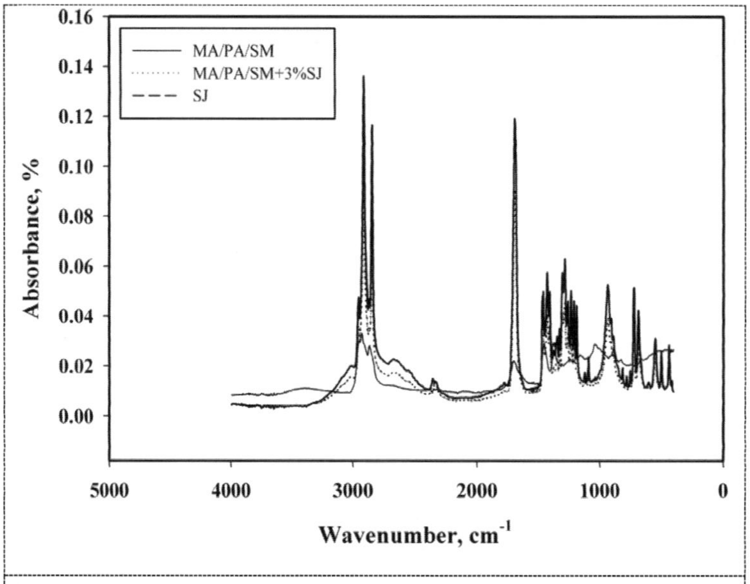

Figure 4. FT-IR curves of MA/PA/SM and MA/PA/SJ.

3.3. Thermal stability of MA/PA/SM/SJ

The thermal stability curve of MA/PA/SM+3%SJ and MA/PA/SL+2%SJ eutectic composites was demonstrated in Fig. 5.a and Fig. 5.b which is shows that the prepared eutectic composites do not show any mass degradation within work temperature 30 C to 160 °C. The weight degradation of MA/PA/SM+3%SJ and MA/PA/SL+2%SJ have appeared once the working temperature at 168.7 °C and 164.28 °C, and thus reaches an optimum weight degradation at work temperature 289.39 °C and 272.64 °C, respectively. Thus, these results indicate that the MA/PA/SM+3%SJ and MA/PA/SL+2%SJ eutectic composites presented a good stability to apply as a CPCM in thermal energy storage application (TES) with operation temperature below of 160 °C.

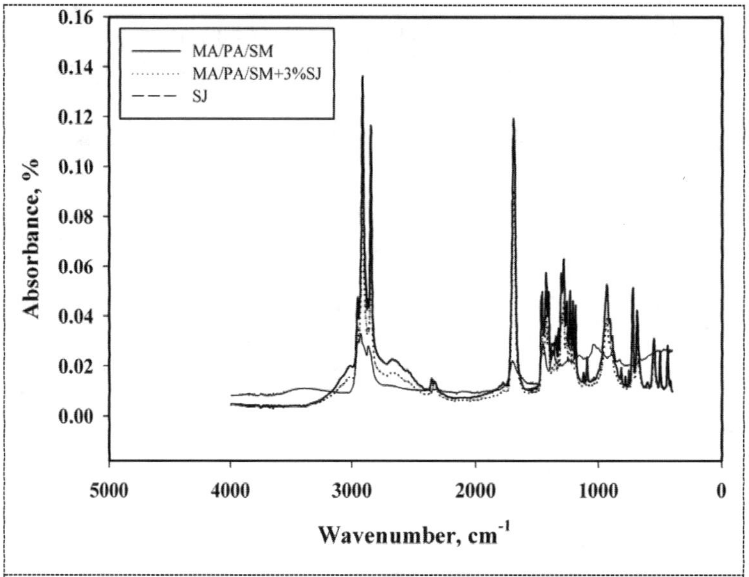

Figure 5. Curve thermal stability of MA/PA/SM+3%SJ.

Thermal stability of composite phase change materials (CPCMs) also been studied by other researchers. Kim and Drza [12], measured thermal stability of paraffin/xGnP composite PCM at range temperature 30 to 600 °C. The mass loss of CPCMs started to decomposition at 200 °C and reach to total decomposition at 280 °C. In the other work, Jeong, Jeon [21] analyzed thermal decomposition of n-octadecane/diatomite CPCMs at range temperature of 30 to 400 °C. In this study obtained that the weight decomposition of n-hexadecane/diatomite composite was 50% lower than pure n-hexadecane PCM at operation temperature 200 °C. According to these studies can be noted that CPCMs have a higher thermal stability compared to the pure PCMs.

4. Conclusions

The preparation and thermal characteristic analysis of two novel eutectic composites phase change material (CPCM) which involves the myristic acid/palmitic acid/sodium myristate (MA/PA/SM) and myristic acid/palmitic acid/sodium laurate (MA/PA/SL) with Shorea Javanica (SJ) have been evaluated in this current study. The thermal conductivity of these CPCMs was simultaneously increased with increasing the composition of SJ as 1, 2, 3, 4, and 5 wt. %, respectively. But, CPCM with composition SJ 3 wt. % and 2% with MA/PA/SM and MA/PA/SL respectively that shows a good improvement on thermal conductivity without significant impact in decreasing the latent heat of fusion of the CPCM mixtures. The eutectic composite of MA/PA/SM+3%SJ and MA/PA/SL+2%SJ also indicate a good thermal performance, no chemical reaction between each component in the mixture and has a good thermal stability without occurs weight degradation within temperature work 30 °C to 160 °C. Therefore, those may conclude that the Shorea Javanica (SJ) is acceptable to use as a porous material to improve the thermal conductivity of composite phase change material (CPCM).

Acknowledgement

The authors acknowledge the University of Malaya for financial support through Institute of Research Management and Services (IPPP) fund.

References

[1] Nomura, T., N. Okinaka, and T. Akiyama, *Waste heat transportation system, using phase change material (PCM) from steelworks to chemical plant.* Resources, Conservation and Recycling, 2010. **54**(11): p. 1000-1006.
[2] Cabeza, L.F., et al., *Materials used as PCM in thermal energy storage in buildings: A review.* Renewable and Sustainable Energy Reviews, 2011. **15**(3): p. 1675-1695.
[3] Al-Abidi, A.A., et al., *Review of thermal energy storage for air conditioning systems.* Renewable and Sustainable Energy Reviews, 2012. **16**(8): p. 5802-5819.
[4] Kuravi, S., et al., *Thermal energy storage technologies and systems for concentrating solar power plants.* Progress in Energy and Combustion Science, 2013. **39**(4): p. 285-319.
[5] Pielichowska, K. and K. Pielichowski, *Phase change materials for thermal energy storage.* Progress in Materials Science, 2014. **65**(0): p. 67-123.
[6] Zalba, B., et al., *Review on thermal energy storage with phase change: materials, heat transfer analysis and applications.* Applied Thermal Engineering, 2003. **23**(3): p. 251-283.
[7] Fauzi, H., et al., *Thermo-physical stability of fatty acid eutectic mixtures subjected to accelerated aging for thermal energy storage (TES) application.* Applied Thermal Engineering, 2014. **66**(1–2): p. 328-334.
[8] Zhou, M., et al., *Heat transport enhancement of thermal energy storage material using graphene/ceramic composites.* Carbon, 2014. **75**(0): p. 314-321.
[9] Zhou, M., et al., *Heat transport enhancement of thermal energy storage material using graphene/ceramic composites.* Carbon, (0).
[10] Zhang, Z. and X. Fang, *Study on paraffin/expanded graphite composite phase change thermal energy storage material.* Energy Conversion and Management, 2006. **47**(3): p. 303-310.

[11] Alkan, C. and A. Sari, *Fatty acid/poly(methyl methacrylate) (PMMA) blends as form-stable phase change materials for latent heat thermal energy storage.* Solar Energy, 2008. **82**(2): p. 118-124.

[12] Kim, S. and L.T. Drza, *High latent heat storage and high thermal conductive phase change materials using exfoliated graphite nanoplatelets.* Solar Energy Materials and Solar Cells, 2009. **93**(1): p. 136-142.

[13] Nomura, T., N. Okinaka, and T. Akiyama, *Impregnation of porous material with phase change material for thermal energy storage.* Materials Chemistry and Physics, 2009. **115**(2–3): p. 846-850.

[14] Fauzi, H., et al., *Phase change material: Optimizing the thermal properties and thermal conductivity of myristic acid/palmitic acid eutectic mixture with acid-based surfactants.* Applied Thermal Engineering, 2013. **60**(1–2): p. 261-265.

[15] Setianingsih, N., *Pemurinian Damar Shorea javanica dengan Menggunakan pelarut Organik dan Bahan Pemucat.* 1992.

[16] Fauzi, H., et al., *Preparation and thermal characteristics of eutectic fatty acids/Shorea javanica composite for thermal energy storage.* Applied Thermal Engineering, 2016. **100**: p. 62-67.

[17] Karaipekli, A., A. Sarı, and K. Kaygusuz, *Thermal conductivity improvement of stearic acid using expanded graphite and carbon fiber for energy storage applications.* Renewable Energy, 2007. **32**(13): p. 2201-2210.

[18] Sarı, A. and A. Karaipekli, *Thermal conductivity and latent heat thermal energy storage characteristics of paraffin/expanded graphite composite as phase change material.* Applied Thermal Engineering, 2007. **27**(8): p. 1271-1277.

[19] Karaipekli, A. and A. Sarı, *Capric–myristic acid/expanded perlite composite as form-stable phase change material for latent heat thermal energy storage.* Renewable Energy, 2008. **33**(12): p. 2599-2605.

[20] Sari, A., A. Karapekl, and K. Kaygusuz, *Fatty Acid/Expanded Graphite Composites as Phase Change Material for Latent Heat Thermal Energy Storage.* Energy Sources, Part A: Recovery, Utilization, and Environmental Effects, 2008. **30**(5): p. 464-474.

[21] Jeong, S.-G., et al., *Optimal preparation of PCM/diatomite composites for enhancing thermal properties.* International Journal of Heat and Mass Transfer, 2013. **62**(0): p. 711-717.

Multivariable model predictive control design of reactive distillation column for Dimethyl Ether production

A Wahid[1,2] **and I G E P Putra**[1]
[1] Chemical Engineering Department, Universitas Indonesia, Depok, Indonesia

E-mail: wahid@che.ui.ac.id; i.gede34@ui.ac.id

Abstract. Dimethyl ether (DME) as an alternative clean energy has attracted a growing attention in the recent years. DME production via reactive distillation has potential for capital cost and energy requirement savings. However, combination of reaction and distillation on a single column makes reactive distillation process a very complex multivariable system with high non-linearity of process and strong interaction between process variables. This study investigates a multivariable model predictive control (MPC) based on two-point temperature control strategy for the DME reactive distillation column to maintain the purities of both product streams. The process model is estimated by a first order plus dead time model. The DME and water purity is maintained by controlling a stage temperature in rectifying and stripping section, respectively. The result shows that the model predictive controller performed faster responses compared to conventional PI controller that are showed by the smaller ISE values. In addition, the MPC controller is able to handle the loop interactions well.

1. Introduction

Dimethyl ether (DME) is a simple ether which is widely used as propellant. It is also well-known as an excellent alternative diesel fuel which has high cetane number with lower NO_x, CO, and particulate emission [1]. Physical properties of DME are similar to liquefied petroleum gas (LPG), DME may hence be used as an alternative of LPG for domestic applications. In addition, DME is a key intermediate of other important chemical, such as dimethyl sulphate, methyl acetate, and light olefins. Due to its wide range of applications, the large scale production of DME is now required [2].

Currently, DME is produced via methanol dehydration in a catalytic fixed-bed reactor, followed by purification of DME in a sequence of two distillation columns [3]. Methanol dehydration in the conventional fixed-bed reactor is an equilibrium-controlled reaction, so that methanol conversion is limited that led the reactor to be operated at a high recycle rate which requires larger capital investment and operating costs [1,2]. This conventional process is costly in terms of operating cost, caused by the energy needed by two distillation columns. A single reactive distillation column can be used to reduced capital and operating cost, instead of one reactor followed by two distillation columns [4]. Bildea et.al. [5] stated that DME production through reactive distillation process led to savings of 30% in capital cost and 6% in energy requirement compared to conventional DME process for the same production rate.

While reactive distillation can have benefit over the conventional process, combination of reaction and distillation on a single column makes reactive distillation process a very complex system with

[2] To whom any correspondence should be addressed.

multiple steady state, high non-linearity of process, and strong interaction between process variables [6]. Wahid and Saputro [4] demonstrated the method of determining control structure of reactive distillation column for DME production. They used sensitivity and rangeability analysis to select a stage temperature as inferred variable to control DME product purity and its corresponding manipulated variable (MV). In these study, they successfully implemented a single-point PI control structure with a stage temperature at rectifying section as controlled variable (CV) and reboiler duty as MV, while the reflux ratio is controlled with ratio control.

Single-point control structure has an advantage on its less loop interaction [7]. However, in the single-point control structure which was proposed by Wahid and Saputro [4], no adjustment is made to control the stripping zone temperature. Therefore, the bottom purity is not controlled. The two-point control structure can maintain both product purities, but it led to strong interaction between CV and MV [6]. This study is focused on the implementation of multivariable model predictive control (MPC) to a reactive distillation column for the DME production based on two-point control structure. Both product purities are maintained by controlling the temperature of two stages which are selected by singular value decomposition (SVD) and rangeability analysis. Dynamic model of the process for the MPC controller is estimated by first order plus dead time (FOPDT) model.

2. Process Simulation

2.1. Reaction kinetics and thermodynamic properties

DME is produced from the methanol dehydration reaction in the liquid phase over an acid-base catalyst. To simplify the simulation, the kinetic for methanol dehydration to DME over ion exchange was modelled by simple power rate law in equation 1 and 2 [2].

$$r'_{\mathrm{DME}} = k'[\mathrm{MeOH}]^m[\mathrm{H_2O}]^n \quad (1)$$

$$k' = A'_0 \exp\left(-\frac{E'_a}{RT}\right) \quad (2)$$

The NRTL model was selected for the liquid phase properties calculation. The vapour phase properties calculated using RK equation of state. There is no azeotrope in the methanol–DME–water system [8].

2.2. Reactive distillation column

Rigorous simulation of the reactive distillation was performed in UniSim Design simulator. A reactive distillation column with capacity of 100,000 tonnes per year DME consist of 7 rectifying stages, 34 reactive stages, and 11 stripping stages [5] was used as base case for this work.

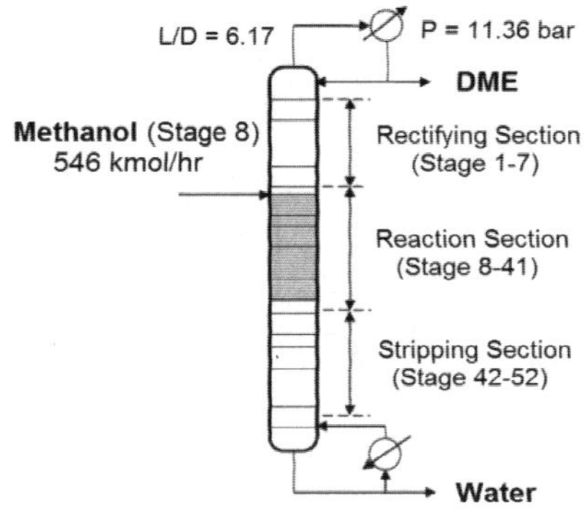

Figure 1. Schematic diagram of reactive distillation for the base case steady-state condition.

2.3. Model predictive control

In this study, a simple FOPDT process model was implemented in a 2 × 2 MPC controller built in UniSim Design. A series of step changes in reflux flow rate and reboiler duty were applied to generate input-output dynamic data. These data were then used to develop the FOPDT process model empirically using Method II [9].

3. Steady-state analysis

3.1. SVD analysis
The primary objective of reactive distillation control is to maintain product quality at its desired value. However, due to a large dead time and high price and maintenance cost, very few distillation columns use online composition analyser. Temperatures are widely used as inferential variable to control composition [10]. The two stages temperatures were selected by Singular Value Decomposition (SVD) analysis [11]. SVD analysis suggest that the stage 5 temperature should be controlled by the reflux flow rate (L) and the stage 47 temperature should be controlled by reboiler duty (Q_r).

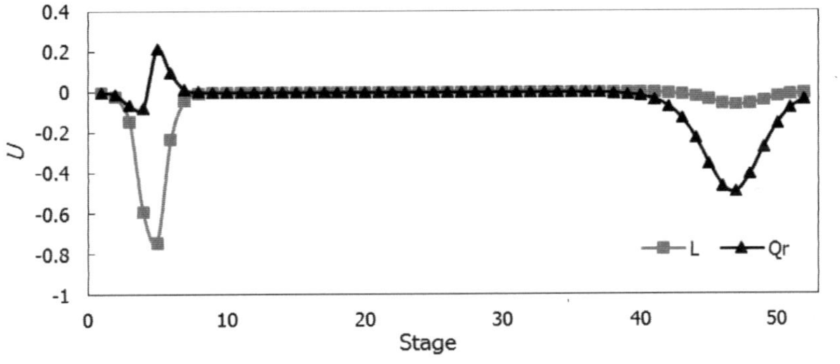

Figure 2. SVD analysis plot.

3.2. Rangeability analysis
Reactive distillation is highly nonlinear and possibly showing input multiplicity behaviour, thus the most sensitive input may have bad rangeability. So, this makes compromise between sensitivity and rangeability is needed for reactive distillation control structure design [4]. The steady-state variation of the sensitive tray temperatures obtained from SVD analysis as the two inputs are varied from its base case is studied in rangeability analysis. These input-output relations are plotted in Figure 3. The 5th and 47th stages temperatures did not show any input multiplicity for the its input variation. This makes 5th and 47th stages temperatures can be the output (CV) and can be paired with its corresponding inputs (MV).

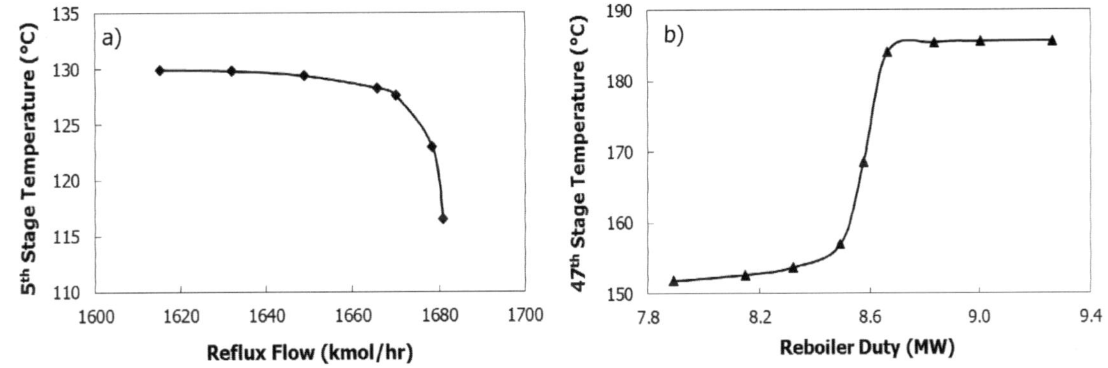

Figure 3. Steady-state input-output relation of a) reflux flow rate to stage 5 temperature, b) reboiler duty to stage 47 temperature.

4. Controller design

4.1. Controller configuration

Figure 4 shows the PI and MPC controller configurations. In the MPC configuration, the stages 5 and 47 temperatures are controlled with the reflux flow rate and the reboiler heat duty in a two-input two-output system. In the two-point PI configuration, the stage 5 temperature is controlled using the reflux flow rate and the stage 47 temperature is controlled using the reboiler duty. As the comparison, this work also investigates the performance of a single-point PI configuration that was proposed by Wahid and Saputro [4]. Temperature at the stage 5 which is the most sensitive stage to the change in reboiler duty, is controlled by reboiler heat duty, while the reflux ratio is controlled by distillate flow rate. The column inventory are controlled by three single-loop PI controllers. The column pressure is controlled by condenser duty and the reboiler level is controlled by bottom flow rate. In the MPC and two-point PI configurations, condenser accumulator level is controlled by distillate flow rate, while in the single-point configuration, condenser accumulator level is controlled by reflux flow rate.

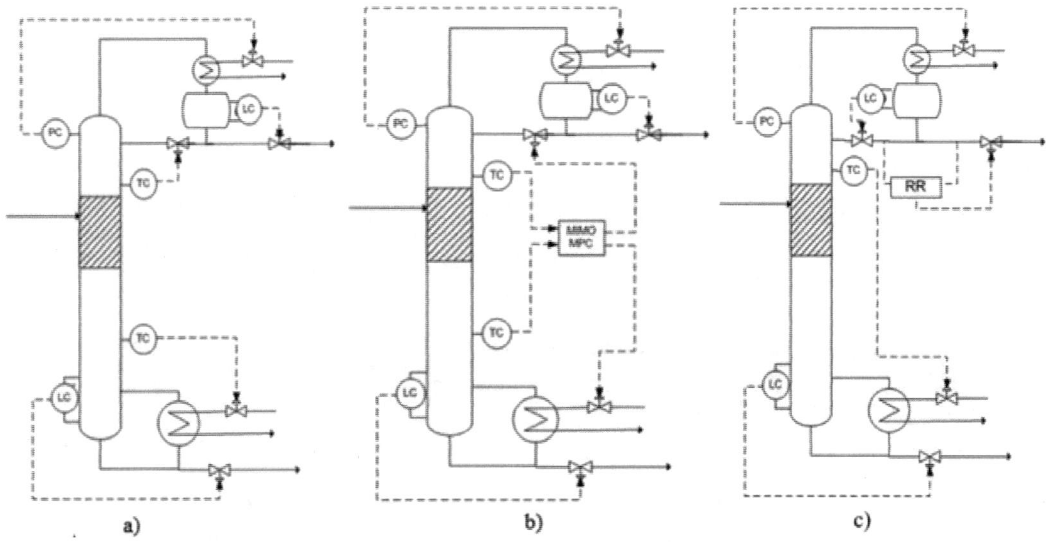

Figure 4. a) Two-point PI control configuration, b) MPC control configuration, c) Single-point PI control configuration.

4.2. Process model identification

In this study, dynamic model of reactive distillation column was modeled as a simple FOPDT model. Model identification was done by changing reflux flow rate and reboiler duty in a step test to generate input-output dynamic data. Table 1 summarizes the obtained transfer functions explaining the relationships of the two inputs (i.e. reflux flow rate and reboiler duty) and the two outputs (stages 5 and 47 temperatures). The units for all time constants are minutes. These model was implemented for the MPC controller.

Table 1. Transfer functions for the process models of DME reactive distillation.

	L	Q_r
T_5	$\dfrac{-6.912e^{-0.035s}}{0.272s+1}$	$\dfrac{1.030e^{-0.077s}}{0.185s+1}$
T_{47}	$\dfrac{-0.537e^{-0.89s}}{0.421s+1}$	$\dfrac{3.351e^{-0.102s}}{0.627s+1}$

4.3. Controller parameter

The two PI controllers were then tuned by using Tyreus-Luyben method, while the MPC controller was tuned by fine tuning. Tables 2 and 3 summarize the controller parameters.

Table 2. Single-point and two-point PI controllers parameters.

Controller	Single-point PI		Two-point PI	
	K_c	T_i (min)	K_c	T_i (min)
Stage 5 temperature	0.921	0.251	0.853	0.145
Stage 47 temperature	-	-	0.055	0.054
Reflux ratio	0.500	0.500	-	-
Column pressure	0.829	0.273	0.829	0.273
Condenser level	1.018	0	1.121	0
Reboiler level	2.450	0	2.450	0

Table 3. MPC controller parameters.

Controller	K_c	T_i (min)	P	M	T
Stages 5 and 47 temperatures	-	-	23	2	1
Pressure column	0.829	0.273	-	-	-
Condenser level	1.121	0	-	-	-
Reboiler level	2.450	0	-	-	-

5. Results and discussion

5.1. Stage 5 temperature control

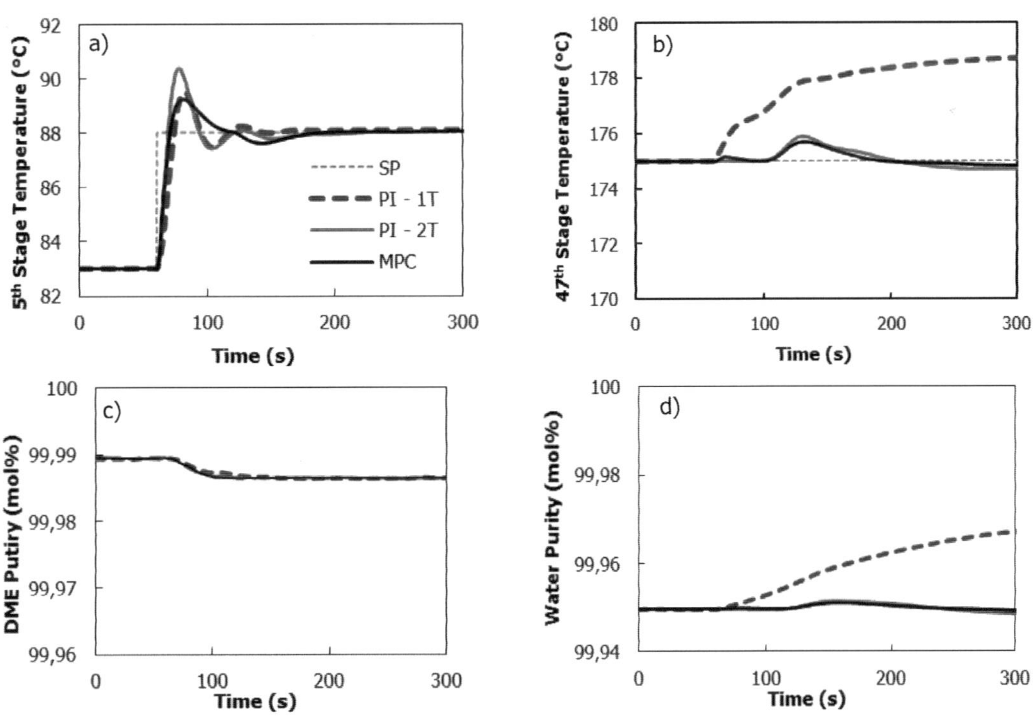

Figure 5. Dynamic responses to a change in stage 5 temperature set-point: a) stage 5 temperature, b) stage 47 temperature, c) DME purity, d) water purity.

Figure 5a shows the dynamic responses in case of the set-point change in stage 5 temperature. The multivariable MPC and single-point PI controllers perform responses with smaller overshoot compared to the response of two-point PI controller. Figure 5b shows the effect of the stage 5 temperature set-point change on the dynamic responses of the stage 47 temperature controller. MPC controller performs the best responses in overcoming the interaction between the stages 5 and 47 temperature control loops. Compared to the two-point PI controller, the single-point PI controller is a better in overcoming the loop interaction because in the single-point PI controller there is only one closed loop to control the temperature. Top and bottom product compositions can be controlled through temperature control of stages 5 and 47 as shown in Figures 5c and 5d. The purity of DME decreases when there is a positive change in stage 5 temperature set-point. The single-point PI controller can not maintain the bottom product because of the uncontrolled stage 47 temperature.

5.2. Stage 47 temperature control

Figure 6b shows that the response of the MPC controller is faster and more stable under a change in the stage 47 temperature set-point. In addition, the undershoot generated by the MPC controller is smaller than the two-point PI controller. A step change in the set-point of the stage 47 temperature causes a disturbance effect on the stage 5 temperature control loop which can be seen from the stage 5 temperature deviation from the set-point as shown in Figure 6a. MPC controller can minimize the disturbance effect on stage 5 temperature and overcome the loop interactions. The purity of the bottom product decreases when there is a decrease in the stage 47 temperature set-point.

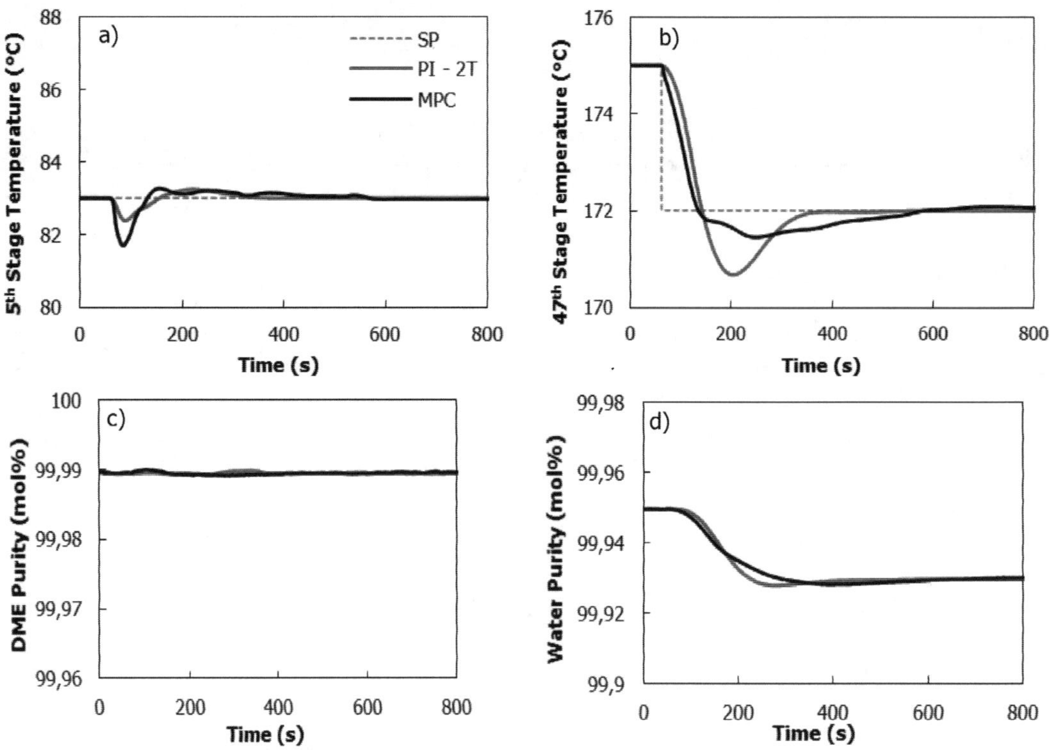

Figure 6. Dynamic responses for to a change in stage 47 temperature set-point: a) stage 5 temperature, b) stage 47 temperature, c) DME purity, d) water purity.

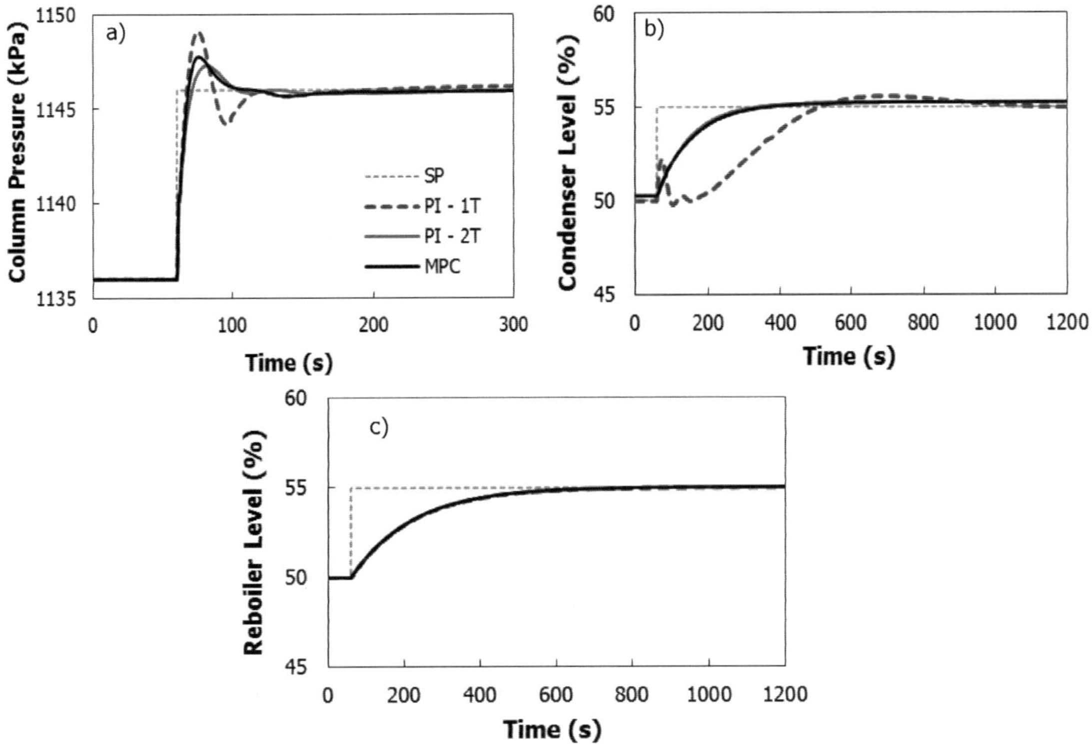

Figure 7. Dynamic responses to a change in the set-point of a) top column pressure, b) condenser accumulator level, c) reboiler level.

5.3. Pressure control

The dynamic responses of pressure controllers under a change in its set-point are shown in Figure 7a. The rise time of the two-point PI and multivariable MPC responses are slightly slower than the single-point PI controller responses. However, the responses of the MPC and two-point PI are more stable with small overshoot.

5.4. Condenser level control

The dynamic responses of the two-point PI and multivariable MPC controller are similar because it use the same configuration and tuning parameters. In the two-point PI and MPC controllers, the condenser accumulator level is controlled by adjusting the distillate flow rate. The dynamic response of the single-point PI controller is different to the other two configurations. In the single-point PI control configuration, there is interaction between the condenser level control and stage 5 temperature control loops. When there is a change in level set-point, the controller initially decreases reflux flow rate to increase the condenser accumulator level. As a consequence, the stage 5 temperature increase, then the stage 5 temperature controller adjusts reboiler duty that makes the vapour flow rate change. The interaction between the condenser level control and stage 5 temperature control loops is shown by the oscillatory response in the Figure 7b.

5.5. Reboiler level control

The three controllers use the same configuration and tuning parameters for the reboiler level control. The dynamic responses of the three controllers show its similar behaviour. Figure 7c shows that there is no interaction between the reboiler level control with the other control loops.

Table 4 shows the ISE (Integral of Squared Error) value of each controller in case of set point tracking. The smaller the ISE value, the better the performance of the controller. Overall, the multivariable MPC controller performs fastest responses with smaller ISE values and less oscillation under set-point step changes. Two-point PI control provides better control performances compared to single-point PI control, regarding to its ability to maintain the stage 47 temperature and the bottom product purity.

Table 4. ISE values in the case of set point tracking.

Controller	Single-point PI	Two-point PI	MPC
Stage 5 temperature	186.0	178.0	134.9
Stage 47 temperature	Uncontrolled	504.0	280.8
Column pressure	301.3	185.5	197.6
Condenser level	4983.0	1000.1	1003.3
Reboiler level	1983.6	1986.8	1986.7

6. Conclusions

In this article, we shown that the controllers with two-point control structure (i.e. two-point PI and multivariable MPC) provide better control performance for DME reactive distillation column, compared to the single-point control structure. The two-point control structures can control both product purity, while the single-point control structure can only control top product purity. The stage 5 temperature in rectifying zone which infers top product purity, is controlled using reflux flow rate and the stage 47 temperature in stripping zone which infers bottom product purity, is controlled using reboiler duty. The multivariable MPC controller performs better responses than the PI controllers as shown by its smaller ISE values. In addition, the multivariable MPC controller is able to handle the loop interactions well.

Acknowledgments

We express our gratitude to the Universitas Indonesia which has funded this research through the scheme of Hibah Publikasi Internasional Terindeks untuk Tugas Akhir Mahasiswa (PITTA) No.2068/UN2.R3.1/PPM.00/2017

References

[1] Azizia Z, Rezaeimanesh M, Tohidian T and Rahimpour M R 2014 *Chem. Eng. Process.* **82** 150–172
[2] Lei Z, Zou Z, Dai C, Li Q and Chen B 2011 *Chem. Eng. Sci.* **66** 3195–3203
[3] Kiss A A and Suszwalak D J P C 2012 *Comput. Chem. Eng.* **38** 74–81
[4] Wahid A and Saputro E F 2016 *Proc. The 1st Int. Joint Conf. on Science and Technology (Bali, Indonesia, 12–13 October 2016)*
[5] Bildea C S, Gyorgy R, Brunchi C C and Kiss A A 2017 *Comput. Chem. Eng.* doi:10.1016/j.compchemeng.2017.01.004
[6] Khaledi R and Young B R 2005 *Ind. Eng. Chem. Res.* **44** 3134–45
[7] Smith C L 2012 *Distillation Control–An Engineering Perspective* (New Jersey: John Wiley & Sons)
[8] Tavan Y and Hosseini S H 2013 *Chem. Eng. Process.* **73** 151–157
[9] Marlin T E 2000 *Process Control–Designing Processes and Control Systems for Dynamic Performance* (Singapore: McGraw-Hill) pp 179–181
[10] Luyben W L 2006 *J. Process Control* **16** 115–134
[11] Moore C F 1992 Selection of controlled and manipulated variables *Practical Distillation Control* ed W L Luyben (New York: Van Nostrand Reinhold) chapter 8 pp 140–177

IOP Conf. Series: Materials Science and Engineering **334** (2018) 012019 doi:10.1088/1757-899X/334/1/012019

Performance analysis of different tuning rules for an isothermal CSTR using integrated EPC and SPC

A H Roslan[1,2], S F Abd Karim[1] and N Hamzah[1]
[1] Faculty of Chemical Engineering, Universiti Teknologi MARA, Malaysia

E-mail: sitifatma07@gmail.com

Abstract. This paper demonstrates the integration of Engineering Process Control (EPC) and Statistical Process Control (SPC) for the control of product concentration of an isothermal CSTR. The objectives of this study are to evaluate the performance of Ziegler-Nichols (Z-N), Direct Synthesis, (DS) and Internal Model Control (IMC) tuning methods and determine the most effective method for this process. The simulation model was obtained from past literature and re-constructed using SIMULINK MATLAB to evaluate the process response. Additionally, the process stability, capability and normality were analyzed using Process Capability Sixpack reports in Minitab. Based on the results, DS displays the best response for having the smallest rise time, settling time, overshoot, undershoot, Integral Time Absolute Error (ITAE) and Integral Square Error (ISE). Also, based on statistical analysis, DS yields as the best tuning method as it exhibits the highest process stability and capability.

1. Introduction

In today's competitive market, many manufacturing industries are seeking ways to optimize process performance and improve product quality in order to maintain a successful operation. For this, two methods are commonly practiced, namely Engineering Process Control (EPC) and Statistical Process Control (SPC) [1].

Variations are undoubtedly among the major concerns for every manufacturing industry. The concept of variation states that no two products will be perfectly identical even when extreme care is taken to ensure the products remain synonymous between one another [2]. Variations can be classified into two categories, namely common cause and special cause [3]. Common cause variation contributes to more than 80% of product defect due to equipment failure, gradual deterioration, wear and tear, irregular scheduled maintenance and many more. These variations are natural and is inherent within the process, which makes it difficult to be completely eliminated. On the other hand, about 20% of the defect is due to special cause variation such as improper tuning, incorrectly adjusted machinery and wrong choice of material [4].

1.1 Statistical Process Control

SPC is a well-known method used to achieve process stability through variability reduction [5]. This is accomplished by monitoring the process condition and detecting the presence of special cause variation. The main reason for the implementation of SPC is to obtain higher and steady product

[2] To whom any correspondence should be addressed.

quality. The term quality varies from one aspect to another, which can be measured based on the level of performance, reliability, durability, serviceability and compliance to standards [6].

One of the primary tools used in SPC is the control charts, which was pioneered by an American physician, engineer and statistician named Walter A. Shewhart in the late 1920s [1]. The purpose of a control chart is to observe the inputs and outputs of a process and measure the parameters against a set of control limits, known as the lower control limit (LCL) and upper control limit (UCL). This is useful in detecting the presence of assignable cause variation and determining whether the process is in-control or out-of-control. Whenever an abnormality is detected, SPC will transmit a signal to the operator whom will carry out investigation and find solutions to rectify the situation. Besides control charts, several other tools available in SPC, which are collectively known as the Seven Quality Tools.

For years, SPC has been successfully implemented within the chemical and process industries as a tool to monitor and maintain the consistency of process systems. The warmth and wide acceptance of SPC is complimented with various publications exploring its use in processes such as chlorine production [5], alkyd polymerization [7], paper and pulp production [8] and many others.

1.2 Engineering Process Control

EPC, or also known as automatic process control (APC) is another technique used to reduce variation and ensure the process output remains at a desired set point. This is accomplished through the use controllers, which perform continuous adjustments to the process variables in order to compensate the effect of disturbance [9].

Controllers play an important role in the implementation of EPC. Various types of controller are available such as PID, fuzzy logic, ANN and others. However, PID is still the most widely used controller in the manufacturing industry due to its advantages such as simple design, cost effective and produce satisfactory control performance [10]. The notations P, I and D are representative of its variables, which are process gain (K_c), time integral (τ_I) and derivative time (τ_D), respectively. Each variable is responsible in governing the stability and response of the process [11]. There are different ways to which a PID controller can be configured into, such as feedback, feedforward and feedback-feedforward, although the former provides the simplest control system.

1.3 Integration of EPC and SPC

Generally, both methods seek to reduce variation as their main objective, albeit in different ways. SPC tracks the process parameters and signals for the presence of assignable cause variation. EPC on the other hand maintains the process output at a specified target by minimizing the effect of disturbance. These are sometimes viewed as a disadvantage as SPC is not able to automatically eliminate the source of variation meanwhile EPC does not address the root cause.

The integration of EPC and SPC initially began in 1988 when Box and Kramer proposed the concept to the SPC research community [9]. Over time, the approach gained momentum and attracted the attention of many researchers as results showed that the use of integrated EPC/SPC yields superior performance compared to the use of either alone. Hence, recent studies have shown increasing interest towards developing an integrated EPC/SPC system to achieve better process performance and product quality.

Within the field of Chemical Engineering, reactors are looked upon as the "heart" of every chemical process. CSTR is among the types of reactors that are frequently used in manufacturing industries as it is simple to construct, operates at low cost and applicable for all types of fluids. However, due to its complex and non-linear characteristics, control of product concentration is difficult and often requires an accurate model [12]. This paper serves to study the integration of EPC and SPC for the control of product concentration for an isothermal CSTR through the following objectives: 1) Analyse the process performance using different tuning method and 2) Determine the most effective type of tuning method by comparing the process stability and capability.

Unfortunately, there has been little effort given to implement an integrated EPC/SPC system for a CSTR. Hence, it is with hope that this study would provide useful information as well as exposure for future researchers to carry out improved trials and discover new findings within this field of study.

2. Methodology

This study was conducted in two stages, namely, process simulation and statistical analysis. Firstly, the process was modelled using SIMULINK MATLAB R2015a as demonstrated by [13] in Figure 1 and Figure 2. The isothermal CSTR has following reaction kinetics, which is known as Van de Vusse reaction:

$$A \xrightarrow{k_1} B \xrightarrow{k_2} C$$
$$2A \xrightarrow{k_3} D$$

Table 1: Parameters for Van de Vusse reaction.

Parameter	Value
k_1	0.83 min^{-1}
k_2	1.66 min^{-1}
k_3	0.166 mol^{-1} min^{-1}
Steady state feed concentration, C_A	10 $gmol^{-1}$

The controller variables (K_c, τ_I and τ_D) were calculated using DS and IMC tuning methods whereas the variables for Z-N method was obtained from [13]. Each process response was evaluated based on the rise time, settling time, overshoot, undershoot, Integral Time Absolute Error (ITAE) and Integral Square Error (ISE). Next, the data from the process output were transferred to Minitab 17 where Process Capability Sixpack was used to analyze the process stability, capability and normality. The number of samples used for this analysis was 125, which is outlined by the Acceptance Quality Limit (AQL).

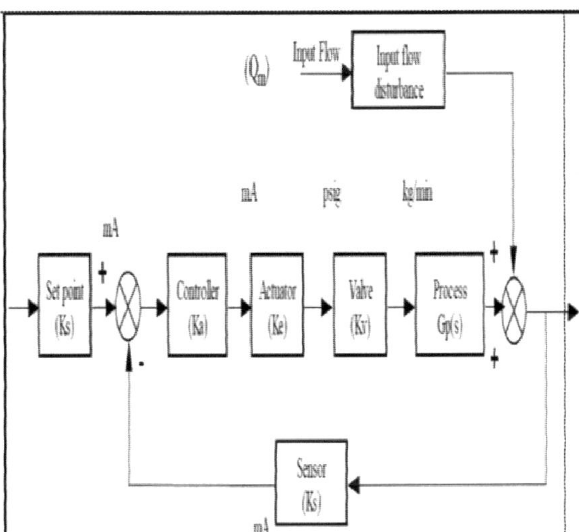

Figure 1. Block diagram based feedback control for product concentration control of isothermal CSTR [13].

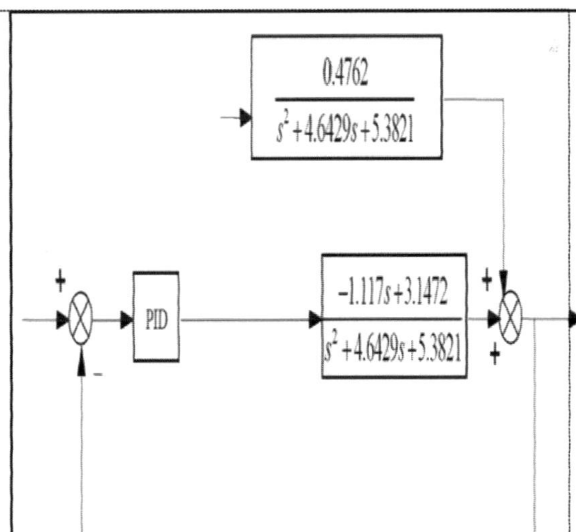

Figure 2. Transfer function based feedback control for control of product concentration of isothermal CSTR [13].

2.1 Ziegler Nichols

According to [10], Z-N is the most widely used method to determine the variables of a PID controller. The principle technique for PID controller tuning are step response and ultimate frequency. The unit step response method is based on the open-loop response of the system and is characterized by two parameters, namely delay (L_1) and time constant (τ). These two parameters can be obtained by drawing a tangent at the point of inflexion, where the slope of the step response is at its maximum. The intersection between the tangent and coordinate axes produce the process parameters, which are then used to calculate the controller variables. The values of K_c, τ_I, and τ_D are 0.2, 0.95 and 0.23, respectively.

2.2 Direct Synthesis

According to [10], DS method for PID controllers are usually based on a time-domain or frequency-domain performance criterion. The controller is designed based on a closed-loop transfer function which is then used to perform analytical calculation and ensure the set point of the closed-loop matches the desired response. One of the advantages of using the DS approach is that performance requirements are directly integrated through specification of the closed-loop transfer function. Also, it is able to provide valuable information between the relationship of the process model and the controller [14]. The calculated values of K_c, τ_I, and τ_D are 0.6880, 1.1501 and 0.5448, respectively.

2.3 Internal Model Control

IMC method has been gaining increasing popularity in controller implementations, although PID controllers are still the standard type of controller used for this method. According to [10], the objective of IMC method is to achieve a fast and accurate set point tracking by removing the effect of disturbance as effectively as possible and also ensuring insensitivity to modelling error. An advantage of IMC method is that it allows uncertainty and trade-offs between performance and robustness to be considered in a systematic approach [14]. The calculated values of K_c, τ_I, and τ_D are 0.2755, 0.8627 and 0.2154, respectively.

2.4 Minitab

Minitab is a software that is widely used within the academic, business and government sectors for its wide collection of statistical tools. [15]. For this study, Minitab is used to produce a report called Process Capability Sixpack. It is consisted of control charts, histogram, normal probability plot and process capability chart. For this study, 125 sample data from the process output were used to produce the Process Capability Sixpack to analyse the stability, normality and capability of each tuning method.

3. Results and Discussion

3.1 Assumptions

Several assumptions were made throughout this study in order to achieve the objectives. Amongst them are:

- Desired product concentration (set point) is equal to the feed concentration at 10 $gmol^{-1}$.
- Flowrate disturbance is represented with “random numbers” in SIMULINK MATLAB.
- Upper Specification Limit (USL) and Lower Specification Limit (LSL) of the process is assumed to be 10.00 ± 0.05 $gmol^{-1}$.
- The significance level (α) for normality test is 1%.

3.2 Process without Control Scheme

Before analyzing the effectiveness between Z-N, DS and IMC, the process was firstly modelled without a control scheme as seen in Figure 3. The purpose of this setup is to observe the output when no corrective action is available to minimize the effect of disturbance. Based on Figure 4, it is

IOP Conf. Series: Materials Science and Engineering **334** (2018) 012019 doi:10.1088/1757-899X/334/1/012019

observed that the process is unable to reach the desired set point of 10 gmol^{-1}, but instead only managed to achieve 5.85 gmol^{-1}. Also, the response displays a large undershoot at the initial stage of the process. From these observations, it is clear that a control scheme is indeed required obtain a desirable and satisfactory process response.

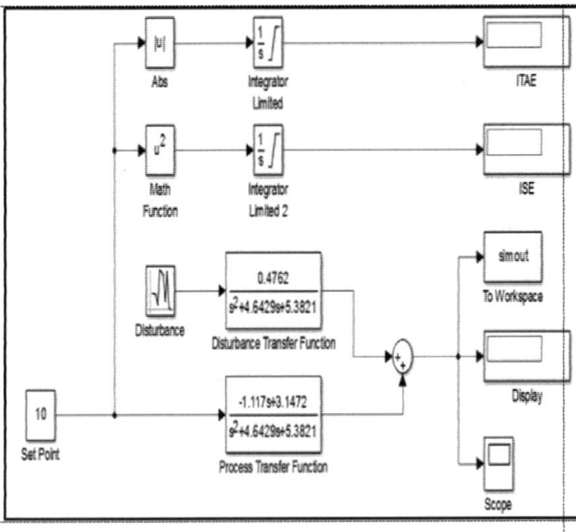

Figure 3. Process model without control scheme.

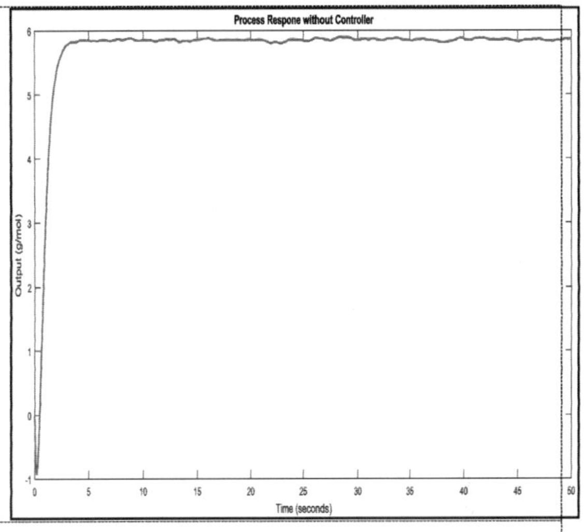

Figure 4. Process response without control scheme.

3.3 Comparison between Z-N, DS and IMC

In this section, comparison will be made for each of the process response obtained from Z-N, DS and IMC methods. The criteria used to evaluate the effectiveness of each tuning method are rise time, (τ_R), settling time (τ_S), overshoot, undershoot, integral square error (ISE) and integral time absolute error (ITAE). The process was modelled as seen in Figure 5 and the response produced from each method are shown in Figure 6.

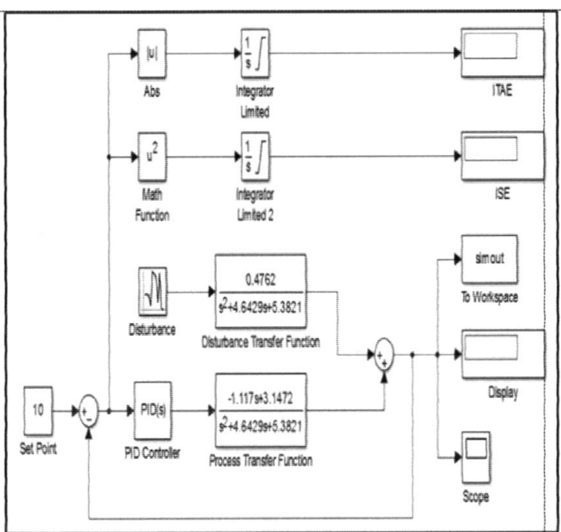

Figure 5. Process model with PID controller.

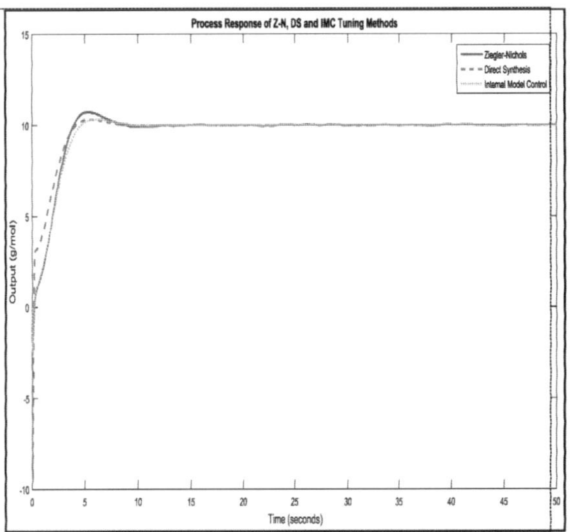

Figure 6. Process response for Z-N, DS and IMC methods.

Based on Figure 6, it is observed that DS method produces the best process response as it has smallest τ_R and τ_S as well as lowest ITAE and ISE. However, it also displays the highest undershoot when compared to other tuning methods. The second-best response is produced from Z-N and lastly, IMC method. Table 2 presents a result summary for each tuning method.

Table 2. Results summary for Z-N, DS and IMC methods.

Method	Rise time (s)	Settling time (s)	Overshoot (%)	Undershoot (%)	ITAE	ISE	Rating
Ziegler-Nichols	0.19	9.23	7.94	28.57	18.64	127	Moderate
Direct Synthesis	0.038	9.04	7.06	96.08	15.62	114.3	Best
Internal Model Control	0.038	10.77	4.03	25.28	22.34	145.4	Worst

3.4 Fine Tuning

A common approach that is practiced in EPC in order to achieve better process response is by performing fine tuning to the controller variables. This is done by adjusting K_c, τ_I, and τ_D through trial-and-error until a satisfactory response is produced. The new controller variables are presented in Table 3 meanwhile Figure 7 presents the process response after fine tuning was performed.

Table 3. Controller variables for Z-N, DS and IMC methods after fine tuning.

Tuning method	K_c	τ_I	τ_D
Ziegler-Nichols	0.6	0.95	0.0144
Direct Synthesis	0.6880	1.15	0.0334
Internal Model Control	0.5510	0.8627	0.0135

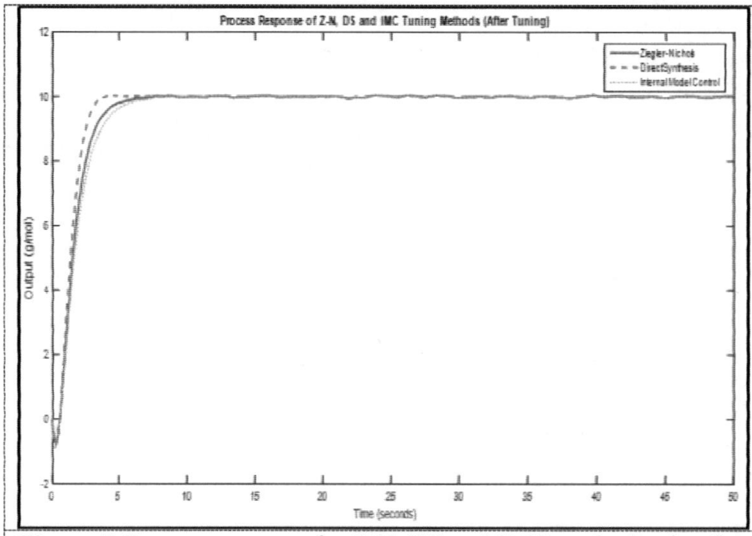

Figure 7. Process response for Z-N, DS and IMC methods after fine tuning.

Based on Figure 7, it can be observed that the process response had improved significantly. The most noticeable difference is that the overshoot was eliminated completely and the undershoot was minimized. Apart from that, the values of ISE and ITAE were also reduced, which indicate that the error within the process was reduced. However, τ_r increased slightly after performing this step, meaning that the process will take a longer time to reach the set point. Apart from that, τ_r decreased which means the process will take a shorter time to stabilize at the set point. Table 4 presents the result summary for each of the tuning method after performing fine tuning.

Table 4. Results summary for Z-N, DS and IMC methods after fine tuning.

Method	Rise time (s)	Settling time (s)	Overshoot (%)	Undershoot (%)	ITAE	ISE	Rating
Ziegler-Nichols	0.38	7.50	-	7.04	18.04	127	Moderate
Direct Synthesis	0.19	5.00	-	8.94	15.62	114.3	Best
Internal Model Control	0.38	8.77	-	7.88	20.42	134.8	Worst

3.5 Performance Tests

Once the optimum controller variables have been calculated, a series of performance tests were conducted to measure its performance in handling changes in set point and disturbance load [11]. To perform set point test, the process was simulated two (2) times, each with different set points – 15 and 20 $gmol^{-1}$. Meanwhile for disturbance test, the process response was studied under two conditions, firstly without disturbance and then with the presence of disturbance.

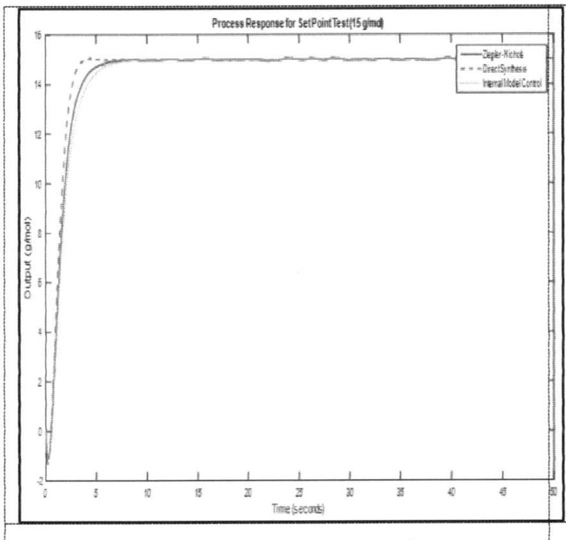

Figure 8. Set point test for 15 $gmol^{-1}$.

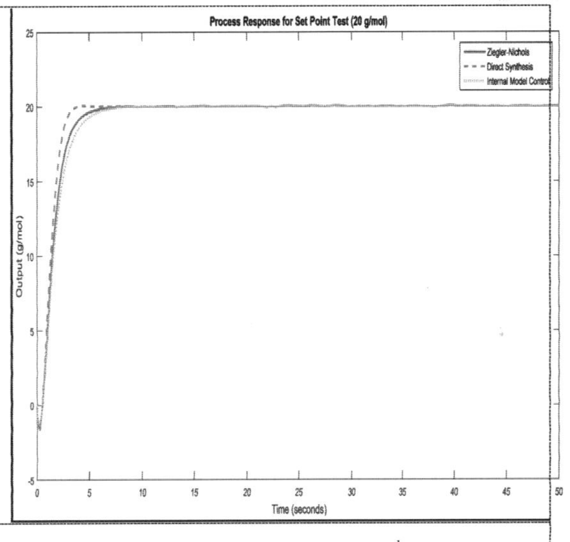

Figure 9. Set point test for 20 $gmol^{-1}$.

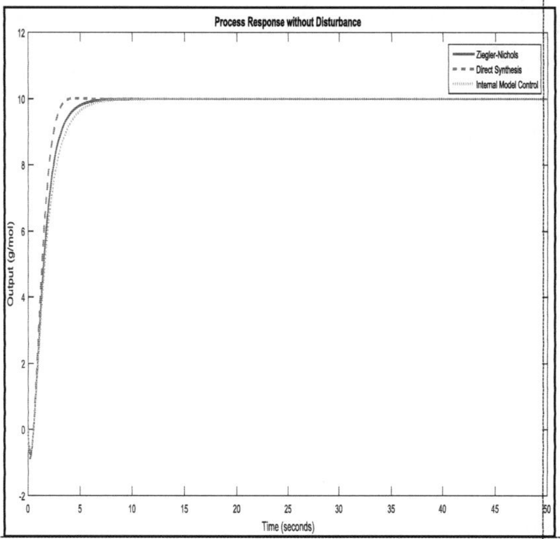

Figure 10. Process response without disturbance.

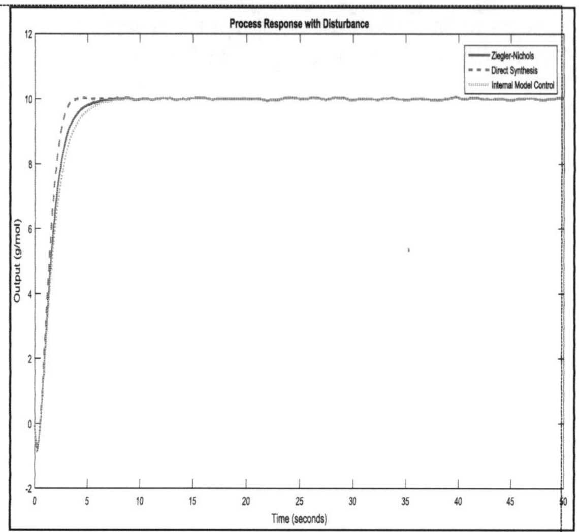

Figure 11. Process response with disturbance.

The results of the set points tests are as shown in Figure 8 and Figure 9. It is observed that the process manages to reach each of the specified target with minimum disturbance. Hence, it is proven that the controller variables used in table 3 pass the set point tests and can be applied for varying set points.

The results of disturbance test are presented in Figure 10 and Figure 11. Based on these graphs, it is seen that the response produced without disturbance is slightly more stable compared to when disturbance is present. However, the latter still manages to achieve the targeted specification of 10 $gmol^{-1}$ with minimal fluctuation. It should be worth considering that a response such as in Figure 10 is unlikely to occur in industrial practice as various disturbance loads will affect the process variables.

3.6 Process Capability Sixpack

Up until this point, the results shown represent the principles of EPC. However as stated by [16], the control of process variables would be significantly improved when SPC is used simultaneously with EPC. Hence, a Process Capability Sixpack will be used to measure the performance of the process with regards to its stability, normality and capability.

IOP Conf. Series: Materials Science and Engineering **334** (2018) 012019 doi:10.1088/1757-899X/334/1/012019

3.6.1 Ziegler-Nichols Method

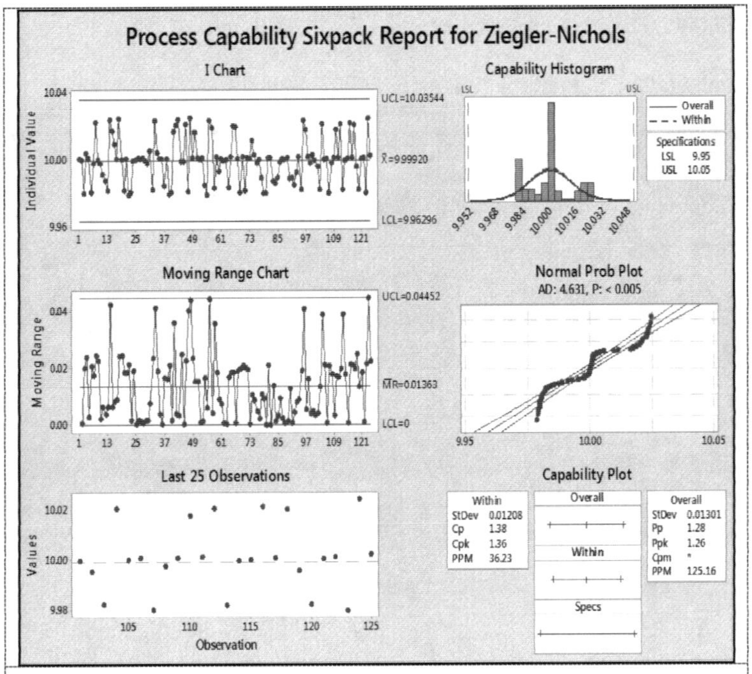

Figure 12. Process Capability Sixpack for Z-N method.

The Process Capability Sixpack report in Figure 12 shows the control charts for I and MR. These charts are used to monitor the presence of special cause variation which is the frequent cause of product defect. It is observed that all of the points lie within the UCL and LCL, indicating that the process is stable and in-control. The average product concentration obtained from the sample is 10.00 $gmol^{-1}$, which is the exact desired specification.

A capability histogram is also included to display the process spread relative to the specification limits. As mentioned previously, the USL and LSL are assumed to be 10.00 ± 0.05 $gmol^{-1}$, which calculates into 10.05 and 9.95, respectively. Based on the chart, it is observed that all of the data fall within the specification limits, suggesting that the process is capable. Additionally, the process spread lies at the center of the specification limits, implying that the process has a high tendency of forming a product that is closest to the average concentration.

The capability of the process can be further analyzed using the process capability plot. Here, various performance indices are provided such as C_p, C_{pk}, P_p and P_{pk}. However, only the value of C_{pk} will be used to measure the process capability as C_p does not consider the location of the process spread (relative to the center of the specification limits) meanwhile P_p and P_{pk} is only applicable when the process is out-of-control [17]. According to [18] the general acceptable value of C_{pk} is 1.33 and any lower than that means the variation is either too wide compared to the specifications or that the process is off-centered. Based on the capability plot, the C_{pk} obtained is 1.36, which is above the acceptable value. This further proves that the process is capable and can indeed achieve the desired specification.

Besides that, a normal probability plot is included to determine the normality of the process. This graph is used to determine whether the data is normally distributed or otherwise. Before conducting the analysis, several parameters need to be clearly understood such as the significance level (α), *p*-value and null hypothesis (H_0). As a brief summary, the null hypothesis is either to be accepted or rejected depending if the *p*-value is greater or lesser than α, respectively.

For this study, the α is assumed to be 1% or 0.01. However, the *p*-value obtained from the graph is <0.005, which is lesser than 0.01. This means that the null hypothesis is rejected and that the data is not normally distributed.

3.6.2 Direct Synthesis Method

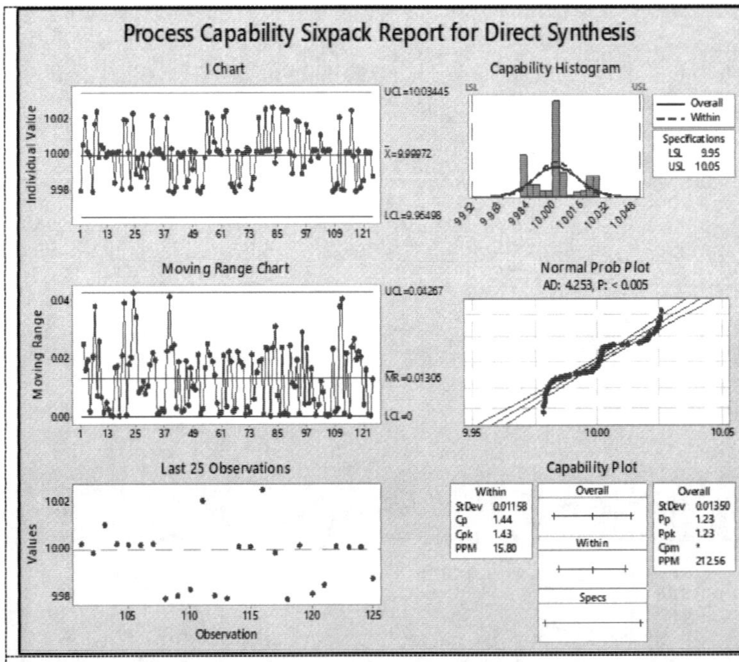

Figure 13. Process Capability Sixpack for DS method.

Figure 13 shows the Process Capability Sixpack for DS method. Based on the I-MR charts, it is observed that all of the points lie within the specification limits, implying that special cause variation is absent from the process and that the process is stable. Similar to Z-N method, the average concentration of the sample is 10.00 gmol^{-1}, which means that the product achieves the desired specification.

Besides that, the histogram shows that the process spread is well-centered between the USL and LSL. This indicates that the process has a high tendency of producing a product that is nearest to the average concentration. Based on the capability plot, the C_{pk} obtained is 1.43, which proves that the process is capable. The C_{pk} obtained from DS method is greater compared to Z-N, meaning DS method is more capable compared to Z-N.

By observing the normal probability plot, it shows that the *p*-value is < 0.005, which is lesser than 0.01. This means that the null hypothesis is rejected and the sample data is not normally distributed.

3.6.3 Internal Model Control Method

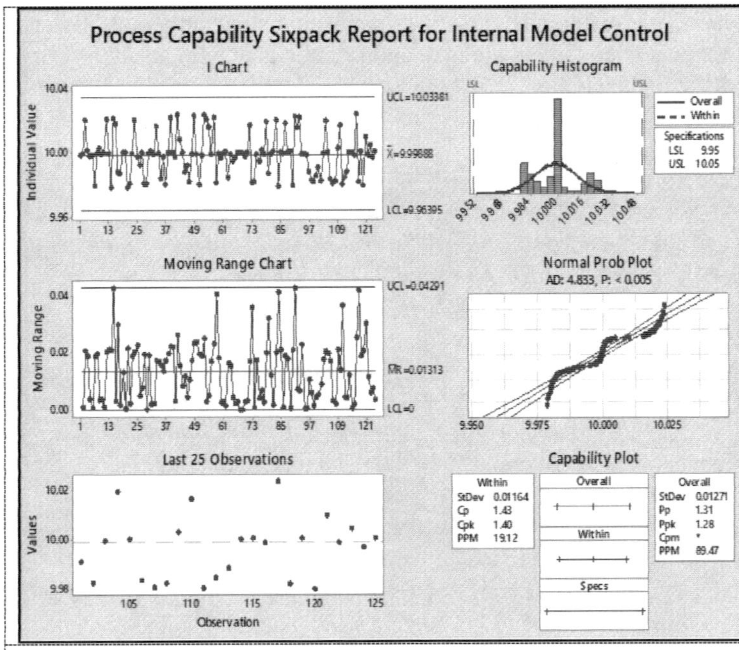

Figure 14. Process Capability Sixpack for IMC method.

The final tuning method that will be analyzed is the IMC as shown in Figure 14. Based on the I-MR charts, all of the points lie within the UCL and LCL, similar to the other two methods. This shows that the process does not contain special cause variation and that it is in-control.

The capability histogram shows that the process spread lie well within the LSL and USL, implying that the process is capable. Similar to Z-N and DS method, the process spread is located at the middle of the specification limits, indicating that the product formed has a high tendency of having the desired concentration.

By referring to the capability plot, the C_{pk} obtained is 1.40, which is greater than 1.33. This further supports the statement that the process is capable and is able to meet the desired specification. When comparing the C_{pk} values obtained from all of the tuning methods, Z-N produces the lowest value, followed by IMC and lastly, DS method.

The normal probability plot shows a *p*-value of <0.005, which is lesser than 0.01. This goes to show that the null hypothesis is rejected and that the data used is not normally distributed.

4. Conclusion

In this paper, control of product concentration for an isothermal CSTR was studied using different PID tuning methods. The tuning methods used were Ziegler-Nichols (Z-N), Direct Synthesis (DS) and Internal Model Control (IMC) as these are among the most commonly used tuning methods within the chemical and process industries. The simulation model was obtained from past literature and was re-constructed using SIMULINK MATLAB. Based on the results, DS produces the best process performance as it displays the shortest rise time and settling time, smallest overshoot as well as lowest ITAE and ISE. This is followed by Z-N method and lastly, IMC method. The process outputs were then used to study the stability, normality and capability of the process using Process Capability Sixpack in Minitab. Based on the results obtained, DS displays the best process performance for having the biggest C_{pk}, indicating that it is the most capable method that can achieve the desired specification. In conclusion, all of the tuning methods used in this study were proven to be able to control the product concentration for an isothermal CSTR, although DS yields as the best tuning method compared to the others. For recommendations, this study can be further expanded and improved upon by applying different tuning methods such as Cohen-Coon, Takahashi, and Lambda tuning methods to obtain a wider scope of result and comparison. Also, other SPC tools such as X-MR chart, EWMA, CUSUM and Hotelling T^2 may serve as an alternative for the statistical analysis.

References

[1] Duffuaa S and Noman S M 2004 *International J. of Production Research.* ***42*** 4109-4118
[2] Mahesh B P and Prabhuswamy M S 2010 *International J. for Quality Research.* **4** 193-203
[3] Jiang W and Tsui K L 2000 *IEEE Transactions.* ***32*** 505-513
[4] Hossain A Choudhury Z A and Suyut S 1996 *IEEE Transactions on Industry Applications.* ***32*** 243-249
[5] Aljebory K M and Alshabeb M 2014 *J. of Mechanical and Industrial Engineering,* ***8*** 243-256
[6] Škulj G. Vrabič R and Sluga A 2013 *CIRP.* *7* 401-406
[7] Hachicha W Moussa I and Kolsi R 2012 *International J. of Control and Automation.* ***5*** 45-62
[8] Matos A S Requejio J G and Pereira Z L 2008 *Integration of Engineering Process Control and Statistical Control in Pulp and Paper Industry* (Lisbon: Elsevier) pp 1-6
[9] Rahim M A Siddiqui Y A and Elshafe M 2014 *Integration of Multivariate Statistical Process Control and Engineering Process Control* pp 472-480
[10] Korsane D T Yadav V and Raut K H 2014 *International J. of Innovative Research in Electrical, Electronics, Instrumentation and Control Engineering.* ***3*** 582-586.
[11] Ishak A A and Abdullah Z 2015 *PID Tuning Fundamental Concepts and Applications* (Kuala Lumpur: UiTM Press)
[12] Kumar S 2012 *International J. of Advanced Research in Computer Science and Software Engineering,* ***2*** 372-375
[13] Vishnoi V Padhee S and Kaur G 2012 *International J. of Scientific and Research Publications.* ***2*** *1-7*
[14] Smith C A and Corripio A 2006 *Principles and Practice of Automatic Process Control* (New Jersey: John Wiley & Sons)
[15] Rahman M N Zain R M Alias A M and Nopiah Z M *International J. of Science and Technology.* ***9*** 193-208
[16] Montgomery D C Keats B J Runger G C and Messina W S 1994 *J. of Quality Engineering,* ***26*** 79-87
[17] Montgomery D C 2009 *Introduction to Satistical Quality Control* (Arizona: John Wiley & Sons)
[18] Wooluru Y Swamy D R and Nagesh P 2014 *International J. for Quality Research.* ***8*** 399-416

IOP Conf. Series: Materials Science and Engineering **334** (2018) 012020 doi:10.1088/1757-899X/334/1/012020

A Comparative study between MPC and PI controller to control vacuum distillation unit for producing LVGO, MVGO, and HVGO

A Wahid[1,2] and A P Prasetyo[1]

[1] Sustainable Energy Research Group, Department of Chemical Engineering, Faculty of Engineering, Universitas Indonesia, Kampus Depok, Depok 16424, Indonesia

E-mail: wahid@che.ui.ac.id

Abstract. This study describes the selection of controllers in the vacuum distillation unit (VDU) between a model predictive control (MPC) and a proportional-integral (PI) controller by comparing the integral square error (ISE) values. Design of VDU in this simulation is based on modified Metso Automation Inc. scheme. Controlled variables in this study are feed flow rate, feed temperature, top stage pressure, condenser level, bottom stage temperature, LVGO (light vacuum gas oil), MVGO (medium vacuum gas oil), and HVGO (heavy vacuum gas oil) flow rate. As a result, control performance improvements occurred as using MPC compared to PI controllers, when testing a set-point change, of feed flow rate control, feed temperature, top-stage pressure, bottom-stage temperature and flow rate of LVGO, MVGO, and HVGO, respectively, 36%, 6%, 92%, 53%, 90%, 96% and 88%. Only on condenser level control PI performs much better than the MPC. So PI controller is used for level condenser control. While for the test of disturbance rejection, by changing feed flow rate by 10%, there is improvement of control performance using MPC compared to PI controller on feed temperature control, top-stage pressure, bottom-stage temperature and flow rate LVGO, MVGO and HVGO 0.3%, 0.7%, 14%, 2.7%, 10.6% and 4.3%, respectively.

1. Introduction

Vacuum Distillation Unit (VDU) is a secondary process unit in refinery. This unit is very important for fractionating atmospheric residue/long residue, bottom product of Crude Distillation Unit (CDU), which has boiling point more than 350°C [1]. Generally, VDU operates in 25-65 mmHg. As the pressure of the columns is reduced, the component's boiling point will be decreased as well. So that, it doesn't need too high temperature to fractionate the atmospheric residue into lighter fraction.

The products of VDU are Light Vacuum Gas Oil (LVGO), Medium Vacuum Gas Oil (MVGO), and Heavy Vacuum Gas Oil (HVGO). These products can be processed further into gasoline and diesel product blending.

VDU has many variables that have a very prominent role to the stability of operation and product quality. Column temperature can stabilize the composition profile within the column. In the other words, temperature column gives indirect composition control [2]. As we know, column pressure affects boiling point of the component. Pressure fluctuations make control more difficult and reduce

[2] To whom any correspondence should be addressed.

unit performance. In addition, pressure variations alter relative volatility and affects fractionation performance [3]. Feed flow rate and temperature will affect the vapor-liquid contact and influence the product's composition [4]. In order to maintain the stability of operation and reject any disturbances, proper controllers are needed.

There are some types of controller that are used in chemical industry, Proportional-Integral (PI) and Model Predictive Control (MPC) are two of them. PI is a controller that works by comparing the error value with the desired response and integrating it so that the difference becomes zero [5].

MPC predicts the future values of the process output using dynamic model and available measurement. The controller outputs are calculated so as to minimize the difference between the predicted process response and desired response [6]. MPC can improve control performance and stabilize unit operation by using optimization [7]. Furthermore, operational costs can be minimized due to more economic process operation [7].

This study focuses on VDU simulation and control to produce LVGO, MVGO, and HVGO. Controller that implemented are MPC and PI. In this study, will be compared the results of MPC and PI controller by calculating the ISE (integral square error) of MPC and PI, respectively, against set-point changes and disturbance. Disturbance that is conducted in this study is feed flow rate increasing by 10%.

2. Methodology

2.1. Simulation Environment

In this simulation, the feed that is going to be fractionated is atmospheric residue which flow rate is 118,021 bpd. The configuration of VDU simulation can be seen in figure 1. In this configuration, feed and products (side streams such as LVGO, MVGO, and HVGO) flow rate are adjusted by flow rate controller. Feed temperature is controlled by controlling heater duty Top stage pressure is controlled by manipulating vapor flow rate. The condenser level is controlled by controlling overhead condensate flow rate. The bottom stage temperature is controlled by manipulating reboiler duty.

In the installation of the controller, the considerations are the control objective, the desired set-point (SP) value and the independent variable that will affect the SP when the dynamic behavior of the vacuum distillation unit system is performed. There are two types of controller which are used in this study, MPC and PI. Furthermore, these two types of controllers will be compared to see which controller provide the best performance.

2.2. FOPDT (first-order plus dead time) Model

MPC and PI are types of controllers which performance is depend on the model that is used. In this study, model which is used is empirical model FOPDT (first-order plus dead time). FOPDT model can be obtained by doing model testing in each controller to get process reaction curve (PRC). The PRC can be seen in figure 2

The empirical model of FOPDT is shown by equation (1)

$$G_p(s) = \frac{K_p e^{-\theta s}}{\tau s + 1} \tag{1}$$

Where, K_P is process gain describing how far PV moves; τ, time constant that describe how fast the PV respond; θ is dead time describing how much delay occurs before the PV first begins to move.

2.3. Controller Tuning

Based on model that is obtained from the PRC, controller tuning is done to get optimal control. For PI controller, tuning parameter is conducted by calculating the parameter using Ziegler-Nichols method to get proportional gain (Kc) and integral time (Ti) [8]. Where, the following is Ziegler-Nichols equation to calculate Kc and Ti for open loop [eq. (2) and (3)] and closed loop control [eq. (4) and (5)]:

$$K_c = \frac{0.9}{K_p}\left(\frac{\theta}{\tau}\right)^{-1} \tag{2}$$

$$T_i = 3.33\theta \tag{3}$$

$$K_c = \frac{Kcu}{2.2} \tag{4}$$

$$T_i = \frac{Tu}{1.2} \tag{5}$$

In the MPC controller the FOPDT model can be used directly. There are two ways how MPC uses this model. MPC uses this model to predict effect of past control moves on future outputs, P (prediction horizon) and using identical model to compute M (control horizon) moves [9].

2.4. *Control Performance*

The performance of the controller is tested by doing set-point tracking and disturbance rejection. The disturbances are the elevation of feed flow rate by 10%. Next is comparing the error value, ISE (integral square error), of the MPC and PI controllers to the set-point (SP) changes and also the disturbance to see which controller has more optimum performance. The smaller ISE values the better the performance of the controller. The equation of ISE can be seen as follow:

$$ISE = \int_0^{\infty} (e)^2 dt \tag{6}$$

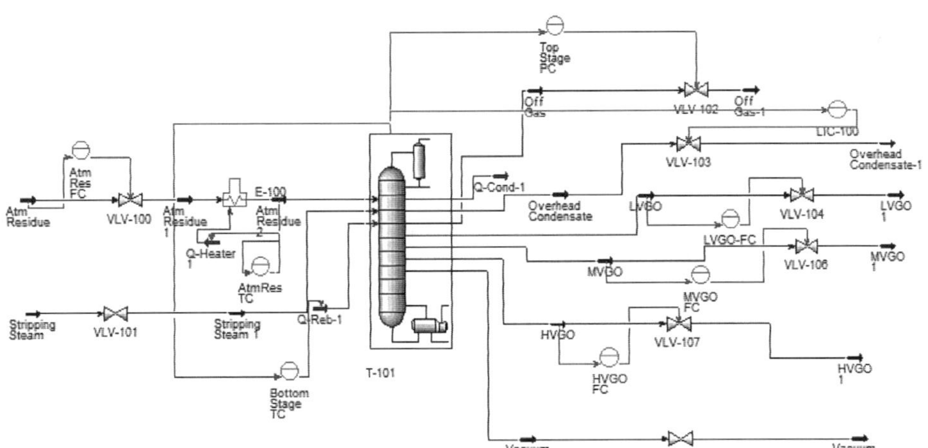

Figure 1. Configuration of VDU simulation

IOP Conf. Series: Materials Science and Engineering **334** (2018) 012020 doi:10.1088/1757-899X/334/1/012020

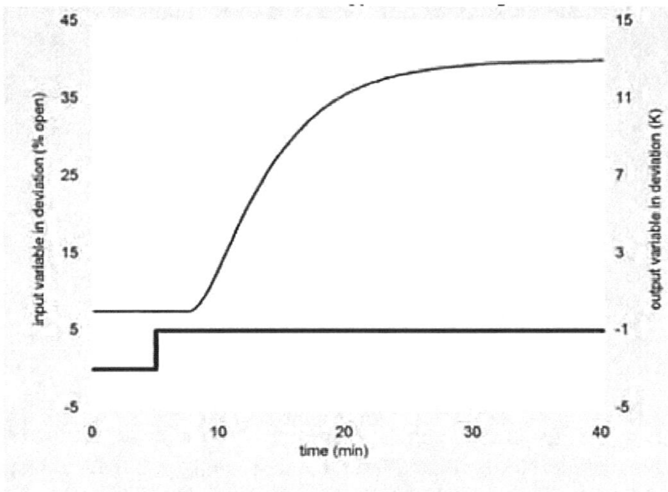

Figure 2. Example of PRC.

3. Result and Discussion

3.1. Set-point Tracking

In this study, set-point (SP) tracking is conducted in order to see the response of the controllers because of SP change. In this test, will be seen how quick MPC is able to respond the SP change compared to PI. The response of the controller due to change of SP of each variable in this test can be seen in figure 3.

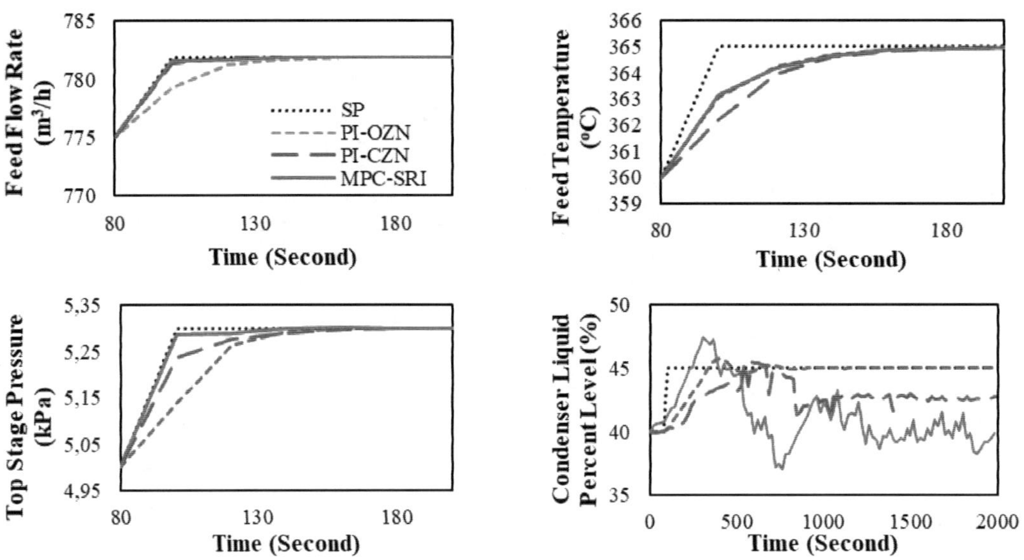

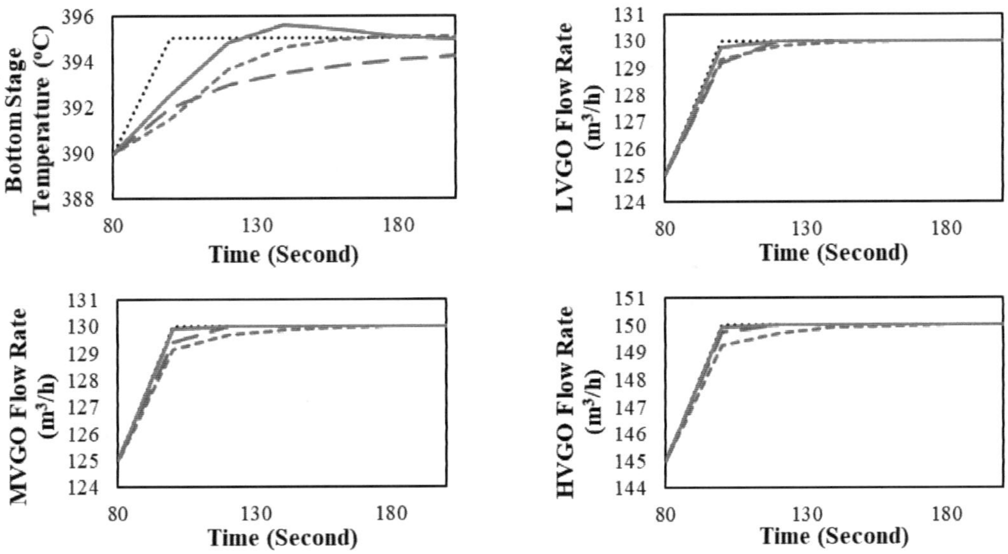

Figure 3. Response of MPC and PI control due to SP change.

The results shown in figure 3 indicate that, generally, MPC can provide better response due to SP change than PI controller. Although MPC is effective, it is not good for level control [7] as it is can be seen in figure 3. So that, PI controller is used for level control of condenser. To find error value of each controller, the ISE value is computed by using equation (4). The result of the calculation is shown in table 1 below

Table 1. ISE of each controlled variable in MPC and PI controller because of SP change.

No.	Controlled Variable	ISE		
		PI	PI (Closed Loop)	MPC
1	Feed Flow Rate	134.6	4.2	2.7
2	Feed Temperature	90	181	84
3	Top Stage Pressure	0.552	0.092	0.007
4	Condenser Level	2129	11069	37840
5	Bottom Stage Temperature	282	385	131
6	LVGO Flow Rate	10.2	13	1.1
7	MVGO Flow Rate	18.29	6.98	0.28
8	HVGO Flow Rate	13.79	1.33	0.16

3.2. Disturbances Rejection

Figure 4 shows the response both MPC and PI controller due to the presence of disturbance.

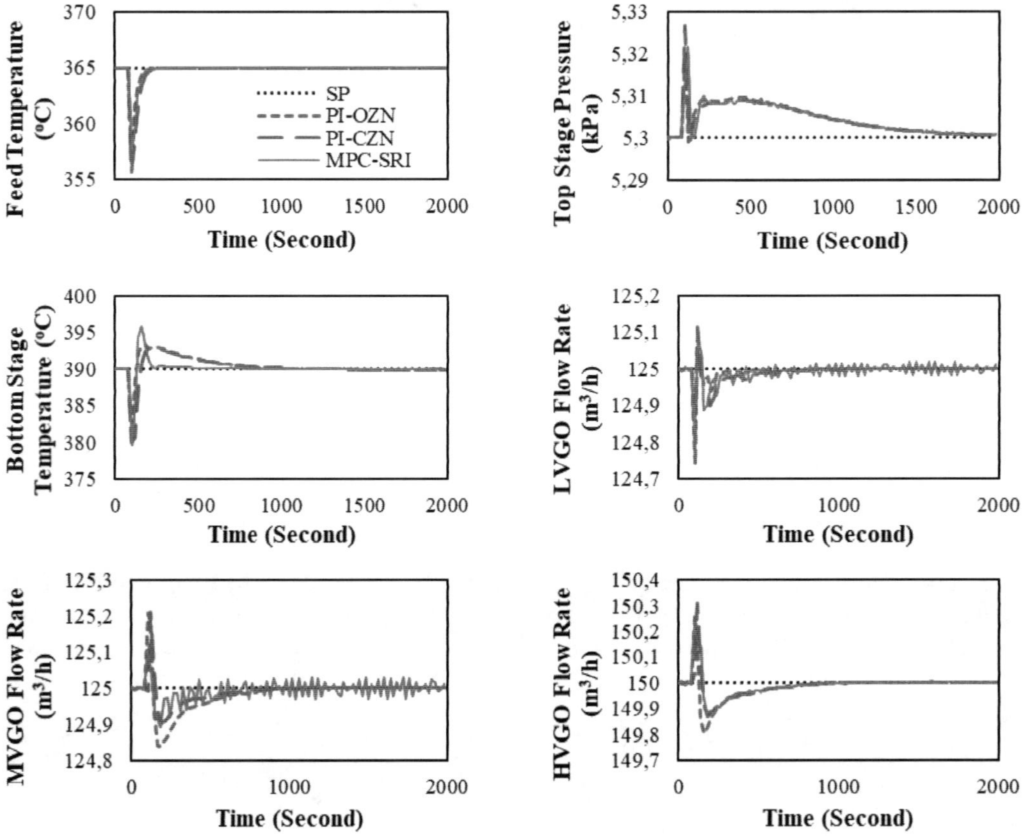

Figure 4. Response of MPC and PI control due to disturbance.

Increased feed flow rate, will cause the temperature of feed and bottom stage drop. Furthermore, higher feed flow rate will lead higher vapor flow rate within the column as well, and will affect the performance of the column and flow rate of the products as shown in figure 4. However, in spite of MPC has better response for maintaining the operation remain stable than PI, because of the presence of disturbance, it only provides slight improvement. The ISE value in disturbance rejection test is shown in table 2 as follow:

Table 2. ISE of each controlled variable in MPC and PI controller because of disturbance.

No.	Controlled Variable	ISE		
		PI	PI (Closed Loop)	MPC
1	Feed Temperature	1731	2883	1726
2	Top Stage Pressure	0.065	0.064	0.063
3	Bottom Stage Temperature	4501	5748	3870
4	LVGO Flow Rate	1.71	1.76	1.67
5	MVGO Flow Rate	4.759	1.995	1.783
6	HVGO Flow Rate	5.387	3.950	3.781

4. Conclusion

In general, MPC can improve control performance because it provide better response than PI when dealing with SP change and disturbance. As a result, control performance improvements occurred as using MPC compared to PI controllers. When testing a set-point change, MPC provide improvement of feed flow rate control, feed temperature, top-stage pressure, bottom-stage temperature and flow rate of LVGO, MVGO, and HVGO, respectively, 36%, 6%, 92%, 53%, 90%, 96% and 88% compared to PI. Only condenser level control that PI performance is much better than the MPC. So, for controlling condenser level in this study, PI controller is used. While for the test of disturbance rejection, by changing feed flow rate by 10%, there is improvement of control performance using MPC compared to PI controller on feed temperature control, top-stage pressure, bottom-stage temperature and flow rate LVGO, MVGO and HVGO 0.3%, 0.7%, 14%, 2.7%, 10.6% and 4.3%, respectively.

Acknowledgments

We express our gratitude to the Universitas Indonesia which has funded this research through the scheme of Hibah Publikasi Internasional Terindeks untuk Tugas Akhir Mahasiswa (PITTA) No. 2068/UN2.R3.1/PPM.00/2017.

References

[1] Chilingarian G V and Yen T F 1994 *Asphaltenes and Asphalts,1* (Amsterdam: Elsevier)
[2] Skogestad S 2007 The dos and don'ts of distillation column control *Chem. Eng. Res. Des.* **85(A1)** 13-23
[3] Sloley A W 2001 *Effectively Control Column Pressure.* (Texas: The Distillation Group Inc.)
[4] Akpa J G and Umuze O D 2013 Simulation of a multi-component crude distillation column. *Am. J. Sci. Ind. Res.* **4(4)** 366-77.
[5] Marlin T E 2000 *Process Control: Designing Processes and Control System for Dynamic Performance. 2nd Edition.* (New York: Mc Graw-Hill)
[6] Himmelblau D M, Edgar T F and Lasdon L S 2001 *Optimization of Chemical Processes, Second Edition.* (New York: McGraw-Hill)
[7] Kano M and Ogawa M 2010 The state of the art in chemichal process control in Japan: Good practice and questionnaire survey *J. Process Contr.* **20** 969-82.
[8] Smith C A and Corripio A B 1997 *Principles and Practice of Automatic Process Control.* (New York: John Wiley & Sons Inc.)
[9] Ahmad A. and Wahid A 2007 Application of model predictive control (MPC) tuning strategy in multivariable control of distillation column. *Reaktor* **11(2)** 66-70.

Design of neural network model-based controller in a fed-batch microbial electrolysis cell reactor for bio-hydrogen gas production

Azwar[1,5], M A Hussain[2], A K Abdul-Wahab[3], M F Zanil[4] and Mukhlishien[1]

[1] Chemical Engineering Department, Faculty of Engineering, University of Syiah Kuala, 23111 Banda Aceh, Indonesia
[2] Chemical Engineering Department, Faculty of Engineering, University of Malaya, 50603 Kuala Lumpur, Malaysia
[3] Biomedical Engineering Department, Faculty of Engineering, University of Malaya, 50603 Kuala Lumpur, Malaysia
[4] Chemical and Petroleum Engineering, Faculty of Engineering & Built Environment, UCSI University, 56000 Kuala Lumpur, Malaysia

E-mail: azwar@che.unsyiah.ac.id

Abstract. One of major challenge in bio-hydrogen production process by using MEC process is nonlinear and highly complex system. This is mainly due to the presence of microbial interactions and highly complex phenomena in the system. Its complexity makes MEC system difficult to operate and control under optimal conditions. Thus, precise control is required for the MEC reactor, so that the amount of current required to produce hydrogen gas can be controlled according to the composition of the substrate in the reactor. In this work, two schemes for controlling the current and voltage of MEC were evaluated. The controllers evaluated are PID and Inverse neural network (NN) controller. The comparative study has been carried out under optimal condition for the production of bio-hydrogen gas wherein the controller output is based on the correlation of optimal current and voltage to the MEC. Various simulation tests involving multiple set-point changes and disturbances rejection have been evaluated and the performances of both controllers are discussed. The neural network-based controller results in fast response time and less overshoots while the offset effects are minimal. In conclusion, the Inverse neural network (NN)-based controllers provide better control performance for the MEC system compared to the PID controller.

1. Introduction

Microbial electrolysis cells (MEC) used for wastewater treatment are a novel and promising renewable energy technology that can produce H_2 while the treatment is being performed. It is a new bio-electrochemical process that is capable of producing hydrogen gas [1]. Currently, several studies of the MEC system have been reported in the literature. One important and interesting phenomenon of the MEC model is a competition between anodophilic and methanogenic microorganisms to consume the substrate in the anode compartment [2, 3]. Competition among these microbial populations has severe

[5] To whom any correspondence should be addressed.

effects on the performance of the MEC bioreactor. An initial study of this system used a model involving competition among anodophilic, methanogenic acetoclastic and hydrogenotrophic methanogenic microorganisms in the biofilm as reported by Pinto et al. [4]. Others have reported improvements in the modeling and simulation of a two-chamber microbial fuel cell [5], conduction-based modeling of the biofilm anode of a microbial fuel cell, analysis of a microbial electrochemical cell using the proton condition in biofilm model [6] and a multi-population model of a microbial electrolysis cell [7]. Picioreanu et al. [8] provided detailed descriptions of the mathematical model for microbial fuel cells with anodic biofilms and anaerobic digestion. The model was based on evaluation of the effect of pH and electrode geometry on microbial fuel cell performance.

Bio-hydrogen production process in the MEC is a nonlinear and highly complex system due to the microbial interaction. Its complexity makes MEC system difficult to operate and control under optimal conditions. However, these problems can be alleviated using an integrated process system engineering approach, which involves process modeling and control simultaneously. Artificial neural networks (ANNs) are one of the most effective techniques and powerful tools to be used in modeling of complex processes and unknown systems. ANNs are able to cope with non-linear process between input and output variables without the requirement of explicit mathematical representation. In the process control system, ANNs have been widely used when conventional control techniques did not give good performance [9, 10].

In the literature, several studies and investigations on the modeling of biohydrogen production using ANN approach is reported. For example, El-Shafie [11], use ANNs to successfully predict hydrogen yield with the following input: initial medium pH, initial glucose concentration and reaction temperature. Sridevi et al. [12] studied the Back propagation NNs modelling of biodegradation and fermentative biohydrogen production using distillery wastewater in a hybrid upflow anaerobic sludge blanket reactor. The hydrogen production profile in batch studies was simulated using ANN by Nasr et al. [13]. The ANN inputs were the initial pH, initial substrate, biomass concentrations, temperature, and time. A model was trained, tested and validated to predict the hydrogen production profile. However, none of these studies in the literature involve neural network based controller for bio-hydrogen gas production in the MEC.

This study focuses on the comparative controller performance in a feedback control approach for the MEC fed-batch reactor, where the comparative study between PID and neural network model-based controller are discussed. The performance and assessment of these PID and neural network controller for tuning and rejection of the load disturbances and noise will be discussed. The requirements for a good controller for optimal H_2 gas production are also highlighted in this work. Where, better control system for optimum bio-hydrogen gas production can be achieved.

2. Mathematical model

The model equations presented here are based on multi-population MEC models, and the following modifications were made to modify the model proposed by Pinto et al. [4] for our case study:

1. The model here is modified for a fed-batch reactor; the Pinto model assumed a continuous system.
2. In this proposed model, the biofilm formation and retention consists of two-phase model biofilm growth, anodic biofilms (Layer 1) and a cathode biofilm population (Layer 2). The Pinto model uses three phases, an outer biofilm layer (Layer 1), an inner biofilm (Layer 2) and cathode biofilms (Layer 3). Using the two-phase model will be more practical and easier to apply in a real plant.
3. This proposed model involves metabolic activities of methanogenic acetoclastic, and methanogenic hydrogenophilic microorganisms without involving the fermentation used in the Pinto model. We assume that it is very difficult to observe two different processes simultaneously in the same reactor, such as fermentation and the bio-electrochemical process [7].

2.1. Mass balances for the MEC system

The dynamic mass balance equations for the components S, Xa, Xm, Xh and Mox in the reactor system are given below as follows:

$$\frac{dS}{dt} = -q_{max,a}\frac{S}{K_{S,a} + S}\frac{M_{ox}}{K_M + M_{ox}}x_a - q_{max,m}\frac{S}{K_{S,m} + S} \quad (1)$$

$$\frac{dx_a}{dt} = \mu_{max,a}\frac{S}{K_{A,a} + S}\frac{M_{ox}}{K_M + M_{ox}}x_a - K_{d,a}x_a - \alpha_1 x_a \quad (2)$$

$$\frac{dx_m}{dt} = \mu_{max,m}\frac{S}{K_{A,m} + S} - K_{d,m}x_m - \alpha_1 x_m \quad (3)$$

$$\frac{dx_h}{dt} = \mu_{max,h}\frac{H_2}{K_h + H_2} - K_{d,h}x_h - \alpha_2 x_h \quad (4)$$

$$\frac{dM_{ox}}{dt} = \frac{\gamma}{V_r x_a}\frac{I_{MEC}}{mF} - Y_M q_{max,a}\frac{S}{K_{A,a} + S}\frac{M_{ox}}{K_M + M_{ox}} \quad (5)$$

$$Q_{H_2} = Y_{H_2}\left(\frac{I_{MEC}}{mF}\frac{RT}{P}\right) - Y_h \mu_h x_h V_r \quad (6)$$

where S is the substrate concentration; x_a, x_m, and x_h are the concentration of the anodophilic, acetoclastic, and hydrogenotrophic microorganisms, respectively; M_{ox} is the oxidized mediator fraction per electricigenic microorganism; and Q_{H_2} is the hydrogen production rate (mL/day). Detailed parameter descriptions are provided in Table 1.

2.2. Electrochemical process

The MEC voltage can be calculated using the theoretical values of electrode potentials by subtracting the ohmic, activation, and concentration losses. Resistance to the flow of ions in the electrolyte and electrode during the MEC operation generates ohmic losses. These partial resistances consist of the counter-electromotive force (E_{CEF}), activation loss (η_{act}), concentration loss (η_{conc}), and ohmic loss (η_{ohm}). Each of these polarizations has different magnitudes for different current density degrees. At low current densities, activation losses are dominant due to reaction-energy barriers at the electrode-electrolyte interface that need to be overcome to start the reaction. At high current densities, the reactant and product diffusion limitations lead to high concentration losses. Finally, ohmic losses increase linearly with current due to electron and ion conduction at the electrodes and electrolytes, contact resistance across each material's interface, and interconnections to electrodes. Therefore the following electrochemical balance can be written as follows:

$$\eta_{ohm} = R_{int}I_{cell} \quad (7)$$

$$-E_{applied} = E_{CEF} - \eta_{ohm} - \eta_{conc} - \eta_{act} \quad (8)$$

$$\eta_{conc,A} = \frac{RT}{mF}\ln\left(\frac{M_{Total}}{M_{red}}\right) \quad (9)$$

$$\eta_{act,C} = \frac{RT}{\beta mF}\sinh^{-1}\left(\frac{I_{MEC}}{A_{sur,A}i_0}\right) \quad (10)$$

$$I_{MEC} = \frac{E_{CEF} + E_{applied} - \frac{RT}{mF}\ln\left(\frac{M_{Total}}{M_{red}}\right) - \eta_{act,C}(I_{MEC})}{R_{int}}\frac{M_{red}}{\varepsilon + M_{red}} \quad (11)$$

$$I_{MEC} = \frac{E_{CEF} + E_{applied} - \frac{RT}{mF}\ln\left(\frac{M_{Total}}{M_{red}}\right) - \eta_{act,C}(I_{MEC})}{R_{int}} \quad (12)$$

With the activation losses at the cathode, the I_{MEC} calculation requires a numerical solution for the nonlinear Eq. (12) as $\eta_{act,C} = f(I_{MEC})$. Because the solution of Eq. (12) could result in negative I_{MEC}

value if $E_{applied}$ is smaller than the sum of η_{conc}, η_{act} and E_{CEF}, only non-negative values of I_{MEC} were considered. The detailed characterization, operation and constant parameters used in these simulation studies are listed in Table 1.

Table 1. Values of kinetic parameters used for the simulation studies.

Symbols	Description	Value
$\mu_{max,m}$	The maximum growth rate of the acetoclastic methanogenic microorganism	0.3 d^{-1}
$\mu_{max,a}$	The maximum growth rate of the anodophilic microorganism	1.97 d^{-1}
$\mu_{max,h}$	The maximum growth rate of the hydrogenotrophic microorganism	0.5 d^{-1}
$q_{max,a}$	The maximum reaction rate of the anodophilic microorganism	13.14 mg-A $mg\text{-}x^{-1}$ d^{-1}
$q_{max,m}$	The maximum reaction rate of the acetoclastic methanogenic microorganism	14.12 mg-A $mg\text{-}x^{-1}$ d^{-1}
$K_{S,a}$	The half-rate (Monod) constant of the anodophilic microorganism	20 mg-A L^{-1} or mg-M L^{-1}
$K_{S,m}$	The half-rate (Monod) constant of the acetoclastic methanogenic microorganism	80 mg-A L^{-1} or mg-M L^{-1}
K_M	Mediator half-rate constant	0.01 mg-M L^{-1}
K_h	Half-rate constant	0.001 mg L^{-1}
Y_{H_2}	The dimensionless cathode efficiency	0.9 [dimensionless]
Y_h	The yield rate for hydrogen consuming methanogenic microorganisms	0.05 [ml-H_2 $mg\text{-}x^{-1}$ d^{-1}
m	The number of electrons transferred per mol of H_2	2 mol-e^- mol-H_2^{-1}
P	The anode compartment pressure	1 atm
M_{ox}	Oxidized mediator fraction	800 mg-M $mg\text{-}x^{-1}$
β	The reduction or oxidation transfer coefficient	0.5 [dimensionless]
$A_{sur,A}$	The anode surface area	0.01 m^2
i_0	The exchange current density in reference conditions	0.005 A m^{2-1}
E_{CEF}	The counter-electromotive force for the MEC	-0.35 V
$E_{applied}$	The electrode potentials	0.80 V
$K_{d,a}$	The microbial decay rates of the anodophilic microorganism	0.04 d^{-1}
$K_{d,m}$	The microbial decay rates of the acetoclastic methanogenic microorganism	0.01 d^{-1}
$K_{d,h}$	The microbial decay rates of the hydrogenotrophic microorganism	0.01 d^{-1}
Y_M	The oxidized mediator yield	34.85 mg-M $mg\text{-}A^{-1}$
γ	The mediator molar mass	663400 mg-M mol_{med}^{-1}
V_r	The anodic compartment volume	10 1

3. Design of Neural Network Controller

The basic direct neural network model used for the controller concept refers to the inverse form of the process. Inverse neural network model with feed-forward structure is used directly as elements within the feedback loop. The diagram of the neural network inverse model based control strategy implementation to control the MEC Current (I_{MEC}) controlled in Fed-batch Microbial Electrolysis Cell Reactor is shown in Fig. 1. From the figure, it can be seen that the inverse model acts as the controller and provides the current control action with respect to certain current and past values of the process variables. In this case, the neural network model is trained to predict the required manipulated variable i.e. Electrode potential $(E_{applied})$ and to bring the process to the set-point i.e. MEC current (I_{MEC}).

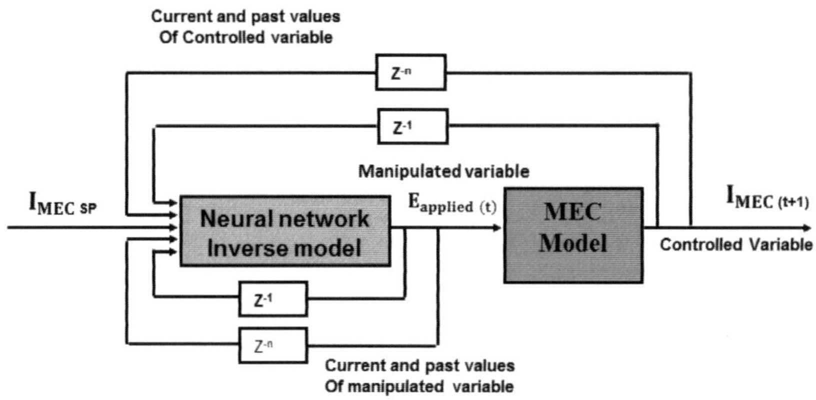

Figure 1. Block diagrams for neural network inverse based model control strategy.

3.1 Training and validation data

The inputs and the outputs of the neural network are fed through a moving window approach. The model is made of 14 input and one output nodes, the input nodes consist of data for substrate (S), anodophilic microorganisms (x_a), acetoclastic microorganism (x_m), ammonium nitrogen (x_h), oxidized mediator fraction (M_{ox}), MEC current (I_{MEC}) and the one output node is electrode potentials ($E_{applied}$) and past and current data for S i.e. $S_{(t)}$, $S_{(t-1)}$; x_a i.e. $x_{a(t)}, x_{a(t-1)}$; x_m i.e. $x_{m(t)}, x_{m(t-1)}$; x_h i.e. $x_{h(t)}$, $x_{h(t-1)}$; M_{ox} i.e. $M_{ox(t)}$, $M_{ox(t-1)}$; I_{MEC} i.e. $I_{MEC(t)}$, $I_{MEC(t+1)}$, $I_{MEC(t-1)}$ and output node is the electrode potentials ($E_{applied}$).

The training of a neural network is done in conjunction with the back-propagation algorithm. The back-propagation pass is used to calculate the derivatives and the errors of every neuron in the network. The back-propagation algorithm learns to recognize and reproduce patterns in an iterative process whereby its weights are adjusted in order to minimize a selected error criterion. Data for training the neural network in the simulation work are obtained by solving the ordinary differential equations (ODE) utilizing the MATLAB 2011a software. Two sets of data have been prepared for training the neural network model and one is used for cross validation purposes in order to test the validity of the trained neural network models. The three training data sets are switched between each other during the training session in order to improve system identification by the neural network models [14]. The validation for neural network inverse based model are shown in Figure 2.

3.2 Forward and inverse modeling

The procedure of training a neural network to represent the dynamics of the system is referred to as forward modeling. Forward modeling refers to training the neural network model to predict the plant output, $I_{MEC(t+1)}$. In this case, connections between layers, or weights, are changed during training in order to minimize the error between the actual plant output and the predicted output from the neural network model. The input to the network consists of present and past value of S, x_a, x_m, x_h, M_{ox}, I_{MEC} and $E_{applied}$ values. The desired network output is the future $I_{MEC(t+1)}$ values. Those input and output values are fed into the network in the moving window approach. From these training exercises, the neural network architecture produced is a 14 nodes input layer, 28 nodes of hidden layer and a 1 output layer system. The activation function utilized is the sigmoidal function in both the hidden and output layer. The forward model can be expressed mathematically as a function of inputs to the model as shown below:

$$I_{MEC(t+1)} = f\begin{pmatrix} S_{(t)}, S_{(t-1)}, x_{a(t)}, x_{a(t-1)}, x_{m(t)}, x_{m(t-1)}, x_{h(t)}, x_{h(t-1)}, Q_{H2(t)}, Q_{H2(t-1)}, M_{ox(t)}, M_{ox(t-1)}, \\ I_{MEC(t)}, \quad I_{MEC(t-1)}, E_{applied(t)}, E_{applied(t-1)} \end{pmatrix} \tag{13}$$

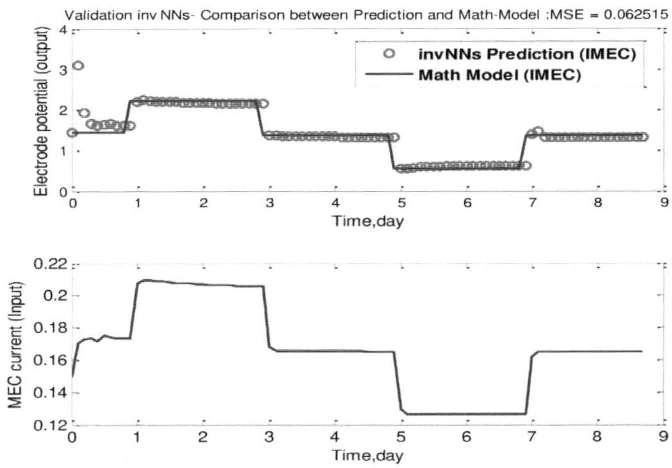

Figure 2. The validation for neural network invers based model.

Inverse modeling refers to the training of the neural network in predicting the input to the plant given past data of the inputs and outputs together with the desired output. The inverse model is used directly as the controller within the feedback loop. Similar to the forward modeling methodology two training data sets were used, those were switch from one to the other during training to improve the identification process. In determining the inverse model to use as the controller, the network architecture and activation functions that were chosen are similar to the forward model. For this inverse model, the two-layered feed-forward network that has 14 input nodes, 24 hidden nodes and 1 output node is used. During training the network is fed with the required future value, $I_{MEC(t+1)}$, together with the present and past input and outputs to predict the current input or control action, $E_{applied(t)}$ as seen in Fig. 3. The inverse model can be expressed mathematically as a function of inputs to the model as shown below:

$$E_{applied(t)} = f\begin{pmatrix} S_{(t)}, S_{(t-1)}, x_{a(t)}, \quad x_{a(t-1)}, x_{m(t)}, x_{m(t-1)}, x_{h(t)}, x_{h(t-1)}, Q_{H2(t)}, Q_{H2(t-1)}, \\ M_{ox(t)}, M_{ox(t-1)}, I_{MEC(t)}, I_{MEC(t-1)}, I_{MEC(t+1)}, E_{applied(t-1)} \end{pmatrix} \tag{14}$$

The data sets generated for the training and validating, transfer functions and method of training the inverse model used are similar to the forward model. The architecture of the inverse model can be seen in Figure 3.

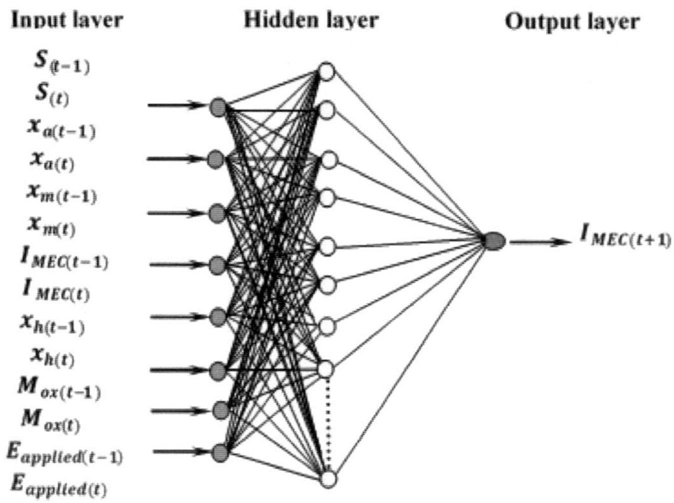

Figure 3. The inverse model architecture for Microbial Electrolysis Cells.

Figure 4 shows the inverse modeling of MEC current to be used as training data. From the result, it can be seen that the artificial neural network accurately tracks the inverse dynamics of the MEC system.

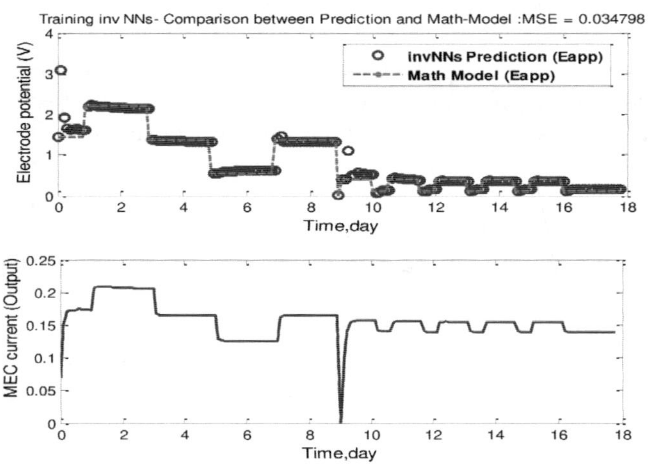

Figure 4. Inverse modeling of MEC current and electrode potential.

4. Neural Network Controller Scheme

In this section, the performance of the basic neural network controller is discussed. The performance of the controller was investigated through studies of nominal operating condition, constant set-point and multiple set-point tracking study for testing the performance of process by loading internal disturbance. The variation in disturbance was generated by changing the counter-electromotive force (nominal value of 0.35 V).

4.1 Constant set-point

In this work, we performed set-point tracking study when the I_{MEC} current was maintained at approximately the nominal optimal operation value of 0.16 A. Fig. 5 shows the process and controller response when using the conventional PID-ZN method and neural network model-based controller for single set-point performance under nominal operating conditions. The controller performs well and is successful in managing the process to follow the given set point. However, the PID (tuned with the Z-N method) gave a higher overshoot and more oscillations than the neural network model-based controller. This is also evident in the behavior of the manipulated voltage of the electrode potential applied for both cases.

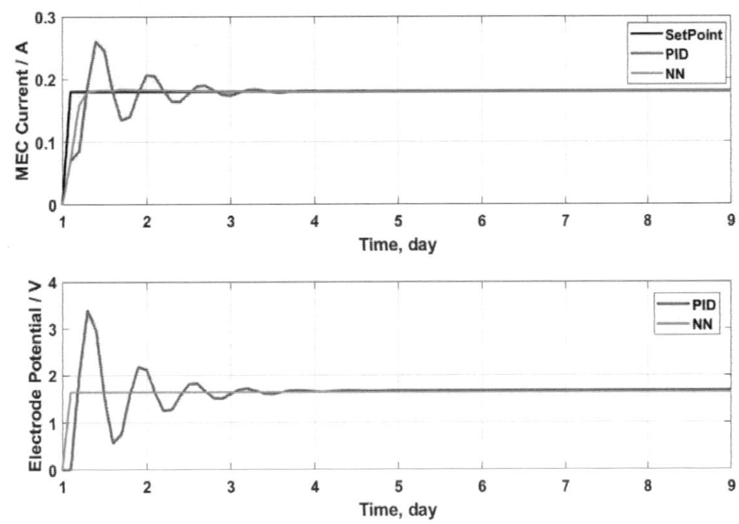

Figure 5. Comparison of PID-ZN and neural network model-based controller for constant setpoint.

4.2 Disturbances rejection

Figure 6 shows the comparison of the control performance of the PID-ZN method and neural network model-based controller for constant setpoint with disturbances in the MEC system. The disturbance considered in this study was generated through changes in the counter-electromotive force (V). Based on these results, the performance of this controller is generally acceptable and the neural network model-based controller gave smoother responses in the MEC system.

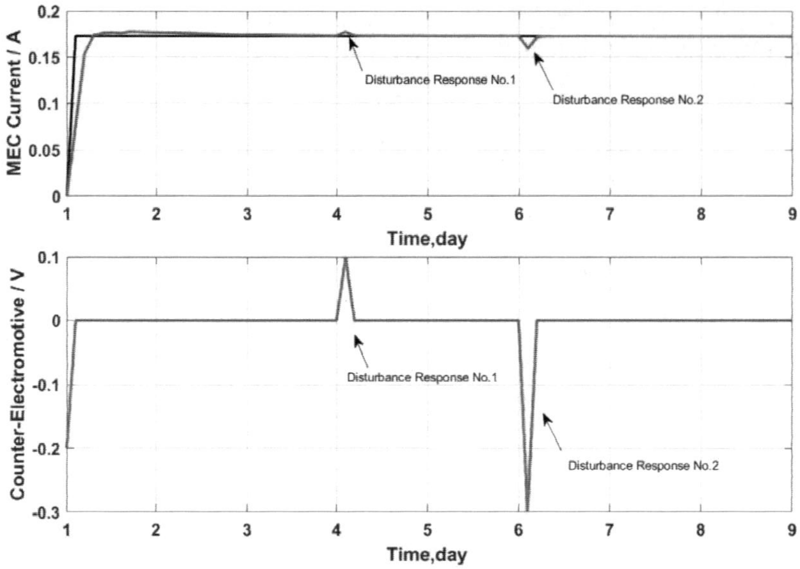

Figure 6. Neural network model-based controller for disturbance rejection

4.3 Multiple set-point tracking study

Figure 7 shows the comparison of PID-ZN and neural network model-based controller for multiple setpoint tracking study. The controller performs well and is successful in managing the process to follow the given set point changes to keep up around at 0.11 A, 0.16 and 0.20 A, respectively. From these figures, we can see that neural network model-based controller can provide better control of the MEC system compared with the Ziegler-Nichols tuning method. It can be seen that the controller performs reasonably well when responding to deviation, but it becomes sluggish when responding to large deviations in the process response. However, the controller is also capable of following the time-varying characteristic of the process in most conditions.

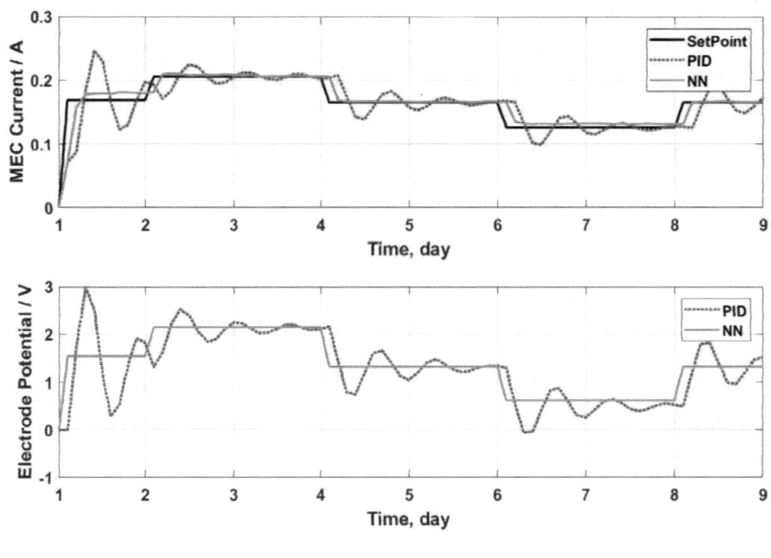

Figure 7. Comparison of PID-ZN and neural network model-based controller for multiple setpoint tracking study

5. Conclusions

In this paper, the optimal control design for MEC with various simulation tests involving single set-point study, multiple set-point changes and disturbance rejection was achieved using the neural network model-based controller. Various simulation tests have been evaluated and the performances of both controllers were discussed. A Ziegler–Nichols tuning method has been used to design the PID controller, while the neural network controller has been designed from the inverse response of the forward MEC model. The neural network-based controller results in fast response time and less overshoots while the offset effects are minimal. In conclusion, the neural network (NN)-based controllers provide better control performance for the MEC system compared to the PID controller.

References

[1] Logan B E, Call D, Cheng S, Hamelers H V M, Sleutels T H J A, Jeremiasse A W, Rozendal R A 2008 Microbial electrolysis cells for high yield hydrogen gas production from organic matter *Environmental Science and Technology* **42** (23) 8630-8640

[2] Wang A, Liu W, Cheng S, Xing D, Zhou J, Logan B E 2009 Source of methane and methods to control its formation in single chamber microbial electrolysis cells *International Journal of Hydrogen Energy* **34** 3653–3658.

[3] Pinto, R P, Srinivasan B, Manuel M F, Tartakovsky B A 2011(a) Unified model for electricity and hydrogen production in microbial electrochemical cells Preprints of the 18th *International Federation of Automatic Control (IFAC) world congress, Milano (Italy), August 28 – Sept. 2*

[4] Pinto R P, Srinivasan B, Manuel M F, Tartakovsky B 2010 A two-population bio-electrochemical model of a microbial fuel cell *Bioresource Technology* **101** (14) 5256–5265

[5] Zeng Y, Choo Y F, Kim B.-H, & Wu P 2010 Modelling and simulation of two-chamber microbial fuel cell *Journal of Power Sources* **195** (1) 79-89

[6] Marcus A K, Torres C I, & Rittmann B E 2011 Analysis of a microbial electrochemical cell using the proton condition in biofilm (PCBIOFILM) model *Bioresource Technology* **102** (1) 253-262

[7] Pinto R P, Srinivasan B, Escapa A M F, Tartakovsky B 2011(b) Multi-population model of a microbial electrolysis cell *Environmental Science & Technology* **45** 5039-5046

[8] Picioreanu C, van Loosdrecht M C M, Curtis T P, & Scott K 2010 Model based evaluation of the effect of pH and electrode geometry on microbial fuel cell performance *Bioelectrochemistry* **78** (1) 8-24

[9] Wang J & Wan W 2009 Application of desirability function based on neural network for optimizing biohydrogen production process *International Journal of Hydrogen Energy* **4** 1253-1259

[10] Shi Y, Gai G, Zhao X, Zhu J, Zhang P 2010 Back propagation neural network (BPNN) simulation model and influence of operational parameters on hydrogen bio-production through integrative biological reactor (IBR) treating wastewater *College of Power and Energy Engineering*, Harbin Engineering University, Hei Long-jiang, China

[11] El-Shafie A 2014 Neural network nonlinear modeling for hydrogen production using anaerobic fermentation Neural Computing and Applications **24** 539-547

[12] Sridevi K, Sivaraman E, Mullai P 2014 Back propagation neural network modelling of biodegradation and fermentative biohydrogen production using distillery wastewater in a hybrid upflow anaerobic sludge blanket reactor *Bioresource Technology* **165** 233-240

[13] Noha Nasr, Hisham Hafez M, Hesham El Naggar 2013 Application of artificial neural networks for modeling of biohydrogen production *International Journal of Hydrogen Energy* **38** 3189-3195

[14] Hussain M A and Mujtaba I M 2001 Application of neural networks and other learning technologies in Process Engineering *Imperial College Press* pg 199-216

Optimization control of LNG regasification plant using Model Predictive Control

A Wahid[1,2] and F F Adicandra[1]
[1] Sustainable Energy Research Group, Department of Chemical Engineering, Faculty of Engineering, Universitas Indonesia, Depok 16424, Indonesia

[2]E-mail: wahid@che.ui.ac.id

Abstract. Optimization of liquified natural gas (LNG) regasification plant is important to minimize costs, especially operational costs. Therefore, it is important to choose optimum LNG regasification plant design and maintaining the optimum operating conditions through the implementation of model predictive control (MPC). Optimal tuning parameter for MPC such as P (prediction horizon), M (control of the horizon) and T (sampling time) are achieved by using fine-tuning method. The optimal criterion for design is the minimum amount of energy used and for control is integral of square error (ISE). As a result, the optimum design is scheme 2 which is developed by Devold with an energy savings of 40%. To maintain the optimum conditions, required MPC with P, M and T as follows: tank storage pressure: 90, 2, 1; product pressure: 95, 2, 1; temperature vaporizer: 65, 2, 2; and temperature heater: 35, 6, 5, with ISE value at set point tracking respectively 0.99, 1792.78, 34.89 and 7.54, or improvement of control performance respectively 4.6%, 63.5%, 3.1% and 58.2% compared to PI controller performance. The energy savings that MPC controllers can make when there is a disturbance in temperature rise 1°C of sea water is 0.02 MW.

1. Introduction

Several ways have been done to transport gas, one of which is by building gas pipes or liquefying natural gas into LNG, to reduce its volume and make the gas in liquid phase so that transportation is easier. Gas pipe is very limited utilization. Because the installation of gas pipelines has a high difficulty, as well as very expensive costs, especially for long distances and through the oceans. For LNG itself, in West Java has been built units FSRU (Floating Storage and Regasification Unit). This unit is useful for returning LNG back into gas phase to be utilized as industrial fuel.

LNG regasification plant should be in optimum design, both the design of the equipment and the control design in order to minimize the energy used. Therefore, in this research we will select the optimal LNG regasification plant design and set the control parameters used to maintain optimum operating conditions in case of process disturbances. The controller used is MPC because MPC is one of the advanced process control that implements the optimization method (cost function) in its control and has been proven to have better control performance compared to conventional controller [1].

2. Methodology Experimental

2.1 FOPDT Model

Figure 1 shows the research flow diagram. The research begins by creating a steady state model developed by Wahid and Ryan [2] and Devold [3]. Then for the selection of models based on energy use and optimization of operating conditions. After that, change the simulation to dynamic mode and install the control system on each equipment according to the control objectives. After that, the test model is performed on each controller. This model test is performed to obtain the PRC curve as shown in Figure 2 to obtain an empirical model.

Empirical model used is first order plus dead-time model (FOPDT) with equation:

$$G_P(s) = \frac{K_p e^{-\theta s}}{\tau s+1} \quad (1)$$

with $K_p = {}^{\Delta}/_{\delta}$ (Δ= magnitude of change of output variable and δ = amount of input variable change affecting output variable), $\tau = 1{,}5\ (t_{63\%} - t_{28\%})$ with $t_{63\%}$ dan $t_{28\%}$ respectively are output response times when it reaches 63% and 28% of the final value of the output variable, and $\theta = t_{63\%} - \tau$ [4].

2.2 Tuning Controller

2.2.1 Proportional-Integral Controller. Based on the parameters of this FOPDT empirical model, the PI controller is adjusted using the Ziegler-Nichols method [5]. The transfer function PI is used equation:

$$G_c(s) = K_c\left(1 + \frac{1}{T_I s}\right) \quad (2)$$

With K_c is the controlling gain and T_i is the integral time. For Ziegler-Nichols method used equation:

$$K_c = \frac{0.9}{K_p}\left(\frac{\theta}{\tau}\right)^{-1} \text{ dan } T_I = 3.33\ \theta \quad (3)$$

2.2.2 *Model Predictive Control.* In MPC a dynamic model is used to predict the future output over the prediction horizon based on a set of control changes. The desired output is generated as a setpoint that may vary as a function of time; the prediction error is the difference between the setpoint trajectory and the model prediction. A model predictive controller is based on minimizing a quadratic objective function in Equation 4 over a specific time horizon based on the sum of the square of the prediction errors plus a penalty related to the square of the changes in the control variable(s) [6].

$$\sum_{i=1}^{P} w_i e^2(k+i) + \lambda \sum_{i=1}^{m} \Delta u^2(k+i-1) \quad (4)$$

where $e(k+i)$denotes the predicted error at time $(k+i), i = 1, \ldots., p,$

$$e(k+i) = r(k+i) - \hat{y}(k+i) \quad (5)$$

$r(k+i)$is the reference value or setpoint at time $k+i$, and $\Delta u(k)$ denotes the vector of current and future control moves over the next *m* sampling instants:

$$\Delta u(k) = [\Delta u(k), \Delta u(k+1), \ldots, \Delta u(k+m-1)]^T \quad (6)$$

Equation 4 contains design parameters that can be used to tune the controller, that is, you vary the parameters (P and M) until the desired shape of the response that tracks the setpoint trajectory is achieved [7].

For the MPC controller, the empirical model parameters obtained from PRC can be directly used on UniSim software R390.1 as a process model. Then re-identified the system, by retesting the model with the close loop method using the default MPC adjustment constants and then the results are compared with the MPC performance by using the previous PRC. The re-identification process is completed if the MPC performance using the previous PRC is better than the performance of MPC with the PRC used. The performance parameter used is ISE. The system identification is performed to obtain the optimum

IOP Conf. Series: Materials Science and Engineering **334** (2018) 012022 doi:10.1088/1757-899X/334/1/012022

empirical model. After that done, tuning parameters MPC Prediction Horizon (P), Control Horizon (M), and Sampling Time (T) by using the fine-tuning method. MPC fine-tuning adjustment is done by trial error and then view system behavior to change adjustment parameters (P, M, T) to obtain stable settings.

2.3 Performance of Controller

Based on these two controller tuning methods, the control performance was calculated using ISE (Integral Square Error) [8]. Setting methods that have smaller ISE values are the best tuning method.

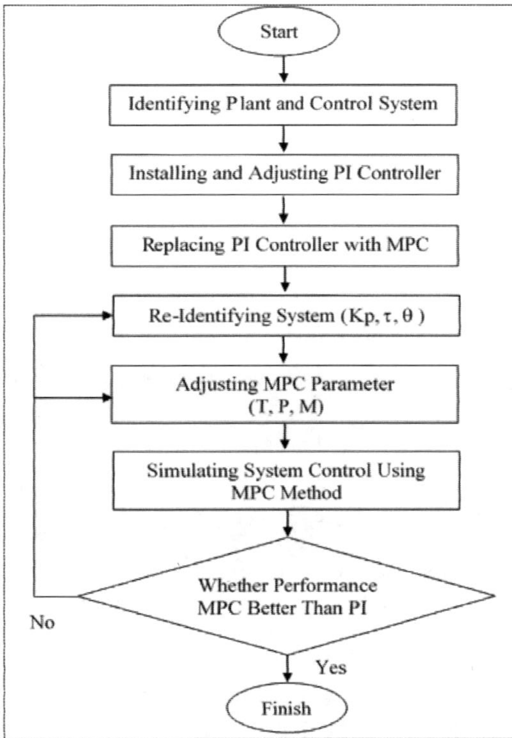

Figure 1. Research Flow Diagram.

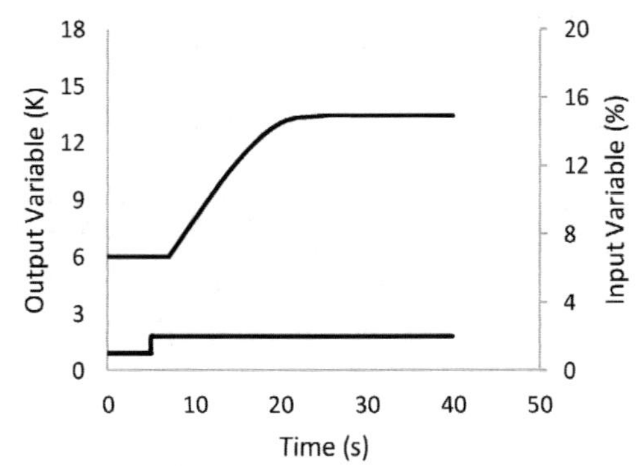

Figure 2. Process Reaction Curve.

3. Result and Discussion

3.1 Optimization of LNG Regasification Scheme

To determine the most optimum regasification process scheme between the two schemes in figure 3 and figure 4, a simulation is performed that can calculate the energy requirements for each process scheme.

IOP Conf. Series: Materials Science and Engineering **334** (2018) 012022 doi:10.1088/1757-899X/334/1/012022

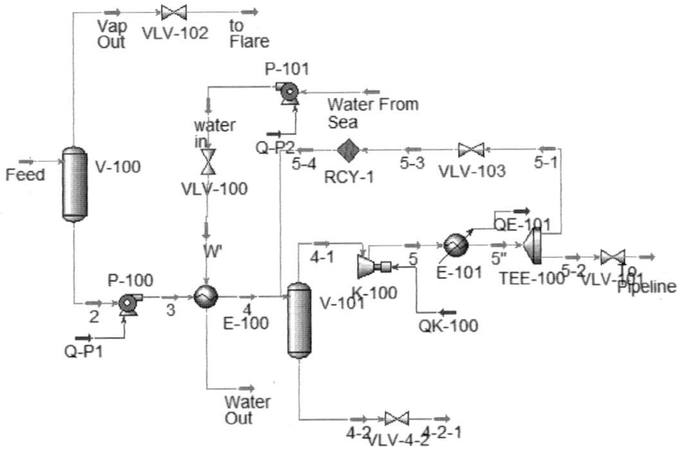

Figure 3. Scheme 1 Regasification Process.

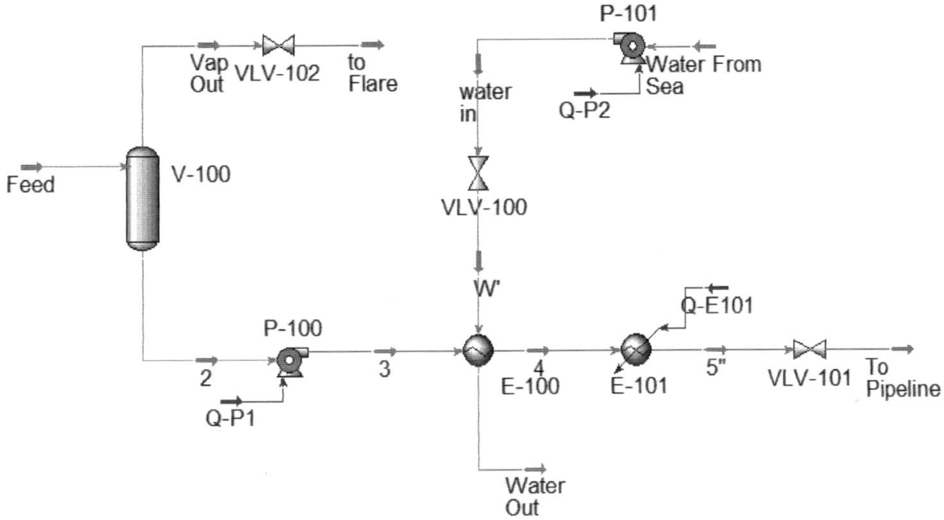

Figure 4. Scheme 2 Regasification Process.

Table 1. Use of Energy at Regasification Plant (MW).

Process Equipment	Scheme 1	Scheme 2
LNG Pump	1.04	2.37
Water Pump	0.79	0.81
Compressor	21.15	-
Cooler	8.87	-
Heater	-	15.97
Total	31.85	19.15

From the simulation results in table 1 and figure 5, it can be seen that the energy in scheme 1 is very large for the compressor process equipment, and small enough for the LNG pump energy needs compared to scheme 2. In scheme 2, the amount of energy required is in the heater unit that functions to increase the gas temperature to achieve the gas pipe specification. In scheme 2, there is no compressor to increase pressure but is replaced by a high pressure LNG pump which requires considerable magnitude compared to scheme 1. If viewed as a whole, scheme 2 requires less energy than scheme 1. Energy savings can be done by replacing the process in order to increase in pressure from the compressor unit to the high pressure LNG pump unit reaches 40%.

3.2 Optimization of LNG Regasification Operating Conditions

At the LNG regasification plant the unit that has an important role is the vaporizer unit. In this unit, there is a phase change from liquid LNG to gas which is the main objective of the regasification process. Therefore, an optimum operating condition is required for this unit to make the plant run efficiently. To obtain optimum condition simulations are performed using Unisim R390.1, to see the effect of vaporizer output temperature with required seawater pump power and duty of heater needed to achieve gas pipe specification.

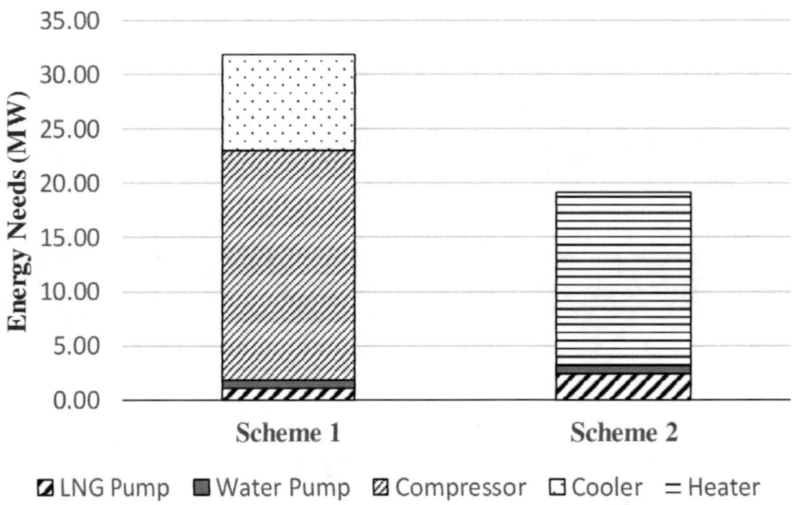

Figure 5. Proportion of Energy Requirement of LNG Regasification Plant.

3.3 Empirical Model and Adjustment of Control

Control system at LNG regasification plant that used in this research shown in Figure 6. The empirical model is obtained from PRC (process reaction curve) resulting from system identification for each controller. From the PRC, FOPDT empirical models were developed as shown in Table 2, which is computed by equation. (1).

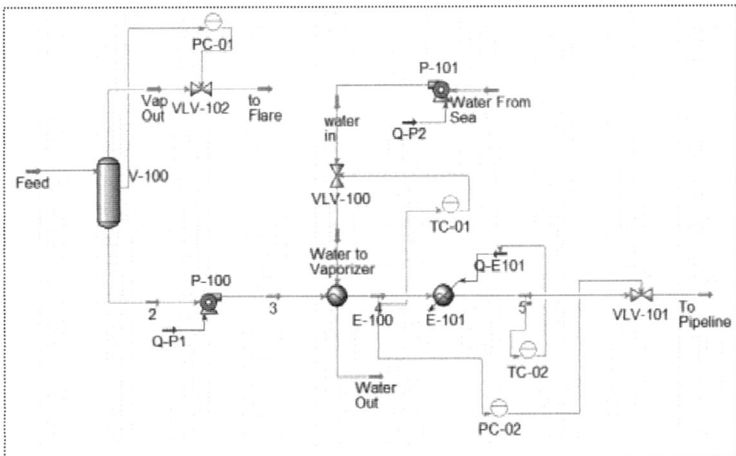

Figure 6. Control System at LNG Regasification Plant.

Table 2. Process Model of Regasification LNG Plant.

Controller	FOPDT
Pressure Control V-100	$\frac{0.069\, e^{-0.0108s}}{0.103s+1}$
Pressure Control Stream 4	$\frac{0.121\, e^{-0.003s}}{0.017s+1}$
Temperature Control E-100	$\frac{0.045\, e^{-0.0182s}}{0.642s+1}$
Temperature Control E-101	$\frac{0.381\, e^{-0.0088s}}{0.0438s+1}$

3.4 Performance of Control

The way to find out the performance of each controller, set point tracking and disturbance rejection are performed and evaluated with the value of ISE. Here are the results of both methods:

3.4.1 Set Point Tracking. In this method, a set point change is made to see how the controller response to achieve process stability. The controller that has the lowest ISE value is the best controller. The result for the set point tracking can be represented by Figure 7. From Figure 7, the tuning with MPC-Fine tuning method produces the best performing control with the total ISE scores in sequence for the vaporizer, heater, storage tank, and stream 4 controllers are 34.88, 7.54, 0.98 and 1792.78.

3.4.2 Disturbance Rejection. In this method, disturbance is given to the system to see the controller response in overcoming the disturbance on the system. The results of the test for disturbance rejection can be seen in Figure 8. From Figure 8, the tuning with MPC-Fine tuning method produces the best performing control with the total ISE scores in sequence for the vaporizer, heater, storage tank, and stream 4 controllers are 7.4, 7.54, 0.35 and 0.016.

3.5 Energy Efficiency

To determine the effect of errors on both controllers on energy to be used can be seen in figure 9 and figure 10. Based on the calculation of energy deviation, the use of control with the MPC-fine tuning method has fewer errors than the proportional-integral method controller. The use of MPC-fine tuning controller can reduce 0.02 MW and MPC controller also reduces error to product quality by 34.25% compared to the PI controller.

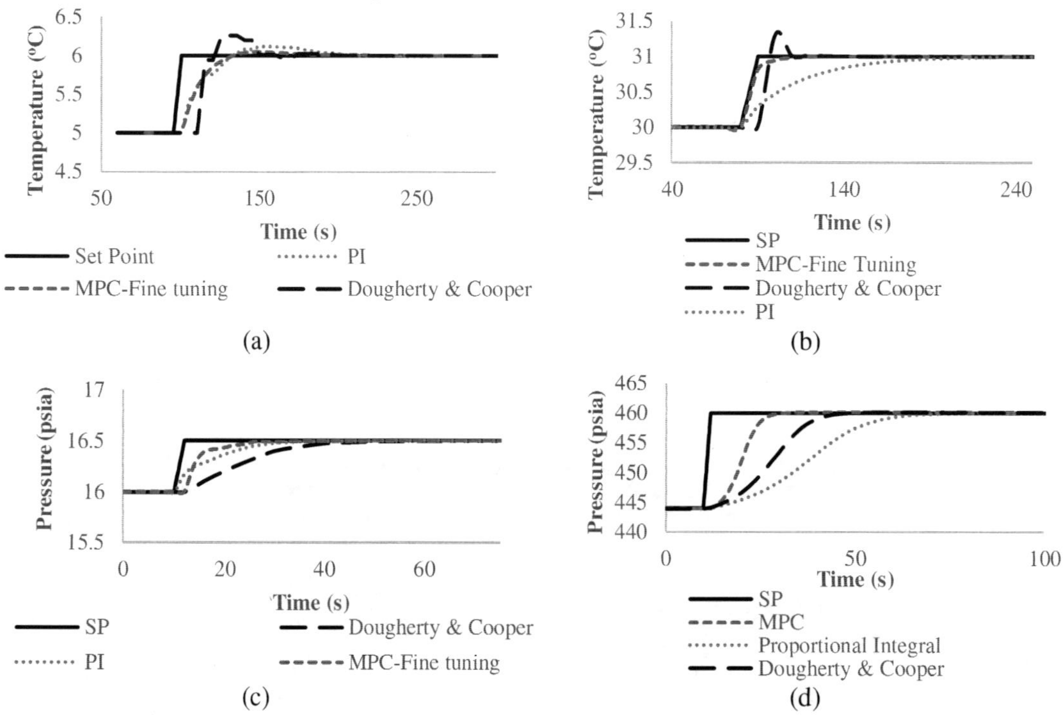

Figure 7. Comparison of Three Controller (Set Point Tracking): (a) Vaporizer (b) Heater (c) Storage Tank (d) Stream 4.

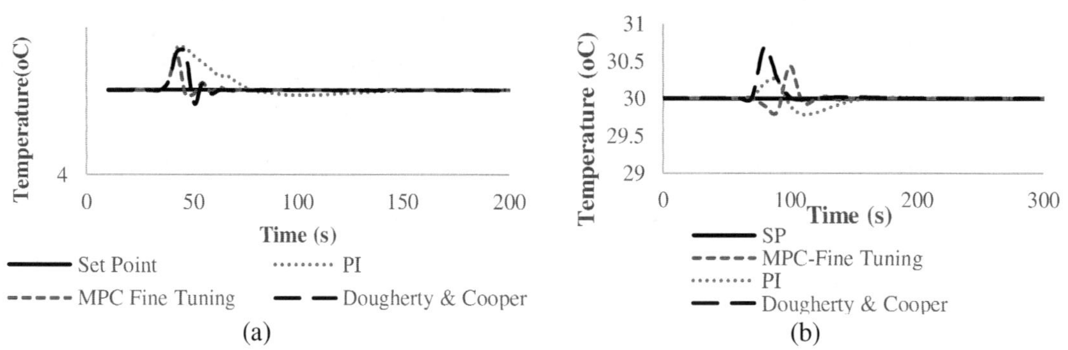

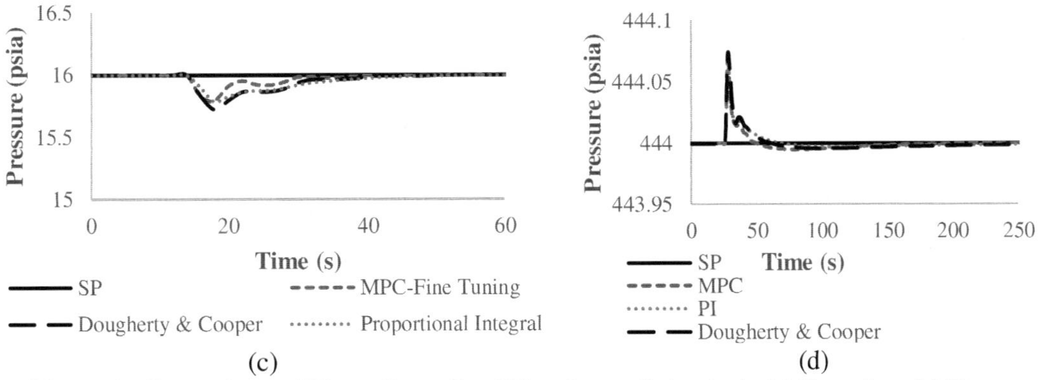

Figure 8. Comparison of Three Controller (Disturbance Rejection): (a) Vaporizer (b) Heater (c) Storage Tank (d) Stream 4.

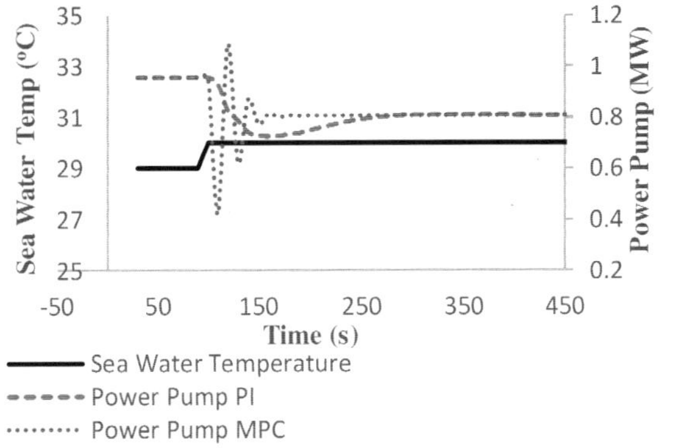

Figure 9. Relationship needs of pump power to disturbance on Sea water temperature.

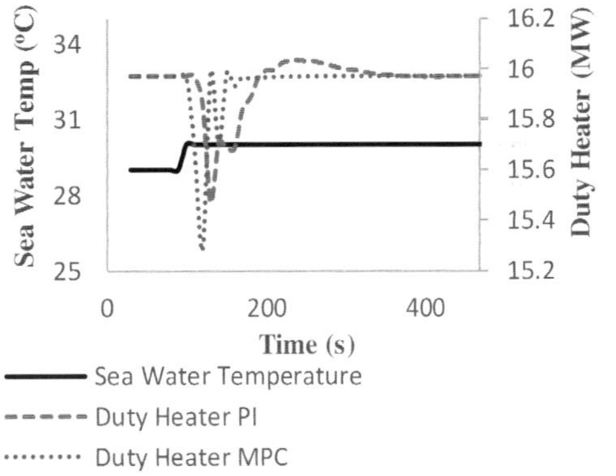

Figure 10. Relationship needs of Duty Heater to disturbance on Sea water temperature.

4. Conclusion

Scheme 2 LNG regasification plant requires less energy than scheme 1, with energy savings reaching 40% from scheme 1, so scheme 2 is chosen. Tuning MPC (P, M and T) that is required as follows: pressure tank control (90, 2, 1), pressure on stream 4 (95, 2, 1), vaporizer temperature (65, 2, 2) and heater temperature: (35, 6, 5) with ISE value in set point tracking respectively 0.99, 1792.78, 34.89 and 7.54, or improved control performance respectively 4.6%, 63.5%, 3.1% and 58.2%, compared to PI controller performance. The energy savings that MPC controllers can perform when a disturbance in temperature rise 1°C of sea water is 0.02 MW.

Acknowledgments

We express our gratitude to the Universitas Indonesia which has funded this research through the scheme of *Hibah Publikasi Internasional Terindeks untuk Tugas Akhir Mahasiswa* (PITTA) No.2068/UN2.R3.1/PPM.00/2017.

References

[1] Wahid, A and Afdal, A 2016 Model predictive control based on system re-identification for methanol and dymethil ether synthesis control. *Proc. Conf. Nasional Teknik Kimia "Kejuangan"*. (Yogyakarta)

[2] Wahid, A and Tanuwijaya, R 2015 Pemilihan Metode Penyetelan Pengendali PI pada Pengendalian

Pabrik Regasifikasi LNG Menggunakan Metode Skor. *Proc. Conf. Nasional Teknik Kimia UNPAR.* (UNPAR: Bandung)

[3] Devold, H 2013 *Oil and Gas Production Handbook: An Introduction to Oil and Gas Production, Transport, Refining and Petrochemical Industry.* (Oslo : ABB Oil and Gas)

[4] Marlin, T 2000 *Process Control: Designing Processes and Control Systems for Dynamic Performance*. (United States: McGraw-Hill Higher Education)

[5] Smith, C and Corripio, A 1997 *Principles and Practice of Automatic Process Control*. (United States: John Wiley & Sons Inc.)

[6] Himmelblau D M, Edgar T F and Lasdon L S 2001 *Optimization of Chemical Processes, Second Edition.* (New York: McGraw-Hill)

[7] Altman, W 2005 *Practical Process Control for Engineers and Technicians*. (IDC Technologies)

Numerical study of influences of crosswind and additional steam on the flow field and temperature of propane non-premixed turbulence flame

Wusnah[1,5], Y Bindar[2], Yunardi[3], F M. Nur[4] and A M. Syam[1]

[1] Department of Chemical Engineering, Malikussaleh University, Lhokseumawe 24355, Indonesia.
[2] Department of Chemical Engineering, Faculty of Industrial Technology, Bandung Institute of Technology, Bandung, Indonesia
[3] Department of Chemical Engineering, Syiah Kuala University, Banda Aceh 23111, Indonesia.
[4] Department of Mechanical Engineering, Malikussaleh University Lhokseumawe 24355, Indonesia.

E-mail: wusnah@unimal.ac.id

Abstract. This paper presents results the process of combustion propane using computational fluid dynamics (CFD) to simulate the turbulent non-premixed flame under the influences of crosswinds and the ratio of fuel (propane) to steam, *S*. Configuration, discretization and boundary conditions of the flame are described using Gambit^{TM} software and integrated with Fluent^{TM} software for calculations of flow and reactive fields. This work focuses on the influence of various crosswind speeds (0–10 m/s) and values of S (0.14–2.35) while the velocity of fuel issued from the nozzle was kept constant at 20 m/s. A turbulence model, k-ε standard and combustion model, Eddy Dissipation model were employed for the calculation of velocity and temperature fields, respectively. The results are displayed in the form of predictive terrain profile of the propane flame at different crosswind speeds. The results of the propane flame profile demonstrated that the crosswind significantly affect the structure velocity and position of the flame which was off-center moving towards the direction of crosswind, eventually affect the temperature along the flame. As the values of S is increasing, the flame contour temperature decreases, until the flame was extinguished at S equals to 2.35. The combustion efficiency for a variety of crosswind speeds decreases with increasing values of S.

1. Introduction

Combustion is a very complex process dealing with interaction of physical and chemical phenomena including flow, turbulence, thermodynamics, chemical reaction, radiation, extinction, and ignition. Almost all burning systems in a variety of equipment such as rocket engines, aircraft engines, kilns and chimneys in industry are implemented in a turbulent environment, with the aim at increasing the rate of combustion. Heat is released from chemical reactions, but the process involves the propagation

[5] To whom any correspondence should be addressed.

of heat transfer and fluid dynamics so that the theoretical interpretation can be derived using an understanding of chemistry, physics, fluid mechanics and applied mathematics [1, 14].

In accordance with the type of fuel, combustion can be divided into three types, gas, liquid and solid; while the flame generated can be classified into premix and no-premix ones. A premix flame occurs when the fuel and oxidizer are mixed before ignition, while the latter when the fuel and oxidizer are not mixed before the combustion process takes place, oxidizer merges with fuel by diffusion. In terms of safety aspects, the use of non-premix flame is better than that of premix flame. As a consequence, the non-premix flame is more widely used for industrial applications such as turbines, aircraft and furnaces [2].

Non-premix jet flames influenced by crossflow occur in several applications, such as gas turbine combustor, refinery flaring operations, and industrial burners. The characteristics of the flames under a crossflow influence are totally different from those of stable ones. Results of research found out that the combustion efficiency decreases when the crosswind speed increases. In addition, the flame will bend towards the wind direction with a smaller size, so that oxygen is diffused less into the flame. When oxygen concentration supply reduces into the flame, burning runs imperfectly and produces more pollutants, which are undesirable [3-5]. However, increased jet exit velocity makes the flame less susceptible to the effects of crosswind. The height of a flame is strongly influenced by the increase in wind speed compared to the flame length, as a result of the rising angle of inclination of the flame [6]. The difference between the combustion characteristics of the species is caused by the difference in mass, surface area, and orientation. Results of the analysis showed that the species and length of the flame are due to the influence of wind speed and humidity. At the time of ignition, a combustion material sample under windy conditions contains higher moisture than the combustion products without the influence of these conditions. The average mass loss on ignition is 13% with the effect of wind and 25% without the influence of wind [7].

A study [8] burnt fuels and removed hydrocarbon gas through a vertical hole with both diameters 0.076 m and 0.305 m. The combustion products were collected, samples were taken and analysed to determine the efficiency of combustion. If the flame is stable (i.e. the effect of exit speed and fuel value of the fuel gas), the efficiency can be greater than 98%. However, based on pilot test, it was found out that high crosswind speed significantly reduced the combustion efficiency. Other workers [9,10] demonstrated that the effect of crosswind with a range of 1–15 m/s can reduce combustion efficiency significantly. The laminar burning rate and adiabatic flame temperature increases with the addition of hydrogen into the fuel flow, thus increasing the mass and thermal diffusivity of hydrogen in the air, making it more reactive. Instead, the addition of steam into the flow of the fuel mixture of methane actually reduces the rate of air and adiabatic flame temperature of combustion [10]. Increasing mole fraction of water vapour in the fuel flow rate will further decrease a flame burning methane [11].

Addition of steam to mix with incoming air thereafter mixed with fuel will affect the combustion process inside the combustor. A research on gas turbine cogeneration systems with steam injection [12] was conducted to study the effect of NOx formation due to steam addition to a methane non-premixed flame. It was found that steam addition stimulates the production of NOx. However, because of the decrease in CH concentration with steam addition, the production rates of HCN and N radicals decrease dramatically. Another study [13] numerically investigated the effect of steam addition on the performance of the combustor. It was reported that steam injection is an effective way to reduce the temperature in the burner keeping other performance like the total pressure loss minimum.

The above review has shown that studies the effect of crosswind and steam additional is done separately. There has been limited studies was conducted to investigate simultaneous influences of crosswind and steam addition on the performance of a flame. It is the objective of this paper are to analyse the effect of various crosswind speeds and the addition of steam to the flame produced, and the combustion efficiency, using computational fluid dynamical approach.

2. Methodology

2.1. Governing Equations

In general, many equations are used in the concept of computational fluid dynamics, because the approach of fluid characteristics attempts to simulate real conditions. The basic equations involved in a laminar and turbulent flow without involving heat transfer and species are shown in Equation (1). This equation is often mentioned as the continuity equation in Cartesian coordinates that is used in computational fluid dynamics. It is also involved in the modeling of turbulent combustion including conservation of mass, momentum, enthalpy and species [1].

$$\frac{\delta p}{\delta t}+\frac{\delta(\rho u)}{\delta x}+\frac{\delta(\rho v)}{\delta y}+\frac{\delta(\rho w)}{\delta z}=0 \quad (1)$$

This equation is the common equation of mass conservation and applies to any compressible and non-compressible flow. The momentum conservation equation defines the fluid motion for the forces on the particles at every certified fluid element in computational fluid dynamics modelling. Momentum conservation equations for Cartesian coordinates are shown in Equations (2)–(4):

$$\rho g_x+\frac{\delta\sigma_{xx}}{\delta_x}+\frac{\delta\sigma_{yx}}{\delta_y}+\frac{\delta\sigma_{zx}}{\delta_z}=\rho\left[\frac{\delta u}{\delta t}+u\frac{\delta u}{\delta x}+v\frac{\delta u}{\delta y}+w\frac{\delta u}{\delta z}\right] \quad (2)$$

$$\rho g_x+\frac{\delta\sigma_{xy}}{\delta_x}+\frac{\delta\sigma_{yy}}{\delta_y}+\frac{\delta\sigma_{zy}}{\delta_z}=\rho\left[\frac{\delta v}{\delta t}+u\frac{\delta v}{\delta x}+v\frac{\delta v}{\delta y}+w\frac{\delta v}{\delta z}\right] \quad (3)$$

$$\rho g_x+\frac{\delta\sigma_{xz}}{\delta_x}+\frac{\delta\sigma_{yz}}{\delta_y}+\frac{\delta\sigma_{zz}}{\delta_z}=\rho\left[\frac{\delta w}{\delta t}+u\frac{\delta w}{\delta x}+v\frac{\delta w}{\delta y}+w\frac{\delta w}{\delta z}\right] \quad (4)$$

These equations are general differential equations of fluid motion. In fact they can be applied to every continuum (solid or fluid) when moving or stationary [1].

When the combustion process involves heat transfer in the fluid or solid, the heat transfer equation needs to be changed into the form of Equation (5).

$$\frac{\delta ph}{\delta t}+\frac{\delta puj^h}{\delta xj}=\frac{\delta p}{\delta t}+u_j\frac{\delta p}{\delta xj}-\frac{\delta j^h j}{\delta xj}+\frac{\delta ui\tau ij}{\delta xj}+\rho w_h \quad (5)$$

Equation (6) is required for systems that involve a chemical reaction of migratory species.

$$\frac{\delta Y\alpha}{\delta t}+\rho u_i\frac{\delta Y\alpha}{\delta xj}=-\frac{\delta j^\alpha j}{\delta xj}+\rho w_\alpha \qquad (\alpha = 1,2,...,N) \quad (6)$$

Solution of Equation 6 requires modelling. Eddy dissipation models (EDM) is one combustion models developed to predict the combustion reaction of gas in a turbulent flow. The combustion reaction is considered as a single reaction that is not reversed at a limited rate. EDM assumes that the reaction proceeds very quickly. When the reactants are mixed at the molecular level, in real-time directly form the product, so that the overall reaction rate is controlled by turbulent flow. The weakness of this model lies in the assumption that the combustion reaction following the one-step reaction that is not reversible. Consequently, it does not enable the combustion reaction kinetics to be completely incorporated into the calculation. Although this model is widely used in CFD commercial software, it is only appropriate for cases where the kinetics do not play an important role, but so far is the use of this model can still be promising for prediction accuracy [1,2].

2.2. Geometry Configuration and Numerical Computation

This study starts with depicting the geometry of the flame being investigated using the Gambit software (Geometry and Building Intelligent Mesh Toolkit). The procedure is carried out drawing a cylinder in three dimensions for the nozzles and domain, followed by the process of meshing or dividing objects into small parts in order to perform the analysis on the computational fluid dynamics program. The diameter of the nozzle to issue the fuel is 2.0 mm, while the diameter of the domain is 0.7 m. Other basic data with regard to fuel are shown in Table 1. Defining the boundary conditions is performed as the velocity inlet, outlet pressure, and the interior wall. Figure 1 shows the geometry of the three-dimensional flame that has a discretized domain.

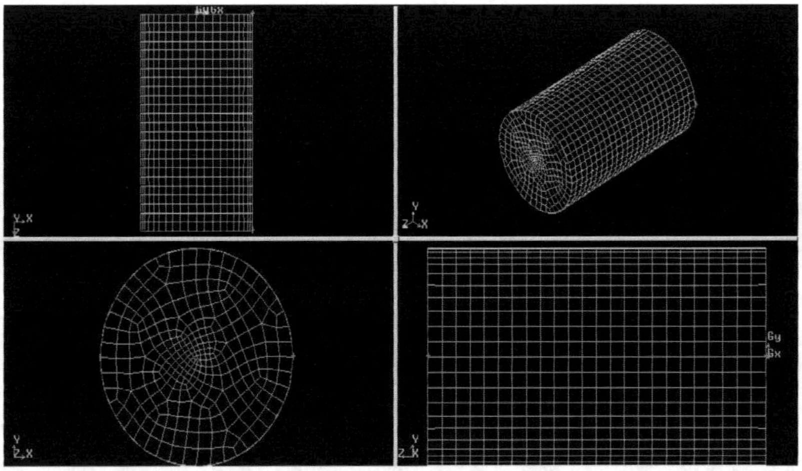

Figure 1. Three-dimensional geometry flame domains.

Calculation of flow and reactive fields, continuity, momentum, energy and species transport equations was solved by commercial CFD code Fluent 6.2.16. With regard to flow field calculation, a standard k - ε turbulence model was employed to solve the momentum transport equation. As the aim of the research to study the influences of crosswind and additional steam to the fuel, the crosswind speed was varied as 0, 1, 3.77, 7.5 and 10 m/s while the mass ratio of steam to fuel (S) was varied as 0.14, 0.25, and 2.35.

Table 1. Characteristics of the flame.

Fuel composition	Propane 100%
Speed Fuel	20 m/s (Re 8800)
Nozel Diameter	2 mm
Domain Diameter	0.7 m

The effects of crosswind speeds and the addition of steam to the flame were analysed to determine the efficiency of the flame combustion process. The combustion efficiency (η_c) is the mass flow rate of carbon in the form of CO_2 produced by flame divided by the mass flow rate of carbon in the fuel, as shown in Equation (7).

$$\eta_C = \frac{\text{Mass flowrate of carbon in the } CO_2 \text{ produced by flame}}{\text{Mass flowrate of carbon in the } C_xH_x \text{ in the fuel gas stream}} \quad (7)$$

Table 2 presented mixture of steam and fuel, starting from pure fuel to give the ratio of steam to fuel, S = 0. The amount of steam to mix with propane fuel was increase as such the maximum S value was 2.35.

Table 2. Species mass fraction flow fuel for various values of S.

S	X_{C3H8}	X_{N2}	X_{H2O}
0	1	0	0
0.14	0.8	0.09	0.112
0.25	0.72	0.1	0.18
2.35	*0.18*	0.39	0.423

3. Results and Discussion

3.1. Flow field profile

Figure 2 illustrates the flow field profile in terms of axial velocity of fuel at various crosswind speeds. The highest fuel speed can be seen at a position close to the surface of the nozzle; the speed decreases with increasing distance towards the flame. Finally, flow rate decreases with increasing distance of flame area. Fluid velocity increases not only at the tip of the nozzle, but also on the downstream flow of the output of the nozzles. This phenomenon takes place because the high cross wind speeds mixing of fuel and air occurs very quickly at the base of the flame, where there is a vortex motion that moves the incoming air into the missing tip of the flame, so that the flame bends towards the direction of the cross wind. The volume of incoming air affects the magnitude of the flame's height.

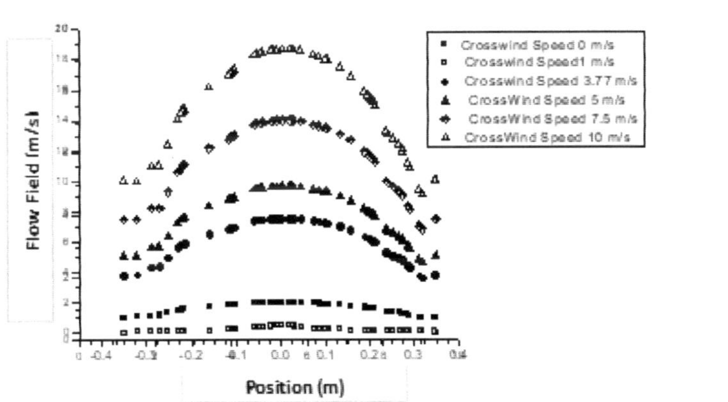

Figure 2. Flame flow field (axial velocity) against axial position for a variety of cross wind speeds.

3.2. Influences of additional steam for a variety of crosswind speeds

Figure 3 shows the relationship between the crosswind speeds and the combustion efficiency at different S values. The combustion efficiency was obtained in accordance with Equation 7. The trend of three curves of different S values is similar, the combustion efficiency tends to decrease with the increase of crosswind speeds. However, at the highest value of S (2.35), the combustion efficiency decreases sharply with respect to the increase of crosswind speeds. This finding support the results of previous researches [8-10]. At lower values of S (0.14 and 0.25), the combustion efficiency was almost stable up to the crosswind 7.5 m/s, and thereafter decreases slightly significant. This finding indicates that at small ration of steam to fuel, combustion efficiency is still high until particular crosswind speed (7.5 m/s). Having a fuel speed of 20 m/s is capable of suppressing the influence of crosswind flow of 7.5 m/s in speed. However, the fuel speed of 20 m/s is no longer able to maintain a stable combustion efficiency at higher crosswind speeds. Results obtained from this study indicate that the mass ratio of steam to fuel and crosswind speed play an important role in calculating the efficiency of the combustion. The highest efficiency of combustion obtained here is 99%.

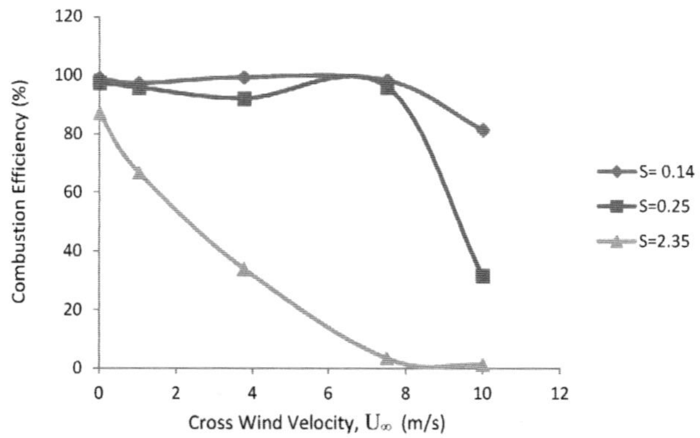

Figure 3. Relationship between crosswind speed and combustion efficiency.

Crosswind speeds also greatly affected the efficiency of combustion, as seen from Figure 3. At higher speeds, the efficiency decreases significantly, which mainly due to incomplete combustion. At higher crosswind, the flame tends to bend towards wind direction and almost extinct. As in the case of value of S = 2.35, imperfect combustion took place, as indicated by low resulting combustion efficiencies even at lower crosswind speeds. At higher crosswind speeds, greater than 7.5 m/s, the flame was extinguished and consequently the combustion efficiency reached zero. This suggests that the addition high portion of steam is not recommended for burning hydrocarbons.

3.3. Flame contour temperature profile

Figure 4 shows the flame temperature contours that are generated at various cross wind speeds and for different mass ratios of steam to fuel. Differences in colour on the contour image reflects the different temperature values generated by each flame. From Figure 4, it can be seen that the value of S at various cross wind speeds greatly affects the combustion temperature. The flame temperature decreases with increasing amount of steam supplied in the fuel flow. At a large value of S (2.35), the flame was almost extinguished, resulting in the release of hydrocarbons into the atmosphere instead of combusting the fuel. Linking this contour temperature with Figure 3, now it is much easier to understand that at high value of S efficiency of the combustion is low.

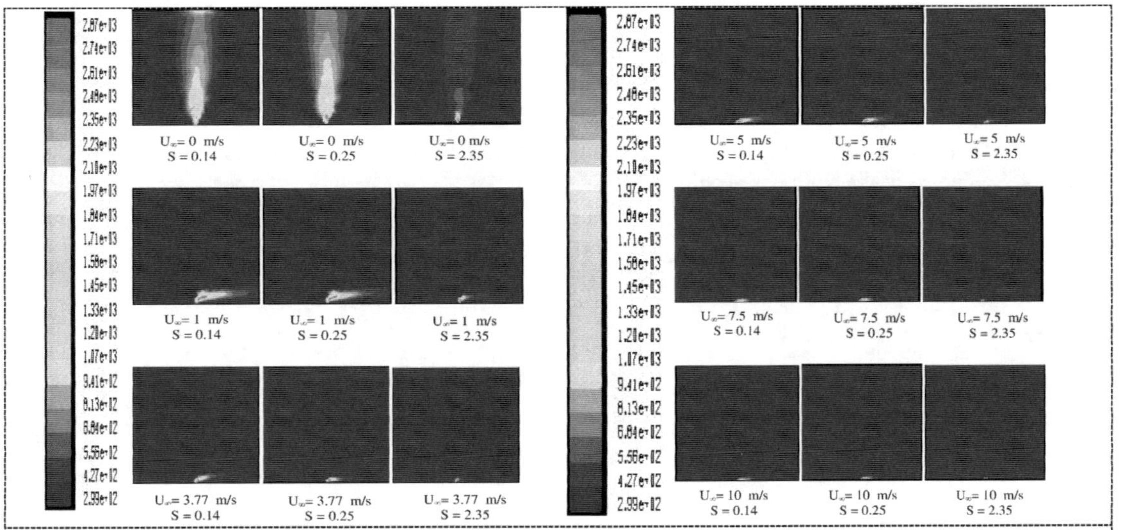

Figure 4. Flame contour temperature at crosswind speed 0 , 1 3.77, 5.0, 7.5 and 10 (m/s) and value S (0.14 , 0.25 and 2.35).

4. Conclusion

A CFD study of a propane non-premixed flame and the impact of crosswinds as well as additional steam have been presented using the k-ε standard for the turbulent model, and eddy dissipation for the combustion model. The results of CFD study showed The addition of a quantity of steam to a combustion system influenced by crosswind lowered the flame temperature. As the temperature getting lower, the combustion efficiency also decreases. The study needs to be enhanced to further investigate the impact of steam addition on soot and NOx production by a flame.

Acknowledgement

The authors would like to acknowledge the Ministry of Research, Technology and Higher Education for financial support.

References

[1] Castineira D and Edgar T F 2008 *Energy Fuel 22*: 1698–1706
[2] Wusnah, Yunardi, Sylvia N, Elwina and Bindar Y 2014 Proceedings of the Annual Conference International, Unsyiah., 109-114
[3] Fairweather M, Jones W P and Lindstedt R P 1992 *Combustion and Flame* 89: 45–63
[4] Sinai Y L and Owens M P 1995 *Fire Safety Journal,* 1–34
[5] Johnson M R, and Kostiuk L W 2000 *Efficiencies of Low-Momentum Jet Diffusion Flames in Crosswinds* (Combustion and Flame 123), 189–200
[6] Nmira F, Consalvi J L, Boulet P and Porterie B 2010 *Fire Safety Journal* 45: 129–141
[7] Cole W J, Dennis M H and Fletcher T H 2011 *International Journal of Wildland Fire* 20: 657–667
[8] Kuipers E W, Jarvis B, Bullman S J, Cook D K and McHugh D R 1996 *Internal Report, Shell Research and Technology Thornton and British Gas Research Centre.*
[9] Bandaru R V and Turns S R 2000 *Combustion and Flame* 121:137–151
[10] Boushaki T, Dhue Y, Selle L, Ferret B and Poinsot, T 2012 *International Journal of Hydrogen Energy*, 37(11) , 9412–9422
[11] Mazas A, Lacoste D, Fiorina B and Schuller T 2009 Proceedings of the European Combustion, 1-6

[12] Zhao D, Yamashita H, Kitagawa K, Arai N and Furuhata T 2002 *Combustion and Flame* 130: 352–360
[13] Xue R, Hu C, Sethib V, Nikolaidis T and Pilidis P 2016 Applied Thermal Engineering 104: 249-257
[14] Yunardi, Wusnah, Darmadi, Hisbullah, Yazid Bindar 2017 International Journal on Advanced Science Engineering and Technology 7: 86-91

IOP Conf. Series: Materials Science and Engineering **334** (2018) 012024 doi:10.1088/1757-899X/334/1/012024

Performance of cellulose acetate membrane with different additives for palm oil mill effluent (POME) liquid waste treatment

N A S Aprilia[1,3], Fauzi[1], N Azmi[1] and N Najwan[1] and A Amin[2]

[1] Department of Chemical Engineering, Faculty of Engineering, Universitas of Syiah Kuala, Darussalam, Banda Aceh, 23111, Indonesia
[2] Department of Mechanical Engineering, University of Abulyatama, Lampoh Keude, Banda Aceh, Indonesia.

E-mail: sriaprilia@unsyiah.ac.id

Abstract. Performance of cellulose acetate membrane for treatment of POME liquid has studied with different additives. Cellulose acetate membranes were prepared with different additive ie formamide and polyethylene glycol and used acetone as solvent. The function of formamide and polyethylene glycol (PEG) is to increase the porosity of the membrane surface. Performance of the membrane were included SEM, FT-IR and coefficient permeability. Membrane performance has been performed for percent rejection of total suspended solid (TSS) and turbidity of POME liquid waste. Cellulose acetate with formamide shows an increased percentage of rejection in removing TSS and turbidity than cellulose acetate with PEG.

1. Introduction

Indonesia still used aerobic siystem for processing liquid waste of palm oil industry. Liquid waste is channeled to the collection unit (fatpit) in the form of a trench and then flowed into the collecting oil (oil quote pool) for the oil to be taken and the temperature drop. After that liquid waste is fed into the aerobic treatment pond. The system is very simple and cheap. But this system has a disadvantage: the land used must be wide, when using a closed system can produce greenhouse gases such as methane and carbon dioxide. Processing this system is not economical because once processed, the waste directly discharged into the water channel.

One of the technologies selected for POME's liquid waste treatment is ultrafiltration membrane technology. This technology does not require a lot of energy and large land. The ultrafiltration membrane is an asymmetric membrane, in which the dense top layer of pore and the lower layer has larger pores which have good mechanical strength. Aprilia (2011) [1] stated that the most important goal in membrane technology is to control the membrane structure that will affect membrane performance. The formation of

[3] To whom any correspondence should be addressed.

asymmetric membrane pores are influenced by several factors, including the use of solvents, non-solvents and additives. A very serious problem in the membrane process is fouling, because the adsorption of the solute on the membrane surface will clog the membrane pores (pore bloking).

Some literature has studied blending a polimer with hydrophilic materials or changed additives. This is considere to be an attractive and simple method to modify the membrane properties in order to minimize fouling. Aryani et al (2016) [2] has used PEG400 and acetone as additives on polysulfone membranes. They investigated that the used PEG400 and aceton in polysulfone membrane will increase the hydophylicity and in a lower fouling will increasing the resistance of organic matter during up to five hours of peat water filtration. To reduce fouling in the membrane system, it is need to add additives as anti fouling. The addition of addtives in membrane hydrophobicity such as in polysulfone, polyether sulfone, polypropylene, polyethylene can be fabricated by increasing the outer surface porosity and removing the inner skin layer during spinning process. Moreover, despite depending on the property of the membrane material, it can also be adjusted by the membrane pore size [3].

The objective of the research was to treat the POME by using performance of cellulose acetate membrane with additive PEG and acetone in removing TSS and tubidity. The characterization of the membrane include SEM, FT-IR and permeability coefficients.

2. Experimental procedure and methodology

2.1. Materials

The type of polymer that has been used in this experiment are cellulose acetate supply by Sigma-Alderich, acetone as solvent, PEG and formamide as additives. As liquid waste was used the POME from PT Surya Panen Subur II Kabupaten Nagan Raya, Aceh Province.

2.2. Membrane preparation

Membrane preparation was carried out by phase inversion method. The cellulose acetate was dissolved into the acetone and then mixed with the additives PEG and formamide at room temperature. The solution was stirred for 8 hours and save in the refrigerator for 2-3 hours until no bubbles were observed. Then, the cellulose acetate solution was casted on a flat glass plate and immediately immersed in a destilled water bath as coagulant. Then the membrane followed by annealing process, and then immersion in hot distillate water (70°C) for 20 minutes. The variations of casting cellulose acetate solution used in this experiment are shown in Table 1.

Table 1. Experimental design of cellulose acetate solution

Membrane code	Composition of cellulose acetate solution			
	Cellulose acetate	Acetone	PEG (%)	Formamide
CA-1	15	85	5	-
CA-2	15	75	-	10

2.3. Characterization of membrane

2.3.1. SEM. The prepared membrane was characterized by scanning electron microscopy (SEM) to obtain visual information of the top surface structure and geometric pore. Surface of the membrane sample was cuted to 0.5 x 0.5 cm and then the sample is attached to the sample plate (brass disk) and deposited with gold using a sputter coater. Then, the sample was imaged using a scanning electron microscope (low vacuum) with a magnification of 5000 times

IOP Conf. Series: Materials Science and Engineering **334** (2018) 012024 doi:10.1088/1757-899X/334/1/012024

2.3.2. FT-IR. Analysis of Fourier Transform Infrared Spectroscopy (FT-IR) Spectrometer employed to characterize the membrane in absorption spectra which provide information about the surface organic structure. The analysis was used by using a FT-IR spectroscopy (Shimadzu FT-IR Spectrometer). The sample cut into 1 x 1 cm then was put into the Nicolet Avatar FT-IR Spectrometer. Spectra were recorded in the range 4000-400cm^{-1}.

2.3.3. Coeficient permeability membrane. Permeability is carried out by permeation experiments with pure water. The membrane flux is obtained from each pressure difference. Membrane permeability was obtained from the graphical slop between pressure and flux.

2.4. Test the performance of membrane

The membrane performance test was performed by passing the POME liquid through the ultrafiltration module at a pressure of 2 bar. POME liquid waste was performed before membrane process, by filtering out suspended particles. The membrane rejection was calculated based on TSS and turbidity.

3. Result and Discussion

3.1. Characterization of the membrane

3.1.1. Effect of additives at SEM analysis. Surface structure of the membrane blengding with PEG and formamide (CA-1 and CA-2) has investigated by SEM in Figure 1. The CA-1 membrane (Figure 1A) provides a uniform pore structure and spread more evenly on the surface compared to the CA-2 membrane (Figure 1B). Membrane CA-2 the structure surface more smooth. This suggests that PEG have a major effect on the formation of pores rather than formamide.

The addition different additives into membranes influent the pores formation in the membranes structure [2]. Ramos-Olmos et al., (2008) [4] also employed that the different additives will increase the voids size. The SEM images indicated that the membranes have asymmetrical structure with a skin layer at the top, intermediate layer and a bottom layer [4]. The different additives displayed improved characteristics of the membranes such as porosity optimum, large pore size and thin membrane thickness coupled with finger-like structure extended from both inner and outer layers of the membrane [5]. All of condition could be observed by SEM cross section of the membranes. But the cross section images does not imply in this study.

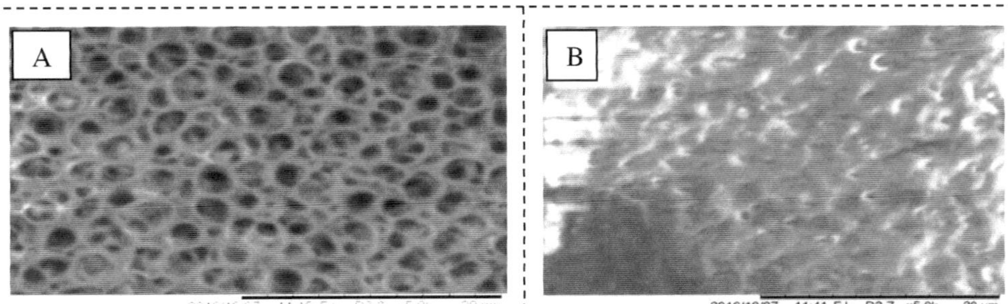

Figure 1. SEM images of the cellulose membrane with additives on the surface membranes (A) PEG and (B) formamide.

Additives in the membrane will provide large pores compared to membranes without additives. This is shown that additing PEG as addtive will increase membrane pore, while maintaining membrane resistance to external factors [6]. Rosnelly (2012) [6] has imployed that high flux on the membrane with the addition of PEG compared to the membrane without PEG. The higher concentration of PEG, the larger pore size form of the membrane. This is because the addition of PEG is homogeneous into the polymer and the solvent. In the process of immersion into PEG coagulation tubs will also diffuse into non-solvent. The more PEG that dissolves into water the larger of pore size is generated and the PEG additive is scattered on the membrane surface and also increases the porosity of the membrane [7].

3.1.2. Effect of additives at FT-IR analysis. FT-IR spectroscopy is used to highlight differences in the functional groups effect of additives PEG and formamida. The effect of different additives on cellulose acetate membrane is used to determine the compounds present in the membrane. The effect of additives on the structure membrane can be seen in Figure 2.

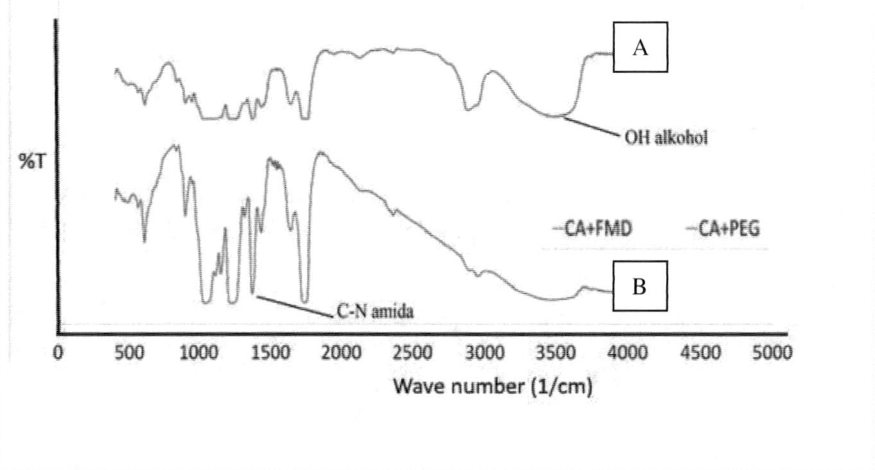

Figure 2. FT-IR spectra of celulose acetate membranes with formamide and PEG.

The presence of carbonyl groups in Figure. 2 shows that cellulose acetate and acetone in the membrane (Figure 2. A and B). In Figure 2A with a wavelength of 3483.44 cm^{-1} representing the absorption of the wavelength contains the OH functional group. In the presence of the group identifies (-OH) the presence of PEG additives on the cellulose membrane. In Figure 2B there is a C-N amide group at wavelength 1180-11360 cm^{-1}. In the presence of CN-amide on the FTIR spectrum identifying the presence of a formamide additive contained in the membrane.

3.1.3. Coefficient permeability membranes. The permeability coefficient (Lp) is the membrane's ability to pass water (aquadest) on the basis of pressure increase [1], where flux is determined by passing aquadest under operating pressure conditions. Lp of the membrane is obtained from the graph slope to the operating pressure [8]. Table 2 show the result of Lp.

Table 2. Coefficient permeability of the membrane.

Type of membrane	Lp ($L/m^2.h.bar$)	Flux ($L/m^2.h$)
CA-1	17.95	23.32
CA-2	9.21	19.40

Base on permeability of the membrane (Lp), ultrafiltration membrane has a Lp range of 10-50 L/m^2.h.bar. and the nanofiltration membrane has a Lp range 1.4-12 L/m^2.h.bar. From Table 2, type of membrane CA-1 with additives PEG in range ultrafiltration membrane but of CA-2 with additive formamide in range nanofiltration membrane.

From the flux it is also known that the CA-1 membrane with PEG additive has a higher flux value than the CA-2 membrane with formamide additives. The second flux value of the membrane can be seen in Table 5. The flux value on the CA-1 membrane is higher than of CA-2. This indicates that the surface pore size is larger [9]. Increased pores result in the flow of water more easily pass through the membrane. Aprilia et al (2013) [10] mentions that membranes with a dense pore structure will have greater mass transfer resistance, so water permeability becomes smaller.

3.2. Performance of the membranes in POME treatment

Flux is one of parameter that is measured in deriving parameters on POME. Table 3 show the fluxes of the two membranes. Flux membrane CA-1 larger than membrane CA-2. The effect of PEG additives on cellulose acetate membranes provides a larger pore size than that of formamide additives in membrane.

Table 3. Fluxes of the membrane in for POME separation.

Membrane Type	Flux (L/m^2.h)
CA-1	8.30
CA-2	7.55

The result of research on membrane performance as an additive influence on decreasing parameter of POME such as turbidity and total suspended solid (TSS) has been done. Turbidity is a parameter used to indicate thelevel turbidity of a waste. In this research, the value of POME turbidity before separation is 4.85 NTU. After the separation process of POME turbidity is successively CA-1 and CA-2 membranes are 3.28 and 2.62 NTU. The rejection calculation obtained turbidity rejection value as in Table 4. From the Table 6 showed that CA-1 membrane capacity gave a greater percentage of yield than the CA-2 membrane.

Table 4. Rejection of TSS and turbidity of the membrane.

Type of membrane	Percent of rejection (%)	
	Turbidity	TSS
CA-1	59.79	38.46
CA-2	63.09	53.85

From the membrane characterization results, the CA-2 membrane has more dense pores, and the ability to pass permeate is better than the CA-1 membrane. So the percent of rejection for TSS and turbidity on the CA-2 membrane is greater than that of the CA-1 membrane. According to Sofyana et al (2011) [11]. The value of rejection is inversely proportional to the value of flux obtained. The higher the value of flux, the value of rejection obtained will be lower. This is because the larger the POME flowing through the membrane, the resistance on the membrane surface becomes smaller.

4. Conclusion

Based on the results of the study can be concluded that SEM analysis shows the additive variation on the cellulose acetate membrane makes the membrane have different pore structure. The addition of PEG additive makes the distribution and pore size of the membrane more evenly compared to the additive of formamide. The CA-1 membrane provides good membrane performance over the CA-2 membrane in terms of flux and the CA-2 membrane performs better than the CA-1 membrane in terms of TSS and turbidity injection percent of POME. Cellulose acetat membrane with PEG additive is better than formamide additives.

References

[1] Aprilia S and Amin A 2011 *Jurnal Rekayasa Kimia dan Lingkungan.* **8**(2): 84-88.
[2] Aryanti P T P, Joscarita S R, Wardani A K, Subagjo S, Ariono D, and Wenten I G 2016 J. Eng. Technol. Sci., **48** (2).
[3] Mansourizadeh A, and Ismail A F 2011 International Journal of Greenhouse Gas Control **5**(2):374-380
[4] Ramos-Olmos R, Rogel-Hernandez E, Flores-Lopez L Z, Lin S W and Espinoza-Gomez H 2008 Journal of the Chilean Chemical Society, **53** (4).
[5] Chong K. C, Lai S O, Lee K M, Lau W J, Ismail A F and Ooi B S 2015 Desalination and Water Treatment **54**, 3218–3226
[6] Rosnelly C M 2012 P *Jurnal Rekayasa Kimia dan Lingkungan.* **9**(1): 25-29.
[7] Utami Z D, Fansuri, Hamzah 2014 *Jurnal Sains dan Seni Pomits.* **2**(1): 1-4.
[8] Lindu M, Puspitasari T and Ismi E 2008 *Jurnal Sains Materi Indonesia.* **4**(4): 107-117.
[9] Murni S W and Sudarmi S 2010 *Jurnal Eksergi.* **10**(2): 36-41.
[10] Aprilia S, Arifin B, Anugerah R R and Safriadi I 2013 *Jurnal Rekayasa Kimia dan Lingkungan.* **9**(4): 173-179.
[11] Sofyana, Mulyati S, Adly F W and Maghfirah Z 2011 *Jurnal Hasil Penelitian Industri.* **24**(1): 28-35.

Adsorption and kinetics study of manganesse (II) in waste water using vertical column method by sugar cane bagasse

H Zaini[1,3], S. Abubakar[2], T Rihayat[1] and S Suryani[1]

[1] Chemical Engineering, Politeknik Negeri Lhokseumawe
[2] Electrical Engineering, Politeknik Negeri Lhokseumawe

Email: halimzaini60@gmail.com

Abstract. Removal of heavy metal content in wastewater has been largely done by various methods. One effective and efficient method is the adsorption method. This study aims to reduce manganese (II) content in wastewater based on column adsorption method using absorbent material from bagasse. The fixed variable consisted of 50 g adsorbent, 10 liter adsorbate volume, flow rate of 7 liters / min. Independent variable of particle size with variation 10 – 30 mesh and contact time with variation 0 - 240 min and respon variable concentration of adsorbate (ppm), pH and conductivity. The results showed that the adsorption process of manganese metal is influenced by particle size and contact time. The adsorption kinetics takes place according to pseudo-second order kinetics with an equilibrium adsorption capacity (q_e: mg / g) for 10 mesh adsorbent particles: 0.8947; 20 mesh adsorbent particles: 0.4332 and 30 mesh adsorbent particles: 1.0161, respectively. Highest removal efficience for 10 mesh adsorbent particles: 49.22% on contact time 60 min; 20 mesh adsorbent particles: 35,25% on contact time 180 min and particle 30 mesh adsorbent particles: 51,95% on contact time 150 min.

1. Introduction

The existence of metals in the environment comes from two sources. First of all natural processes such as chemical weathering, geochemical activities, decaying plants and animals. Other sources are derived from the activities of chemical and other industries. One of the most important metals for humans is manganasse needed as nutrient and the body needs 10 mg of the manganese that can be obtained from food. If the levels are excessive this element is toxic to the respirator. In the water for public consumption problems that often occur also concerns the color of water. Manganasse concentrations greater than 0.5 mg / liter can cause a strange taste in the drink and leave a brown color on the laundry and can also cause damage to the liver.

According to **[1]** the composition of the chemical elements of ground water depends on the layers of land passed. If through lime soil, then the water will become a container because it contains $Ca(HCO_3)_2$ and $Mg(HCO_3)_2$.If through granite rocks then the water is soft and aggressive because it contains gas CO_2 and $Mn(HCO_3)_2$. Sludge Mn will give stains on the material / objects are white. The existence of this element can cause smell and taste in the drink. Besides, the concentration of 0.5 mg / liters of this element is the result of an air barrier that can be achieved [2]. Furthermore, Iron and manganese are dissolved in water in the formof Fe^{2+}, Mn^{2+}, in an insoluble state in the form of $Fe(OH)_3$, $Mn(OH)_4$ and can also be in colloidal form bonded to humic acid [3]

[3] To whom any correspondence should be addressed.

Adsorption generally occurs based on the interaction between metal and functional groups present on the surface of the adsorbent through ion exchange interaction or complex formation, usually occurring on solid surfaces containing functional groups such as -OH, -NH, -SH and COOH. According to **[4]** the components involved in the adsorption process between heavy metals and adsorbents of agricultural waste are the presence of active hydroxyl groups (-OH), carbonyl (C = O), carboxyl (-COOH), amines (-NH2), amides ($-CONH_2$) and thiols (-SH). There are several factors that influence metal adsorption process that is contact time, particle size of adsorbent, flow rate, activation, metal type.

Many methods have been developed to reduce the heavy metal content present in water, such as sedimentation, vaporization, coagulation, electrochemical, electrophoresis and ion exchange, filtration. However, all of these methods are considered inefficient because they require substantial funds and are less effective in their operations [5].

One reasonably good method is that the adsorption method can be adsorption by batching and batch system adsorption with a column system that operates continuously using single, multiple columns. In parallel or in series column system. When compared to these last two methods, the column system is much better than the stirring system. The column system does not require a large space, the surface of the tool used is relatively small, while the stirring system spans the place and the wider space. The stirring system still requires further separation steps such as filtration, without going through the filtration stage of the separation system has not occurred despite adsorption already occurring. In the column system of separation of solids in liquids occurs at once with the process of separation.

In addition there is a difference in the size of the adsorbent particles used. The stirring system of particle size can be the same or smaller than the column system whereas the particle size system is larger. If a larger or smaller particle size is used the absorption process becomes ineffective.

2. Experimental procedure and methodology

2.1. Materials

This research used adsorbate from waste water (industrial waste water) containing Pb, Ca, Mn and Fe metals. The adsorbents are from bagasse with particle size of 10 mesh, 20 mesh and 30 mesh. This type of adsorbentt consists of non-activated adsorbents.

Sugarcane bagasse were collected and used as sorbent for the biosorption of Mn(II) ions. The sugarcane bagasse sample was collected from traditonal market around Lhokseumawe city. Samples were washed several times using reverses osmosis water to remove impurities,extraneous and dust. The samples then dried in sun dried for five days and in an oven at 105°C for 2 h. Finally, the dried bagasse was sieved to get the average adsorbent size of 10 mesh, 20 mesh and 30 mesh. The sieve sugarcane bagasse was then stored in a clean airtight plastic container and keep in desiccator, until the time of use.

2.2. Methods

The adsorption process in the vertical columns with a diameter of 6.35 cm and the height of the blank condition column of 38 cm. Each of the columns was inserted with absorbent 10 mesh, 20 mesh, 30 mesh (50 g).

The study was conducted with fixed variable particle size of 10 mesh, 20 mesh and 30 mesh. Flow rate 7 liter / min, 10 liter adsorbat volume, operating temperature of room temperature 30 °C. Variable free contact time 0; 30; 60s; 90; 120; 150; 180; 210 and 240 minutes. Response variables of remaining adsorbate concentration in solution (ppm), absorbed adsorbate concentration (ppm), adsorption capacity (mg / g),%removal.

In the adsorption process the amount of metal adsorbed in the adsorbent for contact time t is expressed by the equation:

$$q_t = \frac{(Co - C_t)V}{m} \quad (1)$$

Where:
C_o: Initial concentration of adsorbate (mg / l)
C_t: concentration at t (min)
V: the volume of adsorbate (liter)
q_t: adsorption capacity (mg / g)
Percent Removal (%) of metal into adsorbent is expressed by the following equation:

$$\% R = \frac{(Co - Ct)}{C_o} x100\% \quad (2)$$

Sample of analysis of data collected was analyzed using instrumentation method with Atomic Absorption Spectrophotometer (AAS).

3. Result and Discussion

3.1. The concentration of Mn in equilibrium (Ct)

Changes in the concentration of ions Mn^{+2} (Figure 1) over the contact time show the existence of differences with each other. This difference is influenced by the size of the adsorbent particle size may affect the absorption of Mn. The highest absorption occurs in adsorbents whose particle size is 30 mesh, then 10 mesh and last 20 mesh. Small particle sizes such as the 30 mesh adsorbent size in Mn absorption give better results compared to the 10 mesh and 20 mesh sizes. With a column system on 30 mesh particle size provides the best removal ability. Mn content in well water and drilling wells ranged from 0 to 5.26 ppm and drinking water quality standard of 0.5 ppm. Manganese (Mn), whose maximum requirement is 0.015 mg / l for drinking water and 0.5 mg / l for clean water [6].

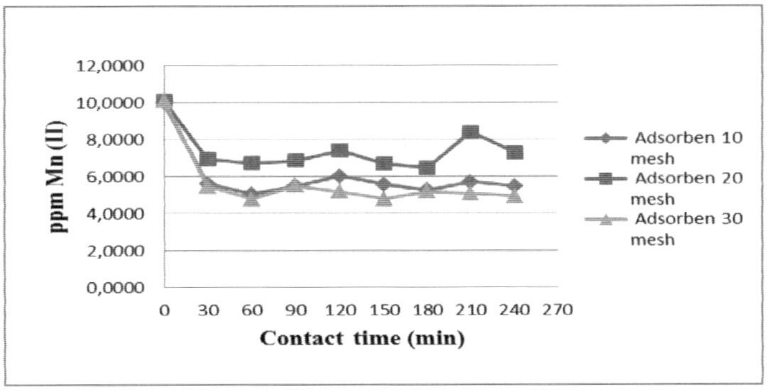

Figure 1. Effect of Contact Time on Mn (ppm) in adsorbate.

Based on the data in table 2 and figure 2 the amount of Mn^{+2} metal ions absorbed into the adsorbent shows that the adsorption equilibrium occurs at 150 minutes, the magnitude of the time of the adsorption capacity is relatively the same. The best adsorption capacity occurs in adsorbents with 30 mesh particle sizes, then for the particle size of 10 mesh and last the particle size of 20 mesh. The difference in absorption occurs between each of the adsorbents aside from the particle size, also due to the porosity of each particle. The adsorbent porosity of a particle size of 30 mesh is better than that of the adsorbent whose particle size is 10 mesh and 20 mesh. Adsorption capacity of highest adsorbent bagasse 1.00 mg / g for 10 mesh adsorbents, 0.7 mg / g for 20 mesh and 1.10 mg / g adsorbents for 30 mesh adsorbents respectively at contact time of 60 minutes

Comparison of bio-adsorbent chemically activated camphor dyers of HCl resulted in the largest adsorption capacity of 1.698 mg/g [7]. The adsorption capacity of maize adsorbent to Mn metal in groundwater is 0.043 mg/g.

3.2. Capacity adsorption Mn on to SCB

The adsorption capacity of bagasse is higher than the capacity of corncob adsorption without activation (**Figure 2**) . The adsorption capacity of the bagasse adsorbent is not activated under the capacity of chemically activated camphor adsorbents with HCl. Biosorption processes tend to occur rapidly, taking form few minutes to a couple of hours and it takes account transfer mass processes and adsorption processes. The biosorption kinetics is controlled mainly by convective and diffusive processes. In a first stage occurs the metal transference from solution to adsorbent surface neighborhood; then in the next step, the metal transference from adsorbent to intraparticle active sites; and finally, the metallic ion removal by the active sites via complexation, adsorption, or intraparticle precipitattion. The first and second steps represent the resistance to convective and diffusive mass transferences and the last one is quick and non-limiting for the overall biosorption velocity.

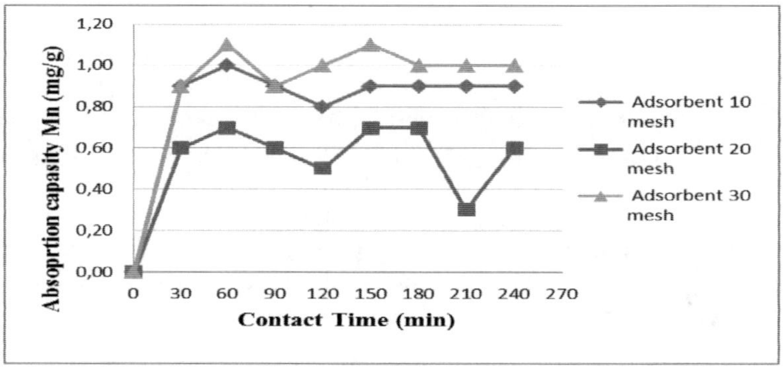

Figure 2. Effect of Contact Time on Capacity Adsorption Mn Onto Sugarcane Bagasse.

Analogously to the biosorption isoterms, the biosorption kinetics in general present hyperbolic shape and they are described by various models. The models more used in biosorption studies are present on Table 3.

Table 3. Kinetics Adsorption Model

Pseudo-first-order	Pseudo-second-order
Differential equation $\frac{dq_t}{dt} = k_1(q_e - q_t)$	$\frac{dq_t}{dt} = k_2(q_e - q_t)^2$
Integral equation $\ln(q_e - q_t) = \ln q_e - k_1 t$	$t/qt = 1/k_2 q_e^2 - t/q_e$
Initial adsorption velocity $v_1 = k_1 q_e$	$v_2 = k_2 q_e^2$

Based on the Figure 3 the Lagergren kinetic pseudo-first order equation for each of the adsorbents yields an equation which does not lead to the form of linear equations. This condition is proved by parameter R^2. The equality of the equation is shown if the value of R^2 equals 1, the closer to the value of 1. The value of R^2 is shown in Figure 3, where the value of R^2 is very far from the value of 1, ie for 10 mesh: 0.0242; For 20 mesh: 0.0270 and for adsorbents of 30 mesh: 0.0745. In addition the value of the adsorption capacity (q_e) of the calculation with the adsorption capacity value of the extrapolation results is different as shown in Figure 3.

3.3. Pseudo-First Order Kinetic

Based on Figure 3 the pseudo-second order Lagergren kinetics equation for each adsorbent yields the equation which leads to the form of the linear equation. This condition is proved by parameter R^2. The equality of the equation is shown if the value of R^2 equals 1, the closer to the value of 1. The value of R2 is shown in table 6, where the value of R2 is very far from the value of 1, ie for 10 mesh adsorbents: 0.9935; For 20mesh adsorbents: 0.6664 and for adsorbents of 30 mesh: 0.9915. In addition, the adsorption capacity value (q_e) of the calculation with the adsorption capacity value of the extrapolation results is relatively different.

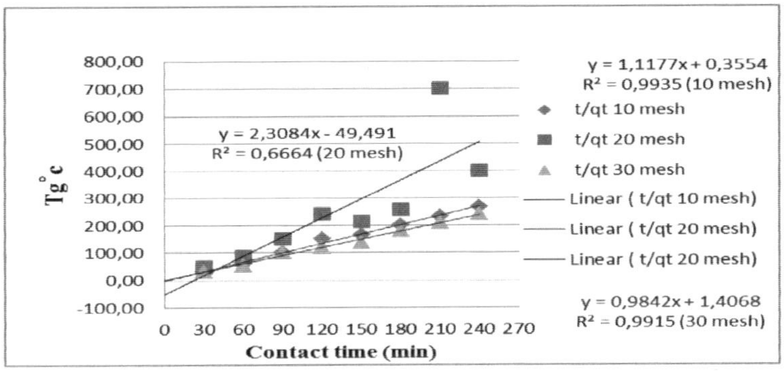

Figure 3. Pseudo-Second Order Kinetic

The rate limiting step of the adsorption was studied by using different kinetic models based on reseach [8], such as pseudo-first-order, pseudo-second-order adsorption equation, elovich equation and the intraparticle diffusion model. Kinetics data was best fitted with the pseudo-second-order type. This method has been successfully applied for reduce the risk of Mn(II) from ground water samples by using SCB adsorbent. Based on Figure 4, the extrapolation results showed significant differences between each adsorbent. This difference is due to differences in particle size. For 10 mesh adsorbents: 0.8947 mg/g; 20 mesh: 0.4332 mg/g and 30 mesh: 1.0161 mg/g.

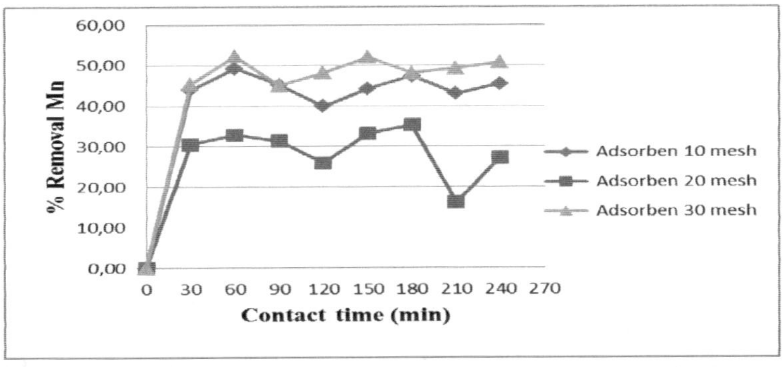

Figure 4. Effect of contact time to efficience removal Mn(II) on to SCB

According to Figure 4, the size of the adsorbant particles on the metal removal of Mn (II) is very influential. The particle size of 30 mesh is the highest separation, then the particle size is 10 mesh and the last is 20 mesh. The highest removal efficiency for each adsorbent was 49.22% for 10 mesh adsorbents at 60 minutes, 35.25% for 20 mesh adsorbents at 180 minutes and 51.95% for 30 mesh adsorbents at 150 minutes.

Esfandiar et al (2014), has studied about sugarcane bagasse (SCB) as biosorbent were used to remove Mn(II). The maximum experimental Mn (II) removal efficiencies is 63%, respectively, were obtained. Additionally, a chemical treatment of SCB had a significant effect on the adsorption, as maximum removal increased up to 99%. The ability to eliminate each of the adsorbents is influenced by the contact time, internal and external properties of the adsorbent.

The use of sugarcane bagasse as adsorbent is considered effective enough to decrease the fatty acid free fat content of frozen vegetable, the biggest decrease 57,3% with particles size 200 mesh particle size 50 g, 2 hours contact time [10, 11, 12]. Adsorption of bagasse which was activated with NaOH for 3 hours to Pb (II) was batched with 76.7% removal ability at 60 minutes. Physicochemical characteristics of waste water.The physicochemical characteristics such as turbidity, alkalinity, hardness, TDS, Chorine and other ions concentration will be measure for waste water.

3.4. Effect of Contact Time to PH Value

The pH of the aquoeus solution is an important operational parameter in the adsorpstion process because it affects the solubility of the metal ions, concentration of the counter ions on the functional groups.

According **[13]** , the potential to remove Zn(II), Cd(II) and Mn(II) form aqueous solution biosorption using maze stalks as an agriculture waste, was investigated ini batch experiments. Different factors influencing metal adsorption such as contact time, initial metal ion concentration, pH, ionic strength and temperature were investigated. The adsorption process was relatively fast and equlibrium was establed after about 90 min. The optimum intial pH for manganase by maize stalks was 5.6

The potential to reduce the risk of Mn(II) from ground water through adsorption using Sugar Cane Bagasse (SCB) as a low cost natural wastes was investigated. The adsorption was taken in batch experiments under different parameters such as pH (2.0-10.0), weight of phase (0.25-2.5 g), contact times (10-180 min), initial concentration (2.0-10.0 ppm). The removal efficiency may be maximum at pH 5 on adsorptsion by using sugarcane bagasse [14].

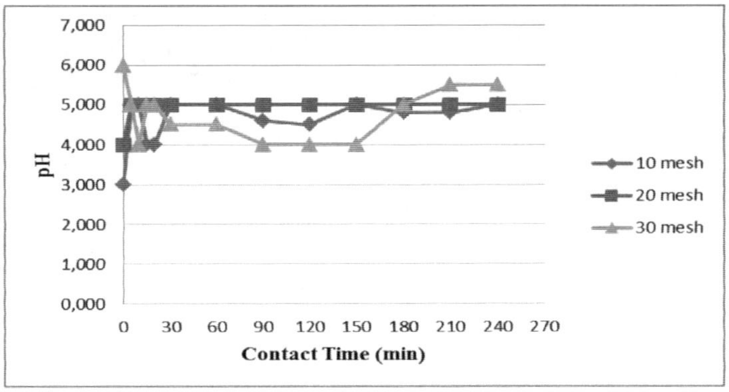

Figure 5. Effect of contact time to pH Value

3.5. Effect of Partikel Size to PH Value

Based on Figure 5, the particle size of the adsorbent affects the pH value of the wastewater solution. However, the pH value that gives a big change is the adsorbent whose particle size is 30 mesh, while for the 10 mesh and 20 mesh particles the pH is relatively stable. In general, the pH value of the solution ranges from 4 to 5.5. This pH value is relatively equal to the pH obtained from manganese adsorption using a corncob adsorbent of 5.6 magnitude.

4. Conclusion

Based on the data of research, data processing, discussion and comparison to previous research can be summarized the following conclusion: The adsorption of manganese (II) metal in wastewater using adsorbent of sugarcane bagasse is affected by contact time and particle size of the adsorbent. Adsorption kinetics of manganese (II) is adsorption kinetics of the pseudo-second order.Adsorption capacity at equilibrium (qe) manganese (II) for particles size 10 mesh adsorbents: 0.8947 mg/g; 20 mesh: 0.4332 mg/g and 30 mesh: 1.0161 mg/g. The highest removal efficiency of manganese (II) metal waste for 10 mesh adsorbents: 49.22% at contact time of 60 minutes; 20 mesh: 35.25% at contact time 180 min and 30 mesh: 51,95% at contact time 150 min.

Acknowledgement

The authors express their gratude and thanks to the Higher Education Ministry for Financial Support through the PPT (Penelitian Produk Terapan) Grant and Ditjen Penguatan Riset dan Pengembangan, Kemenristekdikti

References

[1] Sutrisno,T., and Suciastuti, E. 2010. Water supply technology. Rineka Cipta.
[2] Setiyono, A. 2014.Study of Manganese Content (Mn) In Well Water Dug in Karang Nunggal Village Kabupaten Tasik Malaya. Environmental Health Journal. Vol.10 No.1
[3] Sari, W.K., Karnaningrum, N. 2015. Study of Removal Iron and Manganese Using Cascade Aeroator and Rapid Sand Filter In Well Water Dug. Enveronmental Departement- Institut of Technology Surabaya-Indonesia.
[4] Ahalya, N., Ramachandra, T.V., Kanamadi, R.D.2003.Biosorption of Heavy Metals. Reseach Journal of Chemical and Environment. 7(4),71-79.
[5] Irmawati, A., Ulfin, I. 2013. Utilization of Peanut Skin Biomass For Adsorption of Chromium In Aqueous Solutions By Column Method. Chemistry Departement-Instute of Technology Surabaya.
[6] Munfiah, S., Nurjazuli., Setiani, O. 2013. Physical Quality and Chemical Well Water Drilling and Drilling at Working Area of Puskesmas Guntur II KabupatenDemak.Journal of Environmental Health-Indonesia. Vol 12 No 2, October 2013
[7] Mandasari, S., Purnomo, A. 2016. Removal of Fe Ion and Mn in water with camphor wood sawdust. Technic Journal Instutute of Technology Surabaya. Vol.5 No.1,(2016) ISSN:2337-3539 (2301-9271 Print)
[8] Ahmed,S.,A., El-Roudi,A.,M., Salem, A.,A. 2015. Removal Mn(II) from Gound Water by Solid Wastes of Sugar Industry. http://scialert.net/fulltext/?doi=jest.2015.338.351&org=11
[9] Esfandiar, N., Nasemejad, B., Ebadi, T. 2014. Removal Mn(II) form goundwater by sugarcane bagasse and activated carbon (a comprative study): Application of response Surface Methodology (RSM). Juornal of Industrial and Engineering Chemistry 20 (2014) 3726-3736
[10] Ratno, Mawarani, L.,J., Zulkifli. 2013. The Effect of Sugar Cane As Adsorbent on Jelantah Oil Pretreatment Process on Biodiesel Characteristics.Technic Journal POMITS Vol 2 No 2 (2013) ISSN: 2337-3539 (2301-9271).
[11] Nurhayati, I., Sutrisno, J.2011. Sugar Cane Waste as Heavy Metal Absorbent Pb.National Seminar: Development of Eco-Friendly Technology Toward Environmental Sustainability
[12] Rihayat, T, Saari, M., Suraya, A.R., Mahmood, M.H. , Dahlan, K.Z.H.M. , Yunus, W.M.Z.W. Sapuan, S.M. Synthesis and thermal characterization of polyurethane/clay nanocomposites

based on palm oil polyol. Polymer - Plastics Technology and Engineering, Vol. 45, Issue 12, (2006):1323-1326

[13] El-Sayed, G,O., Dessouki, H., A., Ibrahiem, S., S. 2011. Removal of Zn(II), Cd(II) and Mn(II) From Aqueous Solutions by Adsorption On Maize Stalks. The Malaysion Journal of Analytical Sciences, Vol 15 No 1 (2015): 8-21

[14] Khrisnakumari, B., Isyshawarya, P., Gayathri, R.,G.,R., Sangeetha, N. 2016. 2016. Removal of Iron Content FromDringking Water By Using Coconut Coir And Sugar Bagasse. International Conference on Breakthrough in Engineering, Science & Technology-2016 (INC- BEST'16).

IOP Conf. Series: Materials Science and Engineering **334** (2018) 012026 doi:10.1088/1757-899X/334/1/012026

Comparative study on the removal of COD from POME by electrocoagulation and electro-Fenton methods: Process optimization

A Chairunnisak[1], B Arifin[1], H Sofyan[2], M R Lubis[1] and Darmadi[1,3]

[1] Chemical Engineering Department, Syiah Kuala University, Banda Aceh, Indonesia
[2] Department of Mathematics, Syiah Kuala University, Banda Aceh, Indonesia

E-mail: darmadi@unsyiah.ac.id

Abstract. This research focuses on the Chemical Oxygen Demand (COD) treatment in palm oil mill effluent by electrocoagulation and electro-Fenton methods to solve it. Initially, the aqueous solution precipitates in acid condition at pH of about two. This study focuses on the palm oil mill effluent degradation by Fe electrodes in a simple batch reactor. This work is conducted by using different parameters such as voltage, electrolyte concentration of NaCl, volume of H_2O_2 and operation time. The processing of data resulted is by using response surface method coupled with Box-Behnken design. The electrocoagulation method results in the optimum COD reduction of 94.53% from operating time of 39.28 minutes, 20 volts, and without electrolyte concentration. For electro-Fenton process, experiment points out that voltage 15.78 volts, electrolyte concentration 0.06 M and H_2O_2 volume 14.79 ml with time 35.92 minutes yield 99.56% degradation. The result concluded that the electro-Fenton process was more effective to degrade COD of the palm-oil-mill effluent compared to electrocoagulation process.

1. Introduction

The increased production of palm oil has increased the income of the Indonesian economy, particularly in the industrial field. However, although the benefit of the economic improvement, the impact on the environment emerged are the functional conversion to forest, greenhouse gas production, water, biodiversity, emissions, environmentally friendly process technology, and waste both solid and liquid [1]. Wastewater in particular, in every ton of fruit, will produce 0.65 ton of waste from the process. Palm-oil-mill effluent (POME) is a brownish liquid containing a large amount of water, oil, suspended solids, dissolved solids, and sand. The untreated POME contained some colloidal suspension of 95-96% water, 0.6-0.7% oil, and 4-5% total solids, including 2-4% suspended solids [2]. It has a pH in the range of 3.45 – 4.6, COD concentration of 30,000 – 80,000 mg/l with BOD concentration of 15,000 – 40,000 mg/l, total suspended solid of 1,500 – 50,000 mg/l, and total solid of 16,000 – 95,000 mg/l [2]. One of the popular processes of palm oil wastewater is the pond system (pond). The pond system consists of a sand-and-oil trap, a cooling pond, an acidification pond, an anaerobic pond, a facultative pond, and an aerobic pond. The sand-and-oil trap is a part of wastewater pre-treatment. An aerobic pond has a depth of 1.0 – 1.5 m, whereas the anaerobic pond ranges five to

[3] To whom any correspondence should be addressed.

seven meters. The anaerobic pond could process large amounts of the solid with low-cost need. However, pond system requires a long residence time, large area, huge cost to prepare the pool, and microbes used to produce methane gas that increases the greenhouse effect [3].

Another alternative to overcome problems arising from pond system is to combine this system with other treatment systems. The treatment system used is electrocoagulation and electro-Fenton. Electrocoagulation is a coagulant production method of the in-situ process by applying an electric current through a cathode and an anode that releases metallic ions into solution. The cathode releases hydrogen recovered as an energy source or reactants for other industrial applications. In electrocoagulation, the proper anode material reduces the electrolyte to generate active coagulant species in its original place. Floc formed has a large surface area that can adsorb dissolved organic compounds and captures colloidal particles contained in solution [4]. Metal ions produced further produce hydroxide, poly hydroxide, and metal poly hydroxide compounds that have high affinity against opposite charged ions in solution [5]. In the process of electrocoagulation (EC) stages to degrade or to decrease the pollutant concentration are 1) the dissolving of anode ions in the solution, 2) the generating of OH^- ions and H_2 at the cathode, 3) electrolytic reactions at the electrode surface, 4) adsorption of coagulant in colloidal pollutants and 5), colloid removal by sedimentation or flotation [6]. After the use of the iron electrode, oxidation simplicity and reaction mechanisms that occur are as follows.

Anode: $Fe + 2e^- \rightarrow Fe^{2+}$

Cathode: $2H_2O + 2e^- \rightarrow H_2 + 2OH^-$

Alkali condition: $Fe^{2+} + 2OH^- \rightarrow Fe(OH)_2$

Acid condition: $4Fe^{2+} + O_2 + 2H_2O \rightarrow 4Fe^{3+} + 4OH^-$

The generating of oxygen: $2H_2O + 4e^- \rightarrow O_2 + 4H^+$

The electro-Fenton process requires H_2O_2 addition from the outside and an anode to generate Fe. In electro-Fenton, the processes of electrocoagulation and Fenton combine to increase the organic compound degradation in the wastewater. The electro-Fenton process occurs based on the work of the dissolved anode (Fe). Anodes will give Fe^{2+} ions into the solution, and cathode assists water reduction occurs simultaneously. The efficiency of the Fenton reaction mainly depends on H_2O_2 concentrations, the ratio of Fe^{2+}/H_2O_2, pH, and reaction time. Hydrogen peroxide is useful to give conditions for the Fenton reaction. Besides, Fe^{2+} could continue to regenerate in the cathode depending on the setup of the electrolytic cell [7]. The process uses the iron catalyst both in the form of ferrous (Fe^{2+}) and ferry (Fe^{3+}) salts. Several parameters influence the process of electro-Fenton (EF) such as pH, Fe/H_2O_2 ratio, existence of other anions and the chelating agent [8]. The following chemical reactions occur in the process.

$Fe^{2+} + H_2O_2 \rightarrow Fe^{3+} + OH^{\bullet} + OH^-$

$Fe^{2+} + OH \rightarrow Fe^{3+} + OH^-$

$Fe^{3+} + H_2O_2 \rightarrow Fe^{2+} + H^+ + HO_2^{\bullet}$

$Fe^{3+} + e^- \rightarrow Fe^{2+}$ (in the cathode)

The electro-chemical process had many applications such as the olive mill wastewater processing, cannery [8], surfactants, food processing, semiconductor, mechanical polishing, liquid organic fertilizer [7], electroplating, livestock [9], textile wastewater [10], palm oil mill effluent [11, 12], winery wastewater [13], pulp and paper wastewater.

The purpose of the research is to get the organic compound reduction in wastewater degradation derived from the value of COD analyzed. In this research, the COD reduction process takes place by combining pond system with electrocoagulation and electro-Fenton. The process variables on electrocoagulation process include voltage, operation time, and the electrolyte concentration. For the electro-Fenton process, operation parameters are voltages, operating time, electrolyte concentration and the volume of H_2O_2. Response Surface Methodology (RSM) processes the data by Box-Behnken Design (BBD) to get the optimum condition and related process variables.

2. Experimental procedure and methodology

Wastewater was taken from local palm oil mill (PT Syaukath Seujahtera, Bireuen Regency, Indonesia). The liquid wastewater was from the anaerobic pond I. This wastewater contained significant amounts of organic compounds, had a brown color, and smells bad. The wastewater was then put into a jerry can, and placed in a refrigerator at a temperature of 4°C. The wastewater characteristic after H_2SO_4 addition, such as pH, COD, TSS and TDS were 2; 7,000-11000 mg/L (diluted by 10 times until 700-1100 ppm) ; 4,720 mg/l and 1011 mg/l, respectively. COD in the pond was less than the COD in fat pit that was usually in the range of 30,000 – 80,000 mg/l because the pond was the fourth stage in the wastewater treatment of oil palm factory.

2.1. Materials and research procedure

The main equipment consisted of a reactor, electrodes, and power supply. Reactors EC and EF were made of glass with a size of 14 cm × 12 cm × 15 cm. Electrodes used consisted of two pairs of iron plates with a size of 15 cm × 10 cm × 1.0 cm. The electrode was connected to a DC-power supply (Sanfix 30 A and 30 V). Chemicals used were NaCl, acetone (Merck), and H_2O_2 (technical grade). Research stages were begun by cleaning electrode every time before and after one run. Electrodes were sanded by using sandpaper to make the surface of the electrode slippery and shiny. Cleaning was continued by washing electrode by using acetone and followed by a rinse with aqua distillate.

The next stage was putting wastewater of 1.3 L into a glass reactor. A certain amount of NaCl and H_2O_2 was added to a reactor by BBD design. NaCl and H_2O_2 were added then the electric current was applied to the electrodes. During the research took place, wastewater was stirred continuously by the stirrer speed of 100 rpm. The voltage was adjusted by BBD. All runs were carried out in the room temperature of 28°C. After the run had been completed, wastewater was filtered by using the membrane with a pore size of 0.45 micrometer then the filtrate was collected 100 ml for COD analysis. The electrode was cleaned back by using the stages mentioned above.

2.2. Experimental design

The Box-Behnken Design was useful for adjusting response surface, where a formula referred to a combination of 2^k block factorial with a simple design (Tables 1 – 3). The result of the design was typically very efficient to get the number of experiments conducted. By using RSM, it could be determined the relation between dependent process variables and the result. To find the electrocoagulation variable relation, the dependent variables were set as operation times (A; 15 – 45 minutes), voltage (B; 10 – 20 volt), and electrolyte concentration (C; 0.0 – 1.0 M). In electro-Fenton process, the dependent variables included operation times (A; 15 – 45 minutes), voltage (B; 10 – 20 volts), electrolyte concentration (C; 0.0 – 1.0 M), and volume of H_2O_2 (D; 5 – 15 ml).

Based on the one-factor at an experiment, variables were identified to get strong effects on COD removal. Table 1 shows Box-Behnken design of the research of 17 trials for electrocoagulation and 29 trials for electro-Fenton process. As shown in Table 1, the factors were designed as A, B, C and D. All factors were prescribed into three levels, namely +1, 0, -1 for high, intermediate and low value, respectively.

Table 1. Input variable and code factor of response surface method of Box-Behnken design.

Code Factor	Variable	-1	0	1
For electrocoagulation and electro-Fenton				
A	Time (minute)	15	30	45
B	Voltage (V)	10	15	20
C	Electrolyte concentration (ppm)	0	0.5	1
For electro-Fenton				
D	Volume of H_2O_2 (ml)	5	10	15

Analysis of variance (ANOVA) was carried out as a parametric test to distinguish the average value of more than two data groups by comparing their variances and determining responses to their variables. For predicting the optimum condition, thus the best COD removal had been based on all variables. Based on the relationship between the significant variable and the percentage of COD reduction as response obtained, thus it was obtained a polynomial equation to predict response. The polynomial equation quality was expressed by R^2 coefficient. The function of desirability or Derringer function was the most important and most used multi-criteria method in the analytic procedure optimization.

Chemical oxygen demand was determined by referring to SNI.6989.2:2009 for organic and inorganic compound oxidized by $Cr_2O_7^{2-}$ in closed reflux methods by using UV-VIS spectrophotometer model V-630. The oxidant determination in the sample was by visible light 420 nm.

Based on these results, the percentage of COD reduction was calculated from Equation 1.

$$COD_{reduction}(\%) = \frac{COD_0 - COD_t}{COD_0} \times 100 \quad (1)$$

COD_0 was initial COD, and COD_t was the remaining COD after electrocoagulation.

2.3. Energy consumption

In this study, the energy consumption of electrocoagulation and electro-Fenton referred to the optimum condition. The energy consumption calculation was as follows.

$$\text{Energy consumption}\left(\frac{\text{kWh}}{\text{mg COD}}\right) = \frac{I \times v \times t}{\Delta COD \times V} \quad (2)$$

Where I was a current intensity (A), v was voltage (V), t was the time of electrocoagulation, V was the volume of solution (L), and ΔCOD was the experimental COD reduction in solution [14, 15].

3. Result and Discussion

3.1. Response analysis by Box-Behnken Design

The research applies the processes of electrocoagulation and electro-Fenton to degrade the COD in palm oil mill effluent. The run of the data uses the Box-Behnken design with three levels based on dependent variables for electrocoagulation process such as operation time, voltage, and electrolyte concentration. For the electro-Fenton process, the dependent variables include time, voltage, electrolyte concentration, and volume of H_2O_2. Response variables investigated are the value of COD reduction after finishing the process (Table 2). It gives the COD removal predicted by a quadratic model along with values observed. The agreement between the predicted and experimental data is strong, with a low difference in COD removal.

Experimental data of the electrocoagulation show that the highest COD reduction of POME by electrocoagulation is in 45 minutes, the voltage of 20 V, and with NaCl concentration of 0.5 M. Such conditions result in the percentage of COD reduction of 95.01%. The predicted data of this run for electrocoagulation experimental was 95.86% (Table 2). On the other hand, the lowest COD of 76.85% for 15-minute duration, the voltage of 10 V, and electrolyte concentration of 0.5 M. The predicted data of this run was 76.00%. By electro-Fenton (Table 3), the highest COD reduction in the experiment and predicted data was 99.3% and 99.7%, respectively, at the time of 45 minutes, the voltage of 15 V with the electrolyte concentration of NaCl 0.5 M and the volume of H_2O_2 15 ml. Meanwhile, the lowest COD reduction in electro-Fenton for the experiment and predicted data was 91.00% and 92.20%, respectvely, in 15 minutes of electrolysis time, 15 voltage, 0.5 M NaCl solution concentration and 10 ml volume of H_2O_2. The high COD reduction occurs because of relatively short contact time, relatively short voltage and the H_2O_2 addition. The electro-Fenton uses peroxide that could produce hydroxyl radical. A radical could degrade organic compounds into other simple compounds in wastewater [16].

Table 2. Box-Behnken design with percentage of COD reduction after treatment by electrocoagulation

Run	Time (minute)	Voltage (V)	NaCl Concentration (M)	Actual COD Reduction (%)	Predicted COD Reduction (%)
1	**45**	**20**	**0.5**	**95.01**	**95.86**
2	30	20	1.0	92.50	91.69
3	30	15	0.5	88.50	87.91
4	**15**	**10**	**0.5**	**76.85**	**76.00**
5	15	15	1.0	77.03	78.05
6	30	10	0.0	83.20	84.01
7	45	10	0.5	86.64	86.85
8	45	15	0.0	91.25	90.23
9	45	15	1.0	91.25	91.22
10	15	20	0.5	81.59	81.38
11	30	20	0.0	91.27	91.44
12	30	15	0.5	87.75	87.91
13	30	15	0.5	88.08	87.91
14	15	15	0.0	78.05	78.08
15	30	15	0.5	86.46	87.91
16	30	10	1.0	84.90	84.73
17	30	15	0.5	88.78	87.91

Table 3. Box-Behnken design with percentage of COD reduction after treatment by electro-Fenton.

Run	Time (minute)	Voltage (V)	NaCl Concentration (M)	H_2O_2 Volume (mL)	Actual COD Reduction (%)	Predicted COD Reduction (%)
1	15	15	0.5	5	91.70	92.04
2	45	15	0.0	10	99.30	98.71
3	**45**	**15**	**0.5**	**15**	**99.30**	**99.97**
4	30	15	0.0	5	96.08	96.01
5	45	10	0.5	10	96.73	96.88
6	15	15	0.5	15	92.05	93.15
7	30	10	0.5	15	98.50	97.38
8	30	10	0.5	5	93.34	93.10
9	30	20	0.5	15	98.70	98.41
10	45	20	0.5	5	99.00	99.46
11	45	15	0.5	5	97.00	96.91
12	15	15	1.0	10	93.00	93.04
13	30	15	0.5	10	96.24	96.32
14	**15**	**10**	**0.5**	**10**	**91.00**	**92.20**
15	30	15	0.5	10	97.50	96.32
16	30	20	0.5	5	97.98	98.52
17	30	15	0.5	10	96.34	96.32
18	30	15	0.5	10	95.00	96.32
19	30	10	0.0	10	94.00	95.06
20	15	20	0.5	10	94.50	94.25
21	15	15	0.0	10	93.00	92.72
22	30	20	1.0	5	97.00	99.17

Run	Time (minute)	Voltage (V)	NaCl Concentration (M)	H_2O_2 Volume (mL)	Actual COD Reduction (%)	Predicted COD Reduction (%)
23	30	20	0.0	10	99.00	99.03
24	30	10	1.0	10	95.00	95.98
25	30	15	1.0	15	98.30	98.27
26	30	15	0.5	10	96.20	96.32
27	45	15	1.0	10	99.00	98.73
28	30	15	0.0	15	98.40	98.61
29	30	20	1.0	10	98.50	98.45

An excessive amount of peroxide in wastewater could be other pollutants in wastewater by considering its toxic and harmful properties. Because of peroxide used in electro-Fenton process, corrosion occurs more quickly to electrodes. A part of the applied current not only degrades wastewater, but also deposits electrode, thus it causes corrosion to occur more quickly to the electrode surface.

The response surface method of BBD indicates that the response from COD values is the quadratic model of the processes of electrocoagulation and electro-Fenton. Equation 3 results in the predicted data for COD reduction in electrocoagulation and Equation 4 for COD reduction in electro-Fenton process. The strong relationship between COD removal predicted by Equations 3 and 4, and the experimental results (Tables 2 and 3) show that the model accuracy and ability are good.

$$\text{COD}(\%) = 59.68 + 1.09\,A - 0.03\,B + 1.3\,C + 0.01\,AB + 0.03\,AC - 0.05\,BC - 0.01\,A^2 + 0.01\,B^2 - 1.15\,C^2 \quad (3)$$

$$\text{COD}(\%) = 73.91 + 0.52\,A + 0.80\,B + 1.03\,C + 0.41\,D - 0.004\,AB - 0.01\,AC + 0.007\,AD - 0.15\,BC - 0.04\,BD - 0.10\,CD - 5.32\,A^2 + 5.50\,B^2 + 2.72\,C^2 + 0.02\,D^2 \quad (4)$$

The fit checking of the model is by using the determination coefficient R^2. Based on tabulation data by the response surface method coupled with Box-Behnken design (Tables 2 and 3), both models follow statistically the quadratic model (Table 4). "Model Summary Statistics" focuses on the maximizing the "Adjusted R-Squared" and the "Predicted R-Squared" values to find model reliability. The value of R^2 that approaches one indicates the high correlation degree among observation and predicted values. Similarly, the model also indicates the higher value of *Adj*-R^2 which means that the model is more significant [17]. Based on the experimental data, data agree well with the predicted data (electrocoagulation R^2 0.9829 and the electro-Fenton R^2 0.9448). Furthermore, the R^2 values also show that each model explains 98.29% and 94.48% of the total variation in the COD removal to the operating parameters.

Table 4. Model summary statistics of electrocoagulation and electro-Fenton process.

Response	Source	Std. Deviation	R-squared	Adjusted R-square	Predicted R-square	PRESS
Percentage of COD removal by electrocoagulation	Linear	2.09	0.8819	0.8547	0.7818	104.98
	2 FI	2.31	0.8894	0.8231	0.5524	215.41
	Quadratic	**1.09**	**0.9829**	**0.9608**	**0.8237**	**84.86**
	Cubic	0.9	0.9932	0.9729	-	-
Percentage of COD removal by electro-Fenton	Linear	1.18	0.8124	0.7812	0.7147	51.05
	2 FI	1.21	0.8521	0.7699	0.5484	80.80
	Quadratic	**0.84**	**0.9447**	**0.8894**	**0.7549**	**43.86**
	Cubic	0.98	0.9679	0.8503	-1.1224	379.78

The analysis of variance (ANOVA) observed for response expresses the model reliability, as presented in Table 5. ANOVA is a comparison of the variance because of the treatment and the variance because of random errors inherent in the responses. By this comparison, it is possible to check the regression significance of variables toward responses by considering sources of experimental variance. It is important to test the significance and accuracy of the model through ANOVA.

ANOVA of Table 5 shows that the probability (*p*) values such as time, voltage, electrolyte concentration and volume of H_2O_2. It shows the significant variable determination by applying the *p-values*. As a result, such variables are significant and have a real effect on COD reduction that have a *p-value* lower than 0.05. It is because *p-value* is generally compared to a particular real standard α, usually 0.05 or 5%. Time and voltage variables have the larger interaction effect of degrading COD based on both the same *p-values* ($p < 0.05$). In electro-Fenton, significant coefficients are time, voltage, volume of H_2O_2, and interaction of voltage-volume H_2O_2. In the both processes, electrolyte concentration has a *p-value* above 0.05 which is not a significant variable and indicates that electrolyte concentration has the least interaction among other variables. In addition, the effect of the quadratic terms A^2 on the COD removal is significant.

Table 5. ANOVA results in COD removal by electrocoagulation.

Source	Electrocoagulation		Electro-Fenton	
	P-value	Remark	P-value	Remark
Model	< 0.0001	Significant	< 0.0001	Significant
A – Time	**< 0.0001**		**< 0.0001**	
B – Voltage	**< 0.0001**		**< 0.0001**	
C – Electrolyte concentration	0.5536		0.7314	
D – Volume of H_2O_2			**0.0009**	
AB	0.1384		0.4766	
AC	0.6527		0.8610	
AD			0.2656	
BC	0.8348		0.3875	
BD			**0.0194**	
CD			0.5539	
A^2	**< 0.0005**		**0.0028**	
B^2	0.5403		0.6836	
C^2	0.6043		0.059	
D^2			0.2458	
Lack of Fit	0.2512	not significant	0.6153	not significant

3.2. The effect of variables on COD efficiency

3.2.1. The interaction effect of time and voltage. The percentage of COD removal represents the effect of time (X) and voltage (Y), which was developed by using Design Expert (version 10.0.1) software. Figures of 1 and 2 show the data. To find the time effect on the COD removal, time range varies from 15 to 45 minutes. Figures 1 (a) and 2 (a) indicate that time of the process increases for electrocoagulation and electro-Fenton processes. The contours were circular and become more elongated along the time axis. So a smaller value of time and voltage increased the efficiency of COD removal in palm oil mill effluent. Based on the data, the efficiency of COD removal increases in line with the increase in time in the range of 15 – 45 minutes. The COD removal in electrocoagulation process increases from 80% to 95% (Figures 1 (a) and (b)); likewise, in the electro-Fenton process it also increases from the value of 94% to 100%.

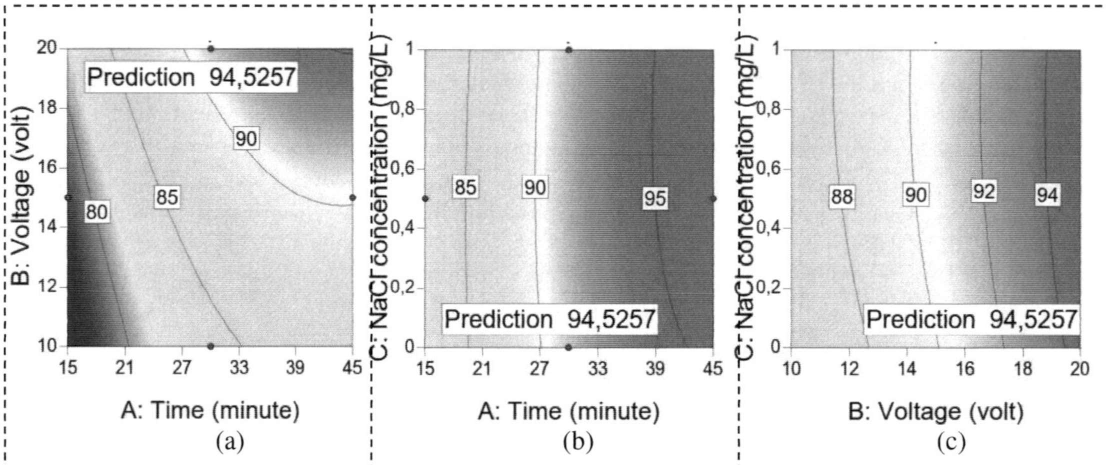

Figure 1. Two-dimensional contour plots express the effect of time, voltage and electrolyte concentration on electrocoagulation.

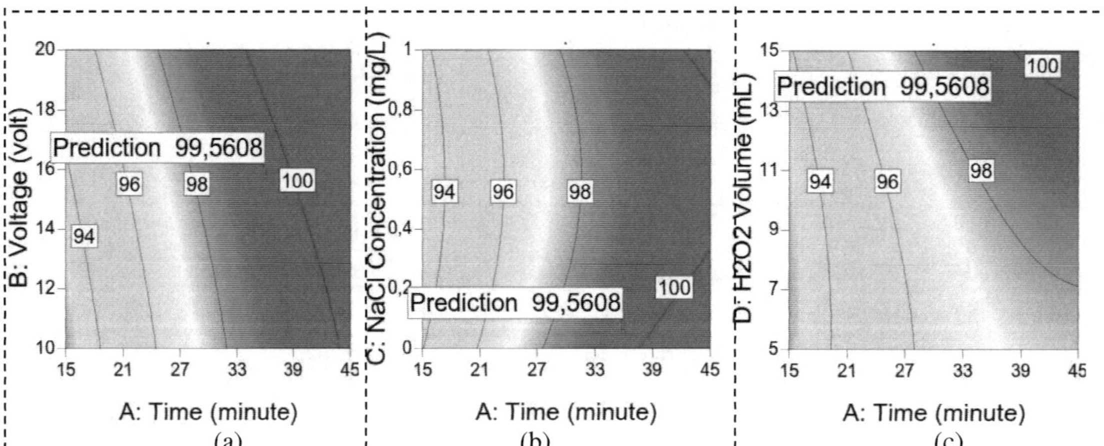

Figure 2. Two-dimensional contour plots express the effect of time, voltage and electrolyte concentration, and H_2O_2 volume on electro-Fenton process.

It indicates that longer time is more favorable in electrocoagulation and electro-Fenton processes of palm oil mill effluent. Time is a variable for controlling the rate of both processes. As a result of the change of time and voltage in the reactor, organic matter and coagulant interact and collide more effectively. The electrocoagulation generates coagulants in its original place during the time. Then, the cathode generates hydrogen bubbles during that time [18]. Pollutant and coagulant interact to form floc. This floc will form bubbles in the top of the reactor by the hydrogen existence. Some heavy flocs will sink to the bottom of the reactor.

3.2.2. The interaction effect of time and electrolyte concentration. The anion concentration on the solution could affect the passive layer stability of electrodes. In the solution, the electrolyte solution generates anions. This research uses a small concentration of the electrolyte into the reactor. This research also considers the effect of supporting electrolytes on the efficiency of an electrocoagulation system by varying the NaCl concentration from 0 to 1 mg/l (Figures 1 (a) and (b)). The electrolyte is various such as NaCl, Na_2SO_4 and others salt. The electrolyte in the research is NaCl. Some earlier works indicated that NaCl was the best electrolyte for electrocoagulation [7, 19]. Chloride ion could induce the passive layer and corrosion (pitting). As a result, anions increased electrical consumption and had a negative effect on EC efficiency [20].

Figures 1 (b) and 2 (b) show the contour plots based on variables of time (min) and electrolyte concentration (mg/L). Based on research data in Figure 1 (b), 1 (c) and 2 (b), the effect of electrolyte concentration was not significant on the removal of COD in both processes. At all level of time, the removal efficiency of palm oil mill effluent was found increased. In electrocoagulation in the range of 39 – 45 minutes and the electrolyte concentration of 0 – 1 M, COD reduction was 95% (Figure 1 (b)). At the shorter time than 39 minutes, the data fluctuated. In the range of 15 – 21 minutes, the COD removal was 85%; however, the removal increased to 90% with the increase in time until 27 minutes. Based on data and theory of electrocoagulation, the electrolyte was not so significant in removing COD in mill effluent POME treatment.

The NaCl addition does not increase conductance and current for both processes. In this research, the removal of organic compounds does not increase with the NaCl concentration addition. The complete removal got is above 39 minutes for both processes. The results of the research show that the degradation occurred efficiently without the NaCl electrolyte presence. It could be stated that the degradation became faster when NaCl electrolyte was not used in this research. It is probably because the use of NaCl causes a side reaction namely the organochloride ion formation, mainly OCl^-.

3.2.3. The interaction effect of time and H_2O_2 volume. This section explains the result of the electro-Fenton process. This process adds several oxidizing solutions. The addition aims to form the radical of hydroxyl which has standard oxidation potential of 2.8 V. This radical can oxidize several organic matters into CO_2, H_2O and inorganic ions by the specific reaction [20]. The optimum volume of H_2O_2 in removal efficiency of COD in this study is in the range of 5 – 15 ml. In 45 minutes, 15 volts, NaCl and 10 ml H_2O_2, the highest removal efficiency is 99.3% (Table 2). However, the increase in the volume of H_2O_2 causes the COD removal is slightly different. It may occur as a result of inactivity of some H_2O_2 because of the voltage applying to the electrode. The highest volume of H_2O_2 indicates the efficiency of COD removal is nearly complete. At the time reaches 45 minutes, such efficiency was almost adjacent. The COD removal efficiency in electro-Fenton depends on H_2O_2 dosage and time. The unlimited presence of the oxidizing agent forms the radical of hydroxyl more effectively to degrade the organic matter [21].

3.3. Energy Consumption

To find the efficiency of both processes, this research investigates the energy consumption/kg COD degraded in the optimum operating variables. Equation 4 indicates energy consumption of electrocoagulation for 39.28 minutes, 20 volts without NaCl is 0,1617 kWh/mg COD degraded. For the electro-Fenton process, the optimum condition in 35.92 minutes, 15.78 V, 3.9 Ampere with 14.79 ml H_2O_2 shows that energy consumption is 0,067 kWh/mg COD degraded. This shows that value of energy consumption of electro-Fenton is lower than that of electrocoagulation. However, further study is necessary to analyze the operating cost of both processes to prove the economic cost which refers to the more reliable value.

3.4. Comparison of removal efficiency for Electrocoagulation and Electro-Fenton

Response surface method of Box-Behnken Design (BBD) determines the optimum condition of this research without adding the experimental cost. BBD allows calculation of COD removal on intermediate levels that are not experimentally investigated. A three-level BBD used in the research is to get the optimum conditions through a minimal investigation number compared to other designs. It uses constraint provided to meet the optimum condition. The experimental and predicted data show that the independent variable setting is different to the optimization by BBD for both processes. For both processes the constraint consists of NaCl concentration, electrolysis time, voltage set in their ranges and for electro-Fenton it also consists of H_2O_2 volume.

The high desirability indicates that the condition becomes optimal (closest to 1). The desirability (Derringer function) is the most important and the most current multi-criteria method used for optimizing analytic rules. This method initially refers to a construction a desirability function of each response [22]. Based on the response surface method of BBD, the optimum COD removal by electrocoagulation is 94.53% at operation time of 39.28 minutes, a voltage of 20 V, and without electrolyte. For electro-Fenton process, it is optimal at 99.56% in operating time of 35.92 minutes, a voltage of 15.78 V, electrolyte concentration of 0.06 M and H_2O_2 volume of 14.79 ml. The optimum COD removal prediction by BBD is close to the experiment data (Tables 2 and 3). Based on optimal prediction and experiment data, the electro-Fenton has the best result in removing the organic matter in palm-oil-mill-effluent treatment by comparing the removal efficiency of the COD.

4. Conclusion

This research investigates electrocoagulation and electro-Fenton performances in palm oil mill effluent treatment in a batch. The efficiency of COD removal is the main response of this research. The research uses response surface method (RSM) with Box-Behnken Design to get the experimental data interaction. As a result, the high R^2 values of COD removal of electrocoagulation and electro-Fenton are 0.98 and 0.95, respectively. ANOVA provides the R^2 values to verify that the processes are quadratic models. The largest removal of COD for electrocoagulation and electro-Fenton are 95.01% and 99.3%, respectively. In electrocoagulation, the optimum COD removal is 94.53% at operation time of 39.28 minutes, a voltage of 20 V and without electrolyte. For electro-Fenton process, it is optimal at 99.56% in operating time of 35.92 minutes, a voltage of 15.78 V, electrolyte concentration of 0.06 M and H_2O_2 volume of 14.79 ml. It is clear that the RSM gives information on interactions among those factors and helps to get optimum values of the investigated factors. Overall, both processes are successful in treating the palm oil mill effluent. This research indicates that the electro-Fenton process is the most successful process that degrades COD. Within all variables tested the COD removal is proportional to the operation time, voltage and electrolyte concentration for electrocoagulation and proportional to H_2O_2 volume for electro-Fenton process. For the application of electro-Fenton process in industries, other electrodes and response could be experimentally designed and tested by response surface method to remove organic pollutants in wastewater.

Acknowledgments

Authors thank for the help or encouragement of colleagues, special work by technical staff and financial support from Magister Grant of Syiah Kuala University in 2016 (Project No. 267/UN11.2/PP/SP3/2016).

References

[1] Hansen S B, Padfield R, Syayuti K, Evers S, Zakariah Z and Mastura S 2015 Trends in global palm oil sustainability research *Journal of Cleaner Production* **100 (1)** 140–49

[2] Tetratech 2013 *Methane Capture and Use Potential at Palm Oil Mills in Indonesia. Methane Expo* (Canada: https://www.globalmethane.org/expodocs/canada13/ag_-11Franco_Miguel_Presentation%20v4.pdf (accessed 10.02.16)

[3] Liew L, Kassim M, Muda K, Loh K, Affam A 2015 Conventional Methods and Emerging Wastewater Polishing Technologies for Palm Oil Mill Effluent Treatment: A Review *Journal of Environmental Management* **149** 222 – 235

[4] Mook W T, Aroua M K and Issabayeva G 2014 Prospective applications of renewable energy based electrochemical systems in wastewater treatment: A review *Renewable and Sustainable Energy Reviews* **38** 36–46

[5] Harif T, Khai M and Adin A 2012 Electrocoagulation versus chemical coagulation: coagulation/flocculation mechanisms and resulting floc characteristics *Water Research* **46 (10)** 3177–88

[6] Khandegar V and Saroha A K 2013 Electrocoagulation for the treatment of textile industry effluent - A review *Journal of Environmental Management* **128** 949–63

[7] Akyol A, Can O T, Demirbas E and Kobya M 2013 A comparative study of electrocoagulation and electro-Fenton for treatment of wastewater from liquid organic fertilizer plant *Separation and Purification Technology* **112** 11–19

[8] Kobya M and Demirbas E 2015 Evaluations of operation parameters on treatment of can manufacturing wastewater by electrocoagulation *Journal of Water Process Engineering* **8** 64–74

[9] Tak B, Tak B, Kim Y, Park Y, Yoon Y and Min G 2015 Optimization of color and COD removal from livestock wastewater by electrocoagulation process: Application of Box-Behnken design (BBD) *Journal of Industrial and Engineering Chemistry* **28** 307–315

[10] Behin J, Farhadian N, Ahmadi M and Parvizi M 2015 Ozone assisted electrocoagulation in a rectangular internal-loop airlift reactor: Application to decolorization of acid dye *Journal of Water Process Engineering* **8** 171–178

[11] Hanum F, Tambun R, Ritonga M Y and Kasim W W 2015 Aplikasi elektrokoagulasi dalam pengolahan limbah cair pabrik kelapa sawit *Jurnal Teknik Kimia USU* **4 (4)** 13–17

[12] Nasution A, Ng B L, Ali E, Yaakob Z and Kamarudin S K 2014 Electrocoagulation of palm oil mill effluent for treatment and hydrogen production using response surface methodology *Pol. J. Environ. Stud.* **23 (5)** 1669–77.

[13] Díez A M, Rosales E, Sanromán M A and Pazos M 2017 Assessment of LED-assisted electro-Fenton reactor for the treatment of winery wastewater *Chemical Engineering Journal* **310 (2)** 399–406

[14] Gumus D and Akbal F 2016 Comparison of fenton and electro-Fenton processes for oxidation of phenol *Process Safety and Environment Protection* 103 (A) 252–58

[15] Popescu M, Sandu C, Rosales E, Pazos M, Lazar G and Sanromán M A 2017 Evaluation of different cathodes and reaction parameters on the enhancement of the electrofenton process *Journal of Electroanalytical Chemistry* In Press Accepted Manuscript

[16] Akyol A, Can O T and Bayramoglu M 2015 Treatment of hydroquinone by photochemical oxidation and electrocoagulation combined process *Journal of Water Process Engineering* **8** 45–54

[17] Zhao Q, Kennedy J F, Wang X, Yuan X, Zhao B, Peng Y and Huang Y 2011 Optimization of ultrasonic circulating extraction of polysaccharides from asparagus officinalis using response surface methodology *International Journal of Biological Macromolecules* **49 (2)** 181–87

[18] Vepsalainen M 2012 Electrocoagulation in the treatment of industrial waters and wastewaters *Dissertation* (Finland: VTT Technical Research Centre of Finland)
[19] Nindeesh P V and Gandhimathi R 2012 Trends in electro-Fenton process for water and wastewater treatment: an overview *Desalination* **299** 1–15
[20] Ma L, Zhou M, Ren G, Yang W and Liang L 2016 A highly energy efficient flow through electro-Fenton process for organic pollutants degradation *Electrochimica Acta*. **200** 222–30
[21] Gümüş D and Akbal F 2016 Comparison of fenton and electrofenton processes for oxidation of phenol *Process Safety and Environment Protection* **103 Part A** 252-258
[22] Bezerra M A, Santelli R E, Oliviera O P, Villar L S and Amelia L 2008 Response surface methodology (RSM) as a tool for optimization inanalytical chemistry *Talanta* **76 (5)** 965–977

IOP Conf. Series: Materials Science and Engineering **334** (2018) 012027 doi:10.1088/1757-899X/334/1/012027

Removal of zinc (II) ion from aqueous solution by adsorption onto activated palm midrib bio-sorbent

F Mulana[1,2,3], Mariana[1], A Muslim[1], M Mohibah[2] and K H Ku Halim[2]

[1] Department of Chemical Engineering, Faculty of Engineering, Syiah Kuala University, Darussalam, Banda Aceh, Indonesia
[2] Faculty of Chemical Engineering, Universiti Teknologi MARA, Shah Alam, andMalaysia

E-mail: faridmln@yahoo.com

Abstract. In this paper, palm midrib that was activated with mixed citric acid and tartaric acid as biosorbent was used to remove Zn (II) ion from aqueous solution. The aim of this research is to activate palm midrib by using a mixed citric acid and tartaric acid and to determine adsorption capacity of activated palm midrib biosorbent on Zn (II) ion uptake from aqueous solution. The effect of several parameters such as contact time, initial Zn (II) ion concentration and activator concentration on the degree of Zn (II) ion removal was examined. Atomic Absorption Spectroscopy method was performed to determine adsorbed amount of Zn (II) ion into activated biosorbent. The result showed that the adsorption process was relatively not so fast and equilibrium was reached after contact time of 120 min. The adsorption capacity of biosorbent reached a maximum when the concentration of mixed citric acid and tartaric acid was 1.6 M. The optimum adsorption capacity was 5.72 mg/g. The result was obtained on initial Zn (II) ion concentration of 80 ppm for 120-min contact time. Langmuir isotherm was found as the best fit for the equilibrium data indicating homogeneous adsorption of metal ions onto the biosorbent surface.

1. Introduction

Recently, the growth of chemical industry moves rapidly so that the disposal of heavy metals into the environment increases accordingly. The heavy metals aqueous wastes have induced many problems for human being and its environment. Electroplating and metal surface treatment processes produce large quantity of wastewater containing heavy metals, which include zinc (II) ion [1].

The presence of zinc (II) ions in aquatic environment, even at low concentrations, is potentially harmful to human health. Zinc may cause some diseases including high blood pressure, infantilism, destruction of red blood cells, impaired wound healing, and kidney damage when it accumulates in the human body [2]. Considering the toxicity and harmful effects of these metals, therefore, before discharging to environment heavy metal has to be treated. There are a lot of treatment processes for heavy metal removal from wastewater such as reverse osmosis, electrocoagulation, ultrafiltration, dialysis/electrodialysis, solvent extraction, precipitation and ion exchange [3,4,5,6].

These separation processes are not entirely economically feasible due to their relatively high costs. Therefore, it is important to study new processes, which are more efficient and economical.

[3] To whom any correspondence should be addressed.

Adsorption is one of the most widely used methods for the removal of metal ions from contaminated water [7].

Many researchers utilized activated carbons as adsorbent for wastewater treatment industries. Because activated carbon is an expensive adsorbent. Therefore, the searching of cheap adsorbents from agricultural solid waste could be an alternative method [8]. Agriculture, forestry and fisheries have been generating large quantities of various biomass wastes, and some of them contain various natural materials with interesting functional groups, such as carboxyl, hydroxy and so on which are as the active site on the adsorbent.

Palm midrib is one of the valuable biomass wastes but it has been partly utilized as cattle feed while the remaining is mostly wasted into the nature. For application of adsorption process, palm midrib that have variety of pore size inside its particles with a proper activation process, those pores size can be enlarged and adapted to the size of molecules of proper liquid waste of heavy metal so that adsorbate can be adsorbed easily into pore of adsorbent.

So far, some researchers have used other biomaterials to sorb heavy metals including coffee waste, orange peel, banana peel, mangosteen peel, hardwood, corn straw and other adsorbents [9, 10, 11, 12, 13, 14]. The adsorption capacity of biomass-based biosorbent on heavy metals can be enhanced by modifying the physicochemical properties of its surface [15, 16, 17]. The modification has been carried out using various reagents such as some acids such as citric acid, tartaric acid and phosphoric acid can increase the adsorption capacity of heavy metal ions [15, 16, 17, 18, 19, 20, 21]. With this activation, some acid groups can occur and form ester groups [17].

As far as we know, there are no previous studies that deal with the chemical modification of palm midrib by using a mixed citric acid and tartaric acid. Thus, the preparation of biosorbent from this agricultural by-product was here proposed. The purpose of this study is to prepare activated palm midrib biosorbent by using mixed citric acid and tartaric acid as activator, and to investigate the adsorption of zinc (II) ions from aqueous solution toward adsorption capacity. The effect of mixed activator concentration at similar ratio (citric acid and tartaric acid) was studied at neutral pH.

2. Experimental procedure and methodology

2.1 Materials and Equipment

Palm midribs, a waste product of palm tree, were used for the removal of Zn (II) ions from aqueous solution. These palm midribs were collected from area of Aceh Besar District, milled, powdered and sieved to get uniform particle size of 40-60 mesh. Some chemicals that used in this research like sodium hydroxide, citric acid 99.9 %, tartaric acid 97.5% and a solution of $Zn(NO_3)_2$ were purchased from chemical supplier with brand of Merck and distilled water was collected from our laboratory. Some equipment that used in this research such as hot plate, glasswares, desiccator, digital scale, oven dryer and sieve shaker. The analysis apparatus was Atomic Absorption Spectrometry (AAS) type AA 7000 (Shimadzu) and SEM TM-3000 (Hitachi).

2.2 Research Procedure

Raw material preparation of palm midrib was performed as following: at the beginning, about 10 mm x 10 mm size of raw palm midrib was washed several times with clean water in order to remove impurities on raw materials and after that was dried under sunlight for 1-2 days. Further washing was performed by soaking of dried palm midrib in hot water for 1 hr, then it was dried again under sunlight for a day and finally was dried in oven dryer at 110 °C for 1 hr. A dried palm midrib was then milled and sieved on the particle size of 40-60 mesh. Lastly, powdered palm midrib was dried again at 110 °C for 4 hr and then was stored in a desiccator for further application.

For preliminary activation process, 10 gr of dried palm midrib was mixed with solution of 0.5 M KOH. The mixing was performed at speed of 80 rpm for 30 min at room temperature. After completion of mixing and filtering, palm midrib biosorbent was neutralized several times using a distilled water. Finally a neutral of palm midrib biosorbent was dried at 50 °C for 2 hr. The purpose of base solution addition is to create some new active sites on the surface of palm midrib so that later would increase adsorption capacity [22].

About 2 gr of netralized palm midrib was mixed with 200 ml of citric acid and tartaric acid (on the same ratio of 0.4 M) at speed of 80 rpm for 2 hr, then filtered and activated palm midrib biosorbent was then washed with distilled water to remove the activator excess until reached a pH of ± 7, followed by re-drying at 50 °C for 24 hr [16]. The similar procedure was repeated for citric acid/tartaric acid concentrations of 0.8; 1.2, and 1.6 M. For adsorption process, 1 gr of activated palm midrib biosorbent was mixed with 100 ml of 20 ppm of $Zn(NO_3)_2$ at speed of 80 rpm for different time (0 to 160 min) at room temperature. After the adsorption process was completed, the filtered solution was analyzed by using AAS. The above treatment was repeated for initial concentrations of Zn (II) ion of 40, 60, and 80 ppm (15, 44, 54 and 73 ppm of AAS actual results).

3. Results and Discussion

3.1 Effect of Contact Time Toward Adsorption Capacity

Based on the literature study it is known that the longer of contact time between biosorbent and adsorbate then the concentration of metal ion that adsorbed will increase until the equilibrium occurs where the largest adsorption capacity occurs at the beginning of the adsorption process. So in order to know the effect of the contacting time on the adsorption capacity of Zn (II) ion by the activated biosorbent, the preliminary experiment has been carried out at different contact time until 160 min. The effect of contact time on the adsorption capacity of Zn (II) ion can be seen in Figure 1.

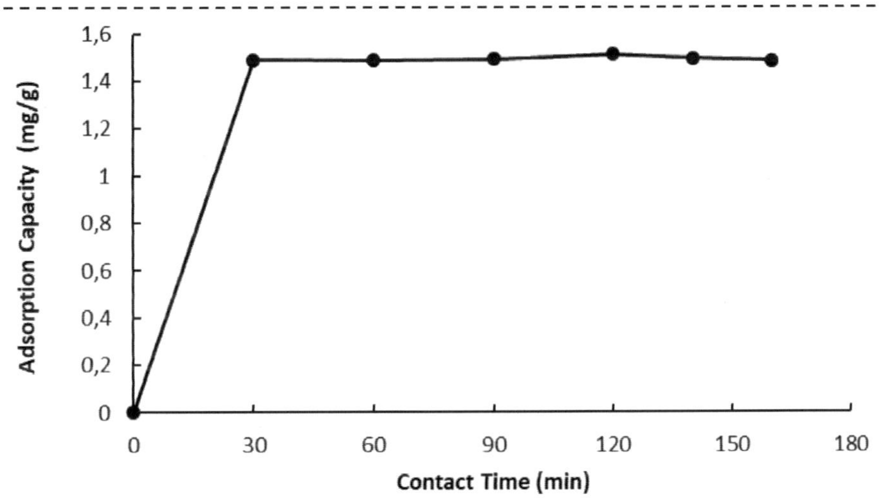

Figure 1. The relationship between contact time (min) and the adsorption capacity (mg/g) at initial concentration of 20 ppm Zn (II) ion after biosorbent activation by using citric acid and tartaric acid concentration of 0.4 M.

Based on Figure 1 we can see that at the beginning of the adsorption process, the adsorption rate increases rapidly until 30 min and later the increasing of adsorption tends to stabilize from the 120 min until reaching equilibrium at the 160 min. The condition of this research was performed by using biosorbent that was activated by citric acid and tartaric acid concentration of 0.4 M while for adsorbate concentration was at initial concentration of 20 ppm Zn (II) ion. The initial rapid uptake of metal ions may be ascribed to the presence of large number of vacant sites available for metal ions, and afterwards the remaining free metal ions are difficult to be occupied, because of repulsive forces between the free and adsorbed metal ions.

The adsorption capacity reached 1.38; 1.48; 1.48, 1.50; 1.49; 1.50 mg/g at different contact time of 0, 30, 60, 90, 120, 140 and 160 min, respectively. Equilibrium time occured after contact time of 120 min, where the adsorption capacity of Zn (II) ion reached 1.5 mg/g on the initial adsorbate concentration of 20 ppm. With increasing of contact time, the capacity of biosorbent to bind to surface of biosorbent will be higher that is enabling more and more bonds to be formed between surface of biosorbent particles and heavy metal ions in solution of adsorbate.

Citric acid and tartaric acid that used as activator during activation process plays an important role on modification of biosorbent surface. The presence of citric acid and tartaric acid during the activation process resulted in degradation of materials on the surface of biosorbents that will form the more pore on particles that later can be also seen on SEM graph in Figure 4. By increasing the concentration of citric acid and tartaric acid, it might be proportional to the increase the pore distribution and surface area of the biosorbent causing a number of binding metal ions of zinc increases and as well as adsorption ability also increases during adsorption process.

Other researcher [23] found that the adsorbed amount of Zn (II) ion on the first 40 min of adsorption process increased significantly and after contact time of 60 min it still remains increasing but reached equilibrium after contact time of 120 min as our finding. The quite fast adsorption of metal ions in the early minutes of adsorption process may be due to at the beginning of the adsorption process many sites and pore of biosorbent is still empty so that the tendency of the metal ion solution to be adsorbed into the biosorbent is much higher with increasing contact time and slow the rate of adsorption began to decline until they reach equilibrium time. Similar to biosorbent that activated by activator of citric acid and tartaric acid in this study, [24] suggested that some natural biosorbent contains cellulose and lignin which is a material with huge potential to be used as biosorbent.

3.2 Effect of Initial Zn (II) Ion Concentration on Adsorption Capacity

To study the effect of initial concentration of Zn (II) ion on the adsorption capacity, the research was done by varying the Zn (II) concentration of 20, 40, 60 and 80 ppm and the adsorption process was done at contact time of 120 min. The effect of the initial concentration of adsorbate on adsorption capacity is shown in Figure 2.

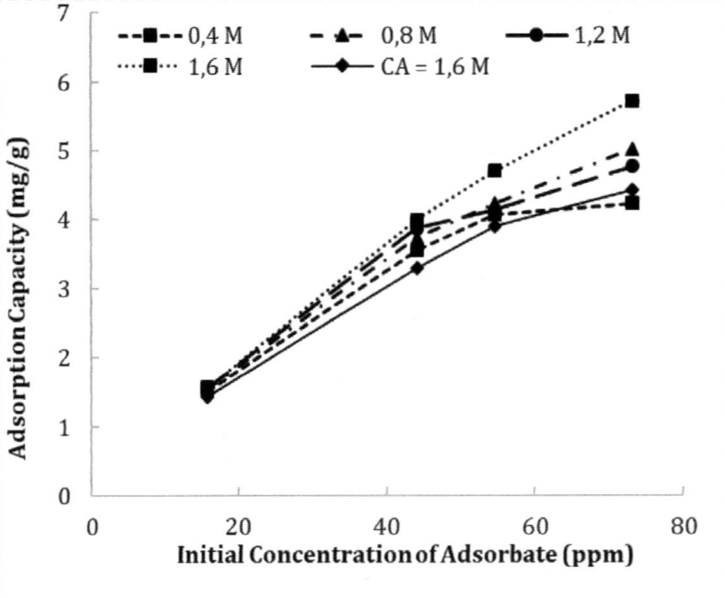

Figure 2. Effect of initial concentration of adsorbate (ppm) and activator concentration (M) on adsorption capacity (mg/g) at the contact time of 120 min (CA = Citric Acid).

Figure 2 showed that the adsorption capacity increases with increase of initial adsorbate concentration. The highest adsorption capacity was 5.72 mg/g that obtained after adsorption process was performed at initial adsorbate concentration of 80 ppm. On the other hand, the adsorption capacity on others initial adsorbate concentration of 20, 40, 60 and 80 ppm using biosorbent that activated by mixed citric acid and tartaric acid of 1.6 M were 1.56,; 3.99; 4.69; 5.72 mg/g, respectively. The utilization of mixed citric acid and tartaric acid as chemical activator to activate the palm midrib biosorbent has improved the adsorption capacity of zinc (II) ion. As comparison, from Figure 2 it can be seen that the adsorption capacity of zinc (II) ion on activated palm midrid by using mixed citric acid and tartaric acid was higher than that of by using only citric acid. Based on the previous study [25, 26, 27], the presence of hydroxyl groups in the biosorbent caused the metal ion interact easily with the hydroxyl group (O-H) from both acids and improving adsorption capacity.
The adsorption capacity on the biosorbent increases linearly with increasing adsorbate concentration. This is because when the initial concentration of the solution was increased, the mass transfer driving force became larger, resulting greater adsorption capacity. It can be clarified also that the more concentration, the more adsorbate molecules and adsorbents interact in the adsorption process. This causes adsorption tends to increase its capacity.

3.3 Effect of Activator Concentration on Adsorption Capacity
The addition of different concentration citric acid and tartaric acid during activation process would affect the adsorption capacity of the adsorbate solution on palm midrib biosorbent. This effect was illustrated in Figure 3.

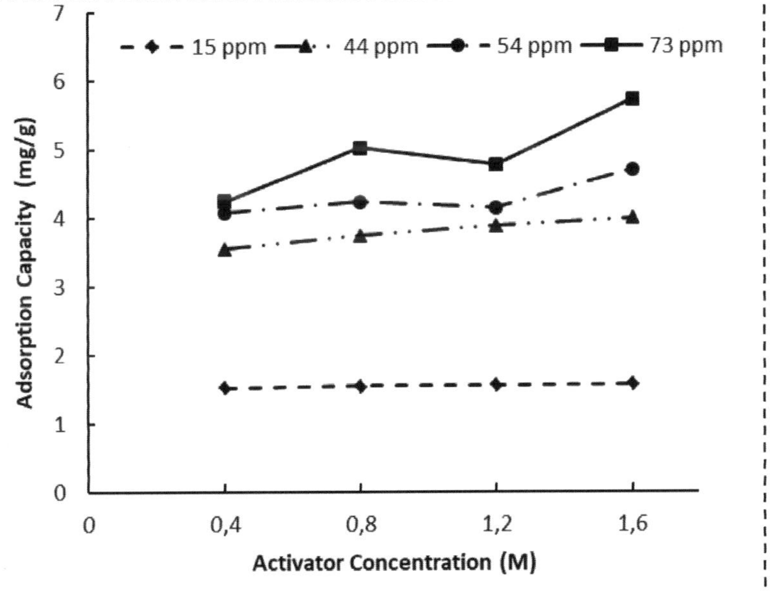

Figure 3. Addition effect of citric and tartaric acid concentration (M) and the initial concentration of adsorbate (ppm) on adsorption capacity (mg/g) at the contact time of 120 min.

The increase of citric and tartaric acid concentration that used to activate biosorbent affected adsorption capacity on each samples with different uptake capacity of Zn (II) ion. The highest adsorption capacity of 5.72 mg/g was obtained when we used biosorbent that activated by using citric acid and tartaric acid concentration of 1.6 M at initial concentration of 80 ppm Zn (II) ion. While the lowest adsorption capacity was obtained at citric acid and tartaric acid concentration of 0.4 M with the uptake capacity of 1.52 mg/g.

The higher concentration of the activator would be the higher the adsorption capacity value of the Zn (II) ion. The increase of citric acid and tartaric acid concentration increases proportionally pore numbers and surface area of the biosorbent, causing the number of Zn ion-binding sites to increase and their adsorption capacity increased as well. Therefore as observed in SEM images analysis, the presence of citric acid and tartaric acid during activation results in the cleaning of impurities materials on the surface of biosorbent to form more pores and active sites.

3.4 Morphological Analysis on Biosorbent Surfaces

To know the morphology on the surface of the biosorbent, SEM analysis of biosorbent has been done and the results can be seen in Figure 4 below.

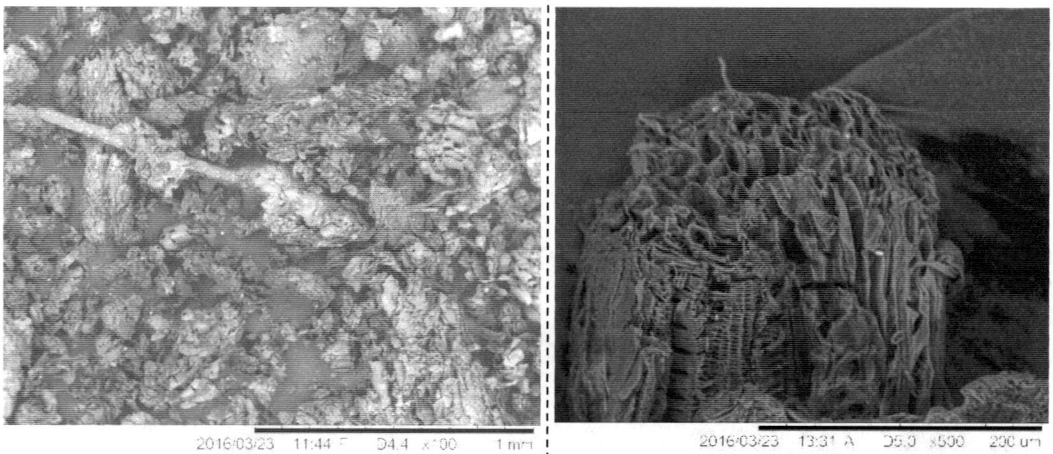

(a)

(b)

Figure 4. SEM photographs of (a) palm midrib biosorbent before activation (b) palm midrib biosorbent after activation with citric acid and tartaric acid of 1.6 M.

Figure 4 (a) showed clearly that the pores and cavities present in the unactivated biosorbent are still covered by many impurities and the pores and cavities are not clearly visible on the surface of the biosorbent. However, after activation by citric acid and tartaric acid as seen in Figure 4 (b), the surface of the biosorbent has been cleaned and broaden and we can see clearly there are many pores and cavities appeared that are some parts look quite regular.

3.5 Adsorpstion Isotherm

Determination of Langmuir adsorption isotherms was obtained by creating a relationship curve between the equilibrium Zn (II) ion concentration in aqueous solution (Ce) with Ce/qe (qe, represents the equilibrium adsorption capacity). In Figure 5 it can be seen that the correlation value of R^2 obtained is 0.998.

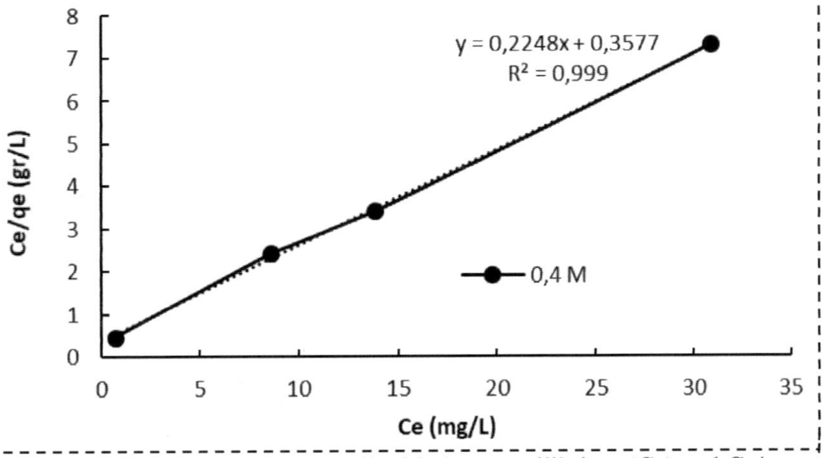

Figure 5. Relationship between liquid phase equilibrium (Ce) and Ce/qe.

The determination of Freundlich adsorption isotherms was obtained by creating a relationship curve between the log Ce and the log qe. Freundlich adsorption isotherm graphs can be seen in Figure 6 where the correlation value obtained is 0.9934.

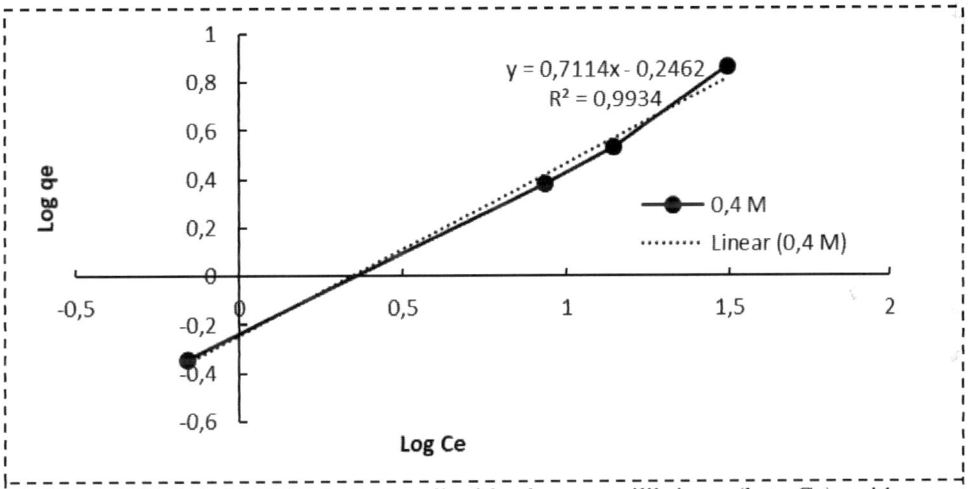

Figure 6. Relationship between liquid phase equilibrium (log Ce) with adsorption capacity on Freundlich adsorption Isotherm (Log qe).

The determination of Langmuir or Freundlich adsorption isotherms is known from the value of R^2 obtained. The adsorption isotherms of zinc (II) ions using palm midrib biosorbent followed the equation having R^2 value close to 1. Based on Figure 5 and Figure 6 it can be seen that Langmuir's isotherm equation has R^2 value close to 1 that is 0.9989, so it can be concluded that adsorption of zinc (II) ions on activated palm midrib follows Langmuir's isotherm equation. The adsorption isotherms show that the initial Zn (II) ion concentration increase is followed by the increase in the amount of adsorbed so as to achieve a balanced state.

The Langmuir equation is used to obtain the maximum adsorption capacity value (qm) and the Langmuir adsorption constant (kL) where by obtaining this value, we can know the amount of adsorbate adsorbed on 1 gr of biosorbent. For this biosorbent the qm value obtained is 4.23 mg/g and the value of kL is 0.62.

4. Conclusions

The adsorption capacity of Zn (II) ion on activated palm midrib reached the equilibrium time at 120 min. The highest adsorption capacity of 5.72 mg/g by using palm midrib that was activated using mixed citric and tartaric acid concentration of 1.6 M was obtained at adsorption condition of initial concentration of 80 ppm zinc (II) ion.

Citric acid and tartaric acid that used as activator during activation process plays an important role on modification of biosorbent surface. The presence of citric acid and tartaric during the activation process resulted in degradation of materials on the surface of materias. The adsorption capacity of biosorbent was influenced significantly by addition of citric and tartaric acid as mixed activator during activation process. The cleaness and opening of pore and cavities would also increase the adsorption capacity significantly. Adsorption of zinc (II) ion on activated palm midrib follows Langmuir's isotherm equation.

References

[1] Malamis S and Katsou E 2013 *J. of Hazardous Materials* **252** 428-461
[2] Anastopoulos I, Bhatnagar A, Bikiaris D N and Kyzas G Z 2017 *Int. J. Mol. Sci.* **18** 114
[3] Kurniawan T A, Chan GY, Lo WH and Babel S 2006 *Chem. Eng. J.* **118 (1)** 83–98
[4] Ali I, Khan T A and Asim M 2012 *Environ. Sci. Poll. Res.* **19(5)** 1668–76
[5] Ghosh G and Bhattacharya P K 2006 *Chem. Eng. J.* **119(1)** 45–53
[6] Khattab I A, Shaffei M F, Hussein H S and El-Rehim S A 2013 *Egypt J. Petrol.* **22(1)** 205–10
[7] Sharma P, Kaur H, Sharma M and Sahore V 2011 *Environ. Monit. Assess.* **183** 151–195
[8] Grassi M, Kaykioglu G, Belgiorno V and Lofrano G 2012 *Emerging Compounds Removal from Wastewater* G. Lofrano (ed.) Springer Science
[9] Rossner A, Snyder S A and Knappe D R U *Water Res.* **43** 3787–3796
[10] Lafi R, Rezma S and Hafiane A 2014 *Desalin. Water Treat.* **1** 12
[11] Ponou J, Wang L P, Matuo S and Fujita T 2014 *Environ. Eng. Manage.* J. **2** 1070–1081
[12] Huang K, Xiu Y and Zhu H 2013 *Environ. Sci. Pollut. Res.* **20** 5930–5938
[13] Chen X, Chen G, Chen L, Chen Y, Lehmann J, McBride M B and Hay AG 2011 *Bioresour Technol* **102** 8877–84
[14] Gisi S D, Lofrano G, Grassi M and Notarnicola M 2016 *Sustainable Mater.* Tech. **9** 10-40
[15] Marshall W E, Wartelle H, Johns M and Toles C A 1999 *Bioresource Technol.* **69** 263–268
[16] Marshall W E, Wartelle L H, Boler D E and Toles C A 2000 *Environ. Technol.* **21** 601–607
[17] Marshall, W E, Chatters A Z, Wartelle L H and McAloon A 2001 *Ind. Crop. Prod.* **14** 191–199
[18] Wong K K, Lee C K, Low K S and Haron M J 2003 *Chemosphere* **50** 23–28
[19] Wafwoyo W, Seo C W and Marshall W E 1999 *J. Chem. Technol. Biotechnol.* **74** 1117–1121
[20] McSweeny J D, Rowell R M and Min S H 2006 *Journal of Natural Fiber* **3** 43-58
[21] Yang J S, Park Y T, Baek K and Choi J 2010 *Sep. Sci. Technol.* **45** 1963-1974
[22] Mulana F, Muslim A, Alam P N and Mariana 2015 Adsorption of Pb (II) Heavy Metals from Wastewater Using Modified Rice Husk as Adsorbent *Proc. of The 5th AIC Unsyiah In conjunction with The 8th ICChESA* 9-11 September, 2015, Banda Aceh, Indonesia
[23] Pratama Y Triandi R and Darjito 2015 *Kimia Student Journal* **1** 741 – 747
[24] Okafor C P 2012 *Inter. J. Electrochemical Science* **7** 12354 – 12369
[25] Mulana F, Mariana, Alam P N and Muslim A 2016 Activation of Palm Midrib by Using Mixed Citric Acid and Tartaric Acid and its Application for Adsorption of Zn (II) Heavy Metals from Wastewater *Proc. of The 6th AIC Unsyiah in conjunction with The 12th ICMSA* 4-6 October 2016, Banda Aceh, Indonesia
[26] Muslim A 2017 *J. Eng. Sci. Technol.* **12(2)** 1823– 4690.
[27] Muslim A, Aprilia S, Suha T A and Fitri Z *2017 J. Korean Chem. Soc.* **61**(3), 89– 96.

IOP Conf. Series: Materials Science and Engineering **334** (2018) 012028 doi:10.1088/1757-899X/334/1/012028

Study on the impact of air quality in agricultural and health sectors

S Chairani[1]
Department of Agricultural Engineering, Syiah Kuala University
Banda Aceh, 23111

E-mail: susi.chairani@gmail.com

Abstract. This study focused on the impact of air quality in agricultural and health sectors. The impact of CO_2 on the agricultural crops was conducted by using literature review and the impact of air quality was conducted using secondary data to calculate the Air Quality Index (AQI), derived from some monitoring stations available in Indonesia. Numerous studies showed that the elevated CO_2 decreased the agricultural productivity. Maize yields decreased by 15 % in areas which used irrigation system and 8 % in areas which used rainfed. Maize yields had already experienced severe losses without increasing CO_2 concentrations. It decreased by 21 % for irrigated maize and 26 % by rainfed maize. In addition, it turned out that other elevated pollutants, such as SO_2, NO_2, SPM, O_3, CH_4, $PM_{2.5}$, PM_{10} and TSP also occurred in the atmosphere. These pollutants' effects might harm human being in term of health concern. The USEPA had developed a tool, called the Air Quality Index (AQI) calculator to calculate the pollutants' concentrations in a daily basis. This tool's function to inform how clean or polluted the air that we breathed was with the health effects based on the concentrations of each pollutant. The AQI also provided the information on sensitive groups, health effects and cautionary statements. Based on the air daily data which derived from Board of Meteorology, Climatology and Geophysics (BMKG) of Indonesia, the AQI in Indonesia varied from good, moderate to unhealthy categories; with level of health concern was respiratory diseases, such as asthma.

1. Introduction

The purpose of this study was to identify the impact of air quality on agricultural and health sectors. These both sectors might be related to each other. Good agriculture made healthy human beings. The sources of agricultural pollutants might be similar with or different from air pollutants. Yet, both agricultural and health sectors got impacted with the same target, human beings.

CO_2 is composed of one atom of carbon and two atoms of oxygen. It is a heavy colorless, odorless and non flammable gas. It is produced during respiration by the decomposition of organic substances; when carbon is burned; or when people and animal breathe out. The US Department of Energy (USDOE) [1] mentioned that the primary sources of CO_2 were divided into two sources: natural CO_2 sources and man made (anthropogenic) CO_2 sources. The greates natural CO_2 sources was derived from the ocean. Other sources came from animal and plant respiration, decomposition of organic matters, forest fires, and emissions from volcanic eruptions. In addition, CO_2 deposits found in rock layers within the Earth's crust. Meanwhile, the anthropogenic CO_2 sources came from our daily

[1] To whom any correspondence should be addressed.

IOP Conf. Series: Materials Science and Engineering **334** (2018) 012028 doi:10.1088/1757-899X/334/1/012028

activities, such as powerplant, transportation, industries, chemical production, petroleum production, and agricultural practices. The USDOE had identified more than 4.000 stationary sources which emit over 3.000 million metric tons of anthropogenic CO_2 per year in USA and some parts in Canada (Figure 1). The blue color represented the electric power which generated the largest emitter of anthropogenic CO_2.

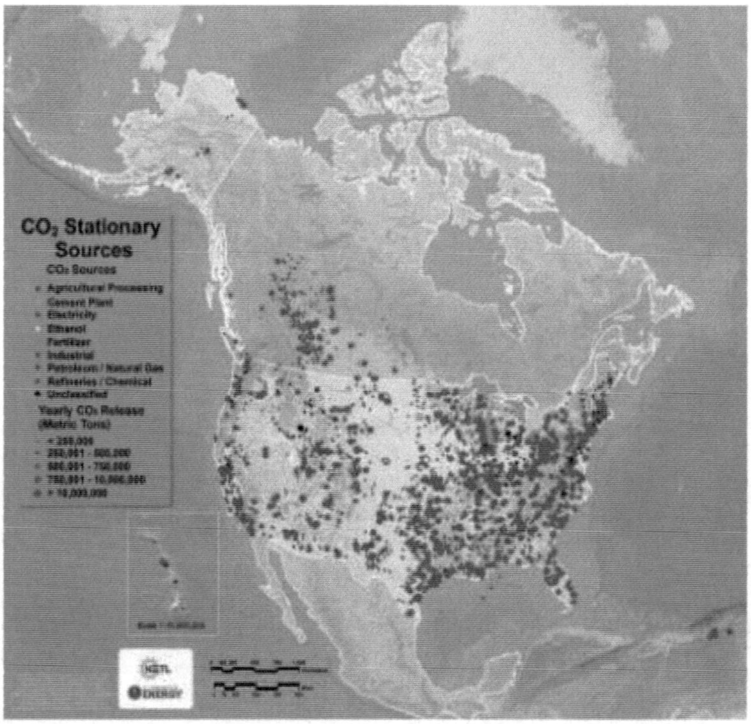

Figure 1. Stationary anthropogenic sources of CO_2 by industry in USA and some parts in Canada [1].

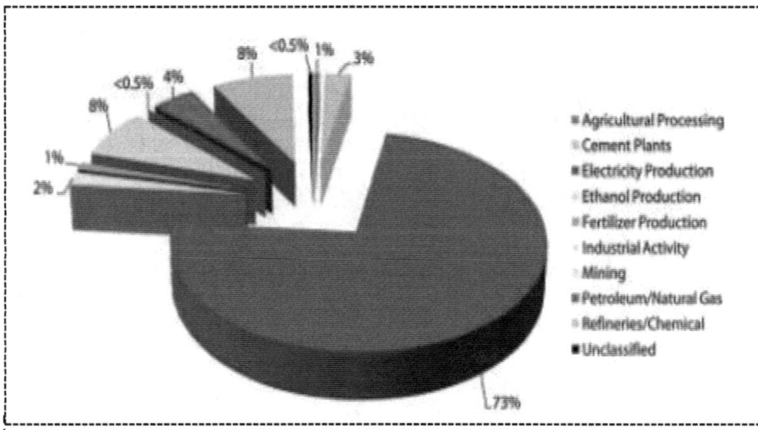

Figure 2. Diagram of stationary anthropogenic of CO_2 emission [1].

Figure 2 above showed the diagram of stationary anthropogenic of CO_2 emission. The largest contributor was the electrical productrion which was about 73%. The elevated CO_2 concentrations in the atmosphere could increase water use efficiency in crops and mitigate yield losses due to climate change [2]. A study reported that yields from wheat, maize, soybean, and rice at levels of CO_2 remained at concentration of 2000 levels would experience severe decline in yields due to higher temperature and drier conditions. But, when all four crops grown at doubled CO_2 concentrations, the crop productivity and yield varied. According to the study, maize experienced yield losses due to the large part of the plant was already greater efficiency in photosynthesis using CO_2 than other crops. Maize yields decreased by 15 % in areas which used irrigation system and 8 % in areas which used rainfed. Maize yields had already experienced severe losses without increasing CO_2 concentrations. It decreased by 21 % for irrigated maize and 26 % by rainfed maize. This study also mentioned that the elevated CO_2 concentrations on crop nutrition could lead to the reduction of nutrients, such as iron, zinc and protein content [2]. Another study which was conducted in Ethiopia [3] revealed that CO_2 emissions had a negative impact on agricultural performance. The impact of CO_2 emissions induced reduction in agricultural factor productivity led to a decrease in production of agricultural traded and non-traded crops, except livestock production.

The impact of elevated CO_2 concentrations on plants in the mid to high latitudes of northern hemisphere showed that the increase of CO_2 concentrations accounted for 98% of water use efficiency on plants. Plants growing at higher altitudes showed higher increase in water use efficiency and more sensitive to the CO_2 increase than plants at lower altitudes. Meanwhile, plants growing at low temperature environments were slightly more sensitive to CO_2 increase than those at higher temperature environments. The research also showed that there was no significant relationship between precipitation and plants' water use efficiency in response to the atmospheric CO_2 increase. In addition, the research reported that altitude and temperature, both impact the water use efficiency in response to the elevated CO_2. Yet, the size of the impact was physically small and could be omitted from ecological models [4].

Agricultural pollutants which important and generally occurring of these pollutants were ethylene, fluorides, ozone, peroxyacyl nitrates (PAN) and sulfur oxides. Their effects upon animals and plants could be best described by the reaction of biologic materials to pollutant concentration and exposure time. The following four criteria were recognized: (1) interference with enzyme systems (2) change in cellular chemical constituents and physical structure (3) retardation of growth and reduction in production from altered metabolism, and (4) acute immediate tissue degeneration [5]. Meanwhile, the sources of air pollutants in the atmosphere were sulfur dioxide (SO_2), nitrogen dioxide (NO_2), Suspended Particulate Matter (SPM), pH of precipitation, ozone (O_3), green house effect, particulate matter (PM_{10}), particulate matter ($PM_{2.5}$) and Total Suspended Particulate (TSP).

2. Methodology

This study focused on literature review on CO_2 impact on agricultural crops and utilized secondary data for air quality which obtained from Board of Meteorology, Climatology and Geophysic (BMKG) of Indonesia to calculate the Air Quality Index (AQI) in Indonesia.

Each of the following pollutant measurement was conducted by BMKG and most of them were analysed at Air Quality Laboratory of BMKG in Indonesia, except for the green house gas which was analysed at NOAA Laboratory. SO_2 measurement was conducted using Passive Sampler with passive gas method. Samples were analysed using an ion chromatography. NO_2 measurement was conducted using Passive Sampler with passive gas method and analysed using a spectrophotometer. SPM sampling was conducted using a High Volume Sampler (HVS) and samples were analysed using an analytical balance. pH of precipitation was measured using an Automatic Rain Water Sampler (ARWS) with Wet and Dry Deposition method. These samples were analysed using an ion chromatograph. O_3 measurement was conducted using an Ozone Analyzer and UV Photometric method. The pollutants measured in Green House Effect were CO_2 and CH_4. Samples were measured using an Analizer Picarro G3010 with Cavity Ring Down Spectroscopy (CRDS). These samples were analysed using an Air Kit Flask Sampling. PM_{10}, $PM_{2.5}$ and TSP were particles in the air with size less than 10 microns, less than 2.5 microns and less than 100 micrometers, respectively.

The quality standard for pollutants was presented in Table 1. The Air Quality Index (AQI) was identified using a tool developed by The US Environmental Protection Agency (USEPA) [7] called the Air Quality Index Calculator which available online on www.airnow.gov. The AQI could be calculated by using equation 1 [8].

$$\mathbf{I_p} = \frac{\mathbf{I_{Hi}} - \mathbf{I_{Lo}}}{\mathbf{BP_{Hi}} - \mathbf{BP_{Lo}}} (\mathbf{C_p} - \mathbf{BP_{Lo}}) + \mathbf{I_{Lo}} \quad (1)$$

Where:
I_p = the index for pollutant p
C_p = the truncated concentration of pollutant p
BP_{Hi}= the concentration breakpoint that is greater than or equal to C_p
BP_{Lo} = the concentration breakpoint that is less than or equal to C_p
I_{Hi} = the AQI value corresponding to the BP_{Hi}
I_{Lo} = the AQI value corresponding to the BP_{Lo}

The AQI was used to report air quality in daily basis. It informed us about how clean or polluted the air was with the health effects which might become our concern. The AQI focused on health effects that we might experience within hour after breathing polluted air. The AQI with level of health concern was classified into 6 categories along with its meanings presented in Table 2 and Table 3, respectively.

Table 1. Quality Standard of Pollutants[a]

Pollutant	Concentration amount	Units
SO_2	0.14	ppb
NO_2	0.08	ppb
SPM	230.00	μ gram/m^3
pH	5.60	-
O_3	120.00	ppb
PM_{10}	150.00	μ gram/m^3
$PM_{2.5}$	65.00	μ gram/m^3
TSP	230.00	μ gram/m^3

[a]BMKG of Indonesia [6]

Table 2. Classification of The Air Quality Index with Level of Health Concern[a]

Air Quality Index	Level of Health Concern	Colors
0 - 50	Good	Green
51 - 100	Moderate	Yellow
101 - 150	Unhealthy for sensitive group	Orange
151 - 200	Unhealthy	Red
201 - 300	Very unhealthy	Purple
301 - 500	Hazardous	Hazardous

[a] USEPA [7]

Table 3. Classification of The Air Quality Index with Level of Health Concern with its meaning[a]

Air Quality Index	Level of Health Concern	Meaning
0 - 50	Good	Air quality is considered satisfactory.
51 - 100	Moderate	Air quality is acceptable; however for some pollutants there may be a moderate health concern for every small number of people who are unsually sensitive to air pollution.
101 - 150	Unhealthy for sensitive group	Members of sensitive groups may experience health effects. The general public is not likely to be effected.
151 - 200	Unhealthy	Everyone may begin to experience health effects; members of sensitive groups may experience more serious health effects.
201 - 300	Very unhealthy	Health alert: everyone may experience more serious health effects.
301 - 500	Hazardous	Health warning with emergency conditions. The entire population is more likely to be effected.

[a] USEPA [7]

3. Results and Discussion

Based on the previous studies, the results showed that the elevated CO_2 could increase crop water use efficiency and reduce crop productivity. Yet, another study showed that there was no significant relatonship between precipitation and plants' water use efficiency with the elevated CO_2.

Table 4. The AQI based on the average pollutant's concentrations from several stations located spreadly in Indonesia in 2016 up to May 2017[a]

Pollutant	Concentration	Unit	AQI[b]	AQI Category[b]	Sensitive Group[b]
SO_2	0.0033	ppb	47	Good	People with asthma are the group most at risk.
NO_2	0.0368	ppb	152	Unhealthy	People with asthma or other respiratory diseases, the elderly, and children are the groups most at risk.
SPM	77.6659	μ gram/m^3	-	-	-
pH	5.1336	-	-	-	-
O_3	17.0000	ppb	16	Good	Children and people with asthma are the groups most at risk.
CO_2	400.0000	ppb	-	-	-

CH_4	1870.0000	ppb	-	-	-
$PM_{2.5}$	18.3623	μ gram/m^3	64	Moderate	People with respiratory or heart disease, the elderly and children are the groups most at risk.
PM_{10}	19.2691	μ gram/m^3	18	Good	People with respiratory disease are the group most at risk.
TSP	0.0000	μ gram/m^3	-	-	-

[a] BMKG of Indonesia [6]
[b] USEPA [7]

Table 5. The AQI with Level of Health Concern in Indonesia[a]

Pollutant	AQI Category[b]	Sensitive Group[b]	Health Effects Statements[b]	Cautionary Statements[b]
SO_2	Good	People with asthma are the group most at risk.	None	None
NO_2	Unhealthy	People with asthma or other respiratory diseases, the elderly, and children are the groups most at risk.	Greater likelihood of respiratory symptoms in active children, the elderly, and people with lung disease, such as asthma; possible respiratory effects in general population.	Active children, the elderly, and people with lung disease, such as asthma, should avoid prolonged or heavy outdoor exertion; everyone else, expecially children, should reduce prolonged or heavy outdoor exertion.
O_3	Good	Children and people with asthma are the groups most at risk.	None	None
$PM_{2.5}$	Moderate	People with respiratory or heart disease, the elderly and children are the groups most at risk.	Unusually sensitive people should consider reducing prolonged or heavy exertion.	Unusually sensitive people should consider reducing prolonged or heavy exertion.
PM_{10}	Good	People with respiratory disease are the group most at risk.	None	None

[a] BMKG of Indonesia [6]
[b] USEPA [7]

The AQI based on the average pollutant's concentrations from several stations located spreadly in Indonesia in 2016 up to May 2017 was presented in Table 4. The results showed that based on its concentrations, the AQI varied from 16 to 152 for each pollutant. The AQI categories varied as well from good, moderate to unhealthy, with sensitive group of people with asthma and respiratory or heart diseases. The AQI only calculated $PM_{2.5}$ (24hr average), PM_{10} (24hr average), CO (8hr average), SO_2 (1hr average), SO_2 (24 hr average), O_3 (8hr average), O_3 (1hr average), and NO_2 (1hr average). The AQI also provided the health effects and cautionary statements presented in Table 5.

4. Conclusion

This study identified the impact of air quality in agricultural by conducting some literature reviews and the impact of air quality in health sectors by calculating the AQI using AQI Calculator in Indonesia. Studies showed that the elevated CO_2 decreased the agricultural productivity. Meanwhile, the air pollutants in the atmospheresuch as SO_2, NO_2, SPM, O_3, CO_2, CH_4, $PM_{2.5}$, PM_{10} and TSP had an impact on human's health. The air quality for health sector used the Air Quality Index (AQI) Calculator to inform how clean or polluted the air that we breathed was with the health effects based on the concentrations of each pollutant. The AQI in Indonesia was in good, moderate to unhealthy categories.

References

[1] USDOE National Energy Technology Laboratory 2017 Carbon dioxide 101 U S DOE available online on www.netl.doe.gov

[2] NASA 2016 NASA Study: Rising carbon dioxide levels will help and hurt crops available online on www.nasa.gov

[3] Mulatu D W, Eshete Z S, Gatiso T S 2016 The impact of CO_2 emissions on agricultural productivity and household welfare in Ethiopia: A Computable general equilibrium analysis *Environ. for Develop.* EfD DP 16-08

[4] Wang G, Feng X 2012 Response of plants' water use efficiency to increasing atmospheric CO_2 Concentration *Environ. Sci. Technol.* 16 8610-20

[5] Middleton J T, Emik L O, Taylor O C 2012 Air quality criteria and standards for agriculture *J Air Pollution Control Assc.* 10 476-480

[6] Badan Meteorologi, Klimatologi dan Geofisika (BMKG) Indonesia 2017 Kualitas udara available online www.bmkg.go.id

[7] USEPA 2017 Air quality index calculator available online on www. airnow.gov

[8] USEPA 2017 Technical Assistance Document for the Reporting of Daily Air Quality – the Air Quality Index (AQI) EPA-454/B-16-002 May 2017

IOP Conf. Series: Materials Science and Engineering **334** (2018) 012029 doi:10.1088/1757-899X/334/1/012029

Activation and characterization of waste coffee grounds as bio-sorbent

Mariana[1,2,4], Marwan[1], F Mulana[1], Yunardi[1], T A Ismail[3] and M F Hafdiansyah[3]

[1] Department of Chemical Engineering, Syiah Kuala University, Banda Aceh 23111, Indonesia
[2] Environmental and Natural Resource Research Center, Syiah Kuala University, Banda Aceh 23111, Indonesia
[3] Undergraduate Program of Chemical Engineering, Syiah Kuala University, Banda Aceh 23111, Indonesia

E-mail: mariana_hasyim@yahoo.com

Abstract. As the city well known for its culture of coffee drinkers, modern and traditional coffee shops are found everywhere in Banda Aceh, Indonesia. High number of coffee shops in the city generates large quantities of spent coffee grounds as waste without any effort to convert them as other valuable products. In an attempt to reduce environmental problems caused by used coffee grounds, this research was conducted to utilize waste coffee grounds as an activated carbon bio-sorbent. The specific purpose of this research is to improve the performance of coffee grounds bio-sorbent through chemical and physical activation, and to characterize the produced bio-sorbent. Following physical activation by carbonization, a chemical activation was achieved by soaking the carbonized waste coffee grounds in HCl solvent and carbonization process. The activated bio-sorbent was characterized for its morphological properties using Scanning Electron Microscopy (SEM), its functional groups by Fourier Transform Infra-Red Spectrophotometer (FTIR), and its material characteristics using X-Ray Diffraction (XRD). Characterization of the activated carbon prepared from waste coffee grounds shows that it meets standard quality requirement in accordance with Indonesian National Standard, SNI 06-3730-1995. Activation process has modified the functional groups of the waste coffee grounds. Comparing to natural waste coffee grounds, the resulted bio-sorbent demonstrated a more porous surface morphology following activation process. Consequently, such bio-sorbent is a potential source to be used as an adsorbent for various applications.

1. Introduction

Raw materials derived from organic materials can be converted into activated carbon since they contain carbon. Activated carbon is a porous solid produced from carbonaceous materials by heating at high temperatures. The more surface area of activated carbon the higher the adsorption capacity [1]. Previous research mentioned that the activated carbon from coffee grounds was able to adsorb iron ion up to 99,43% and was able to adsorb 99% mercury metal [2].

[4] To whom any correspondence should be addressed.

The Province of Aceh, in the northern tip of Sumatera Island, Indonesia has been well known as the biggest Arabica coffee production in the country. In addition to coffee production, the region is also considered as the one with high consumption of coffee in the country. It is easily indicated by large number of traditional and modern coffee shops scattered in the capital city of Aceh Province, Banda Aceh. Every day, those shops produce a large quantity of spent coffee grounds as waste without any effort to convert them to various valuable products. In fact, since the used coffee grounds are wasted as trash, new environmental problems, such as odor and esthetics issues arise. Converting the waste coffee grounds into valuable products will not only provide economic potential, but also at the same time reduce the environmental problem.

In the Province of Aceh, particularly in Banda Aceh, a coffee drink is prepared in traditional way by pouring boil water through a small textile filter bag containing a bed of fresh roasted coffee grounds. A more concentrated coffee can be obtained by pouring the filtrate again and again through the filter bag. During the traditional dripping coffee making, water infiltrates into the coffee-ground and absorbs the liquid compounds and essence from the coffee, subsequently establishing the final, textural micro- or macro-pores within the coffee-grounds. Consequently, waste coffee grounds represent a porous material that has a potential for the use as a bio-sorbent to remove particular compounds from liquid phase.

Natural bio-sorbent of waste coffee grounds requires improvement in terms of their adsorption capacity by increasing their accessible pores and adsorption surface area. Such improvement could be achieved through both physical activation and chemical activation. Physical activation can be done by heating the natural porous materials at high temperatures, while chemical activation can be completed by adding a chemical solution [1]. Previous research conducted by Adhitiyawarman [3] shows that activated carbon from shell activated by chemical solution has water content, ash content, surface area, surface acidity and better adsorption capacity than activated carbon. Other study conducted by Li [4] on the production of activated carbon from seaweed by carbonization and chemical activation processes. It was noted that the purpose of carbonization is to obtain an initial pore on the structure of the activated carbon layer for next activation process. While the purpose of activation is to create new pores and expand the specific surface area of the produced activated carbon. Therefore, in this study we used two processes to obtain activated carbon that is the carbonization pstage following by chemical activation using HCl. This study will compare the characteristics of bio-sorbent before and after the activation process. The chemical composition of the coffee grounds is shown in Table 1.

Table 1. Chemical composition of coffee grounds

Composition	Percentage
Ether extract	0.48
Crude fiber	21.40
Raw protein	10.10
Ash	1.50
Free nitrogen extract	31.30
Tannin	7.80
Peptic	6.50
Non producing sugar	2.00
Low sugar	12.40
Chlorogenic acid	2.60
Caffeine	2.30
Total caffeine acid	1.60

Source: Murthy K N, 200

2. Experimental procedure and methodology

2.1. Materials and equipment
The equipment and tools used in this research were a muffle furnace (Isuzu EPTR-26R), mechanical sieve of 100 mesh, multi-purpose balance (Acis-BC 500), hot plate & stirrer (Cimarec), desiccator, centrifuge, pH meter, electric oven (GE-171), shaker, *Scanning Electron Microscope* (SEM) JEOL-JSM 6510 LA, UV-Vis Spectrophotometer, FTIR Spectrophotometer Prestige-21 (Shimadzu) and X-Ray Diffraction (Shimadzu 6000). While the materials used in this study are the coffee grounds, obtained from the Solong coffee shop located in Lampineung Jl. Panglima Nyak Makam, Banda Aceh, Indonesia.

2.2. Activation of Waste Coffee Grounds
Activation of coffee grounds was performed physically and chemically. Physical activation was conducted through the following stages: the waste coffee ground powder was brewed with hot water for 10 minutes and then filtered. Subsequently, the material was dried under the sun then carbonized in a muffle furnace at 600 °C for 4 h. The carbonized waste coffee grounds then were cooled and weighted to compare their weight before and after carbonization. Furthermore, the carbonized coffee grounds were sieved with the size of 100 mesh [7,8]. The sieved activated coffee grounds were then subjected to chemical activation by soaking in a solution of HCL 0.1 M for 48 h. It then was drained and placed in an oven at 105 °C for 24 h. The dried materials were stored in the desiccator.

2.3. Characterization of Bio-sorbent
The characterized parameters in this study were performed by the method of Scanning Electron Microscopy (SEM), X-Ray Diffraction (XRD), X-Ray Fluorosence (XRF) and Fourier Transform Infrared (FTIR).

2.3.1. Water Content. The calculation of water content was performed according to Equation (1) where the weight of the activated carbon before and after heating at 105°C for 3 h was noted.

$$Water\ Content\ (\%) = \frac{(a-b)}{a} x100\% \tag{1}$$

where:
a = initial carbon weights (g)
b = final carbon weight (g)

2.3.2. Ash Content. The calculation of ash content was performed according to Equation (2) where the weight of the dried activated carbon before and after heating at 600°C for 4 h was noted.

$$Ash\ Content(\%) = \frac{a}{b} x100\% \tag{2}$$

where:
a = ash weight (g)
b = initial dry carbon weight (g)

2.3.3. Absorbability of Iodine. An amount of 2 g activated carbon together with 50 ml of 0.1 N iodine solvents was inserted into a dark and closed bottle, then shaken for 15 minutes. The solution was separated using a centrifuge and a 10 ml of filtrate that was then titrated with 0.1 N sodium thiosulfate solutions. If the yellow color of solvent is close to disappearing, an indicator of 1% amilum was added. The titration was continued until the right blue color disappeared. The capability of the bio-sorbent to adsorb Iodine is given by Equation 3.

$$Adsorbability\ \ Iod\ \left(\frac{mg}{g}\right) = \frac{\left\{10 - \left(\frac{N \times V}{0,1}\right)\right\}}{S} \times 12,69 \times 5 \qquad (3)$$

where:

V = required sodium thiosulfate solvent
N = normality of sodium thiosulfate solvent
S = weight of activated karbon (g)
12.69 = the amount of Iodine needed for a 1 mL sodium thiosulphate 0.1 N

3. Results and Discussion

3.1. Characterization of activated carbon

The analysis results of the activated carbon characteristics are shown in Table 2. Table 2 shows the activated carbon characteristics that including water content of 6.38%, ash content of 1.05% and iodine absorption of 856.58 mg/g. From those activated carbon characteristics, it can be concluded that the prepared activated carbon meets the permitted quality standard as in accordance with the quality requirements of active carbon as outlined in SNI No. 06-3730-1995.

Table 2. Characteristics of activated carbon from coffee grounds.

Parameter	Result analysis	Quality requirement of active carbon (SNI No. 06-3730-1995)
water content	6.38	Maximum 15 %
ash content	1.05	Maximum 2,5 %
absorbability of iodine	856.58	Minimum 750 mg/g

3.2. Morphology of activated carbon

Morphological structure of coffee grounds before activation and after activation was analysed using Scanning Electron Microscope (SEM). The characterization results from SEM were shown in Figure 1.

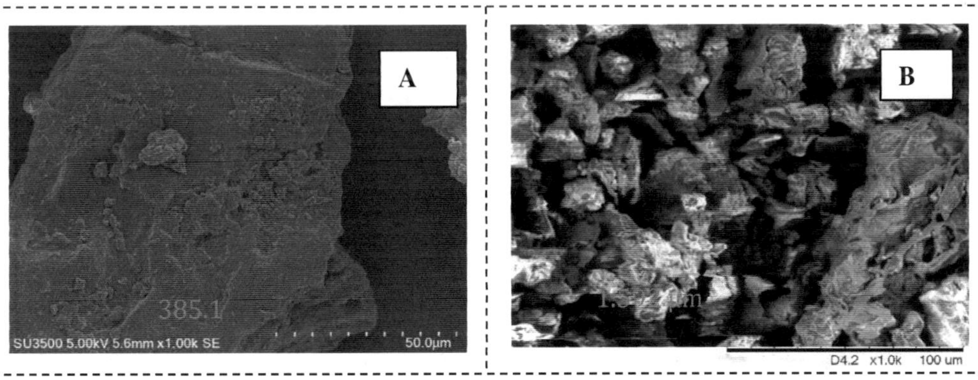

Figure 1. SEM of coffee grounds: (A) before activation, and (B) after activation.

Figure 1 (A) shows that the bio-sorbent coffee grounds that have not been activated yet have irregular pores with an average pore size of only 385.1 µm. The results of this measurement are also in accordance with the results of research that has been done by other researchers [9]. They stated that the adsorbents that have not yet been activated have irregular pores so that the cavity on the pore is still rough and loose. Figure 1 (B) shows that the sorbent that has been activated both physically and chemically has a pore size that is regular and porosity is arranged with the average pore size of 1.592

µm. The above results indicate that bio-sorbent activation can form the distribution of the pores over the entire surface of the adsorbent which can improve the performance of the sorbent.

3.3. Functional groups analysis of bio-sorbent
The functional groups of bio-sorbent of coffee grounds before and after activation were identified using fourier transformation infrared (FTIR) in the range of 400 to 4000 cm^{-1} as shown in Figure 2 and Table 2.

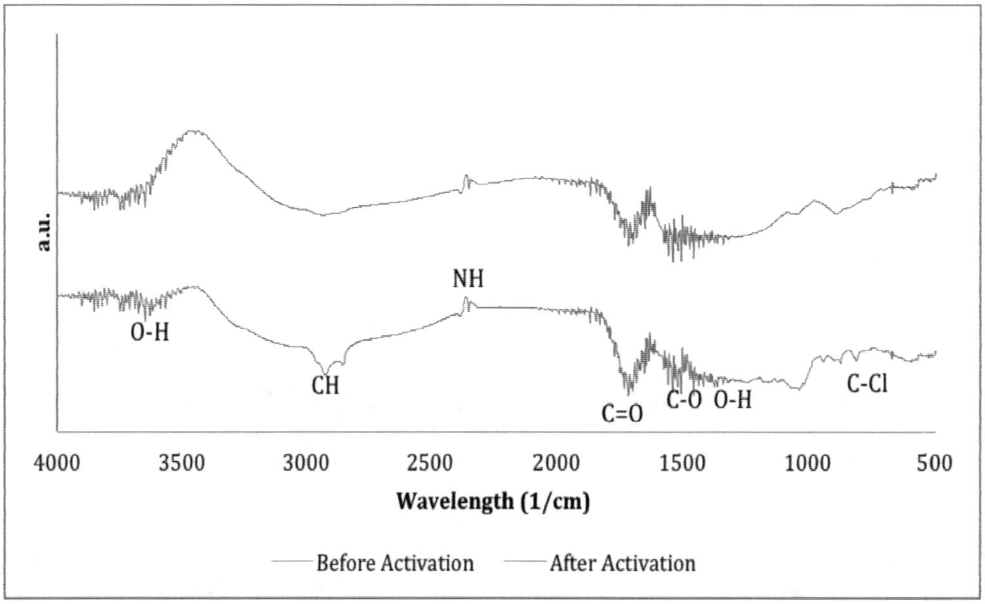

Figure 2. Pattern of FTIR of coffee grounds before activation and after activation.

Table 3. FTIR spectrum results of coffee grounds before activation and after activation.

Functional Cluster	Bonded	wavelenght / cm^{-1}	
		before activation	after activation
Alkanes (-CH-)	CH	2905-2875	2905
Carboxylic Acids, Free OH	O-H	3550-3500	3550
Charged Amines (C=NH$^+$)	NH$^+$	2500-2325	2400
α-amino acid hydrochloride	C=O	1755-1730	1755-1730
-COOH	C-O str plus O-H def	1440-1395 1320-1310	1440-1395
Polyclorinated	C-Cl	800-700	790

Table 3 shows the FTIR analysis of coffee grounds before activation and after activation. Table 3 shows that the spectra before activation and after activation of samples have different patterns. The occurrence of this pattern change is caused by the effect of activation which results in the presence of

O-H group which is hydroxyl group at peak of 1320 - 1310 cm^{-1}. The infrared spectra of the activation sorbent also show the presence of a specific absorbance band of C - O carbons at peak region around 1440 - 1395 cm^{-1}. This phenomenon proves that the bio-sorbent after activation has an R-COOH functional group of carboxylic acids group which can improve the performance of the bio-sorbent when used in the adsorption process.

3.4. X-Rays diffraction (XRD) and X-Ray Fluorosence (XRF) analysis

Figure 3 shows the XRD patterns of both prepared bio-sorbent samples before and after activation. A very significant difference can be seen in Table 3 (A) and (B) where the Cl_2 compound increases when the sample of coffee ground has been carbonized and activated. Figure 3 shows a graph of the analysis results where a very significant difference can be seen in Table 2 (A) and (B). SO_3 compound increases when the sample of coffee ground has been carbonized. It could be happen due to the evaporation of the water content at the time of physical activation, where the space that was occupied by the free and bonded water molecules becomes empty so that the sorbent adsorption capacity increases [11].

The activation process with HCl will decrease the levels of Ca, K and Na that are impurities which bind to the surface of the adsorbent that act as the cation exchange material. With the decreasing levels of Ca, K and Na inside the bio-sorbent there will be a lot of free space on the surface of the bio-sorbent so that the ability of the bio-sorbent to adsorb would also increases.

Performance of the bio-sorbent after activation would increase compared to without the activation process. This is because the bio-sorbent that has been treated by the activation process is generally dominated by Cl_2 about 33.2% of which is amorphous and porous solid and has an inert, neutral, and large surface area so it has large adsorption capacity properties. In addition, the activated bio-sorbents are also dominated by SO_3 of about 32% of which alumina can adsorb cations and anions with a cation adsorption sequence by alumina [10].

Table 4. Data of XRF.

Compounds	Content (%)	
	Before Activation (A)	After Activation (B)
SO_3	27.7	32
$PbCl_2$	7.5	7.9
Cl_2	20.6	33.2
KNa_2	18.9	13.9
N_2	9.4	9.5
Ca	16	3.4

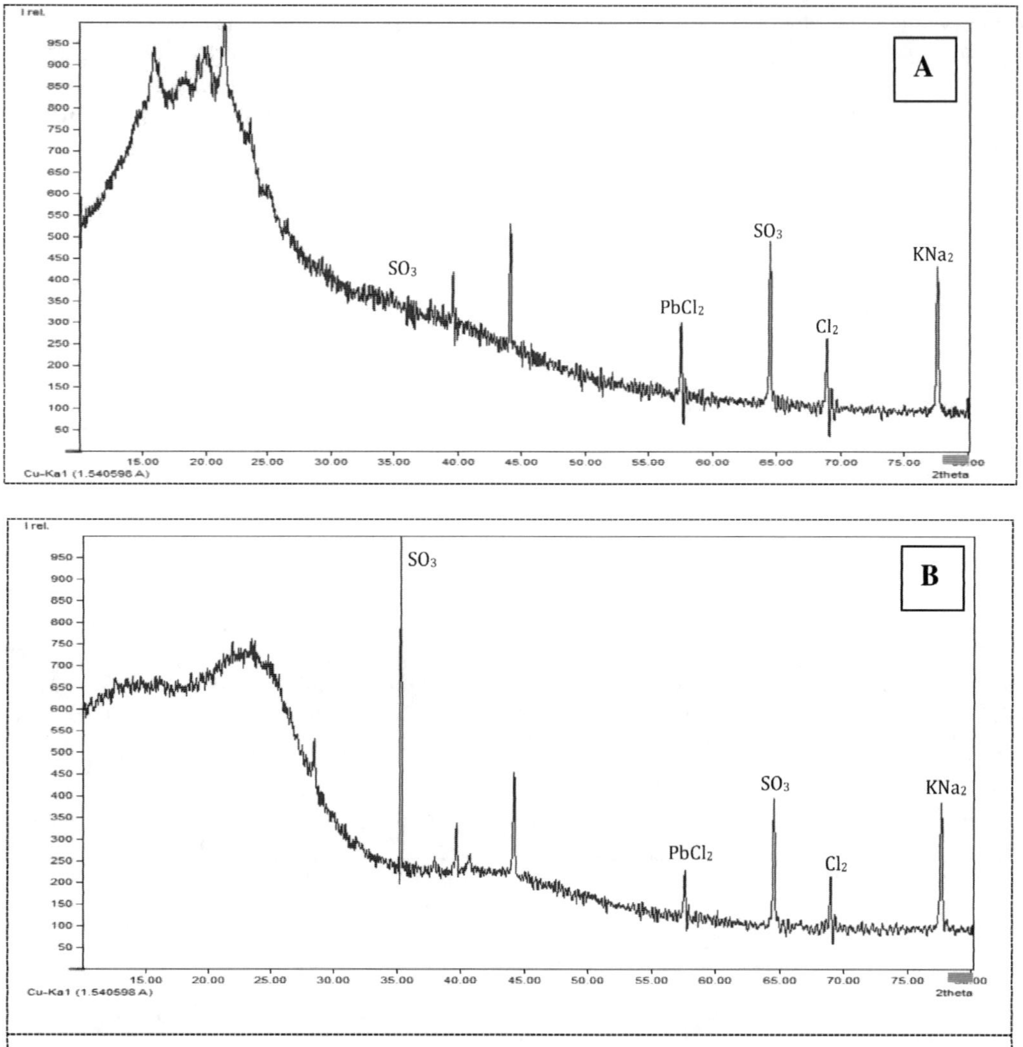

Figure 3. Graph of the XRD pattern: (A) before activation, and (B) after activation.

4. Conclusions

The characteristics results of activated carbon prepared from coffee grounds shows that the activated carbon meets the permitted quality standard according to SNI 06-3730-1995. SEM analysis shows that the pore size of bio-adsorbent of coffee grounds before activation is only about 385.1 μm, while after activation it increases to a pore size of 1.592 μm. FTIR analysis shows the appearance and disappearance of functional groups on bio-sorbent from coffee grounds after activation. The appearance of O-H bond at peaks of 3550 – 3500 cm^{-1} indicated that the surface of the bio-sorbent of activated coffee grounds allows to bind ions. X-Ray Diffraction (XRD) analysis show that Cl_2 and SO_3 intensity increased after activation process that would also increase adsorption capacity of activated bio-sorbent. The impurities contained in the coffee grounds can be minimized through the activation process and carbonization process.

References

[1] Sembiring M T and Sinaga T S 2003 *Arang Aktif*
[2] Lubis S and Nasution R 2002 *Jurnal Natural* **2** 2 12
[3] Adhitiyawarman A I 2015 *Jurnal Kimia Khatulistiwa* **4** 2

[4] Li S, Han K, Li J, Li M and Lu C 2017 *Microporous and Mesoporous Materials* **243** 291-300.
[5] Murthy K N, D'Sa A and Kapur G 2004 *Arabica*, **167** 121-050.
[6] Rizki A P 2015 *Paper presented at the Seminar Nasional Teknik Kimia Kejuangan*
[7] Irmanto I and Suyata S 2010 *Molekul* **5** 1 22
[8] Alfiany H Bahri S and Nurakhirawati N 2013 *Journal of Natural Science FMIPA* **2** 3
[9] Wang Man 2017 *Ecotoxicology and Environmental Safety* **139** 36
[10] Utari Tresye 1994 *Pembuatan Adsorben Alumina Dari Kaolin*. Diss. **27164**
[11] Sari, Wulan T I, Muhsin M and Wijayanti H 2016 *Konversi* **5.2** 20-25

Adsorption performance of fixed-bed column for the removal of Fe (II) in groundwater using activated carbon made from palm kernel shells

N Sylvia[1,4], L Hakim[1], N Fardian[2] and Yunardi[3]

[1] Department of Chemical Engineering, Malikussaleh University, Lhokseumawe, 24351, Indonesia
[2] Department of Nutrition, Medical School, Malikussaleh University, Lhokseumawe, 24351, Indonesia
[3] Department of Chemical Engineering, Syiah Kuala University, Banda Aceh 23111, Indonesia.

E-mail: nxsylvia@gmail.com

Abstract. When the manganese is under the acceptable limit, then the removal of Fe (II) ion, the common metallic compound contained in groundwater, is one of the most important stages in the processing of groundwater to become potable water. This study was aimed at investigating the performance of a fixed-bed adsorption column filled, with activated carbon prepared from palm kernel shells, in the removal of Fe (II) ion from groundwater. The influence of important parameters such as bed depth and the flow rate was investigated. The bed depth adsorbent was varied at 7.5, 10 and 12 cm. At a different flow rate of 6, 10 and 14 L/minute. The Atomic Absorb Spectrophotometer was used to measure the Fe (II) ion concentration, thereafter the results were confirmed using a breakthrough curve showing that flow rate and bed depth affected the curve. The mathematical model that used to predict the result was the Thomas and Adams-Bohart model. This model is used to process design, in which predicting time and bed depth needed to meet the breakthrough. This study reveals that the Thomas model was the most appropriate one, including the use of Palm Kernel Shell for processing groundwater. According to the Thomas Model, the highest capacity of adsorption (66.189 mg/g) of 0.169-mg/L of groundwater was achieved with a flow rate of 6 L/minute, with the bed depth at 14 cm.

1. Introduction

The need of clean water is increasing along with the development of people and technology. This requires various facilities including clean water, whereas most people use groundwater. Groundwater contains relatively high Fe. A small amount of Fe is needed to develop red blood cell, but higher amounts can cause dangerous ailments to human health and the environment. Thus, a further processing step is required in groundwater [1]. One method to eliminate Fe in groundwater is adsorption. This is performed by activated carbon as the adsorbent. The most important part of the process was using effective and cheap adsorbent, for instance using solid waste of crude palm oil plant

[4] To whom any correspondence should be addressed.

in North Aceh, namely palm kernel shell [2, 3, 4]. In crude palm kernel oil, palm kernel shell only a small portion of which is used as fuel, the rest is thrown away and ultimately impacts the environment such as odor and pollution on the ground if it accumulates in large quantities. Okoroigwe et al [5] mentioned that palm kernel shell was effective to be used as active carbon for water treatment.

There are two types of adsorption, namely the batch and the continuous (column) systems. The continuous (column) system is considered more efficient compared to a batch system, and more appropriate to be used in larger scales. Two methods used in a column system down flowed and up flow [6, 7, 8, 9]. In this research, down flow was used due to its easier operation on flow rate. The success of adsorption by column system was determined by several factors such as flow rate, effluent fluid concentration and the height of bed adsorbent.

Mathematical methods have been developed to describe pollution mass in the column-shaped adsorbent. As mentioned by Messaoudi et al [10] performed research on Congo Red bio-sorption (CR) in fixed bed column of fluid using shell jujube.. The highest bio-sorption capacity was 80.49 mg/g) out of 100 mg/ L CR fluid, reached in the flow rate of 2.8 mL/min, bed depth 4 cm and JS particle sizes of 50-100 μm. Data gained was coherent to Thomas Model.

Mohamed [11] conducted research using Spent Tea Leaves (STL) in adsorption crystal violet (CV) from textile plant waste by fixed bed column. Fourier transform infrared (FT-IR) spectroscopy and thermal analysis (TGA) were used for analysis. Based on the breakthrough curve, the optimal result was reached in 30 mg/l of concentration, 20 cm of bed height and 5 ml/min of flow rate.

Nassar et al [1] conducted research on Fe and Mn adsorption on Palm Fruit bunch and maize cob on fixed bed column. He analysed the effect of the process variable of bed height, flow rate, and percentage breakthrough. The results showed that the adsorbent used was effective in adsorbing metal ion.

Most studies of metal ion wells in ground water are conducted using a batch system that is generally very easy to apply in laboratory studies and is limited to the processing of waste water in small quantities. Data obtained under batch conditions are generally not appropriate for most treatment systems. The weakness of the batch system between the contact times in the batch system is too short to reach the equilibrium state in the column system. While the continuous system is inexpensive in operating costs and is able to adapt to the ongoing process [10]. Based on previous research to analyze the performance of adsorbent, this research focused on the effectiveness of adsorption column using palm kernel shell as active carbon, in this matter was down flow fixed-bed column [2,3,4,5].

2. Mathematical description

Various simple mathematical models have been developed to describe and possibly predict the dynamic behaviour of the bed in column performance [12]. The mathematical models such Thomas and Bohart-Adams are the most used to describe the dynamic behaviour of pollutants removal in fixed-bed column [10].

Thomas model is most one the most general and widely used in the column performance theory. This model has been applied for biosorption progress where the external and internal diffusion limitations are absent [13, 14], the maximum adsorption capacity, where q_o of an adsorbent is necessary in the design of an adsorbent column. The Thomas model is commonly applied to determine q_o, where the model is given in equation 1.

$$\frac{C_t}{C_0} = \frac{1}{1+\exp\left[\frac{k_T}{Q}(q_0 m - C_0 t)\right]} \quad (1)$$

where k_T is the Thomas rate constant (L/mg.min), C_0 is the initial lead concentration (mg/L), C_t is the equilibrium concentration (mg/L) at time t (min), q_o is the maximum column adsorption capacity (mg/g), Q is the volumetric flow rate through column (L/min), m is the mass of adsorbent in the column (g).

The Bohart-Adams model assumes that the adsorption rate is proportional to both the residual capacity of the adsorbent and the concentration of the adsorbing species [15]. The Adams–Bohart model is used for the description of the initial part of the breakthrough cure, this model is limited to range of the condition used. Where the model is given in equation 2:

$$\frac{C_t}{C_o} = \exp(k_{AB} C_0 t - k_{AB} N_0 \frac{Z}{u}) \quad (2)$$

From this equation, values describing the characteristic of the operational parameters of the column Z is the bed depth (cm), N_o is the column adsorption capacity in BDST model (g/L), u (cm/min) is the linear velocity calculated by dividing the flow rate by the column sectional area, k_{AB} is the adsorption rate constant (L/mg·min).

3. Experimental procedure and methodology

3.1. Preparation of palm kernel shell activated carbon

Palm kernel shell was obtained from a palm oil company, PT. Syaukat Sejahtera, Kuta Blang, Gerugok, North Aceh, Indonesia. The raw palm kernel shell was washed with deionized water several times to remove all traces, such as oil and dirt. The material was dried in an oven having a temperature of 100 oC until its weight is constant. Carbonization of the material was achieved by heating at 400 °C in a furnace for 3 hours under air. The carbonized material was ground into fine particles and sieved to a particle size of 50 mesh. Activation of the carbonized particles was performed by physical heat treatment in a furnace at a temperature of 800 °C for one hour under air. The activated carbon was then cooled to a room temperature before it was used as an adsorbent.

3.2. Activated carbon analysis

The results of the activated carbon quality test prior to activation and after activation are listed in Table 1. The characteristic of kernel shell palm before activation and after activation is done by using scanning electron microscopy SEM (JSM-6510LA). The functional groups of palm kernel shell before and after activated were identified by Fourier transform infrared FTIR (Jasco 4100).

3.3. Fixed Bed column studies

The groundwater sample was taken the local groundwater well in the Faculty of Chemical Engineering, Malikussaleh University, Lhokseumawe, Indonesia, consisting of Fe (II), Ca (II) and Mg (II). Ca (II) and Mg (II) content are relatively small at 0.01 mg/L. But the content of Fe (II) is quite large at 0.169 mg/L. The concentration of Fe (II) is made influent (C0) 0.169 mg/l, pH 7.09, the flow rate was varied among 6; 10; and 14 L/min. The model of flow was down flowing fixed-bed column continuously as shown in Figure 1. Palm kernel shell adsorbent used was placed in a column of 30 cm and 6.4 cm in diameter, whereas the height of adsorbent in a column was varied to 7.5; 10 and 12.5 cm. For several minutes, samples were collected as much as 5 ml to be analysed using AAS (AA-6300 Shimadzu) as effluent Fe (II) ion concentration (C_t). The flow was stopped once the column was fully marked by $C_t = C_0$. In this model, fluid was absorbed on to the superficial layer of Palm Kernel Shell. The top layer of adsorbent was a place of direct contact to fluid at its highest concentration (C_0), while the lower layer of adsorbent will absorb the liquid in lower concentration. After a certain period, the top layer will be full and be less efficient in absorbing than the lower layer will replace it. This layer is named the adsorption zone. The relation of C_t/C_0 and t,

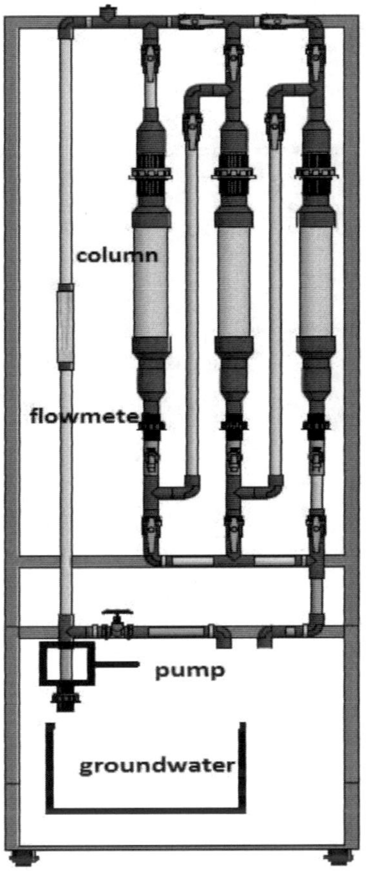

Figure 1. Laboratory scale fixed bed adsorption Column.

4. Result and discussion

The characterization of the adsorbent used in this study includes the quality functional group test and the active carbon texture analysis before and after the activation.

4.1. Element analysis of activated carbon, FTIR and SEM analysis

The determination of moisture content in Table 1 aims to determine the hygroscopic nature of activated carbon. The amount of water vapour in the air as well as the length of the cooling, grinding and sieving processes influence the water content contained in the activated charcoal. High water content in activated carbon can reduce the adsorption power. Measurement of ash content on activated carbon aims to determine the percentage of mineral content. The higher the mineral content, the higher the ash content other than that, the ash can interfere with the adsorption process because of the excessive ash content can lead to blockage of the pores of activated carbon so as to decrease the ability of adsorption.

The volatile substances are substances other than water, i.e., non-carbon compounds and ash contained in the activated carbon. The high volatile substances caused by imperfect decomposition of non-carbon compounds such as CO2, CO, and H2. High volatile substances in activated carbon can reduce the adsorption ability. Carbon content will be higher after physically activated. This indicates that the palm kernel shell fits into the active carbon.

IOP Conf. Series: Materials Science and Engineering **334** (2018) 012030 doi:10.1088/1757-899X/334/1/012030

Table 1. Element analysis of activated carbon.

No	Testing type	Raw palm kernel shell	Palm kernel shell Activated carbon (pysical Teratment)
1	Water	7.64	4.81
2	Ash	1.79	2.53
3	Vaporized steam (950°C)	30.39	14.18
4	Carbon	60.18	78.48

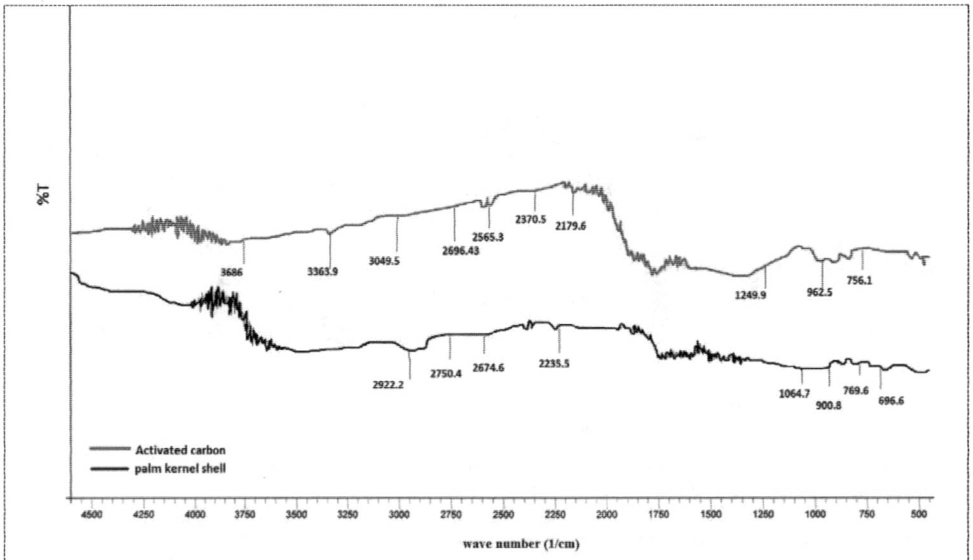

Figure 2. FTIR spectra palm kernel shell and activated carbon.

The activated carbon activation function of the palm kernel shell is analyzed using FTIR (IRPrestige-2, Shidmazu). The functional group analysis was performed to find out the functional group changes occurring on activated carbon, prior to activation and after physics activation. The FTIR spectrum of the activated carbon and palm kernel shell in Fig. 2 has an absorption band at the 4000-2500 cm^{-1} wave number which is a vibration of C-H, O-H, and 2500-2000 cm^{-1} indicating the presence of C = C vibration. The absorption is at the wave number 1550 - 650 cm^{-1} which are the C-O. The resulting activated carbon has an absorption pattern with the O-H (3686, 3363.9 cm^{-1}), C-H (3049.5 cm^{-1}) , C-O (1249, 962.5 and 756.1 cm^{-1}), and C = C (2565.3, 2370.5 and 2179.6 cm^{-1}) bond types [10]. The presence of O-H and C-O bonds indicates that the activated carbon produced tends to be more polar. Thus the resulting activated carbon can be used as an adsorbent.

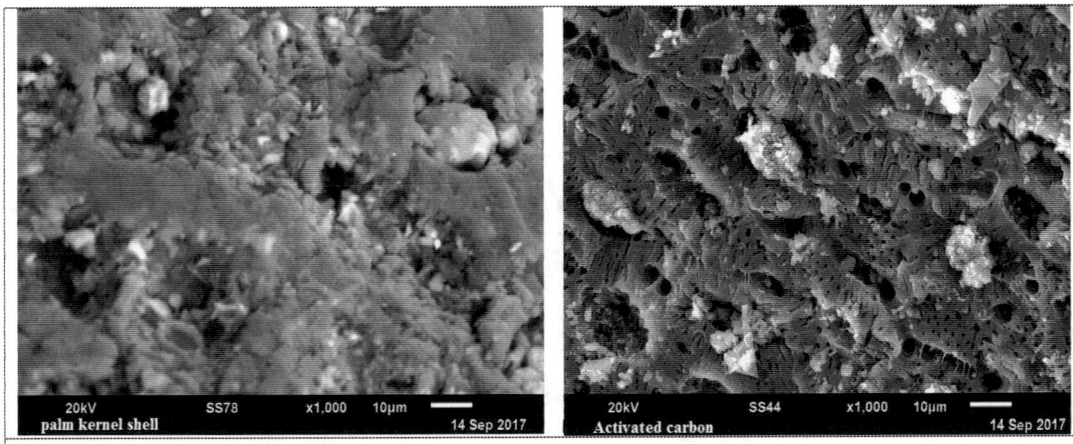

Figure 3. FTIR spectra palm kernel shell and activated carbon.

The SEM observation in the palm kernel shell before activation shown in Figure 3 on the left has a non-uniform particle size and the pores are not yet formed. This is because the process of activation has not been done so that the shape of the surface is still tightly bound to each other causing the morphology and topography of carbon do not form pores. SEM observation of palm kernel shell after physical activation Figure 3 right side of the surface of the pores seen more open pores spread over the surface and walls of active carbon cavity of palm kernel shell. If the activated carbon is cleaner than the impurities the pores is more and the surface area is greater.

4.2. Effect of flow rate on breakthrough curve.

Breakthrough curve for column was determined by plotting the ratio of C_t/C_0 against time, as shown in Figure 4. The column was found to perform effectively in the lowest rate of 6 L/min. Previously, breakthrough and exhaustion was reached at a flow rate of 6 to 14 L/min. The column breakthrough time (C_t/C_0 = 0.05) was reduced from 1000 minutes to 500 minutes, by flow rate was increasing between 6 to 14 L/min. This was due to the decline of contact time, limiting the contact of Fe (II) against activated carbon. On a higher flow rate, Fe (II) had no sufficient time to diffuse into the pores of activated carbon and Fe (II) was out of the column prior to equilibrium. Similar results were found in the adsorption of Fe and Mn of the fixed-bed system using Fruit Bunch and maize cob as adsorbing media in the column [1].

4.3. Effect of different bed depths on breakthrough curve

Breakthrough curve in various bed depths of 7.5 (50 g), 10 cm (150 g) and 12.5 cm (200 g) is shown in Figure 5, at a constant flow rate of 6 L/min, influent Fe (II) ion concentration of 0.169 mg/L and particles size of 50 μ m. The results indicate that by increasing the bed height from 7.5 to 12.5 cm the adsorption capacity of the column increased from 66.189 to 75.685 mg/g (Table 1).

Observations show that the highest adsorption capacity happened in the higher bed. This was caused by a wider surface of activated carbon, making more sites of adsorption [16]. Nevertheless, the conclusion is that the higher the bed depth, the better the q_e.

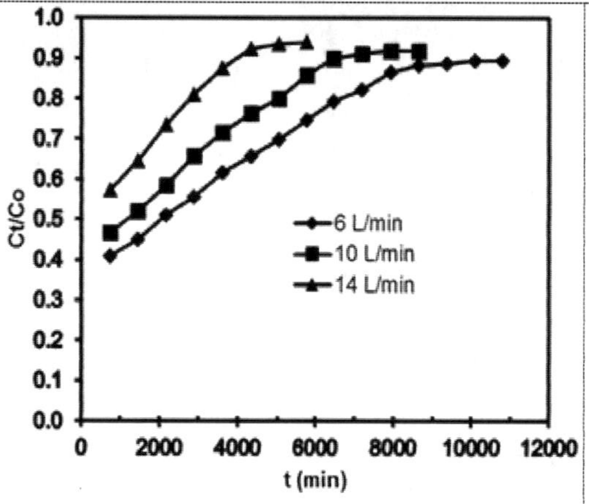

Figure 4. Breakthrough curves: the effect of flow rate on Fe (II) adsorption (C_0 = 0.169 mg/L, Z = 7.5 cm).

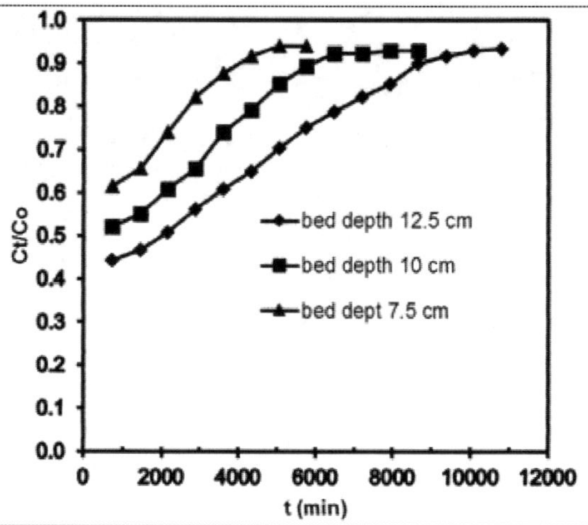

Figure 5. Breakthrough curves: the effect bed depth on Fe (II) adsorption (C_0 = 0.169 mg /L, and flow rate 6 L/min).

4.4. Thomas Model

The result in Table 2 was the capacity of adsorption in various flow rates, measured by the equation above (1). The Thomas model showed Constanta k_T and q_0 was measured by the plot ln $[C_0/C_t-1]$ versus t on various flow rates. Column data using Thomas Model was aimed to determine the Constanta Thomas (k_T) and capacity adsorption (q_e). Measurement coefficient (R^2), Constant relative was measured by using linear regression and error less than 0.15 also shown in Figure 1, as the range of R^2 were between 0.977-0.982. So the correlation C_t / C_0 and t based on the equation was significant. The higher the flow rate, the lower the q_e but the k_T number was increasing. As comparison, q_e was formulated from calculation and research showed that the model was close to the experimental data. The Thomas model was appropriate for the adsorption process where internal and external diffusion are not limiting the process. [16].

4.5. Bohart-Adams Model

The initial part of the breakthrough curve contained experimental data, which used the Bohart- Adams adsorption model for its description. This approach focused on the estimation of characteristic parameters such as maximum adsorption capacity (N_0) and kinetic constant (k_{AB}) from the Adams–Bohart model. For all breakthrough curves, respective values of N_0, k_{AB} was calculated and is presented in Table 3, together with the correlation coefficients (R^2 >0.906). From Table 2, the values of N_0 at all conditions have no significant difference. The values of k_{AB} increased with both initial Fe concentration and flow rate increase, but it decreased with bed depth increase. This shows that external mass transfer in the initial part of adsorption in the column has a large influence on the overall system kinetics [17].

Table 2. The Thomas model parameters for the adsorption iron, C_0 = 0.169 mg/ L

Q, (L/ min)	Z, (cm)	k_T (L /mg.min)	q_e (mg /g)	R^2
	7.5	0.095	66.189	0.981
6	10	0.070	70.875	0.982
	12.5	0.050	75.685	0.982
10	7.5	0.124	35.442	0.979
14	7.5	0.189	5.577	0.977

Table 3. The Thomas model parameters for the adsorption iron, C0 = 0.169 mg/ L

Q, (L/ min)	Z, (cm)	k_{AB} (L /mg.min)	N_0 (mg /L)	R^2
6	7.5	0.005	965.707	0.906
	10	0.004	967.871	0.908
	12.5	0.002	970.658	0.906
10	7.5	0.030	961.986	0.910
14	7.5	0.027	736.021	0.914

5. Conclusion

On the basis of the experimental results of this investigation, activated carbon made of palm kernel shell is applicable for wastewater treatment to remove Fe (II) ion from groundwater. The adsorption of Fe was dependent on the flow rate and bed depth. The Thomas model adequately described the adsorption of Fe (II) ion onto activated carbon in column mode.

Acknowledgement

The authors express their sincere gratitude to Malikussaleh University through the Ministry of Research, Technology and Higher Education for financial support of this research.

References

[1] Nassar M M, Ewida K T, Ebrahiem E E, Magdy Y H and Mheaedi M H 2003 Fixed-bed Adsorption for the Removal of Iron and Manganese onto Palm Fruit Bunch and Maize Cob *Adsorpt. Sci. Technol.* **21** 161

[2] Jalani N F, Aziz A A, Wahab N A, Hassan W H and Zainal N H 2016 Application of Palm Kernel Shell Activated Carbon for the Removal of Pollutant and Color in Palm Oil Mill Effluent Treatment *J.Earth. Environ.Health. Sci.* **2** 15

[3] Abechi S E, Gimba C E, Uzairu A and Dallatu Y A 2013 Preparation and Characterization of Activated Carbon from Palm Kernel Shell by Chemical Activation *Res. J. Chem. Sci.* 3 54

[4] Evbuomwan B O, Agbede A M and Atuka M M 2013 A Comparative Study of the Physico-Chemical Properties of Activated Carbon from Oil Palm Waste (Kernel Shell and Fibre) *I.J.S.E.I.* **2** 75

[5] Okoroigwe E C, Ofomatah A C, Oparaku N F and Unachukwu G O 2013 Production and evaluation of activated carbon from palm kernel shells (PKS) for economic and environmental sustainability *Int. J. Phys. Sci.* **8** 1036

[6] Afroze S, Sen T K and Ang H M 2016 Adsorption performance of continuous fixed bed column for the removal of methylene blue (MB) dye using Eucalyptussheathiana bark biomass *Res. Chem. Intermed*. **42** 2343

[7] Karthikeyan G, Anbalagan K and Andal N M 2004 Adsorption Dynamics and equilibrium Studies of Zn(II) onto Chitosan *Indian J. Chem. Sci.* **116** 119

[8] Chen N, Zhang Z Y, Feng C P, Li M, Chen R Z and Sugiura N 2011 Investigations on the batch and fixed-bed column performance of fluoride adsorption by Kanuma mud *Desalination* **268**, 76

[9] Wu J and Yu H Q 2008 Biosorption of 2,4-dichlorophenol from aqueous solutions by immobilized Phanerochaete chrysosporium biomass in a fixed-bed column *Chem. Eng. J.* **138** 128

[10] Noureddine El Messaoudi, Mohammed El Khomri, Abdellah Dbik, Safae Bentahar, Abdellah Lacherai, Bahcine Bakiz 2016 Biosorption of Congo red in a fixed-bed column from

aqueous solution using jujube shell: Experimental and mathematical modeling. *J.E.C.E* **4** 3848

[11] Mohamed Sulyman 2014 Fixed-bed Column Packed with Low-Cost Spent Tea Leaves for the Removal of Crystal Violet from Aqueous Solution, *I.C.E.S.T* **69** 110

[12] Aksu Z and Gonen F 2004 Bisorption of phenol by immobilized activated sludge in a continuous packed bed: prediction of breakthrough curves *Process. Biochem.* **39** 599

[13] Thomas H C 1944 Heterogeneous ion exchange in a flowing system *J. Am. Chem. Soc.* **66** 1664

[14] Rozada F, Otero M, García A I and Morán A 2007 Application in fixed-bed systems of adsorbents obtained from sewage sludge and discarded tyres *Dyes. Pigm.* **72** 47

[15] Bohart G and Adams E Q 1920 Some aspects of the behaviour of charcoal with respect to chlorine *J. Am. Chem. Soc*. **42** 523

[16] Aksu Z and Gonen F 2004 Biosorption of phenol by immobilized activated sludge in a continuous packed bed: prediction of breakthrough curves *Process. Biochem* **39** 599

[17] Zulfadhly Z, Mashitah M D and Bhatia S 2001 Heavy metals removal in fixed-bed column by the macro fungus Pycnoporus sanguineus, *Environ. Pollut.* **112** 463

Analysis the potential gas production of old municipal solid waste landfill as an alternative energy source: Preliminary results

A P Hayati[1], N Emalya[1], E Munawar[1,3], T Schwarzböck[2], J Lederer[2] and J Fellner[2]

[1] Department of Chemical Engineering, Syiah Kuala University, Jl. Syech Abdurrauf No. 7, Banda Aceh, Indonesia
[2] Institute for Water Quality, Resource and Waste Management, Vienna University of Technology Karlsplatz 13, 1040 Vienna, Austria

E-mail: edi.munawar@unsyiah.ac.id

Abstract: The MSW landfill produces gas which is represent the energy resource that lost and polluted the ambient air. The objective of this study is to evaluate the potential gas production of old landfill as an alternative energy source. The study was conducted by using 10 years old waste in landfill simulator reactor (LSR). Four Landfills Simulator Reactors (LSR) were constructed for evaluate the gas production of old MSW landfilled. The LSR was made of high density poly ethylene (HDPE) has 50 cm outside diameter and 150 cm of high. The 10 years old waste was excavated from closed landfill and subsequently separated from inorganic fraction and sieved to maximum 50 mm size particle prior emplaced into the LSR. Although quite small compare to the LSR containing fresh waste has been reported, the LRS containing 10 years old waste still produce much landfill gas. The landfill gas produced of LSR operated with and without leachate recirculation were about 29 and 21 litter. The composition of landfill gas produced was dominated by CO_2 with the composition of CH_4 and O_2 were around 12.5% and 0.2 %, respectively.

1. Introduction

The Indonesian government enacted Act 18 on Waste Management in May 2008. This acts requires MSWM authorities to implement integrated waste management by considering the waste hierarchy of reduce, reuse and recycle (3Rs) [1, 2]. MSWM includes the collection, sorting and disposal at landfill site. A landfilling of MSW should be carried out in an environmentally sound manner in order to avoid impacts on humans and the environment. In addition, the act obliges local governments to close landfills that operate as an open dump no later than five years after enactment of this act. Within the same time frame, new landfills to be constructed to replace the open dump ones. The construction, operation and closure of the open dump should be carried out in accordance with the standards and regulations in force [2].

Although there are no exact number how many landfills were operated in Indonesia, the State Ministry of Environment (MoE) estimated that all districts and municipalities have at least one

[3] To whom any correspondence should be addressed.

landfill. In 2006, the amount of MSW generated and collected was estimated to be about 38.5 million tons [3]. Most of it ended up at landfill sites spread across the 126 districts and municipalities, while only 7% of the waste generated and collected was recycled or composted. Furthermore, about 37% of the total MSW generated arose from major cities such as Jakarta, Bandung, Surabaya, Medan, etc.[3].

The landfilling of municipal solid waste is associates with the environment problems, such as aesthetic problems, air pollution and ground water contamination. Soil contamination is caused by leachate, a liquid which is produce of degradation organic matter including the precipitation that has percolated through a landfilled waste and leached out any soluble materials such as salt, organic and nitrogen compounds. In addition to the leachate, the MSW landfill also produces gas which is mainly methane gas (CH_4) and carbon dioxide (CO_2). These gases are classified as greenhouse gas that caused the increase of atmospheric temperature so called global warming effect. According to Pohland and Harper, 1985 [4], the quantity and composition of landfill gas formed was significantly influenced by the age of landfilled waste. At the same time, the landfills gas is represent the energy resource that lost and polluted the ambient air since most landfill does not equipped landfill gas capture facilities and subsequently use as energy source. The objective of this study is to investigate the landfill gas production of closed landfill, particularly in tropical climate region which identified by high precipitation rate by using landfill simulator and evaluate its potential as energy source.

2. Materials and Methods

2.1. Experimental setup

Four Landfills Simulator Reactors (LSR) were constructed for evaluate the gas production of old MSW landfilled. The LSR was constructed using high density polyethylene (HDPE) pipe of 50 cm outside diameter (OD), 2.5 cm wall thickness and 150 cm of high (total volume of LSR approximately 0.57 m3). The both end of LSR is equipped with lids made of similar material to LSR body. As during the degradation organic matter process, the biogas and leachate are generates, biogas and leachate outlet ports were installed at the top and bottom of the reactor, respectively. Additionally, a showerhead were installed for fresh water to simulate rainfall or leachate recirculation and distribution were also installed at the top of LSR.

The biogas outlet port was connected to the eudiometer to measure the volume of biogas generate subsequently determine its composition by using gas analyzer (Polytector G750, GFG. Inc). Whilst, the leachate outlet port was connected with U-type silicone tube to prevent ambient air enter to the LSR after the leachate drain out. An external plastic reservoir (10 liter capacity) were placed to collect the leachate produced subsequently re-circulate into the LSR. Four thermocouples have 20 cm of length were installed at 40, 65, 90 and 115 cm of bottom LSR to measure the internal temperature of LSR. A two layers blanket made of synthetic wool and aluminum foil was coverage the entire LSR surface to reduce heat transfer from

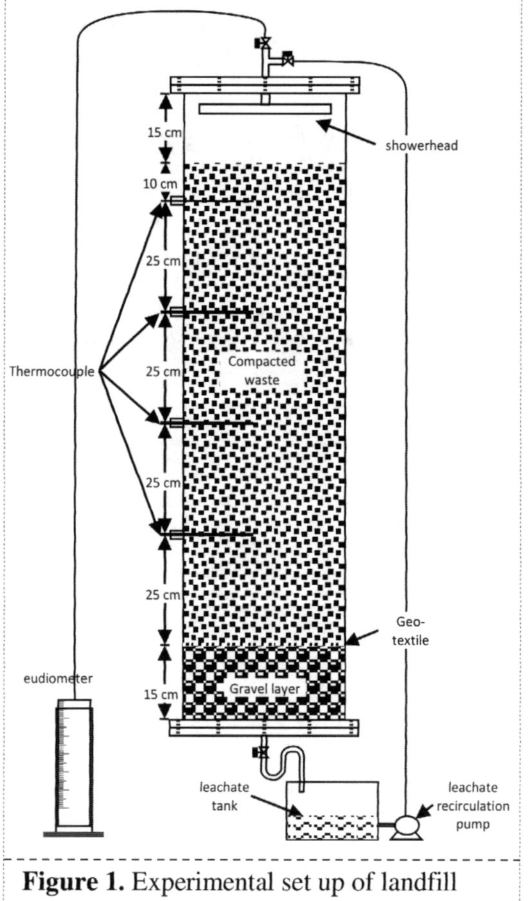

Figure 1. Experimental set up of landfill simulation reactor.

inside to the ambient air. No an electrical device has been use to maintain the internal temperature.

The LSR were filled with a 15 cm layer of sand (particle size ranged 1-2.5 cm) at the bottom followed by geo textile sheet to prevents the pore of gravel layers to be clogged by waste materials. Approximately the last 15 cm at the top of LSR was kept empty as headspace. The dimensions of the reactors are given in Figure 1.

2.2. Waste collection and preparation

The aged waste was collected from Cot Padang Nila landfill. This landfill was located at Tunong Tanjong village, Padang Tiji, Pidie district (5°26'24.3"N 95°49'10.9"E), around 86 km southeast east of Banda Aceh city. The Cot Padang Nila landfill was landfill constructed to received MSW generated from Pidie district area. This landfill has total area around 2.2 ha equipped 3 landfilling cells, treatment pound, and other supporting facilities (see Fig. 2). The Cot Padang Nila landfill was built also by UNDP after tsunami disaster in 2004 and operate immediately after completion. The aged waste was excavated from the first cell which was closed in 2012 after received around 50,000 m^3 of waste. More than 3 ton of waste was excavated to have 1 ton waste with particle size around 1 to 5 cm. Most excavated waste was inorganic matter such as plastic sheet, rubber, textile, and wood mixed with the soil. The inorganic matter was subsequently removed manually of excavated waste to easier sieving the process. The sieving of waste was carried out in two steps, first by using sieving with screen size 5 cm and followed by using sieving with screen size 1 cm. Then the sample was transported to the laboratory for emplaced into the LSR.

Figure 2. Cot Padang Nila landfill, Padang Tiji, Pidie district.

Prior emplaced into the LSR, the collected and sieved waste with maximum particle size of 5 cm was mix to homogenously and divided into four to ensure each LSR received the similar waste. The waste was placed by layers of approximately 30 cm and compacted with hydraulic press up to an apparent density of 375 kg/m^3. Finally, all LSRs was seeded with approximately 20 liters of mature leachate in order to stimulate degradation process and leaching conditions as suggested by Vroon et al., 1999 [5] and Oonk and Woelders, 1999 [6].

Two LSR (A1-A2) was operate according to the standard operating procedure described in Andreas et al., (1998) for LSR under anaerobic conditions. The fresh water was added to simulate the landfill operation with a permeable top-cover and thus the infiltration of rainwater occurred. The other two LSR (B1-B2) was also be operated under anaerobic conditions. However, contrary to LSR A1 and A2, the leachate was re-circulate to simulate landfill operation with a impermeable top-cover and thus the landfill gas capture for use as energy source. The aim this experiment is to determine the residual emission loads from the aged wastes consist of mainly moderately degradable organic compounds as the easily degradable ones have already been degraded. Hence, the landfill gas produced is resulted of the degradability of moderately degradable organic compounds, which will help to determine the ideal way to dealing with the MSW landfill that has been closed.

2.3. Sampling and Analysis

During the experiment, the volume and composition of gas produced were conducted daily basis. The gas was transferred from headspace of the LSR into eudiometer to determine the volume produced gas. The gas subsequently extracted by using portable gas analyzer (Polytector G750, GFG Germany) to determine the gas composition CH_4, CO_2, O_2 and H_2S). In addition to gas, the volume and composition of leachate produced was also determined. The measurement volume, pH and conductivity of leachate samples were carried out every day by using graduated cylinder (Iwaki, Pyrex USA), pH meter (HI9126, Hanna Instruments Inc.USA) and conductivity meter (Pinnacle 541, Corning USA), respectively. COD, NH_4, NO_3 and NO_2 content of the leachate samples were analyzed once a week according to standard methods (APHA, 2005). The heavy metals content (Fe, Cd, Cu, Pb) of the leachate was analyzed with a Atomic Absorption Spectrometer equipped with flame detector (AAS-6300, Shimadzu corp. Japan).

3. Result and discussion

3.1. Landfill gas production

Although four LSR was operated two for each operation method, the data reported here in was one for each. The two other LSR were has some leak in the beginning operation which caused the anaerobic condition has been interrupted for some days before recovered. However, after the leak was fixed, the gas production followed similar trends of other two LSR. The data reported in paper was data collected until 18 day after filling of waste.

The landfill gas was started produce one day after waste filling, this probably due to the microorganism already exist and there was no significant different the environmental condition between the landfill site and LSR. The wastes were excavated and transported from site to laboratory without any adjustment except the particle size. Hence, there was no such long time for bacteria to adapted and restart activity. However, the landfill gas produce quite low compare to the LSR containing fresh waste. The landfill gas produced was around 1.7 l/day, this amounts about a ten of landfill gas produced from LSR containing fresh waste [7]. This was caused by the aged wastes consist of mainly moderately degradable organic compounds as the easily degradable ones have been degraded. The total landfill gas produced of LSR operated with and without leachate recirculation were about 19 and 26 l, respectively (see Figure 3).

The composition of landfill gas produced was dominated by CO_2, with average CH_4 and O_2 contents were 3.7 and 0.7, respectively. The landfill gas composition was lowest compare to the typical landfill gas was reported of experiments with bioreactor landfill simulators; nevertheless biogas composition depends strongly on composition of degradable organic matter and the pH of the system [8]. The low CH_4 content in landfill gas usually occurs in the beginning of anaerobic degradation stage, where the acidogenic process take place. This condition was indicated by decreasing of pH leachate.

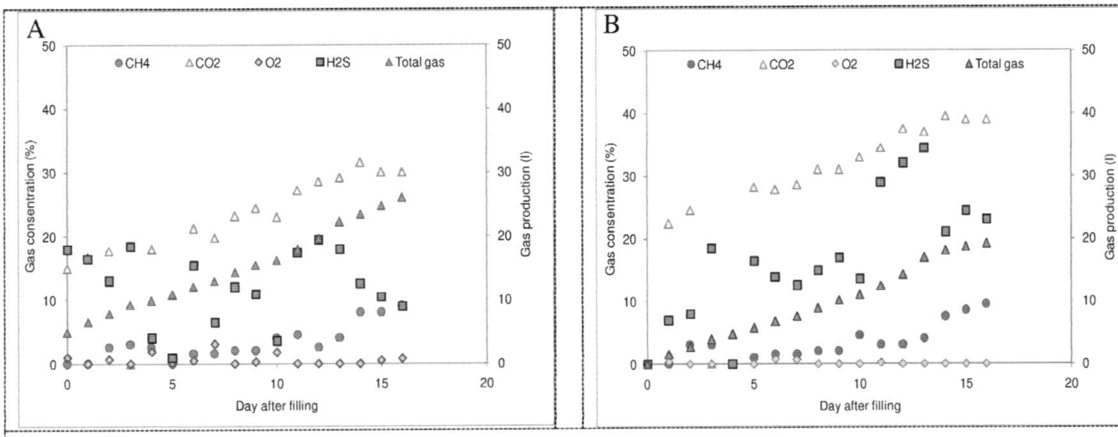

Figure 3. Landfill gas production and composition, (A) LSR operated without leachate recirculation and (B) LSR operated with leachate recirculation.

3.2. Physical-chemical parameters

In addition to volume and composition of landfill gas, the physical-chemical parameters also measured. The temperature inside LSR was measured by using digital thermometer which connected to the attached thermocouple has 20 cm of length. The temperature of LSR was relative constant at about 30 (± 2)° C. There was no significant change compare to the initial condition. This might due the easily degradable organic compound has been completely degraded and the degradation of moderately degradable organic compounds was running very slow. Thus the Heat generated from the degradation of organic compounds does not increase the temperature of the LSR. The LSR temperature around the 18 day operation was about 30 (± 2)° C. The similar trend was found to the pH value of leachate. The initial pH was around the initial pH was about 6.9 then increase to about 7.1 at day 19. For all reactors, the conductivity reached its lowest level 6.93 mS/cm at day 10 and reached its highest level 7.7 due to salt dissolved into the leachate.

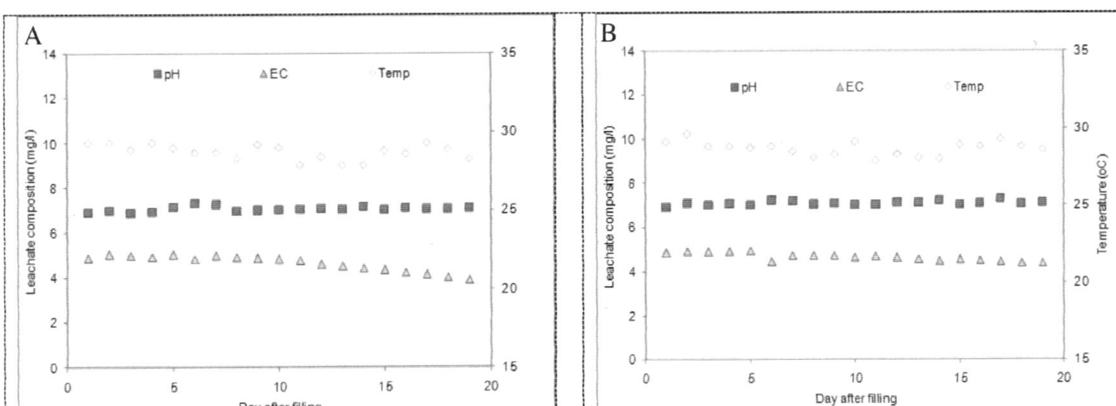

Figure 4. Physical-chemical parameters of LSR, (A) LSR operated without leachate recirculation and (B) LSR operated with leachate recirculation.

3.3. Biological parameters

As the leachate analysis was carried out once a week, only few data have been collected. Based on these data collected, the COD concentration has trend to decrease week to week. The COD concentration of LSR A was decreased from 799 mg/l at first week to 342 mg/l after three weeks operation, while COD concentration LSR B was decrease from 480 mg/l to be 320 mg/l (see Fig. 5). On contrary to the COD, the concentration of NO_3 of both LSR tend to increase. The NO_3 concentration of LSR A was increase from 17 mg/l to 31 mg/l, while NO_3 concentration LSR B was increased dramatically from 8 mg/l to 40 mg/l. Although the NO_3 concentration has increase tendency, but the NH_4 concentration in both LSR was observed stable. The NH_4 concentration of both LSRs were around 12.6-14.2 mg/l.

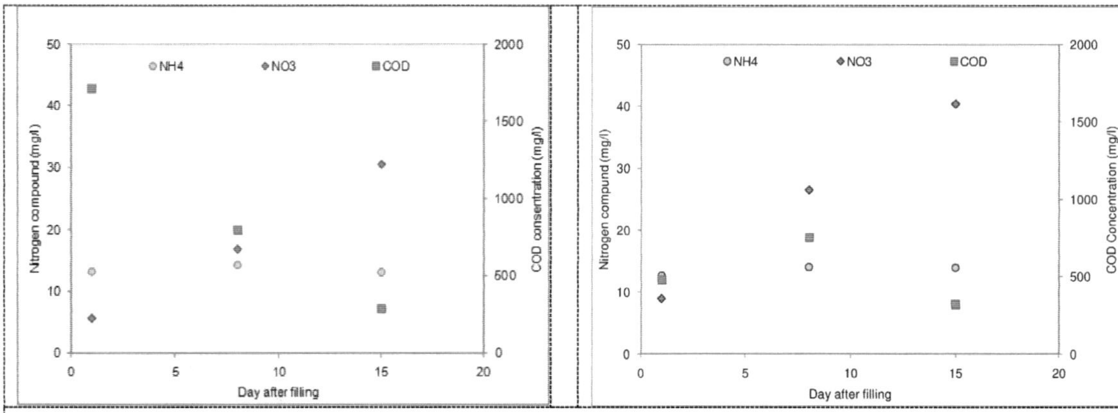

Figure 5. Biological parameters of leachate produce from landfill simulator, (A) LSR operated without leachate recirculation and (B) LSR operated with leachate recirculation.

4. Conclusion

The landfilling of municipal solid waste is associates with the environment problems, such as aesthetic problems, air pollution and ground water contamination. In addition to the leachate, the MSW landfill also produces gas which is mainly methane gas (CH_4) and carbon dioxide (CO_2) which classified as greenhouse gas that caused the increase of atmospheric temperature so called global warming effect, the landfills gas is represent the energy resource that lost and polluted the ambient air since most landfill does not equip landfill gas capture facilities and subsequently use as energy source. The aim of this study is to investigate the landfill gas production of closed landfill, particularly in tropical climate region which identified by high precipitation rate by using landfill simulator and evaluate its potential as energy source.

The landfill gas was started produce one day after waste filling, this probably due to the microorganism already exist and there was no significant different the environmental condition between the landfill site and LSR. The highest methane gas composition was found at reactor B that is about 8.5% and O_2 0.2%. Along with the length of time the waste remains in the reactor, the methane gas produced increases each day on all reactors. Leachate circulation served to keep the moisture of waste and not only can reduce water and soil pollution but also can accelerate the degradation of organic compounds and increase the methane gas production.

References

[1] GoI, *Act of The Republic of Indonesia No. 18 year 2008 Regarding Waste Management*. 2008, Ministry of Environmental: Jakarta.

[2] GoI, *The Goverment Regulation No. 81 of 2012 on Municipal Solid Waste Management*. 2012: Jakarta, Indonesia.

[3] MoE, *Indonesian Domestic Solid Waste Statistic Year 2008*. 2008, State Ministry of Environment the Republic of Indonesia: Jakarta, Indonesia.

[4] Pohland, F.G., *Leachate Recycle as Landfill Management Option.* Journal of the Environmental Engineering Division, 1980. **106**(6): p. 1057–1069.

[5] Vroon, R., H. Oonk, and W. van Marwijk, *A laboratory-scale exploration of the long-term behaviour of mechanically separated organic residue in a flushing bioreactor.* Waste Management & Research, 1999. **17**(6): p. 527-534.

[6] Oonk, H. and H. Woelders, *Full-scale demonstration of treatment of mechanically separated organic residue in a bioreactor at VAM in Wijster.* Waste Management & Research, 1999. **17**(6): p. 535-542.

[7] Valencia, R., et al., *The effect of hydraulic conditions on waste stabilisation in bioreactor landfill simulators.* Bioresource Technology, 2009. **100**(5): p. 1754–1761.

[8] Garcia–Heras, J.L., *Reactor Sizing, Process Kinetics and Modelling of Anaerobic Digestion of Complex Wastes*, in *Biomethanization of the Organic Fraction of Municipal Solid Wastes*, J. Mata–Alvarez, Editor. 2002, TJ International Ltd., Padstow: Cornwall, UK.

IOP Conf. Series: Materials Science and Engineering **334** (2018) 012032 doi:10.1088/1757-899X/334/1/012032

Assessment of groundwater vulnerability using DRASTIC Model and GIS : A case study of two sub-districts in Banda Aceh city, Indonesia

I Machdar[1,3], T Zulfikar[1], W Rinaldi[1] and Y Alfiansyah[2]

[1] Chemical Engineering Department, Faculty of Engineering, Syiah Kuala University, Jalan Tengku Syech Abdur Rauf No. 7, Darussalam, Banda Aceh 23111, Indonesia
[2] Civil Engineering Department, Faculty of Engineering, Syiah Kuala University

E-mail: machdar@unsyiah.ac.id

Abstract. This present study assessed the groundwater vulnerability to protect aquifer in part of Banda Aceh City (the sub-district of Banda Raya and Lueng Bata), Indonesia. The study provides an additional tool for local planner and manager as for managing and protecting groundwater resources. The study area covers 1,164 ha and total population was estimated around 50,000 inhabitants. DRASTIC model in a GIS (Geographic Information System) environment was used in this study to generate vulnerability maps. The maps were created by applied seven criteria as standard in DRASTIC approach, i.e. depth to groundwater, recharge, aquifer type, soil properties, topography, impact of the vadose zone, and hydraulic conductivity. The vulnerability maps provides five categories of vulnerability, i.e. less, low, medium, high, and very high. It was found that the village areas, labelled with the high groundwater pollution potential, are mainly in the area of Lamlagang and the part of Geuce Kaye Jatoe and Geuce Komplek (Banda Raya sub-district) and the part of Batoh and Suka Damai (Lueng Bata sub-distric) This study prompts that the DRASTIC approach is helpful and efficient instrument for assessing groundwater vulnerability. The generated map can be an effective tool for local administrators in groundwater management as well.

1. Introduction

Surface water and groundwater are both essential sources of water supply and agriculture. Although surface water is used frequently as source of water supply due to its capacity and continuity, to date, groundwater is still a main source in several areas especially in developing countries. Issues related to groundwater mostly address to ecological degradation, salinization, and massive overuse [1, 2]. Groundwater pollution is problematic in Indonesia, for example, in Jakarta 45% of groundwater is already contaminated by faecal coliform [3]. Mainly, the sources of the contaminants originate from septic tank overflow [4], landfill leachate [5], industrial effluent [3].

Assessments and actions for groundwater pollution prevention are more effective ways than remedial ones. It is because a contaminated groundwater is very expensive to recover and take a few years back to origins. The investigation of potential for groundwater contamination or groundwater vulnerability has been intensively studied with a number of approaches during the past year [5-10].

[3] To whom any correspondence should be addressed.

The DRASTIC method has been most frequently applied for mapping vulnerability in porous aquifer [11].

The objective of this study was to elaborate groundwater vulnerability in a shallow aquifer using DRASTIC model and a GIS software. Vulnerability means the sensitivity of an aquifer system to deterioration due to an external action [12].

2. Study Area

The study region in which the DRASTIC model and GIS was adapted is located in the Banda Aceh City, in the northwestern tip of the Sumatra island (Figure 1). The city of Banda Aceh is the provincial capital of Aceh Province, Indonesia, and is also center of education, commerce, and culture of the province. This region is characterized by a very flat with average elevation around 2 meters above sea level. The city comprises 9 sub-districts, i.e. Baiturrahman, Banda Raya, Jaya Baru, Kuta Alam, Kuta Raja, Lueng Bata, Meuraxa, Syiah Kuala, and Ulee Kareng.

In this study, the sub-district of Banda Raya and Lueng Bata was selected since it has been subject to inter-city terminal, warehousing, permanent settlement and business center since the development of Banda Aceh's spatial planning in 2006. The sub-district of Banda Raya and Lueng Bata comprises 10 and 9 villages, respectively, and has a total surface of 1,164 ha. At present, the total population of the two districts is approximately 50,000 inhabitants. It is around 20% of the total population in the Banda Aceh City.

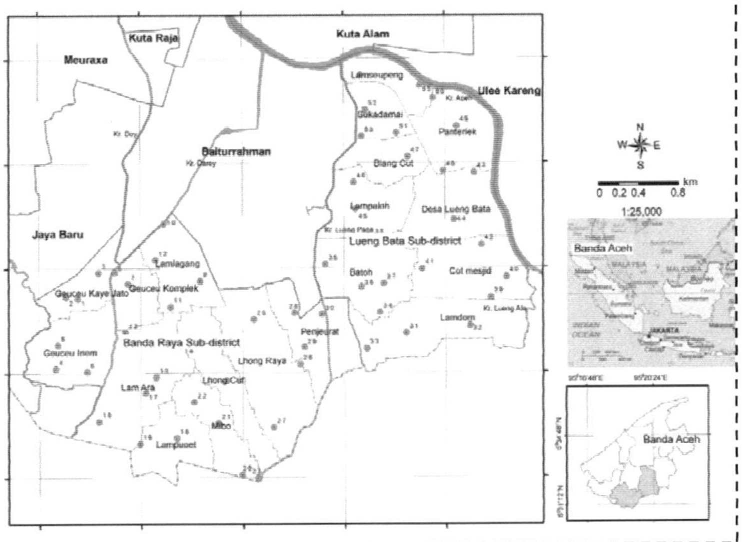

Figure 1. Location Map of Banda Aceh City, boundary of the study area and monitoring wells.

3. Experimental procedure and methodology

3.1. The DRASTIC index method

The DRASTIC is an approach developed by the US Environmental Protection Agency (USEPA) to provide a standardized evaluation of the potential for groundwater contamination [11]. The DRASTIC index or vulnerability rating present numerical ranking collective description of all the primary hydrological and geological considerations that influence groundwater flow through the vertical profiles of an area. In this approach, spatial collective data on depth to groundwater (D), recharge (R), aquifer type (A), soil properties (S), topography (T), impact of the vadose zone (I), and the hydraulic conductivity (C) are collected. For a defined area being investigated, each criterion is assigned a typical range and a rating relative value scale (r) from 1 to 10. In this scale, the higher values represent more sensitive area for contamination. These index criteria further are combined in an additive equation. Moreover, each criterion is decided a weight factor (w), from 1 to 5, reveal the relative importance of each factor. In the end, the linear equation of the total impact criterion score as the DRASTIC index or vulnerability rating has the following formula:

$$\text{DRASTIC index (vulnerability rating)} = D_r D_w + R_r R_w + A_r A_w + S_r S_w + T_r T_w + I_r I_w + C_r C_w \quad (1)$$

where w is weight and r is rank.

In this formula, it applies four assumptions, i.e. (i) the contaminant is introduces at the ground surface; (ii) the contaminant is leached into the groundwater by precipitation; (iii) the contaminant is soluble; and (iv) the evaluated area is 40 ha or larger [11]. Obtained data is simplified by a GIS model interface. The GIS software was used in this study (ArcGIS) to prepare seven maps represent the seven criteria of DRASTIC. Individual map was classified and designated ratings and weighted according to DRASTIC standards [13] as shown in Table 1 and Eq. (1) was rewritten as follows:

$$\text{DRASTIC index (vulnerability rating)} = 5D_r + 4R_r + 3A_r + 2S_r + T_r T_w + 5I_r + 3C_r \quad (2)$$

Table 1. DRASTIC standard weight.

Criteria	Weight
Depth to groundwater (*D*)	5
Recharge (*R*)	4
Aquifer type (*A*)	3
Soil properties (*S*)	2
Topography (*T*)	1
Impact of the vadose zone (*I*)	5
Hydraulic conductivity (*C*)	3

4. Results and Discussion

4.1. Depth to groundwater (D)
Depth to groundwater is crucial criterion because it assigns the length that contaminant have to deliver prior to getting the groundwater. A deeper groundwater (low water table) is less vulnerable than high water table. In this study, the depth to groundwater was monitored from existing inhabitant shallow wells that scatter in the sub-district of Banda Raya and Lueng Bata. The total sampling comprises 55 shallow wells, 30 wells located in Banda Raya and 25 wells in Lueng Bata Sub-District. It was found that, depth to ground water was very shallow varies from 0.23 to 3.49 m._Accordingly, DRASTIC rating values were 9 and 10 for shallow and deep ground water, respectively [14] and affected by a weight of 5 (Figure 2).

4.2. Recharge (R)
Groundwater recharge is essential parameter for vulnerability because it is a route transportation of contaminant into aquifers. Net recharge criterion of this study is the sum of precipitation without taking in account of surface runoff and evapotranspiration. The source of recharge is the annual rainfall data taken from a local climatology station for 10-year period. The average annual recharge value was 130 mm, correspond to DRASTIC index of 8 (Figure 3).

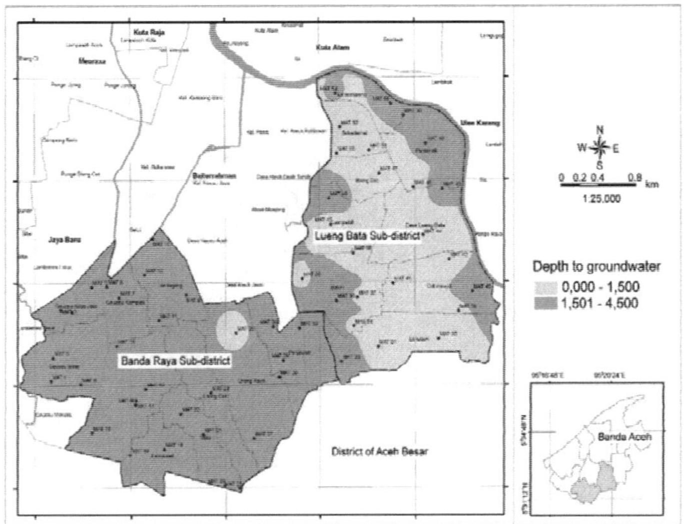

Figure 2. Depth to groundwater of the study area.

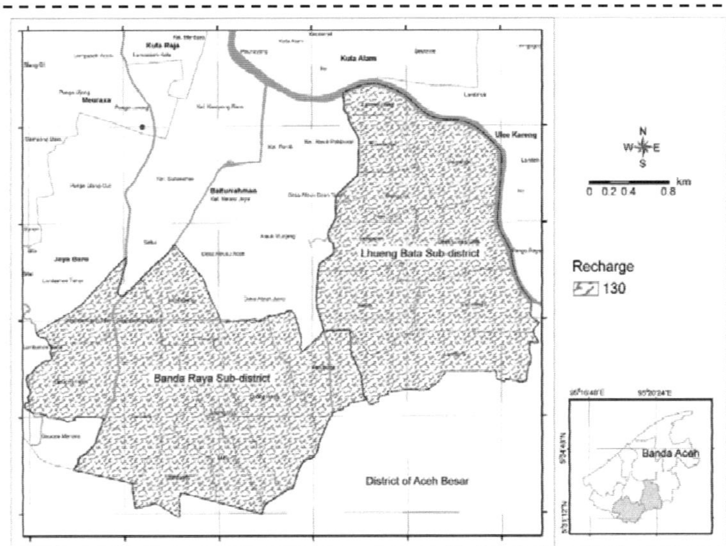

Figure 3. Recharge of the study area.

4.3. Aquifer type (A)

The aquifer layer acts a main function over the pathway and leaching of contaminant. Normally, the larger the particle size distribution or the more permeable rock within the aquifer giving the higher the permeability and the lower the attenuation capacity. Aquifer layer information of this study was prepared using research data from the BGR Germany survey [15]. The main aquifer material layers in the study area are sand, gravel, and lime stone. The layers were assigned rating value from 2 to 8 and affected by a weight of 3 (Figure 4). A higher aquifer layer rating value implies higher permeability of the aquifer media and higher vulnerable potential.

IOP Conf. Series: Materials Science and Engineering **334** (2018) 012032 doi:10.1088/1757-899X/334/1/012032

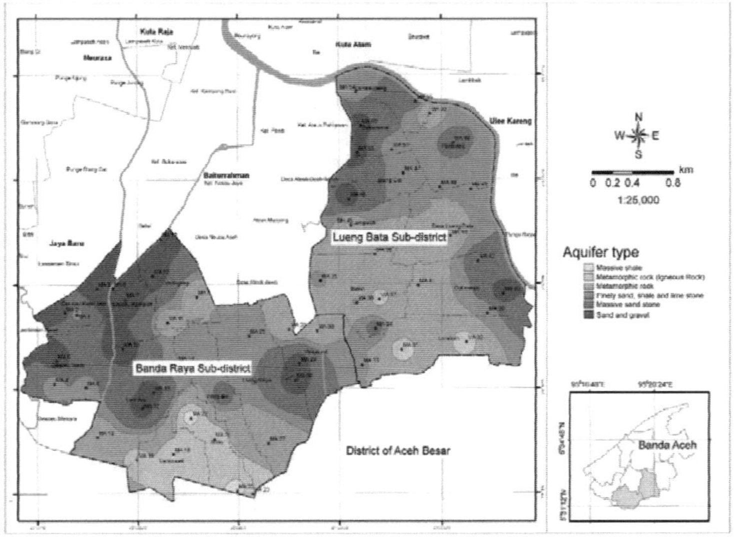

Figure 4. Aquifer type of the study area.

4.4. Soil properties (S)

The soil layer is the upper part of the vadose zone which averages to a depth of 2 m or less from the ground surface. Soil properties affect groundwater pollution potential as they control the movement of air and contaminant [16]. Soil property information was evaluated from the map gathered from the BGR Germany survey [15]. The primary soil of the district of Banda Raya and Leung Bata is clay and sandy clay. DRASTIC standard ratings for this criterion vary between 1 and 5 (Figure 5).

4.5. Topography (T)

Topography (slope of an area) offers a high possibility for pollutant to infiltrate. It will be related to higher groundwater contamination. Moreover, topography controls the direction of contaminant flow. Areas with low land slope tend to hold water for longer time and enable a greater infiltration or recharge to generate a greater potential contaminant transportation. The topography map was obtained from the BGR Germany survey [15]. Slope variation in the study area is very small and is divide into two classes i.e. 2-6% and 6-12%. Areas of steadily low topography have higher vulnerability as contaminant can pool and more contaminant percolation into the groundwater. Topography ratings, according to standard DRASTIC for this area study are 5 to 9 [14] (Figure 6).

4.6. Impact of the vadose zone (I)

The vadose zone is an unsaturated layer above the water table. The impact of vadose zone on groundwater potential contamination is in principle identical to that of characteristic soil cover. The impact of vadose zone is a difficult phenomenon, depending on aquifer layers and topographic characteristics [17]. A higher impact of vadose zone if it is composed of porous soils. The data for impact of vadose zone was retrieved from identification and classification of the sub-surface geology map. The vadose zone were assigned rating value from 1 to 6 and affected by a weight of 5 (Figure 7). Higher rating value means higher vulnerability.

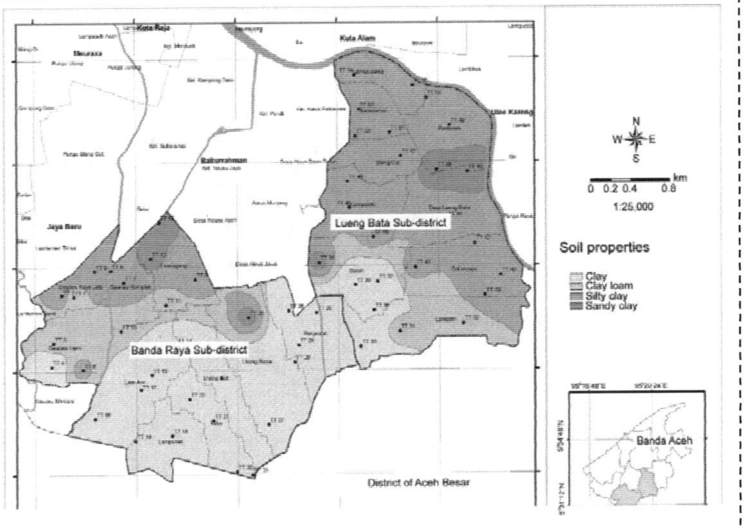

Figure 5. Soil properties of the study area.

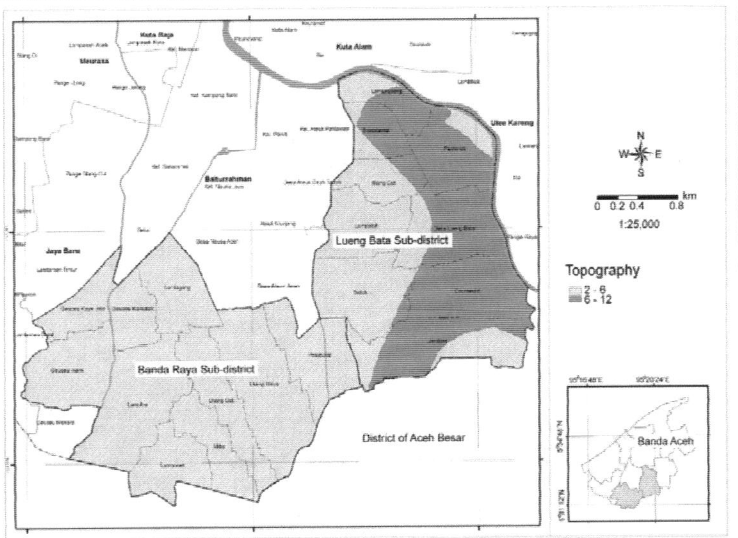

Figure 6. Topography of the study area.

4.7. Hydraulic conductivity (C)

Hydraulic conductivity is the rate of water flow through horizontally an aquifer layer. Hydraulic conductivity of an aquifer is the ability of the aquifer to spread water depends on its permeability and the amount of saturated zone. As the conductivity increase, dispersion of contaminants in aquifer also increases. Accurate values of the hydraulic conductivity are not easy to obtain and it is also reflected as a drawback of the DRASTIC approach [8, 9]. In this study, the values of aquifer hydraulic conductivity was calculated based on pumping tests of existing inhabitant shallow wells in around study areas. The hydraulic conductivity values in the area were assigned rating value of 1 (0 – 4.08 m/d) and 2 (4.08 – 12.24 m/d) (Figure 8). An area with high conductivity is vulnerable to considerable contamination of groundwater. However, the contaminant can immigrate immediately through the aquifer. This phenomenon is distinct from an aquifer which has an impermeable media, it can still water content in the presence of fractures [6].

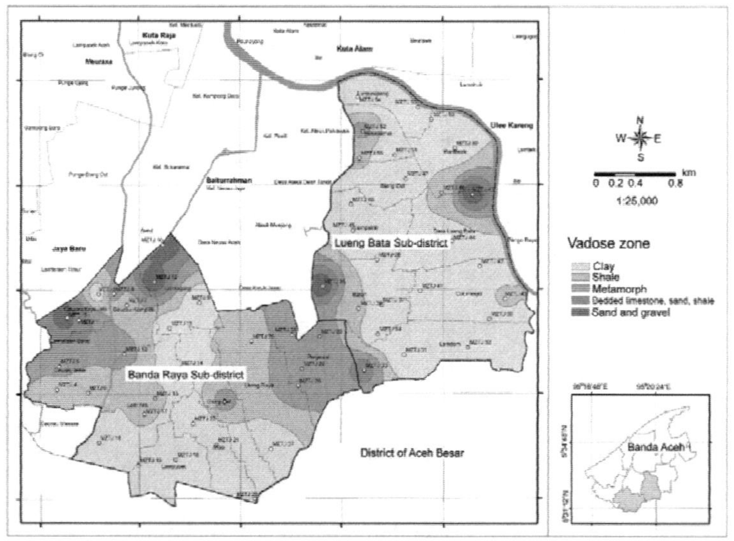

Figure 7. Vadose zone of the study area.

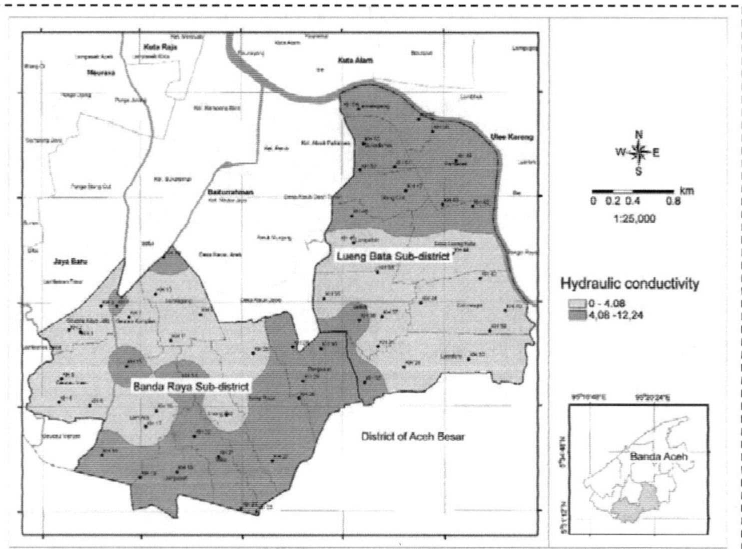

Figure 8. Hydraulic conductivity of the study area.

4.8. Land use (L)

Land use information was integrated with DRASTIC standard criteria as an additional parameter to evaluate potential vulnerability in around of the study area. A weighting value equal to 5 was multiplied to the land use parameter [10, 18]. If the DRASTIC index in (Eq. 2) equal to DI, the resultant index includes land use parameter (MD) [19]:

$$MD = DI + L_r L_w \quad (3)$$

where $L_r L_w$ is the land use index.

In order to attain detail vulnerable assessment, the land use parameter was included population density factor in the area investigated. The factor was assigned value of 5 (low, 487 – 1,945 people), 7 (medium, 1,956 – 3,405 people), and 9 (higher than 3,404 people) (Figure 9).

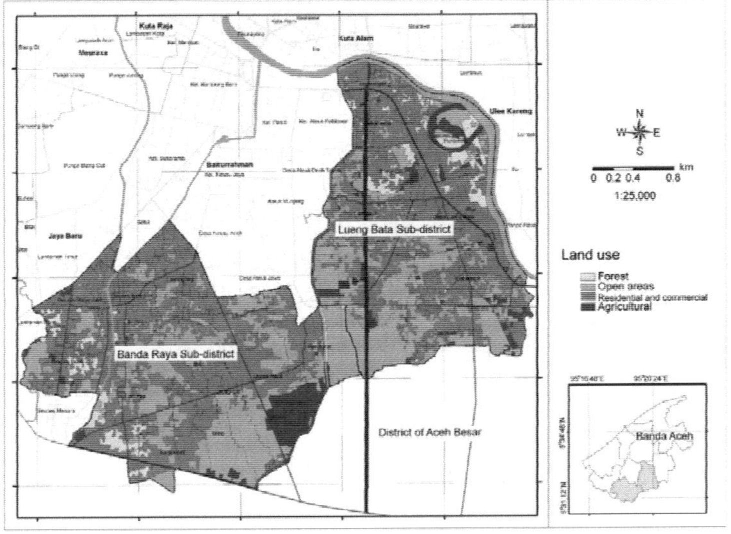

Figure 9. Land use of the study area.

4.9. The DRASTIC values

The DRASTIC values describe the whole groundwater vulnerability and its individual criterion is useful to identify the contaminant pathway through aquifer in a study area. In the criterion of depth to groundwater rating layer, the study area with rating value of 9 and 10 cover 67.5% and 32.5%, respectively. The whole study area has a high net recharge rate, i.e. 8. For the aquifer layer result, only 33.8% of the study area contributes to groundwater pollution potential with rating value of 8. The prominent soil properties in the study area were covered by clay and sandy clay with rating value of 3 (95.3%) with low groundwater pollution potential and only 4.7% of the land area investigated contributes to contaminant pollution. In the criterion of topography rating layer, the low-lying area with the rating value of 9 was 75.3% coverage. The vadose zone media with the rating value 1 was 94.3% of the land area with meaning mainly low groundwater pollution potential. In hydraulic conductivity parameter, the rating layer of 1 covers 48.8% of the area study.

4.10. Groundwater pollution vulnerability

The groundwater vulnerability of the study area is present in Figure 10 based on overlay map of the DRASTIC map index (static vulnerability). The index values lay between 97 and 148. This range is classified on five categories of vulnerability, i.e. less (97-107), low (108-117), medium (118-127), high (128-139), and very high (140-148). The study illustrates that of the total 1,164 ha of the study area, it has 355 ha (30%), 411 ha (35%), 254 ha (22%), 115 ha (10%), and 30 ha (3%) categorized as less, low, medium, high, and very high groundwater pollution potential, respectively. The village areas, labelled with the high groundwater pollution potential, are mainly Lamlagang and part of Geuce Kaye Jatoe and Geuce Komplek (Banda Raya sub-district) and part of Batoh and Suka Damai (Lueng Bata sub-district).

Domestic wastewater discharge from inhabitant and small-scale business activities including schools, government offices, and mosques contribute to groundwater pollution potential as in the area is not installed an appropriate sewerage system. Analysis groundwater vulnerability of the area study using seven parameters of the DRASTIC approach combine with land use as additional parameter will give more clear picture of the area study sensitivity. Figure 11 presents map sensitivity analysis

IOP Conf. Series: Materials Science and Engineering **334** (2018) 012032 doi:10.1088/1757-899X/334/1/012032

(dynamic vulnerability) that was performed by adding land use parameter. The index values of sensitivity lay between 101 and 188. This range is classified on five categories of vulnerability, i.e. less (101-118), low (119-134), medium (135-150), high (151-167), and very high (168-184). Of the total area, it was found that 11% and 2% of the area was high and very high category, respectively.

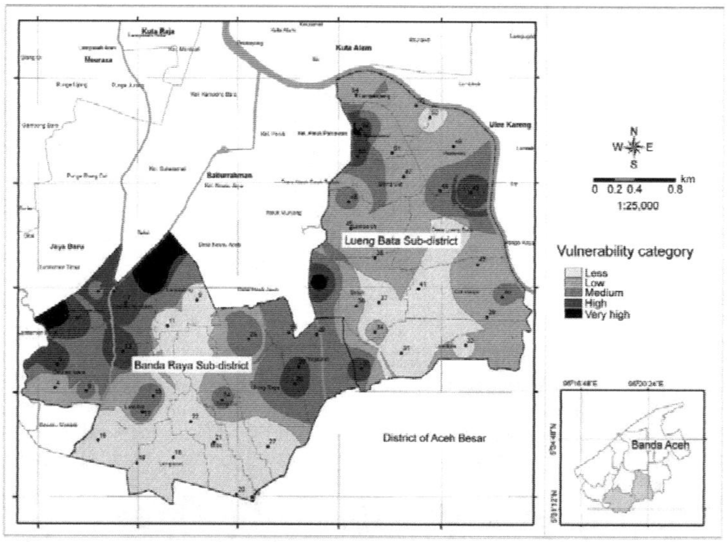

Figure 10. Static vulnerability map of the study area.

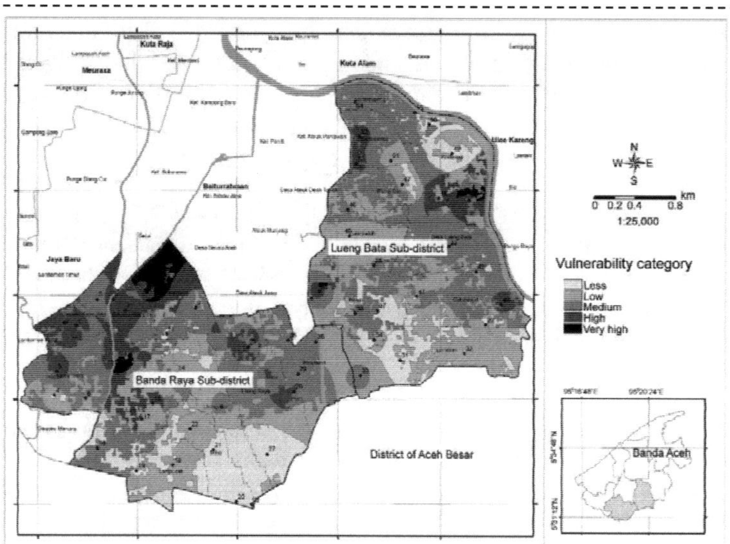

Figure 11. Dynamic vulnerability map of the study area.

5. Conclusions

The vulnerability maps of groundwater pollution potential based on the DRASTIC approach in the part of Banda Aceh, Indonesia were produced in an ArcGIS environment. These investigations can assist the Banda Aceh government official as an instrument for development of Banda Raya and Lueng Bata sub-district. The vulnerability map generated to this area study presents five categories. The map shows that only 115 ha or 10% and 30 ha or 3% of the total area study has high and very high vulnerability category, respectively. It was found that the village areas, labelled with the high

IOP Conf. Series: Materials Science and Engineering **334** (2018) 012032 doi:10.1088/1757-899X/334/1/012032

groundwater pollution potential, are mainly in the area of Lamlagang and the part of Geuce Kaye Jatoe and Geuce Komplek (Banda Raya sub-district) and the part of Batoh and Suka Damai (Lueng Bata sub-distric). Actually, the DRASTIC approach shows some weakness (the difficulty to validate vulnerability), however these generated maps could be use an additional tool by local planner.

Acknowledgement
This study was supported by research Grants from the Ministry of Research, Technology, and Higher Education, Indonesia.

References

[1] Kagabu M, Shimada J, Delinom R, Nakamura T and Taniguchi M 2013 *Hydrological Processes*, **27** 2591
[2] Kooy M 2014 Water Alternatives, **7** 35
[3] Kees B 2016 Indonesia: Country water assessment *Asian Development Bank* ISBN: 978-92-9257-361-4 (Metro Manila, Philippines)
[4] Machdar I, Matsura N, Kodera H and Ohashi A 2014 *J. of Water & Env. Tech* **6** 459
[5] Sholichin M 2012 *Int. J. of Civil & Env. Eng.* 12 74.
[6] Fritch T G, McKnight C L, Yelderman J C and Arnold A L 2000 *Environ. Management* **25** 337.
[7] Srinivasamoorthy K, Chidambaram S, Sarma V S, Vasanthavigar M, Vijayaraghavan K, Rajivgandhi R, Anandhan P and Manivannan R 2009 *Res J. of Environ and Earth Sci.* **1** 22
[8] Bastida M J J, Arauzp M, Valladolid, M 2010 *Hydrogeology Journal* **18** 681
[9] Lathamani R, Janardhana M R and Mahalingamc B S 2015 *Aquatic Procedia* **4** 1031
[10] Kihumba A M, Vanclooster M and Longo J N 2017 *J. of African Earth Sci.* **126** 13
[11] Aller L, Bennet T, Lehr J H, Petty R J and Hackett G 1987 US EPA Report 600/287/035, U.S. Environmental Protection Agency
[12] Al-Zabet T 2002 *Environmental Geology* **43** 203
[13] Aller L, Lehr J H and Petty R 1985 US, USEPA Report 600/02-85/018 U.S. Environmental Protection Agency
[14] Zhou Y 2008 *Groundwater Monitoring*, UNESCO-IHE, Delft, The Netherlands.
[15] Mangeonad 2007 *Project Management of Georisk in NAD* BGR Germany and Mining and Energy Department (NAD Province, Indonesia)
[16] Kiely G 1998 Environmental Engineering (Malaysia: McGraw-Hill)
[17] Rahman A 2008 *Applied Geography* **28** 31
[18] Panagopoulus G, Antonakos A and Lambrakis N 2006 *Hydrogeol J* **14** 894
[19] Secunda S, Collin M and Melloul A J 1998 *J Environ Manage* **54** 39

Prevention of Major Accident Hazards (MAHs) in major Hazard Installation (MHI) premises via land use planning (LUP): a review

M Q Khudbiddin[1], Z A Rashid[1,2], A F M S Yeong[1], A B Alias[1], M F Irfan[1], M Fuad[1] and H Hayati[1]

[1] Faculty of Chemical Engineering, Universiti Teknologi MARA (UiTM), 40450 Shah Alam, Selangor, Malaysia.

E-mail: zulmas06@yahoo.com.my

Abstract. For a number of years, there is a concern about the causes of major hazards, their identification, risk assessment and the process of its management from the global perspective on the activities of the industries due to the protection of the environment, human and property. Though, industries cannot take pleasure in their business by harming the nature of the land, there are a number of measures that need to be put into consideration by the industries. Such measures are in terms of management and safety for the businesses, lives, properties, as well as the environment. The lack of consideration in the selected appropriate criteria can result in major accidental hazards (MAHs). This paper will review the land use planning (LUP) methods used in the past and present to prevent major accident hazards at major hazard installation (MHI).

1. Introduction

Major Accident Hazards (MAHs) is defined as an incident involving the loss of life that occurs within or without the site area with ten or more injuries and/or one or more injuries outside or release of toxic chemical or explosion or fire of spillage of hazardous chemical resulting in 'on-site' or 'off-site' emergencies or damage to equipments leading to the stoppage of process or having an adverse effect(s) on the environment [1]. A major accident is also defined as an "unexpected, sudden occurrence including, in particular, a major emission, fire or explosion, resulting from abnormal developments in the course of an industrial activity, leading to a serious danger to workers, the public or the environment, be it in immediate or delayed, inside or outside the installation and involving one or more hazardous substances" [2].

Several accidents such as the Bhopal tragedy on December 2nd, 1984, Mexico City on November 19th, 1984, and more recently, the Enschede fireworks disaster on May 13th, 2000, and the chemical explosion in Toulouse on September 21st, 2001, have obviously showed that how the outcomes of the industrial accidents may perhaps need to be critically reviewed to find the best solution. Thus, industries cannot take the pleasure in their business by harming the nature of the land. There are many considerations that need to be put in place by the industries, particularly in terms of its management and safety for the businesses, lives, properties, and environment. The lack in such considerations could

[2] To whom any correspondence should be addressed.

very well cause major accidental hazards. Even though the industries are making various efforts, there is still much to do to enhance the policies for the purpose of reducing the dangers (impact) of major accident hazards. Therefore, this paper will review brief discussions on the issues related to the implementation of the LUP since it was introduced.

2. Literature Review

The major accident that had taken place in Bhopal (1984) arising from the leakage of methyl isocyanate (MIC) gas was the worst tragedy ever happened in the industrial process world. It was learnt that in this accident, the government of Madhya Pradesh, 2006 reported that 500,000 people were exposed to MIC gases, thus sacrificing a total of 3,787 people [3]. According to Bowonder [4], the plant was located very close to residential areas and was only 2 to 3 km from the Bhopal Railway. According to Crowl and Louvar [5], originally, the nearest civilian was 1.5 miles (2.41 km) away when the plant was constructed. However, the settlement of the population around the plant had increased so drastically that it grew very close to the plant, from 2 to 3 km to less than just 100 m away [4, 5]. Population growth in the nearby area is likely due to the fact that the plant became the most dominant job attraction in the area. The same scenario applies to the PEMEX LPG plant, Mexico in 1984, when explosions occurred from 11,000 m^3 of LPG until the LPG facility was destroyed and it had devastated the local town of San Juan Ixhuatepec, with 500-600 people killed and 5,000-7,000 others suffering severe burns [6]. According to Lees [7] and most reports related to the PEMEX incident, it shows that the cause of this massive loss of life is due to the siting factor of the LPG facility that is very close to the residential area. The view posed by Lees [7], that is the unreasonable plant siting factor, has been identified as one of the causes for severe impact, or the industrial accidents may perhaps need to occur as a result of the hazardous environments that had originated from the industrial locations close to the residential sites, and high-density population areas have actually been discussed from past research reports and researches through past industrial incidents such as Feyzin in 1964 and Flixborough in 1974, but there are still weaknesses in the implementation of the recommendations. Thus, this reinforces the need of land use regulations in the hazardous installations resulted in revising of Directive 82/501/EEC (the Seveso Directive) [8] that is developed under the new Directive 96/82/EC [8], to prevent MAH or minimize the consequences of accident.

The approach to having the LUP as a requirement is very reasonable since most industrial facilities are able to cause major accidents under certain circumstances with consequences extended outside the borders of the establishment and can harm human beings and the environment. Therefore, separation distance that is adequate (i.e. safe distance) to ensure the safety of human beings and the environment from hazardous activities between MHI facilities and residential and commercial areas should be taken seriously. However, the issue of separation distance needs to be carefully and reasonably refined so as not to be misinterpreted and misused, for example, if there are investors who are interested in building a chemical plant in a particular area, the authorities concerned in maintaining public safety cannot arbitrarily defend the investors on having an absurdly long separation distance without reasonable excuse(s). Therefore, a very detailed technical study that provides information relating to the potential major accidents in the form of risk assessment that may occur on MHI facilities is particularly important, so that the exact separation distance can be determined, thus avoiding extensive unexploited areas as well as economic loss.

In order to realize the element of the LUP in Seveso II Directives 1996, the European Working Group, on land use planning, composed by representatives of the Competent Authorities and appointed by the Joint Research Centre of the European Commission [5] has been established, which require that the Member States should take measures towards this direction. More specifically, Article 12 of the Directive (see Appendix V. 1) foresees: -

- That the Member States shall ensure that the objectives of preventing major accidents and mitigating the consequences of such accidents are taken into account in their land use policy and especially through controls on the siting of new establishments, the modifications to existing ones, and new developments (residential areas, areas of public use, transport links, etc.) in the vicinity of existing establishments;

- That their land-use policy takes account of the need to establish and maintain appropriate separation distances between the establishments covered by the Directive and residential areas, areas of public use and areas of particular natural sensitivity or interest;
- That the land-use policy takes account of the need for additional technical measures so as not to increase the risk to people;
- That all Competent Authorities and planning authorities shall set up appropriate consultation procedures to facilitate the implementation of the policies mentioned above (i.e. the land-use policies that take into account the major-accident hazards).

However, the above General Directives do not state clearly and in detail the exact value of separation distance. Instead, they submit to Member States and the competent authorities to determine the appropriate distance to allow the construction and operation of MHI facilities.

As a result of the process to realize these Directives, there are several constraints, among which, based on the present expansion of the European Community, only a restricted number of European countries have developed specific general guideline for LUP in relation to major accident hazards after several years from which the instructions have been introduced and enforced among Member States. Majority of European Communities still performed the control of land use planning in the vicinity of hazardous installations by non-specific legislation, and the risk is not explicitly considered within the land use policies [10-13]; it is related to another policy for the plan Economic strategic and etc. This is due to the difficulty in determining acceptability criteria of the appropriate LUP depending on the cultural background, socio-economic and the special characteristics of each region. For example, there are countries where a certain level of risk is acceptable and the LUP aims at the greater minimization of risk, as their country has very limited land area with high population rates per km^2 such as Hong Kong, Singapore and the likes, whereas for others, the "zero risk" concept is applied, not allowing any development which poses risk to the population, for instance, a country with very low population density.

Another factor that slowed the implementation of Seveso II Directives, 1996, was a lack in the understanding of the objective of the European Union (EU) Directive 96/82 / EC (Seveso-II Directive) among Member States, which deals with two key features of LUP; which are the separation among hazardous installations and residential, and other sensitive areas (i.e. safety distance) as well as the systematic technical framework for its assessment and scrutiny. For instance, there is no explanation from the Directive on how LUP policy should be implemented by the EU Member States into their National LUP laws, because in addition, the technical elements consist of a number of other aspects that need to be considered, such as technological, social, cultural, and economic, among others. [14]. This kind of attitude has contributed to the development of LUP hence, affecting the policy made for the protection of accident hazards in environment, while the policy is meant to be abided by all the respective countries.

Another issue related to LUP is the "domino effect", where in this issue, the interaction between neighbouring process units and plants is largely related to the MAH's risk severity chain impact, which can be avoided by adequately siting the relevant installations and planning the uses of land around them. In general, the scope and objective of LUP in the vicinity of hazardous installations is to ensure the consequences of the potential accidents are taken into consideration when decisions are made concerning siting of new installations:-

- Extension (or modifications) of existing installations;
- Determination of uses of land in the vicinity of establishments;
- Proposal for new developments in the vicinity of establishments.

To implement the Seveso Directives 1982 and Seveso II Directives 1996, the HSE of the UK implemented several guidelines such as CIMAH Regulations, 1984. The guidelines said are as follows [8]:-

- Identify all the hazards involved in their undertaking, which may pose risks to personnel;
- Evaluate the risks arising from those hazards;
- Identify and put in place the means to manage the hazards and reduce the risks;
- Make a demonstration that those risks are as low as reasonably practicable.
- Under this guideline, it is also required for the risk attached to the hazardous environment be informed to persons who are within the area measured for the avoidance of the accident threat.

Nevertheless, in 2015, [14-16,18] the Control of Major Accident Hazards (COMAHs) [9], was introduced for the purpose of preventing major accidents that consist of dangerous substances and reducing the outcomes to the people and the environment. The COMAH 2015 is the realization of the main Seveso III Directive (2012/18/EU). The regulations the COMAH came into force as of June 1st, 2015 and its purpose on the other part is to assist duty holders in abiding by COMAH in the health and safety management in the most fundamentals to MAH industries. The government of Malaysia adopted the CIMAH regulations and promulgated the Regulations in February 1996, under the Department of Occupational Safety and Health (DOSH), Malaysia. Under this CIMAH 1996, Malaysia regulations, should there be any premises that process, handle, store, and manufacture the chemicals listed in Schedule 1 and 2, CIMAH 1996 in this study that exceeds the threshold equivalent quantities specified as in Schedule 2, CIMAH 1996, then the premise is required to present CIMAH Safety report from a CIMAH competent assessor according to Schedule 6 (Subregulation 14 (1) and 15 (1)) i.e. "Information To be Included In The Report on Industrial Activity". It contained a total of four components and the list presented as follows [23]:

(a) Information relating to every hazardous substance involved in the activity in relevant quantity as listed in Schedule 2.
   i) The name of the hazardous substance as given in Schedule 2 or for a hazardous substance included in either of those Schedules under a general designation, the name corresponding to the chemical formula of the hazardous substance;
   ii) A general description of the analytical methods available to the manufacturer for determining the presence of the hazardous substance, or references to such methods in the scientific literature;
   iii) A brief description of the hazards, which may be created by the hazardous substance;
   iv) The degree of purity of the hazardous substance, and the names of the main impurities and their percentages.

(b) Information relating to the installation, namely
   i) A map of the site and its surrounding areas to a scale large enough to show any features that may be significant in the assessment of the hazard of risk associated with the site;
   ii) Scale plan of the site showing the locations and quantities of all significant inventories of the hazardous substance;
   iii) A description of the processes or storage involving the hazardous substance and an indication of the conditions under which it is normally held;
   iv) The maximum number of persons likely to be present on site;
   v) Information about the nature of the land use and the size and distribution of the population in the vicinity of the industrial activity to which the report relates;
   vi) Information on the nearest emergency services (fire station, hospital, police station, community hall, etc.)

(c) Information relating to the system of management for controlling the industrial activity, namely:
   i) The staffing arrangement for controlling the Industrial activity with the name of the person responsible for safety on the site and the names of those who are authorized to set emergency procedures in motion and to inform outside authorities

ii) The arrangements made to ensure that the means provided for the safe operation of the industrial activity are properly designed, constructed tested, operated, inspected, and maintained.
iii) The arrangements for training of persons working on the site.

(d) Information relating to a potential major accident in the form of risk assessment, which contains the following:
i) A description of the potential sources of a major accident and the conditions or events, which could be significant in bringing one about.
ii) A diagram of the plant in which the industrial activity is carried on, sufficient to show the features which are significant as regards the potential for a major accident or its prevention or control.
iii) Information about prevailing meteorological conditions in the vicinity of the site;
iv) A estimate of the number of people on site who may be exposed to the hazards considered in the report;
v) The consequences to the surrounding areas in the form of risk assessment.

The same was done by other countries such as Austria, Belgium, Denmark, Finland, France, Germany, Greece, Ireland, Italy, Luxembourg, Holland, Portugal, Spain, Sweden, Singapore, Hong Kong, and United Kingdom, that is, creating specific guidelines for assessing the effects of a worst-case scenario of an incident, should it occur. This information will be used to determine the hazard boundary zone that may be caused by the MHI incident and determine the acceptability site, whether the presence of the MHI facility is safe or not, as well as to determine the appropriate emergency response method.

Nonetheless, all of the above-mentioned countries have a different approach or method of risk assessment approach, namely consequences-oriented approach, risk-oriented approach, land use planning criteria and generic safety distance based on Environmental Impact Assessment (EIA). Countries such as France, Holland, United Kingdom, Germany, and Luxembourg use the land use planning criteria approach. For Malaysia, our approaches are a combination of consequences-oriented approach and LUP criteria. However, there are several new approaches, one of them being Alessandro et al. [14, 16, 20] who examined the LUP in the industrial environment as the occurrence of major accidents. The LUP was used in identifying the criteria in the accident scenarios. Generally, the appropriate choice of area in the chosen criteria of major accident would result into a critical effect in terms of assessment and the follow-up policy. The model used in developing the LUP is an improved version of MIMAHs (Methodology for the Identification of Major Hazard Accidents).

In an attempt to provide safety to the human and environment, Kyusang, Seonghyun and En Sup Yoon [21] discusses the process that can ensure the protection to land and human through the minimization of the risk and dangers of chemical process plant attached to human beings through the plants. The article further focuses on the process of providing safety measures and therefore, provides the steps that can be used at the earliest period in processing the designated plant.

In addition, John and Richard [22] provides an important insight on the chemical reaction in hazards environment as the key factors to the incidents occurred in manufacturing and processes of chemical substance around the world.

Christou et al. [14-17, 19] provides the key factor summaries associated to the LUP and the role of risk management. The chapter provides two major operational scenarios that are fundamental to the LUP, that is, firstly, fixed site risk which covers storage and manufacturing facility, and secondly, transport risks which covers bulk road, rail tankers of chemicals, petroleum products, gas, and shopping. The literature provides a comprehensive explanation between the factors and their connection to each other. Furthermore, it explains the routine activities that are involved in each scenario by making reference to the itemized issues under the factors. As part of the efforts of the activities, it clearly shows that the LUP is mandatory in ensuring the stakeholders discharging their obligations. It further identifies the relevance of business and industrial facilities in LUP, where it provides the activities that show the conceptual framework of the business and industry as in Figure 1. This is seen as an interaction between the risk and business impact, where business and residential businesses are impacted on low impact. When the risk impact increases, special attention should be taken to minimize the impact.

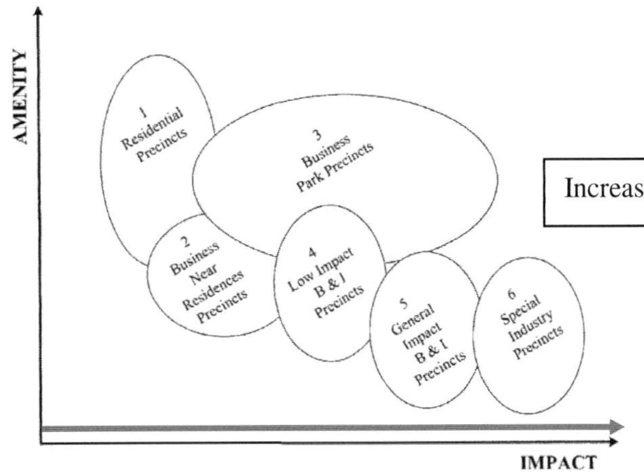

Figure 1. Business and Industry Precincts - A Conceptual Framework

A number of recognized, published articles had provided methodological concept in determining the LUP as the method of identification of accident scenario, whereby such concept will be adopted in this review paper. The regulations will also be used in evaluating the effectiveness of the provisions in determining high risk and safety of the entire public and environment. In a proposed procedure in the current research, the target is at the identification of the accident scenarios, which is relevant in determining LUP in the area of around major hazard institutions as classified in Seveso sites without the help of specific risk-based or consequence-based approaches that had been shortly adopted in the decisional phase of LUP (Cozzani et al., 2006) [8]. The proposed procedure is also applicable to the existing and new plants. As such, it would involve the usual effort through information required for risk assessment studies (CCPS, 2000; Mannan, 2005; Uijt de Haag & Ale, 2005). A sound basis is required in the approach to LUP. As such, it is of importance that the accident scenarios are carefully selected via a validated procedure that aims at efficient identification process of reference accident scenarios for LUP purposes. A draft table (Table 1) of various accident scenarios based on an enhanced version of the MIMAH methodology is created. Then, HAZOP analysis is utilised in validating the relevance of the draft table produced as well as other techniques applied. Subsequently, the possible accident scenarios are assessed for LUP relevance against criteria accounting for expected severity, frequency and time scale of the scenarios said [6].

Furthermore, in the case of adopting this method, an updated version of the MIMAH methodology was developed to identify possible accident scenarios, as mentioned beforehand. MIMAH is a step-by-step methodology for the identification of accident scenarios; this is usually carried out with the development of generic fault and event trees. The methodology is based on categorization of utensils and of properties of the hazardous substances, and includes a database of reference fault and event trees. The adoption of the MIMAH method approach is justified based on the rational representative of the current state-of-the-art in accident scenarios identification, since it was originally developed within the EU FP6 ARAMIS project [7]. The second stage provided a systematic procedure applied to the draft table of possible accident scenarios obtained by MIMAH to identify the relevant scenarios that should be applicable for LUP. In the first stage, the draft table is revised and integrated with the results obtained from a layered approach based on specific identification techniques. Then, the second stage is the validation process, where practical rules are provided to select the accident scenarios relevant in the LUP context from the general list obtained in Stage I.

Table 1. Reference customization criteria introduced for the identification of accident scenarios

Task	Criteria	Example
Identify hazardous equipment category (EQ)	For equipment not included in original EQ classes of MIMAH, select EQ based on geometrical and functional similarity.	Loading/unloading facilities, hoses and connection arms are considered as the likeliest geometrical class "EQ10 – pipe network".
Customize the generic tree	Exclude causes due to material properties.	Eliminate causes like "internal combustion/explosion" for non-flammable/reactive materials.
	Exclude causes due to equipment characteristics.	Eliminate causes like "leak/rupture of internal high pressure source" if utilities such as HP steam coil are not present.
	Exclude causes due to operation characteristics.	Not consider scenarios like "filled beyond normal level" if material is in gas/vapour state.
Customize the generic tree	Exclude non-relevant events due to properties of the contained material.	Eliminate fire scenarios for non-flammable materials.
	Propose conditional exclusions depending upon the verification of specific additional information.	Check if initial concentration and temperature of a water ammonia pool may yield toxic concentrations in air.
	Mark with warning notes the specific scenarios that may occur only under particular enabling conditions.	Only the release of LNG over water can give raise to a Rapid Phase Transition scenario.

3. Conclusions

The MAH is one of the major incidents that keep affecting the development of the environment in a destructive way. Though the MAH is always monitored, supervised and regulated by institutions meant for that, accidents remain inevitable at times. The major concern is the application of the land use planning (LUP) procedure for effective use in securing the safety of the environment, people and property. The MAH is still a threat due to its effects on the environment; this was resulted from the use of chemicals such as biological, physical and mechanical. These accidents were generally referred to as industrial sites and more often than not, some industries are in the habit of violating the land planning policies. These policies were made to ensure the safety of the operation of industries. The industries are expected to fully comply with the respective regulations in enhancing the safety of the environment, people and their properties. Usually, the result of the increasing incidence leads to injuries, loss of property and environmental degradation. Some of the accident that happens around the globe had affected the loss of a reasonable number of lives, property and environment. It had since become essential for operators to ensure the compliance of the effective measures and its proper utilisation for security purposes. I believe that the LUP plays a vital role in increasing the effectiveness of the strategies applied in curbing the ever-existing risks, even in a properly regulated site. Although each country's approach to controlling the risk of MAH is different, the effectiveness of each of the approaches practiced by the affected State is not known. Accordingly, in the authors' view, all the MAH control approaches need to be taken into account and validated to see fit before a new approach is to be introduced, not counting if the approach in the form of criteria, guideline, code of practice or the likes of a country is to be adopted by different geographical boundaries. Thence, the process of screening all of the parameters should be done more carefully in terms of safety, economic and sustainability of the future.

Acknowledgement
We, the authors, would like to extend our sincerest gratitude to the Ministry of Higher Education (MoHE) for administering the provision to us in a form of research grant with the grant number (reference code) of FRGS/1/2016/TK02/UITM/02/13. We would also like to thank the Research Management Institute (RMI) of Universiti Teknologi MARA (UiTM) for managing our affairs in attending this conference. We appreciate all the hard work and dedication provided by the staff and all those involved.

References

[1] Disaster Management Institute Bhopal, accessed on 6/3/2017, available at http://www.hrdpidrm.in/e5783/e26901/
[2] International Labour Organization, "Prevention of major industrial accidents", Geneva, International Labour Office, 1991, at p. 3. An ILO contribution to the International Programme on Chemical Safety of UNEP, the ILO and the WHO (IPCS), Geneva, Switzerland.
[3] "Madhya Pradesh Government: Bhopal Gas Tragedy Relief and Rehabilitation Department, Bhopal". Mp.gov.in. Retrieved 28 May 2017.
[4] B Bowonder, "An analysis of the Bhopal accident", Prajeci Appraisal, volume 2, number 3, September 1987, pages 157-168. Beech Tree Publishing, 10 Watford Close, Guildford, Surrey GUl 2EP, England.
[5] Crowl D. A. And Louvar J.F, "Chemical Process Safety: Fundamental with Applications" 2nd edition, 2002, Prentice hall.
[6] Arturson, G., "The tragedy of San Juanico--the most severe LPG disaster in history.", Burns Incl Therm Inj. 1987 Apr;13(2):87-102.
[7] Lees, F. P., " Loss Prevention in the Process Industries," Butterworth Edited by Sam Mannan, 2005.
[8] V. Cozzani et al. "Application of land-use planning criteria for the control of major accident hazards: A case-study", Journal of Hazardous Materials A136 (2006), at p. 170.
[9] Council Directive 96/82/EC of 9 December 1996 on the control of major accident hazards involving dangerous substances, Official Journal of the European Communities, L 10/13, Brussels, 14.1.97. [5] O'Leary S K, Foutz B E, Shur M S, Bhapkar U V and Eastman L F 1998 *J. Appl. Phys.* **83** 826
[10] [5] M. Smeder, M.D. Christou, S. Besi, Land use planning in the context of major accident hazards—an analysis of procedures and criteria in selected EU member states, Report EUR 16452 EN, Institute for Systems, Informatics and Safety, JRC Ispra, Ispra (I), 1996.
[11] M.D. Christou, A. Amendola, M. Smeder, The control of major accident hazards: the land-use planning issue, J. Hazard. Mater. 65 (1999) 151–178.
[12] M.D. Christou, M. Mattarelli, Land-use planning in the vicinity of chemical sites: risk-informed decision making at a local community level, J. Hazard. Mater. 78 (2000) 191–222.
[13] C. Hamilton, R. De Cort, K. O'Donnel, Report on land use planning controls for major hazard installations in the European Union, EUR 15700 EN, European Commission, 1994.
[14] A. Tugnolia, Z. Gyenes, L. V. Wijk, M. Christou, G. Spadoni, V. Cozzani, "Scenario Selection for Land Use Planning Purposes around "Seveso" Establishments", Chemical Engineering Transactions, Vol. 26, 2012, p. 417-422.
[15] M. Smeder, M.D. Christou, S. Besi, Land use planning in the context of major accident hazards—an analysis of procedures and criteria in selected EU member states, Report EUR 16452 EN, Institute for Systems, Informatics and Safety, JRC Ispra, Ispra (I), 1996.

[16] A. Tugnoli et al., “Reference criteria for the identification of accident scenarios in the framework of land use planning “, Journal of Loss Prevention in the Process Industries, 26, 2013, at p. 614.

[17] Graham Dalzell, “Understanding Major Accident Hazards –The Cutting Edge Of Common Sense”, Symposium Series No. 149, 2003 Icheme, at P 502

[18] Health and Safety Executive, “The Control of Major Accident Hazards Regulations 2015”, Guidance on Regulation. L111 (Third edition) Published 2015.

[19] M. Smeder, M.D. Christou, S. Besi, Land use planning in the context of major accident hazards—an analysis of procedures and criteria in selected EU member states, Report EUR 16452 EN, Institute for Systems, Informatics and Safety, JRC Ispra, Ispra (I), 1996.

[20] Tugnoli, Alessandro, Zsuzsanna Gyenes, Lorenzo Van Wijk, Michalis Christou, Gigliola Spadoni, and Valerio Cozzani. "Reference criteria for the identification of accident scenarios in the framework of land use planning." *Journal of Loss Prevention in the Process Industries* 26, no. 4 (2013): 614-627.

[21] Han, Kyusang, Seonghyun Cho, and En Sup Yoon. "Optimal layout of a chemical process plant to minimize the risk to humans." *Procedia Computer Science* 22 (2013): 1146-1155.

[22] John Barton and Richard Rogers, Chemical Reaction Hazards, A Guide to safety, second edition, 1997.

[23] Occupational Safety and Health Act 1994(Act 514) and Regulations, 1996, ILBS.

IOP Conf. Series: Materials Science and Engineering **334** (2018) 012034 doi:10.1088/1757-899X/334/1/012034

Focussing on recycling attitudes of engineering students at UiTM Shah Alam - towards zero discharge

A S A K Sharifah[1,2], P F M Khamaruddin[1], N N Mohamad [1] and M Q Saharuddin [1]

[1] Faculty of Chemical Engineering, Universiti Teknologi Mara
40450 Shah Alam, Selangor Darul Ehsan, Malaysia

E-mail: drsharifah@salam.uitm.edu.my

Abstract. The generation of municipal solid waste (MSW) in Malaysia over the past 10 years has increased by 95 percent due to rapid development in the urban areas. 16.76 million tonne of waste is expected to be generated by Malaysian in the year 2020. Presently, there are about 70 percent of waste produced reported to be collected and 95 percent of it are disposed in landfills with only 5 percent left are being recycled. A 40 percent reduction of waste disposed to landfill by year 2020, through recycling and intermediate treatments such as waste to energy, composting and material recovery is very much needed as opening new landfills are not socially attractive. It clearly shows the awareness in reducing waste through recycling is still at infancy among Malaysians. Hence, a proper solid waste management should be enhanced through education among youngsters to ensure an integrated solid waste management towards zero discharge is achievable. The purpose of the research is two-fold. Firstly, to determine the recycling practices awareness among engineering students through manual and on-line survey. Secondly, a study on the effectiveness of recycling bins at the engineering students' centre. The data collected were analyzed using statistical analysis. Results show that there is significant relationship ($p<0.05$) between gender and knowledge on recycling using Chi square test. However, there is an insignificant relationship ($p>0.05$) between knowledge and awareness of students towards recycling using partial correlation. Nevertheless, it is important to note that recycling practices through the provision of well-designed recycling bins and continuous education will ensure a sound society upholding a Zero Discharge attitude in the future.

1. Introduction

Malaysia presently is experiencing rapid growth of economy and urban transformation since the last decade. With respect to this, the amount and types of solid waste have increased gradually corresponding to the economic growth and improving living standard. Solid waste is one of the major environmental problems faced by most municipalities in Malaysia [1]. The generated amount of solid waste increases from 17,000 per day to 19,100 tons in 2005, an average of 0.8 kilogram per capita per day [2]. From the statistics obtained, it is expected that in year 2020, the amount of solid waste generated is 30,000 tons per day [1]. The waste in Malaysia consists of 45% food waste, 23% plastic, 7% paper and 6% iron. It is approximated that 95-97% of waste collected is taken to landfill for disposals. Meanwhile, the remaining waste is diverted to recyclers or being dumped illegally [1].

[2] To whom any correspondence should be addressed.

Solid waste management program is one of the greatest challenges to achieve institutional sustainability [3]. There are few methods of handling municipal waste which are reduce, reuse, recover, treatment and disposal to the landfills. In order to achieve an effective solid waste management, the society should be educated and preferably students are the first target. One of the methods to reduce waste generated is through recycling. The waste management in the campus can be initiated from the effective recycling program that is being implemented by the university [4]. Recycling can be defined as a recovery process of waste from used products through their reuse in which it may be used for other purposes or for their original purpose [5]. A successful recycling programmed is affected by few factors which are the technology, participation of public, the development and maintenance of environmentally responsible behaviour [6].

In 1987, the Action Plan for a Beautiful and Clean Malaysia (ABC Plan) was implemented which aims to minimize the waste generated but unfortunately, there was no legal and fiscal instrument to regulate the plan which leads to the first and second recycling program in 1993 and 2000. Ever since, National Recycling Day has become an annual event in Malaysia celebrated on every November 11th [4]. Nevertheless, based on the world recycling rate data in 2015, Malaysia holds the fifth lowest rank in the world where the recycling rate is at 17.5% [8] which is unimpressive compared to her neighbour, Singapore which holds the top rank where 59% of the Singaporean practices recycling. To date, the awareness of the recycling practices is still lacking and targeting 22% in the year 2020 [9]. Table 1 represents the methods of waste disposal in Malaysia in year 2002, 2006 and the targeted year of 2020 of waste disposal [10]. There are six treatment of waste disposal comprising of recycling, composting, incineration, inert landfill, sanitary landfill and other disposal sites. Back in year 2002 and 2006, the most popular method of disposal is by treating the waste to other disposal sites. Other disposal sites refer to open dumpsites which have illegal dumping or have accumulated a huge quantity of waste. However, Malaysia has targeted to reduce to zero percent of waste being sent to other disposal sites in 2020. Instead, sanitary landfill has been targeted to have the highest percentage followed by recycling and incineration in year 2020. To emphasise the importance of waste segregation, the government has imposed the "Waste Segregation Bill" effective 1st September 2016 to homeowners and fines to be imposed to those who ignore the call.

Table 1. Method of waste disposal in Malaysia.

Treatment Methods	Percentage		
	2002	2006	2020 (Targeted)
Recycling	5.0	5.5	22.0
Composting	0.0	1.0	8.0
Incineration	0.0	0.0	16.8
Inert landfill	0.0	3.2	9.1
Sanitary landfill	5.0	30.9	44.1
Other disposal sites	90.0	59.4	0.0
Total	100.0	100.0	100.0

It is crucial to start educating the young ones about the importance and benefits of recycling as well as encouraging them to participate in recycling to care for the environment. It is believed that the current and future waste generators will respect and conserve natural resources through education by making informed waste prevention choices [11]. Hence the authors wish to emphasise that University Teknologi Mara (UiTM) with a total number of students of more than 150,000 and being the biggest university in Malaysia, plays a vital role to subscribe to waste management education to be embedded in all the fields she offers. UiTM Shah Alam is the biggest branch and has a total population of some 42,000 students where 25% of them study engineering. Although a recycling campaign was organized by the facility department of the university in year 2013-2014, where recycle bins were placed at the centre of engineering building, the lack of response and support from the students made the effort seemed futile. Thus, continuous education on the importance of recycling practice among the students is required to be done and provision of recycling bins at the students' centres are the main focus of this research. Issues regarding the operational maintenance and placement of recycling bins at selected location were not particularly attended to, but, these were in no way jeopardise the clear intention of why sound waste management particularly recycling should be performed by the students.

2. Experimental procedure and methodology

The study used questionnaires to elicit responses, with specific focus on awareness and behaviour of recycling among students and effectiveness of recycling bins at engineering students' centre. Engineering students are being used in the study because they contribute to 25% of total population of students in UiTM Shah Alam. The following method of data collection was selected in order to fulfil the objectives of the research.

The students' centres are located at the beautifully designed Kompleks Kejuruteraan Sultan Abdul Halim which comprises of two tower buildings that houses the lecturers' rooms and five blocks of lecture halls, tutorial rooms and laboratories. The students' centres for the civil engineering students are located at level 5 of Block 1, for mechanical engineering is located at Level 3 of Block 3, and for both the electrical engineering and chemical engineering students centres located at Block 4 and 5 respectively. Besides these locations, students seating area are also available at the base of the tower buildings close to the library. The study is being conducted at the students' centre/ lecture venues since it is the strategic and the most frequent place for the students to conduct group discussion or self-study. Thus, indirectly, the behaviours of students towards sustaining recycling practices can easily being observed.

The respondents are randomly selected from all the four branch of engineering fields. However, it is to be noted there are limitations while conducting the research. Limitations are factors than cannot be controlled by the researcher meanwhile, delimitations are restrictions that the researcher deliberately imposes on the study to narrow the scope. The first limitation of the study was that some of the respondents received may well be of the same respondents but at the same time there are those approached that do not want to continue to participate in the survey conducted after the first or second survey. Another crucial limitation is the respondents' degree of honesty when filling out the questionnaire. In this case, the respondents would answer the questions based on what they think is the right ones or selection when the fact is it opposes to what they actually do.

There are three (3) phases of questionnaires delivered gradually to fulfil the objectives to study the awareness and perceptions of engineering students towards recycling practices as well as to determine the effectiveness of recycling bins provided in the campus. In the first phase of questionnaire, the questions asked focuses on the waste management in the campus and to identify the awareness of respondents on recycling practices. It was conducted in November 2016. Besides the questionnaires, to complement the survey, there are five (5) recycling bins being placed at the students' centres previously described. These recycling bins were supplied at the faculty at the beginning of Phase 2. Hence, the second phase of the questionnaire that was conducted on March 2017 were aimed to study the effectiveness of the new recycling bins at engineering students' centres. The final phase of questionnaires in May 2017 seeks further respondents understanding on how waste segregation helps to conserve the environment. A total of 1129 respondents were received from all sets of questionnaires. The data collected is then analyzed using statistical analysis (Minitab) to determine the significant of the selected variables.

3. Results and Discussion

The questionnaires were conducted in stages with a total of 1129 respondents. Table 2 shows the demographic data of the students involved in the first and second phase of the study. No demographic data was needed in the third phase of the questionnaire since it only focuses on how waste segregation helps to conserve the environment. In the first phase, the questions asked focused on the awareness of respondents on the knowledge and awareness of recycling at the faculty. The questions were purposely asked to determine the knowledge of respondents towards recycling practices and to determine if they are aware on the environmental issues surround their place of study.

Table 2. Demographic data.

Demographics data	Percentage
Gender	
Male	54%
Female	45%
Age	
19-21	18%
22-24	74%
25-27	7%
>27	1%
Field of engineering	
Electrical	21%
Chemical	44%
Civil	18%
Mechanical	17%

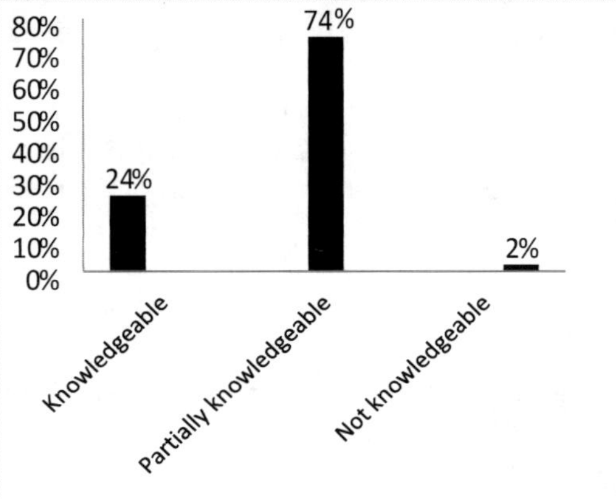

Figure 1. Knowledge on recycling.

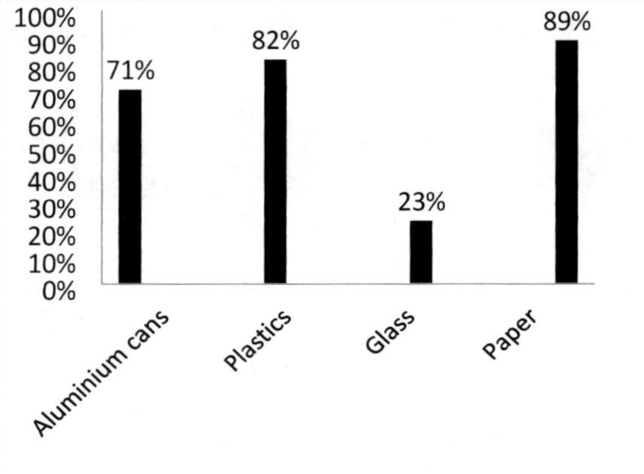

Figure 2. Potential recyclable items within the campus

Regarding the knowledge of recycling among the students, most of the respondents are partially knowledgeable about recycling, 24% are knowledgeable and a minority of 2% knows nothing at all (figure 1). A statistical analysis was done using partial correlation between respondents' knowledge and the awareness level towards recycling. From the analysis obtained, it was found that an insignificant relationship of ($p>0.05$) with a correlation of 0.81 was obtained. It has shown that there is no significant relationship between knowledge and awareness level. A Chi Square Test was conducted which is associated with gender and the knowledge on recycling. From the statistical analysis obtained, the Pearson Chi Square, gives a p-value less than significance level ($p<0.05$). Thus, it is concluded that there is a relationship between gender and knowledge on recycling. Specifically, this approach is appropriate because the sampling method consists of simple random sampling.

To complement the survey, a waste composition study was performed to know what are the general items of the waste discarded by the students. As the research is focussing on recyclable materials, the waste is categorised only as plastics, paper, food waste, aluminium cans and those that are of multicomponent such as battery, cartridge ink, motherboard or any electrical circuit and wires are termed as miscellaneous (table 3). However, it must be noted that the results intend to show the typical composition of students' as generators at the university taken at specific time and days of the week during the period of study. Besides, other waste composition such as textile, rubber and metals are not included.

Table 3. Waste composition at faculties of engineering, UiTM.

Composition of Waste \ Date/Day	24/4/17 Mon	3/5/17 Wed	8/5/17 Mon	15/5/17 Mon	22/5/17 Mon
Paper	15.20	20.91	26.50	30.26	25.26
Plastic	34.80	27.71	29.85	26.24	28.81
Aluminium Can	1.82	0.71	1.70	1.78	1.30
Food Waste	28.04	30.35	34.09	30.42	29.97
Miscellaneous	14.10	15.23	19.18	16.80	13.99
Total	94.26	94.91	111.32	105.50	99.33

As can be observed in table 3 the waste collected varies between 94 – 111 kg depending on the students' activities or presence at the centres. The waste collected on Monday 8/5/17 is the highest because it was the start of the examination week where many students are present at the centres. This results also serve as the basis for the question posed related to the potential recyclable items within the campus. The purpose of the question asked was to identify if the respondents are aware on the potential of recycling activities at their surroundings. From the results as depicted in figure 2, the respondents agreed that the most potential items to be recycled are paper followed by plastics, aluminium cans and glass. This confirmed the earlier observation made (figure 2) whereby a large proportion of the waste generated at complex of engineering can be recycled or is potentially recyclable. In fact, the program for segregation and recycling is feasible in the faculty. It is noteworthy that the results obtained is similar to the study conducted in Campus Mexicali I of the Autonomous University of Baja California (UABC) where paper represents the largest percentage of recyclable waste in the faculty [12]. Pearson correlation factor was conducted for age of respondents and their awareness on potential recyclable items at the faculty. It was found that the p-value is more than the significant level ($p>0.05$). This shows that correlations between the two variables are statistically insignificant. So, there is in conclusive evidence about the significant of the association between variables.

On the question about what might cause them not to practice recycling, variety of reasons was given. As shown in figure 3, the major excuse is on the availability of the recycling bins. This also suggests that there is still recycling potential among them. Because of the absence of recycling bins, the students throw their rubbish in the available bins provided. However, 47% of the respondents said that the bins are full, 23% are unsure if the products are recyclable, 13% do not even bother to care about recycling and only 10% believe that recycling is too time consuming.

In the second phase of questionnaire, the questions asked related to the effectiveness of recycling bins in the faculty. As previously mentioned, in year 2013/2014, recycling bins placed at the library lobby seating area provided by the facility department of the university. Due to lack of support from the students, the recycling bins are left ignored by the students. Therefore, as mentioned earlier, there are five (5) new recycling bins placed at the complex of engineering as the alternative, were made and placed at each engineering centres as shown in figure 4 during phase 2 of the study. This approach was made to encourage the engineering students to practice recycling in the environment of study. As can be observed in figure 5, 91% of respondents are aware on the existence of the bins and a minority of 9% are not aware. The recycling bins provided are mainly for the segregation of aluminium cans, plastic bottles and papers. These are clearly specified on the recycling. However, they are unable to differentiate and understand what the objectives of the placement of the bins. Even with a clear signage, the students are unable to perform the act of recycling properly. This strongly recommends, the education on recycling is very important to increase the students' awareness on recycling in future.

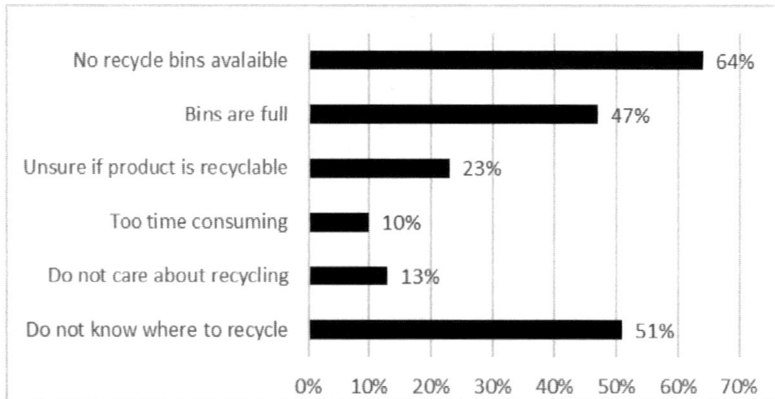

Figure 3. Obstacles preventing students from recycling.

Figure 4. Recycling bins at engineering complex.

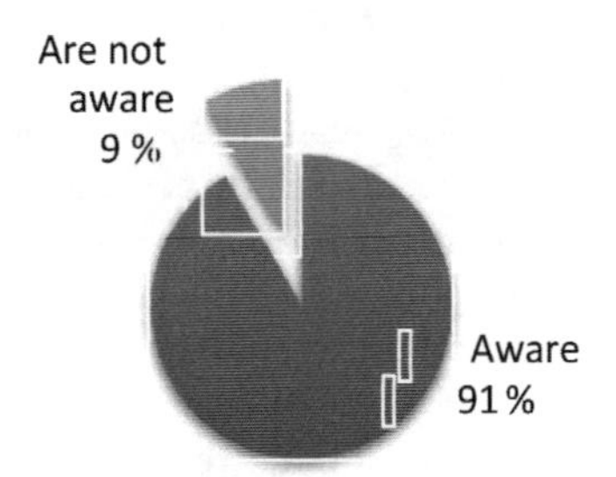

Figure 5. Awareness on the existing recycling bins in the campus.

Recycling is good for the environment in the sense that old waste products of no use are then converted into same new products. In order to encourage students to recycle more, education should start at home. On the question whether, recycling is continued at home, from the survey conducted, 56% of respondents practice recycling at home and 44% respondents do not recycle at home (figure 6). The result also indicated that 38% of female practice recycling and only 18% of the male do. This result agrees with the findings conducted in Dalhousie University in which the statistical calculation of t-tests is not statistically significant and female tend to practice recycling more compared to male.

The sustainability of the waste management should be enhanced through higher educational institution [13]. Strategies on how to improve recycling should be implemented among youngsters. The results show that 73% respondents suggested to provide more recycling bins and make them accessible (57%) followed by education on recycling (48%), visible signage (46%) and impose fine on those who neglect recycling (15%) (figure 7). This result agrees to the study conducted in the University of the West Indies, Cave Hill Campus, Barbados where the data suggested that signage and the accessibility of recycling bins are the potential key strategies that could be used to increase awareness about recycling.

Figure 6. Students who practice home recycling.

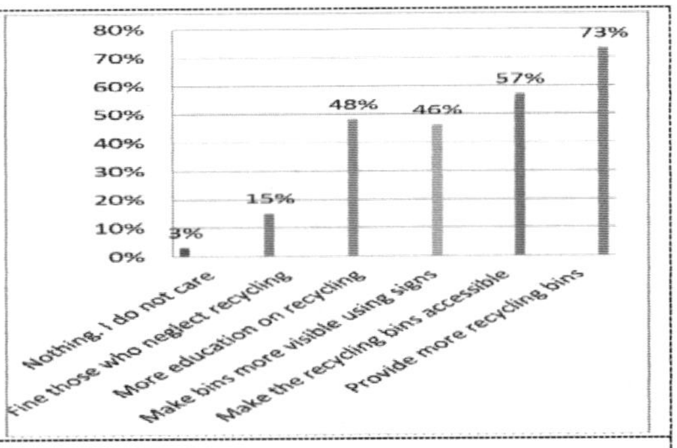

Figure 7. Strategies for improving recycling.

Recycling helps to reduce the amount of waste being sent to the landfill in which it helps to conserve raw materials and protects natural habitats for the future. In the last phase of questionnaire distribution, the questions delivered focused on how waste segregation helps to conserve the environment. From table 4, majority of 52%, said they will start practice waste segregation in everyday life. Also, most of the respondents want to take advantage of the concept of waste to wealth (W2W) that converts waste into profitable end products. The results finally capture respondents' awareness on the importance of conserving the environment and shows positive attitudes towards the ideas.

Table 4. Opinions on contribution to the conservation of environment.

Respondents opinions on contribution to the conservation of the environment	% of respondents
I will start practicing waste segregation in everyday life	52
I will help in promoting the importance of waste segregation	27
I will promote recycling products	22
I would be more interested to know about recyclable items	24
I want to aspire to be entrepreneurs from recycled products	12
I want to take advantage of the concept of waste to wealth (W2W) that converts waste into profitable end products	39

4. Conclusion

The research conducted among engineering students focus on the knowledge and awareness of recycling practices. The results obtained shows that the students' awareness and knowledge on recycling are unfavourable at the beginning. Since the research has shown that most of the students' knowledge on waste segregation are still low, it reflects the urgency to educate all students on zero waste management. Most of the students believe that the obstacles that prevent them to recycle are due to the absence of recycling bins or the waste bins are full. Results shows that there is significant relationship ($p<0.05$) between gender and knowledge on recycling using Chi square test and gives an insignificant relationship ($p>0.05$) between knowledge and awareness of students towards recycling using partial correlation. As the recycling facilities (recycling bins) are provided, majority of students are aware on the existence of the new recycling bins. It was also observed that waste segregation at home are also been practised more by the female than the male students. Hopefully, with the new existing facilities provided in the faculty (and many more are needed), these will enable to help increase the awareness of students towards the importance of waste segregation. Further, the authors felt the study is most rewarding when a major portion showed their willingness to continue participation in conserving the environment and even more exciting that a small percentage would become eco-entrepreneurs in the future. As an overall conclusion, it is necessary to develop students' attitude and willingness through providing a firm understanding on the importance to conserve the environment and to practice recycling to prevent problems related to solid waste management in order to guarantee a sustainable future in the country.

Acknowledgement
Thank you to Universiti Teknologi Mara (UiTM) Shah Alam for the grant support 600-RMI/MyRA/10/3/12 and the participation of the students during the research study.

References

[1] Desa, A., N.B.Y.A. Kadir, and F. Yusoff 2013, *A Study On The Knowledge, Attitudes, Awareness Status And Behaviour Concerning Solid Waste Management.* Jornal of Social and Behavioral Sciences, 2010: p. 643-648.
[2] Poor H, Negara, J.P.S.P. 2015, *Kitar Semula*, P.D.K.T. Kementerian Kesejahteraan Bandar, Editor.
[3] Smyth, D.P., A.L. Fredeen, and A.L. Booth 2010, *Reducing Solid Waste InnHigher Education: The First Step Towards'greening' A University Campus.* Journal of Resources, Conservation & Recycling: p. 1007-1016.
[4] Moh, Y.C. 2016, *Solid Waste Management Transformation and Future Challenges of Source Separation and Recycling Practices in Malaysia.* Journal of Resources, Conservation and Recycling: p. 1-14.
[5] Zain, S.M., et al. 2011, *Focusing on Recycling Practice to Promote Sustainable Behavior.* Journal of Social and Behavioral Sciences: p. 546-555.
[6] Kelly, T.C., et al. 2005, *University Community Responses to On-Campus Recycling.* Journal of Resources, Conservation & Recycling: p. 42-55.
[7] Peavy, H.S., D.R. Rowe, and G. Tchobanoglous, *Environmental Engineering*. 1985, Singapore: McGraw Hill International Editions.
[8] Bank, W. 2017, *Atlas Of Sustainable Development Goals*, ed. W.D. Indicators. 2017, Washington.
[9] Ramayah, T., J.W.C. Lee, and S. Lim 2012, *Sustaining the environment through recycling: An empirical study.* Journal of Environment Management: p. 141-147.
[10] Shin, T.J., et al. 2013, *Current Practice Of Waste Management System In Malaysia: Towards Sustainable Waste Management*, in *Faculty of Technology Management and Business Faculty of Science, Technology and Human Development Faculty of Electrical and Electronic Engineering*., University of Tun Hussein Onn Malaysia.
[11] Mrema, K. 2008, *An Assessment of Students' Environmental Attitudes and Behaviours and The Effectiveness of Their School Recycling Programs*, in *School of Resource and Environmental Studies*., Dalhousie University.
[12] Vega, C.A.d., S.I. Benitez, and M.E.R. Barreto 2008, *Solid Waste Characterization and Recycling Potential for a University Campus.* Journal of Waste Management, **28**: p. 521-526.
[13] Bailey, J., M. Pena, and T. Tudor 2015, *Strategies For Improving Recycling At A Higher Education Institution: A Case Study Of The University Of The West Indies, Cave Hill Campus, Barbados.* Journal of Waste Management. **8**: p. 1-11.

IOP Conf. Series: Materials Science and Engineering **334** (2018) 012035 doi:10.1088/1757-899X/334/1/012035

Production and characterization refuse derived fuel (RDF) from high organic and moisture contents of municipal solid waste (MSW)

P Dianda[1,3], Mahidin[2] and E Munawar[2]

[1] Graduate Program Department of Chemical Engineering, Syiah Kuala University, Banda Aceh 23111, Indonesia
[2] Department of Chemical Engineering, Syiah Kuala University, Banda Aceh 23111, Indonesia

E-mail: edi.munawar@unsyiah.ac.id

Abstract. Many cities in developing countries is facing a serious problems to dealing with huge municipal solid waste (MSW) generated. The main approach to manage MSW is causes environmental impact associated with the leachate and landfill gas emissions. On the other hand, the energy available also limited by rapid growth of population and economic development due to shortage of the natural resource. In this study, the potential utilized of MSW to produce refuse derived fuel (RDF) was investigate. The RDF was produced with various organic waste content. Then, the RDF was subjected to laboratory analysis to determine its characteristic including the calorific value. The results shows the moisture content was increased by increasing organic waste content, while the calorific value was found 17-36 MJ/kg. The highest calorific value was about 36 MJ/kg obtained at RDF with 40% organic waste content. This results indicated that the RDF can be use to substitute coal in main burning process and calcinations of cement industry.

1. Introduction

Indonesia and other developing countries are facing serious problem to handle the huge amount municipal solid Waste (MSW). The MSW generated increase with the population growth and economic development. The State Ministry of Environment (MoE), the Republic of Indonesia was estimated the total MSW generated in 2006 approximately 38.5 million ton [1]. This amount is predicted increase to five times by 2020 [2]. Thus this condition lead Indonesia to the serious environmental problems. There are many method can be used to dealing with the MSW, for instant thermo chemical treatment by using incineration, combination mechanical and biological treatment, and bio–gasification, and landfilling. All of these methods has its advantages and disadvantages [3]. For instant, although incineration has been widely used to destroy the MSW due to its capability to reduce the volume of MSW up to 90% of original volume and the energy resulted from the MSW burning can be recovered to generate an electricity or district heating [4, 5]. However, this method is potentially to harmful the environment through the concentration heavy metals content in the bottom ash, as well as the dioxins and furans in flue gas [6, 7].

[3] To whom any correspondence should be addressed.

Unlike to the incineration, the capability of mechanical and biological treatment to reduce volume of MSW relative small as the recyclable fraction in MSW much low compare to the organic fraction. However, during the biological treatment the organic carbon content of MSW can be reduced, which will be reduce the greenhouse gas emission from the next process (i.e. landfilling). Bio–gasification is biochemical process to convert an organic fraction of MSW to produce biogas, which include anaerobic digestion and fermentation. This method only suitable for homogenous wastes having high percentage of organic biodegradable matter and high moisture content. Whilst the landfilling of MSW is associated with the environmental problems mainly with the leachate and biogas emissions [8].

Although all above method have been successful widely used to treat the MSW, such as technologies are applicable in high GDP countries. In city at the emerging countries such as Banda Aceh, Indonesia, the landfilling still the main method for waste management. Moreover, the landfilling of waste was carried out without any pre–treatment, since the waste management alternatives mentioned above are associated with costs that are beyond economic capacities [9]. In addition, Banda Aceh also faced with a problem of available the energy due to lack of natural resource for energy production. Therefore, the refuse derive fuel (RDF) is a potential MSW treatment options which is can solve both, the waste management problem and energy shortage. Thus the objective of this study is to produce RDF from high organic and moisture contents of MSW, characterize its properties, and to investigate composition of organic waste is suitable for RDF production. Finding of this study is very important for waste management development, particularly in developing countries.

2. Material and Method

2.1. Study area description

Banda Aceh or Banda Aceh is the capital and largest city in the province of Aceh, Indonesia. Banda Aceh city is located in northwestern tip of Sumatra Island coverage an area 61.36 square kilometer which divided into nine sub–districts (shown in Fig. 1) including Baiturrahman, Banda Raya, Jaya Baru, Kuta Alam, Kuta Raja, Lueng Bata, Meuraksa, Syiah Kuala and Ulee Kareng. According to the Statistic Bureau report, the total of populations of Banda Aceh city in 2014 is 249,499 inhabitant spread across these sub–district as shows in Table 1 . Banda Aceh was not frequently the subject of discussion before December 26, 2004, when the earthquake at magnitude 9.3 in Indian Ocean was triggered tsunami waves and hit the western coast of Sumatra island. Banda Aceh was the closest city to the earthquake's epicenter, and severely affected after tsunami wave sweep this city. Based on the central government report, approximately 126,602 people have been killed, 93,638 people was missing, and another 514,150 people was displacement [10].

Table 1. Population, area and population density of nine administrative sub–districts within the Banda Aceh city.

No.	Sub–district	Population (persons)	Area (km^2)	Population density ($person/km^2$)
1	Baiturrahman	35,249	4.54	77.64
2	Banda Raya	22,961	4.79	47.94
3	Jaya Baru	24,481	3.78	64.76
4	Kuta Alam	49,545	10.05	49.30
5	Kuta Raja	12,831	5.21	24.63
6	Lueng Bata	24,581	5.34	46.03
7	Meuraksa	18,979	7.26	26.14
8	Syiah Kuala	35,702	14.24	25.07
9	Ulee Kareng	25,170	6.15	40.93
	Banda Aceh	249,499	61.36	40.66

Source: [11]

Approximately 130.7 ton of a MSW is collected within the Banda Aceh city and transported to the landfill site daily. The MSW collection and transportation is undertaken by the Sanitation and Park Department (DKP) of Banda Aceh municipality. The collection was carried out by two method, first door–to–door collection of waste from variety of sources in within the city area by using 26 ordinary trucks, 4 compactor trucks and 10 pick–ups. The frequency of waste collection is 1–4 trips per day depend on the population density and volume of waste generated. Second, MSW was collected from temporary disposal station. There is 85 temporary disposal stations across of Banda Aceh area made of container or brick–box located close to the residential and commercial area, hence the community can be disposed their waste into it any time. The DKP will be transported the waste to the landfill site by using arm–roll truck or ordinary trucks 1–7 times per week. At the landfill site, the waste was directly disposed onto the landfill without any preliminary treatment.

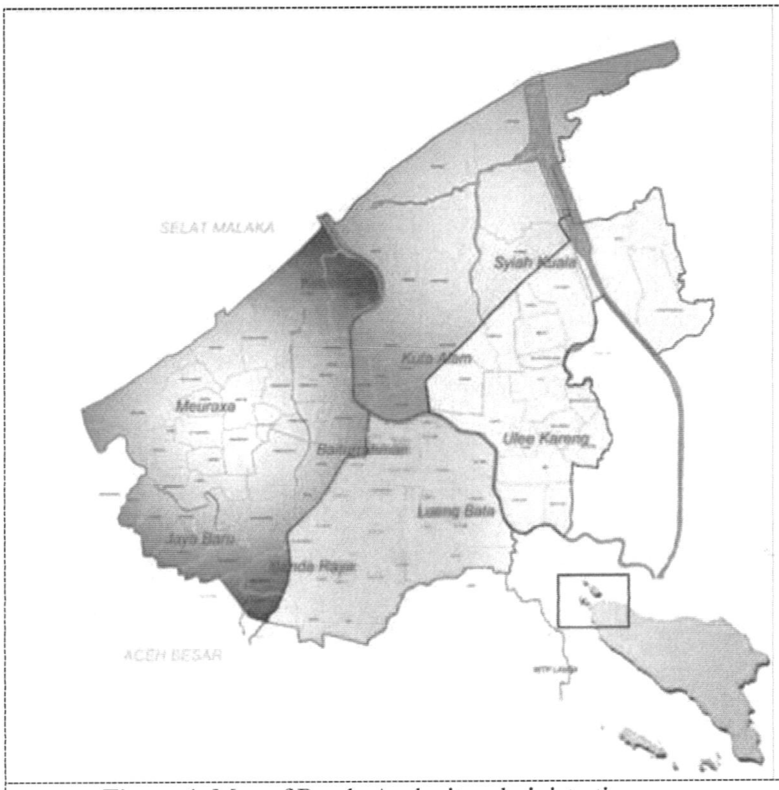

Figure 1. Map of Banda Aceh city administrative area.

3. Results and Discussion

3.1. Composition and volume MSW generated

Result of waste sampling and measuring found that each Banda Aceh people was generated approximately 0.52 kg/capita/day. This amount is small compare to MSW generated in other city such as Kuala Lumpur for about 1.62 kg/capita/day [12, 13], Beijing 0.85 kg/capita/day [14], in Macao 1.52 kg/capita/day [15]. It is can be understood since the waste generation rate was influenced by numerous factors, such as gross domestic product (GDP), norm and culture, waste management policies, consumption habits and lifestyle. In this case, the GDP of Banda Aceh also lower compare to other cities mentioned. In term source of waste, the residential was the larger source of MSW in Banda Aceh. The residential area was contribute for 71.23% of 130.7 ton/day was MSW generated from

Banda Aceh. The second large contributor was shop (e.g. copying and printing shop, restaurant, etc.) which is contribute for about 23.54%, while the schools, markets, and offices were contribute for about 4,38%, 0,53%, 0,24% (see Fig. 2(a)), respectively.

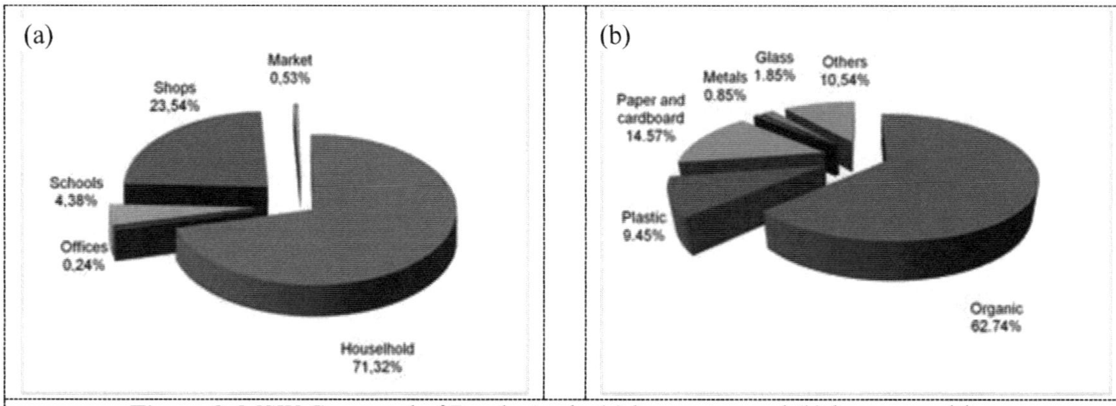

Figure 2. MSW Generated of Banda Aceh (a) by source, and (b) by composition.

As in other developing country, the organic wastes was the major content of MSW generate in Banda Aceh. The composition of organic waste was comprising to 62.74% of total MSW collected. This figure a bit small compare to other study was conducted by Meidiana and Gamse [16], where the composition of organic wastes generated from three classified as big to small cities in java for about 64.5% to 78.1%. On the national level, an average organic waste fraction was about 65% [17]. However, the composition of organic waste found in this study was typical waste in developing countries which is around ±70% organic content [18]. The composition of plastic, paper and cardboard, metals, and glass were 9,45%, 14,57%, 0,85%, and 1,85%, respectively. Whilst, around 10,54% of waste was classified as unidentified waste. Fig. 2(b) is present the composition of Banda Aceh MSW based on wet weight of waste.

3.2. Characteristic of RDF

The characteristic of RDF is greatly influence by the proximate (e.g. moisture, ash and volatile contents) and ultimate (e.g. C, H, O, N and S contents, as well as HHV) analysis. These information is very useful to represent the RDF quality. Therefore, the RDF was analyzed proximately and ultimately as presented in Table 2 and compared to the low rank coals from Nagan Raya district, Aceh and the charcoal made of its as reported by Khairil, Irwansyah [19].

Table 2. Proximate and ultimate analysis of RDF and other fuel.

Fuel type	Proximate Analysis (%wt)				Ultimate Analysis (% wt)					Calorific value (MJ/kg)
	Moisture content	Volatile content	Fixed carbon	Ash content	C	H	N	O	S	
Coal[1)]	11.82	46.61	38.03	3.54	58.02	5.5	1.26	35.01	0.21	22.38
Charcoal[1)]	6.8	48.13	41.16	3.91	64.55	5.23	1.41	28.59	0.22	23.91
RDF[2)]	6.4	8.82	75.58	9.2	49.75	9.04	10.77	28.55	1.89	24.96

Source: [19] and this study.

Table 2 shows that the moisture and volatile contents of RDF is lower than coals and its derived charcoal. In contrary, the ash content of RDF is higher than that fuels. The high content of ash may come from dust of street waste, whilst in the coals and charcoals the ash content is mostly impurities contained in the pore of surface coals. The impurities of coals should be smaller than in MSW, since the coals is quite homogeneous compare to the MSW. The low moisture and volatile contents in RDF probably due to low moisture and volatile contents in its origin wastes, thus apart of them have been released during the drying process. Whereas in the coals and charcoal the moisture and volatile matters were trapped in the pore space.

Table 2 also present the comparison of ultimate analysis and HHV of RDF from Banda Aceh MSW to coals and charcoals prepared in laboratory furnace under temperature maintained at 250 °C [19]. The data in that table shows that the carbon content in RDF quite small compare to the coals and charcoals, however this figure a bit higher than the carbon content of RDF have been report by Nobusuke, Yoshinori [20]. Nobusuke was produced seven type of RDF from various wastes, including industrial waste, MSW, and synthetic waste. The carbon content of RDF prepare from MSW was ranged 42–44% dry weight. The carbon content of RDF from Banda Aceh MSW also slightly higher than the commercial RDF produced in Italia. The carbon content of RDF in Italia was around 49% with the moisture content for about 25–30% [21]. The similar comparison have been found for the other substances. In contrary to the carbon content, the calorific value of RDF from Banda Aceh MSW is higher to coals and charcoal from Nagan Raya district. Moreover, the calorific value RDF from Banda Aceh MSW about 25–42% high than the minimum calorific value of RDF in Finland, United Kingdom, and Italia. The minimum calorific value of RDF in these countries are about 3,107, 4,469, and 3,585 kcal/kg, respectively [22, 23]. The RDF from Banda Aceh MSW also already meet the minimum calorific value of RDF use as co–fuel for main burning process in cement industry which is required calorific value of 4,780 kcal/kg [24]. This finding indicated that the RDF from Banda Aceh MSW is potentially produce in commercial scale and use to substitute apart of coals consumption in cement industry was located next district. This effort will not only can reduce the green house gas emissions from waste management sector and cement industry, but also can solve the MSW management problem faced by cities in developing country, such as Banda Aceh as well.

3.3. Effect of organic waste content on the RDF properties

As the nature of MSW in developing contain high organic waste and moisture contents may be affect to the RDF quality. Thus, the effect of organic waste content to RDF properties was investigate. Fig. 4. presents the effect of organic waste content on proximate analysis and ultimate analysis of RDF.

Fig. 4 (a) shows the increase of organic content in RDF was not effect significantly to the proximate analysis (i.e. moisture, volatile, fixed carbon and ash contents). This is can be understand since the organic matter mostly comprising carbohydrate (e.g. cellulose, hemicelluloses and lignin) in some waste types (e.g. wood, branch of three), which is completely combustible. However, the moisture content was much increase by increasing the organic waste content. The moisture content of RDF was increase from 3% for RDF without organic waste to be around 10% at RDF containing organic waste for 80%.This was caused by high content of water in the organic waste.

In contrast, the carbon content and calorific value were increase by increasing organic waste content up to 40% of total weight of RDF. The carbon content and calorific value at this composition was about 62 % dry weight and 36 MJ/kg, respectively. This figure 48–59% higher than low rank coal from Nagan Raya district [19] and 90–145% higher than the minimum calorific value of RDF in Finland, UK and Italia [22, 23]. However, carbon content and calorific value was dropped for more organic waste content. The similar trend has been seen for hydrogen content, but it was in lower magnitude. The oxygen content was shows on the opposite response to the carbon content and calorific value, while the nitrogen content tend to increase by putting more organic waste. This may caused by the oxygen content in organic waste decrease during the drying process as the oxygen content tend to deplete by temperature increase. In contrast, the nitrogen content not effect by temperature. This phenomena also found when [19] increase the carbonizing temperature from 250 to 450 °C.

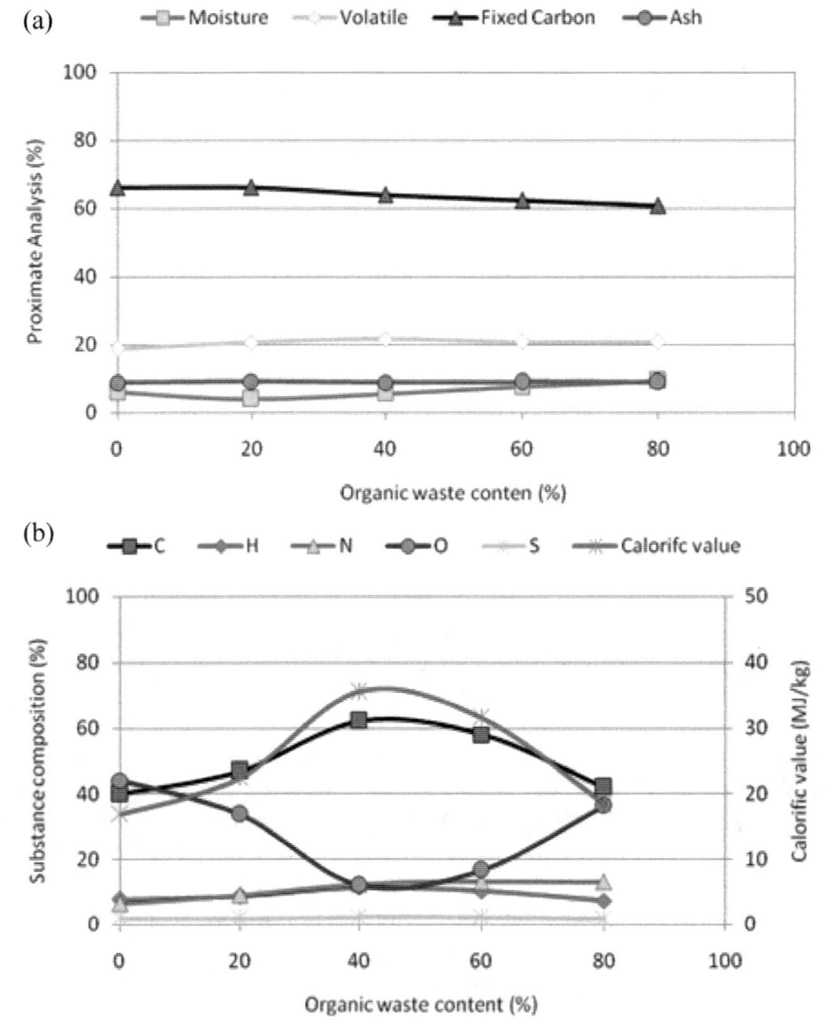

Figure 3. Effect of organic waste content on RDF properties, (a) proximate analysis and (b) proximate analysis.

4. Conclusion

The intention to alternative waste management in developing countries is getting higher due to the impact traditional waste management practice (e.g. landfilling) to the environment become more serious by rapid growth population and economic development. The landfilling of MSW associated with the leachate and landfill gas emissions. Although landfill gas emissions will become negligible within decades, while leachate emissions will be continue to polluted the environmental for hundreds of years. This paper present a study of production and characterization RDF from high organic waste and moisture content of MSW, which nature MSW in the developing countries. The use of MSW to produced RDF can be an option waste management problem and simultaneously the energy shortage due to lack of natural resource for energy. The aim this study is to produce RDF from high organic waste contents, characterization and exploring the best composition of organic waste use for RDF production.

The RDF was produced contain various percentage of organic waste and particle size. Although the organic waste much influenced to the RDF characteristic, the experimental data clearly show that MSW of cities in developing countries, such as Banda Aceh is can use for RDF production. Moreover, the calorific value of RDF was higher than the low rank coal from Nagan Raya district, Aceh provinces and comparatively to RDF produce from different sources and district in Ulsan city, Korea. The calorific value of RDF was resulted heating value ranged from 17 to 36 MJ/kg, which the highest heating value was obtained from RDF with organic waste content 40%. Furthermore, according to the minimum calorific value requirement, the RDF can be use to substitute apart of the coal utilized in cement industry, both in main burning process and calcinations. These can be provides by RDF with organic waste content 40%, which has the calorific value for about 36 MJ/kg.

Acknowledgment

This study was funded by Syiah Kuala University through the post graduate research grant with the contract number 654/UN11/S/LK–PNBP/2015. The authors are very grateful to Mr. Purnama Arkha and Mr. Iqbal Manani for their contribution in this work.

References

[1] MOE, Indonesian Domestic Solid Waste Statistic Year 2008, State Ministry of Environment the Republic of Indonesia, Jakarta, Indonesia. 2008.
[2] Kardono. Integrated Solid Waste Managemen in Indonesia. in International Symposium on Eco Topia Science, ISETS07. 2007. Nagoya, Japan.
[3] Lee, B.-K., M.J. Ellenbecker, and R. Moure-Eraso, Analyses of the recycling potential of medical plastic wastes. Waste Management, 2002. **22**(5): p. 461-470.
[4] Cheremisinoff, P.N. and A.C. Morresi, Energy from Solid Wastes. 1976: Marcel Dekker Inc.
[5] Kušar, H.M.J., A.G. Ersson, and S.G. Järås, Catalytic combustion of gasified refuse-derived fuel. Applied Catalysis B: Environmental, 2003. **45**(1): p. 1-11.
[6] Sell, N.J., Industrial Pollution Control: Issues and Techniques, 2nd Edition. 1992.
[7] Lee, B.-K., M.J. Ellenbecker, and R. Moure-Ersaso, Alternatives for treatment and disposal cost reduction of regulated medical wastes. Waste Management, 2004. **24**(2): p. 143-151.
[8] Fellner, J., G. Döberl, and P.H. Brunner. Impacts of Leachate Flow Conditions on the Duration of the Aftercare Period. in Kalmar ECO-TECH′03. 2003. Kalmar, Sweden.
[9] Fellner, J. and D. Laner, The potential of nitrogen assimilation in aerated municipal solid waste landfills. chemistry, 2011. **33**: p. 37.
[10] Bappenas, Master Plan of Rehabilitation and Reconstruction Aceh and Nias, North Sumatera Regions. 2005, National Planning Development Board, the Republic of Indonesia.
[11] BPS, Statistic of Banda Aceh Population 2014. 2015, Bureau Statistic of the Republic of Indonesia: Banda Aceh, Indonesia.
[12] Periathamby, A., F.S. Hamid, and K. Khidzir, Evolution of solid waste management in Malaysia: impacts and implications of the solid waste bill, 2007. Journal of Material Cycles and Waste Management, 2009. **11**(2): p. 96-103.
[13] Saeed, M.O., M.N. Hassan, and M.A. Mujeebu, Assessment of municipal solid waste generation and recyclable materials potential in Kuala Lumpur, Malaysia. Waste Management, 2009. **29**(7): p. 2209-2213.
[14] Zhen-shan, L., et al., Municipal solid waste management in Beijing City. Waste Management, 2009. **29**(9): p. 2596-2599.
[15] Jin, J., Z. Wang, and S. Ran, Solid waste management in Macao: Practices and challenges. Waste Management, 2006. **26**(9): p. 1045-1051.
[16] Meidiana, C. and T. Gamse, The new Waste Law: Challenging opportunity for future landfill operation in Indonesia. Waste Management & Research, 2011. **29**(1): p. 20-29.

[17] Helmy, M., T.B. Laksono, and D. Gardera, 3R Implementation in Indonesia, in Senior Official Meeting on the 3R Initiative. 2006, Japan International Cooperation Agency (JICA): Tokyo. Japan.
[18] Diaz, L.F., et al., Composting and recycling municipal solid waste. 1993: Lewis Publishers.
[19] Khairil, et al., Fundamental Study on the Carbonization Characteristics of Low Rank Coal Under Low Temperature and its Application on Traditional Blacksmith. Advanced Materials Research, 2012. **576**: p. 615-618.
[20] Nobusuke, K., et al., The behavior of flue gas from RDF combustion in a fluidized bed. Powder Technology, 2005. **151**(1–3): p. 87-95.
[21] Galvagno, S., et al., Steam gasification of tyre waste, poplar, and refuse-derived fuel: A comparative analysis. Waste Management, 2009. **29**(2): p. 678-689.
[22] Gendebien, A., et al., Refuse Derived Fuel, Current Practice and Perspectives. 2003, European Commission – Directorate General Environment.
[23] Rotter, V.S., et al., Material Flow Analysis of RDF-Production Processes. Waste Management, 2004. **24**: p. 1005–1021.
[24] Ibbetson, C. and K. Wengenroth. Optimisation of Fuels from MBT processes. in Optimisation of Fuels from MBT processes. 2007. Wasteconsult international.

IOP Conf. Series: Materials Science and Engineering **334** (2018) 012036 doi:10.1088/1757-899X/334/1/012036

Utilization of wastewater on seed germination and physiological parameters of rice (*Oryza sativa L.*)

V Huy[1,3] and C B Iwai[1, 2]

[1] Department of Land Resources and Environment, Faculty of Agriculture, Khon Kaen University, Khon Kaen, 40002 Thailand
[2] Integrated Water Resource Management Research and Development Center in Northeast Thailand, Khon Kaen University, Khon Kaen, 40002 Thailand

E-mail: chuleemas1@gmail.com

Abstract. Due to increasing world population and demand, fresh water availability is becoming a limited resource. Reusing wastewater for agriculture has received attention since it contains nutrients, which are beneficial for growing crops. Even though wastewater can be used as the nutrient source for the plant, the toxicity of wastewater can still be a cause for concern and investigation. The main objective of this paper was to assess the effect of different sources of wastewater on the germination of Jasmine rice (KDML105), White rice (Phatum Thani 1), and Sticky rice (RD6) under laboratory conditions. Petri dish cultures were used with various concentrations (0, 50, and 100%) of wastewater collected from swine farm, aquaculture activity, and domestic. Seed germination, root length, shoot length, seed vigor index, fresh weight and dry weight were measured after each experiment. The results have shown that domestic wastewater and aquaculture activity wastewater did not decrease performance of Jasmine rice (KDML105), White rice (Phatum thani 1), and Sticky rice (RD6) while the germination of Jasmine rice (KDML105), White rice (Phatum thani 1), and Sticky rice (RD6) decreased when irrigated with swine farm wastewater. Therefore, using domestic and aquaculture activity wastewater for irrigation are suitable for growth of these crop.

Introduction

Due to increasing population numbers in the developing countries, demand for natural resources have been put under pressure and stress from excess waste. Access to water has become an essential and limited resource since industrial, agricultural and human activities keep rising to fulfill economic demands. Many countries in Africa, Latin American, Southern Asia, Middle East and more isolated island countries in Oceania may have problems with the supply and disposal of water because of the diverse requirement including irrigation of crops [1]. Increasing human demand for rice has focused attention on increasing agricultural activities by fertilizer use, irrigation and value adding food processes and this requires further access to water resources. In the same way, wastewater continues to increase due to these activities and is a major main problem for the environment. Wastewater is considered one

[3] To whom any correspondence should be addressed.

of the worst kinds of pollution because of chemical and waste material from agricultural and food production. The environment can be affected by the additional heat, high or low pH level, and chemical contamination.

However, wastewater may contain various amount of organic matter (OM) and plant nutrients [nitrogen (N), phosphorus (P), potassium (K), calcium (Ca), sulfur (S), copper (Cu), manganese (Mn), and zinc (Zn)] which benefit agriculture by increasing crop production [2]. A major crop in Asia is rice which consumes a lot of water during its life cycle. Additionally, wastewater from livestock production, aquaculture, and domestic have high nutrient levels and have a biochemical oxygen demand (BOD). After proper dilution, wastewater can be used as a potential nutrient source for plant growth in agricultural production, especially with rice. Therefore, reusing wastewater in agriculture has received much attention for the exploitation of its nutrients. For instance, using wastewater for irrigation can cause an increase in OM, N, P, and K to the soil and can provide significant benefit to farming communities [3]. For example, an experiment on seed germination using jute mill wastewater indicated that jute mill wastewater plays a profound role on germination characteristics and carbohydrate content of *Cicer arietinum* and *Pisum sativum* [4]. An experiment on the effects of treat and untreated wastewater on the seed *Avena sativa L* reported that treated wastewater can improve the seeding growth, and its amylase and lipase activities [3]. An experiment on seed performance of pea, lentil and gram from textile wastewater showed that wastewater can be utilized for irrigation propose [5]. It was also reported that municipal wastewater had an effect on germination, seedling performance, nutrient uptake, and increased chlorophyll content of rice [2]. A study on wastewater treatment on the seed *Abelmoschus Esculentus L* showed that the wastewater is an important nutrient source for seed germination and plant growth [6]. An experiment on rice germination from domestic wastewater reported that with the proper dilution domestic wastewater could be used as a potential source for seed germination [7]. There are many researchers studying the effect of wastewater on seed germination, but most of the studies focus on a single source of wastewater to test seed germination. There is limited information about comparing the different sources of wastewater on seed germination. Therefore, the aim of this study is to investigate the applicability of different sources and different concentrations of wastewater on selected rice varieties in order to improve germination and yield.

Experimental procedure and methodology

1.1. Wastewater sampling

Domestic wastewater (DW) was collected from Khon Kaen University wastewater, Khon Kaen province, Thailand. Swine farm (SW) and aquaculture activity wastewatr (AW) were collected from animal farms, also located in Khon Kaen University. Wastewater was stored in plastic containers and put in the refrigerator below 5 ^{0}C until use. Wastewater was diluted to 0, 50 and 100 % (pure wastewater without dilution).

1.2. Physical-chemical characteristic of wastewater

Physical-chemical parameters of wastewater were analyzed: pH, Temperature, Electrical Conductivity (EC), Dissolved Oxygen (DO), Biochemical Oxygen Demand (BOD), Chemical Oxygen Demand (COD), Total Dissolved Solids (TDS), Total Solids (TS), Suspended Solids(SS), Hardness, Fats, Oil, and Grease (FOG), Nitrogen (N), Nitrate (NO_3^-), Phosphorus (P), Potassium(K), Copper (Cu), Cadmium (Cd), Lead (Pb), Chromium (Cr) and Zinc (Zn) following the Standard Method for Examination of Water and Wastewater (APHA, 2013) [8].

1.3. Seed germination experiment

A seed germination experiment was conducted for 5 days at Ecotoxicology Laboratory, Department of Land Resource and Environment, Faculty of Agriculture, Khon Kaen University, Khon Kaen, Thailand. The experiment followed a Completely Randomized Design (CRD) with three replications. Rice seeds (*Oryza sativa L.*) from three different varieties comprising Jasmine rice (KDML105),

White rice (Phatum Thani 1), and Sticky rice (RD6) were sterilized with 1% solution of hydrochloride for 15 minutes to remove microbes. Seeds were then washed with distilled water. Whatman Grade 181 No. 1 filter papers were put into petri dishes to which 10 seeds from each representative rice variety were added. Each petri dish was irrigated with 5 mL of different concentration of wastewater, and then incubated at 25 °C in an Aqualytic Tc 455s incubator. After completion of seed germination, seeding from each treatment was measured for germination percentage, root length, shoot length, seeding vigor index, fresh weight, and dry weight.

1.4. Germination percentage estimation
Seed germination percentage was calculated from each treatment based on the primary root emergence from the seeds and the results were expressed in percentage following the calculation as in Equation 1 [4].

$$\text{Germination (\%)} = \frac{\text{Total number of seed germinated in particular treatment}}{\text{Total number of seed treated in particular treatment}} \times 100 \qquad (1)$$

1.5. *Seeding Length estimation*
The shoot length was determined based on the primary seed plumule and root length was determined by the primary seed radical. Seeding length results were expressed as centimeter.

1.6. Fresh weight and dry weight
Fresh weight was measured using an Ohaus PA 2102 electronic balance. Dry weight was measured after drying the seedling in hot air oven at 80 ^{0}C for 24 hours and was then recorded using electronic balance [4].

1.7. Seed vigor index
Seed germination, root length, and shoot length were used to calculate seed vigor index, which followed the formula as in Equation 2 [9].

$$\text{Vigor index} = (\text{Mean root length} + \text{Mean shoot length}) \times \text{Percentage germination} \qquad (2)$$

1.8. Statistical analysis
The Statistic 10 software (version 10, USA) was used to analyze the data including the analysis of variance (One Way ANOVA). Treatment means were compared using least significance difference (LSD) at $P< 0.05$.

Results and Discussions
1.9. Wastewater characteristics
The chemical and physical characteristics of swine farm, domestic, and aquaculture activity wastewater are presented in Table 1. The different sources of wastewater express difference characteristics. Swine farm, domestic, and aquaculture activity wastewater are slightly alkaline with pHs of 7.78, 7.37 and 7.8, respectively. Swine farm wastewater shows the highest electrical conductivity (EC), Total Dissolved Solids (TDS), Chemical Oxygen Demand (COD), Fats, Oil, and Grease (FOG) when compared with the other wastewaters. The concentration of Total Solids (TS), Suspended Solids (SS), Biological Oxygen Demand (BOD), Hardness and Dissolved Oxygen (DO) in swine farm, aquaculture, and domestic wastewater show a similar order. The concentration of Nitrogen (N), Nitrate (NO_3^-), Phosphorus (P), and Potassium (K) in swine farm, domestic and aquaculture activity wastewater are also similar. Copper (Cu) and Zinc (Zn) are higher in swine farm wastewater when compared with other wastewater. It is reported that Copper (Cu) and Zinc (Zn) are found in higher quantity in swine farm wastewater because of supplement added to the pig's feed with higher amounts of Copper (Cu) and Zinc (Zn) to improve feed efficiency and the health of swine [10].

Lead (Pb) concentration is higher in domestic wastewater compared with other wastewater (Table 1). It is reported that the higher concentration of Lead (Pb) in domestic wastewater is due to the use of leaded petrol when undertaking domestic activities [11]. Zinc (Zn) and Lead (Pb) are not detected in aquaculture activity wastewater (Table1).

Table 1. Physical and chemical characteristic of different sources of wastewater

Characteristics	Swine farm	Domestic	Aquaculture
pH	7.78	7.37	7.81
Electrical conductivity (µs/cm)	3081.7	477.3	494.7
Total dissolve solids (mg/L)	1521.7	238.7	247.3
Total Solids (mg/L)	4600	5000	5800
Suspended solids (mg/L)	1500	1500	2000
Biochemical oxygen demand (mg/L)	205	167	140
Chemical oxygen demand (mg/L)	7200	91.33	24
Hardness (mg/L)	72	30	26
Fats, Oil, and Grease (mg/L)	6802.5	1500	2005
Temperature (^{0}C)	28.33	29.30	26.29
Salt (mg/L)	1.54	0.22	0.23
Dissolve Oxygen (mg/L)	1.63	1.83	1.36
Nitrate as NO_3^-(mg/L)	0.05	0.09	0.072
Nitrogen as N (mg/L)	2.24	1.49	1.12
Phosphorus as P (mg/L)	1.34	1.52	0.1
Potassium as K (mg/L)	148.3	80.27	70.95
Copper as Cu (mg/L)	0.098	0.003	0.001
Zinc as Zn (mg/L)	0.308	0.031	ND
Lead as Pb (mg/L)	0.002	0.164	ND

Note: ND is not detected

3.1. Germination results

The results of the seed experiment of Jasmine rice (KDML105), Sticky rice (RD6), and White rice (Phatum Thani 1) are showed in Table 2, 3 and 4.

Table 2. Effect of wastewater on Jasmine rice (KDML105) germination (GM), roots length (RL), shoots length (SL), seeding vigor index (SVI), fresh weight (FW) and dry weight (DW)

Sources of WW	Concentration (%)	Jasmine rice (KDML105)					
		GM (%)	RL (cm)	SL (cm)	SVI	FW (g)	DW (g)
DI	0	100^a	3.42^a	2.07^b	550.28^a	0.605^b	0.20^a
SW	50	0^b	0^b	0^c	0^b	0^c	0^c
SW	100	0^b	0^b	0^c	0^b	0^c	0^c
DW	50	90^a	3.77^a	2.32^{ab}	548.92^a	0.75^{ab}	0.20^a
DW	100	95^a	4.03^a	2.57^a	628.30^a	0.87^a	0.21^a
AW	50	90^a	3.52^a	2.10^b	506.25^a	0.73^{ab}	0.19^a
AW	100	95^a	3.90^a	2.33^{ab}	593.60^a	0.78^a	0.205^a

Values in same letters in the columns are not significantly different ($p \leq 0.05$), WW= wastewater, DI=distill water,
SW=swine farm wastewater, DW= domestic wastewater, AW= aquaculture activity wastewater.

Table 3. Effect of wastewater on Sticky rice (RD6) germination (GM), roots length (RL), shoots length (SL), seeding vigor index (SVI), fresh weight (FW) and dry weight (DW)

Sources of WW	Concentration (%)	Sticky rice (RD6)					
		GM (%)	RL (cm)	SL (cm)	SVI	FW (g)	DW (g)
DI	0	100^{a}	4.695^{a}	1.98^{b}	667.50^{a}	0.57^{a}	0.205^{a}
SW	50	10^{b}	0.25^{b}	0.13^{c}	3.80^{b}	0.2^{b}	0.05^{b}
SW	100	0^{b}	0b	0^{c}	0^{b}	0^{c}	0^{b}
DW	50	90^{a}	4.85^{a}	2.29^{ab}	642.60^{a}	0.58^{a}	0.21^{a}
DW	100	95^{a}	4.915^{a}	2.49^{a}	703.475^{a}	0.61^{a}	0.23^{a}
AW	50	90^{a}	4.58^{a}	2.24^{ab}	613.80^{a}	0.545^{a}	0.205^{a}
AW	100	95^{a}	4.78^{a}	2.41^{ab}	683.05^{a}	0.61^{a}	0.215^{a}

Values in same letters in the columns are not significantly different ($p \leq 0.05$), WW= wastewater, DI=distill water,
SW=swine farm wastewater, DW= domestic wastewater, AW= aquaculture activity wastewater.

Table 4. Effect of wastewater on White rice (Phatum Thani 1) germination (GM), roots length (RL), shoots length (SL), seeding vigor index (SVI), fresh weight (FW) and dry weight (DW)

Sources of WW	Concentration (%)	White rice (Phatum thani 1)					
		GM (%)	RL (cm)	SL (cm)	SVI	FW (g)	DW (g)
DI	0	100^{a}	4.075^{b}	2.45^{b}	653^{a}	0.815^{b}	0.195^{a}
SW	50	20^{b}	0.15^{c}	0.08^{c}	9.20^{b}	0.225^{c}	0.075^{b}
SW	100	0^{b}	0^{c}	0^{c}	0^{b}	0^{c}	0^{c}
DW	50	90^{a}	5.46^{ab}	2.64^{b}	729.45^{a}	1.04^{ab}	0.205^{a}
DW	100	95^{a}	6.07^{ab}	3.14^{a}	874.65^{a}	1.175^{a}	0.215^{a}
AW	50	85^{a}	5.35^{ab}	2.54^{b}	671.85^{a}	0.97^{ab}	0.20^{a}
AW	100	95^{a}	5.75^{ab}	2.81^{ab}	816.60^{a}	1.03^{ab}	0.205^{a}

Values in same letters in the columns are not significantly different ($p \leq 0.05$), WW= wastewater, DI=distill water,
SW=swine farm wastewater, DW= domestic wastewater, AW= aquaculture activity wastewater.

The result of seed germination from 0, 50, and 100% of swine farm, domestic, and aquaculture activity wastewater on Jasmine rice (KDML105), Sticky rice (RD6) and White rice (Phatum Thani 1) are significantly different (Table 2, 3 and 4). However, seed germination of Jasmine (KDML105), Sticky rice (RD6) and White rice (Phatum Thani 1) irrigated with 50% and 100% of domestic wastewater, aquaculture wastewater and control are not significantly different and the germination of all rice varieties are lower when compared with the control. The results are similar to report from Quetta City on the effects of wastewater on seeding lettuce growth which determined that seed growth was decreased because of high soluble salt levels within the wastewater [12]. It was reported that seed germination of *Capsicum annun* was delayed because the high level of salinity and toxicity of high osmotic pressure found in wastewater, which can decrease water uptake, by the seed [13]. It was also indicated that the high osmotic pressures of seed germination solution makes inhibition more difficult and restarts seed germination. The ability of seeds to germinate under high osmotic pressure differs with variety as well as species [14]. All the rice varieties were less germinated with 50% and 100 % swine farm wastewater because of the effect of high EC (3081.7 μs/cm) which is presented in Table 1. The report on the effect of tannery effluent on seed germination and growth of two sunflower cultivars indicated that water successfully used for irrigation has EC less than 2250 µs/cm [15]. If electrical conductivity is higher than the permissible limit, it will be attributed to the relatively high salinity, which can decreased of the crop germination percentage [16]. Or,it can also be attributed to the toxicity of high concentration of chloride ion in wastewater which could be reduce seed germination [17]. The shoot length of Jasmine (KDML105), Sticky rice (RD6) and White rice (Phatum 1) which irrigated with 100% domestic wastewater is significantly increased when compared with control. An experiment on the impact of different dilutions of sewage water on seed germination characteristics and carbohydrate content of seeding *Vigna unguiculata* indicated that when high nutrients are applied to seed, the shoot length would be increased [4]. In the same way, by increasing wastewater concentration, seeding height was increased since wastewater contains essential nutrients, so it can promote plant growth [18]. The fresh weight of Jasmine rice (KDML105) is significantly increased when irrigated with 100% domestic and aquaculture wastewater when compared with other concentrations. The fresh weight of White rice (Phatum Thani 1) is also significantly increased with 100% domestic wastewater when compared with other treatments. It was already found that when the root and shoot length were healthier, the plants were heavier in weight [15]. Plants irrigated with 100 % wastewater can have a greater seed weight when compared with tape water because wastewater application can be contributed plant appropriated nutrition and increase photosynthetically active area during seed filling period [15]. There is no significant difference on the fresh weight of sticky rice (RD6). It was reported that different plant varieties have different abilities to absorb the different range of the nutrients, which can affect length, thickness, surface area, and weight [19]. For instance, modern genotypes of rice are more tolerance than the traditional genotype and different varieties have different capacities in root absorption, translocation and shoot demand [19]. The 100% domestic wastewater was very effective for seed performance when compared with other wastewater. It might be the plant nutrients presenting in domestic more than other wastewater, which are shown in Table.1. However, the 50% domestic and aquaculture wastewater were less effective for seed performance, possibly because of diluted concentrations, which can reduce the plant nutrient.

4. Conclusions

From this study, it can be concluded that 100 % domestic and aquaculture activity wastewater were influence seed performance of Jasmine (RDML105), Sticky rice (RD6) and White rice (Phatum Thani 1) and 100% domestic wastewater was given the best result. However, Swine farm wastewater was not suitable for irrigation purpose since it might be contained high EC (3081.7 µs/cm) which was higher than permissible limit (2250 µs/cm) for irrigation seed germination and plant. It could be suggested that using domestic wastewater directly for irrigation purpose is appropriated. The use of wastewater for plant nourishment would be beneficial for alternative resources to fresh water. Wastewater contains essential nutrients, which is important for soil as well as crop; moreover, it is necessary to minimize pollution and to assist the environment. However, toxic contamination such as heavy metal and health safety should be a concerned. Further, study on food safety needs to be investigated.

Acknowledgements

The author thanks the Integrated Water Resource Management Research and Development Centre in Northeast Thailand, The Research Developing and Learning Centre on Earthworm for Agriculture and Environment, and the KKU scholarship for ASEAN and GMS Countries for financial support. We would like to thank Professor Dr. Barry Noller for his kind help with editing the English.

References

[1] Naddafi K, Jaafarzadeh N, Mokhtari M, Zakizadeh B and Sakian M R 2005 Effects of wastewater stabilization pond effluent on agricultural crops *Int J Environmental Science Technology* **1(4)** 273–7
[2] Gassama U M, Puteh A B, Abd-Halim M R and Kargbo B 2015 Influence of municipal

wastewater on rice seed germination, seedling performance, nutrient uptake, and chlorophyll content *J Crop Science Biotechnology* **18(1)** 9–19
[3] Fendri I, Ben S R, Khemakhem B, Ben H N, Gdoura R and Abdelkafi S 2013 Effect of treated and untreated domestic wastewater on seed germination, seedling growth and amylase and lipase activities in Avena sativa L *J Sci Food Agriculture* **93(7)**1568–74
[4] Manisha P and Angoorbala B 2015 Impact of different dilutions of sewage water on seed germination characteristics and carbohydrate content of vigna unguiculata *Int J Agriculture Innovation Research* **4(2)** 330–1
[5] Khan M G, Daniel G, Konjit M, Thomas A, Eyasu S S and Awoke G 2011Impact of textile wastewater on seed germination and some physiological parameter in pea(*Pisum sativum* L.), lentil(*Lens eculentum* L.) and gram (*Cicer arietinum* L.) *Asian Journal* 1–5
[6] Khaleel R I, Ismail N and Ibrahim M H 2013 The impact of waste water treatments on seed germination and biochemical parameter of Abelmoschus Esculentus l *Social Behavioral Science* **91(2012)** 453–60.
[7] Dash A K 2012 Impact of domestic waste water on seed germination *Biologicial* 280–6.
[8] APHA (American Public Health Association) 2013 *Standard Method for the Examination of Water and Wastewater* (Washington DC: APHA)
[9] Abdul-Baki A and Anderson J D 1973 Vigor determination in Soybean seed by multiple criteria *Crop Science*
[10] Mavrogianopoulos G, Vogli V and Kyritsis S 2002 Use of wastewater as a nutrient solution in a closed gravel hydroponic culture of giant reed (*Arundo donax*) *Bioresource Technolology* **82(2)**
[11] Shamuyarira K K and Gumbo J R 2014 Assessment of heavy metals in municipal sewage sludge *Int J Environ Res Public Health* **11(3)** 2569–79.
[12] Bazai Z A and Achakzai A K K 2006 Effect of wastewater from Quetta city on the germination and seedling growth of lettuce (*lactuca sativa l.) J Apply Science* **6(2)** 380–2.
[13] Khan S S and K H Sheikh 1976 Effect of different level of salinity on seed germination and growth of capsicum annum *Biologicial* **22** 15-25
[14] Ayyasamy P M, Yasodha R, Rajakumar S, Lakshmanaperumalsamy P, Rahman P K S Ma and Lee S 2008 Impact of sugar factory effluent on the growth and biochemical characteristics of terrestrial and aquatic plants *Bull Environ Contam Toxicol* **81(5)** 449–54.
[15] Hussain F, Malik S, Athar M, Bashir N, Younis U and Hassan M 2013 Effect of tannery effluents on seed germination and growth of two sunflower cultivars *African J Biotechnology* **9(32)** 5113–20
[16] Panaskar D B and Pawar R S 2011 Effect of textile mill effluent on growth of Vigna unguiculata and Pisum sativum seedlings *Indian J Sci Technol* **4(3)** 266–72.
[17] Hayward H.E and C.H Wadleight 1949 Plant *growth on the salin and alkali solids advances in agronomy* (New York, Acad:press)
[18] Moradi S, Heidari H, Saeidi M and Nosratti I 2016 Effect of sewage-contaminated water on seed production, heavy metals accumulation and seedling emergence in OAT *Glob Nest J* **18(2)** 329–38.
[19] Baligar V C, Fageria N K and He Z L 2001 Nutrient use efficiency in plants *commun soil science plant anal* **32(7)** 921–50

IOP Conf. Series: Materials Science and Engineering **334** (2018) 012037 doi:10.1088/1757-899X/334/1/012037

Application of banana peels waste as adsorbents for the removal of CO_2, NO, NO_x, and SO_2 gases from motorcycle emissions

V Viena[1,3], Elvitriana[1] and S Wardani[2]

[1] Department of Environmental Engineering, Faculty of Engineering, The University of Serambi Mekkah, Jl. Tgk Imum Lueng Bata Banda Aceh, Aceh, 23245, Indonesia
[2] Department of Farm, Faculty of Agriculture, University of Abulyatama, Jl. Blang Bintang Lama, KM. 8,5, Lampoh Keude, Kabupaten Aceh Besar, Aceh 24415, Indonesia

E-mail: veraviena@serambimekkah.ac.id

Abstract. The aims of the study were to investigate the application of banana peels as adsorbent for the removal of CO, NO, NOx and SO_2 gases from motorcycles emissions. The effect of differents thermal activation on the characteristics of banana peels adsorbent (BPA) such as moisture content, ash content, volatile matter and fixed carbon has been studied using proximate analysis. The study of Iodine adsorption capacity of BPA was obtained at 952 mg/g adsorbent. Structure and morphology of BPA were characterized by Fourier transform infrared (FTIR) and field emission scanning electron microscopy (SEM). The results showed that BPA could significantly adsorbed the CO and SO_2 gases emissions from motorcycles, but not applicable for NO, NOx gases. After 10 minutes of flue gas analysis at idle mode using BPA adsorption tube, CO gas could be totally removed, from initial 19618 ppm to 0 ppm, while SO_2 gas could also be totally removed from 24523 ppm to 0 ppm. SEM test showed that temperature of activation had significant effect on the size of pores of BPA formed. BPA was suitable for application in removing CO and SO_2 gases emissions from motorcycles and it helps to reduce the green house gas effects of fossil fuel to the environment.

1. Introduction

Industrial development in the past five decades has brought the planet to a very worrying collapse from the environmental point of view because of the uncontrolled increase in the use of fossil fuels as the main energy source; the average global temperature will increase and thereby alter the ecosystem. In particular, the excess emission of gases from the internal combustion engine and industry has become one of the major causes of environmental pollution, where gases such as CO_2, CO, hydrocarbons, SO_2, and NOx generate volatile organic compounds (VOC) and particulate matter. These compounds have led to the greenhouse effect, acid rain, photochemical smog, stratospheric ozone depletion, and the aforementioned global warming. These events impose considerable negative effects on human and animal health due to the high toxicity of the compounds mentioned [1][2].

[3] To whom any correspondence should be addressed

Studies in a number of cities (Bangkok, Cairo, Jakarta, Santiago and Tehran, to name five) have assigned priority to controlling lead and particulate matter concentrations, which present the greatest hazard to human health. Where photochemical ozone is a problem, control of ozone precursors (nitrogen oxides and volatile organic compounds) is also important both because of the damaging effects of ozone itself and because of the secondary particulate matter formation resulting from atmospheric reactions with ozone. Carbon monoxide and toxic air contaminants have been assigned lower priority for control at the present time, but measures to reduce volatile organic compounds exhaust emissions will generally reduce carbon monoxide and toxic substances as well [3].

To reduce these emissions, several studies have used porous materials such as mesoporous materials, modified mesoporous materials, aluminas, silicon compounds, compounds derived from coal as both chemically and physically activated powders, granular materials, and fabrics modified with TiO_2 [4][5]. But most of it uses porous materials for UV irradius gaseous emission adsorptions modified with TiO_2.

Until now, there is no available literatures on the use of banana peels waste for the removal of gaseous emission from motorcycles. Only one study reported on the use of banana peels for CO_2 removal in the atmostphere[6]. Most of the studies were used agro-wastes for the removal of heavy metals from water and wastewater as it has a long term effects on the environment. Different adsorbents are developed from agro-wastes and used for heavy metals removal such as rice straw [7], seaweed [8], wood and bark [9], tea-waste [10], maize corn cob, jatropha oil cake, sugarcane bagasse [11], tamarind hull [12], sawdust [13], rice husk [14], saltbush [15], marine algal biomass [16], olive pomace [17], activated sludge [18], sugar beet pulp [19], wool, olive cake, sawdust, pine needles, almond shells, cactus leaves, and charcoal [20], seafood processing waste sludge [21] and pine bark [22].

The activated carbon (AC) is a well known adsorbent that can be used efficiently for removal of a broad spectrum of pollutants from air, soil and liquids. The adsorbents are usually porous solids, and adsorption occurs mainly on the pore walls inside particles. Among them, AC is more efficient adsorbent for elimination of many pollutants (organic, inorganic, and biological) of concern in water and wastewater treatment [23]. Activated carbon (AC) is carbon material having a high porosity, high adsorption capacity, high mechanical strength, and a very large surface area. Activated carbon (AC) are widely used in industry such as to separate the various types of pollutants such as dyes and metal ions from industrial waste, as gas storage and others. Comercial ACs are generally derived from coal which is non-renewable carbon source with relatively expensive price. As an alternative, it has been extensively researched and developed ACs from biomass, such as coconut shell, palm and bamboo. [24].

In the recent study, the used of banana peels waste as the potencial adsorbent for the removal of motorcycles gas emission is important, because the public mayor concern on air pollution dispersion in Indonesia. The research to find the most efficient, low cost, environmental friendly adsorbent from agrowastes was keep on going. Banana as one of the biggest fruit commodity contributors in Aceh Province, Indonesia, in the amount of 504.850 quintals [25]. Along with the high productivity, the amount of waste from banana peels also increased. When the harvesting time, banana fruits are used only flesh while the peels is discarded to the environment. Nowadays banana peels is only used as fertilizer or as animal feed. This piles of banana peels has potency as one of raw materials for making natural activated carbon from biomass waste [26]. These wastes can be potentially used as a low-cost bioadsorbent. In addition, using these wastes in various other potential applications will eliminate them from the environment and reduce solid-waste handling, which will add some value to these wastes [27][28]. Identified functional groups of organic compounds from banana peel which act as an adsorbent were -OH, -COO, and -NH groups, which had abundant potency to be made as one of low cost biomass activated carbon through physical activation. [29].

Using environmental friendly materials for reducing gas emissions from motorcycles was our main aims. Along with this study, we hope that BPA not only could become one of alternatives most reliable adsorbent for reducing green house gas effect to the human and environment, but also could be used in wider range of application for treating water, wastewater and gaseous pollutions in the future.

2. Experimental procedure and methodology

2.1. Preparation of Samples

Banana peels were obtained from local market in Banda Aceh, Indonesia. The peels were washes troughly and then cut into pieces of ± 3 cm and dried in the oven at 105 °C for 24 hour to remove the moisture content. The dried peels is carbonized at a temperature of 400 °C for 1.5 hours to become charcoal, then ground and sieved at 50 mesh. The banana peels carbon is activated thermally using a furnace at operating temperature e.i. 500 °C, 600 °C and 700 °C, respectively for 45 minutes. Banana peels activated carbon (BPA) prowered then used for the characterization studies.

2.2. Chemicals

All chemicals reagents used in the titration were supplied by Merck and were of analytical grade.

2.3. Instrumentation and characterization of BPA

Proximate analysis of BPA were carried out to determine the ash percentage, and contents of moisture, volatile matter and fixed carbon. Properties of BPA particles such as Iodine absorption capacity was determined. BPA were further characterized by Fourier Tranform Infrared Spectra (FTIR) using Shimadzu IR Prestige and Scanning electron microscope (SEM) using JSM Series for morphology of BPA.

2.4. Proximate analysis

Proximate analysis of BPA was done according to standard procedure (SNI No. 06- 3730-1995)[30]. Moisture content of BPA was calculated as given in eq. (1). BPA sample was kept in a hot-air oven at 105 ± 2 °C till constant weight was obtained. Then the sample was kept in a desiccator till it reached room temperature and finally the weight of the sample was taken to determine the loss in weight due to the removal of moisture. Volatile matter was determined by heating the air dried sample of BPA in muffle furnace at 900 ± 10 °C for 7 min and calculated using eq. (2). To determine percentage ash content, the air-dried BPA sample was heated in a muffle furnace at 500 °C for 30 min; the temperature was further increased from 500 °C to 815 °C during the next 30 min and maintained until constant weight was obtained. The percentage ash was then calculated using eq. (3). Fixed carbon was obtained after deducting the sum of moisture, ash and volatile matter contents (expressed as per cent) from 100 and calculated using eq. (4). The experiment was carried out in N_2 atmosphere.

$$\%\ \text{Moisture} = \frac{100\ (Loss\ in\ wight\ after\ drying)}{Total\ weight\ of\ sample\ taken} \quad (1)$$

$$\%\ \text{Volatile matter} = \frac{100\ (Loss\ in\ weight\ after\ heating)}{Total\ weight\ of\ sample\ taken} - \%\ \text{Moisture} \quad (2)$$

$$\%\ \text{Ash} = \frac{100\ (Weight\ of\ ash\ obtained\ after\ heating}{Total\ weight\ of\ sample\ taken} \quad (3)$$

$$\text{Fixed carbon} = 100 - (\%\ \text{Moisture} + \%\ \text{Volatile matter} + \%\text{Ash} \quad (4)$$

2.5. Gasses Emission Adsorption Process

Process of gases emissions adsorption study was accomplished within 2 steps; (1) making of the adsorption tube or knalpot model for filling the adsorbents, that has been formed in cubicles size of 1x1x1 cm; (2) Gases emissions adsorption study was done on motorcycle Honda Astrea year 2000.

Measurement of CO, NO, NOx dan SO2 gases emission from motorcycle was determined using Flue gas Analyzer 6000 at idle condition, after 0, 5 and 10 minutes analysis. Removal of CO, NO, NOx dan SO_2 gases emission were determined by comparing the analysis value recorded before and after adsorbents filled in the model knalpot, while the motorcycles being accelerated for 10 minutes.

3. Results and Discussion

3.1. Proximate Analysis

The results of proximate analysis on BPA prepared was shown on Table 1. The moisture content greatly affect the quality of activated carbon. The moisture content obtained ranging between 2-6%. Low water levels can be due to fewer active charcoal surface containing polar functional groups so that the interaction between polar group is small [31].

Table 1. Proximate analysis of banana peels adsorbent (BPA) at various temperature

Temperature (°C)	Moisture (%)	Volatile (%)	Ash (%)	Fixed Carbon (%)	Iodine Adsorptin (mg/gr)
400	4,00	14,00	9,00	73,01	793,19
500	6,00	11,00	4,00	79,05	824,92
600	2,00	12,00	2,01	84,03	951,83
700	5,00	19,00	3,02	73,01	751,94

The ash content is influenced by the amount of silica content, the greater the amount of silica then the resulting ash content greater. Levels of volatile substance is the result of the decomposition of substances - compilers of charcoal due to heating during the authoring process and not the composition of charcoal [31]. The results of volatile matter of activated charcoal banana peels ranges between 11-19%. Low levels of substances that evaporate due to the evaporation of the compound - non-carbon compounds that are volatile at the carbonization process.

The BPA characteristics of fixed carbon range between 73,01-84,03%. Levels of carbon bonded carbon is the fraction bound in charcoal besides fraction of water, substances evaporate and ash. High and low levels of carbon bound in charcoal influenced by the value of the ash content, content of volatile substance and hydrocarbons that are still attached to the surface of charcoal, and is also influenced by the content of cellulose and lignin materials that can be converted into a carbon atom [31].

Lower moisture content (2-9%) means that the fruits and vegetable adsorbent could be storage for longer time periods because it is resistant to mould growth. The organic nature of Fruits and vegetable adsorbents is confirmed by higher volume of volatilee matter (up to 85 %) along with raised in temperature of activation. [32] This organic matter mainly included proteins, carbohydrates and lipids, all of which are good source of nutrients. The ash content (2-9%) from fruit and vegetable adsorbents could be a good source of minerals and micronutrients.

The parameters that can precisely indicate the quality of low cost activated carbon made of banana peels biomass is the adsorption capacity of the iodine solution. The larger the iodine number of activated carbon, the greater the ability to absorb the solute in the environment. The results of analysis iodine absorption capability has ranged from 752 - 952 mg/g. Low cost carbon made of banana peels has a tend to change before and after activated, which caused by activating agent which can cause a shift based on their chemical environment. Images can be viewed on the catchment area in the position of 2850 - 2960 cm-1, which contains C-H functional groups.

The processes of adsorption, which involves the transference of a solid mass in a fluid phase to the surface of a solid component, are being largely studied due to their high efficacy and for being economically viable in the treatment of wastewater effluents. The utilization of low cost material in the production of activated carbon is an alternative pathway to food industry waste, which is usually disposed and end up losing its value [33].

3.2. Analysis of Functional Groups using FTIR (Fourier Transform Infra Red)

Figure 1. showed the comparison of the wavelength number of carbon absorption band shell Before activation frequency of C-H group is in the intensity of 81.23, while the increased intensity after activation of C-H group that is equal to 92.50. The intensity level of C-H functional groups value, were influenced by the content of H_2O. The higher the content of H_2O, the intensity of C-H group is getting low. Activation temperature has highly effect on the absorbed area of Iodine and the functional group on the surface of banana peels activated carbon.

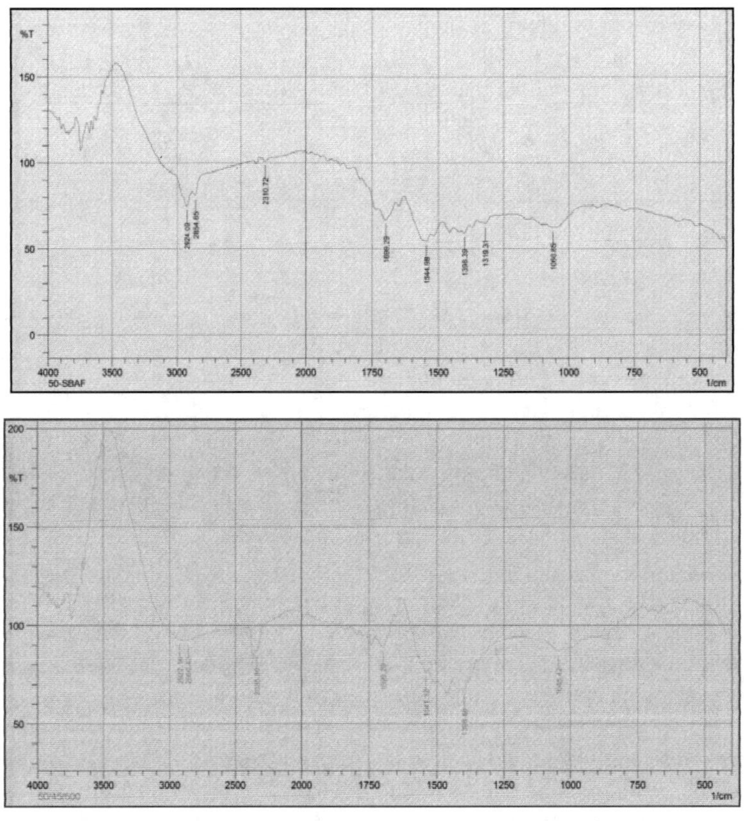

Figure 1. FTIR analysis banana peels activated carbon before thermal activation (left) and after thermal activation (right), Adsorbent size of 50 mesh, T = 600 °C.

3.3 Analysis of surface morpholgy using SEM

The surface morphology of the BPA was analized using SEM (Scanning Electron Microscopy) at 1000 - 3000X. From figure 2. it could be shown that the adsorbent pores surface has more opening after being activated by thermally/physically compare to before activation process.

Figure 2. showed that before activation the BPA tend to have small pores, but after the thermal activation the pores become wider and hollower. This was mainly caused by the temperature used for carbonization, which is 400 °C. A temperature of 300°C to 400°C led to incomplete carbonization of banana peel and the cell structure is maintained. There are seemed some small pores due to the release of some volatile compounds. The study of Mendez et al. [34] showed that the main important weight changes are produced due to the light compounds volatilization and pyrolysis transformations at 450°C. This stage is primary pyrolysis (in the 300°C to 500°C range) with evolution of most gases and tars with formation of the basic structure of the char [35].

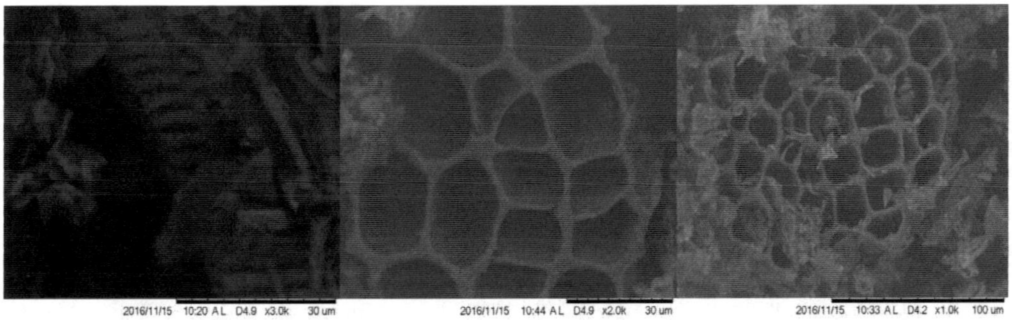

(a) (b) (c)

Figure 2. Surface morphology of BPA (a) before thermal activation, (b) after themal activation, and (c) after gas emission analysis.

3.4. Gas Emission Removal

The object of our research was the use of BPA as adsorbent for the removal of gaseous emission from automotive/motorcyles which has been used since year of 2000. The Motorcycle gas emission analysis was accomplished by making a model of adsorbant tube from knalpot that filled with BPA media that has formed into small cubicles shapes of 1x1x1cm. Modified shape of BPA were aims to minimize the destruction of adsorbents pores during the emission test. The BPA formed were arrange layer by layer, which allow the empty space between the adsorbent as the pathway to adsorption process flows during the acceleration of motorcycles movement. These empty space between the activated carbon meant to give better contact and adsorption of gaseous NO, SO, NOx, SOx, CO, HC. The results of the emission analysis was showed in Table 1.

Tabel 2. Results of gas emission analysis before and after filled with BPA media

Type of gas emission	Gas Emission without activated carbon (ppm)	Gas Emission adsorption using BPA media (ppm)			Adsorption efficiency (%)	
		0 min	5 min	10 min	5 min	10 min
CO	14385	19618	10230	0	47,854	100
NO	24	17	33	33	NA	NA
NOx	25	17	67	145	NA	NA
SO2	2187	24523	13788	0	43,775	100

From Table 2, it could be explained that BPA was a potencial media for adsorbing CO and SO2 gases emission, but not for NO and NOx gases, as the parameter keep raising up during the emission test. Meanwhile, after 10 minutes of flue gas analysis at idle mode using BPA adsorption tube, CO gas could be totally removed, from initial 19618 ppm to 0 ppm, while SO_2 gas could also be totally removed from 24523 ppm to 0 ppm. These was caused by the total pores size of BPA particles didn't have capability to adsorbed NO and NOx gases, as the BPA only has limited funtional group that could removed motorcycle gases emission. Ones the pores were filled with gas, then there're only limited pores left for further adsorption. This phenomena also could be explained from the SEM analysis above (Fig.7). After the gas emission analysis using motorcycle idle position for 10 minutes, the pores were filled with dirt and other pollutant and it blocked the adsorption pathways and it made the pores become saturated. That's why in the future reserach we suggested for modification of banana peels adsorbents with other nano particles to enhance the porosity and adsorption area of each particles produced.

The results of this study provide an understanding of the proximate analysis and ultimate value of BPA produced and it suggested that the material can be used as a low-cost adsorbent for adsorbing gaseous emission and its applicable for reducing the air pollution in Banda Aceh or other Indonesian area. The BPA also could be applied as one of biofilter technology along with knalpot for reducing green house gas effect from fossil fuels to the environment.

4. Conclusion
In conclusions, the proximate analysis showed that the best thermal activation temperature was at 600 ^{o}C, and 45 minute activation, as it resulting a low moisture content, higher fixed carbon percentage and best Iodine adsorption capacity. The BPA prepared showed best adsorption capacity on CO and SO_2 gaseous emissions removal, but not for NO and NOx gases. The BPA has potency to be used as one of low cost adsorbent for removing gaseous pollution or wastewater treatment.

Acknowledgments
Special gratitute to the Indonesian Ministry of Research, Technology and Higher Education for providing financial support through DIPA 2016 (Applied Product Grant).

References

[1] Ao, C.H.; Lee, S.C.; Mak, C.L.; Chan, L.Y. 2003 *Appl.Catal. B*, ***42***,119-129
[2] Haywood, J.M.; Cooper, C.D. 1998. *J. Air Waste Manage. Assoc.* ***48,*** 238–246
[3] Faiz, A. Christopher S Weaver and Walsh, M.P, 1996, *Air pollution from Vehicles: Standard and Technologies for Controlling Emissions*, ISBN 0-8213-3444-1
[4] Sekiguchi, K.; Saito, M.; Lee.Y.; Kim, Y.; Sakamoto, K. *Chem. Eng. J.* 2006, ***118***, 127–130
[5] Castro, T.; Madronich, S.; Rivale, S.; Muhlia, A.; Mar, B. *Atmos. Environ.* 2001, ***35***, 1765–1772
[6] Borhan, A., Thangamuthu, S., Taha, M.F., Ramdan, A.N., (2014), *AIP Conference Proceeding*, Vol 1674, Issue 1, 10.1063/1.4928819.
[7] Gao H, Liu Y, Zeng G, Xu W, Li T and Xia W, *J. Hazard. Mater.*, vol.**150**, 2008, pp.446-452.
[8] Basha S, Murthy ZVP and Jha B, *Chem. Eng. J.*, vol.**137**, 2008, pp.480-488.
[9] Mohan D, Pittman Jr. CU, Bricka M, Smith F, Yancey B, Muhammad J, Steele PH, Alexandre-Franco MF, Gomez-Serrano V and Gong H, *J. Colloid Interface Sci.,* vol.**310**, 2007, pp.57- 73.
[10] Malkoc E and Nuhoglu Y, *Sep. Purif. Technol.,* 2007, vol.**54**, pp.291-298.
[11] Garg UK, Kaur MP, Garg VK and Suda D, *J. Hazard. Mater.*, vol.**140**, 2007, pp.60-68.
[12] Verma A, Chakraborty S and Basu JK, *Sep. Purif. Technol.*, vol.**50**, 2006, pp.336-341.
[13] Memon SQ, Bhanger MI and Khuhawar MY, *Anal. Bioanal. Chem.*, vol. **383**, 2005, pp.619-624.
[14] Kumar U and Bandyopadhyay M, *J. Haza. Materials,* vol.**129**, 2006, pp. 253-259.
[15] Sawalha MF, Gardea-Torresdey JL, Parsons JG, Saupe G and Peralta-Vide JR, " *Microchem. J.*, vol.**81**, 2005, pp.122-132.
[16] Sheng PX, Ting Y, Chen JP and Hong L, *J. Colloid Interface Sci.*, vol. **275**, 2004, pp.131-141.
[17] Pagnanelli F, Mainelli S, Veglio F and Toro L, *Chem. Eng. Sci.*, vol.**58**, 2003, pp.4709-17.

[18] Kim DW, Cha DK, Wang J and Huang CP, *Chemosphere,* vol.**46**, 2002, pp.137-142.
[19] Reddad Z, Gerente C, Andres Y and Le Cloirec P, *Environ. Sci. Technol.*, vol. **36**, 2002, pp.2242-2248.
[20] Dakiky M, Khamis M, Manassra A and Mer'eb M, *Adv. Environ. Res.,* vol**. 6**, 2002, pp. 533-540.
[21] Lee SM and Davis AP, *Water Res*., vol.**35**, 2001, pp. 534-540.
[22] Al-Asheh S, Banat F, Al-Omari R and Duvnjak Z, *Chemosphere,* vol. **41**, 2000, pp.659-665.
[23] Mohammad-Khah, A.; Ansari, R. 2009. *International Journal of ChemTech Research*, **4**:859-864.
[24] DNK Putra Negara *et al* 2017. Chemical, strength and microstructure characterization of Balinese bamboos as activated carbon source for adsorbed natural gas application. *IOP Conf. Ser.: Mater. Sci. Eng.* **201** 012033.
[25] BPS, 2013, *Acheh Fruits-production Comodities*, Central Statistics Board (Aceh Province).
[26] Nasir W., N., S., Nurhaeni dan Musafira (2014), *Online Journal of Natural Science,* Vol **3**, No. 1, hal 18 – 30, ISSN 2338 – 0950.
[27] Midha, S., *Horticulture Statistics in India*, 2012, pp. 148–161.(India)
[28] Palma, C., Contreras, E., Urra, J. and Martınez, M. J., *Waste Biomass Valorization*, 2011, **2**, 77–86.
[29] Castro, G., Caetano L, Ferreira G, Padilha P, Margarida J, Zara F., and Antonio M, 2011, *Journal of American Chemical Society*, **50** : 3446-3451
[30] SNI,1995, *Technical Charcoal, National Indonesian Standard*, No. 06-3730-1995, (Indonesian republic of Industrial Department): p 29 – 57.
[31] Fauziah, N .2009, *Departement Hasil Hutan*, Fakultas Kehutanan, (Institut Pertanian Bogor).
[32] Pathak, P.D, Mandavgane. S.A, Kulkarni, B.D. 2016. *J. Current Science,* Vol **110**, No.11.
[33] Junchen, P.T.; Gobi, F.P.; Honorio, J. F.; Veit, M.T.; Gonçalves, G.C.; Klein, M.R.F. State University of West Paraná, Center for Engineering and Exact Sciences. Federal Technological University of Paraná, Chemical Processes, *the V SEQ Anais, Chemical Engineering Symposium*, (Maringá – Paraná, Brazil)
[34] Mendez A, Gasco G, Freitas MMA, Siebielec G, Stuczynski T, et al. 2005. *Chem Eng J* **108**: 169-177.
[35] Girgis BS, El-Hendawy AN. 2002. *Micropor Mesopor Mater* **52**: p 105-117.

IOP Conf. Series: Materials Science and Engineering **334** (2018) 012038 doi:10.1088/1757-899X/334/1/012038

Comparison of H_2S adsorption by two hydrogel composite (HBC) derived by Empty Fruit Bunch (EFB) biochar and Coal Fly Ash (CFA)

N H Meri[1,2], A B Alias[1], N Talib[1], Z A Rashid[1] and W A W A K Ghani[1]
[1] Faculty of Chemical Engineering, Universiti Teknologi MARA, 40450 Shah Alam, Selangor, Malaysia

E-mail: norhidayahmeri@yahoo.com

Abstract. This study are covered the adsorption performance of two adsorbent Empty Fruit Bunch Hydrogel Biochar Composite (EFB-HBC) and Coal Fly Ash Hydrogel Composite (CFA-HC) on hydrogen sulphide. The EFB biochar were produce by pyrolysed and heated from room temperature to 550°C at 10°C/min under the Nitrogen flow. Meanwhile, coal fly ash collected from a power plant located in Selangor, Malaysia. Both of the materials is a waste from different industries and became the precursor to our adsorbents. EFB biochar and coal fly ash has been synthesized to become hydrogel by polymerization process with acrylamide (AAm) as monomer, N,N'-methylene bisacry lamide (MBA) as cross linker and ammonium persulfate (APS) as initiator. In addition, because of the speciality of hydrogel itself, which is has high ability in storing water, the effect of H_2O wetness on EFB-HBC and CFA-HC were investigate in adsorption of H_2S. EFB-HBC gave a longest breakthrough time and highest adsorption capacity compared with CFA-HC in both condition (dry/wet). The result also indicated that, the increased the bed height, increased the adsorption capacity.

1. Introduction

Hydrogen Sulfide (H_2S) is defined as hazardous gas produced from the natural process such as microbial decomposition and human-made process especially from wastewater treatment plant [1]. H_2S gas is extremely toxic to living organism and even corrosive to the concrete and steel [2]. In the H_2S treatment sector, activated carbon was chosen as the medium in removing H2S gas due to its advantages in the aspects of high surface area and porosity for sorption process [3]. However, since there are a few flaws in activated carbon application for this purpose such as it demanded high temperature, high pressure and an activation process in its manufacturing, the researchers had turned their focus on more environmental friendly and less costly materials [4].

Empty fruit bunch is chosen as raw material because of its availability in Malaysia as abundance biomass waste as this country is the largest producer of palm oil [5]. Thermal conversion of raw EFB will produced carbonaceous material called biochars. Besides, Pyrolysis is suitable process to produce chars that used as precursor to activated carbon especially by slow pyrolysis around less than 700 °C [6]. Meanwhile, CFA produced as a waste product during the combustion process of coal in generation of electricity. They are cluster of spherical particles with a diameter ranging from 1 to 100 μm and typically grey in colour, quite abrasive and mostly alkaline in nature [7]. CFA are amorphous and its

[2] To whom any correspondence should be addressed.

high silica content with some alumina present makes it a suitable precursor as zeolites with application as adsorbent in wastewater treatment [8].

In this study, the empty fruit bunch biochar and coal fly ash will go through polymerization process by using Acrylamide (AAm) monomer to produce hydrogel biochar. The presence of the hydrogel coating of the biochar is known to help improves the adsorption efficiency by increasing the intake and holding up of water. Due to the high water content of the hydrogel plus porous structure networks, more solute will be trap by the hydrogel component [9].

2. Experimental procedure and methodology

2.1. Raw Material and Chemical

Raw Empty fruit bunch collected from palm oil mill in Negeri Sembilan, Malaysia. Meanwhile coal fly ash collected from a power plant located in Selangor, Malaysia. For the preparation of the hydrogel composite, acrylamide (AAm) used as monomer, N,N'-methylen ebisacry lamide (MBA) as cross linker and ammonium persulfate (APS) as initiator are supplied from R&M company. Distilled water used to remove unreacted monomer and to adjust the pH value of the biochar and coal fly ash.

2.2. Pyrolysis

Raw EFB dried at temperature of 100°C for 24 hours in an oven to reduce moisture content. The dried EFB pyrolysis in a fixed bed reactor and heated from room temperature to 550˚C at 10˚C/min under Nitrogen flow. Then the biochar has washed with distilled water to remove unwanted material. Then, the biochar dried in an oven with set temperature of 40°C for 24 hours.

2.3. Polymerization process to hydrogel composite

1.0g of acrylamide (AAm) monomer dissolved in 1mL of distilled water. 0.6g of the EFB biochar was mixed with 0.001g of MBA cross linker. The mixture of the EFB biochar and the cross linker then added to the AAm solution. 0.2mL of 0.1g aqueous solution of APS initiator added to the solution to initiate the polymerization. The precursor solution immediately place into plastic mould and place in an oven at 40^0C for 30 minutes to speed up the polymerization and cross-linking process. Then the precursor solution left for 24 hours to let the polymerization and cross-inking process completed. After 24 hours, the EFB-HBC was taken out from the plastic mould and cut to desired size and washed with distilled water to remove unreacted monomer and low molecular weight matter from theEFB-HBC. Then the EFB-HBC dried in vacuum oven at 40 °C for 24 hours. After 24 hours, the EFB-HBC was stored in desiccator until further used. The procedures repeated for coal fly ash to produced CFA-HC. This hydrogel formation method are following to procedure used by Karakoyun et al. (2011) with some modification [9].

2.4. Adsorption Performance

2.4.1. Effect of Bed Height. The adsorption studies were carried out in three layers stainless steel column with 3 inch height each of layer and 1.85-inch diameter. This experiment was carried out at constant temperature and flowarate at 27 °C and 60 L/hr respectively. Every layers were supported with filter trays and 1 inch height of glass wool to prevent the fine particles pass through and damage the system. Then, the outlet stream were attached with Crowcon Gasman H_2S gas detector. The system was illustrated in Figure 1. Experiment were conducted to study the effect of bed height and H_2O wetness of EFB-HBC (Empty Fruit Bunch-Hydrogel Biochar Composite) and CFA-HC (Coal Fly Ash Hydrogel Composite). A comparison study of adsorption performance between EFB-HBC and CFA-HC were included in this study.

This study were conducted with different bed height of each trays. All the details of experiment were tabulated in Table 1, Table 2, Table 3 and Table 4. For the first experiment, in order to study the effect of bed heights in adsorption of H_2S, each trays were filled with 0.5 inch height of adsorbent followed by next run were filled with 1 inch height of adsorbent for each trays. Finally, the maximum bed heights is 2 inch height of adsorbent were filled. Next, the experiment was repeated by using CFA-HC in order to study the comparison in adsorption of H_2S.

In preparation of wet adsorbent, water has added to adsorbent until the adsorbent fully swelled and the mass of total wet adsorbent became double of its initial weight. Then, the wet adsorbent with 100 % swelled has placed in column and compressed until achieved the required bed height, 0.5 inch, 1.0 inch or 2.0 inches.

The adsorption capacity q_e (mg/g), was calculated by using Eq. 1 [10] ;

$$q_e = \frac{(C_o - C_e)V}{W} \quad (1)$$

Where C_e (mg/L) is the concentration of H_2S at equilibrium, C_o (mg/L) is the initial concentration of H_2S supplied. Meanwhile, V is the volume of the adsorbent (L) and W is the mass of adsorbent used (g).

To understand the adsorption behaviour, breakthrough curves Cf/Ct against time were analysed in this study.

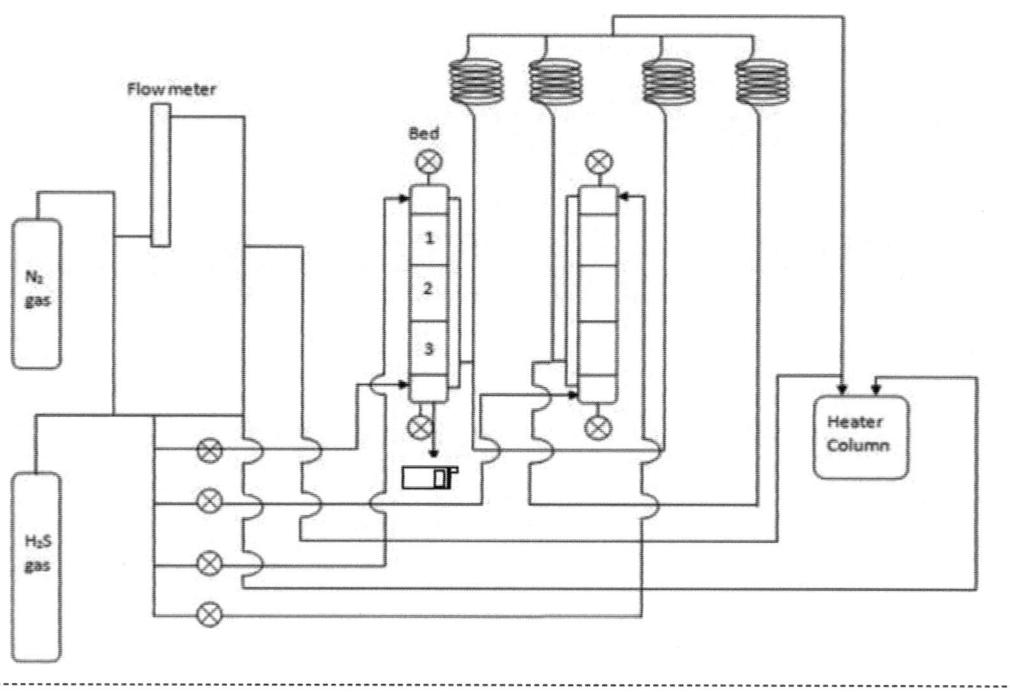

Figure 1. Multilayers adsorption column.

Table 1. Experiment's details for effect of bed height in adsorption of H2S by using dry EFB-HBC.

	Bed height for each layers (Inch)	Total height used (Inch)	Total height used (cm)	Volume (L)	Total Mass (g)
Dry 0.5 EFB-HBC	0.5	1.5	3.81	0.066	12 g
Dry 1.0 EFB-HBC	1.0	3	7.62	0.132	18 g

Dry 2.0 EFB-HBC	2.0	6	15.24	0.264	36 g

Table 2. Experiment's details for effect of water wetness on EFB-HBC in adsorption of H_2S.

	Bed height for each layers (Inch)	Total height used (cm)	Total height used (cm)	Volume (L)	Total Mass (g)
Wet 0.5 EFB-HBC	0.5	1.5	3.81	0.066	24 g
Wet 1.0 EFB-HBC	1.0	3	7.62	0.132	36 g
Wet 2.0 EFB-HBC	2.0	6	15.24	0.264	72 g

Table 3. Experiment's details for adsorption of H2S by using dry CFA-HC.

	Bed height for each layers (Inch)	Total height used (Inch)	Total height used (cm)	Volume (L)	Total Mass (g)
Dry 0.5 EFB-HBC	0.5	1.5	3.81	0.066	15 g
Dry 1.0 EFB-HBC	1.0	3	7.62	0.132	20 g
Dry 2.0 EFB-HBC	2.0	6	15.24	0.264	42 g

Table 4. Experiment's details for adsorption of H2S by using Wet CFA-HC.

	Bed height for each layers (Inch)	Total height used (Inch)	Total height used (cm)	Volume (L)	Total Mass (g)
Dry 0.5 EFB-HBC	0.5	1.5	3.81	0.066	30 g
Dry 1.0 EFB-HBC	1.0	3	7.62	0.132	40 g
Dry 2.0 EFB-HBC	2.0	6	15.24	0.264	84 g

3. Result and Discussion

3.1. Effect of bed height

Figure 2 shows the breakthrough profile for effect of bed height in removal of H_2S. In the beginning of experiment, no H_2S detected in the outlet gas just after the 25 ppm H_2S was supplied. Then after a few moment, the concentration started to increased gradually. After plotted a breakthrough curves by Cf/Ci (Final concentration/Initial Concentration) against Time (s), breakthrough times were calculated. As shown in Figure 2, the higher the bed heights, the longer the time taken to reach breakthrough. The highest bed height 6-inch has a longer breakthrough time, which is 80 s, followed by 3-inch bed height with 60 s, and the shortest is 50 s for 1.5-inch bed height. This is because increasing in bed heights will increasing the amount of adsorbent used. Then, as stated by Choo et al., 2013, the longer the breakthrough time is because of the portion of adsorbent in front of the mass transfer zone had not yet been exposed to H_2S hence the H_2S adsorption capacity increased [11]. By using Eq.1 the adsorption capacity were calculated. The adsorption capacity for 1.5-inch, 3.0-inch and 6.0-inch bed height of Dry EFB-HBC are 1.2485, 1.914 and 2.2513 respectively. Then, the adsorption capacities (mg/g) were increased with bed height increased.

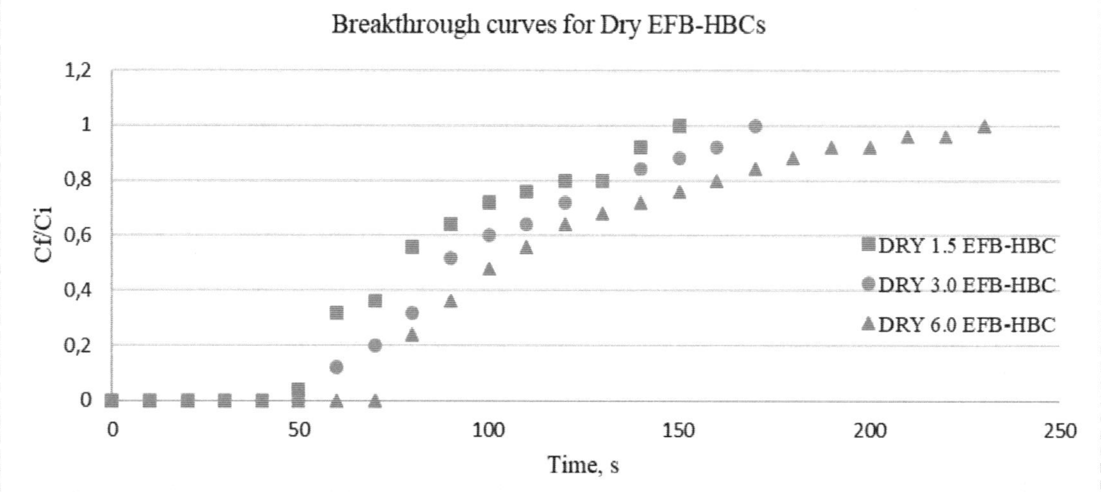

Figure 2. Effect of adsorbent's amount on breakthrough profile for adsorption of Hydrogen Sulphide.

3.2. Effect of water wetness in adsorbent

The specialty of hydrogel is able to absorbed water in the same time good in storing water without spilled. In addition, H_2S is soluble in water and it was predicted that the wet EFB-HBCs have significant adsorption performance. Then an experiment were proceed with wet adsorbent in removal of H_2S. As shown in Figure 3, wet EFB-HBCs have longer breakthrough time compared with dry EFB-HBCs which are 340 s, 360 s and 420 s for 1.5 inch, 3.0 inch and 6.0 inch bed height respectively. In the same time the adsorption capacity were increased by bed height and were compared with dry EFB-HBC shown in Figure 4. It shows that, the adsorption capacity (mg/g) for 1.5 inch, 3 inch and 6 inch bed heights of wet EFB-HBC are 2.7399, 2.7913 and 3.2551 respectively. Previous study were conducted by Wang, 2010 in investigation the effect of moisture content of adsrobent in removal of H_2S, stated that, the higher the moisture content the higher the capability of H_2S adsorption. This explained that the chemical reaction between H_2S and water were provide a good adsorption performance [12].

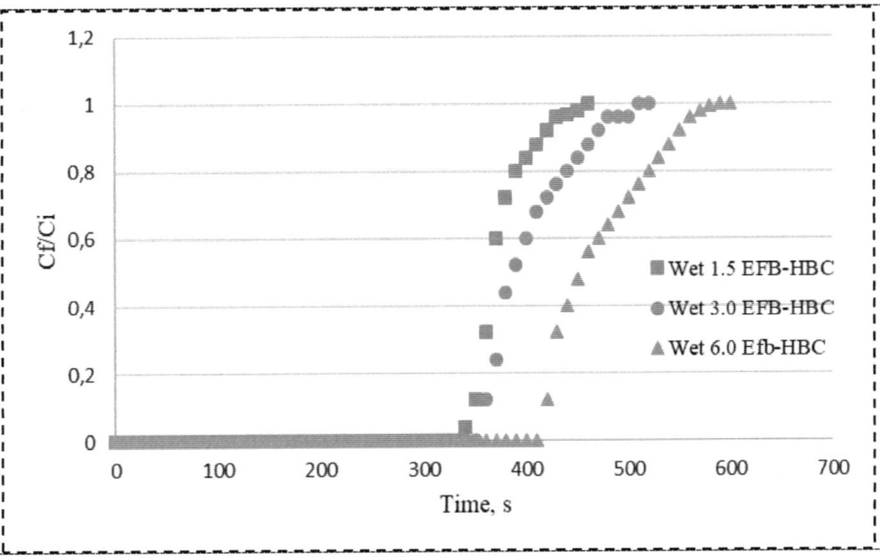

Figure 3. Breakthrough profile for wet EFB-HBCs in adsorption of Hydrogen Sulphide.

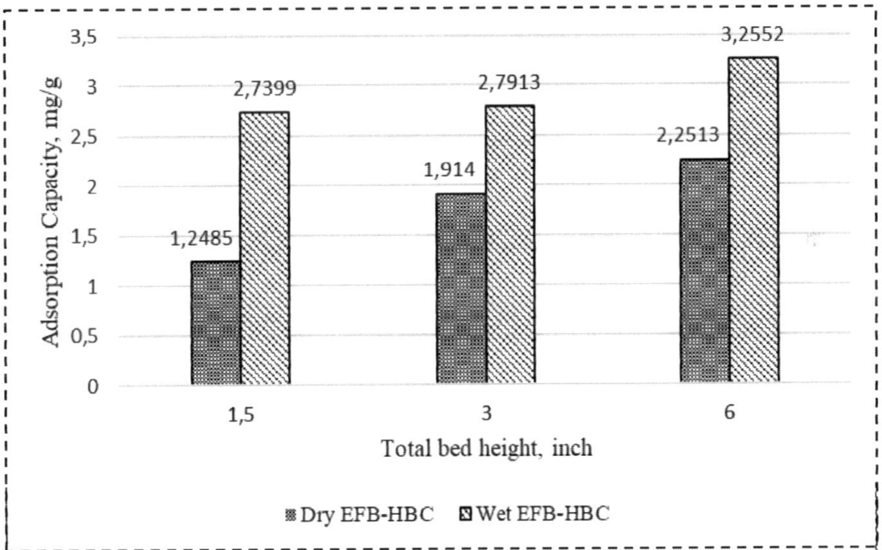

Figure 4. Comparison study between dry EFB-HBC and wet EFB-HBC.

3.3. Adsorption Performance for Dry EFB-HC and Wet EFB-HC

The next experiment is to investigate another industrial waste Coal fly ash which were upgraded to hydrogel composite and used as adsorbent in removal of Hydrogen Sulphide. Figure 5 and Figure 6 shows the breakthrough curves for dry CFA-HC and wet CFA-HC respectively. Breakthrough time for dry CFA-HC at 1.5 inch, 3 inch and 6 inch of bed heights are 20 s, 40 s, and 60 s respectively. Meanwhile, wet CFA-HCs have a longer breakthrough time compared to dry CFA-HC, which are 100 s, 170 s and 230 s for 1.5 inch, 3.0 inch and 6.0 inch bed height respectively. The adsorption capacity also increased when the bed height were increased for both condition dry and wet CFA-HC. Adsorption capacity of dry CFA-HC and Wet CFA-HC has compared and plotted by a bar chat shown in Figure 7. It shows that, Coal Fly ash is a low surface area material compared to biochar [13]. Since then, by upgraded the CFA became hydrogel composite can improved the adsorption performance especially in wet condition.

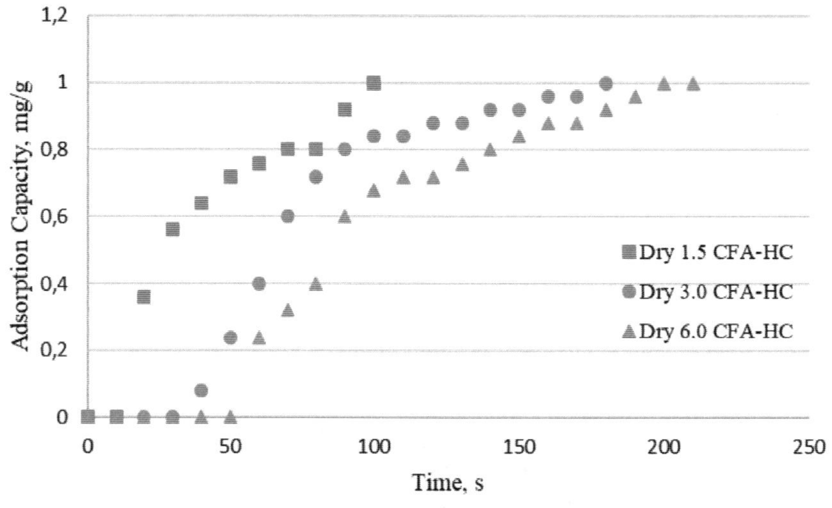

Figure 5. Breakthrough Curves of adsorption of H2S by Dry CFA-HC.

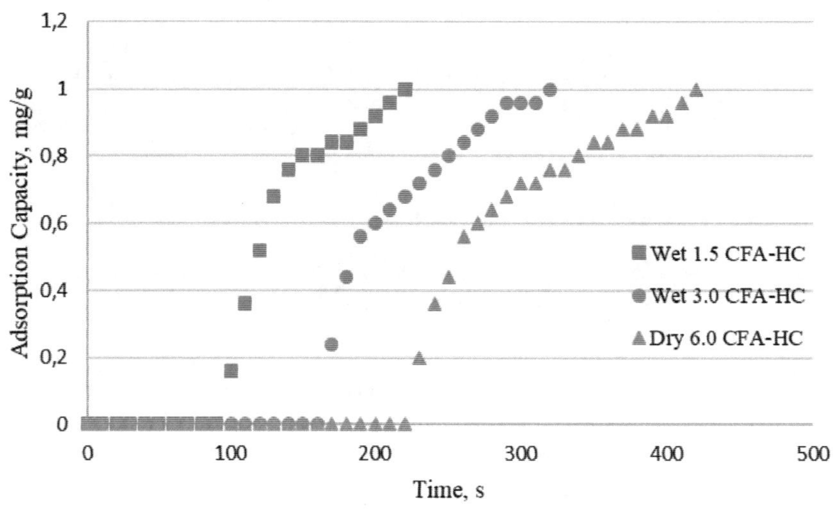

Figure 6. Breakthrough Curves for adsorption of H2S by Wet CFA-HC.

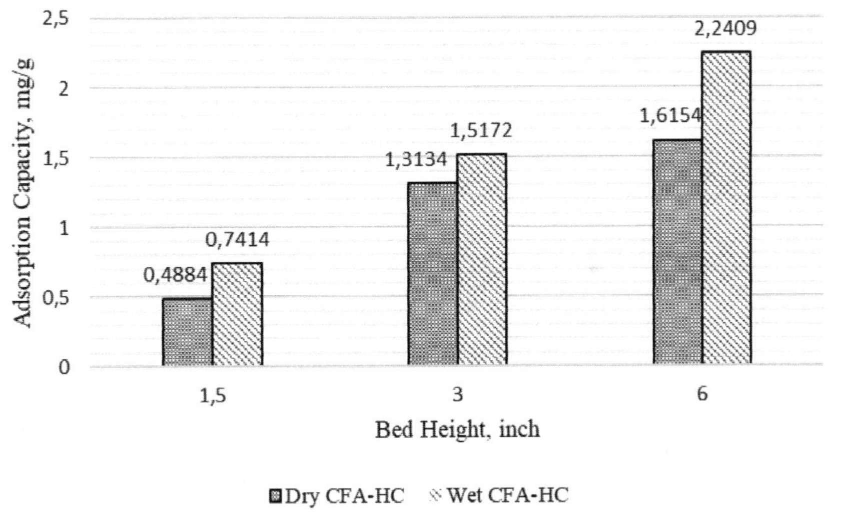

Figure 7. Comparison srudy between Dry CFA-HC and Wet CFA HC.

4. Conclusion

In conclusion, the idea for wet hydrogel composite in adsorption of Hydrogen Sulphide increased the adsorption performance. The results indicated that, hydrogel composite that derived from Empty Fruit Bunch biochar have a good adsorption performance especially in wet condition. This is because of the EFB biochar is a carbonaceous and porous material which good in adsorption, meanwhile the presence of water inside the hydrogel composite were helped in removal of Hydrogen Sulphide. The upgraded a non-porous material such as Coal Fly Ash became hydrogel composite gave a good result in adsorption performance. It proved that, the formation of hydrogel composite as binding agent between particle increased their capability in adsorption. Based on this study, increased the bed height, gave a longer breakthrough time and have a higher adsorption capacity. In this study, complete adsorption modelling for the adsorption process wa not shown in in this study because it will study and reported in subsequent publication.

Acknowledgment

This research was supported by a grant 600-IRMI/DANA 5/3/BESTARI (0008/2016). We would like to thank to Ministry of Higher Education (MOHE), Universiti Teknologi Mara (UiTM) and Faculty of Chemical Engineering for the grant sponsorship, facilities and motivation.

Refferences

[1] Dennis Darylmple and Bryan Petrinec, "New Process for H2S Management Now Commercially Available," *Tartan Engineering*, Canada, pp. 2–4, 2007.

[2] Pushpak Bhandari, "Synthesis and Evaluation of Novel Biochar-Based and Metal Oxide-Based Catalyst For Removal of Model Tar (Toluene), NH3, and H2S From Simulated Producer Gas," University of Pune, 2012.

[3] X. Xu, X. Cao, L. Zhao, and T. Sun, "Comparison of sewage sludge- and pig manure-derived biochars for hydrogen sulfide removal," *Chemosphere*, vol. 111, pp. 296–303, 2014.

[4] G. Shang, G. Shen, L. Liu, Q. Chen, and Z. Xu, "Kinetics and mechanisms of hydrogen sulfide adsorption by biochars," *Bioresour. Technol.*, vol. 133, pp. 495–499, 2013.

[5] M. T. M. N. and R. M. Muhammad Afif Ariffin,Wan Mohd Faizal Wan Mahmood, "Potential of Oil Palm Empty Fruit Bunch (EFB) Biochar from Gasification Process," *Aust. J. Basic*

Appl. Sci., vol. 8, no. 19, pp. 149–152, 2014.
[6] Aino Siirala, "Comparison of gasification , pyrolysis and combustion," *Introd. to biorefineries biofuels*, 2013.
[7] Y. Lee, J. Thet, S. Zhang, J. Ahn, M. Bum, and W. Ahn, "Synthesis of nanoporous materials via recycling coal fly ash and other solid wastes : A mini review," *Chem. Eng. J.*, vol. 317, pp. 821–843, 2017.
[8] V. K. Jha, M. Matsuda, and M. Miyake, "Sorption properties of the activated carbon-zeolite composite prepared from coal fly ash for Ni 2 + , Cu 2 + , Cd 2 + and Pb 2 +," vol. 160, pp. 148–153, 2008.
[9] N. Karakoyun, S. Kubilay, N. Aktas, O. Turhan, M. Kasimoglu, S. Yilmaz, and N. Sahiner, "Hydrogel-Biochar composites for effective organic contaminant removal from aqueous media," *Desalination*, vol. 280, no. 1–3, pp. 319–325, 2011.
[10] G. Vijayakumar, R. Tamilarasan, and M. Dharmendirakumar, "Adsorption , Kinetic , Equilibrium and Thermodynamic studies on the removal of basic dye Rhodamine-B from aqueous solution by the use of natural adsorbent perlite," vol. 3, no. 1, pp. 157–170, 2012.
[11] H. S. Choo, L. C. Lau, A. R. Mohamed, and K. T. Lee, "Hydrogen sulfide adsorption by alkaline impregnated coconut shell activated carbon," *J. Eng. Sci. Technol.*, vol. 8, no. 6, pp. 741–753, 2013.
[12] N. Wang, "An investigation of H2S adsorption mechanisms on Tire Derived Rubber Particles (TDRP TM)," 2010.
[13] S. Zhong, G. Yin, H. Peng, J. Xiang, and Q. Lin, "Preparation , characterization and sludge conditioning performance of," vol. 0, pp. 1–8, 2017.

IOP Conf. Series: Materials Science and Engineering **334** (2018) 012039 doi:10.1088/1757-899X/334/1/012039

Adsorption of mercury from aqueous solutions using palm oil fuel ash as an adsorbent - batch studies

M S Imla Syafiqah[1,2] and H W Yussof[1]

[1] Faculty of Chemical Engineering & Natural Resources, Universiti Malaysia Pahang, 26300 Gambang, Pahang, Malaysia

E-mail: imlasyafiqah@gmail.com

Abstract. Palm oil fuel ash (POFA) is one of the most abundantly produced waste materials. POFA is widely used by the oil palm industry which was collected as ash from the burning of empty fruit bunches fiber (EFB) and palm oil kernel shells (POKS) in the boiler as fuel to generate electricity. Mercury adsorption was conducted in a batch process to study the effects of contact time, initial Hg(II) ion concentration, and temperature. In this study, POFA was prepared and used for the removal of mercury(II) ion from the aqueous phase. The effects of various parameters such as contact time (0- 360 min), temperature (15 – 45 °C) and initial Hg(II) ion concentration (1 – 5 mg/L) for the removal of Hg(II) ion were studied in a batch process. The surface characterization was examined by scanning electron microscopy (SEM) and particle size distribution analysis. From this study, it was found that the highest Hg(II) ion removal was 99.60 % at pH 7, contact time of 4 h, initial Hg(II) ion concentration of 1 mg/L, adsorbent dosage 0.25 g and agitation speed of 100 rpm. The results implied that POFA has the potential as a low-cost and environmental friendly adsorbent for the removal of mercury from aqueous solution.

1. Introduction

Malaysia produces about 47% of the world's palm oil supply and can be considered as one of the world's largest producers and exporters of palm oil. Due to the intensive planting and milling operations, Malaysia generates huge quantity of oil palm biomass including oil palm fibres, oil palm kernel shells, empty fruit bunches (EFB) as residues from harvesting and processing activities. POFA is a waste material generated in boiler due to the burning of palm oil industry waste as fuel to generate electricity [1]. The conventional method to discard these wastes is by incineration in the boiler as fuels for steam production in the oil palm mill [2]. However, the issue regarding palm oil wastes management has led to concerns. This is because, after processing and extraction of oil, solid residues and liquid wastes will be generated from the fresh fruit bunches and resulted in varying by-products including empty fruit bunches fibre, shell, and effluent. As a result, air, river, sea, and groundwater pollutions have increased in recent years due to the large amount of waste produced [3]. Therefore, prevention should be taken to manage the agriculture by-products for sustainable future. To avoid environmental pollution, the by-products can be reused and recycled for other purposes. Thus, the study of POFA as a low cost adsorbent has been done to tackle the waste disposal problem.

[2] To whom any correspondence should be addressed.

One of the major environmental concerns is the pollution of water bodies by heavy metals [4]. The World Health Organization (WHO) considers mercury as one of the top ten chemicals or groups of chemicals of major public health concern. Mercury is one of the most harmful environmental contaminants, which is harmful even at low concentrations. The main sources of mercury emissions besides its natural origins are ore mining, the burning of fossil fuels, and the industrial production processes [5]. The Hg(II) ions can cause respiratory failure, kidney injury, chronic diseases, central nervous system disorders, brain damage and severe environmental pollution [6].

Many conventional techniques have been utilized to reduce Hg(II) ion concentrations in water, including solvent extraction, precipitation, ion-exchange, reverse osmosis, membrane separation, coagulation and adsorption [7]. Among the various water treatment techniques described, adsorption is generally viewed as the most reliable wastewater treatment containing Hg(II) ion due to its convenient operation, low cost, and high efficiency. Several adsorbents have been used for mercury removal from wastewater such as coal fly ash [5], coconut pith [8], eucalyptus wood [9] and fir wood sawdust [10]. Among these adsorbents, activated carbon has become the most popular and widely used adsorbents for wastewater treatment. Activated carbon is a form of porous carbon with a high surface area, used extensively as an adsorbent in the removal of organic and inorganic pollutants [10]. Ash materials are significant sources for activated carbon preparation as they contain appropriate carbon fractions [5]. Ash is a good candidate to produce activated carbon, because it is often used as an adsorbent to remove Hg(II) ion, for example, [11] investigated the different fly ash samples on mercury speciation in simulated flue gas. Also, [12] prepared activated carbon from palm ash as an adsorbent to remove dyes from aqueous solution. As a result, Hg(II) ion removal from water using adsorption techniques proved to be the most applicable on an industrial scale [4].

The objective of this study is to evaluate the adsorption of Hg(II) ions by POFA from aqueous solutions in batch process. The effects of operating variables namely temperature, contact time and initial Hg(II) ion concentration were investigated under batch experimental conditions. Experimental results indicated that POFA has higher adsorption rate for the removal of mercury (Hg(II) ion) in an aqueous solution.

2. Experimental procedure and methodology

2.1. Materials

POFA was obtained from Kilang Lepar Sawit in Kuantan, Pahang. It was cleaned with distilled water several times to remove dust and impurities and dried in an oven at 110 °C for 24 hr to remove any surface moisture. 50 g of POFA was activated with 0.5 M H_2SO_4 for 24 hr. It was sieved through 100 μm and stored in an airtight container.

2.2. Adsorbate

Analytical grade mercury standard solution obtained from Merck was used to prepare the stock solutions containing 1,000 mg/L of mercury metal ions which were further diluted with deionized water to obtain the required concentrations. The prepared solution was used for batch adsorption experiments.

2.3. Characterizations of POFA

Particle sizes distributions of POFA was analysed using Mastersizer 2000 (Malvern Instruments Ltd, UK). Mastersizer EVO 50 SEM (Carl Zeiss, Germany) instrument was used to study the morphology of the POFA. POFA sample was sprinkled on double-sided carbon tape mounted on a SEM stub and the images of 1000X magnifications were captured.

2.4. Batch experiments

The batch adsorption experiment was done by using 50 ml of initial Hg(II) ion concentration of solution in a conical flask for each experimental run. The pH of the solution was adjusted using 0.1 M sodium hydroxide (NaOH) and 0.1 M hydrochloric acid (HCl) as required. The adsorbent was added as specified in the experimental design and specified time accordingly. The mercury removal process was analysed using the effect of initial Hg(II) ion concentration, contact time and temperature. In the first set of experiment, the contact time was varied in the range of 0 - 360 min at a fixed pH (pH 7). The influence of initial Hg(II) ion concentration on the removal efficiency was studied in the range 1 - 5 mg/L. The effect of temperature was studied in the range of 15 - 45 °C using an incubator shaker for 4 hr. The temperature was controlled using temperature-controlled incubator shaker operated at 100 rpm. Equations (1) was used to obtain the removal efficiency.

$$R\,(\%) = \frac{C_0 - C_e}{C_0} \times 100 \quad (1)$$

where R is the percentage of the removal of mercury, C_0 and C_e are the initial and residual Hg(II) ion concentrations (mg/L), respectively. The adsorption capacity (q_e) was calculated according to the following equation (2):

$$q_e = \frac{(C_0 - C_e) \times V}{m} \quad (2)$$

where m is the mass of adsorbent used (g), V is the total volume of Hg(II) ion solution (ml), C_0 is the
initial Hg(II) ion concentration (mg/L), and C_e is the residual Hg(II) ion concentration (mg/L).

2.5 Analysis of the metal

Samples of solutions were filtered through vacuum filter and the filtrates were analysed using the direct mercury analyser RA-3310 (Nippon Instrument Corporation, Japan).

3. Results and discussion

3.1. Characterization of the adsorbent

Figure 1(a)(b) shows the characteristic morphology of the treated POFA and POFA after adsorption with mercury. Figure 2(a) illustrated that treated POFA possesed regular spherical shape and has a porous surface. Previous studies reported by [13] stated that ash consists of a series of spherical clear particles of different sizes. Figure 2(b) showed that the shape of the particles to be regular spherical shapes to completely irregular shapes. This is because POFA surface are covered with smaller particles and irregular deposits. Similar conclusions were made by [14] while studying the characterization of fly ash from coal combustion and found that the morphological features of the smallest fraction of fly ash are irregular in shape. This is evidence that porous structures with several smaller particles are covering the porous surface.

Figure 2 shows the particle size distribution of POFA. The bell-shaped curve for the particle size distribution were obtained. The graphical representation shows that there is a growing trend from 1 μm up to 100 μm and the highest peak of the function is observed in the area of 100 μm. The maxima occur at a relatively smaller range (90 - 100 μm) of particle diameters with the volume percentage of the particles is 5.2 %. It can be stated that between the range of 1 μm and 100 μm, have the largest volume percentage of the POFA and quantitatively negligible when it reached 600 μm.

3.2. Effect of initial Hg(II) ion concentration

Figure 3 presented the impact of initial Hg(II) ion concentration in the range of 1 - 5 mg/L. It is clear that the adsorption capacity of mercury gradually increased with the increase of initial Hg(II) ion concentration. In this experiment, the highest adsorption capacity of Hg(II) ions was 0.98 mg/g at the initial Hg(II) ion concentration of 5 mg/L. When all the active sites are involved in the adsorption process, the saturation and the maximum adsorbent capacity is achieved. Similar trend has been reported by [15] where the adsorption capacity of clay increased with the increase of initial metal ion concentration before attaining a plateau profile.

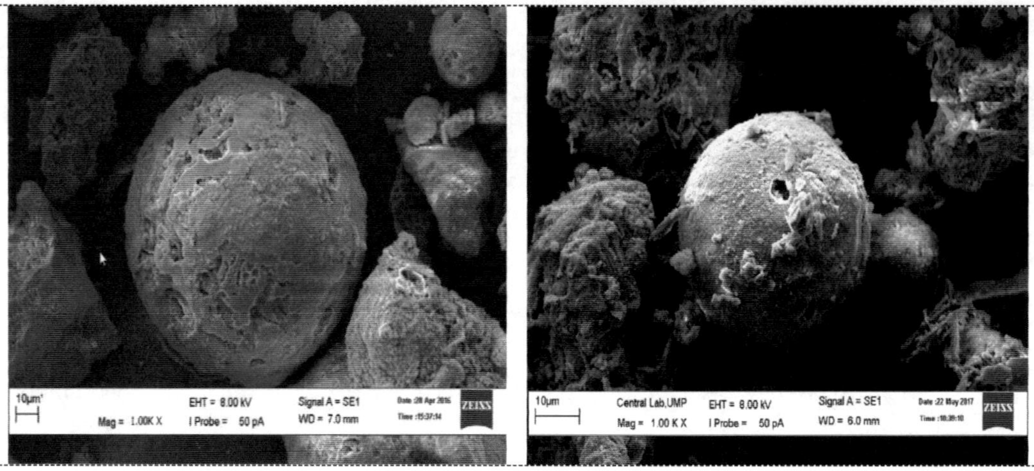

(a) (b)

Figure 1. SEM image of a) Treated POFA b) POFA after adsorption with mercury.

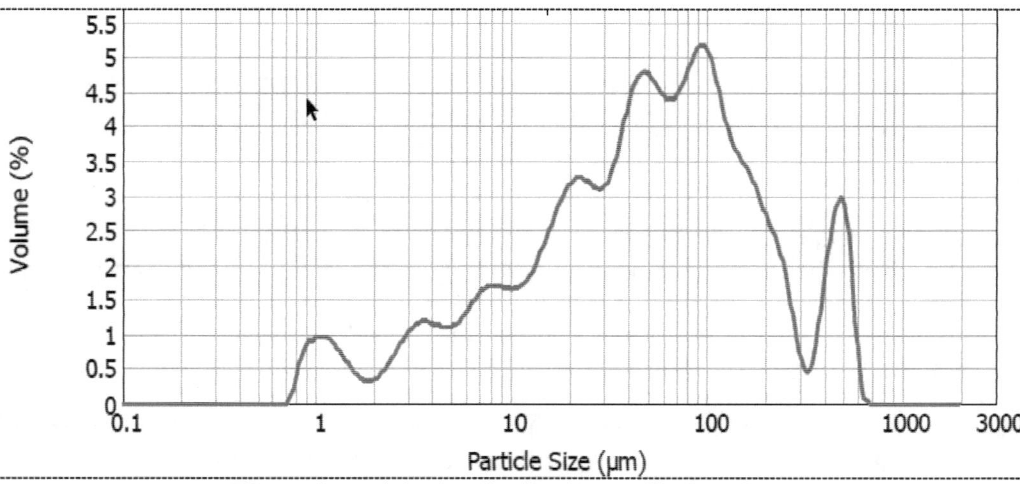

Figure 2. Particle size distribution profile for POFA.

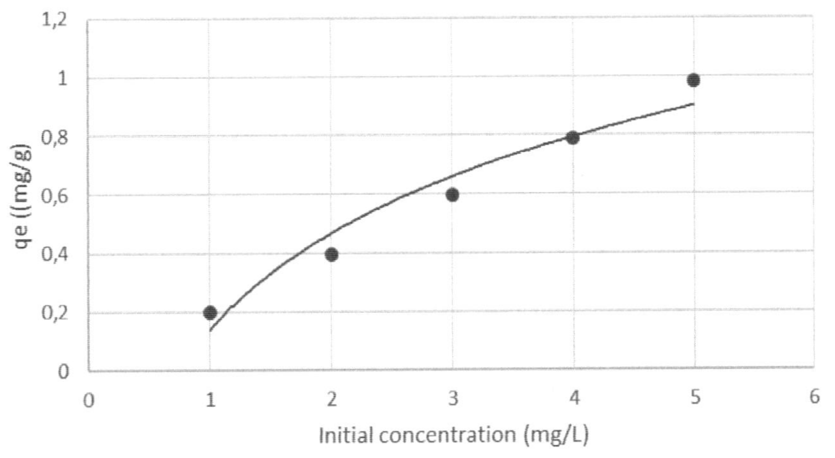

Figure 3. Effect of initial Hg(II) ion concentration on the adsorption of mercury. by POFA. Conditions: time = 4 h; temp. = 25 °C; agitation rate = 100 rpm; pH 7; adsorbent dosages = 0.25 g.

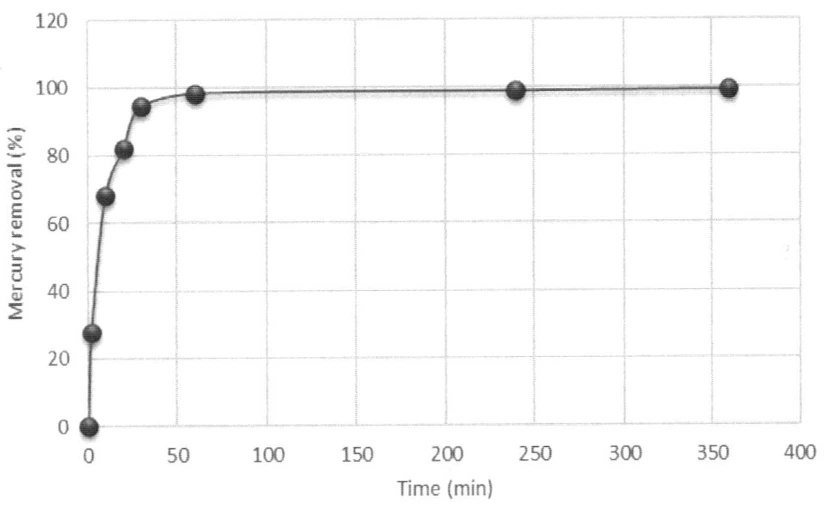

Figure 4. Effect of contact time on the adsorption of mercury by POFA. Conditions: C_0 = 5 mg/L; temp. = 25 °C; agitation rate = 100 rpm; pH 7; adsorbent dosages = 0.25 g.

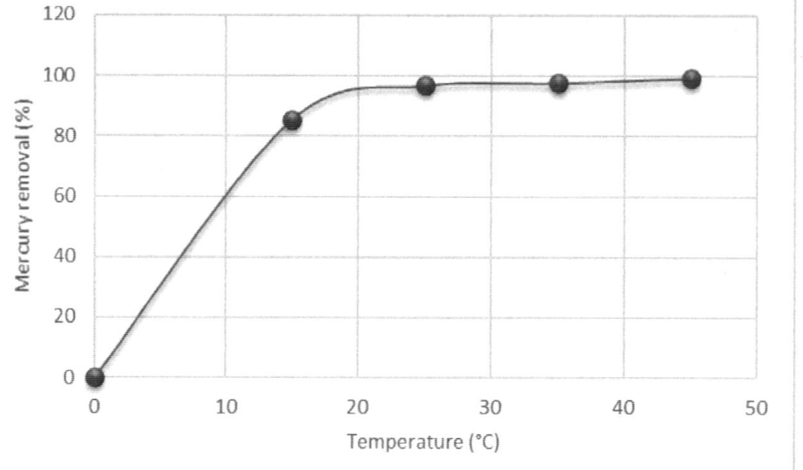

Figure 5. Effect of temperature on the adsorption of mercury by POFA. Conditions: time = 4 h; C_0 = 5 mg/L; agitation rate = 100 rpm; pH 7; adsorbent dosages = 0.25 g.

3.3. Effect of contact time
The effect of contact time on the adsorption of Hg(II) ion are shown in Figure 4. It can be observed that the adsorption of Hg(II) ion removal increased significantly with time from 0 to 60 min. The contact time needed for Hg(II) ion solutions to reach equilibrium was 60 min and reached a plateau at 60 - 240 min with 98 % of Hg(II) ion is adsorbed. According to [6] the binding of Hg(II) ions is high due to the existence of a greater number of functional sites available for Hg(II) ions adsorption in the initial stage. Then, when the available functional sites decreased, the adsorption efficiency slowed down due to the formation of repulsive forces between the Hg(II) ions on the POFA surface.

In fact, [10] reported that the adsorption of mercury using fir wood sawdust shows the same general trend when the adsorption of Hg(II) ion removal increased noticeably over a short period of contact time, then decreased gradually while approaching the equilibrium.

3.4. Effect of temperature
The effects of temperature are one of the most important factors for the adsorption of mercury. The effect of temperature on the mercury removal was studied in the range of 15 - 45 °C with an initial Hg(II) ion concentration 5 mg/L, pH 7, adsorbent dosages of 0.25 g and agitation rate 100 rpm. The result is shown in Figure 5. The mercury removal increased significantly to 97 % as the operating temperature was raised to 25 °C. When the temperature rises, the pores enlarge and more surface will be available for the adsorption of Hg(II) ions. Hence, there is an increase in the removal of Hg(II) ion. The mercury removal increased with the increase in temperature due to the endothermic nature of this process [15]. By increasing the temperature of Hg(II) ion solution, the attractive forces in the aqueous phase is weaken and consequently, more Hg(II) ions becomes free and reach the solid surface. Similar conclusions were made by [16] while studying the effect of temperature on the mercury adsorption using eucalyptus bark and found that a decrease in Gibbs free energy, ΔG, as well as a positive enthalpy value, ΔH, by raising the temperature revealed that the adsorption process is endothermic. Additionally, the increase of temperature can enhance the rate of intraparticle diffusion and the adsorption increases accordingly. [17] reported a similar result for the adsorption of Hg(II) ion removal by MnO_2/CNTs nanocomposite adsorbent. They also showed that the intraparticle diffusion plays an important role in the adsorption of Hg(II) ion removal.

4. Conclusion

In this study, the removal of Hg(II) ion from aqueous solutions using POFA was examined using various parameters namely the initial Hg(II) ion concentration, temperature, and contact time. The highest mercury removal was 99.60 % when the parameters were at 100 rpm of agitation speed, 4 hr of contact time, initial Hg(II) ion concentration of 1 mg/L, adsorbent dosage 0.25 g of adsorbent dosage and pH 7. Overall, it can be concluded that POFA has the potential to be used as a suitable adsorbent for the removal of mercury from aqueous solution as it can be obtained at low cost.

Acknowledgment

This research was funded by Universiti Malaysia Pahang under the internal grant RDU170350. The authors would like to thank the Faculty of Chemical and Natural Resources Engineering, Universiti Malaysia Pahang for providing the infrastructure facilities for this study.

References

[1] Ranjbar N, Behnia A, Alsubari B, Birgani P M and Jumaat M Z 2016 Durability and mechanical properties of self-compacting concrete incorporating palm oil fuel ash. *Journal of Cleaner Production.* **112** 723–730.

[2] Acquah C, Sie Yon L, Tuah, Z, Ling Ngee N and Danquah M K 2016 Synthesis and performance analysis of oil palm ash (OPA) based adsorbent as a palm oil bleaching material. *Journal of Cleaner Production.* **139** 1098–1104.

[3] Mujah D (2016). Compressive strength and chloride resistance of grout containing ground palm oil fuel ash. *Journal of Cleaner Production.* **112** 712–722.

[4] Mubarak N M, Sahu J N, Abdullah E C, Jayakumar N S and Ganesan P 2015 Microwave assisted multiwall carbon nanotubes enhancing Cd(II) adsorption capacity in aqueous media. *J Ind Eng Chem.* **24** 24–33

[5] Attari M, Bukhari S S, Kazemian H and Rohani S 2017 A low-cost adsorbent from coal fly ash for mercury removal from industrial wastewater *J Environ Chem Eng.* **5(1)** 391–9.

[6] Awual R., Hasan M, Eldesoky G E, Khaleque A, Rahman M M, and Naushad M 2016 Facile mercury detection and removal from aqueous media involving ligand impregnated conjugate nanomaterials. *Chemical Engineering Journal.* **290** 243–251.

[7] Kokkinos E, Simeonidis K, Pinakidou F, Katsikini M and Mitrakas M 2017 Optimization of tetravalent manganese feroxyhyte's negative charge density: A high-performing mercury adsorbent from drinking water. *Sci Total Environ.* **574** 482–9.

[8] Johari K, Saman N, Song S T and Mat H 2015 Adsorption equilibrium and kinetics of elemental mercury onto coconut pith. *Journal of Environmental Science and Technology.* **8(2)** 74–82.

[9] Silva H S, Ruiz S V, Granados D L and Santángelo J M 2010 Adsorption of Mercury(II) from Liquid Solutions Using Modified Activated Carbons. *Materials Research.* **13(2)** 129–134.

[10] Kazemi F, Younesi H, Asghar A, Bahramifar N and Heidari A 2015 Thiol-incorporated activated carbon derived from fir wood sawdust as an efficient adsorbent for the removal of mercury ion : Batch and fixed-bed column studies. *Process Safety and Environmental Protection.* **100** 22–35.

[11] Bhardwaj R., Chen X., Vidic R D, Bhardwaj R., Chen X and Vidic R D 2016 Impact of Fly Ash Composition on Mercury Speciation in Simulated Flue Gas Impact of Fly Ash Composition on Mercury Speciation in Simulated Flue Gas. *Journal of the Air & Waste Management Association.* **2247(59)** 1331-1338.

[12] Ahmad A A., Hameed B H and Aziz N (2007) Adsorption of direct dye on palm ash: Kinetic and equilibrium modeling. *Journal of Hazardous Materials.* **141(1)** 70–76.

[13] Al Bakri A M M, Kamarudin H, Bnhussain M, Nizar I K, Rafiza A R and Zarina Y 2012 The Processing, Characterization, and Properties of Fly Ash Based Geopolymer Concrete. *Rev. Adv. Mater. Sci.* **30** 90–97.

[14] Stoch A 2000 Fly ash from coal combustion – Characterization. *Tecnico Lisboa*. 1–10.
[15] Rajamohan N and Al-sinani J 2016 Removal of boron using clay-effect of process parameters, kinetic and isotherm studies. *Journal of Engineering Science and Technology*. **11(3)** 311–326.
[16] Ghodbane I and Hamdaoui O 2008 Removal of mercury(II) from aqueous media using eucalyptus bark: Kinetic and equilibrium studies. *Journal of Hazardous Materials*. **160** 301–309.
[17] Moghaddam H K and Pakizeh M 2015 Journal of Industrial and Engineering Chemistry Experimental study on mercury ions removal from aqueous solution by MnO_2 / CNTs nanocomposite adsorbent. *Journal of Industrial and Engineering Chemistry*. **21** 221–229.

IOP Conf. Series: Materials Science and Engineering **334** (2018) 012040 doi:10.1088/1757-899X/334/1/012040

Enhancement of polyethersulfone (PES) membrane performance by modification with rice husk nanosilica for removal of organic matter in water

S Mulyati[1,3], M A Armando [1], H Mawardi[1], F A Azmi[1], W P Pratiwi[1], A Fadzlina[1], R Akbar[1], and Syawaliah[2]

[1] Chemical Engineering Department, Syiah Kuala University, Banda Aceh, Indonesia, 23111
[2] Doctoral School of Engineering Science, Syiah Kuala University, Indonesia

E-mail: sri.mulyati@unsyiah.ac.id

Abstract. This paper reports the effects of rice husk nanosilica addition on the performance of polyethersulfone (PES) membrane. Polyethersulfone membrane (PES) was fabricated by using N-methyl-2-pyrolidone (NMP) as a solvent and rice husk nanosilica as a modifying agent. The influence of the rice husk nanosilica additive on the characteristics and performance of the membrane has been studied. Scanning Electron Microscopy (SEM) analysis confirmed that the manufactured membrane has an asymmetric morphological structure consisting of two layers. The upper part of the membrane is a thin layer, meanwhile in the bottom side is a porous layer. The addition of 5% nanosilica resulting a PES membrane to have a bigger porous than that of pristine PES. The pure water flux of nanosilica-modified membranes were greater in comparison to the pure water flux of unmodified PES membrane. The performance of all membranes were evaluated on humic acid removal. The highest selectivity was showcased by pure PES membrane. The introduction of rice husk nanosilica additive to the membrane declined the selectivity of the membrane to humic acid in the feed solution. This is caused by the pores enlargement and enhanced hydrophilicity of the membrane after modification with rice husk biosilica.

1. Introduction

Water is a fundamental necessity for human. Clean water is used in daily life for drinking, cooking, washing and so on. Drinking water comes from a variety of sources including common sources (rivers, lakes), wells, or bottled mineral water. Water used as drinking water must meet the quality requirements that have been determined, both physically, chemically and bacteriologically. The consumption of water with poor quality will induce negative impacts on public health. Therefore, the proper treatment is needed in order to produce a pathogen and chemicals free clean water [1]. Some of the most common contaminants existing in raw water are hydrophobic organic matters, for instance humic acid, fulvic acid, and humin [2].

The removal of organic compounds in water can be carried out by various methods, including adsorption [3], electrocoagulation [4], anaerobic granular sludge [5], ozonisation [6], Flotation [7],

[3] To whom any correspondence should be addressed.

integrated ion exchange - bioelectrochemistry [8], electrocoagulation and dead end microfiltration hybrid process [9]. In addition to those mentioned techniques, membrane technology can also be used as an alternative. A number of researchers have applied membrane technology to reduce organic compounds in water such as using carbon nanotube composite membrane [10], bioreactor membrane [11], and poliamide nanofiltration membrane [12]. Membrane technology owns several advantages namely simple of use, integratable with other processes, chemical-free, and resistant to a wide range of operation temperature [13]. Polyethersulfone (PES) is one of the most commonly used polymers in the production of ultrafiltration membranes, it is due to the characteristic of PES which can work at high temperatures and have outstanding mechanical resistance [14]. In addition, the selection of PES as a membrane material is also due to its affordable availability, ease of utilization, favorable selectivity, excellent permeability, high mechanical and stability to high temperature [15]. However, this polymer has a weakness because it is hydrophobic. This property causes the membrane performance to decrease due to the occurrence of fouling or concentration polarization on the membrane surface.

To reduce the impact of polymerization and fouling on PES membranes, a number of authors have performed a modification on the membranes for instance by irradiation with gamma rays [16], introducing cobalt into the system to produce a nanocomposite membrane [17], blending with polysulfide-amide (PSA) [18]. All those endeavours were aimed to improve the hydrophilicity nature of PES membrane thus the performance could be enhanced by hindering the fouling or concentration polarization phenomenon. The addition of nanoparticles such as SiO_2, Fe_3O_4, Al_2O_3, ZrO_2, TiO_2, and ZnO have a significant influence on the characteristics of the resulted membranes, including the membrane top layer, pore size, thickness, hydrophilicity, and other parameters associated with membrane structures such as porosity and macrovoid [19].

In this study, the modification on the PES membrane was done by adding silica nanofiller from rice husk to improve the performance of the membrane. Besides enhancing the separation performance, the utilization of rice husk as an additive is also a green approach because it encourages the reusability of agricultural waste. To produce the nanosilica, the silica was initially isolated of rice husks then it was added to the membrane casting solution as an additive. PES membrane was fabricated by phase inversion techniques. The variables used in this study were the concentration of rice husk nanosilica additive (0, 3 and 5%). The morphological structure of resulting silica and membranes was analyzed by means of Scanning Electron Microscopy (SEM) and Fourier transform infrared spectroscopy (FTIR) was used to evaluate the functional groups of silica and membranes. The pure water permeation experiments of the fabricated membranes were also carried out. Humic acid was employed as a model compound to study the performance of the membranes in terms of flux and rejection against organic matter in feed water sample.

2. Experimental procedure and metodhology

2.1. Preparation and Characterization of Rice Husk Biosilica

The raw rice husk was obtained from rice mills around Banda Aceh city. The bio silica was isolated of rice husk by the extraction technique using KOH as a solvent. The rice husk was dissolved with KOH 37% at temperature of 30ºC. To precipitate the bio silica, HCl 37% was then added into the solution. Following that, the formed bio silica was separated from the solution by filtration method and followed by the drying process. The particle size of resulting bio silica was reduced by means of ball mill until the smallest possible size was obtained. Furthermore, the bio silica was characterized by using Scanning Electron Microscopy (SEM). This characterization was conducted to examine the morphological structure of the bio silica. In addition, the evaluation of functional groups present in rice husk bio silica was also carried out by using the Fourier transform infrared spectroscopy (FTIR).

2.2. The fabrication of PES-rice husk biosilica modified membrane

The casting solution was prepared by dissolving PES polymer into the NMP solvent. The solution was stirred until homogeneous. The rice husk nanosilica additive of a certain concentration was added slowly into the solution until completely dissolved. The solution was then poured on a glass plate and casted by using a casting knife. Subsequently, the casted glass plate was dipped into a bath containing non-solvent (water). At that moment, the phase inversion process began and after a few minutes the membrane was detached from the glass plate. After that, the membrane was washed and stored in distilled water before use.

2.3. Filtration Process
The ultrafiltration experiment was performed using a dead-end module with Trans Membrane Pressure (TMP) as a driving force. The TMP used were 3, 3.5 and 4 bars. A membrane sheet with an area of 1.5 x 10-3 m^2 was placed in a membrane module with an active surface layer in the direction of feed inlet. The pure water permeability of the membrane was evaluated. This test was carried out using distillate water as a feed solution. Furthermore, the experiment of organic matter removal process was done by using a solution of humic acid as a model sample. The volume of permeate which passed through the membrane was accommodated, the flux and selectivity to humic acid were afterward calculated.
Flux was calculated by using equation (1).

$$L_P = \frac{J}{\Delta_P} \qquad (1)$$

Here, Lp indicates the permeability coefficient (L/m^2.hour.atm), J is flux (/m^2.hour), and Δp is pressure (atm).

Selectivity test of the membrane was conducted by using humic acid (HA) solution with concentration of 10 mg/L. The ability of membrane to reject the macromolecules in the feed solution can be calculated by equation (2).

$$R\% = \left(1 - \frac{C_P}{C_f}\right) x100 \qquad (2)$$

Where R is coefficient of rejection (%), Cp is humic acid concentration in permeate, and Cf is the concentration of humic acid in the feed solution.

3. Results and Discussion

3.1. Characterization of Rice Husk Nano Silica
The morphological and FTIR analysis results of nano silica can be seen in Figures 1 and 2. From the results of SEM analysis it appears that the obtained bio silica from rice husk has a size in the nanometer range (10^{-9}). The silica has a non uniform shape of surface, some of the parts are agglomerated but the nanometer particle size is seen dominating.

From the FTIR analysis it can be observed that the 1099 cm^{-1}, 950 cm^{-1}, and 468 cm^{-1} represent the functional group of siloxane (Si-O-Si) [20] Whereas OH group is typified at a wave number of 3305 cm^{-1} [21].

3.2. Morphology of the Membrane
The surface and cross-sectional morphology of pure and silica-modified PES membranes are presented in Figure 3 and 4. As seen from Figure 3, it is evidencing that there is the appearance of nanosilica deposition on the surface of the silica-modified PES membrane. The result of cross-sectional analysis in Figure 4 shows that the membrane prepared by phase inversion technique has an asymmetric structure, consisting of two layers. The top layer consists of a dense layer and the bottom layer is a porous layer. The diffusion of non-solvent into the membrane (the demixing process) causes the membrane to have a macrovoidal structure. From the figure it is confirmed that all the resulting membranes have a finger-like structure on the sublayer and different macrovoidal forms. The changed in membrane structure may be caused by the addition of the nanofiller, which may have an impact on the growth or suppression of macrovoid [19]. The addition of additive to the membrane has an effect

on the pore size on the bottom layer. The increase in additive concentration results in the enlargement of membrane pore size.

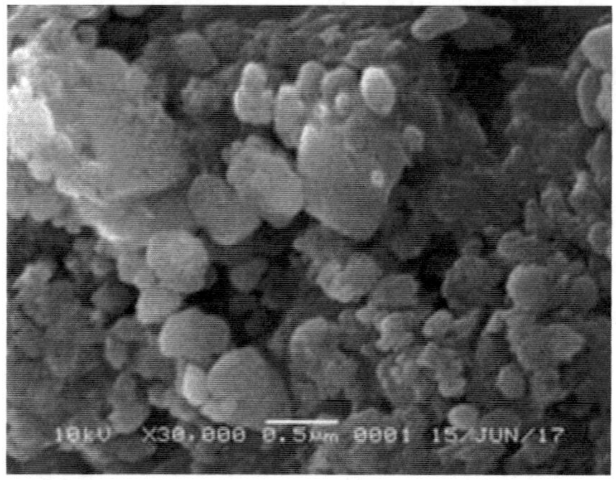

Figure 1. Morphology of Silica From Rice Husk

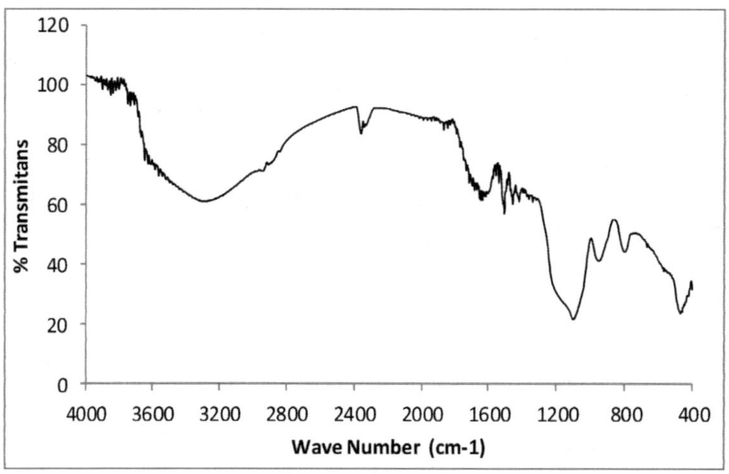

Figure 2. Spectra FTIR of Biosilica Rice Husk

(a) PES (b) PES+ aditif 3 % (c) PES+ aditif 5 %

Figure 3. SEM images of the outer surface of PES membrane with and without additive of biosilica

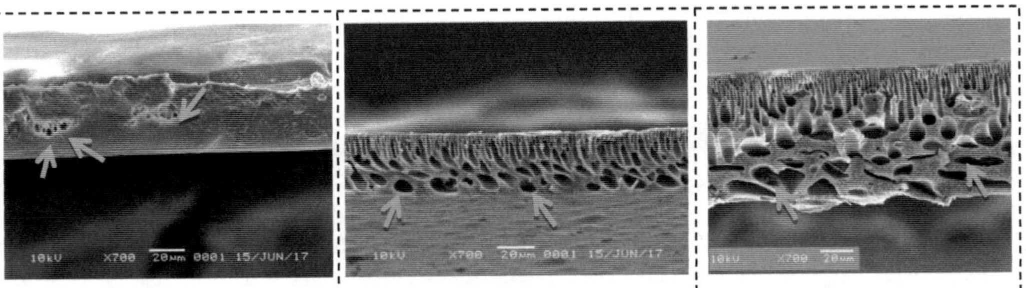

(a) PES (b) PES + silica 3% (c) PES + silica 5%

Figure 4. SEM images of the cross section of PES membrane with and without additive.

3.3. Pure Water Permeation (PWP).

Permeability is the ability of the membrane in letting a particular substance to pass through. Pure water permeability is one of the most important characteristics for membranes when applied to the industry. Membrane permeability is determined by measuring pure water flux. Pure water flux reflects pore size, pore size distribution and membrane morphology [22]. The relationship between pure water flux and Trans Membrane Pressure (TMP) is shown in Figure 5.

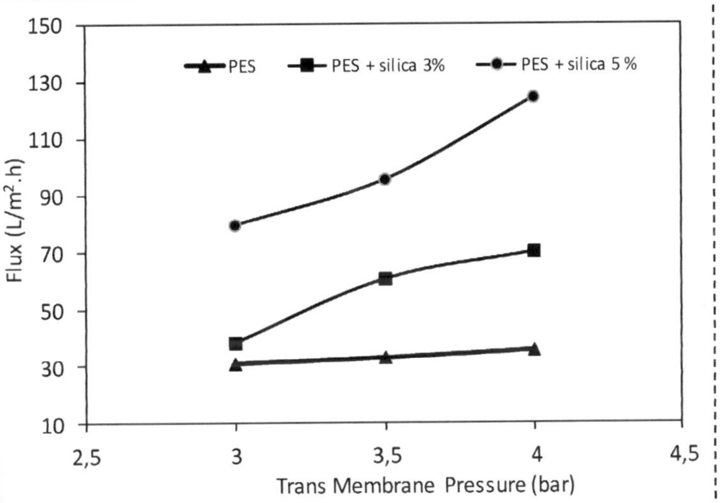

Figure 5. Pure Water Permeation for PES membranes with and without additive

The filtration experiment in this research was conducted based on pressure-driven, therefore the increase of pressure (TMP) causes the pure water flux passing through the membrane to be larger for all membranes. From the Figure 5, it is clearly seen that the pure water flux is influenced by the addition of rice husk nano silica additive into the membrane casting solution. The pure water flux obtained from experiment using original PES was 35.5 L/m^2.hour, 0 L/m^2.hour for PES + 3% additive and 124 L/m^2.hour for PES + 5% additive. Pure water flux improved with increasing concentration of additives added to the casting solution. The presence of the additive causes the membrane pore size to become larger hence permeate can pass through the membrane more easily.

3.4. Membrane Performances

The removal rate (rejection or selectivity), and permeation are important parameters in determining membrane performance. In this study, water feed containing humic acid with a concentration of 10 ppm was passed through the membrane with a TMP of 3 bar. Flux and rejection of pure PES, PES + 3% additives and 5% additives are shown in Fig. 6. The resulting fluxes show that the addition of the nano silica additive into the membrane affected the permeation rate. The membranes which modified with nanosilica from rice husk showed an enhanced flux in comparison to that of without modification. The addition of 5% rice husk silica has successfully double-augmented the flux of the PES membrane. This is due to pore size enlargement and improved hydrophilicity of the membrane after the silica addition. The percentage of rejection obtained in the experiment is in contrast to the resulting flux. The addition of the additive decreases the selectivity or rejection of the membrane to the humic acid in the sample solution. However, the addition of 3% nanosilica resulted in a rejection which comparable to that of pure PES membrane. The resulting rejection for pure PES membrane, PES + 3% additive and PES + 5% additive were 75.19 , 74.39 and 62.25 % respectively.

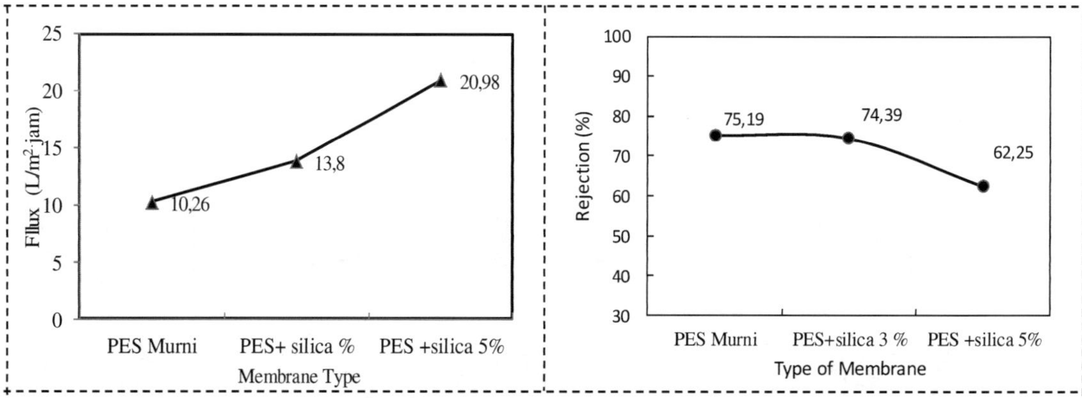

Figure 6. Filtration Performance for PES membranes with and without additive.
Left : Humic acid flux Right : Humic acid rejection

4. Conclusion

PES membrane has been prepared by phase inversion technique and the effect of addition of biosilica from rice husk as additive on membrane characteristics and performance has been studied. SEM analysis showed enlargement of macrovoid size on the silica-modified membranes which leads to increasing of pure water permeation. Filtration performance on humic acid enhanced in terms of flux, meanwhile the solute rejection declined with increasing concentration of biosilica.

References

[1] Bergamasco R, Konradt-Moraes L C, Vieira M F, Fagundes-Klen M R, Vieira A M S 2011 Chem. Eng. J. **166** 483-489
[2] Matilainen A, Vepsäläinen M, Sillanpää M 2010 Adv. Colloid Interfac. **159** 189-197
[3] Bhatnagar A, Sillanpää M 2017 Chemosphere **166** 497-510
[4] Hakizimana J N, Gourich B, Vial Ch, Drogui P, Oumani A, Naja J, Hilali L 2016 Desalination **393** 90-101
[5] Tomar S, Gupta SM, Mishra BK 2015 Bioresource Technol. **197** 171-177
[6] Cai Mp, Lin Yp Chemosphere 2016 **151** 332-338
[7] Xu Zl, Alsalhy Q F 2004 J. Membrane Sci. **233** 101-111
[8] Lim Sj, Kim Th 2015 Bioresource Technol. **189** 107-112
[9] Ben-Sasson M, Zidon Y, Calvo R, Adin A, 2013 Chem. Eng. J. **232** 338-345
[10] Lee J, YeY, Ward A J, Zhou C, Chen V, Minett A I, Lee S, Liu Z, Chae Sr, Shi J 2016 Sep.

Purif. Technol. **163** 109-119
[11] Zhuang H, Hong, X, Han, H, Shan S 2016 Bioresource Technol.**221** 262-269
[12] Zhao Y, Li N, Xia S 2016 Compos Sci. Technol.**132** 84-92
[13] Mulder M 1996 Basic Principles of Membrane Technology ISBN: 978-0-7923-4248-9 (Print) 978-94-009-1766-8 (Online)
[14] Sun W, Chen Cx, Li Jd, Lin, Yz 2009 Chinese J. Polym. Sci. **27** 165-172
[15] Arthanareeswaran G, Starov, Victor M 2011 Desalination **267** 57-63
[16] Zhang X, Niu L, Li F, Yu S, Zhao X, Hu H 2016 Radiat Phys Chem **127** 127-132
[17] Gzara L, Ahmad R Z, Khan S B, Alamry K A, Albeirutty M H, El-Shahawi M S, Rashid M I, Figoli A, Drioli E, Asiri A M -2016 J Taiwan Inst Chem E **65** 405419
[18] Jalali A, Shockravi A, Vatanpour V, Hajibeygi M 2016 Micropor. Mesopor. Mat. 228 1-13
[19] Lin, J., Ye W, Zhong K, Shen J, Jullok N, Sotto A, Van der Bruggen B 2016 Chem Eng Process **107** 194-205
[20] Amutha K., Sivakumar G.2013 Spectrochim Acta A **112:** 219-222
[21] Yusmaniar S B 2007 Jurnal Sains Meteri Indonesia, Edisi Khusus 115-117
[22] Tao M-m, Liu F, Xue, L-x 2013 Ultrason. Sonochem. **20** 232-238

Preparation and characterization compatible pellets for immobilization of colloidal sulphur nanoparticles

M Adlim[1,2], F Zarlaida[1], I Khaldun[1] and R Dewi and M Jamilah [1]
[1] Chemistry Department, FKIP Syiah Kuala University, Darussalam Banda Aceh 23111, Indonesia

E-mail: adlim@unsyiah.ac.id

Abstract. Mercury pollution in atmosphere is dominated by mercury vapour release from coal burning and gold-amalgam separation in gold mining. The initial steps in formulating a compatible mercury absorbent for mercury stabilization was fabrication of pellet supported colloidal sulphur. Sulphur is used to stabilize mercury vapour by formation of metacinnabar that has much lower toxicity. The sulphur reactivity toward mercury vapour can be enhanced by using colloidal sulphur nanoparticles immobilized on compatible pellets. Clay pellets would have heat resistance but in fact, they were less stable in aqueous solution although their stability increased with inclusion of rice husk ash and sawdust or pineapple leaf fibre in the composite. Pellets made of rice husk ash and polyvinyl acetate were stable in water at least for 24 hours. Sulphur from thiosulfate precursor that immobilized onto surface of pellet using chitosan as the stabilizer and the binding agent gave lower sulphur content compared to sulphur from other precursors (sulphur powder and sulphur-CS_2). Sulphur from thiosulfate precursor was in form of colloid, has nanosize, and disperse particles on the surface of rice husk ash pellets. Sulphur immobilization methods affect on sulphur particles exposure on the pellet surface.

1. Introduction

Elemental mercury and the compounds are among the most toxic heavy metal groups [1-2]. The main source of mercury pollution was from burning coal and burning amalgam in gold mining [3]. Mercury vapour backs to the earth and forms some water-soluble compound that eventually pollutes food chain [4].

Sulphur powder was commonly used to stabilize mercury vapour pollution since it will form meta cinnabar that is much lower toxicity [5-7]. Sulphur powder has large particle size and it is ineffective to stabilize mercury vapour in continuous gas reactor since the sulphur particles will block the vapour stream. To overcome this issue, sulphur powder was immobilized in a solid support such as granular carbon and coal [8-9]. Reactivity of impregnated sulphur toward mercury vapour was considered low especially operated at low temperature as reported previously. Mullett and co-workers used sulphur immobilized on granular carbon to stabilize mercury vapour 25°C for 8 weeks exposing and found low mercury uptake [10]. Mercury uptake was much higher at higher temperature (140°C) as reported by Reddy and colleagues [11]. Operating condition at above sulphur melting point will cause sulphur melt out from the support, aggregated, block the vapour steam and finally reduce the reactivity. Thereby finding more reactive sulphur that feasible operated at moderate temperature is preferable.

[2] To whom any correspondence should be addressed.

Attempting to improve sulphur reactivity is by reducing the particle size into nanometer scale. This can be prepared in colloidal form by using a stabilizer [12]. The colloidal sulphur requires stabilizer and a support to disperse the particles and to avoid particle aggregation. Since colloidal sulphur was usually prepared from sodium thiosulfate-HCl, which is polar [13-14], then it needs compatible pellets to immobilize the sulphur particles.
Screening the best material and the formula for preparation of compatible pellets, the effect of sulphur precursors and the immobilization colloidal sulphur technique toward the characteristic of sulphur particles are the main discussion in this papers.

2. Material and Methods

2.1. Materials
Rice husk ash, dried clay, sawdust, sago starch that pass through 200-mesh sifter. Banana leaf and pineapple leaf fibres. $Na_2S_2O_3.5H_2O$ (p.a, Merck), $Na_2S.9H_2O$ (ACS, Fischer, 98%), Chitosan medium molecular weight (Fluka), CS_2 (ACS, Merck), mercury metal (>99.6%, Merck), acetic acid glacial (CH_3COOH), sodium cyanide (KCN), iron (III) chloride ($FeCl_3.6H2O$), methanol (CH_3OH, Pure Analysis), hydrochloric acid (HCl), and acetone, sulphur powder purified (Merck) were purchased without further purification.

2.2. Equipment
Several analytical instrumentations were used for collecting data. Light microscope (optilab). UV-Vis Spectrophotometer. Hitachi SU-3500 SEM and Horiba EDX equipment were used for pellet surface analysis. The measurement was at Vacc 20 kV and spot intensity of 50. The test was based on point analysis in four points for inorganic element searching using Back Scatter Electron (BSE) with 500-2500 times magnification. Torontech TT-EDXPRT. XRF was used to analyse the chemical content of the samples. Philip CM 12 Electron Microscope was used for TEM micrograph to measure particle size.

2.3. Methods
2.3.1. Screening for pellet formula
(a) Pellet made of rice husk ash (RHA) and sago starch as the adhesive
Dried RHA and sago starch that pass through 200 mesh particles were blended and water was added carefully until with weight proportion of 70 g (RHA): 30 g (SS): 70 mL (water) as shown in Table 1. While mixing, the mixture was heated 80°C and a few drops of water were added until the mixture rigid. The mixture was inserted into pellet dies hole (4 mm length and 3 cm diameter) and pressed at 100 Nm^2 to fabricate the pellets. Pellets were dried in oven at 105°C until constant weight.

(b) Pellet made of clay and natural fibre composite
The dried clay (earthenware type) in form of powder, rice husk ash (RHA) powder and pineapple leaf chopped fibres (2 mm length) that all pass through 200 mesh sieve were mixed using mixer and water was added drop wise until from a sticky and coagulated mixture as shown in Table 1. The weight proportion was 90% (clay): 10% (RHA) with and without additional 5% of natural fibres. The mixture was inserted into pellet dies hole (4 mm length & 3 mm diameter) and pressed with hydraulic press at 100 Nm^2. Dried pellets were soaked for 48 hours in 200 mL distilled water to test the decomposition resistance in water.

Each type of pellet was divided into two porcelain crucibles, one was heated in a furnace at 200^{0}C, and the other was heated up 400°C for 2 hours. The sulphur immobilization described in procedure point (a) was repeated for these pellets. Dried pellets were soaked for 48 hours in 200 mL distilled water to test the decomposition resistance in water.

(c) Pallets made of RHA with Polyvinyl Acetate as Adhesive

RHA as described above was mixed with polyvinyl acetate (PVAc) as the adhesive material using a blender with weight ratio of RHA:PVA:H_2O was 7:2:1. The mixture then inserted into holes of pellet dies (3 mm diameter holes, 4 mm height). It was pressed with a hydraulic press machine at 0.5 tones before heating at 50°C and dried for 24 h. Dried pellets were soaked for 48 hours in 200 mL distilled water to test the decomposition resistance in water.

Dried pellets were dipped in chitosan solution made of (2% wt chitosan in 5% of aqueous-acetic acid) for 2 minutes until all pellets were covered with chitosan solution. The pellets were separated and were dried with hot air using hair drier for 2 minutes. The hair dryer was place 30 cm from the sample to avoid excessive heating.

Table 1. Pellet composition

Pellet	Clay (g, %)	RHA, (g)	(PF, g)	Sawdust (g)	H_2O (mL)	SS (g)	PVAc	Temp. (°C)
A	-	140 (70%)	-	-	140	60 (30%)		100
B1	150(100%)	-	-	-	35	-		200
B2	225(90%)	25(10%)	-	-	63.5	-		200
B3	112.5(90%)	12.5(10%)	-	-	35.5	-		200
B4	135(90%)	15(10%)	-	-	35	-		200
C1	150(100%)	-	-	-	35			400
C2	225(90%)	25(10%)	-	-	63.5	-		400
C3	112.5(90%)	12.5(10%)	-	-	63.5	-		400
D1	112.5 (88%)	12.5 (11%)	2.5(2%)		63.5			400
D2	112.5 (88%)	12.5 (11%)	-	2.5(2%)	63.5			400
D3	0	70 (77%)	0	0	10		20 (22%)	50

PF = Pineapple fibre; RHA = Rice Husk Ash; SS = sago starch

2.3.2. Sulphur immobilization and chitosan coating. Chitosan coating and sulphur immobilization was divided into methods that were stepwise and simultaneous methods. Stepwise methods: pellet was coated chitosan film yang dipping chitosan pellets on viscos chitosan solution. Dried chitosan films on surface of pellets were swelled by dipping them into 1.5% aqueous acetic acid solution before sinking into $Na_2S_2O_3.5H_2O$ (2 M) solution and HCl (4 M). The simultaneous methods: 10 mL of $Na_2S_2O_3.5H_2O$ (2 M) was added into 50 mL of chitosan solution and 10 ml of HCl (4 M) was added dropwise until form slurry solution. The pellets were dipped into slurry solution and dried until constant weight. This immobilization method was modified from Chaudhuri and Paria [13]. Finally, the pellets were separated and dried with hot air using a hair drier until constant weight. Dried pellets were soaked for 48 hours in 200 mL distilled water to test the decomposition resistance in water.

2.3.3. Effect of sulphur precursor on the RHA-PVAc pellets-immobilized sulphur. Three types of sulphur precursors were used that were sulphur powder, sulphur powder in CS_2 and $Na_2S_2O_4$-HCl. Pellet'surface covered with S from CS_2 solution (P-S_{CS2}): dried pellets (50 g) were dipped 2 minutes in saturated cold solution of sulphur-CS_2 (100 mL) then; the pellets were filtered and air-dried for 2 hours. Pellet covered with S powder (P-S_{powder}): This chitosan coating techniques was modified from previous methods [12]. $P_{chitosan}$ (50 g) were placed in a plastic sifter, dipped in 50% methanol-water for one minute, filtered and transferred into a boiling flask containing 100 g sulphur-powder. The flask was rotated with a rotary evaporator until all pellets' surfaces were covered by sulphur powder. Pellet covered with S from sulphur colloid (P-$S_{colloid}$): Another half portion of $P_{chitosan}$ (50 g) was placed in a plastic sifter and dipped in water-methanol (50:50) before shook gently and sprayed with 2 M of $Na_2S_2O_4.5H_2O$ solution until all part of pellets were wet. These pellets were placed on a plastic strainer then were carefully sprayed with 4 M HCl solution until all were wet and changed to white, then dried again with hot air using hair dryer until pellets reached constant weight. The hair dryer was placed 25 cm from the sample to avoid excessive heating. All types of pellets (a, b & c) were sprayed with 1% of Na_2S solution (Antony 1962) and dried with a hair dryer before store in oven at 70°C for 24 h.

2.3.4. Sulphur determination on the surface of pellets. Sulphur content within the pellets was determined using XRF, EDX and gravimetric analysis. In gravimetric analysis, the pellets was dried in oven at 105°C until constant weight. Then, 10 gram of the pellets was extracted using acetone. The filtrate was dried in water bath until constant weight. The residue was weight as sulphur content.

3. Result and Discussion

3.1. Screening for pellet formula

A number of pellets with different material and composition was denoted Pellet A up to Pellet D3 in methodology sections (Table 1) were tested on the stability in water by counting the percentage of pellets that crack or decomposed in water for 24 hours soaking and the interaction in in chemical environment as shown in Table 2. The pellet shape was cylindrical with dimension of 4 mm height and 3 cm diameter as displayed in Figure 1 (a-d).

It is found that all pellets made of clay only (B1) and calcinated at 200°C was not stable in water as shown by 100% of pellets decomposed. The stability increase when they were prepared at higher temperature (C1) although these pellets did not last in acid solution. The stability of clay pellets increased if RHA is part of the composite (B3 & B4). The calcination temperature also affect the pellet endurance as shown in case of pellet C2 & C4 that calcinated at 400°C. The addition of pineapple fibre and sawdust in clay composite make even more better stability for the pellets (D1 & D2) in water and chemical environment but they still did not last long in acidic solution. Pellets made of RHA and PVAc (D3) show better characteristic, there were no decomposition in water at least 24 hours and stand for acidic interaction. The further study on these pellets was carried out on immobilization of sulphur from various precursors.

Some explanations for this research funding was deduced from literatures. Since most clay containing kaolinite, reducing temperature would remove water between the particles, shrinked the particles and increase the stiffness a long with the temperature escalation. Elimbi et al reported the best geopolymer cement formation was between 400-700°C [15]. Addition RHA into the clay matrix reduced the water absorption but the revise properties were reported at higher proportion of RHA (> 10%), consequently less recrystallization then subsequently reduced the pellet strength (Fernando, 2017) [16]. The compatible fibres on the composite aimed to improve the flexibility of composite. Husen et al reported the improvement in tensile strength and modulus clay composite by approximately 8 % and 15 % respectively due to inclusion of raw jute fibres [17]. The objective of using clay in this study was to prepare the compatible pellet at high temperature and aqueous solution but in fact it is found the weaknesses, then clay was excluded from pellet matrix and substituted with PVAc as the adhesive agent. The later composition was stable in water solution but it has low heat resistance. It would decompose above PVAc boiling point that 112°C.

Sulphur immobilization was carried out with and without chitosan that function as stabilizer and adhesive to avoid sulphur particle aggregation and to bind sulphur particles on the surface of pellets. In utilizing chitosan, we used two comparative methods that were stepwise or simultaneous sulphur colloid-chitosan immobilization. In stepwise method, pellets were coated with chitosan and subsequently sulphur was impregnated on the surface of chitosan-coated pellets. The chitosan coating on the surface the pellet was displayed in Figure 2. The mean thinness of coating was 26-89 micron. Colloidal sulphur was immobilized on the surface of pellets that previously coated by chitosan. Chitosan function as binding agent for sulphur particles onto pellet surface and also functioned as stabilizer to disperse sulphur particles and to avoid sulphur particle aggregation. Using chitosan as stabilizer and binding agents has been reported previously in preparation of colloidal Nobel metal nanoparticles [18-20].

IOP Conf. Series: Materials Science and Engineering **334** (2018) 012041 doi:10.1088/1757-899X/334/1/012041

Table 2. Pellets made of various compositions and the stability in water and in chemical environment

Pellets	Number of pellets decomposed (%)				Remarks	
	In water	In chitosan solution	In $Na_2S_2O_3$ solution	In acid solution		
A	0	0	25	30	instable	
B1	100	0	-	-	instable	
B2	0	10	-	-	instable	
B3	0	70	-	-	instable	
B4	0	25	-	-	instable	
C1	0	0	0	100	instable	
C2	0	25	-	-	instable	
C3	0	0	100	-	instable	
D1	0	0	0	50	instable	
D2	0	0	0	10	instable	
D3	0	0	0	0	Stable	continued for further study

Figure 1. Shape and dimension of pellets; (a) & (b) clay pellets, (c) RHA pellet, (d) pellet size.

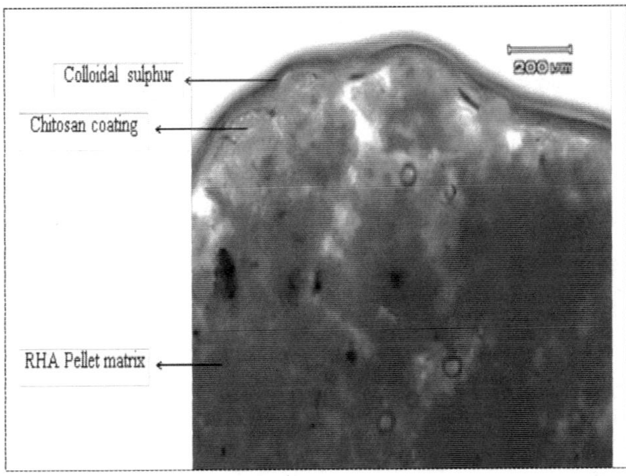

Figure 2. The immobilized colloidal sulphur on the surface of RHA pellets using chitosan as stabilizer and adhesive.

3.2. Effect of sulphur precursor on the RHA-PVAc pellets-immobilized sulphur.

Pellets made of RHA-PVAc were stable in water at least for 24 hour and relatively stand in chemical environment. The effect of different sulphur precursor on the sulphur characteristics was studied using RHA-PVAc pellets. Sulphur immobilized on the pellet surface were from three precursors that were sulphur powder (P-S_{powder}), sulphur dissolved in CS_2 (P-S_{CS2}) and colloidal sulphur from sodium thiosulphate-HCl (P-$S_{colloid}$). It is found that different sulphur precursors gave different colour sulphur-immobilized pellets. P-S_{powder} was yellow, P-S_{CS2} and P-$S_{colloid}$ were yellowish and grey-white respectively.

XRF displayed the sulphur content that follows the trend of P-S_{powder}> P-S_{CS2}> P-$S_{colloid}$ but the weight percentages (wt%) were much lower than found in gravimetric and in EDX methods as shown in Table 3. Probably this is due to sulphur sublimation as reported by Chinchon et al [21]. From the EDX spectrum (Figure 3) randomly scanned on several points, the relative sulphur content was the lowest in P-$S_{colloid}$, which was only 2.55 wt% compared to P-S_{powder} and P-S_{CS2} that were 20.75 and 25.21 wt% respectively. This trend was also consistent with gravimetric data although slightly different in terms of wt% sulphur content. EDX data showed the variance of wt% sulphur content on pellet surface, which was smallest in P-$S_{colloid}$(s^2 = 9.5). The larger data variances was found in P-S_{powder} (s^2 = 1462.7) and smaller variance was in P-S_{CS2} (s^2 = 738.8). This suggested that sulphur particles dispersed better on the surface of P-$S_{colloid}$ compared to the others.

Table 3. XRF data for pellet and coated pellet chemical composition

Pellet chemical components	P-control (%wt)	P-S_{powder} (%wt)	P-S_{CS2} (%wt)	P-$S_{colloid}$ (%wt)
SiO_2	96.6	10.59	90.3	91.72
K_2O	1.51	0.25	1.50	1.54
P_2O_5	0.55	0.10	0.59	0.61
SO_3	0.10	88.44	5.67	2.93
CaO	0.39	0.06	0.34	0.37
MgO	0.29	0.06	0.25	0.29
Na_2O	0.19	0.41	1.56	2.14
Others	0.37	0.09	0.34	0.40

The sulphur powder (S-powder) has bulky particles and such characteristics were brought into pellet surface. Dissolved sulphur in CS_2 have find particles but they aggregated easily on pellet surface soon after CS_2 evaporated. These were different from $Na_2S_2O_3$ sulphur precursor, sulphur was in aqueous colloidal form on the pellet surface. The aggregation slowly occurred since water was not as volatile, therefore sulphur particles were finest compare to other preparation.

Having determined that P-S-$_{colloid}$ gave best characteristic pellets among other sulphur immobilized pellets. Then, the immobilization techniques were compared between simultaneous and stepwise immobilization. The effect of immobilization technique was deduced from the TEM images of sulphur particles in chitosan matrix. To obtain the contrast images of sulphur particles, the mercury vapour was flown through the pellets and it formed black particles of HgS. In TEM sample preparation, the coating layer (chitosan & HgS) was swelled by suspending in 1.5% of aqueous acetic acid to obtain a slurry solution containing those particles. TEM analysis gave different TEM images as displayed in Figure 4. TEM images Figure. 4(a) represents particles in on the surface PS_{powder}. It showed that the diameter of HgS was around 100 nm on the surface of P-S_{powder} but they were 25 nm and 10 nm on the surface of P-$S_{colloid-chi-simultaneous}$ and P-$S_{colloid-chi-setpwise}$ respectively.

In P-$S_{colloid-chi-simultaneous}$ system, it shows that most of HgS particles were covered by chitosan film. This is different from the stepwise techniques that apparently HgS were smaller and more accessible on the surface. These phenomena confirmed the previous publication that chitosan stabilized palladium catalyst prepared by simultaneous immobilization has lower catalytic activity than the stepwise one due to shielding effect of chitosan layer [20]. Illustration of the immobilization technique that affected on the sulphur particles exposure on pellet surface is presented in Figure 5.

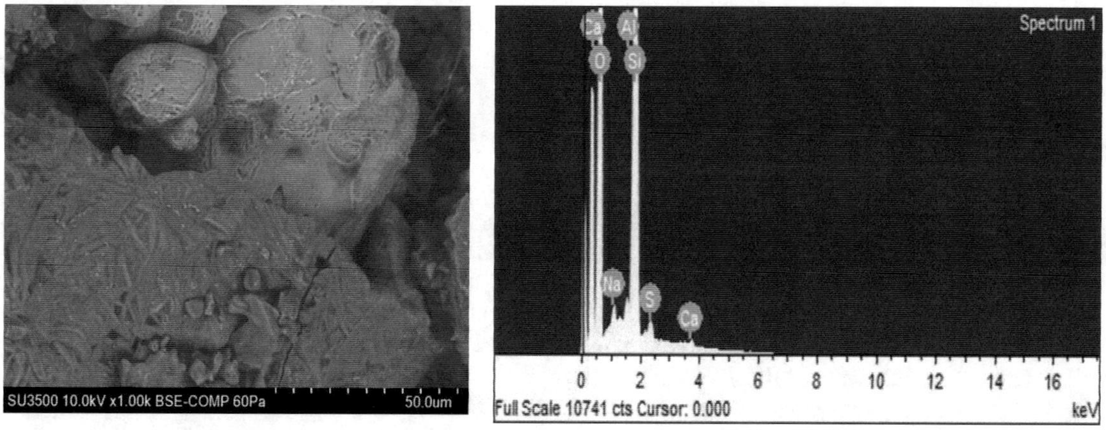

(a)

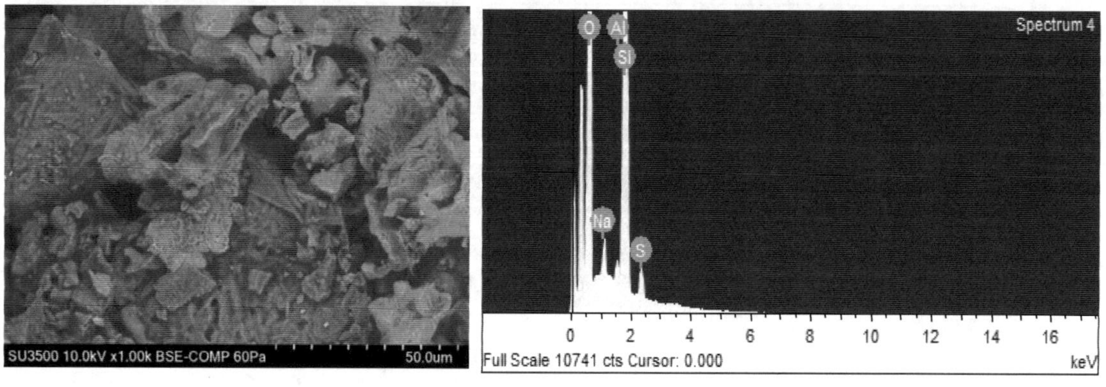

(b)

IOP Conf. Series: Materials Science and Engineering **334** (2018) 012041 doi:10.1088/1757-899X/334/1/012041

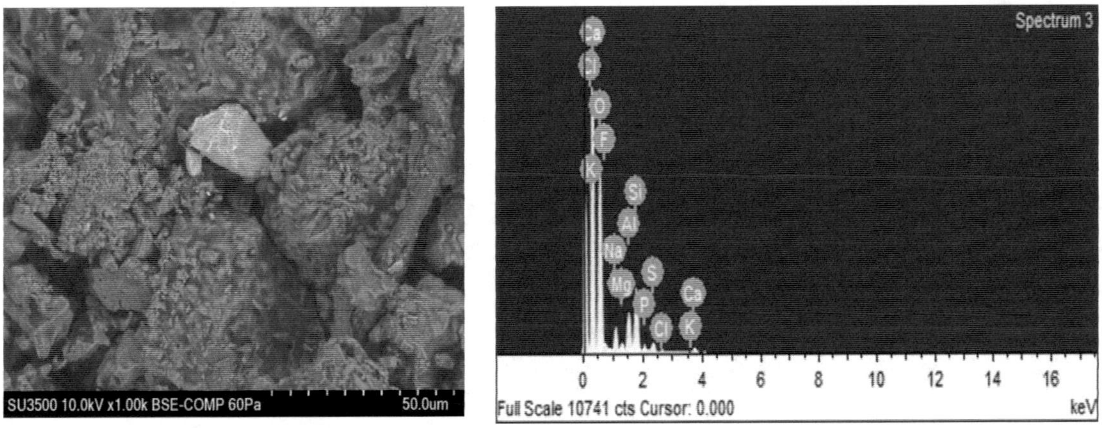

(c)

Figure 3. SEM images and EDX spectra for (a) P-S_{powder}; (b) P-S_{CS2} and (c) P-$S_{colloid}$

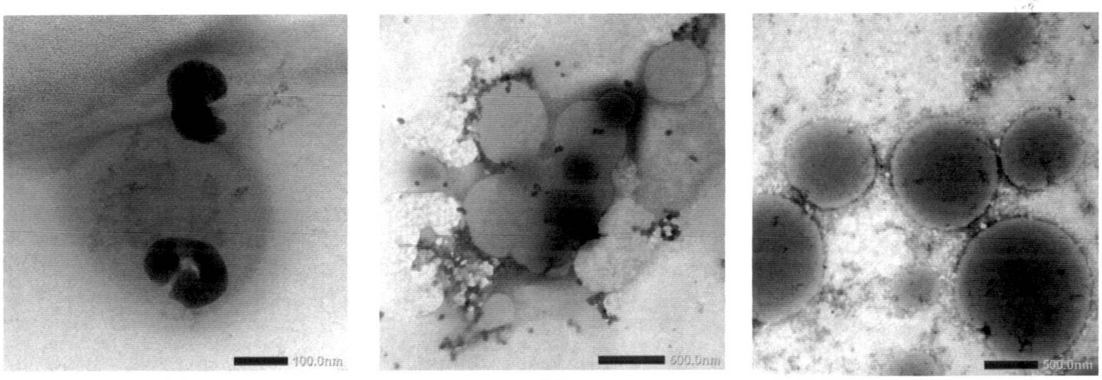

(a) (b) (c)

Figure 4. Representative TEM images comparison of (a) P-S_{powder} , (b) P-$S_{colloid\text{-}simultaneous}$ and P-$S_{colloid\text{-}setpwise}$.

(a) Simultaneous immobilization

$Na_2S_2O_3$(aq) + Chi(aq) + HCl(aq) → immobilized onto pellet →

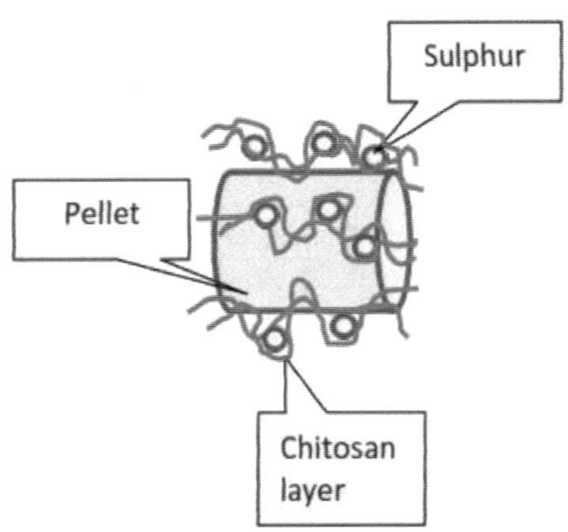

(b) Stepwise immobilization

$Na_2S_2O_3$(aq) + HCl(aq) → immobilized onto chi coated pellet →

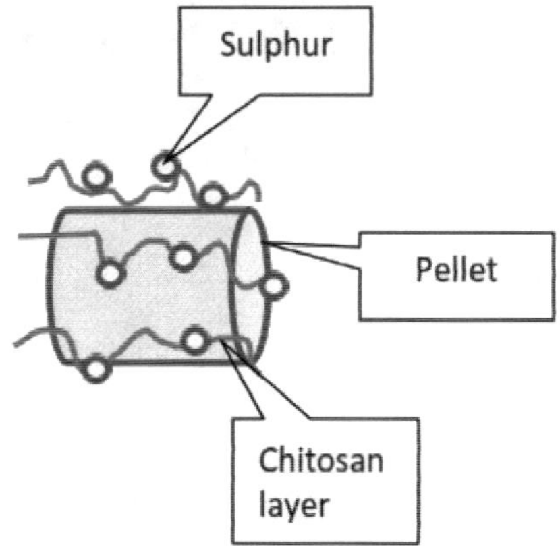

Figure 5. Ilustration of immobilization effect on the sulphur particles explosure. (a) simultaneous immobilization let to sulphur particles covered up by chitosan layers, (b) stepwise immobilization made surface of suphur particles more excessable for futher chemical reaction.

4. Conclusion

Calcination temperature and type of the filler affected on the stability of clay pellets. High calcination and the present of sawdust or pineapple leaf fibres as the filler in pellet matrix enhanced the stability pellets in aqueous solution. Pellets made of rice husk ash and sago starch produced more stable pellets but the best composite in this study was rice husk ash-polyvinyl acetate since it gave much more stable pellets. Type of sulphur precursor has correlation with particle size of immobilized sulphur on the surface of pellets. Sulphur fabricated from thiosulfate-HCl formed colloidal nanosize particles. Colloidal sulphur immobilized on the pellets that previously coated by chitosan showed better dispersion of sulphur particles compared to those that prepared by simultaneous immobilization of colloidal sulphur-chitosan onto pellet surface.

Acknowledgements

This research funded by research grant from Ministry of Research, Technology and Higher Education, Republic of Indonesia with grant registration number of "Penelitian Unggulan Perguruan Tinggi" 121/UN11.2/PP/SP3/2016. We would like to thank several institutions including Laboratories at Syiah Kuala University, Physics laboratory at Universiti Sains Malaysia, TEM Lab of Gadjah Mada University and Physics Laboratory at Indonesian Science Institute (LIPI).

References

[1] Adlim M, Fitri ZA 2015 *AACL Bioflux* 8 656.
[2] Li Y, Wang Y, Li Y, Li T, Mao H, Talbot R, Nie X, Wu C, Zhao Y, Hou C, Wang G, Zhou J, Qie G 2017 *Sci. Total Environ* **574** 1424.
[3] Dziok T, Strugala A, Rozwadowski A, Macherzyn M 2015 *Fuel* **159** 206.
[4] Zarlaida F, Adlim M 2017 *Microchim. Acta* 184 45
[5] Son HY, Lee S, Park SB, Kim MS, Choi EJ, Singh TS, Bae Y, Kwack SJ, Kang TS, Shin HI, Baek MC, Kim SH 2010 *Immunopharmacol Immunotoxicol* 32 277.
[6] Lee KJ, Lee TG 2012 *J. Hazard. Mater* **241** 1.
[7] Rodriguez O, Padilla I, Tayibi H, Lopez-Delgado A 2012 *J. Environ. Manage* **101** 197.
[8] Feng W, Borguet E, Vidic RD 2006a *Carbon* **44** 2998.
[9] Feng W, Kwon S, Feng X, Borguet E, Vidic RD, Asce M 2006b *J. Environ. Eng* **132** 292.
[10] Mullett M, Pendleton P, Badalyan A 2012 *Chem. Eng. J* **211** 212.
[11] Reddy KSK, Shoaibi AA, Srinivasakannan C 2014 *J. Ind. Eng. Chem* **20** 2969.

[12] Adlim M 2006. *Indo. J. Chem* 6 1
[13] Chaudhuri RG, Paria S 2011 J. Colloid Interface Sci **354** 563.
[14] Chaudhuri RG, Paria S 2010 J. Colloid Interface Sci **343** 439
[15] Elimbi A, Tchakoute HK, Njopwouo D 2011 Construct. Build. Mater. **25** 2805.
[16] Fernando PR 2017 Adv. Recycl. Waste Manage. **2** 120.
[17] Hossen MF, Hamdan S, Rahman MR, Rahman, MM, Liew FK, Lai JC 2015 Fibers Polym.**16** 479.
[18] Adlim M, Bakar MA, Liew KY, Ismail J 2004 J. Mol. Catal. A Chem 212 141.
[19] Adlim M, Bakar MA 2013 Kinet. Catal 54 586.
[20] Adlim M 2006 *J. Mat. Sains* 11 125.
[21] Chinchon JS, Lopez-Soler A, Traveria A, Vaquer R 1988 Spectrom **17** 217.

IOP Conf. Series: Materials Science and Engineering **334** (2018) 012042 doi:10.1088/1757-899X/334/1/012042

Investigation properties of superparamagnetic nanoparticles and magnetic field-dependent hyperthermia therapy

Z Hedayatnasab[1], F Abnisa[1,2,3] and W M A Wan Daud[1]

[1] Department of Chemical Engineering, Faculty of Engineering, University of Malaya, 50603 Kuala Lumpur, Malaysia
[2] Department of Chemical Engineering, Faculty of Engineering, University of Malikussaleh, Lhokseumawe 24351, Indonesia

E-mail: faisal.abnisa@gmail.com

Abstract. The application of superparamagnetic nanoparticles as heating agents in hyperthermia therapy has made a therapeutic breakthrough in cancer treatment. The high efficiency of this magnetic hyperthermia therapy has derived from a great capability of superparamagnetic nanoparticles to generate focused heat in inaccessible tumors being effectively inactivated. The main challenges of this therapy are the improvement of the induction heating power of superparamagnetic nanoparticles and the control of the hyperthermia temperature in a secure range of 42 °C to 47 °C, at targeted area. The variation of these hyperthermia properties is principally dependent on the magnetic nanoparticles as well as the magnetic field leading to enhance the efficiency of magnetic hyperthermia therapy at targeted area and also avoid undue heating to healthy cells. The present study evaluates the magnetic hyperthermia therapy through the determination of superparamagnetic nanoparticles properties and magnetic field' parameters.

1. Introduction

Hyperthermia has been considered as a therapeutic method for the treatment of cancerous tumors when a sarcoma had been disappeared after a very high fever [1, 2]. This finding had been the reaction of the immune system towards a bacterial infection. Based on this observation, the fact of the cancer cells is vulnerable to high temperatures and their growth can be terminated at temperatures ranging from 42 °C to 47 °C has been established [3, 4]. Meanwhile, local hyperthermia has been used for such cases that cancers have not spread the whole body in order to avoid side effects as well as enhance the treatment efficiency. However, this technique has not much more effective for deep-seated tumors due to the distribution of heterogeneous temperature and also incapability in preventing overheating nearby healthy cells consequently, superparamagnetic nanoparticles (SPMNPs) have been applied. As a matter of fact, SMNPs have been attracted attention for this cancer treatment due to easy controllability, biological compatibility, physicochemical properties, and superior magnetic properties to enhance the hyperthermia efficiency. Technically, SPMNPs can be injected locally or through the intravascular region within the vicinity of external alternating current magnetic field (ACMF). This technological procedure produces focus-generated heat by the stable colloidal suspensions of the NPs

[3] To whom any correspondence should be addressed.

on the affected cells which is called magnetic hyperthermia or magnetic nanofluid hyperthermia therapy (MNFHT) [5, 6].

This cancer treatment was first experimentally performed in 1957 [7], to cure carcinoma in lymphatic nodes through maghemite (γ-Fe_2O_3) NPs, with 20-100 nm particle size heated between 43 °C to 47 °C by 1200 kHz ACMF. Subsequently, MNFHT has been well established for the application of the SPMNPs in treating prostatic and esophagus cancers and brain and neck tumors beside the investigation of the clinical prophase and animal-based experiment stages of different cancers [8]. The proper functionality of MNFHT depends on two crucial factors; heat power of SPMNPs known as specific absorption rate (SAR) as well as heat temperature called hyperthermia temperature (T_H) which can be achieve by applying the appropriate stable SPMNPs and the ACMF' parameters (frequency and amplitude) [9]. These considerations can provide SPMNPs with high magnetic capability to destroyed cancerous cells while the temperature controlled in the secure range of 42 °C to 47 °C to protect healthy cells being overheated. This present paper tries to investigate the crucial properties of SPMNPs as well as ACMF' parameters that have direct effects on the factors of MNFHT to improve its efficiency.

2. Magnetic nanofluid hyperthermia therapy

Hyperthermia therapy commonly remedies cysts, inflammation and arthritis pains through the raising of the blood flow to deliver nutrition to the affected cells then muscle spasms are released. In regards to the sensitivity of cancerous cells towards the high temperature (above 42 °C) which derive from deficiency of oxygen and subsequently blood flow in their own cells; hence, they are incapable to dissipate the extra heat via the heat transfer phenomenon resulted in being destroyed. Indicating, the viability of cancerous cells will be reduced by increasing the body temperature [10]; however, such increased-temperature is inadequate to destroyed serious cancers. Therefore, strong magnetic mediates required to not only produce satisfactory heat but even focus on the affected area to prevent disorders for patients. Magnetic nanofluid hyperthermia therapy applys SPMNPs as magnetic mediators in the presence of ACMF through the attachment of SPMNPs on the cancer cells while stimulating the body's active immunity.

Figure 1 experimentally shows, magnetic nanofluids are placed in a micro-centrifuge tube in the center of a circular coil exposed to ACMF to measure the temperature changes over the time (dT/dt) via a thermocouple connected to a data processing system in order to record T_H [3, 4]. The desirable case in this procedure is achieving the T_H between 42 °C and 47 °C since SPMNPs within this secure range are prone to burn cancerous cells while the healthy cells are unaffected. Meanwhile, the induction heating power of SPMNPs plays effectively to burn and eradicate and cancerous cells. Hence, determination the factors affect these magnetic hyperthermia parameters will assure to perform a promising cancer treatment such as particle size and shape of SPMNPs, the surface modification of SPMNPs, frequency and amplitude of ACMF.

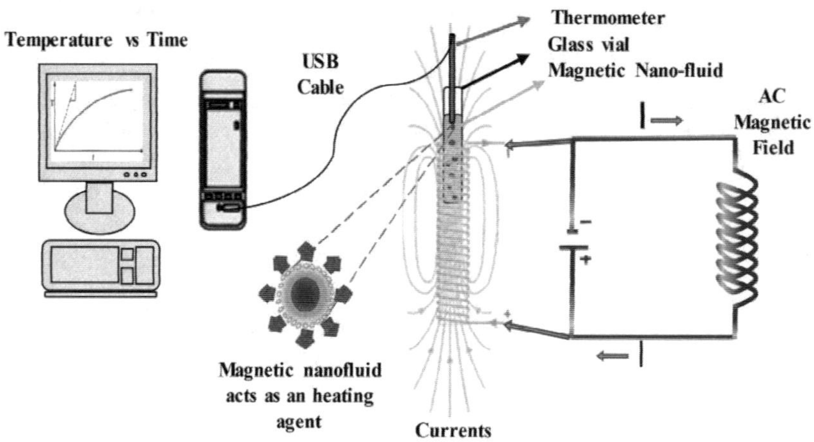

Figure 1. Schematic of experimental magnetic hyperthermia setup [6].

2.1. The parameters of ACMF

The variation of frequency (f) and amplitude (H) of ACMF have directly effects on SAR and T_H values. Generally, the SAR values rise by increasing these parameters and can be beneficial until the temperature being stable in the hyperthermia range; otherwise, disorders may be occurred. Atkinson et al. [11], suggested that there was a criterion for induction field' parameters that should not be exceeded value of 4.85×10^8 A/(ms). This value obtained by the multiplication of the coil diameter of 30 cm, amplitude (A/m) and the frequency of 200 kHz to protect healthy cells from overheating by eddy currents. However, they ignored to consider some measurement in this heating process which can diminish the eddy currents including smaller coils with inhomogeneous fields and off-axis field directions and also SPMNPs as magnetic mediators with Neel and Brownian heating mechanism which are varied from the ones that Atkinson used. This criterion cannot be acceptable; thus, a new criterion proposed [12], that $H \times f = 5 \times 10^9$ A/(ms) considering SPMNPs with small coil diameter.

According to the latter criterion, polyethylene glycol-coated rod-shaped $NiFe_2O_4$ NPs were synthesized with 16 nm length and 4.5 nm diameter by Iqbal et al. [13]. T_H values of coated $NiFe_2O_4$ (8.7 mg/mL) increased by the increase of the amplitude which were 35, 42 and 47 °C in accordance with 2.3×10^3, 3.9×10^3 and 5.5×10^3 A/m with the frequency of 260 kHz. Although coated $NiFe_2O_4$ with 5.5 KA/m amplitude was appropriate to protect the normal cells, low value of SAR, 18 W/g, indicated incapability of heating power to destroyed cancerous cells. Different values of amplitudes (20, 30, 40, 50 and 60× 103 A/m) were performed on 1.5 mg ml−1 chitosan-coated MnFe2O4 NPs at 307 kHz [14]. The SAR and T_H values improved by increasing the amplitude from 57.2 to 278.69 W/g and from 44.10 to 65.38 °C respectively; however, two SAR values of 57.2 and 97.5 W/g that were acceptable for MNFHT measured in 20 and 30 × 103 A/m amplitudes at 307 kHz which were beyond than allowable criterion value.

Sakellari et al. [15], opposed these criteria and believed that those values were inadequate for the biological tissues in clinical. They compared the SAR values of nanorods magnetite NPs in low (12×10^3 A/m and 210 kHz, $H \times f = 2.5 \times 10^9$ A/(ms)) and high (24×10^3 A/m and 765 kHz, $H \times f = 18.3 \times 10^9$ A/(ms)) magnetic field. These values were measured of 10W/g and 759W/g, respectively. The large values of ACMF parameter made nanorod MNPs appropriate candidate for MNFHT which beyond the presented criteria. Similarity, comprehensive experiments performed have indicated that the permissive value of $H \times f$ have been considerably larger than both criteria ranging between 1.8×10^9 A/m to 18.7×10^9 A/m [16, 17]. Consequently, a new criterion of $H \times f$ value is required to perform the promising clinical MNFHT without heating disorders for patients. Interestingly, Salunkhe et al. [18], mentioned the key point that besides the amplitude and frequency, the concentration of SPMNPs plays important role in the variation of the magnetic hyperthermia factors. They observed that the SAR values decreased with increasing concentrations, the trend of T_H values was in the opposite of SAR values due to the enhancement in exchange coupling energy derived from dipole–dipole interactions.

2.2. Coating of SPMNPs
Uncoated SPMNPs display commonly low capability to retain the stability in a colloidal suspension and also make some difficulties during their synthesis and then hyperthermia therapy; consequently, their surface modification is acquired to provide a proper coater to

- Reduce the sensitivity of the SPMNPs surface toward air and moisture and make more appropriate surface for further functionalization and absorbption of proteins.
- Create the hydrophilic molecules on the surface to improve the dispersity of SPMNPs (prevent agglomeration then control the particle size and reduce the risk of blood capillary obstruction) and enhance blood circulation (transport SPMNPs to targeted area).
- Preserve the physiochemical properties of SPMNPs.
- Prevent the opsonization of SPMNPs by reticuloendothelial system causes their fast clearance from the blood stream before reaching the targeted
- Make a biocompatible and nontoxic shield around SPMNPs because their surface is directly in contact with blood and tissues [19-21].

Besides, if the coating layer of the SPMNPs is thin, it can increasingly affect the magnetic as well as hyperthermia properties by forming a new material with enhanced magnetic properties. The main difference between coated and uncoated SPMNPs is saturation magnetization (Ms) which derived from their surface areas. In the latter case, due to their large surface-area-to-volume ratio, the attractive force among them increased and become aggregated (particle size increased), thereby larger block temperature occurred and subsequently Ms and SAR values decreased. Whereas, the block temperature of the former is forced to the lower temperature, leading to boost Ms and then SAR values [21]. Therefore, coated SPMNPs (with high SAR) have stronger heating power to burn cancer cells with less settling time in the human body with lower concentration considered the desirable case for clinical trials. Clinically, acceptable doses of SPMNPs (magnetite) are between 0.56-3 mg Fe/kg human body weight that is greatly less than the normal blood iron concentration (≈33 mg Fe/kg body weight) as well as total body iron (≈3,500 mg) [22].

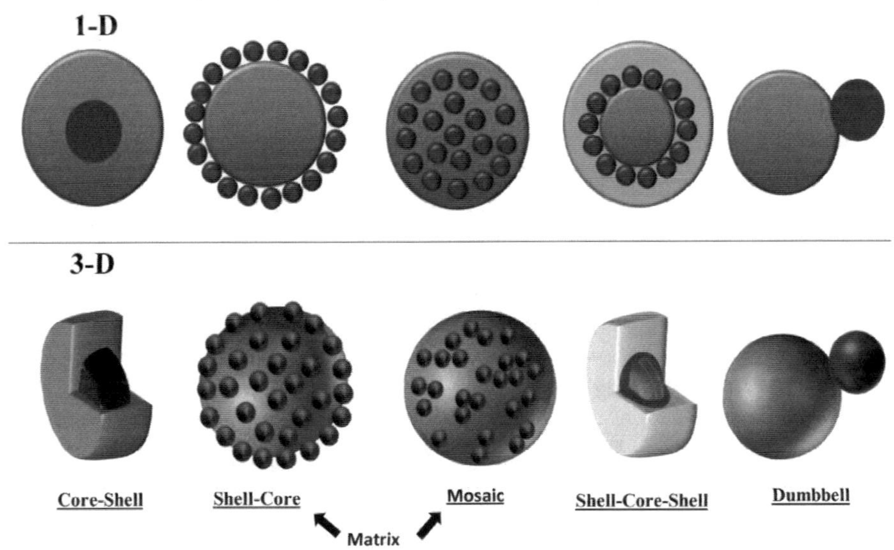

Figure 2. Different structures of coated SPMNPs [6].

A wide variety of materials are applied as a coating agent for SPMNPs can be categorized into two main groups, namely, organic and inorganic. Inorganic coating renders the SPMNPs' surface to bind biological ligands while maintaining the stability of particles to make SPMNPs target specific [19]. The former group functionalizes SPMNPs by functional reactive groups leading to three main supposed structures, core shell [23], matrix (mosaic, shell-core), and shell$_x$-core-shell$_y$ [24], (as illustrated in figure 2). Among various organic coatings, if the shell of such structure is polymer, it may render more benefits like better dispersion of particles, high colloidal stability, solubility, and drug loading on the shell for further treatments. Moreover, polymeric coating promotes biocompatibility, circulation of MNPs in the blood and reduce the toxicity and risk of blood capillary obstruction [25]. Recently, in situ reduction method applied for core-shell Fe_3O_4 with the Ag and polyacrylic acid. SPMNPs Fe_3O_4@Ag and Fe_3O_4-polyacrylic acid showed Ms values of 75.1 and 82.4 emu/g and SAR values of 76 and 87 W/g, respectively. Indicating, the thin polymer coating layer had a great influence in enhancing of MNFHT efficiency [26]. Shete et al. [27], investigated the effects of chitosan on Fe_3O_4 SMNPs. Polymer coating did not significantly influence in the reduction of magnetization. The Ms values were 49.96 and 51.68 emu/g for the particle size of 21.8 ± 5.3 and 15.1 ± 5.0 nm, respectively. This finding implies well dispersition and less agglomeration degree of SPMNPs with thin polymer layer coating. Interestingly, the SAR value of coated SPMNPs boosted to 118 W/g at 26.67 × 10^3 A/m and 265 KHz indicating, in addition to coating, other parameters affect hyperthermia properties. The coated SPMNPs become also saturated in less time as compared with the uncoated ones.

2.3. Particle size and particle shape of SPMNPs

A small particle size leads to high magnetization, as well as high induction heating power, whereas a particle size below 5 nm makes some disorders on the surface spin of NPs consequently, Ms and SAR values decreased. SPMNPs with the optimal particle size of $5<d<20$ nm can influence greatly in increasing Ms and SAR values which related to the Neel and Brownian heating mechanism in single-domain state. Muller et al. [28] compared the single-domain SPMNPs with different particle sizes (10.9, 12.6, and 20.9 nm). It showed that by increasing particle size in single-domain, Ms rose from 64.9 emu/g to 79.9 emu/g, and coercivity and residual magnetism appeared gradually, implying that magnetic behavior is changing from superparamagnetic to ferromagnetic with hysteresis heating mechanism. Moreover, they realized that wide size distribution can negatively influence the magnetic properties because of the particles' statistical orientation. A wide range of particle sizes from 5 nm to 110 nm investigated on the SAR value [29]. While particle size increased moderately from 24 nm to 110 nm, the SAR value decreased from 137 W/g until it become null 1 W/g by reaching the multi-domain state. As matter of the fact when the dispersity of MNPs changes from mono- to poly-, the SAR value decreases due to the reduction in homogeneous particle distribution, which helps in the increase of the total heat generation.

Furthermore, shape anisotropy (spherical, cubic, rod, and facet irregular) is another important parameter in enhancing magnetic as well as hyperthermia properties. Fe_3O_4 SPMNPs with rod and spherical shapes showed high magnetic and hyperthermia; however, the rod-shaped SPMNPs had higher magnetization as compared to the counterpart with the same material and volume [30]. It can be said that this difference in their magnetization related to the shape anisotropy of nanorods avoided them becoming magnetized in directions apart from magnetization along their easy magnetic axis. Spherical shaped-MNPs as compared to the cubic shaped-ones exhibited lower Ms due to the interface coupling energy of spherical shaped was larger than cubic shaped resulted in larger shape anisotropy [31]. This trend was observed in SAR values of the spherical and cubic shaped-MNPs because of different particle morphologies [29]. The variation of magnetic and hyperthermia properties of spherical and facet shaped-$CoFe_2O_4$ NPs with different particle sizes were studied by Joshi et al. [32]. They observed that with increasing particle size in both shapes, the Ms and SAR values increased; whereas, these values of spherical-shaped were significantly smaller than their counterparts which due to the partial attachment of the magnetic moments of the facet shaped-$CoFe_2O_4$ and also severity in their orientation in the direction of ACMF. Hyperthermia temperatures followed the similar trend of above magnetic properties that rose by increasing the particle size of both shapes, while T_H of the facet not only was lower than spherical but also could not reach the allowable hyperthermia temperature range 42 °C to 47 °C.

3. Conclusions

Hyperthermia therapy is considerably developed by applying SPMNPs as heating nano-mediators. Magnetic hyperthermia properties, SAR and T_H, act crucially in the destroying of cancerous cells without damage to nearby healthy cells if the temperature controlled in secure temperature range. The consideration of SPMNPs' properties as well as magnetic field' parameters not only can improve the efficiency of this cancer therapy on the affected cells but also avoid undue heating to healthy cells. Application the coating layer on the SPMNPs' surface meanwhile outweighs the uncoated SPMNPs due to the provided improved properties if the coating layer of SPMNPs is thin. However, a great research is still required in the case of magnetic field' parameters to present a reliable criterion.

Acknowledgments

The authors appreciate the University of Malaya for financial support through the PPP project number PG065-2015B.

References

[1] Busch W 1867 Aus der sitzung der medicinischen *Berl Klin Wochenschr* **13** 137
[2] Coley W B 1891 Ii. Contribution to the knowledge of sarcoma *Annals of surgery* **14** 199-220
[3] Chiriac H, Petreus T, Carasevici E, Labusca L, Herea D-D, Danceanu C and Lupu N 2015 In vitro cytotoxicity of fe–cr–nb–b magnetic nanoparticles under high frequency electromagnetic field *Journal of Magnetism and Magnetic Materials* **380** 13-9
[4] Moroz P, Jones S K and Gray B N 2002 Magnetically mediated hyperthermia: current status and future directions *International Journal of Hyperthermia* **18** 267-84
[5] Khot V M, Salunkhe A B, Thorat N D, Ningthoujam R S and Pawar S H 2013 Induction heating studies of dextran coated mgfe2o4 nanoparticles for magnetic hyperthermia *Dalton Trans* **42** 1249-58
[6] Hedayatnasab Z, Abnisa F and Daud W M A W 2017 Review on magnetic nanoparticles for magnetic nanofluid hyperthermia application *Materials & Design* **123** 174-96
[7] Gilchrist R K, Medal R, Shorey W D, Hanselman R C, Parrott J C and Taylor C B 1957 Selective inductive heating of lymph nodes *Annals of surgery* **146** 596-606
[8] Pang C L K 2015 *Hyperthermia in Oncology*: CRC Press)
[9] Hergt R, Dutz S and Roder M 2008 Effects of size distribution on hysteresis losses of magnetic nanoparticles for hyperthermia *Journal of Physics: Condensed Matter* **20** 385214
[10] Jordan A, Scholz R, Wust P, Fähling H and Roland F 1999 Magnetic fluid hyperthermia (MFH): Cancer treatment with AC magnetic field induced excitation of biocompatible superparamagnetic nanoparticles *Journal of Magnetism and Magnetic Materials* **201** 413-9
[11] Atkinson W J, Brezovich I A and Chakraborty D P 1984 Usable Frequencies in Hyperthermia with Thermal Seeds *IEEE Transactions on Biomedical Engineering* **BME-31** 70-5
[12] Hergt R and Dutz S 2007 Magnetic particle hyperthermia—biophysical limitations of a visionary tumour therapy *Journal of Magnetism and Magnetic Materials* **311** 187-92
[13] Iqbal Y, Bae H, Rhee I and Hong S 2016 Control of the saturation temperature in magnetic heating by using polyethylene-glycol-coated rod-shaped nickel-ferrite (NiFe2O4) nanoparticles *Journal of the Korean Physical Society* **68** 587-92
[14] Oh Y, Lee N, Kang H W and Oh J 2016 In vitro study on apoptotic cell death by effective magnetic hyperthermia with chitosan-coated MnFe(2)O(4) *Nanotechnology* **27** 115101

[15] Sakellari D, Brintakis K, Kostopoulou A, Myrovali E, Simeonidis K, Lappas A and Angelakeris M 2016 Ferrimagnetic nanocrystal assemblies as versatile magnetic particle hyperthermia mediators *Materials Science and Engineering: C* **58** 187-93
[16] Kossatz S, Ludwig R, Dähring H, Ettelt V, Rimkus G, Marciello M, Salas G, Patel V, Teran F J and Hilger I 2014 High Therapeutic Efficiency of Magnetic Hyperthermia in Xenograft Models Achieved with Moderate Temperature Dosages in the Tumor Area *Pharmaceutical Research* **31** 3274-88
[17] Mamiya H 2013 Recent Advances in Understanding Magnetic Nanoparticles in AC Magnetic Fields and Optimal Design for Targeted Hyperthermia *Journal of Nanomaterials* **2013** 17
[18] Salunkhe A B, Khot V M, Ruso J M and Patil S I 2016 Water dispersible superparamagnetic Cobalt iron oxide nanoparticles for magnetic fluid hyperthermia *Journal of Magnetism and Magnetic Materials* **419** 533-42
[19] Salunkhe A B, Khot V M and Pawar S H 2014 Magnetic hyperthermia with magnetic nanoparticles: A status review *Current Topics in Medicinal Chemistry* **14** 572-94
[20] Laurent S, Dutz S, Häfeli U O and Mahmoudi M 2011 Magnetic fluid hyperthermia: Focus on superparamagnetic iron oxide nanoparticles *Advances in Colloid and Interface Science* **166** 8-23
[21] Gupta A K and Gupta M 2005 Synthesis and surface engineering of iron oxide nanoparticles for biomedical applications *Biomaterials* **26** 3995-4021
[22] Huang J, Li Y, Orza A, Lu Q, Guo P, Wang L, Yang L and Mao H 2016 Magnetic Nanoparticle Facilitated Drug Delivery for Cancer Therapy with Targeted and Image-Guided Approaches *Advanced Functional Materials* **26** 3818-36
[23] Ghosh Chaudhuri R and Paria S 2012 Core/shell nanoparticles: classes, properties, synthesis mechanisms, characterization, and applications *Chem Rev* **112** 2373-433
[24] Chomoucka J, Drbohlavova J, Huska D, Adam V, Kizek R and Hubalek J 2010 Magnetic nanoparticles and targeted drug delivering *Pharmacological Research* **62** 144-9
[25] Zhang Q, Lee I, Joo J B, Zaera F and Yin Y 2013 Core–Shell Nanostructured Catalysts *Accounts of Chemical Research* **46** 1816-24
[26] Ding Q, Liu D, Guo D, Yang F, Pang X, Che R, Zhou N, Xie J, Sun J, Huang Z and Gu N 2017 Shape-controlled fabrication of magnetite silver hybrid nanoparticles with high performance magnetic hyperthermia *Biomaterials* **124** 35-46
[27] Shete P B, Patil R M, Thorat N D, Prasad A, Ningthoujam R S, Ghosh S J and Pawar S H 2014 Magnetic chitosan nanocomposite for hyperthermia therapy application: Preparation, characterization and in vitro experiments *Applied Surface Science* **288** 149-57
[28] Müller R, Dutz S, Neeb A, Cato A C B and Zeisberger M 2013 Magnetic heating effect of nanoparticles with different sizes and size distributions *Journal of Magnetism and Magnetic Materials* **328** 80-5
[29] Gonzalez-Fernandez M A, Torres T E, Andrés-Vergés M, Costo R, de la Presa P, Serna C J, Morales M P, Marquina C, Ibarra M R and Goya G F 2009 Magnetic nanoparticles for power absorption: Optimizing size, shape and magnetic properties *Journal of Solid State Chemistry* **182** 2779-84
[30] Mohapatra J, Mitra A, Tyagi H, Bahadur D and Aslam M 2015 Iron oxide nanorods as high-performance magnetic resonance imaging contrast agents *Nanoscale* **7** 9174-84
[31] Salazar-Alvarez G, Qin J, Šepelák V, Bergmann I, Vasilakaki M, Trohidou K N, Ardisson J D, Macedo W A A, Mikhaylova M, Muhammed M, Baró M D and Nogués J 2008 Cubic versus Spherical Magnetic Nanoparticles: The Role of Surface Anisotropy *Journal of the American Chemical Society* **130** 13234-9
[32] Joshi H M, Lin Y P, Aslam M, Prasad P V, Schultz-Sikma E A, Edelman R, Meade T and Dravid V P 2009 Effects of Shape and Size of Cobalt Ferrite Nanostructures on Their MRI Contrast and Thermal Activation *The Journal of Physical Chemistry C* **113** 17761-7

IOP Conf. Series: Materials Science and Engineering **334** (2018) 012043 doi:10.1088/1757-899X/334/1/012043

Synthesis characterisation series of newly fabricated type II CdSe CdSe/CdTe nanocrystals and their optical properties

A S Ahmed[1,3] and W Christopher[2]

[1] Institute of Chemistry, Shah Abdul Latif university Khaipur (Mirs), Pakistan
[2] The School of Chemistry and Material Science, The University of Manchester, Oxford Road, Manchester, M13 9PL, UK.

Email: ashafique2010@gmail.com

Abstract Nanocrystalline semiconductors exhibit different properties due to two basic factors. They possess high surface to volume ratio and the actual size of particle can determine the electronic and physical properties of the material. The small size results in an observable quantum confinement effect, defined by the increasing bandgap accompanied by the quantization of the energy levels to discrete values. In present work we have synthesized the series of cadmium selenide/cadmium telluride (CdSe/CdTe) core/shell and CdSe/CdTe/CdS core/shell/shell to investigate the biexciton energy through transient absorption measurements. These structures are type II nanocrystals are with a hole in the shell and the electron confined to the core. We specifically investigate the effect of nanoparticle shape on the electronic structure and ultrafast electronic dynamics in the band-edge exciton states of CdSe quantum dots, nanorods, and nanoplatelets. Particle size was chosen to enable straightforward comparisons of the effects of particle shape on the spectra and dynamics without retuning the laser source. In our results the Uv-vis showed only a mild redshift in the first excitonic an elongated tail with increasing shell thickness. High resolution Transmission Electron Microscopy (HRTEM) shows the slight agglomeration of the nanocrystals but still the size distribution was calculate able. Spherical small crystals ranging from 5.9 nm to 10 nm are observed. CdTe/CdSe structures were quasi spherical with a rough diameter 6 nm with some little agglomerated structure. . The spherical nanocrystals could be peanut shaped oriented along the c axis or the spherical only, which could explain the two peak emission. p-XRD results indicate the predominant wurtzite structure throughout.

1. Introduction

In last 20 years the quantum dots (QDs) which are also known as the crystals at the nano-scale are major focus of the research in the material science [1],[2],[3]. The luminescence properties of the nano-crystals semiconductor have given the edge over the traditional glowing dyes due to size tuneable emission [4], wide range of the photoexcitation properties [5],[6]. When these properties combine together these produce very good properties which significantly enhance the properties such as photoemission, high fluorescence and considerable battle to photo bleaching [7].

[3] To whom any correspondence should be addressed.

The Quantum Dots have the potential to confine the imagination when the laser light is being thrown, however it depends on the size of nanocrystals because the smaller size NCs can glow more than the bigger size one. The role of NCs becomes more important because of its unique internal heterostructure, these properties make them use in the wide variety of the fields among them are telecommunication, LED, optoelectronics [9] and impressive application in bio labelling [10]. Peculiar properties of semiconductor particles when these are enough smaller than 10 nm in diameter size , these behave someway or other just like the single atom, in spite of being made up of thousands of atoms. These type of the semiconductors are highly applicable in computer chips and lasers, these can be define as the quantum dots [10],[11]. The semiconductor particles ranging from 1-10 nm are refereed as quantum dots and from 10-100 nm are termed as nanomaterials. Any material above to 1000 nm is termed as bulk materials, these does not have properties like the nanomaterials.

The nanocrystals have been changed to water soluble to make them useful for many biological, biochemical and industrial applications. Further the nanoparticles have made its way in the application as photovoltaic cell to increase its efficiency of converting occurrence beam energy into electrical energy. The other major fact which makes the semiconductor NPs useful in the light emitting diodes (LED) is in fact its band gap, which make these materials to transform electrical generated power into light with 30% efficiency [12]. In addition to above described properties there are huge applications in electronic devices like computers to cellular phones [21].

Interestingly, the semiconductor nanomaterial properties are entirely different than their bulk materials. In the bulk materials the electrons have range of the energies which are foud in different energy levels [22],[23]. This is all due to their packing pattern that almost infinite lattice of molecule can be formed leading them to hold the countless number of energy levels. Such type of the energy levels formulates the valance and conduction bands, these are split by the forbidden electron energy known as an energy or band gap.

2. Experimental/Synthesis

2.1 Synthesis of CdSe NCs:

Cadmium Selenide (CdSe) quantum dots (QDs) were grown at relatively high temperature using colloidal synthesis method [13]. CdSe was syntheized using Cadmium Oxide in the mixture of Oleylamaine and Octadecylamine (ODE) and Selenium powder in the elemental form. The summary of the synthesis is given below in detail.

15 mL of oleylamine was degassed under vacuum for 20 minutes at 100 °C. Under the N_2, 0.45 mL of 0.45 Se stock solutions was then added degassed for 20 minutes under vacuum. Under N_2 the temperature was increased to 300 °C, 3 mL of Cd Oleate solution (0.3 m) loaded in syringe was then injected quickly into the reaction flask. The temperature was then set at 280 °C for 25 minutes. Cleaned using 1:1 butanol : methano (degassed anhydrous). Re-dispersed in small amount of hexane (degassed anhydrous).

The CdSe nanoparticles have different diameter ranging from 4 to 12 nm depend on the injected amount of selenium into the solution, obviously this will affect the order to vary the emission wavelengths. A varying range of photoluminescence from 500 to 610 nm can be achieved by varying the TOPSe used and the reaction time as these were two major factors which affects the size of the resultant nanoparticles [14].

2.2 Synthesis of core/shell and core/shell/shell Naocrystals

The synthesis of core /shell CdSe/CdTe and core/shell/shell CdSe/CdTe/CdS were also synthesized and their optical properties investigated using Uv – vis and photoluminescence techniques. The aim of these investigations was to determine the biexciton energy through transient absorption measurements. In this synthesis the CdSe was used a core and the CdTe was grown using both the cadmium and tellurium precursors which were added separately. The summary of synthesis is given below [15].

IOP Conf. Series: Materials Science and Engineering **334** (2018) 012043 doi:10.1088/1757-899X/334/1/012043

2.3. Synthesis of Cadmium Stock solution

Cd stock solution- CdO 0.8 g added to already degassed 20 ml of Oleic acid and 20 ml ODE heated to 240 °C under N_2 to from a clear solution. Tellurium stock solution – Te powder 0.125 g dissolved in already degassed 2.9 ml of TBP (tributyl phosphine) and 9.5 ml of ODE.

The Te stock solution was injected hot to a Cd solution at 300 °C in less than a 2 second. As soon as the Te-solution added to Cd-solution, the colure turned dark brown. It was then cooled to 250 °C for growth of nanoparticles for 20 minutes. Then cooled to 60 °C and added 10 ml of dry methanol, re-dispersed in toluene – methanol cycles to remove any un-reacted solvents. It was finally dispersed in toluene for further use. The CdTe was also shelled from 1 ml thickness to 5 ml thickness to investigate the stability and optical properties of these materials [16].

3. Result and Discussion

3.1 Optical Properties

The Uv – vis and photoluminescence spectra are given in figure (1-3). Figure 1 show the absorption data of CdSe core and CdSe/CdTe core shell with increasing shell thickness. Figure 2 explain the elongated tail with increasing shell thickness and weakening of the first excitonic absorption features.

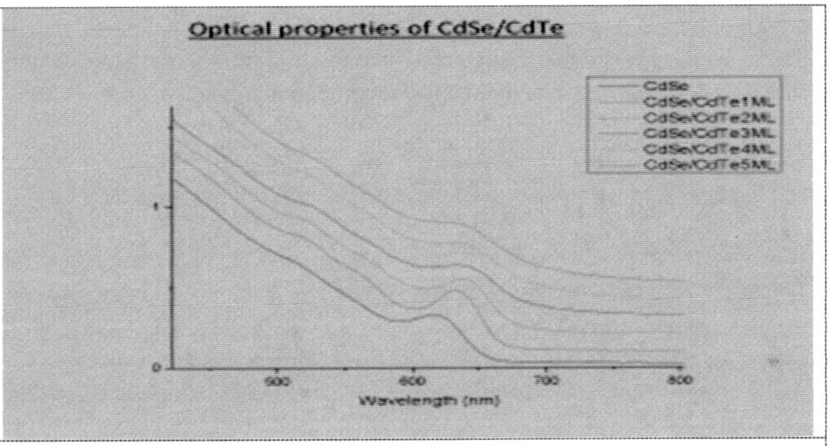

Figure 1. Absorption data of CdSe core and CdSe/CdTe core shell with increasing thickness.

This trend is clear in both figure (1 & 2) the effect of thickness of CdTe shelling from 1 ml to 5ml, the weakening of excitonic peak which predominant in first case.

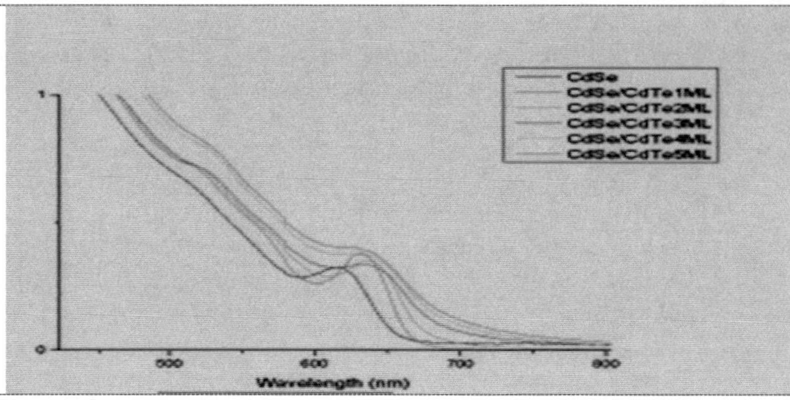

Figure 2. Absorption data of CdSe core and CdSe/CdTe core shell highlighting the elongated tail with increasing shell thickness and the weakening of first excitonic feature.

Figure (3) show the photoluminescence of CdSe core and CdSe/CdTe core shell nanoparticles with increasing shell thickness. It was observed during these photoluminescence studies that when there was low concentration of selenium used there was very low or no emission found. This shows that selenium has important role in luminescence properties of the reported materials [18].

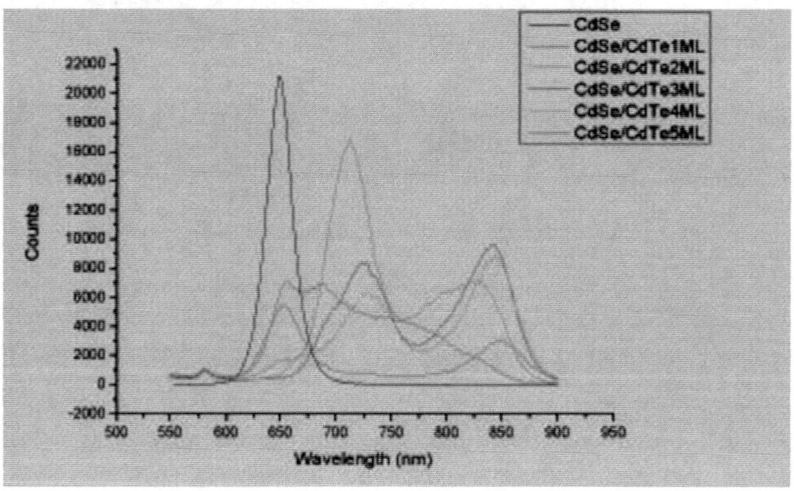

Figure 3. The photoluminescence of CdSe core and CdSe/CdTe core shell.

In UV studies it was further confirmed that in case of CdSE core, CdSe/CdTe core shell with the increasing thickness of CdTe shelling and increasing concentration of selenium the redshift has been observed which show the tendency of these particles to grow larger. By increasing the concentration of selenium, but keeping the concentration time constant, the red shift observed was greater than 70 nm highlighting that this has a larger effect on the size of the nanoparticles when compared with the length of the reaction. Fluorescent intermittency is also known as blinking in emission is common but at the same time it is not process confined to quantum dots [17]. Fluorescent intermittency is observed in dye molecules, fluorescent proteins, single polymer segments, porous silicon and high harvesting complex [17].

3.2 HRTEM and XRD Studies

High resolution Transmission Electron Microscopy (HR-TEM) images are shown figure (4-5). Figure 4 show the low and high images showing CdSe nanocrystals with an average diameter of 5.9 nm. The reaction time is still an important factor in the determination of particle size by TEM.

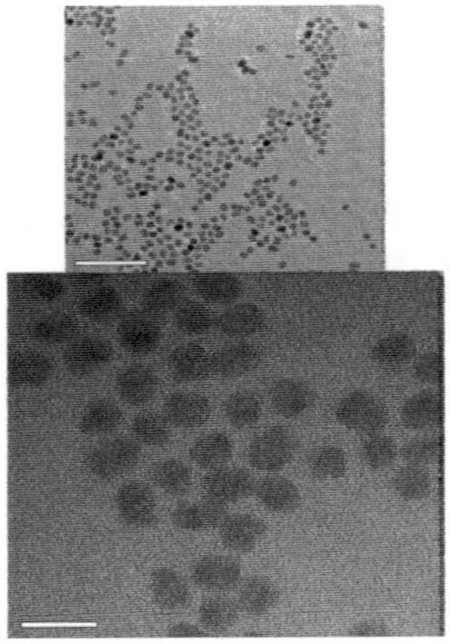

Figure 4. Low and high Resolution Images showing CdSe nanocrystals with an average diameter 0f 5.9 nm.

To ensure that the reaction finished after the desired time, the flask was cooled using a water bath or compressed air. As expected with a short reaction time the particle synthesized was small and luminescent to words the blue region of the visible spectrum [19]. It can be seen in the high resolution image of CdSe the crystals are spherical and peanut size and all the noncoastal are separated from each other.

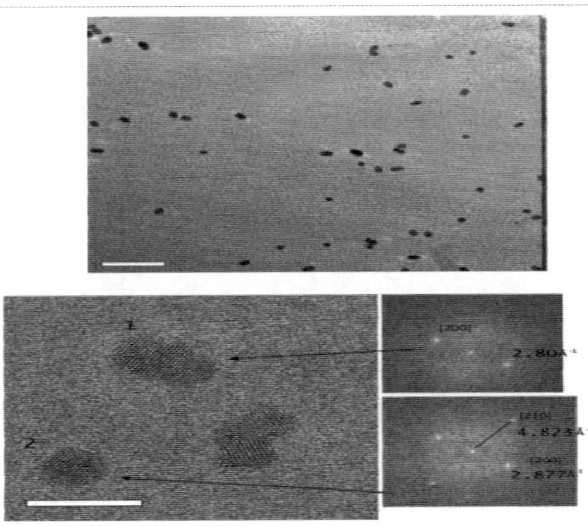

Figure 5. Low and High Resolution images of CdSe/CdTe having peanut and spherical nanocrysts 1.

Shows a probable wurtzite structure showing the [100] lattice 2 shows a probable wurtzite structure There is little chance of agglomeration of these particles. The lattice fringes are very clear which correspond to XRD pattern in figure 6. TEM image of the CdSe/CdTe core shell 5ml thickness are shown in figure 5. These images confirm the spherical and peanut shaped nanocrystals for the core shell nanocrystals. The crystal lattice marked as 1 show probable wurtzite structure showing the [100] and lattice 2 also showing the same type of wurtzite structure. FFT of the high resolution images indicated a wurtzite structure for both the spherical and peanut shaped nanocrystals.

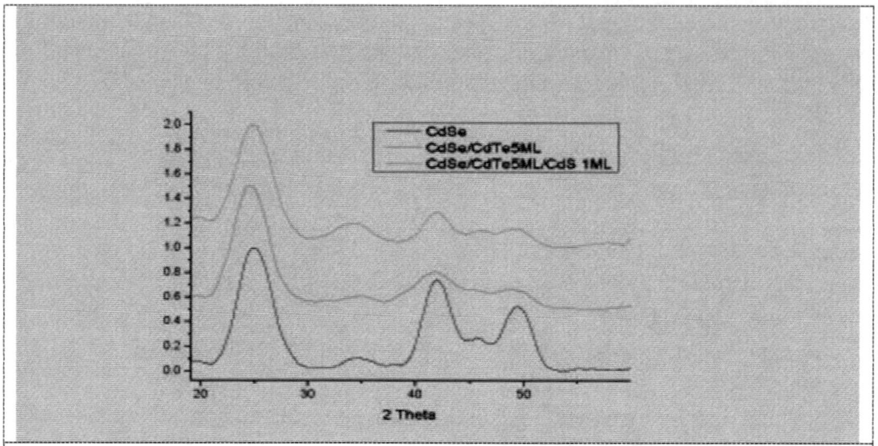

Figure 6. XRD data of CdSe CdSe/CdTe CdSe/CdTe/CdS. All showing the wurtzite crystal structure.

The spherical nanocrystals could be peanut shaped nanocrystals oriented along the c axis or the presence of spherical only, which could explain the two peak emission. The structural pattern was determine using powder X-Ray diffraction (p-XRD) shown in figure 6. CdSe nanocrystal exist in two different crystalline lattice structures can be either wurtzite (w, hexagonal) or cubic [20-21]. The XRd data of CdSe , CdSe/CdTe 5ml/ CdS 1ml all show the wurtzite crystal throughout. The CdS shelling was carried out to protect the CdTe from oxidation.

4. Conclusion

Absorption and photoluminescence measurements showed a redshift of the exctonic peak and PL emission. The PL emission red shifted from a core PL of 595 nm to 650 nm for the shelled structure. The HRTEM images of CdTe showed the nanocrystals to be peanut shaped with a wurzite lattice. The wurzite peanut shaped structures are predominant in p-XRD results. Absorption and Photoluminescence measurements showed a slight blue shift from a core emission of 650 to 642 nm. The blue shift is typically indicative of alloying. Therefore it could be concluded that this structure could be producing quasi type –II structure whilst also having strain alloying effects to actually increase the conduction band-gap of the core.

High Resolution Transmission Electron Microscopy (HRTEM) was conducted on previously dried solutions and showed little agglomerations of nanocrystals howver it was not difficult made getting a size distribution. There were predominantly small spherical crystals 5.9 nm for the CdTe core but quite a few large nanocrystals possibly agglomerated nanocrystals in excess of 10 nm.

The CdTe/CdSe structures were quasi spherical with a rough diameter 6 nm with some little agglomerated structure. The spherical nanocrystals could be peanut shaped oriented along the c axis or the spherical only, which could explain the two peak emission. p-XRD results indicate the predominant wurtzite structure throughout.

5. Summary

Photoluminescence properties, size and shape of the nanocrystals play an important role in deciding nanomaterial for different applications and uses. Core/shell, core/shell/shell nanocrystals of CdSe, CdTe and CdS synthesized, characterized by Uv – Spectroscopy, Photoluminesnce Spectroscopy (PL), Transmission Electron Microscopy (TEM) and X-Diffraction (XRD) techniques. Result shows with increasing thickness of CdTe shell, the redshift is observed in Uv –studies which is indicative of tendency of these particles to grow larger. Most of the nanaocrystals synthesized are ranging in between 5.9 nm to 10 nm having peanut shaped wurzite structures.

IOP Conf. Series: Materials Science and Engineering **334** (2018) 012043 doi:10.1088/1757-899X/334/1/012043

References

[1] A. D. Yoffe, Adv. Phys. 42, 173 (1993)
[2] A. L. Efros, M. Rosen, Phys. Rev. Lett., 1997, 78, 1110 – 113.
[3] A. P. Alivisator, J. Phys.Chem. 100,13 226 (1996)
[4] A.L. Efros, M.Rosen, M. Kunro, M. Nirmal, D.J. Norris, M.G. Bawendi, Phys, Rev. B., [5]1996, 54, 4843 – 4856. **(51)**
[6] Blackman et al. chemMater. 2007, 19, 3815 -3821.
[7] D. Patidar, R. Sharma, N. jani, T. P. Sharma, N. S. saxena. Bull. Mater. Sci. 29, 21 (2006)
[8] E.L. Bentzen, F. House, T. J. Utley, J. E. Crowe, D.W. Wright, Nanolett., 2005, 5, 591 -595.
[9] H. Weller, Angew, Chem. Int, ed. Engl.32,41 (1993)
[10] J. W. Choi, A. Bhupathiraju, M. A. Hassan, J. M. Iannon, J. Cryst. Growth 255, 1 (2004).
[11] Jennifer M. Zemike and Justin Franz, Journal of Chemical Education, 2016, 93 (4) 747 – 752.
[12] Mohammad et al. J. Phys, Chem, B, 2005, 2005, 109, 10533 -10537.
[13] Paul O' Brien, N. L. Pickett, T. Trindad, Chem . Mater., 2001, 13, 3843 – 3858.
[14] Priciipales and Application of Thermal analysis, Paul Gobbot, John Wiley and sons, Oxford – UK, 2007.
[15] R. W. Brikmire, B. E. McCandless, S. S. Hegedus, Solar Energy 12, 15 1992.
[16] R.J Cook, H.Kimble, Phys, Rev, Lett,. 1985, 54, 1023 – 1026.
[17] R S. Agilan, M. Thambidurai, N. Murugan, Chalcogenide Letters Vol. 6, No. 4, 171 (2009)
[18] S. K. Das, Sol. Energy Mat. Sol. Cells, 29, 277 (1993) (Both are the same)
[18] S. Kundu and L. C. Olsen, Thin Solid films, 471, 298 (2005).
[19] Tunable visible emission of TM – doped ZnS quantum dots, Eur. Phys. J. Appl. Phys, 2012 , 59 10404.
[20] U. Gangopadhyay, K. kim, D. Mangalaraj and Yi, Appl. Surf. Sci., 230, 364 (2004).
[21] W. I. Danaher, L. E. Lyons, and G. C. Morris, Sol. Energy Mater. 12, 137 (1985).
[22] Y. Wang and N. Herron, J. Phys. Chem. 95, 525 (1991).
[23] Zong et al., J. Phys. Chem. C, 2007, 111, 526 -531.

IOP Conf. Series: Materials Science and Engineering **334** (2018) 012044 doi:10.1088/1757-899X/334/1/012044

Identification of Carbon loss in the production of pilot-scale Carbon nanotube using gauze reactor

P P D K Wulan[1,2] , W W Purwanto[1], N Yeni and Y D Lestari[1]
[1] Chemical Engineering Department, Universitas Indonesia, Depok 16424, Indonesia

E-mail: wulan@che.ui.ac.id

Abstract. Carbon loss more than 65% was the major obstacles in the Carbon Nanotube (CNT) production using gauze pilot scale reactor. The results showed that the initial carbon loss calculation is 27.64%. The calculation of carbon loss, then, takes place with various corrections parameters of: product flow rate error measurement, feed flow rate changes, gas product composition by Gas Chromatography Flame Ionization Detector (GC FID), and the carbon particulate by glass fiber filters. Error of product flow rate due to the measurement with bubble soap gives calculation error of carbon loss for about ± 4.14%. Changes in the feed flow rate due to CNT growth in the reactor reduce carbon loss by 4.97%. The detection of secondary hydrocarbon with GC FID during CNT production process reduces carbon loss by 5.14%. Particulates carried by product stream are very few and merely correct the carbon loss about 0.05%. Taking all the factors into account, the amount of carbon loss within this study is (17.21 ± 4.14)%. Assuming that 4.14% of carbon loss is due to the error measurement of product flow rate, the amount of carbon loss is 13.07%. It means that more than 57% of carbon loss within this study is identified.

1. Introduction

The broadening use of Carbon Nanotube (CNT), especially within the purpose of research and commercial leads to the need of producing CNT on a mass scale. CNT is applied as one of the components in nano-technological electronic equipments, sensor, electrodes, semiconductors, fuel cells, within the sector of health, and research. Seeing the big potential for CNT applied in various fields, research and investment continue to be developed to be able to synthesize CNTs worldwide [1]. In Indonesia, the mass production of CNT is considered to be in the development process meaning that within a few years it will become a new promising business opportunity in which the needs of CNT for the purpose of research and industry can be met from the domestic production. Related to the CNT production, the Department of Chemical Engineering of Universitas Indonesia has successfully developed the study as well as the production of CNT on a pilot scale [2], [3], and [4].

CNT is synthesized using a gauze-structured catalyst reactor through decomposition reaction of catalytic methane. Methane is chosen as a reactant due to high ratio of hydrogen/carbon [5]. This reactor system is designed with the capacity of 1kg/day CNT production which is scaled up based on a gauze-structured laboratory-scale catalyst reactor [6]. On a small scale, CNT produced is far from the

[2] To whom any correspondence should be addressed.

target of 1 kg per day; as a matter of fact, the production of CNT is only 240 gram per day [2]. This is due to the instability of feed flow rates of the reactor which is prone to leakage. Reevaluation as well as improvement regarding the pilot-scale reactor has been held [3]. The increase of the CNT production, however, does not show any significance (307 gram/day). CNT production is sometimes restrained due to catalyst factor [7].Using the same reactor, the replacement of the catalyst with Fe/MgO and Ni/MgO is conducted in order to obtain CNT with better quality and quantity [4]; nevertheless, the production of CNT decreases compared to the one using Ni-Cu-Al as the catalyst. Further evaluation shows that carbon loss during the synthesis of CNT using the system of gauze-structured catalyst reactor is considerably high, more than 65% of the carbon mass is unidentified (Figure 1). This leads to the structured catalyst reactor designed for the production of CNT 1 kg/day is not applicable commercially.

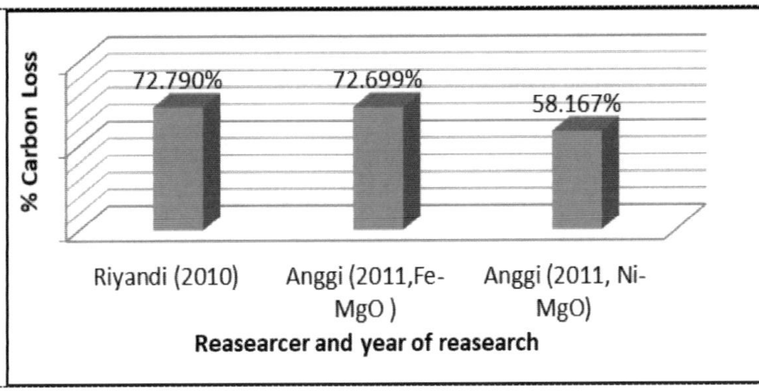

Figure 1. Carbon loss on pilot scale CNT production using gauze reactor.

The cause of the high number of carbon loss during the synthesis of CNT remains unknown. An error during the process of collecting data, especially during the process of measuring the feed flow rates is allegedly considered to have caused the carbon loss; notwithstanding, further study regarding that is yet to be conducted. On the other hand, the change on the feed flow rate due to the increase of resistance for the presence of catalyst and the growth of CNT in the reactor need further examination as it affects the calculation of carbon loss. Another possibility is that the release of CNT as particulates during the process of production and handling becomes another issue causing the magnitude of carbon loss [8]. The uncertain conversion of methane into other compounds such as ethane, ethylene, and other hydrocarbons becomes another possible factor. A number of studies report that site-products in the form of volatiles of organic compounds (VOC) and polycyclic aromatic hydrocarbon (PAH) are formed during the synthesis of CNT with ethylene as the feedstock [9]. As for methane with other hydrocarbon concentration as the feedstock, namely ethylene and ethane which are much smaller than H_2 and CH_4; however, this gives an impact towards the mass of carbon converted while the reaction takes place. The unidentified compounds during the synthesis of CNT result in the considerably high number of carbon loss.

Therefore, this study aims at identifying the carbon loss during the synthesis process of CNT using a pilot-scale structured catalyst reactor. Carbon loss is identified by analyzing the error measurement of the product flow rate on the bubble soap and the change of the feed flow rate because of the presence of catalyst and growth of CNT within the reactor, the analysis regarding whether or not carbon particulates exist in the product stream using gravimetric method where carbon particulates are filtered using filters made out of glass fiber, and the analysis of the composition of the reactor output product using DC-FID and GC TCD. The success of this study will become a stepping stone in increasing the production of CNT using gauze-structured catalyst reactor. As the identification of carbon loss during the synthesis of CNT brings to light, it means that the improvement of the reactor in order to achieve the production target of 1 kg/day is able to be conducted. Such process is ready for the commercial purpose on an industrial scale so that it is able to mass produce CNT in Indonesia so that it meets the domestic needs of CNT for research and industrial purpose.

2. Experimental Method

This research will be conducted in Chemical and Natural Product Engineering Laboratory (CNPE), Department of Chemical Engineering, UI. The first step of the research is analyzing the error measurement of flow rates using bubble soap. Nitrogen gas is flowed using a mass flow controller (SIERRA Smart-trak series) with the flow rate set at 2.5L/min, then conducting a gas sampling to bubble soap. The next step is to measure the actual flow rate of the gas for 9 times using different volumes of bubble soap: 100 ml, 150 ml, and 200 ml, 3 times of sampling on each bubble soap. Further step is to repeat the above steps for the flow rate set at 1 L/min, 2 L/min, and 3 L/min and then analyzing the error during collecting the data of the flow rate using bubble soap.

For CNT production, 18.29 gram of $Fe(NO_3)_3.9H_2O$, 0.9 gram $(NH_4)_6Mo_7O_{24}.4H_2O$, 150 gram $Mg(NO_3)_2.6H_2O$, and 76.29 gram of citric acid is dissolved in 100ml of deionized water; then being sonicated for 40 minutes and being stirred for 3 hours until forming into gel. The catalyst is calcined for an hour at 800 ^oC; then the catalyst powder is grounded and dissolved in isopropanol to be the coat of the wiremesh 304 which is in the form of structured configuration using the method of spray coating. Wiremesh 304 is type of non-magnetic stainless steel that contain in the range of 18-20% chrome, and nickel in the range of 8-12% with additional elements [10]. The catalyst substrate is loaded in to a quartz tube equipped with a digital atmospheric furnace. The series of the research is available in Figure 2. Glass fiber filter (Pall Corp. A/E type with the diameter of 4.7 cm), which has been weighed, is installed in-line within the circuit in order to filter carbon particular carried within the gas flow.

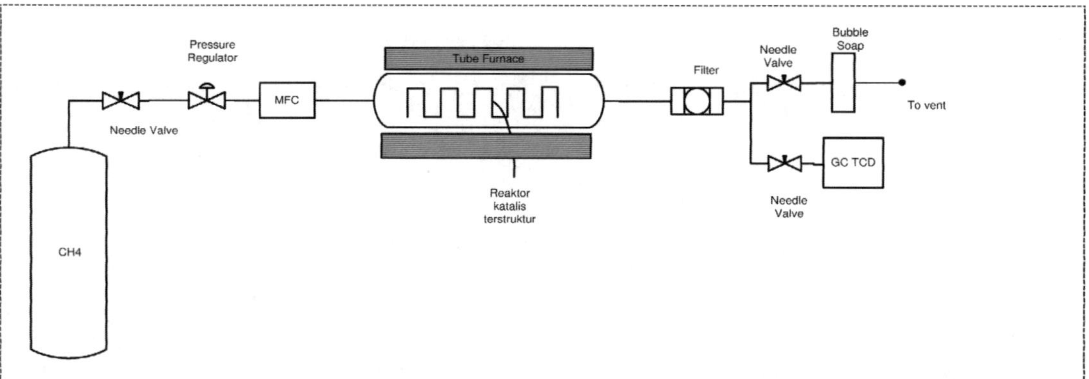

Figure 2. Schematic reactor equipment.

Prior to the synthesis of CNT, calibration is conducted between gas chromatography with Thermal conductivity detector (GC TCD), gas chromatography with Flame ionization detector (GC FID) and massflow controller. GC calibration is performed in order to determine the retention time and response factor (RF) of each gas which is going to be identified, while the flow meter calibration aims to adjust the listed values of the flow rates in the flow meter with the actual gas flow rate measured by bubble soap. Leakage check is performed by draining nitrogen gas to the circuit.

The catalyst, then, is heated up to 850 ^oC. Once the heated catalyst meets the temperature, H_2 gas with the flow rate of 240 L/h is discharged for 10 minutes in order to reduce the oxides of the catalyst. The reaction begins to take place by canalizing methane gas (purity level 99.9%) into the reactor. Methane flow rate set in the mass flow controller is 2,414 L/min. During the reaction process, the product flow rate is measured using bubble soap, while the output gas composition is analyzed using GC TCD (8A type Shimadzu), Porapak Q column packing with Argon gas carrier to determine the composition of CH_4, H_2, CO, and CO_2. While, in order to find out the presence and the composition of secondary hydrocarbon, GC FID (Shimadzu 2014), Porapak N column packing with Helium gas carrier are used. After the process of reaction, the result of CNT is weighed using a digital scale. Glass fiber filters are also weighed and analyzed using SEM EDX (Scanning Electron Microscopy Energy Dispersive X-Ray) in order to find out regarding particulate compositions. The calibration of feed rate (methane) is conducted for three times in order to comprehend the effect of catalyst and CNT growth in the reactor towards the feed flow rate. The first one is conducted when the reactor is empty, the second one is done when the catalyst is loaded into the reactor, and the last one is conducted after the reaction process in which CNT will have grown within the reactor. The calibration is held using methane gas set at the same set point with the mass flow controller (2,414 L/min) and under the same condition (P = 1 atm, T = 27 °C).

3. Result and Disscusion

3.1 Initial Calculation of Mass Balance and Carbon Loss

Several studies have been conducted as shown in Figure 1. The result showed that the average carbon loss reached 68%. This is due to the instability of feed flow rates of the reactor which is prone to leakage. The calculation of the initial mass balance and carbon loss only involves the raw data obtained during the reaction process: input feed rate, product rate measured using bubble soap, and product gas composition identified by only Gas Chromatography Thermal Conductivity Detector (GC TCD) before any correction using identification method undertaken within this study. The input of feed flow rate is 1,266.84 mL/min for the first 7 minutes then constantly set at 2,919.61 mL/min until the reaction process lasts for 70 minutes. The actual CNT mass obtained is 19.6 grams. Based on the experimental data, the input carbon mass rate, the output carbon mass rate, and the theoretical CNT mass formed can be seen in Figure 3.

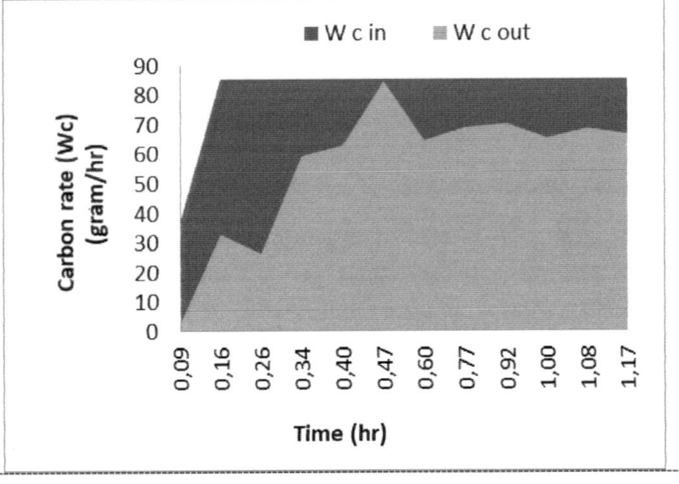

Figure 3. The carbon mass rate in and out during the reaction.

$M_{CNTtheoretical}$ is the same as dark area in Figure 3 which is 27.09 grams. $M_{CNT\ actual}$ obtained from the experiment is 19.6 grams. The initial carbon loss can be evaluated using equation 1.

$$\%\ \text{Carbon loss} = \frac{M_{CNT\ Theoretical} - M_{CNT\ actual}}{M_{CNT\ Theoretical}} \times 100\% \quad (1)$$

IOP Conf. Series: Materials Science and Engineering **334** (2018) 012044 doi:10.1088/1757-899X/334/1/012044

Based on equation (1) the initial carbon loss is 27.64%. These result showed that the initial carbon loss requires correction using aforementioned correction factors.

3.2 The effect of product flow rate measurement error using bubble soap towards carbon loss

Based on the result of the study, the value of relative error towards flow rate measurement using bubble soap is available in Figure 4.

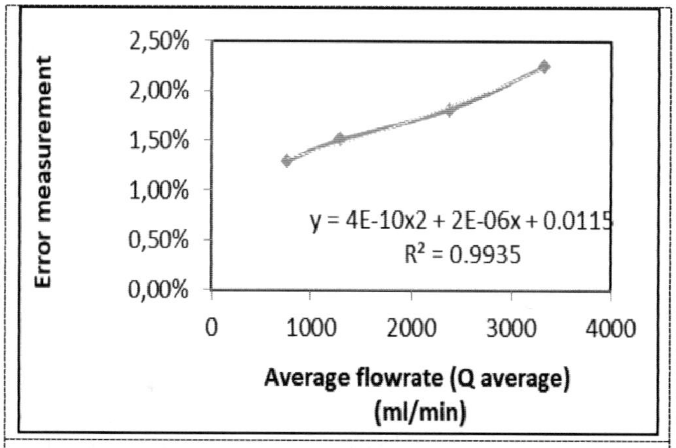

Figure 4. Relative measurement error of flow rate.

Errors during the measurement of product flow rate cause further errors in the calculation of carbon loss. From the calculation in Table 1, the error commonly occurs in the calculation of carbon loss since the measurement of product flow rate using bubble soap is 4.14%, so the value of carbon loss is formulated as follows:

$$\% \text{ Carbon Loss} = \frac{\int_{t_0}^{t}\left[W_{C_{in}} - \left(Q_{product}\cdot y_C\cdot \frac{P.MR_C}{R.T}\right)\right]dt - M_{CNT\ actual}}{\int_{t_0}^{t}\left[W_{C_{in}} - \left(Q_{product}\cdot y_C\cdot \frac{P.MR_C}{R.T}\right)\right]dt} \text{x}100\% \qquad (2)$$

From equation (2) the amount of carbon loss is $(27.64 \pm 4.14)\ \%$

Table 1. Result of error calculation of product flow rate to carbon loss

Time (minute)	**$Q_{product}$ (mL/min)**	***relative error***	***absolut error* (ml/min)**	***error Carbon loss***
0.00	0.000	0.00%	0.00	
5.48	2,188	1.78%	38.94	3.56%
9.42	3,856	2.52%	97.03	4.50%
15.46	3,102	2.16%	66.89	4.71%
20.42	3,084	2.15%	66.26	4.30%
24.00	3,044	2.13%	64.83	4.28%
28.00	3,017	2.12%	63.92	4.25%
36.00	2,798	2.02%	56.62	4.14%
46.00	2,839	2.04%	57.94	4.06%
55.00	2,855	2.05%	58.47	4.09%
60.00	2,602	1.94%	50.54	3.99%
65.00	2,704	1.98%	53.64	3.93%
70.00	2,602	1.94%	50.54	3.93%

Carbon loss Error (average)	4.14%

It can be seen from Table 1, the calculation is merely based on errors caused during the measurement of product flow rate. The error occurred during the process of weighing the actual mass of CNT and the measurement of product gas composition can be removed since the value is smaller compared to the error during the measurement of product rate.

3.3 The effect of feed flow rate changes due to CNT growth in the reactor towards carbon loss

In order to determine the effect of catalyst and CNT growth towards the change of feed flow rate. Feed flow rate calibration was done when the reactor is empty and reactor load with catalyst substrate. Changes occurred in the feed flow rate affect the amount of carbon loss. The result of the feed flow rate calibration is shown in Figure 5. Catalyst substrate inside the reactor will affect carbon loss value that can be calculated using equation 3.

$$\%\ \text{Carbon Loss} = \frac{\int_{t_0}^{t}\left[\left(Q_{CH_4\,feed}\cdot\frac{P.MR_c}{RT}\right) - W_{cout}\right]dt - M_{CNT\ actual}}{\int_{t_0}^{t}\left[\left(Q_{CH_4\,feed}\cdot\frac{P.MR_c}{RT}\right) - W_{cout}\right]dt} \text{x } 100\% \qquad (3)$$

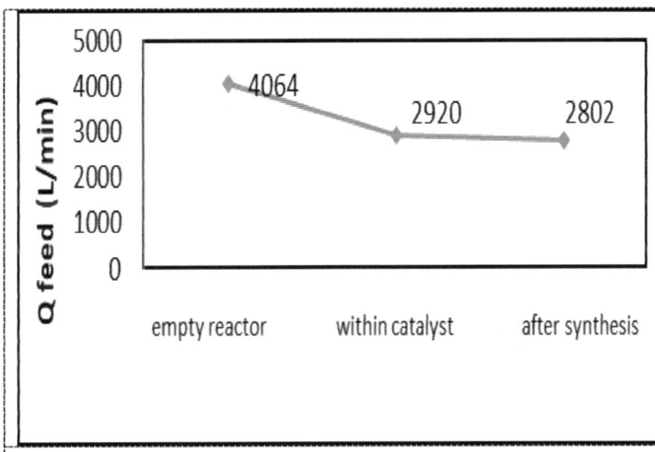

Figure 5. Calibration of feed flowrate in 1 atm and 27 °C condition.

Figure 5 shows that the addition of catalyst substrate within the reactor results in the decrease of the feed flow rate to 1,144.5 mL/min or 28.16% compared to the flow rate in the empty reactor. The decrease of flow rate is caused by the resistance of wire mesh that leads to the rising of feed gas pressure. The increasing pressure of the feed gas then leads to a greater mass of the gas that the gas flow rate turns smaller. The result of calibration after synthesis shows a decrease in the feed flow rate of 118 mL/min. The decrease of the flow rate is caused by the resistance due to CNT growth in the catalyst substrate. The large amount of the resistance as a result of CNT aligns to the growing mass of CNT. The existence of CNT, will increase the resistance that leads to the decreasing of feed flow rate to 4.04%.

Within this research, if the feed flow rate used is the calibration with empty reactor, the theoretical mass of CNT is 63.51 gram, while the actual mass of CNT is only 19.60 gram meaning that the accumulation of the carbon loss is 69.14%. This result is in line with the previous studies which have average carbon loss of 68%. The result of carbon balance calculation after correction of feed flow rate reduce the initial carbon loss from 27.64% to 22.66% or decrease by 4.97% without any correction.

Accordingly, calibration using empty reactor is not applicable for this study. Therefore, it is required to calculate the initial mass balance and carbon loss using the feed flow rate after correction within catalyst inside the reactor.

3.4 Analysis of the reactor output gas with GC FID and its effect towards carbon loss calculation

From the experimental results, in the first few minutes, it is only hydrogen and methane gas detected by GC TCD in which the hydrogen concentration is much greater than methane. In 28th minute, CO and CO_2 emerge with high concentration: 36.43% for CO and 46.22% for CO_2. The presence of these two gases during the reaction process is caused by the imperfect catalyst reduction. The reaction is as follows [11]:

$$6CH_4 + 6Fe_2O_3 \rightarrow 12Fe + 3C_{(solid)} + 3CO_2 + 12H_2O$$
$$6CH_4 + 5Fe_2O_3 \rightarrow 10Fe + 3C_{(solid)} + 3CO + 12H_2O$$

Moreover, at the 28th minute, hydrogen loss becomes considerably high compared to other times. This is due to the fact that H_2O is not detected by both GC TCD and GC FID. The following from top to bottom shows peaks appearing in the minute of 5.48, 36, and 70 when product gas characterization with GC FID:

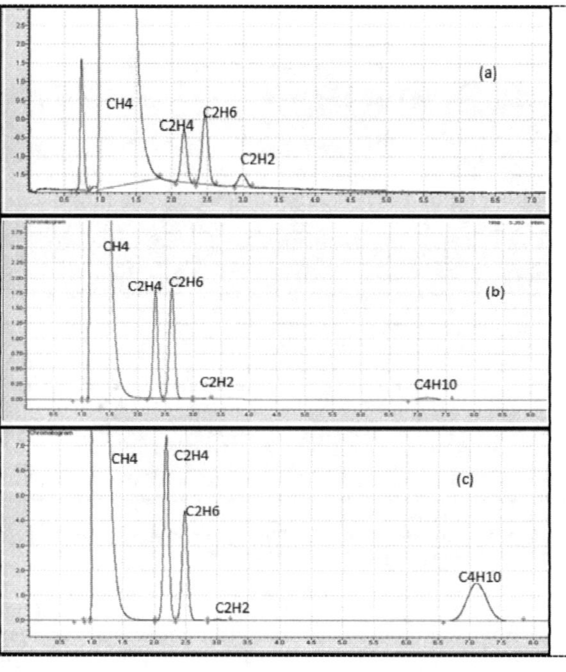

Figure 6. GC FID Chromatograph of gas product (a)6 minute, (b)36 minute, and (c) 70 minute.

The result of characterization with GC FID in Figure 6, shows that during the process of methane decomposition, secondary hydrocarbon which is not detected by GC TCD is formed. Hydrocarbons detected by GC FID are CH_4, C_2H_2, C_2H_4, C_2H_6, and C_4H_{10}. In the first few minutes, there is no C_4H_{10}. Only after the 20th minute, C_4H_{10} begins to form in which its concentration keeps increasing. The formation of such secondary hydrocarbon is due to the reaction process of coupling (merging) on the surface of the catalyst [11]:

$$2CH_4 \rightarrow C_2H_4 + 2H_2 \rightarrow 2C_{(solid)} + 4H_2$$
$$2CH_4 \rightarrow C_2H_2 + 3H_2 \rightarrow 2C_{(solid)} + 4H_2$$

$2CH_4 \rightarrow C_2H_6 + H_2 \rightarrow 2C_{(solid)} + 4H_2$
$C_2H_6 + C_2H_6 \rightarrow C_4H_10 + H_2 \rightarrow 4C_{(solid)} + 8H_2$

The concentration of secondary hydrocarbon continues to increase as the time of reaction process continues. However, the number of secondary hydrocarbon formed is not very high. The average concentration is 1.44%; the highest concentration of secondary hydrocarbon is during the 28th minute which is 15.93%. It is the moment when CO and CO_2 are formed. The detection of secondary hydrocarbon with GC FID affects the composition of product gas and reduces the value of carbon loss. The relation between carbon loss and the product gas composition is expressed by equation 2. Based on the calculation the identification of secondary hydrocarbon on GC FID reduces the number of carbon loss by 5.4% compared to the one merely using GC TCD. It means that 5.41% of carbon loss is the result of methane conversion into a secondary hydrocarbon which was not detected using merely GC TCD. Carbon loss which was initially 27.64% now becomes 22.23%.

3.5 Analysis of particulates in the output of the reactor and its effect on carbon loss
The result of the study shows that the addition of filter mass after the 70-minute reaction process is 0.0127 gram. Based on FESEM result in Figure 7 (a) and (b), there is a black bundle on the filter after being characterized using EDX in Figure 7 (c) and (d). The filtered substance using fiber filter is carbon. It is also supported by comparing the initial composition of the filter with the composition after the filter is used for the reaction process. Initially, the carbon composition on the filter was 9.35% and after the reaction process, the carbon composition rises to 51.64%. It means that there is carbon particulates in the gas stream which is very little compared to those found in the catalyst subtract.

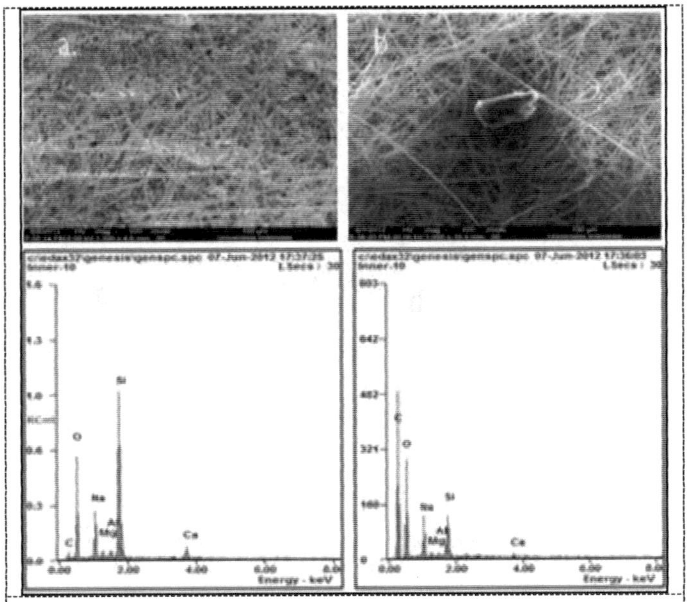

Figure 7. SEM EDX of filter (a) & (c) before CNT production; (b) & (d) after production.

In this study, particulates carried by the product stream do not provide any significant changes towards carbon loss identification. Based on equation 4 the carbon loss during the reaction process after the particulates analysis is 27.59%. This means that 0.05% of carbon loss is the result of carbon carried by product gas as particulates.

IOP Conf. Series: Materials Science and Engineering **334** (2018) 012044 doi:10.1088/1757-899X/334/1/012044

$$\% \text{ Carbon loss} = \frac{M_{CNT\ Theoretical} - M_{CNT\ actual} - M_{particulate}}{M_{CNT\ Theoretical}} \text{x}100\% \quad (4)$$

3.6 Carbon mass balance and total carbon loss

If all the factors mentioned in the previous subchapters are considered, the carbon loss equation is shown by the equation 5.

$$\% \text{ Carbon loss} = \frac{\frac{P.MR_C}{RT}\int_{t_0}^{t}[Q_{CH_4\ feed} - Q_{product} \cdot y_c]dt - M_{CNT\ actual} - M_{particulate}}{\frac{P.MR_C}{RT}\int_{t_0}^{t}[Q_{CH_4\ feed} - Q_{product} \cdot y_c]dt} \text{ x } 100 \quad (5)$$

Accordingly, the calculations of mass balance and carbon loss within this subchapter have taken all the factors of carbon analysis. The calculations are performed using a calibrated feed flow rate with the presence of catalyst, adding the correction of the feed flow rate due to CNT growth, correction of gas composition based on gas product analysis of GC TCD and FID, and particulates. Generally, the conversion of carbon mass during the reaction process can be seen in Figure 8.

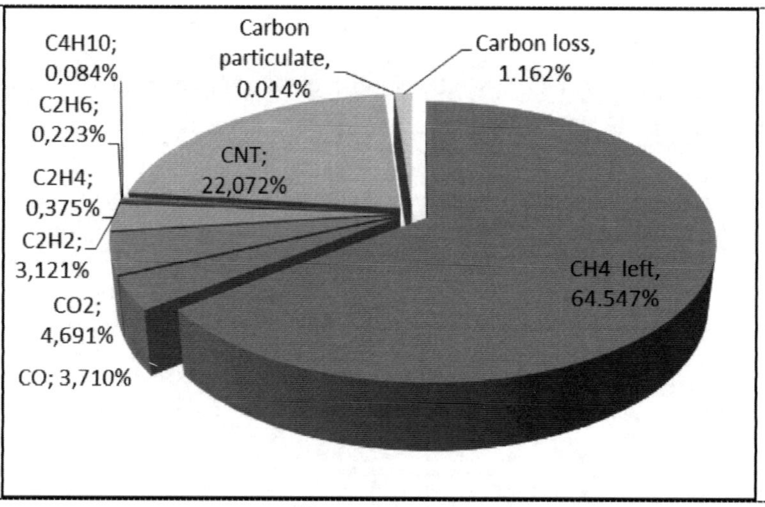

Figure 8. Carbon feed converted into other products.

Based on the discussion in 3.2, the number of error in the calculation of carbon loss is 4.14%. Hence, from the entire analysis of all factors causing the high amount of carbon loss during the pilot-scale CNT production, the following value is obtained: Carbon loss = (17.21 ± 4.14) %. Based on the calculation above, if 4.14% of carbon loss is due to the error measurement of product flow rate using bubble soap, then the amount of carbon loss is 13.07%. Such amount is smaller than the amount of carbon loss before any correction by the studied factors which is 27.64%. The amount is much smaller compared with carbon loss during empty reactor calibration which is 69.14%. The comparison and variation of carbon loss identification techniques results is available in Figure 9.

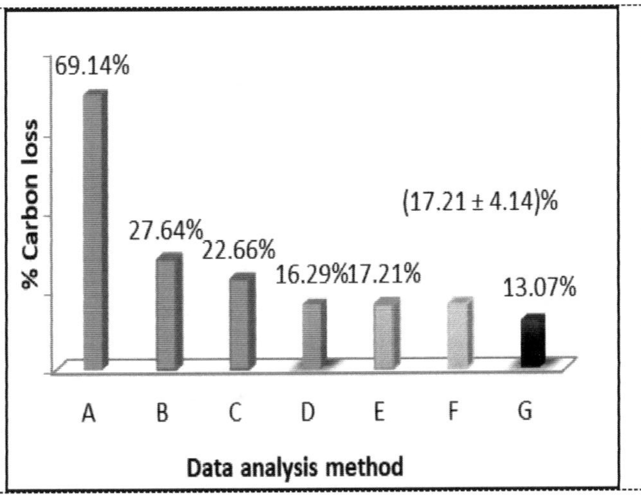

Figure 9. Effect of data analysis method to carbon loss calculation result.

Figure 9 shows that the use of the feed rate during reactor calibration containing catalyst gives the most significant impact in carbon loss identification. This is due to the difference in calibrated feed rate without and with catalyst is 28% which means that the theoretical mass of CNT is comparatively larger than it should be. Furthermore, the use of GC FID and feed flow rate correction due to CNT growth also has considerable influence. With the presence of GC FID, methane conversion to C_2 and C_4 is detected although the concentration is rather small. However, since the carbon in the compounds has more than one mole coefficient ratio (2 for C_2 and 4 for C_4), it means that the 0.01 mol loss of gas will become proportional to the 0.02 mol loss of carbon at C_2 and 0.04 mol loss of carbon at C_4 so that the carbon mass loss becomes high. Within the study, carbon loss of ± 13.07% is left unidentified. The possibility is that the formation of other compounds such as poly aromatic hydrocarbon (PAH) or other in situ compounds during CNT production with CVD (chemical vapor deposition) process contributes to the occurrence of carbon loss. Such compound is commonly identified as GC-MS, yet it is not used within the study.

4. Conclusion

By this study, we have successfully identified carbon loss during the synthesis process of CNT using a pilot-scale structured catalyst reactor. The result of initial carbon loss is 27.64%. The value of initial carbon loss decrease from 27.64% to 22.66% by using correction of feed flow rate. Error of product flow rate due to the measurement with bubble soap gives calculation error of carbon loss for about 4.14%. It is also can be identified that the detection of secondary hydrocarbon with GC FID during CNT production process reduces carbon loss by 5.14%. Particulates carried by product stream are very few and merely correct the carbon loss about 0.05%. Taking all the factors into account, the amount of carbon loss within this study is 13.07%. The success of this study will become a stepping stone in increasing the production of CNT using gauze-structured catalyst reactor. As the identification of carbon loss during the synthesis of CNT brings to light, it means that the improvement of the reactor in order to achieve the production target.

Acknowledgments

The experimental work of this study was supported by Directorate of Research and Community Services, University of Indonesia (PITTA Project 2017).

IOP Conf. Series: Materials Science and Engineering **334** (2018) 012044 doi:10.1088/1757-899X/334/1/012044

References

[1] Wulan P P D K, Silaen T P 2017 *AIP Conference Proceedings*, 1840 080003.
[2] Francy. 2009. Scale-up of Structured Catalyst Reactor Gauze Pilot Scale for Hydrogen and Nanocarbon Production Through Methane Reaction Catalytic Decomposition. *Thesis Chemical Engineering Department University of Indonesia.*
[3] Siregar R C. 2010. Reevaluation of Gauze Pilot Catalyst Reacted Catalyst Reactor for Hydrogen and Nanocarbon Production through Methane Catalytic Decomposition Reaction. *Thesis Chemical Engineering Department University of Indonesia.*
[4] Manggiasih. 2011. Performance Comparison of Structured Fe and MgO-Ni Structures For Carbon Nanotube Synthesis Through Methane Catalytic Decomposition Reaction. *Thesis Chemical Engineering Department University of Indonesia.*
[5] Wulan P P D K, Purwanto W W, Sudibandriyo M 2015 *International Journal of Technology* 1119-1127.
[6] Yulianti. 2008. Structured Catalyst Reactor Design for CNT and Hydrogen Production through Cobalt Catalytic Decomposition Process. *Thesis Chemical Engineering Department University of Indonesia.*
[7] Wulan P P D K, Wijardono S B 2017 *International Journal on Advanced Science Engineering Information Technology* **7** No 2.
[8] Thai S J, Hofman M, Hallock M, Ada E, Ellenbecker M 2009 *Sci.Technol* **43** 6017-6023
[9] Schmitt T C, Biris A S, Miller D W, Biris A R, Lupu D, Trigwell S, Rahman Z U 2006 *Carbon* **44** 2032–2038
[10] Wulan P P D K, Cendana K D 2016 *International Journal Sustainable Future for Human Security* **4** 16-21.
[11] Harutyunyan, Avetik R , Oleg A K, Christopher J B, Elena M, and Gugang C A 2009 *ACS Nano* **3** 379–385.

IOP Conf. Series: Materials Science and Engineering **334** (2018) 012045 doi:10.1088/1757-899X/334/1/012045

Experimental investigation on thermal conductivity and viscosity of maghemite (γ –Fe_2O_3) water-based nanofluids

I Nurdin[1,3], M R Johan[2] and B C Ang[2]

[1] Chemical Engineering Department Lhokseumawe State Polytechnic, 24301 Aceh, Indonesia
[2] Nanomaterials Engineering Research Group, Advanced Materials Research Laboratory, Department of Mechanical Engineering, Faculty of Engineering, University of Malaya, 50603 Kuala Lumpur, Malaysia

Email : irwan@pnl.ac.id; irwan_nurdina@yahoo.com

Abstract. Thermal conductivity and kinematic viscosity of maghemite nanofluids were experimentally investigated at a small volume fraction of maghemite nanoparticles and temperatures. Maghemite nanofluids were prepared by suspending maghemite nanoparticles in water as base fluids. Results show that the thermal conductivity of maghemite nanofluids linearly increase with increasing particle volume fraction and temperature, while kinematic viscosity increase with increasing particle volume fraction and decrease with increasing temperature. The highest enhancement of thermal conductivity and kinematic viscosity are 18.84% and 13.66% respectively, at particle volume fraction 0.6% and temperature 35.

1. Introduction

Nanofluid is a new class of engineering material consisting of solid nanoparticles with sizes smaller than 100 nm suspended in a base fluid. It provides useful applications in industrial fluids system, including as heat transfer fluids, magnetic fluids, and lubricant fluids [1].

Maghemite (γ-Fe_2O_3) nanoparticles are promising magnetic materials that are intensively explored in recent years due to their superparamagnetic behavior [2]. The suspended of these magnetic nanoparticles in a liquid medium generate a new class of liquid called "magnetic fluids." The uniqueness of this fluid is its superparamagnetic and biocompatibility behavior, and it can be controlled by external magnetic fields.

Due to its unique characteristics, the magnetic fluids act as smart fluids. They have potential applications in a variety of areas including microelectronics as in audio speaker, magnetic seal in motors, energy supply, HVAC, magnetic ink for banking cheques, magnetic recording media, magnetic resonance imaging media, drug delivery or therapeutic agents in cancer therapy and thermal engineering applications [3-7].

For such utilizations, the thermal conductivity and viscosity of magnetic fluids are critical parameters that influenced the performance of the process. Therefore, the measurement of these properties and understanding of these transport parameters become important in the application of the magnetic fluids.

[3] To whom any correspondence should be addressed.

It is well-known that the thermal conductivity and viscosity of the nanoparticle in liquids depends on the particles concentration and temperatures. Most research works related to the measurement of thermal conductivity of non-magnetic nanoparticle suspensions show a good thermal conductivity enhancement [8-11]. However, only a few of research associated with the thermal conductivity of magnetic nanofluids, particularly maghemite nanofluids were reported.

Many researchers focus on the measurement of thermal conductivity of magnetite (Fe_3O_4) nanofluids with varying of particle volume fraction and temperature without applying magnetic fields [12-14]. They found that the thermal conductivity of magnetite nanofluids increased with the increase of the particle volume fraction and temperature.

At the same time, lack of data regarding the viscosity of nanofluids measurements particularly magnetic nanofluids in the literature. Most of the investigations available in the literature on the viscosity of nanofluids, concerning the effect of volume fraction agree upon the fact that the viscosity of nanofluids increases with increasing the particle volume fraction. These phenomena also occurred in magnetic nanofluids.

Although significant investigations have been done concerning this material, the amount of data remains very limited. Efforts are still being conducted primarily to characterize the thermal and physical properties of magnetic fluids.

In this study, thermal conductivity and viscosity of water-based maghemite nanofluids were measured at a low particle volume fraction and temperatures.

2. Methodology

The Maghemite nanoparticles used in this experiment have been synthesized using a chemical co-precipitation method and characterized using several characterization techniques.

2.1. Nanofluid preparation

Maghemite nanofluids were prepared by dilution of a known concentration of a stock solution in deionized water at a different percent of the particle volume fraction of 0.1, 0.2, 0.3, 0.4, 0.5, and 0.6%.

The concentration of the nanofluids in percent volume fraction can be estimated by Eq. (1).

$$\phi = \frac{\frac{m_p}{\rho_p}}{\frac{m_p}{\rho_p}+\frac{m_f}{\rho_f}} \times 100\% \qquad (1)$$

where ϕ is particle volume fraction, m_p is mass of nanoparticles, ρ_p is density of nanoparticles m_f is mass of base fluids and ρ_p is density of base fluids.

2.2. *Thermal conductivity measurement*

Thermal conductivity measurements were conducted by KD2 Pro thermal property analyzer equipment by Decagon Devices Company, USA, which is according to transient hot-wire method. It consists of a handheld controller and sensors that inserted in the medium. The KS-1 single-needle sensor with 1.3 mm diameter and 60 mm length was used. The sensor integrates with its interior a heating element and a thermo-resistor, and it connected to a microprocessor for controlling and conducting the measurements. The measurements were performed using 30 ml sample in a cylindrical glass tube bottle. The needle probe inserted into the sample bottle then placed and exposed to a jacketed vessel that the temperature kept constant with a JEO circulating and refrigerating water bath. The samples were allowed in the bath for 30 minutes to get the desired temperature. The measurements conducted at different temperatures and particle volume fractions. Before measurement, the meter was calibrated using the standard verification solutions of known thermal conductivity. Thermal conductivities data were recorded five times with an interval of 15 minutes for each concentration at given temperature. The average of data measurements was reported.

2.3 Viscosity measurement

The Viscosity of the maghemite nanofluids were measured by Cannon Fenske calibrated glass capillary viscometer. The measurement conducted at different particle volume fractions and temperatures. The temperatures were kept constant with a JEO circulating and refrigerating water bath. The samples were placed in the viscometer and allowed for immersion for 30 minutes to get the desired temperature. Viscosity measurements were conducted five times at each particle volume fraction of the maghemite nanofluids at a given temperature. The average of data measurements was reported.

3. Results and Discussion

The sizes and form of the maghemite nanoparticles from TEM image is shown in figure 1. It observed that the shape of the particles is spherical and the size around 9.5 nm.

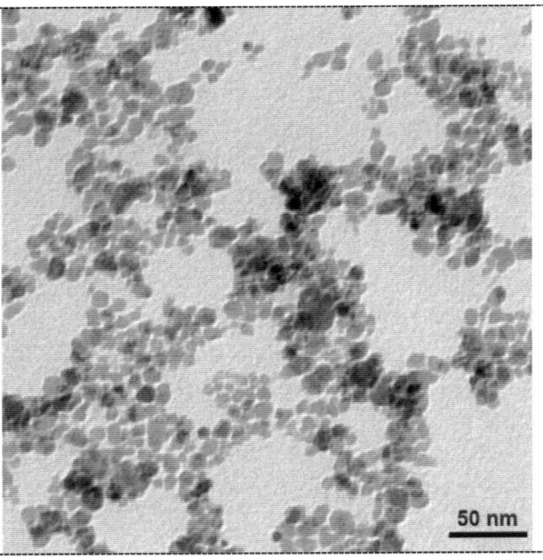

Figure 1. TEM image of maghemite nanofluids.

Figure 2 shows the thermal conductivity of maghemite nanofluids as a function of volume fraction of maghemite nanoparticles. Overall, the thermal conductivities of the maghemite nanofluids increase linearly with particles volume fraction and significantly higher than that of the base fluids. The mixture of the nanofluids contains three components, namely nanoparticle, interfacial layer, and liquids [8, 15, 16]. When the nanoparticles suspended in base fluids, interaction occurred between nanoparticles, the interfacial layer, and base fluids. These interactions caused by a random movement of the particles in the base fluids due to collisions with the molecules of base fluids [17]. Besides that, there is also consideration towards the effect of the interfacial layer between fluids and particles as the increase of layer thickness increases the thermal conductivity of nanofluids [18].

At the same time, the conductivity increased as the temperature increases due to the rise of maghemite nanoparticles Brownian motion. The highest enhancement of thermal conductivity of maghemite nanofluids is 18.84% that obtained at a particle volume fraction of 0.6% and temperature of 35 °C. These results are in good agreement with other's work [12].

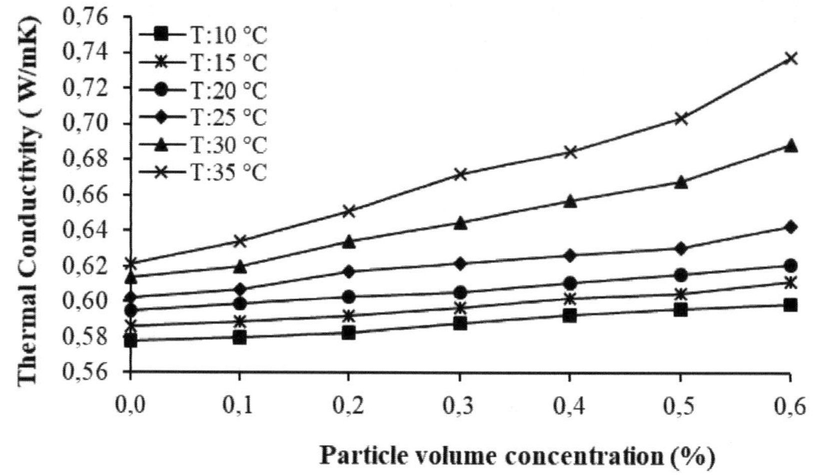

Figure 2. Thermal conductivity of maghemite nanofluids as a function of particle volume fraction at different temperatures.

The effect of temperature on the thermal conductivity of maghemite nanofluids shown in figure 3. Generally, the thermal conductivities are nonlinearly increased with the rise of temperature. When the temperature rises, the intensity of Brownian motion augments significantly, which leads to a corresponding enhancement of thermal conductivity. The thermal conductivity of high volume fraction of maghemite become larger than that with lower volume fraction due to the difference in intensity of Brownian motion at elevated temperature [8].

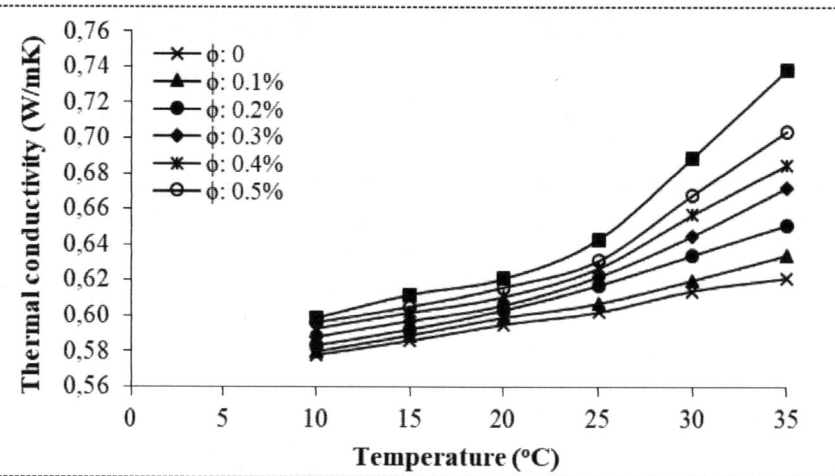

Figure 3. Temperature dependent thermal conductivity of maghemite nanofluids with the different volume fraction of maghemite nanoparticles.

The effect of the particle volume fraction of maghemite nanoparticles on the kinematic viscosity of maghemite nanofluids is presented in figure 4. It can be seen that the increase of a particle volume fraction of maghemite nanoparticles improve linearly with the kinematic viscosity of maghemite nanofluids.

As a nanofluids is a mixture of nanoparticle and base fluids, there is an interaction between particles and its surrounding. The flow of nanofluids created frictions between nanofluids and its surrounding that resist the flow of nanofluids. The more concentrated of nanoparticles in the fluids will enhance the resistance of nanofluids flow, and this will increase the viscosity of nanofluids. The highest enhancement of kinematic viscosity of maghemite nanofluids is 13.66% for a particle volume fraction of 0.6% and temperature of 35 °C.

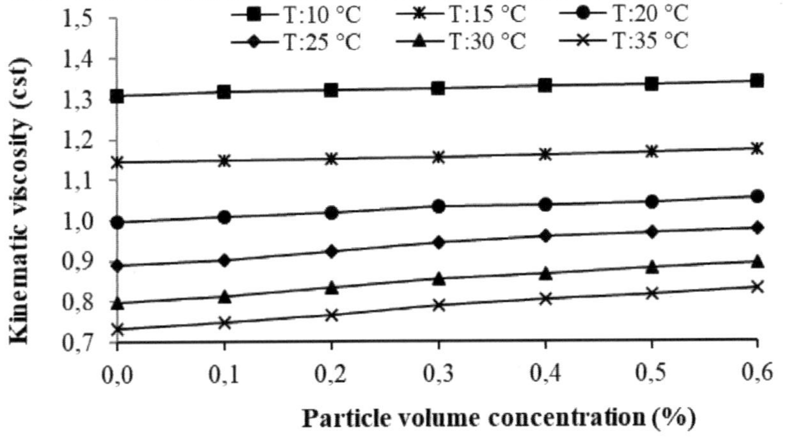

Figure 4. Kinematic viscosity of maghemite nanofluids as a function of maghemite nanoparticles volume fraction at different temperatures.

The influence of temperature on kinematic viscosity of maghemite nanofluids is presented in figure 5. It can be seen that the kinematic viscosity of maghemite nanofluids decreased with increasing temperature. The liquid molecules are tightly bonded together by attractive intermolecular forces (e.g. Van der Waal forces) at low temperature. These attractive forces are responsible for the viscosity since the individual molecules are difficult to move because they are firmly bound to their neighbors. Therefore, the kinematic viscosity is larger at low temperature.

The impact of increasing temperature will be weakening of inter-particle and inter-molecular adhesion forces hence accelerate the movement of the particles in liquids [19]. The increase in temperature causes the enhancement of kinetic or thermal energy of particles in the solutions that make the molecules more mobile. Therefore, the viscosity is reduced.

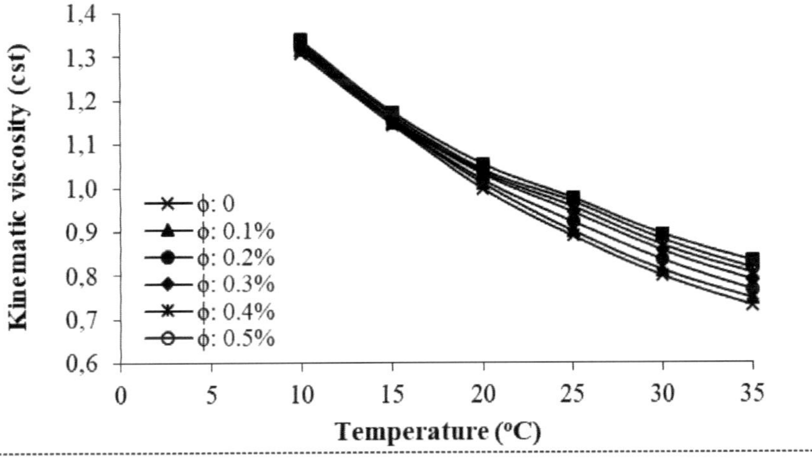

Figure 5. Kinematic viscosity of maghemite nanoparticle suspensions at various temperature.

4. Conclusions

The influence of low particle volume fraction of maghemite nanoparticles and temperature have been studied on the thermal conductivity and viscosity of maghemite nanofluids.

The thermal conductivity of maghemite nanofluids increases linearly with increasing of a particle volume fraction of maghemite nanoparticles and temperature. The highest enhancement of thermal conductivity is 18.84% at a particle volume fraction of 0.6% and temperature of 35 °C. The kinematic viscosity of maghemite nanofluids increase linearly with increasing of particle volume fraction and decrease with increasing temperature. The highest enhancement of kinematic viscosity of maghemite nanofluids is 13.66% that obtained at a particle volume fraction of 0.6% and temperature of 35 °C.

References

[1] Correa J R, Canetti D, Castillo R, Llópiz J C, Dufour J 2006 *Mater. Res. Bull.* **41** 703-13.
[2] Nurdin I, Johan M R, Yaacob I I, Ang B C 2014 *Sci. World J.* 2014.
[3] Rosensweig R, Hirota Y, Tsuda S, Raj K 2008 *J. Phys-Condens. Mat.* **20** 204147.
[4] Nkurikiyimfura I, Wang Y, Pan Z 2013 *Renew. Sust. Energ. Rev.* **21** 548-61.
[5] Wang Y, Cao X, Liu G, Hong R, Chen Y, Chen X, et al. 2011 *J. Magn. Magn. Mater.* **323** 2953-9.
[6] Li Q, Xuan Y 2009 *Exp. Therm. Fluid Sci.* **33** 591-6.
[7] Huminic G, Huminic A, Morjan I, Dumitrache F 2011 *Int. J. Heat Mass Tran.* **54** 656-61.
[8] Murshed S, Leong K, Yang C 2008 *Int. J. Therm. Sci.* **47** 560-8.
[9] Jeong J, Li C, Kwon Y, Lee J, Kim S H, Yun R 2013 *Int. J. Refrig.* **36** 2233-41.
[10] Yang J-C, Li F-C, Zhou W-W, He Y-R, Jiang B-C 2012 *Int. J. Heat Mass Tran.* **55** 3160-6.
[11] Phuoc T X, Massoudi M, Chen R-H 2011 *Int. J. Therm. Sci.* **50** 12-8.
[12] Abareshi M, Goharshadi E K, Zebarjad S M, Fadafan H K, Youssefi A 2010 *J. Magn. Magn. Mater.* **322** 3895-901.
[13] Philip J, Shima P, Raj B 2007 *Appl. Phys. Lett.* **91** 203108--3.
[14] Sundar L S, Singh M K, Sousa A C 2013 *Int. Commun. Heat Mass Tran.* **44** 7-14.
[15] Tillman P, Hill J M 2007 *Int. Commun. Mass Tran.* **34** 399-407.
[16] Fu H L, Gao L 2012 *Int. J. Therm. Sci.* **61** 61-6.
[17] Godson L, Raja B, Mohan Lal D, Wongwises S 2010 *Renew. Sust. Energ. Rev.* **14** 629-41.
[18] Mintsa H A, Roy G, Nguyen C T, Doucet D 2009 *Int. J. Therm. Sci.* **48** 363-71.
[19] Nguyen C, Desgranges F, Roy G, Galanis N, Mare T, Boucher S, et al. 2007 *Int. J. Heat Fluid Fl.* **28** 1492-506.

IOP Conf. Series: Materials Science and Engineering **334** (2018) 012046 doi:10.1088/1757-899X/334/1/012046

Nanocrystal cellulose as drug excipient in transdermal patch for wound healing: an overview

S A Mohd Zuki[1], N Abd Rahman[1,2] and N F Abu Bakar[1]
[1] Faculty of Chemical Engineering, Universiti Teknologi MARA, 40450 Shah Alam, Selangor, Malaysia

E-mail: noraz695@salam.uitm.edu.my

Abstract. Wound must be carefully treated to avoid serious infection that needs costly treatment. Method to enhance the recovery of the wound is crucial to have effective wound treatment. One of the technologies in wound treatment is transdermal patch that has the benefits of being non-invasive, easy to handle and permits constant drug dosage. In order to obtain a good controlled drug release, drug excipient needs to be investigated. Recently, natural Nanocrystal Cellulose (NCC) which can be synthesized from animal, algae, microorganism or plant has been actively used in drug delivery system as excipient. The application of NCC is advantageous due to its large surface area, biodegradable, non-toxic and abundance source.

1. Introduction

Wound is a common infection that happens to human body and animal. It can occur due to mechanical, chemical or thermal injury [1]. The injury ranges from a simple skin damages or it can be dangerous damages which involves deep cut. Serious injury can penetrates deep into subcutaneous tissue and harm other structures such as tendons, muscles, vessels, nerves, parenchymal organs and bones [1]. Wound injury to critical patients such as diabetic patients, really need a good care and rapid healing rate to aid in their recovery. There are two basic principles in wound management which are; removing the impediments, and provide and maintain clean and conducive environment to heal the wound [2]. The wound has to be kept clean and the wound surface has to be insulated and protected [2].

One of the drug delivery system (DDS) for wound healing that has been used for long period of time is transdermal patch. In December 1979, the United States Food and Drug Administration has approved the first prescription patch which is scopolamine for motion sickness. The drug delivery technology for dermatology disorder has evolved from ointment and cream to the transdermal patch because of its advantages such as; it allows constant dosing which prevent the fluctuation of drug level and it is also non-invasive. Besides, it is a better alternative to be used instead of the drug delivery through the stomach which sometimes becomes problematic when there is a difficulty for the drug to be absorbed through the gastrointestinal track. The steps for transdermal permeation of drug begins

[2] To whom any correspondence should be addressed.

with the sorption by stratum corneum, followed by the penetration of drug through viable epidermis and continued with the uptake of the drug in the dermal capillary layer.

There are four types of transdermal drug delivery system which are membrane permeation controlled, matrix diffusion-controlled, adhesive dispersion-controlled and microreservoir/microsealed dissolution controlled system. The transdermal patch is not only used for wound healing but also has been used in other treatments as shown in table 1.

Table 1. Literature review on transdermal patch.

Author	Treatment	Drug	Remarks
[3]	Wound	Ciprofloxacin (antibiotic)	-Electrospinning technique -A sustained and controlled drug release have been obtained
[4]	Wound	Vitamin B12, Curcumin and Diclofenac (anti-inflammatory, analgesic, antipyretic)	-Electrospinning technique -The resulted nanomembrane of the drug and excipient were stable -The vitamin B12 was nearly 100% -The curcumin release was up to 70% The Diclofenac release was 80%
[5]	Schizophrenia	Blonanserin (antipsychotic)	-Investigation on effect of permeation enhancer, fisopropyl myristate -The release rate of blonanserin increased with an increasing concentration of fisopropyl myristate
[6]	Pain killer	Methadone	-Film casting method -The problem of drug release onto skin has been solved by blending the methadone with the enhancer (Dimethylsulfoxide) -The permeation of the methadone was increase up to 70 %
[7]	Pain killer	Fentanyl	-Fentanyl containing geopolymer granules results in a better resistance to tampering
[8]	-	Eserine and pralidoxime chloride	-The patch resulted in a complete drug release after 72 hours - The transdermal patches were stable for 6 months at $40^0C/75\%$ of Relative Humadity
[9]	Attention deficit/Hyperactivity Disorder (ADHD)	D-threo-methylphenidate (D-threo-MP)	-Solvent evaporation technique -Acrylic pressure sensitive adhesive used as excipient -No significant of enhancer on drug permeation -The best drug loading is resulting from 15% weight percentage of D-threo-MP
[10]	Muscle relaxant	Cyclobenzaprine	-Solvent evaporation technique -Excipient used is Cotran™ 9700 - Cotran™ 9700 give significant effect on drug release -The cumulative drug release decreasing after 7 days

From table 1, transdermal patch shows great potential to be studied as drug delivery technique and can be applied for wound healing treatment. The drug release of the antibacterial drug in the patch can be sustained and controlled. Usually, the transdermal patch consists of outer liner and matrix membrane formed by active material and excipient. Besides the active material, it is also important to study the drug excipient because it has its own role in delivering the drug into targeted cell, tissue or organ.

2. Drug Excipient

Typically, the highest content in a drug tablet, capsule, patch or liquid is not the active pharmaceutical ingredient (API) but the excipient. Excipient is an inactive material that is purposely included in formulation of drug. Usually it has been properly tested for safety and it is considered essential to include the excipient in drug formulation to improve the manufacturability and stabilization of the API. There are many type of drug excipients such as binder (povidones, polysaccharide), filler or diluent (calcium phosphate, lactose), disintegrant (sodium starch glycolate, crospovidones), lubricant (magnesium stearate, glycerides), glidant or anticaking agent (talc, colloidal silicon dioxide), colorant (titanium dioxide; food, drug and cosmetic (FD&C) colours), capsule shell (gelatin, hypromellose), coating agent (hypromellose, shellac), flavour and fragrance (peppermint, berry), release modifier (ethylcellulose, guar gum), pH modifier (citric acid and its salts, salts of phosphoric acid), wetting or solubilizing agent (sodium lauryl sulfate, polysorbates), antimicrobial preservative (glycerin, benzyl alcohol), chelating or complexing agent (ethylenediaminetetraacetic acid salts, cyclodextrins), antioxidant (ascorbic acid, butylated hydroxyanisole) and sweetening agent (sucrose, saccharin) [11]. Among the type of drug excipient that has been introduced before, polysaccharide was actively studied as an excipient in transdermal patch [12-16]). Starch, glycogen and dextrans are some of the example of polysaccharide used as an excipient and all of them are converted into energy in liver and muscles for a later use. Nowadays, there are numerous researches regarding cellulose (natural polysaccharide) as drug excipient which is discussed in next section.

3. Nanocrystal cellulose (NCC)

Nowadays, natural polysaccharide has become popular to be studied as an excipient in drug delivery system because of its properties such as highly safe, non-toxic, has abundant resources in nature, low cost in its processing, biodegradable, biocompatible, high water solubility and bioactivity [17-20]. There are many natural sources that supply polysaccharide i.e animal (chitosan, Chondroitin sulphate), plant (guar gam, pectin, cellulose, mannan), algae (alginate) and microorganism (dextran, pullulan) [21-23]. Besides, large number of reactive functional groups havehydroxyl, amino and carboxylic acid groups on their backbone which make the polysaccharide structure to be easily derived. This characteristic contribute to their structural and functional diversity [24]. From multi-functional group, it can be biochemically or chemically modified into many types of polysaccharide derivatives such as cellulose [25].

Rod-shaped cellulose with typical 10–100 nm in length and 1–100 nm in diameter is called nanocrystal cellulose (NCC) and it has potential to be investigated as drug excipient [26, 27]. There are many studies that have been done in drug delivery field using NCC as shown in table 2. NCC derived from plant has been actively studied in polymer reinforcement as a potentially new material [28]. Commonly, the plant fibre contains of cellulose (40–50%), hemicelluloses (20–30%) and lignin (10–18%). The advantages of NCC are due to its large surface area and excellent colloidal stability [29].

Table 2. Literature review on nanocrystal cellulose as drug excipient.

Author	NCC	Drug	Drug Delivery System	Remarks
[30]	Chitosan	Procaine	Transdermal	-Drug binding and release

	oligosaccharide	hydrochloride, imipramine hydrochloride	patch	are allied with the interaction nature and types that exist between the NCC and drug molecules
[31]	Carboxymethyl -β-cyclodextrin	1-Ethyl-3-(3-dimethylaminopropyl) carbodiimide, N-hydroxysuccinimide	Injection	-Drug binding resulting in stronger inhibition rate and enhanced cellular uptake through folate mediated internalization
[32]	Folic acid-conjugated NCC	Folic acid	Oral intake	-Cellular binding of the conjugate higher than non-targeted NCC
[33]	Chitosan	Doxorubicin hydrochloride, curcumin	Oral intake	-Stable microcapsules were formed after the five-layer deposition of NCC -The surface of microcapsules have the same morphologies as the thin film
[34]	Chitosan	Metformin	Injection	-The swelling of the formulated films were decrease in acidic, neutral and alkaline medium respectively - The formulated film showed potential applications as light emitting materials
[35]	Chitosan	Antofloxacin	Transdermal patch	-Drug release showed pH sensitivity - Drug release can be controlled by manipulating pH system and amount of drug encapsulated
[36]	Cotton wool	2-dimethylamino ethyl methacrylate	Transdermal patch	-Needle-like shape of NCC improve the transfection efficiencies and low cytotoxicities of drug binding
[37]	β-cyclodextrin	Polyvinyl alcohol (PVA)-styrylpyridinium	Transdermal patch	-FTIR spectra showed a successful binding between β-cyclodextrin and the drug -Average fiber diameter increased because of the swelling of nanofiber (interactions between hydroxy of PVA or β-cyclodextrin and aldehyde group)
[38]	Mucilage (Taro	Diltiazem	Transdermal	-Folding endurance and

	corms), hydroxypropyl methylcellulose	hydrochloride	patch	tensile strength of formulated patch increase as the mucilage concentration increase -The drug release controlled by addition of mucilage in formulation -Skin test shows free of potentially hazardous skin irritation
[39]	Pectin, gelatin	Testosterone	Transdermal patch	-Pectin shows important effect on rheological characteristic of formulated patch -The drug release controlled by the formulation of the patch matrix

Most of the results from previous researches showed that NCC give positive effects on the quality of the drug release. The drug release has been significantly affected by pH of solution and formulation ratio. Since it shows the potential in controlling the drug release, it will result in constant drug permeation into the targeted area. Hence, this shows that NCC is a promising drug carrier to be further studied.

4. Conclusion

In this overview, transdermal patch shows a great potential to be further studied as drug delivery system to deliver the active ingredient onto the targeted treatment area especially skin because it can give constant drug release and fulfill the patient compliance since it is a non-invasive treatment. It is easy for the patient to use the patch to treat the targeted area such as wound. Besides, other ingredient in the patch that gives important effect on drug release and permeation is the excipient. The excipient shows positive result on controlling the drug release. Nanocrystal cellulose (NCC), a type of polysaccharide extracted from natural source such as plant has been actively studied to be used as excipient and from the results, it shows that NCC has great potential to be used in improving the control of the drug release for transdermal patch application.

Acknowledgment

The authors would like to thank the financial support received from Research Management Institute of Universiti Teknologi MARA through Geran Inisiatif Penyeliaan (600-IRMI/GIP 5/3 (0015/2016)), Ministry of Higher Education, Malaysia, Faculty of Chemical Engineering, UiTM and Universiti Teknologi MARA, Malaysia.

References

[1] Mohil R "Classification of wounds" in *Principles and Practice of wound care* V. T. S. Sarabahi Eds. New Delhi, India Jaypee Brothers Medical Publishers 2012

[2] Dressings and More: Guidelines for Topical Wound ManagementDoughty D 2005 Dressings and More: Guidelines for Topical Wound Management *Nursing Clinics of North America* **40** 2 217-231

[3] In vivo wound healing performance of drug loaded electrospun composite nanofibers transdermal patchKataria K, Gupta A, Rath G, Mathur R B, and Dhakate S R 2014 In vivo wound healing performance of drug loaded electrospun composite nanofibers transdermal patch *International Journal of Pharmaceutics* **469** 1 102-110
[4] Hybrid electrospun chitosan-phospholipids nanofibers for transdermal drug deliveryMendes A C, Gorzelanny C, Halter N, Schneider S W, and Chronakis I S 2016 Hybrid electrospun chitosan-phospholipids nanofibers for transdermal drug delivery *International Journal of Pharmaceutics* **510** 1 48-56
[5] Effect of isopropyl myristate on the viscoelasticity and drug release of a drug-in-adhesive transdermal patch containing blonanserinZhao C, Quan P, Liu C, Li Q, and Fang L 2016 Effect of isopropyl myristate on the viscoelasticity and drug release of a drug-in-adhesive transdermal patch containing blonanserin *Acta Pharmaceutica Sinica B* **6** 6 623-628
[6] Design, development and characterization of transdermal patch of methadoneMuñoz M D, Castán H, Ruiz M A, and Morales M E Design, development and characterization of transdermal patch of methadone *Journal of Drug Delivery Science and Technology*
[7] Development and evaluation of a tampering resistant transdermal fentanyl patchCai B, Engqvist H, and Bredenberg S 2015 Development and evaluation of a tampering resistant transdermal fentanyl patch *International Journal of Pharmaceutics* **488** 1–2 102-107
[8] Accelerated stability testing of a transdermal patch composed of eserine and pralidoxime chloride for prophylaxis against (±)-anatoxin A poisoningBanerjee S, Chattopadhyay P, Ghosh A, Bhattacharya S S, Kundu A, and Veer V 2014 Accelerated stability testing of a transdermal patch composed of eserine and pralidoxime chloride for prophylaxis against (±)-anatoxin A poisoning *Journal of Food and Drug Analysis* **22** 2 264-270
[9] Transdermal patches for D-threo-methylphenidate (free base): Formulation aspects and in vivo pharmacokineticsZhang C, Luo H, Lin G, Zhu Z, Zhang F, Zhang J, Wu Y, Luo J, and Wang H 2016 Transdermal patches for D-threo-methylphenidate (free base): Formulation aspects and in vivo pharmacokinetics *Journal of Drug Delivery Science and Technology* **35** 50-57
[10] Effect of backing films on the transdermal delivery of cyclobenzaprine patchLv S, Quan P, Liu X, and Fang L 2016 Effect of backing films on the transdermal delivery of cyclobenzaprine patch *Asian Journal of Pharmaceutical Sciences* **11** 6 780-783
[11] Jyllian K. (2014, 3 February 2014) Eyes on Excipients : Inactive but Essential *Chemical & Engineering News*. 9-11.
[12] Effectiveness of an Oxidized Cellulose Patch Hemostatic Agent in Thyroid Surgery: A Prospective, Randomized, Controlled StudyAmit M, Binenbaum Y, Cohen J T, and Gil Z 2013 Effectiveness of an Oxidized Cellulose Patch Hemostatic Agent in Thyroid Surgery: A Prospective, Randomized, Controlled Study *Journal of the American College of Surgeons* **217** 2 221-225
[13] Fibrin Sealant Patch (TachoSil) vs Oxidized Regenerated Cellulose Patch (Surgicel Original) for the Secondary Treatment of Local Bleeding in Patients Undergoing Hepatic Resection: A Randomized Controlled TrialGenyk Y, Kato T, Pomposelli J J, Wright Jr J K, Sher L S, Tetens V, and Chapman W C 2016 Fibrin Sealant Patch (TachoSil) vs Oxidized Regenerated Cellulose Patch (Surgicel Original) for the Secondary Treatment of Local Bleeding in Patients Undergoing Hepatic Resection: A Randomized Controlled Trial *Journal of the American College of Surgeons* **222** 3 261-268
[14] Cellulose-pectin composite hydrogels: Intermolecular interactions and material properties depend on order of assemblyLopez-Sanchez P, Martinez-Sanz M, Bonilla M R, Wang D, Gilbert E P, Stokes J R, and Gidley M J 2017 Cellulose-pectin composite hydrogels: Intermolecular interactions and material properties depend on order of assembly *Carbohydrate Polymers* **162** 71-81
[15] Study of the keratin-based therapeutic dermal patches for the delivery of bioactive molecules for wound treatmentNayak K K and Gupta P 2017 Study of the keratin-based therapeutic dermal

patches for the delivery of bioactive molecules for wound treatment *Materials Science and Engineering: C* **77** 1088-1097

[16] Preparation, characterization, and in vitro diffusion study of nonwoven electrospun nanofiber of curcumin-loaded cellulose acetate phthalate polymerRavikumar R, Ganesh M, Ubaidulla U, Young Choi E, and Tae Jang H 2017 Preparation, characterization, and in vitro diffusion study of nonwoven electrospun nanofiber of curcumin-loaded cellulose acetate phthalate polymer *Saudi Pharmaceutical Journal*

[17] Synthesis of water dispersed nanoparticles from different polysaccharides and their application in drug releaseAyadi F, Bayer I S, Marras S, and Athanassiou A 2016 Synthesis of water dispersed nanoparticles from different polysaccharides and their application in drug release *Carbohydrate Polymers* **136** 282-291

[18] Polysaccharide based nanogels in the drug delivery system: Application as the carrier of pharmaceutical agentsDebele T A, Mekuria S L, and Tsai H-C 2016 Polysaccharide based nanogels in the drug delivery system: Application as the carrier of pharmaceutical agents *Materials Science and Engineering: C* **68** 964-981

[19] Millifluidic synthesis of amorphous drug-polysaccharide nanoparticle complex with tunable size intended for supersaturating drug delivery applicationsTran T-T, Nguyen M-H, Tan Y Z, Chew J W, Khan S A, and Hadinoto K 2017 Millifluidic synthesis of amorphous drug-polysaccharide nanoparticle complex with tunable size intended for supersaturating drug delivery applications *European Journal of Pharmaceutics and Biopharmaceutics* **112** 196-203

[20] Preparation and application of micro/nanoparticles based on natural polysaccharidesYang J, Han S, Zheng H, Dong H, and Liu J 2015 Preparation and application of micro/nanoparticles based on natural polysaccharides *Carbohydrate Polymers* **123** 53-66

[21] Polysaccharide biological response modifiersLeung M Y K, Liu C, Koon J C M, and Fung K P 2006 Polysaccharide biological response modifiers *Immunology Letters* **105** 2 101-114

[22] Polysaccharide-based nucleic acid nanoformulationsRaemdonck K, Martens T F, Braeckmans K, Demeester J, and De Smedt S C 2013 Polysaccharide-based nucleic acid nanoformulations *Advanced Drug Delivery Reviews* **65** 9 1123-1147

[23] Polysaccharides in colon-specific drug deliverySinha V R and Kumria R 2001 Polysaccharides in colon-specific drug delivery *International Journal of Pharmaceutics* **224** 1–2 19-38

[24] Nurunnabi M, Revuri V, Huh K M, and Lee Y-k "Chapter 14 - Polysaccharide based nano/microformulation: an effective and versatile oral drug delivery system A2 - Andronescu, Ecaterina" in *Nanostructures for Oral Medicine* A. M. Grumezescu Eds. Elsevier 2017 409-433

[25] Heterogeneous 'organoclick' derivatization of polysaccharidesHafrén J, Zou W, and Córdova A 2006 Heterogeneous 'organoclick' derivatization of polysaccharides *Macromolecular Rapid Communications* Article **27** 16 1362-1366

[26] Cellulose nanocrystals: Chemistry, self-assembly, and applicationsHabibi Y, Lucia L A, and Rojas O J 2010 Cellulose nanocrystals: Chemistry, self-assembly, and applications *Chemical Reviews* Article **110** 6 3479-3500

[27] Chemistry and applications of nanocrystalline cellulose and its derivatives: A nanotechnology perspectivePeng B L, Dhar N, Liu H L, and Tam K C 2011 Chemistry and applications of nanocrystalline cellulose and its derivatives: A nanotechnology perspective *Canadian Journal of Chemical Engineering* Review **89** 5 1191-1206

[28] Cellulose nanocrystals from acacia bark–Influence of solvent extractionTaflick T, Schwendler L A, Rosa S M L, Bica C I D, and Nachtigall S M B 2017 Cellulose nanocrystals from acacia bark–Influence of solvent extraction *International Journal of Biological Macromolecules* **101** 553-561

[29] Recent advances in the application of cellulose nanocrystalsGrishkewich N, Mohammed N, Tang J, and Tam K C 2017 Recent advances in the application of cellulose nanocrystals *Current Opinion in Colloid & Interface Science* **29** 32-45
[30] Comparative release studies of two cationic model drugs from different cellulose nanocrystal derivativesAkhlaghi S P, Tiong D, Berry R M, and Tam K C 2014 Comparative release studies of two cationic model drugs from different cellulose nanocrystal derivatives *European Journal of Pharmaceutics and Biopharmaceutics* **88** 1 207-215
[31] Carboxymethyl-β-cyclodextrin conjugated nanoparticles facilitate therapy for folate receptor-positive tumor with the mediation of folic acidSu C, Li H, Shi Y, Wang G, Liu L, Zhao L, and Su R 2014 Carboxymethyl-β-cyclodextrin conjugated nanoparticles facilitate therapy for folate receptor-positive tumor with the mediation of folic acid *International Journal of Pharmaceutics* **474** 1–2 202-211
[32] Synthesis and Cellular Uptake of Folic Acid-Conjugated Cellulose Nanocrystals for Cancer TargetingDong S, Cho H J, Lee Y W, and Roman M 2014 Synthesis and Cellular Uptake of Folic Acid-Conjugated Cellulose Nanocrystals for Cancer Targeting *Biomacromolecules* **15** 5 1560-1567
[33] Layer-by-Layer Assembled Thin Films and Microcapsules of Nanocrystalline Cellulose for Hydrophobic Drug DeliveryMohanta V, Madras G, and Patil S 2014 Layer-by-Layer Assembled Thin Films and Microcapsules of Nanocrystalline Cellulose for Hydrophobic Drug Delivery *ACS Applied Materials & Interfaces* **6** 22 20093-20101
[34] Controlled release of metformin from chitosan–based nanocomposite films containing mesoporous MCM-41 nanoparticles as novel drug delivery systemsShariatinia Z and Zahraee Z 2017 Controlled release of metformin from chitosan–based nanocomposite films containing mesoporous MCM-41 nanoparticles as novel drug delivery systems *Journal of Colloid and Interface Science* Article **501** 60-76
[35] Preparation of a novel chitosan-microcapsules/starch blend film and the study of its drug-release mechanismHuo W, Xie G, Zhang W, Wang W, Shan J, Liu H, and Zhou X 2016 Preparation of a novel chitosan-microcapsules/starch blend film and the study of its drug-release mechanism *International Journal of Biological Macromolecules* **87** 114-122
[36] Redox-Responsive Polycation-Functionalized Cotton Cellulose Nanocrystals for Effective Cancer TreatmentHu H, Yuan W, Liu F-S, Cheng G, Xu F-J, and Ma J 2015 Redox-Responsive Polycation-Functionalized Cotton Cellulose Nanocrystals for Effective Cancer Treatment *ACS Applied Materials & Interfaces* **7** 16 8942-8951
[37] Preparation and characterization of electrospun polyvinyl alcoholstyrylpyridinium/β-cyclodextrin composite nanofibers: Release behavior and potential use for wound dressingZhou J, Wang Q, Lu H, Zhang Q, Lv P, and Wei Q 2016 Preparation and characterization of electrospun polyvinyl alcoholstyrylpyridinium/β-cyclodextrin composite nanofibers: Release behavior and potential use for wound dressing *Fibers and Polymers* journal article **17** 11 1835-1841
[38] Taro corms mucilage/HPMC based transdermal patch: An efficient device for delivery of diltiazem hydrochlorideSarkar G, Saha N R, Roy I, Bhattacharyya A, Bose M, Mishra R, Rana D, Bhattacharjee D, and Chattopadhyay D 2014 Taro corms mucilage/HPMC based transdermal patch: An efficient device for delivery of diltiazem hydrochloride *International Journal of Biological Macromolecules* **66** 158-165
[39] Hydrogel blends with adjustable properties as patches for transdermal deliveryMazzitelli S, Pagano C, Giusepponi D, Nastruzzi C, and Perioli L 2013 Hydrogel blends with adjustable properties as patches for transdermal delivery *International Journal of Pharmaceutics* **454** 1 47-57

IOP Conf. Series: Materials Science and Engineering **334** (2018) 012047 doi:10.1088/1757-899X/334/1/012047

Synthesis and innovation of PLA/clay nanocomposite characterization againts to mechanical and thermal properties

S Salim[1,2,3], H Agusnar[1], B Wirjosentono[1], Tamrin[1], H Marpaung[1], T Rihayat[2], Nurhanifa[2] and Adriana[2]

[1] Department of Chemistry, Faculty of Mathematics and Natural Science, Universitas Sumatera Utara, 20155, Medan, Indonesia
[2] Departement of Chemical Engineering, Politeknik Negeri Lhokseumawe, 24301, Aceh Indonesia

E-mail: suryanisalim@yahoo.com

Abstract. Plastic polymer is one of the most dominant materials of daily human activities because of its multifunctional nature, light and strong and anti-corrosion so it is easy to apply in various equipment. Plastic is generally derived from petroleum material so it is nonbiodegradable. Therefore, this study aims to create a breakthrough of natural and biodegradable biodegradable plastic materials from plant starch (pisok kepok starch) with the help of 3 types of acid (HNO_3, HCl and H_2SO_4) called Poly Lactid Acid (PLA). PLA is enhanced by mixing with a clay material with a variation of 1, 3 and 5% composition to form a PLA / Clay Nanocomposite material which is expected to have superior properties and resemble conventional plastics in general. Several types of characterization were performed to see the quality of the resulting material including tensile strength test with UTM tool, thermal endurance test with TGA tool, morphological structure test using SEM tool and additional test to see filler clay quality through X-RD tool. Based on the characterization of tensile and thermal test, 5B nanocomposite with addition of 5% clay and HCl acid aid showed the best tensile strength of 36 Mpa and the highest stability was 446,63 oC. Based on the results of morphological analysis of the best samples (5B) showed good interface ties. Meanwhile, based on the results of filler analysis, the opening of clay layer d-spacing occurred at 0.355 nm.

1. Introduction

Currently research on environmentally friendly products is of great concern by many researchers and industry players especially in the field of polymers [1]. One of the interesting biopolymers studied is PLA (Poly Lactid Acid) due to its biodegradable, biocompatible, recycleble and flammable properties that are safe for the environment [2]. Unlike typical synthetic polymers derived from petroleummaterials, PLA comes from renewable sources from a variety of plants that have high starch content such as bananas, cassava, sweet potatoes, potatoes, corn and so on that help support the movement of go green and Reduce environmental waste.

PLA can be formed by esterification process of lactic acid obtained by fermentation by bacteria by using starch substrate or simple sugar [3]. These polymers may be amorphous and may also be

[3] To whom any correspondence should be addressed.

crystalline. The PLA is applied usually to plastic bags, plastic packaging, drink bottles, plastic cups, hard plastic materials, tableware, car components, shelves, buckets and so on. The latest applications from the PLA, one of them in the medical field, among others, is used as artificial leather, sewing thread surgery, drug capsules and also for tissue engineering because it can be adsorbed by the body. However, this biopolymer also has a weakness that is lower melting point than other plastic polymers so that heat stability is not as good as other polymers [4]. In addition, PLA has mechanical properties that are also lower than synthetic polymers. Therefore, to overcome these weaknesses, PLA can be improved its properties by the addition of nano-sized filler material to form nanocomposite.

Based on some of the most prominent references, clay is a subject of considerable attention based on its ability to spread between layers and its ability to expand. Therefore, the clay can be processed into a composite with a polymer. Clay is a mineral decomposition of volcanic ash where its use as an economical filler to modify the creation and performance of the material. The mineralogy of bentonite is defined as clay consisting of 80% montmorillonite having a chemical formula ($Al_2O_3.4SiO_2xH_2O$) [3].

There are many types of materials used as fillers, but clay (bentonite / clay) has a long record as the most inorganic material added as a filler into the polymer as a matrix [5]. This is potentially very large when compared to the original material without mixing. The purpose of adding fillers is to improve and improve the properties of polymeric materials to be mechanically stronger and more heat resistant.

Interest in research on nanocomposite polymers among researchers now shows an increasing trend. In addition, the material is also easy to find and economical. Plants rich in starch content are available in large quantities in Indonesia. Moreover, banana crops in the local area, especially in the North Aceh region are available in large quantities have great potential to be utilized as a natural raw material for the manufacture of plastic products (PLA). So far the utilization of these fruits is limited only as fruits and other processed foods. However, by doing this research is expected to be able to increase the value of use and also the selling value of banana with end product in the form of biodegradable plastic. Clay as a filler that is abundant in Indonesia especially in North Aceh region is also expected can be utilized to increase the superiority of plastic polymer product (PLA) in terms of strength and resistance to heat so it has value more than other conventional plastics.

Based on these reasons, this study aims to produce PLA (Poly Lactic Acid) as an environmentally-friendly bipolimer made from banana kepok starch. Additionally clay as a filler used is modified to improve biopolymer performance. The effect of acid variables (HNO3, HCl and H2SO4) and the effect of adding clay amounts to the PLA polymer on the resultant characterization of the resulting PLA / Clay Nanocomposite was also identified. The final product is tested by performing some of the main karaterization including tensile and thermal tests to determine its quality. In addition, morphological testing is also performed to support the results of the main characterization, as well as special tests on the quality of filler through d-spacing clay test also became one of the targets in this study.

2. Methodelogy/Experimental

2.1 Samples

The material used in this study is banana kepok originating from local area of North Aceh, aquades, glycerol, acidic substance (HNO_3, HCl, H_2SO_4), ethanol, reactant benedict, Lactobacillus Plantarum bacteria, zinc oxide catalyst, TBAB surfactant, dispersant $(NaPO_3)_6$ and clay from Aceh Utara as filler.

2.2 PLA Preparation

2.2.1 Starch Extraction. Banana kepok peeled, cut into pieces then mixed with aquades at a ratio of 1:3 and ground to form a slurry. Then the slurry was filtered and the filter result was precipitated for 8 hours. Furthermore, the precipitate is dried in an oven with a temperature of 60 °C and saturated into starch flour.

2.2.2_Hydrolysis Starch Becomes Glucose. The amount of starch flour was dissolved in aquades, adjusting the pH to ± 2 with various acid catalysts (5% HNO_3, 5% HCl, 2% H_2SO_4) for each sample. Further hydrolysis is done by heating the solution in the two neck flask and stirring done for 3 hours with a temperature of 100 °C. The resulting glucose was then tested using a Benedict test.

2.2.3 Glucose Fermentation Becomes Lactid Acid. Fermentation of glucose into lactic acid using Lactobacillus plantarum bacteria. Fermentation was carried out for 4 days on three hydrolyzed glucose samples with acid variation.

2.2.4 *Polymerization of Lactic Acid forms the PLA.* Acid produced by fermentation is then polymerized by ROP (Ring Opening Polymerization) method. This process consists of three stages: prepolymerization, depolymerization, and ring opening polymerization. In general the process is heating lactic acid as much as 100 ml under vacuum under pressure of 300 mmHg with a temperature of 100 °C for 6 hours and with the help of ZnO (Zinc Oxide) catalyst at 0.25 gr in the first 4 hours, and 0.75 gr at 6 Hour and added 1.75 ml glycerol as plasticizer.

2.3 Modifiying of Filler

Each 100 grams of bentonite is weighed and then pounded with pestle. Performed sizing on 250 mesh. The sieved bentonite was introduced into 1200 ml of aquades and added $(NaPO_3)_6$ by 1 g and then dispersed by stirring for 6 hours using a medium-spin magnetic stirrer. The dispersion results are then allowed to form bentonite precipitate and separated by water containing impurities. The precipitate is washed with aquades to remove excess residue $(NaPO_3)_6$ and filtered using filter paper. Cleaned bentonite was put back into 1,250 ml of aquades and added 1.24 g of TBAB then dispersed with a magnetic stirrer for 2 hours. The dispersion result is then disenstrifugated for 2 minutes at a rate of 700 rpm to be separated from the liquid. Bentonite is dried in an oven with a temperature of ± 105 °C to dry. Furthermore, an interlayer opening analysis was used X-Ray Difraction (XRD) tool for raw bentonite and purified bentonite.

2.4 Preparation of Nanocomposite

The formation of nanocompsoite was done by mixing the PLA of the synthesized product with 3 kinds of acid catalyst variations of ± 100 ml and the modified filler clay with variations of 1, 3 and 5 % wt, then heated over hot plate at 65 °C for 20 min. The thickened solution is formed and then poured into a plastic pan and soaked in the oven to dry up to form a nanocomposite sheet of clear plastic. PLA / clay nanocomposites formed are expected to have mechanical and thermal properties that resemble conventional plastics in general.

2.5 Characterization Techniques

2.5.1 X-Ray Diffraction. The nanoclay interlayer spacing was determined by X-ray Diffraction (XRD) measurements using Cu Ka radiation (k = 0.1542 nm), while the generator was set up at 50 kV and 40 mA. The data was collected over a range of scattering angles (2θ) of 1–10.

2.5.2 Universal Tensile Machine (UTM). Dog-bone-shaped specimens for nanocomposite of 150 mm in total length, with a gage section of 12 mm wide by 4 mm thick and 80 mm long, type I based on standard ASTM D638, were prepared by compression molding at 150 °C and a pressure of 25 MPa and subsequent cooling down to 5 °C to avoid sample shrinkage. An Universal Tensile Machine was used to measure the tensile strength of pure PLA adn PLA/Bentonite nanocomposite, according to standard ASTM D638. All tests were carried out under ambient conditions using a cross-head speed of 5 mm/min.

2.5.3 Thermogravimetric Analyzer (TGA). A Thermogravimetric analyzer (TGA) Seiker Extar 6300, Japan) was used to measure the thermal weight loss of pure PLA, and PLA/Bentonite nanocomposite in the temperature range between 30 °C and 900 °C with a heating rate of 10 °C/min under nitrogen steam.

2.5.4 Scanning Elctron Microscope (SEM). The morphology of the disk-shaped nanocomposites was investigated by Electron Microscopy (SEM) at room temperature. A high resolution Hitachi S-4700 microscope operated at 2 kV accelerating voltage was employed for SEM to observe surface of nanocomposite such nanofiller agglomeration and distribution.The SEM samples surfaces were cut using an ultramicrotome equipped with a diamond knife at 100 °C, and then coated with platinum vapor.

3. Result and Discussion

3.1. Characteristic of Filler

The process of clay modification as an initial stage before mixing with the PLA polymer in this study consists of the cleaning and opening stage of the d-spacing layer. The crushed clay filler is cleansed using Sodium Hexametafosphate $(NaPO_3)_6$ which acts as a dispersant. The dispersant time spent for 6 hours and the dispersant quantity of 1% of the treated clay weight is the best value [3]. The purge stage aims to eliminate the impurity content of the clay so as to produce pure montmorillonite. Montmorillonite has the ability to experience swelling to allow the polymer to be intercalated into the mineral gallery and to form a nanocomposite. This swelling power will affect whether or not the quality of nanocomposite produced at the time of mixing. Tetra n-Butyl Ammonium Bromide (TBAB) of 1.24 g as a surfactant used in the opening stage of clay-spacing layer aims to increase the distance between layers of the clay so as to maximize the filler interaction with the polymer to form the nanocomposite. The following table shows the results of the clay layer opening test (d-spacing) to compare the swelling clay power before and after the modification stage.

Table 1. X-RD Result.

Clay	**2θ (degree)**	**d-spacing layer**
Raw Clay	7,2	1,142 nm
Modified Clay	4,52	1,511 nm

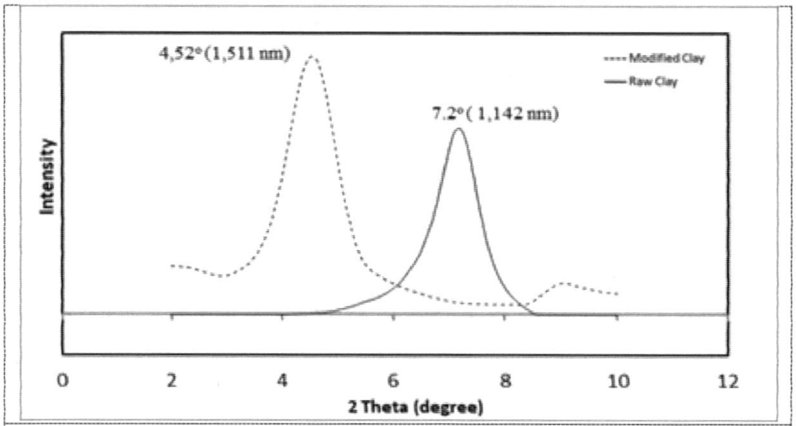

Figure 1. Graph of X-RD Analysis Results.

Based on Figure 1, d-spacing layer clay before and after modification yields values that are not too significantly different. The X-RD graph plots data between the angle of reflection on the x-axis and the intensity on the y-axis to automatically generate the clay d-spacing layer size. Using the peak 2θ setting which yields data at the reflection angle from 2° to 10° indicates that, clay North Aceh prior to modification has a maximum reflection angle peak at 7.2° with a d-spacing value of 1.142 nm, and an increase in the size of its d-spacing After being modified to 1.511 nm at the maximum reflection angle of 4.52°. If the calculated opening of the d-pacing layer on the clay is 0.355 nm. The increased d-spacing layer is caused by the Na + ion composition that affects the swelling properties. This proves that the use of Tetra n-Butyl Ammonium Bromide (TBAB) as a surfactant has succeeded in opening the d-spacing layer in bentonite in the presence of an organic compound exchanging between the two cations (ammonium) of surfactants with ions in montmorillonite (Na+). The opening of d-spacing is also due to the use of $(NaPO_3)_6$ as a cleansing agent which has worked effectively because it is capable of removing impurities that can block the interaction between surfactant and montmorillonite causing clay modification not to occur.

3.2 Mechanical Properties

PLA / Clay nanocomposites which have been mixed according to the variation of the composition are tested for their mechanical properties of the level of tensile strength through the axial forces provided by the tensile test apparatus until it reaches the maximum limit until interrupted. Here is a graph of tensile test results from nine nanocomposite samples and pure PLA samples without mixing with filler clay as a comparison.

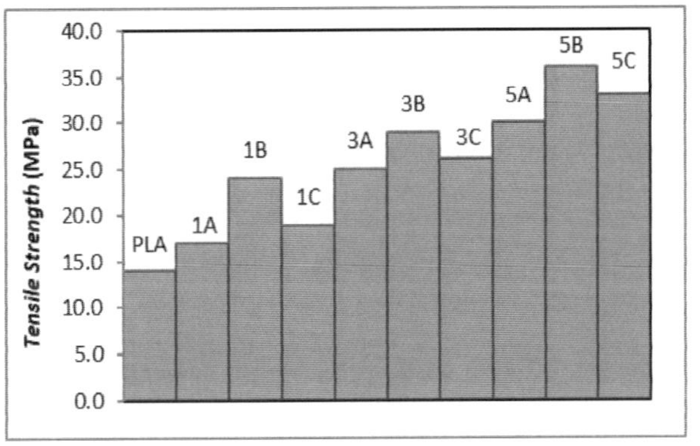

Figure 2. Graph of Tensile Test Results.

Figure 2. above shows the magnitude of the tensile strength value of each sample on the different compositions of PLA and filler and the different acid types. Based on the graph, it can be seen that the addition of a filler showed improved tensile strength to the polymer when compared to the original polymer without mixing. The tensile strength value of the nine samples is 17 to 36 Mpa, while the pure PLA tensile strength value is only 14 MPa. The graph shows that the tensile strength value of the 5B sample is PLA with the addition of 5% filler clay and the hydrolysis result using HCL acid is the maximum value of 36 MPa. Meanwhile, the 1A sample which is an PLA with 3% clay addition and hydrolysis result with 5% HNO_3 acid has the minimum tensile strength value of nanocomposite formed at 17 MPa. Based on the graph, it can be seen that the value of tensile strength of nanocomposite is directly proportional to the amount of filler. The more the amount of clay that is mixed into the PLA matrix, the greater the value of the resulting tensile test. That is, the amount of montmorillonite content as an amplifier causes the polymer and filler to bind well so that it has a higher stretch resistance to its peak. In retrospect, the best average tensile test results are shown by the hydrolysis resultant hydrolysis with 5% HCL acid at each filler concentration. This is because this type of acid has a better level of effectiveness where the rate of hydrolysis reaction of starch more quickly occurs so that the resulting glucose is more optimal than using other types of acids.

3.3 Thermal Stability

Thermal degradation test using TGA (Thermo Gravimetric Analysis) tool is aimed to know qualitatively the thermal stability of the nanocomposite based on the change of sample weight due to heating from room temperature to high temperature is usually hundreds of degrees Celsius, so the sample will be degradation as it burns at a certain temperature. This happens because of the decomposition of chemical bond termination. Here is a table of TGA test results of nine samples of nanocomposites that are formed.

Table 2. TGA Result.

Sampel	1A	3A	5A	1B	3B	5B	1C	3C	5C
Onset Temperature	317,61	328,45	348,09	398,62	442,59	446,63	325,15	351,12	365,04

Table 2 above presents the thermal stability test results in terms of the temperature value on set ie the temperature at which the sample is degraded (thermally decreased) thermally by the increase in temperature. The degradation temperature of nanocomposite samples ranges from 300-450 oC. According to Table 3, PLA / Clay nanocomposites formed from hydrolysis results with three types of acid catalysts (5% HNO_3, 5% HCl and 2% H_2SO_4) have different onset values. Three nanocomposite samples with the best values of these test results (1B, 3B, 5B) began to degrade at a temperature of 398.62 °C; 442,59 °C; And 446.63 °C. The highest degradation temperature was shown by the 5B nanocomposite hydrolysis with HCl acid and the addition of 5% filler clay. As has been reported in the previous carrarization results that this type of acid is the best to be used as a catalyst because the resulting glucose product is more leverage. The results also show that the concentration of clay added in the PLA polymer affects the thermal stability of the nanocomposite. The higher the filler concentration is added, the better the defensive material is characterized by increased degradation temperature. This increase in degradation temperature is caused by bonding on the polymer and the fused filler is stronger so it is difficult to break and the material decomposition becomes slower[8].

3.4 Morphology

The SEM test as an additional test aims to support the main test results of tensile test and thermal degradation test. The samples tested were samples with maximum results (5B) and terminimum (1A) from previous test results. Here is a picture of the results under an electron microscope (SEM).

Figure 3. SEM Test Result of 5B.

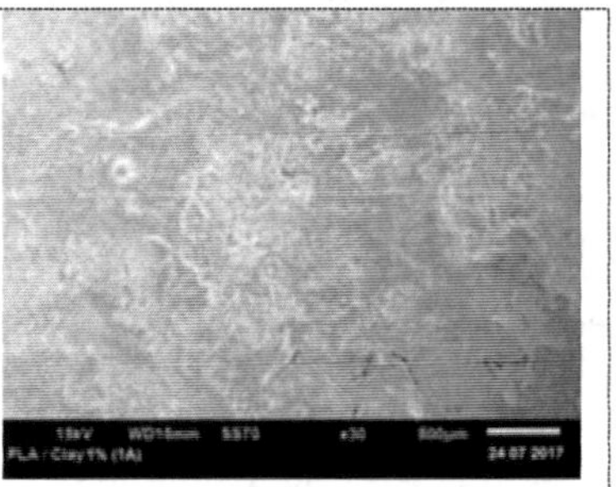

Figure 4. SEM Test Result of 1A.

Figure 3 is the result of SEM test to the 5B sample which is the best resultant nanocomposite and Figure 4 is the result of SEM test to sample 1A which is the nanocomposite with the least result from the result of the previous characterization. When compared, with x50 magnification the two structures look almost identical but the surface of the 5B sample tends to be smoother and forms a good interfacial bond between the matrix and the filler than the sample 1A which shows some aggregate shapes that clump and the cracked surface. This is due to the amount of filler added. Clay fillers as fillers function to bind polymers by the nature of swelling that allows the intercalation of polymers into the clay mineral gallery so as to experience exfoliation or good unification. When correlated with tensile and thermal tests, these SEM test results are correlated where the best nanocomposite surface structure has resulted in good tensile strength and thermal stability[9].

4. Conclusion

This research has generally obtained results in the form of PLA / Clay nanocomposite shaped clear plastic sheet with a thin, flexible and smooth texture that almost resembles conventional plastic in general from the physical aspect. The natural clay has been successfully modified so that it becomes compatible as a PLA filler based on the results of the d-spacing layer test. Based on tensile and thermal test results, tensile strength and thermal stability of nanocomposites increased with increasing amount of added filler and starch hydrolysis using HCL acid showed the best result on final product formation compared to other acids. Samples with maximum values were in 5B nanocomposite Hydrolysis with HCL acid and addition of 5% filler clay with value of tensile strength equal to 36 Mpa and degradation temperature equal to 446,63 ^{o}C. The SEM analysis results show that the best sample nanocomposite surface structure of 5B shows a uniform and supportive surface structure over the main test results.

Acknowledgment

The authors express their gratude and thanks to the Ministry of Research, Technology and Higher Education for Financial Support through the Stranas Grant, 2017 and Ditjen Penguatan Riset dan Pengembangan, Kemenristekdikti and the Directorate of Student Affairs and Education, on funding through grants Student Creativity Program (PKM) at 2017.

References

[1] Rihayat, T., Suryani., 2010. Synthesis and properties of biobased polyurethane/clay nanocomposites. *World Academy of Science, Engineering and Technology*. **65**, 918-922

[2] Averous, L. 2008. Synthesis, Properties and Applications, *In* Monomers, Polymers and Composites from Renewable Resources. Poly Lactid Acid. **21**, 223-225.

[3] Fukusshima, K, Z., Tabuani, D., Camino, G. 2012. Poly (lactid acid)/Clay nanocomposite : Effect of Nature and Content of Clay on Morphology, Thermal and Thermo-Mechanical Properties. *Material Science and Engineering C,*1790-1795.

[4] Gong, Z., Liao, L., Lv, G., Wang, X. 2012. A Simple Method For Physical Purification Of Bentonite. *Applied Clay Science,***119**, 294–300

[5] Kristian, R., Tehrany, EA., Imran, M., Jaquot, M., Desobry, S. 2010. Poly-Lactid Acid Production Application, Nanocomposites And Release Studies.Comprehensive Reviews In Food Science and Food Safety. Vol. **9**, 560.

[6] Mansa, R., Huang, C-Te., Quintel, A., Rocha, F., Detellie, C. 2015. Preparation And Characterization Of Novel Clay/Pla Nanocomposites. *Applied Clay Science*, **115**, 87-96.

[7] Najafi, N., Heuzey, M.C., Carreau, P.J., Adam, M.W. 2012. Control Of Thermal Degradation Of Poltlactide (PLA)-clay nanocomposites using chain exttenders. *Polymer Degradation and Stability*, **97**, 554-565

[8] Rihayat, T., Suryani, S., Zaimahwati, Z. 2014. Effects of heat treatment on the properties of polyurethane/clay nanocomposites paint. *Applied Mechanics and Materials*. **525**, 97-100.

[9] Rihayat, T., Suryani. 2013. Morphology properties of polyurethane/clay nanocomposites based on palm oil polyol paint. *Advanced Materials Research*. **647**, 701-704.

Inhibition effect of fatty amides with secondary compound on carbon steel corrosion in hydrodynamic condition

I M Ibrahim[1,3], J Jai[1], M Daud[2] and Md A Hashim[1]

[1] Faculty of Chemical Engineering, Universiti Teknologi MARA (UiTM), 40450 Shah Alam, Selangor, MALAYSIA.
[2] Industrial Technology Division, Malaysian Nuclear Agency (Nuclear Malaysia) Bangi, 43000 Kajang Selangor, MALAYSIA

E-mail: iznimariah@yahoo.com

Abstract. The inhibition effect demonstrates an increase in the inhibition performance in presence of a secondary compound in the inhibited solution. This study introduces fatty amides as corrosion inhibitor and oxygen scavenger, namely, sodium sulphite as a secondary compound. The main objective is to determine the synergistic inhibition effect of a system by using fatty amides together with sodium sulphite in hydrodynamic condition. The synergistic inhibition of fatty amides and sodium sulphite on corrosion of carbon steel in 3.5 wt% sodium chloride solution had been studied using linear polarization resistance method and scanning electron microscope (SEM) with energy dispersive X-ray spectroscopy (EDX). Electrochemical measurement was carried out using rotating cylinder electrode at different flow regimes (static, laminar, transition and turbulent). Linear polarization resistance experiments showed the changes in polarization resistance when the rotation speed increased. It found that, by addition of fatty amides together with sodium sulphite in test solution, the inhibition efficiency increased when rotation speed increased. The results collected from LPR experiment correlated with results from SEM-EDX. The results showed inhibition efficiency of system was enhanced when fatty amides and oxygen scavengers were present together.

1. Introduction

Presently, the use of organic inhibitor has become a known practice for protection of internal surface made of carbon steel that are in contact with aggressive environment [1]. Amide-based organic compounds have an extensive utilization in industry especially as corrosion inhibitor [2]. An oxygen scavenger, namely, sodium sulphite was introduced as a secondary substance in a test solution containing fatty amides to increase the inhibition efficiency. In flow velocities (hydrodynamic), one would expect that the corrosion mechanism to be more aggressive compared to static condition. For example in oil and gas pipeline system, the transportation of fluid in the pipeline involved the hydrodynamic factor. The fluid flow causes an increase in the transportation rate of chemical species to/from the metal surface, thus resulting in an increase of the internal corrosion rate. Some literature reported on pipeline leaking in Tioga, North Dakota in September 2013 was due to severe internal

[3] To whom any correspondence should be addressed.

corrosion [3]. Such corrosion problems have cost billions of dollars every year and in the middle of 2013 this cost has reached about 3-5% of Gross Domestic Product (GDP) in the United States [4].

According to previous studies, the addition of a secondary substance has increased the inhibition efficiencies due to synergistic effect [5], [6]. Most of the investigations on synergistic effect were under static condition. There are a few studies in the literature on the hydrodynamic effects due to synergism. Some authors have investigated the synergistic effect on inhibition efficiencies in hydrodynamic using different chemicals. For example, Carrillo, et al (2012) [7] evaluates the effectiveness of sodium molybdate with nitrite, phosphate and silicate. Another authors used silicate and phosphonate combination to control corrosion in transmission pipeline [8] and a previous study used ternary inhibitor formulation containing folic acid, a phosphonate and zinc ions [9]. Thus, the main objective of this study is to determine the inhibiton efficiencies when fatty amides are used together with sodium sulphite in different rotation speeds at room temperature (25°C).

2. Experimental

The test solution used 3.5 wt% sodium chloride (NaCl) solution. The fatty amide was synthesized by following the reaction path as reported previously [10]. The working electrode (cylinder electrode) used is made of carbon steel with chemical composition C 0.04 %, Al 0.23 %, Mn 0.74 %, Ni 0.056 %, Cu 0.059 %, Cr 0.060 %, Zn 0.54 %, Pb 0.02 % and the remaining iron analysed by using Inductively Coupled Plasma (ICP) with Optical Emission Spectrometer (OES). The details dimension of the cylinder electrode was reported in our previous study [11]. The cylinder electrode was first mechanically grind and polished using sand paper in grades from 220 to 1000. Then it washed with distilled water and rinsed with acetone. Similar preparation for cylinder electrode was reported by the previous study [12]. Another two electrodes used are saturated calomel and graphite rod as reference and graphite electrode. The electrochemical tests carried out in a rotating cylinder electrode, RCE (Model 636) connected to a Gamry potentiostat. The experiments were performed with rotation speed at 1, 10 and 100 rotation per minute (RPM) which classified as laminar, transition and turbulent respectively at room temperature (25°C). The rotation speed was calculated from Reynolds number (Re) as the following Equation (1) [13]:

$$\mathrm{Re} = \frac{\rho \times u_{cyl} \times d_{cyl}}{\mu} \quad (1)$$

Where ρ, u, dcyl and μ are solution density (1.023 g/cm3), linear velocity (cm/s) of cylinder electrode, diameter of cylinder electrode and solution viscosity (0.00959 g/cm.s) respectively. The measurement of linear polarization resistance test was performed in one hour at a scan rate of 0.166 mV/s for a range of voltage ±20 mV. Constant concentration of fatty amides (20 ppm) with addition of different concentrations of sodium sulphite (500, 1000, and 2000 ppm) were tested in the 3.5 wt. %, NaCl solution. The effect of different concentrations of sodium sulphite was studied. The selection of concentration is based on the preliminary result obtained previously. Then, the inhibition efficiency, IE (%) was calculated using corrosion rate (C.R) as shown in the following Equation (2):

$$I.E(\%) = \frac{C.R_{blank} - C.R_{inh}}{C.R_{blank}} \times 100\% \quad (2)$$

Where $C.R_{blank}$ and $C.R_{inh}$ are the corrosion rate in the absence and presence of the inhibitors respectively. The surface morphologies of the polarized samples were investigated using scanning electron microscope equipped with an Energy dispersive X-ray spectrometer for chemical analysis.

3. Result and discussion

3.1 Electrochemical measurements

In general, the flow is considered under laminar in using RCE provided that Re is less than 50. When Re is greater than 200, the flow is considered under turbulent flow. The transition regime from laminar to turbulent flow occurs when Re between 50 and 200 [14]. The calculated Reynolds number values presented in Table 1 shows three different flow regimes were investigated namely laminar, transition and turbulent flow. The linear polarization resistance (LPR) test was conducted to obtain corrosion rate, C.R (mm/yr). LPR measurement was used because it is an accurate and a rapid method to measure the corrosion rate [15].

Table 2 shows the formulations and the indicators used in this study. The rotation speed at different flow regimes influenced the corrosion rate. The corrosion rate against rotation speeds showed in Figure 1 while its numerical value was tabulated in Table 3. The solutions with addition of 20 ppm fatty amides together with different concentrations of sodium sulphite were tested in different flow regimes. The individual substance of fatty amides or sodium sulphite was also tested to illustrate the comparison. From Figure 1, two patterns were observed describing the trend of corrosion rate. Firstly; for combination system of fatty amides with sodium sulphite, the corrosion rate decreased from laminar (1 RPM) to transition (10 RPM) and increased in turbulent (100 RPM). Secondly; for most of the individual substance (individual fatty amides and individual sodium sulphite), the corrosion rate increased when the rotation speed is increased.

Table 1. Variation in Reynolds number for different rotational speeds in NaCl solution at room temperature (25°C) with laminar, transition and turbulent.

Rotation speed (Revolution per minute, RPM)	1	10	100
Reynolds number (Re)	8	80	804
Anticipated flow regimes	Laminar	Transition	Turbulent

Table 2. The formulation and indicator of chemicals added into the test solution.

Formulation	Indicator
Blank solution	F_o
Individual 20 ppm fatty amides	F_1
Individual 500 ppm sodium sulphite	F_2
Individual 2000 ppm sodium sulphite	F_3
20 fatty amides with 500 ppm sodium sulphite	F_4
20 fatty amides with 2000 ppm sodium sulphite	F_5

Table 3 shows that, the highest inhibition efficicency is F_5 in each rotation which are 94, 51, 61 and 74% in 0, 1, 10 and 100 RPM respectively. They provided dual protection to the carbon steel surface. The fatty amides consist of mixture of amides molecules including stearamide, linoleamide, erucamide, palmitamide and oleamide. The difference in chain lengths of amides provided significant advantages to play a role as an effective corrosion inhibitor to protect the surface of carbon steel. More various chain-lengths of fatty amides compounds formed more protection layers on the surface. In fact, the long chains of carbons can lead to difficulties and prolong the diffusion path for aggressive molecule in solution especially oxygen and water molecule to penetrate onto steel surface. Besides, fatty amides molecules contain nitrogen element (N) that is known to have a pair of electron (lone pair) from sp^2 electron configuration that promotes surface adsorption. The lone pair can adsorb onto carbon steel surface then, forming a protective layer. This finding was also supported by previous studies [16], [17]. In addition, the presence of sodium sulphite in the inhibited solution containing fatty amides reduced the culprit molecule in the aqueous solution namely dissolved oxygen. Sodium sulphite reacted with the dissolved oxygen and then reduced the concentration of dissolved oxygen in the solution. The reaction between sodium sulphite and dissolved oxygen is as following the Equation (3):

$$2Na_2SO_3 + O_2 \rightarrow 2Na_2SO_4 \quad (3)$$

The corrosion rates are greatly decreased in all tested rotation speeds when fatty amides and sodium sulphite (F_4 and F_5) were used together as shown in Table 3. It could be that, the combination of fatty amides with sodium sulphite that contribute to synergistic effect to enhance the inhibition efficiency compared to merely individual fatty amides or sodium sulphite. The presence of sodium sulphite in the inhibited solution suppressed the cathodic reaction in corrosion mechanism and indirectly helped the film formation of fatty amides. The sodium sulphite indirectly supported film formation by scavenging the dissolved oxygen in the solution and then reduced the chances of dissolved oxygen to attack the metal surface. The reaction of sodium sulphite and dissolved oxygen produced sodium sulfate, Na_2SO_4. However, the sodium sulfate do not contribute any consequences to the corrosion process and was proven in the previous study [18]. When concentration of sodium sulphite increased (500, 1000 and 2000 ppm) in the inhibited solution, more dissolved oxygen molecules in the solution reacted and depleted in the solution which results in more protected film formation of fatty amides. Another highlight is the corrosin rate and inhibition efficiency achieved by both individual fatty amide and sodium sulphite (F_1, F_2 and F_3). The corrosion rate of specimens increased in addition of individual substances when tested in the different rotation speeds. Thus, the calculated inhibition efficiency of individual substance (F_1, F_2 and F_3) in Table 3 shows much lower values compared to the inhibition by combination of both in all tested flow velocities. In fact, there are some negative values in the inhibition efficiency of individual fatty amides and individual sodium sulphite when tested in moving condition. It suggested that both individual fatty amide and sodium sulphite able to promote the corrosion reaction instead of inhibited the corrosion process. When facing different flow velocities (laminar, transition and turbulent) the individual substance increased the corrosion rate due to insufficient protection. Thus, the corrosion inhibition provided by formulation of F_1, F_2 and F_3 unable work properly, particularly in the rotation speeds. It also indicated that the presence of both fatty amides and sodium sulphite are needed to maintain the inhibition effeciency even in turbulent condition.

The influence of rotation speeds on the corrosion performance is an important factor to be considered in this study. According to Jiang et al (2005) [19], the rotation speeds have two opposite effects on the inhibition performances. First, it increased the mass transport of inhibitor molecules to the specimen surface and it was beneficial to improve the inhibition efficiency. On the other hand, the rotation speeds results in high shear stress that will remove the inhibitor layers from the specimen surface. This second effect of rotation speed had an adverse effect to inhibition performance. In this current study, the formulation of combination substances (F_4 and F_5) decreased the corrosion rate from 1 RPM (laminar) to 100 RPM (turbulent) as illustrated in the Figure 1 and numerical values as tabulated in Table 3. The increase in mass transfer of fatty amides onto specimen surface at higher rotation speeds provided the maximum protection in this study. Larger amount of fatty amides were either physically or chemically adsorb, then forming protection layers. At the same time, large amount of sodium sulphite in the inhibited solution reacted with dissolved oxygen in the solution making the specimen surface indirectly well protected. In the combination system (F_4 and F_5), the formulations can perform better in the following order $F_5 > F_4$ in most of flow velocities tested. The corrosion rate increased when tested in the different rotation speeds for formulation of individual substances (F_1, F_2 and F_3). It showed the inhibition process of individual substances during rotation speeds was not working properly. In fact, as shown in the Table 3, the inhibition efficiency of individual substance reached negative values. It indicates that, the addition of individual substance in difference rotation speeds promoted corrosion reaction instead of inhibited the corrosion process.

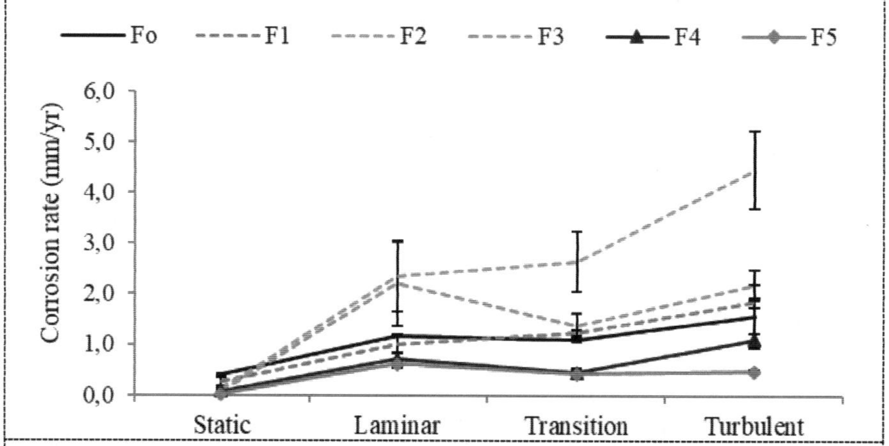

Figure 1. Influence of rotation speed on corrosion rate values of steel at different flow regimes; static, laminar (1 RPM), transition (10 RPM) and turbulent (100 RPM) using linear polarization resistance (LPR) test.

Table 3. The values of corrosion rate, C.R (ohm.cm^2) and inhibition efficiency, IE (%) of systems in different rotation speeds (0, 1, 10 and 100 RPM).

Solutions	0 RPM (Static)		1 RPM (Laminar)		10 RPM (Transition)		100 RPM (Turbulent)	
	C.R (mm/yr)	I.E (%)	C.R (mm/yr)	IE (%)	C.R (mm/yr)	IE (%)	C.R (mm/yr)	IE (%)
F_0	0.396	-	1.178	-	1.095	-	1.572	-
F_1	0.257	35	1.003	2	1.234	-3	1.844	-16
F_2	0.119	70	2.355	-108	2.652	-121	4.466	-131
F_3	0.066	83	2.202	-55	1.388	-11	2.190	-28
F_4	0.080	80	0.709	33	0.443	59	1.097	45
F_5	0.025	94	0.613	51	0.426	61	0.468	74

3.2 Surface morphology and surface analysis by SEM-EDX

Figure 2 shows the Scanning Electron Microscope, SEM images of the selected carbon steel specimen under critical condition of turbulent flow. Figure 2(a) shows a selected area of specimen tested in the presence of 20 ppm fatty amides individually and Figure 2(b) shows specimen in the presence of 20 ppm fatty amides with 2000 ppm sodium sulphite (F_5). It is observed that the specimen surface tested in solution containing individual 20 ppm fatty amides in turbulent condition was severely damaged with the emerging of cracks and pores as observed in Figure 2(a). It was suggested that the dissolution of metal and penetration of aggressive species in the test solution such as chloride ions (Cl^-) onto the metal surface was occurred. The steel surface was not well protected by individual fatty amides during turbulent flow. The adsorption of fatty amides molecule were detached from the surface due to high shear stress applied. It was supported by analysis of chemical composition which indicated the absence of nitrogen element from the surface. Meanwhile, figure 2(b) shows a better surface morphology with the presence of both fatty amides and sodium sulphite, F_5. In addition, the EDX evaluations show a higher percentage of elemental oxygen can be observed in Figure 3(a) compared to Figure 3(b) due to the higher formation of oxide products in specimen tested in individual fatty amide during turbulent condition.

4. Conclusion

The inhibition property can be enhanced with the presence of both fatty amides and sodium sulphite at different flow velocities. The enhancement of inhibition actions are due to binary protections provided by surface adsorption of fatty amide and reaction of sodium sulphite with dissolved oxygen molecules. This shows a synergistic effect between fatty amides with sodium sulphite present in the inhibition system even when facing different flow velocities.

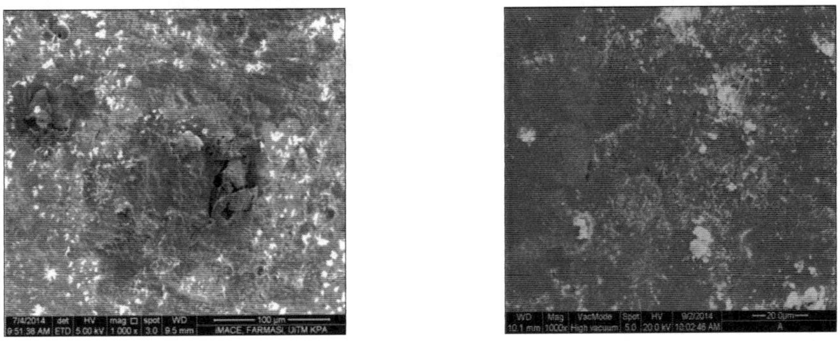

(a) (b)

Figure 2. SEM images for specimens surface immersed in test solution in turbulent condition containing (a) only 20 ppm fatty amides (b) 20 ppm fatty amides with 2000 ppm sodium sulphite.

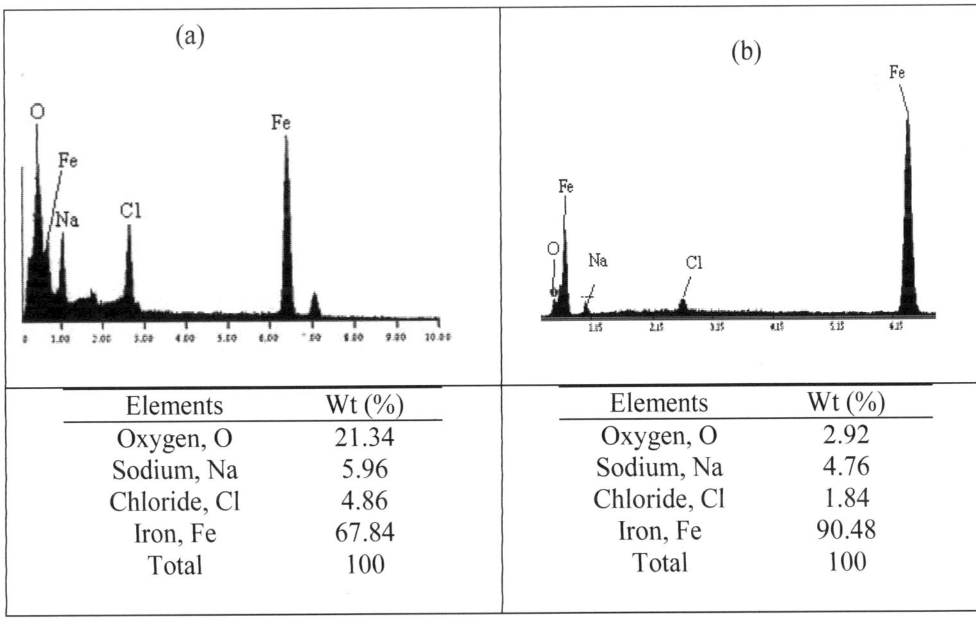

Elements	Wt (%)
Oxygen, O	21.34
Sodium, Na	5.96
Chloride, Cl	4.86
Iron, Fe	67.84
Total	100

Elements	Wt (%)
Oxygen, O	2.92
Sodium, Na	4.76
Chloride, Cl	1.84
Iron, Fe	90.48
Total	100

Figure 3. EDX analysis of polarized samples in turbulent condition containing (a) 20 ppm fatty amides (b) 20 ppm fatty amides with 2000 ppm sodium sulphite.

Acknowledgement
The authors would like to thank and gratefully acknowledged use of services and research facilities available at the Faculty of Chemical Engineering, Universiti Teknologi MARA, Malaysia and MTeG, Malaysian Nuclear Agency (Nuclear Malaysia).

References

[1] G. Khan, K. M. S. Newaz, W. J. Basirun, H. Mohd Ali, F. Faraj Lafta, and G. M. Khan, "Application of Natural Product Extracts as Green Corrosion Inhibitors for Metals and Alloys in Acid Pickling Processes- A review," *Int. J. Electrochem. Sci.*, vol. 10, pp. 6120–6134, 2015.
[2] A. Yildirim, S. Öztürk, and M. Çetin, "Novel amide-based cationic surfactants as efficient corrosion inhibitors for carbon steel in HCl and H2SO4 media," *J. Surfactants Deterg.*, vol. 16, no. 1, pp. 13–23, 2013.
[3] P. Hoffman and W. Bryan, "Year-in-Review : 2013," 2014.
[4] M. P. H. Gerhardus H. Koch and and N. G. T. Y. P. V. J. H. P. Brongers, "Corrosion costs and preventive strategies in the United States," 2002.
[5] S. A. Umoren and U. F. Ekanem, "Inhibition of mild steel corrosion in H2SO4 using exudate gum from pahylobus edulis and synergistic potassium halide additives," *Chem. Eng. Commun.*, vol. 197, no. 10, pp. 1339–1356, 2010.
[6] A. Y. Musa, A. B. Mohamad, A. A. H. Kadhum, M. S. Takriff, and L. T. Tien, "Synergistic effect of potassium iodide with phthalazone on the corrosion," *J. Corros. Sci.*, vol. 53, 2011.
[7] I. Carrillo, B. Valdez, R. Zlatev, M. Stoytcheva, M. Carrillo, and R. Bäßler, "Electrochemical study of oxyanions effect on galvanic corrosion inhibition," *Int. J. Electrochem. Sci.*, vol. 7, pp. 8688–8701, 2012.
[8] M. Salasi, T. Shahrabi, and E. Roayaei, "Effect of inhibitor concentration and hydrodynamic conditions on the inhibitive behaviour of combinations of sodium silicate and HEDP for corrosion control in carbon steel water transmission pipes," *Anti-Corrosion Methods Mater.*, vol. 54, no. 2, pp. 82–92, 2007.
[9] D. S. Kalyani, S. R. S, S. B. M, and B. Sreedhar, "Synergistic effect of folic acid on corrosion inhibition of carbon steel in presence of a phosphanet and zinc ions," *Int. J. Recent Sci. Res.*, vol. 6, pp. 4361–4365, 2015.
[10] I. M. Ibrahim, J. Jai, and M. A. Hashim, "Comparative inhibition effect of synthesized fatty amides mixture, pyridine and pyrrole on carbon steel in saline environment," *Appl. Mech. Mater.*, vol. 575, pp. 206–209, 2014.
[11] I. M. Ibrahim, M. Daud, J. Jai, and A. Hashim, "Effect of secondary species on inhibition efficiency of fatty amide mixtures in dynamic environment," *Int. J. Eng. Technol.*, vol. 7, no. 3, pp. 844–851, 2015.
[12] I. M. Ibrahim, S. Yunus, and M. A. Hashim, "Relative performance of isoproopylamine, pyrrole and pyridine as corrosion inhibitors for carbon steels in saline water at mildly elevated temperatures," *Int. J. Sci. Eng. Res.*, vol. 4, no. 2, 2013.
[13] Pine Research Instrumentation, "Study of mass-transport limited corrosion using pine rotating cylinder electrodes-An overview of theory and practise," pp. 1–6, 2006.
[14] A. Y. Musa, A. H. Kadhum, and A. B. Muhamad, "Corrosion inhibitor film forming in aerated and deaerated solutions," *Int. J. Electrochem. Sci.*, vol. 5, pp. 1911–1921, 2010.
[15] ASTM G59-97 (Reapproved), *Standard test methods for conducting potentiodynamic polarization resistance measurement*. 2009, pp. 1–4.
[16] B. E. A. Rani and B. B. J. Basu, "Green inhibitors for corrosion protection of metals and alloys: An overview," *Int. J. Corros.*, vol. 2012, no. i, 2012.
[17] I. M. Ibrahim, J. Jai, and M. A. Hashim, "Inhibitive effect of fatty amide and secondary species on the corrosion of carbon steel," *Adv. Mater. Res.*, vol. 1133, pp. 366–370, 2016.

[18] T. I. Wu and J. K. Wu, "Effect of sulfate ions on corrosion inhibition of AA 7075 aluminum alloy in sodium chloride solutions," *Corrosion*, vol. 51, no. 3, pp. 185–190, 1995.
[19] X. Jiang, Y. G. Zheng, and W. Ke, "Effect of flow velocity and entrained sand on inhibition performances of two inhibitors for CO2 corrosion of N80 steel in 3% NaCl solution," *Corros. Sci.*, vol. 47, no. 11, pp. 2636–2658, 2005.

Recovery of Cobalt from leach solution of spent oil Hydrodesulphurization catalyst using a synergistic system consisting of Versatic™ 10 and Cyanex®272

Yuliusman[1,2], I T Ramadhan[1] and M Huda[1]
[1] Chemical Engineering Department, Faculty of Engineering, Universitas Indonesia, Kampus UI Depok 16424

E-mail: usman@che.ui.ac.id

Abstract. Catalyst are often used in the petroleum refinery industry, especially cobalt-based catalyst such as CoMoX. Every year, Indonesia's oil industry produces around 1350 tons of spent hydrodesulphurization catalyst in which cobalt makes up for 7%wt. of them. Cobalt is a non-renewable and highly valuable resource. Taking into account the aforementioned reasons, this research was made to recover cobalt from spent hydrodesulphurization catalyst so that it can be reused by industries needing them. The methods used in the recovery of cobalt from the waste catalyst leach solution are liquid-liquid extraction using a synergistic system of Versatic™ 10 and Cyanex®272. Based on the experiments done using the aforementioned methods and materials, the optimum condition for the extraction process: concentration of Versatic™ 10 of 0.35 M, Cyanex®272 of 0.25 M, temperature of 23-25°C (room temperature), and pH of 6 with an extraction percentage of 98.80% and co-extraction of Ni at 93.51%.

1. Introduction

Petroleum consists not only of hydrocarbons, but also contains impurities such as sulfur, nitrogen, oxygen, and metals (mainly Ni and V). The presence of these contaminants results in damage to equipment, catalysts, poor product quality, and environmental pollution [9], therefore they should be removed. In petroleum processing, hydrodesulphurization process is the way to remove all of these contaminants. The oil and oil fraction hydrodesulphurization process are universally carried out by the catalytic reaction between hydrogen and the sulfur components present in the oil and its fractions to produce hydrogen sulfide (H_2S) which can be easily separated from the oil. In naphtha pretreatment, this is done by using a cobalt-molybdenum catalyst to avoid contact between sulfur and precious metal catalysts used in the oil reform process, which cannot tolerate the presence of sulfur.

The cobalt (CoMoX) catalyst used in the hydrodesulphurization process is usually solid. CoMoX $(\gamma - Al_2O_3 - CoO - MoO_3)$ is the most commonly used catalyst for hydrofining. CoMoX takes on the appearance of a dull blue solid. In general, the active constituents present in the catalyst are cobalt oxide in the form of CoO 6.00%wt. and molybdenum oxide in the form of MoO_3 20.2%wt. [3]. However, The Cobalt Development Institute also mentioned that its active constituents consist of cobalt oxide in the form of Co_3O_4 in the amount of 3-5%%wt. and molybdenum oxide in the form of MoO_3 in the amount of 14%%wt [8]. The cobalt catalyst is used in the industry continuously in a

[2] To whom any correspondence should be addressed.

continuous process. As the catalyst continues to be used and regenerated, the catalyst activity will decrease until it is uneconomical to regenerate it, leaving the catalyst disposed of into waste. On average, the lifespan of a cobalt catalyst is 3 to 5 years.

Based on calculations made using Schuman & Shalit method [7] using the latest data from international energy statistics from US Energy Information Administration [10], given that the aggregate production of oil refineries in Indonesia reaches 889.3 thousand barrels/day, it will produce spent catalyst waste of 1,350 tons/year. Thus, the wasted cobalt metal is about 7% of that amount [3], which is 70 kg of cobalt/1 ton of catalyst waste. This amount is very significant considering that cobalt is a metal that has a high selling value. The price of cobalt in mid-2016 is $27,000 per ton or about Rp 356,000,000 per ton, so we can assume that the oil industry can recover around 34 billion rupiahs per year.

Based on the aforementioned reasons, cobalt recovery from spent catalyst weight is necessary. One commonly used method is liquid-liquid extraction. Cobalt is generally extracted using a single extractant (SX) such as Cyanex®272. Cyanex®272 can extract cobalt with an efficiency of >98.6% but this process requires two stages and produces large excess of organic phase [4]. Thus, some institutions also used Versatic™ 10 because this extractant has a low percentage of extraction on impurities such as magnesium and calcium metals. However, the use of this extractant is inefficient because it requires extraction and stripping of two stages to achieve the desired level of purity of cobalt extraction. It can be concluded that single extractants are generally not selective to cobalt when compared with other impurities found in leach liquor, such as copper, zinc, iron (II), cadmium, calcium, magnesium, and manganese. Therefore, a technique often used in the field of extractive hydrometallurgy to obtain higher separation efficiency and metal concentration is to use a synergic mixture of multiple extractants (SSX). When Versatic™ 10 is used as a component in a synergistic extractant mixture with Cyanex®272, the percentage of cobalt extraction is increased to 97.48% [6]. Mubarok and Hanif did an extraction experiment with a synergistic mixture between Versatic™ 10 20%vol. and Cyanex®272 20%vol. at pH of 6.95 and temperature of 40°C.

2. Materials and Methods

2.1. Materials

In the extraction process, the extractants used are Versatic™ 10 and Cyanex®272. Versatic™ 10 is a synthetic tertiary carboxylic acid that has many branches and has a trivial name of neodecanoic acid. Versatic™ 10 has 10 carbon atoms on the main chain and is liquid at room temperature. In this experiment, Versatic™ 10 was imported from Haihang Industry Co., Ltd. Cyanex®272 ($C_{16}H_{34}PO_2H$) is an extractant that belongs to the acid extractant class, in which its main component is bis-(2,4,4-trimethylpentyl) phosphinic acid compound that has POOH as an active group [11]. In this experiment, Cyanex®272 was given by BASF Indonesia. The reason why this extractant mixture is chosen to recover the cobalt metal is:

- The synergistic mixture of Versatic™ 10 and Cyanex®272 meet the fundamental criteria that include low solubility properties in aquatic phases, have good stability to survive several months of recycling on the extraction process without degradation, will not form a stable emulsion with the aquatic phase when stirred, has a high-loading capacity, not flammable, non-volatile, relatively inexpensive, soluble in organic solvents, and has good extraction kinetics.
- Versatic™ 10 and Cyanex®272 synergistic mixture is a commonly used mixture for cobalt separation of leach solution in sulfuric acid or chloride media by liquid-liquid extraction method. In addition, this synergistic mixture can also extract cobalt metal ions up to high concentrations at optimum pH.

In general, Versatic™ 10 and Cyanex®272 extractants are soluble in aromatic and aliphatic solvents, stable against heat and hydrolysis so the solvent used to dilute them is kerosene. Selection of kerosene as an organic solvent is due to:

IOP Conf. Series: Materials Science and Engineering **334** (2018) 012049 doi:10.1088/1757-899X/334/1/012049

- Kerosene density is quite low at 0.6 g/L thus reducing possible errors at two phase separation as the boundary between two phases is visible.
- When compared with other organic solvents having a dielectric constant above 2.2, kerosene is the best solvent ever tested with a 2-2.2 dielectric constant. Increasing the dielectric constant will increase the solvency of the solvent.
- Low kerosene viscosity makes the value of surface tension also low. With a low surface tension the extractant can pull the metal from the aqueous phase more easily.

2.2. Methods

The samples used in this experiment were from the leach solution of spent catalyst waste from Hydrogen Plant UP VI Balongan that contained 1.3263 mg/g of Co and 156.17 of Ni. In this study, extraction experiments were performed using the Versatic™ 10 and Cyanex®272 extractant synergic mixtures dissolved in kerosene. In the process of extraction of cobalt metal, the variations being tested are aquatic phase pH and extractant concentration. Both of these variables can affect the extraction efficiency.

In this paper, two studies were conducted with different types of test variables. The first tested variable is the extractant concentration variable, then after the optimal concentration of extractant is known, a second research is done with pH variation of the leaching solution. In the experiment of concentration variation, the extraction was done three times (single stage SX using Versatic™ 10, SSX using Versatic™ 10 and Cyanex®272 where the concentration of Versatic™ 10 is varied and Cyanex®272 is at 0.25 M, and SSX using Versatic™ 10 and Cyanex®272 where the concentration of Cyanex®272 is varied and Versatic™ 10 is at 0.25 M) with variation of extractant concentration of 0,25 M, 0,35 M, 0,4 M and 0,5 M. In the experiment of pH variation, pH variation was performed by NaOH into the solution. The addition of NaOH changes the pH of the acid to a less acidic state. The pH variations were 4, 4,5, 5, 5,5, and 6. The extraction process for each variation was performed using a magnetic stirrer, with a volume ratio O:A (organic:aquatic) of 1:1, with the volume of each solution being 20 ml. Stirring was done for 60 minutes per sample. The next process is the separation of organic and aqueous phase. The two phases naturally separated into two layers after the extraction process was completed. The separation of the two phases is done using a separation funnel. AAS (Atomic Absorption Spectroscopy) test is used to view the results of the liquid-liquid extraction process for the aqueous phase, so that the metal content of the aqueous phase can be known. The AAS test is done in Laboratorium Kimia Terpadu IPB Baranangsiang.

3. Results and Discussion

3.1. Extractant Concentration Variation Results

The extraction process was carried out using a synergistic mixture of Versatic™ 10 and Cyanex®272 dissolved in an organic solvent called kerosene. This synergistic mixture is mixed with the leach solution that has its pH converted to 0.5 beforehand – following the minimum operating conditions proposed by Cheng, et al. [2]. The process of changing the pH is done by adding NaOH solids gradually until the pH reaches 0.5. The mixture is then filtered using a filter paper to remove the Na_2SO_4 salt content which can interfere with the extraction process and the separation process later in the separatory funnel. The volume ratio between the organic phase and aquatic phase is 1:1. After that, the solution is stirred at 750 rpm (3 mot) for 30 minutes at room temperature. After stirring, the mixture is transferred to and kept at the separatory funnel for 15-30 minutes until there is a separation between the organic phase and the aqueous phase. Result of AAS analysis of aquatic phase after extraction and calculation of efficiency of cobalt extraction process shown in Table 1 and Figure 1.

Table 1. Results of Extractant Concentration Variation.

Extractant Concentration (M)	Co Concentration (mg/g)	Extraction Efficiency (%)
Versatic™ 10 (V.) 0.25 M	0.159	85.28
V. 0.35 M	0.155	87.26
V. 0.4 M	0.162	86.71
V. 0.5 M	0.158	85.42
Cyanex®272 (C.) 0.25 M + V. 0.25 M	0.167	87.45
C. 0.25 M + V. 0.35 M	0.131	90.14
C. 0.25 M + V. 0.4 M	0.134	89.97
C. 0.25 M + V. 0.5 M	0.167	87.48
V. 0.25 M + C. 0.35 M	0.163	87.45
V. 0.25 M + C. 0.4 M	0.207	87.77
V. 0.25 M + C. 0.4 M	0.206	84.49

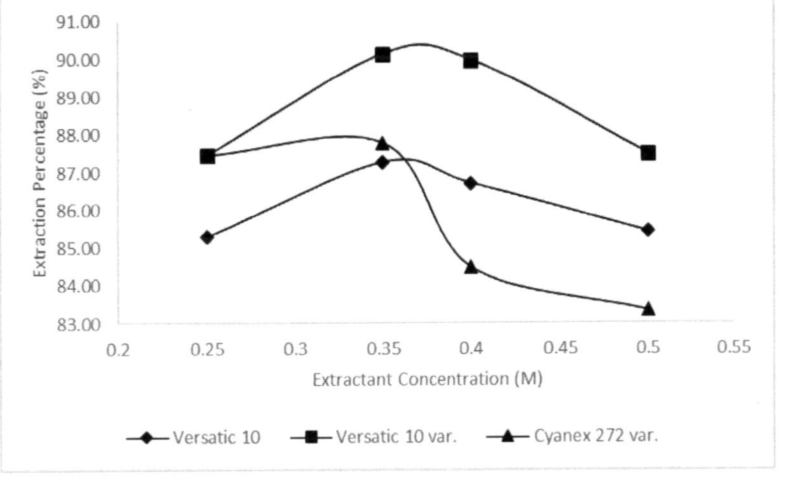

Figure 1. Results of Extractant Concentration Variation.

The AAS test is carried out on the aqueous phase - not the organic phase because kerosene is a flammable solution - after the extraction process so that the amount of cobalt that remains in the organic phase can be known and the amount will be compared with the amount of cobalt metal prior extraction. The smaller the number of cobalt metal on the AAS results indicates that more cobalt metal is successfully extracted, whereas on the contrary if there is a lot of cobalt metal in the AAS results, then the amount of cobalt metal extracted is small so that much remains in the extraction solution.

3.1.1. Single Stage SX Using Versatic™ 10

Le Chatelier principle states that changes in any factor in the system will affect the system equilibrium. If one part is added then the equilibrium will shift to the side which is not added so adding the amount of extractant theoretically will add the product that is produced. When ionized, the active group COOH in Versatic™ 10 will form a COO- group and will bind to the cobalt ion to form the extracted metal compound. The more COOH active groups that is present in the mixture the higher the percentage of cobalt metal extraction. This is evidenced by the increase in the percentage of extraction when the concentration is increased from 0.25 M to 0.35 M.

The amount of bonded metal will always be lower than the theoretical amount because the extractant will form a dimer bond where the neodecanoic acid compounds will bind together because of the hydrogen bond. The formation of dimers occurs because of the presence of hydrogen bonds at a high enough concentration where the collision between the extractants is greater. The dimer effect affects the extraction process if the metal content in the leach liquor is low. Further increases in the concentration of the extractant will not increase the percentage of significant extraction - it can be seen that the percentage of extraction decreases as the extractant concentration is above 0.35 M. So, the addition of extraction becomes inefficient from an economic point of view.

3.1.2. Single Stage SSX Using Versatic™ 10 and Cyanex®272

The mechanism of the Co^{2+} extraction process is where two molecules of Versatic™ 10 bind one M^{2+} molecule. The increase in the percentage of cobalt extraction is quite large with the increased concentration of Versatic™ 10. However, the percentage of extraction that can be achieved by Versatic™ 10 can still be improved by using the SSX method. Therefore, Cyanex®272 is added to increase the percentage of cobalt extraction. Figure 1 shows that the addition of Cyanex®272 increases the percentage of cobalt extraction. The image also shows that the percentage of extraction higher when the concentration that is being varied is Versatic™ 10. This indicates that Cyanex®272 is a synergist in this synergistic mixture. This result is consistent with the hypothesis given by Wilson [11].

The existence of two different types of ligand complexes leads to an increase in the hydrophobic character of the metal complexes produced between the cobalt and the extractant thus permitting the formation of more stable compounds or the binding of higher quantities of metal ions. In general, the reaction occurring in this process is the condition where the more acidic extractant ligand (A) has not been able to meet the coordination capacity of the MA_z chelate so the synergist (S) will trigger the formation of MA_zS_x adduct which has a higher extraction ability than the MA_z parent chelate. The extraction mechanism that occurs is the coordination between the extractant ligand and the metal ion and if the coordination number of the metal ion is not fulfilled then the water molecule will enter into the inner sphere. Furthermore, the water molecule will be displaced by a ligand from a donor (synergist) that has a higher affinity to the metal ion (reaction 1). Another mechanism may also be the expansion of the coordination spheres of the metal ions, thus increasing the probability of the ligand of the synergist to bind to the complex (reaction 2). In addition, donor ligands may also bind to a metal-chelate complex by hydrogen bonding with water molecules present in the inner sphere (reaction 3).

$$M(L)_2(H_2O)_2 + (\text{donor}) \rightarrow M(L)_2(\text{donor})(H_2O) + H_2O \quad (1)$$

$$M(L)_2(H_2O)_2 + (\text{donor}) \rightarrow M(L)_2(H_2O)_2(\text{donor}) \quad (2)$$

$$M(L)_2(H_2O)_2 + (\text{donor}) \rightarrow M(L)_2(HO - H - \text{donor}) \quad (3)$$

in which the reactions above take place with the assumption that the metal ions are bound in the form of hexa co-ordination in simple aqua complex $M(L)_2(H_2O)_2$ and L is the bidentate ligand.

However, the addition of excessive amount of Cyanex®272 will decrease the percentage of cobalt extraction. This phenomenon can be explained by three possibilities. First, this phenomenon occurs due to the increased viscosity that occurs to the Cyanex®272 extractant when its concentration is increased. Ayanda, et al. declares that if Cyanex®272 is saturated up to 12-20 gram/dm^3 Co^{2+}, then its viscosity will rise to >150 cst [1]. In this condition, the ability of the magnetic bar to mix the mixture becomes reduced so that the percentage of extraction also becomes reduced. Secondly, there may also be antagonistic extraction happening in which Versatic™ 10 and Cyanex®272 extractants are joined together with each other thus inhibiting MA_z formation. Thirdly, just like Versatic™ 10, Cyanex®272 can also form dimers if their concentration is higher than metal concentrations. This phenomenon can be seen in Figure 1 where the percentage of extraction decreases when the concentration of Cyanex®272 is more than 0.35 M to the extent that the extraction percentage is even lower than the extraction using single stage Versatic™ 10 SX.

3.2. Leach Solution pH Variation Results

pH of aquatic phase is a highly influential variable on the extraction process. This is because in the separation process using extraction method, the optimum pH value is needed to be known where the ionic compounds to be taken is bound to the extractant while the impurities do not because their optimum extraction pH are in a different range. The pH of the leaching solution was varied with a pH range of 4-6 wherein the percentage of extraction using a Versatic™ 10 0.35 M/Cyanex®272 0.25 M synergistic mixture at each point of variation is shown in table 2 and figure 2.

Table 2. Leach Solution pH Variation Results.

pH	Co Concentration (mg/g)	Extraction Efficiency (%)
0.5	0.131	90.14
4	0.0153	96.62
4.5	0.0133	97.06
5	0.0083	97.90
6	0.0055	98.80

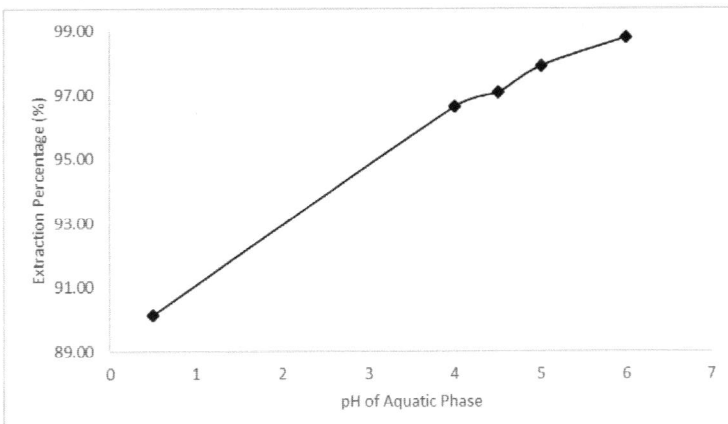

Figure 2. Leach Solution pH Variation Results.

In accordance with the figure above, it can be seen that the greater the pH of the waste the greater the percentage of cobalt extraction. The highest value of extraction efficiency (98.80%) was achieved at the pH of 6. Increase in pH causes an increase in the percentage of extraction. But, with higher value of pH, the extraction percentage will gradually decrease. pH system will affect the metal ions and extractants. At low pH, the ability of extraction decreases due to the effect of proton donation to the extractant. The proton donation causes imperfect dissociation of the extractant molecules so that the metal ion extraction process is not running perfectly. The cobalt extraction process increases with increasing amount of proton concentration up to a certain concentration.

The extraction process begins with the extractant dissociation process. After that, the activated group of the ionized extractant binds the metal ion so that the metal compound will bind to the extractant. The cobalt ion will move from the aqueous phase into the organic phase. The reaction is a reversible reaction, so that if the process reaches the equilibrium point, then the reaction will shift and the percentage of extraction will not increase anymore. After the metal ions are bonded with the extractant, the H^+ and SO_4^{2-} ions will be released, and will return to the aquatic phase. In theory, the release of the ion will affect the pH. However, the addition of these ions has no effect on the actual percentage of extraction. This is due to the presence of Co^{2+} ions in the mixture so that it can still bind to phosphate ions or carboxylates and also because the based extractant already underwent perfect ionization so that the pH change does not decrease the extraction efficiency.

At high pH, the extraction ability decreases due to the hydrolysis of the ion. The hydrolysis reaction will result in a soluble metal hydroxide compound. This causes the metal that is not in ionic form to be difficult to extract. The hydrolysis reaction is the reaction of the breakdown of the O-H bond on the hydrate ion $[M(H_2O)_6]^{n+}$ formed when the salt of the metal cation is hydrated in water. This reaction occurs because the tensile forces between the ion and the water molecule are so large that the O-H bond on one or more water molecules will break. The dissociation constant (K_b) for the hydrolysis reaction of Co^{2+} is 14.12×10^{-11} in which the greater the dissociation constant for the hydrolysis of the metal ions indicates the pH at which the metal ion is hydrolyzed lower. Ma, et al. proves that the cobalt metal will be hydrolyzed upon reaching a pH of 7.5 [5]. Therefore, a decrease in the percentage of extraction will occur when the system pH is >7.5.

3.3. Optimum Condition of the Extraction Process

In this experiment, it was concluded that the optimum condition of the cobalt metal extraction process from spent catalyst waste was by using an initial leach solution pH of 6, Versatic™ 10 concentration of 0.35 M, and Cyanex®272 concentration of 0.25 M resulting in an extraction percentage of 98.80 %. Unfortunately, this was accompanied by co-extraction of Nickel at 93.51%, thus additional treatment is needed to separate cobalt from the organic phase.

4. Conclusion

Based on the results of research and analysis conducted, it can be concluded that the operating condition of liquid-liquid extraction process that yields optimum cobalt extraction by using the Versatic™ 10/Cyanex®272 extractant synergic mixture:

- Leach solution pH of 6.
- Versatic™ 10 concentration of 0.35 M.
- Cyanex®272 concentration of 0.25 M.

Reference

[1] Ayanda, O. S., Baba, A. A., Ximba, B. J. & Fatoki, O. S., 2013. Application of Cyanex® extractant in Cobalt/Nickel separation process by solvent extraction. *International Journal of Physical Sciences,* 8(3), pp. 89-97.

[2] Cheng, C. Y., Barnard, K. R., Zhang, W., Zhu, Z. & Pranolo, Y., 2016. Recovery of nickel, cobalt, copper and zinc in sulphate and chloride solutions using synergistic solvent extraction. *Chinese Journal of Chemical Engineering,* 24(2), pp. 237-248.

[3] Gour, P. K., Upadhyay, S. N., Tiwari, J. S., Ghosh, P. K., Bhattacharyya, N. B. & Sen, S. P., 1977. Studies on CoMoX Catalyst - PART 1: A Study on Fresh γ-Al_2O_3-CoO, γ-Al_2O_3-MoO_3, and γ-Al_2O_3-CoO-MoO_3 Systems by ESR Technique. *Journal of the Research Institute for Catalysis Hokkaido University,* 25(2), pp. 91-98.

[4] Jones, D. L., McCoy, T. M., Mayhew, K. E., Cheng, C. Y., Barnard K. R. & Zhang, W., 2010. *A New Process for Cobalt - Nickel Separation.* 1st Edition. Perth: ALTA.

[5] Ma, L., Nie, Z., Xi, X. & Han, X., 2013. Cobalt recovery from cobalt-bearing waste in sulphuric and citric acid systems. *Hydrometallurgy,* Volume 136, pp. 1-7.

[6] Mubarok, M. Z. & Hanif, L. I., 2016. Cobalt and Nickel Separation in Nitric Acid Solution by Solvent Extraction Using Cyanex 272 and Versatic 10. *Procedia Chemistry,* Volume 19, pp. 743-750.

[7] Schuman, S. C. & Shalit, H., 1971. Hydrodesulfurization. *Catalysis Reviews,* 4(2), pp. 245-318.

[8] The Cobalt Development Institute, 2016. *About Cobalt.* [Online] Accessible at: http://www.thecdi.com/general.php?r=12ENJWIVAD [Accessed on 29 September 2016].

[9] Topsøe, H., Clausen, B. S. & Massoth, F. E., 1996. Hydrotreating Catalysis. *Catalysis: Science and Technology.* Berlin: Springer Berlin Heidelberg.

[10] U.S. Department of Energy, 2015. *Total Petroleum and Other Liquids Production 2015.* [Online] Accessible at: http://www.eia.gov/beta/international/rankings/#?prodact=53-1&cy=2015 [Accessed on 27 September 2016].

[11] Wilson, A. M., Bailey, P. J., Tasker, P. A., Turkington, J. R., Grant, R. A. & Love, J. B., 2014. Solvent extraction: the coordination chemistry behind extractive metallurgy. *Chemical Society Reviews,* Volume 43, pp. 123-134.

Development of black ink for calligraphy purpose in the production of Al-quran

F F A Rahim[1,2], J Jai[1], F Hamzah[1], and N F A Bakar[1]
[1] Faculty of Chemical Engineering, Universiti Teknologi MARA 40450 Shah Alam, Selangor Darul Ehsan, Malaysia

E-mail: ffatiharahim@gmail.com

Abstract. Commercialised calligraphy inks are currently formulated for general purposes and is not to be practiced solely for the writings of Al-Quran. Thus the usage on Al-Quran is uncertain due to the unknown ingredients used. The virtue of this work is to develop halal and genuine formulations of black inks for calligraphy purpose in the production of Al-Quran manuscript. The black ink produced is required to have few properties; rich solid black, soft handling, fast drying time, non-lifting, non-feathering and waterproof. Pigment used include graphene, charcoal and lampblack. The binders are shellac and acrylic emulsion while solvent used are glycerol and ethanol. From the chosen pigments, binders and solvent, 19 formulation of inks were developed by manipulating the type and amount of each element. Evaluation of all 19 inks produced were done by an expert calligrapher using a professional scale. From the evaluation, one ink was chosen to have the best characteristics of an ink (ink code S9) with formulation of 0.3 g lampblack and 13 mL shellac. The produced black ink has fulfilled the requirement by the calligraphy expert.

1. Introduction

The writings of Al-Quran require the best quality of black inks which has the properties of; rich solid black, waterproof base, non-feathering, soft handling when written, drying time and non-lifting as accordance to an expert calligrapher. Table 1 explains the properties of inks to be produced. At present, commercialized inks are primarily utilized with the purpose of general writings but not specialized for the writings of Al-Quran. Therefore, the ingredients used are uncertain to be applied on it. In order to preserve the purity of Al-Quran, few researches has been developed by using natural resources for instance lampblack, mangosteen charcoal and salt [1]. However, the establishment from local researches are yet still in development. The purpose of this paper is to develop formulations of black inks for calligraphy purpose which is halal and pure suitable to be used on Al-Quran manuscript.

Halal prescription often related to foods and drinks. In terms of inks, standard practical guidelines emphasize on the source of raw material used, preparation, handling, storage, processing aids and manufacturing process based on Islamic law. In terms of raw material, Islam has clarify few important guidelines; it must be clean and safe without anything dirty that is considered to be lawful according to Islamic law.

[2] To whom any correspondence should be addressed.

Raw materials from plant origin and plant products that are poisonous or hazardous to health are considered 'haram' unless it has been removed completely [2]. Arabic calligraphy is a form of artwork which is closely associated with the Quran. The Quranic calligraphy came to be recognized as highly regarded form of art in the nineteenth century. There are six types of Arabic calligraphy scripts which are well known to be as *the kufic, thuluth, nasakh, tauqi, muhaqqaq and riq'a* [3]. Black ink is extensively used in the writings of Al-Quran manuscript by using the stated styles of script.

Inks generally are made up of four basic components; pigments, binders, solvents and additives. Pigments are colored, organic or inorganic solid powder acting as the source of coloristic properties of an ink. Hiding power and coloring power of an ink will depend on the particle size of a pigment; the smaller the particle size, the darker the pigment will become [3]. Binders has a specific role to bind all ingredients of the ink together ink onto the paper. Manipulating the type and amount of binder used will influence the penetration of ink towards paper surface. Therefore, the amount of binder should be sufficient enough so that the pigment has an adequate adhesion to the paper surface [5].

Solvents are used to keep the ink liquid flow from when it is applied to papers. Factors affecting good solvents are the evaporation rate of the solvents [6]. The lower the boiling point of solvent, the faster the drying rate of an ink [7]. Finally, the additives are used to alter the final properties of an ink. The most common types of additives are rheology, wax, driers, chelating agents and surfactants. However, the additives element will not be included yet in this work so as to ensure the effectiveness when using pigment, binder and solvent into formulations.

Table 1. Ink Properties and Explanation.

Ink Properties	Explanation
Feathering	Feathering is when the ink spreads out, creating blurry indistinct lines. It can be caused by the paper, nib, and the nature of the ink.
Darkness	Darkness of black inks are not the same depending on the formula created.
Drying Time	An ink that dry fast means high resistance to smearing and smudging.
Lifting	Lifting is when an ink becomes visibly lighter due to erasing. When half of the ink has come off, this shows a bad ink produced.
Waterproof	Able to maintain the permanency of ink when water is test on it.
Soft Handling	The flow of ink when writing on paper.

2. Experimental procedure and methodology

2.1. Materials

Charcoal was obtained in a bulk of wood approximately about 3 kg and was rich black in colour. First, the material was crushed using a hammer before grinding it by using a dry blender. Then, the course powdered charcoal was further grinded using mortar and pestle. In order to form the smallest particles, the material was sieved using a sieve shaker to obtain 45μm particle size. The fine powdered of charcoal was then stored in plastic bottles for further used. Graphene is obtained in the finest particle of size about 0.142 nm and is black in colour. It can be directly used after received.

Lampblack was obtained homemade from the combustion of kerosene oil, burnt with insufficient supply of oxygen in a cracker container. The particle size of lampblack ranging from 9 nm to 25 nm. The fine powdered of lampblack was then kept in glass bottles for further used.

2.2. Preparation of the Inks

The pigment powder; either single pigment or combined pigment, were measured by using a weighing balance. Shellac binder was measured by using a measuring cylinder while acrylic binder used a weighing balance and the solvent were used a measuring cylinder. Firstly, the binder and the pigment were brought into a beaker and stirred with stirring rod until it has fully dispersed. Then the solvent was added and stirred for 2 hours with magnetic stirring bar on a hot plate at 450 rpm. No specified temperature was set because the usage of hot plate was only to keep the sample inks stirred constantly

IOP Conf. Series: Materials Science and Engineering **334** (2018) 012050 doi:10.1088/1757-899X/334/1/012050

for 2 hours. The finished inks were stored into a glass bottle for storage and kept away from direct contact with any form of lights. This is because the exposure to lights will cause the inks to fade and cause damages to physical and chemical structure of an ink [8].

Inks produce will be varied in terms of the formulation and amount of materials used however the procedure to produce one shall remain the same.There were also few inks set to be filtered. This method was hypothesized to improve soft handling and darkness.

2.3. *Ink Formulation*

Nineteen trials of producing black ink varies in the amount of pigment, solvent and binder. As one of the subjects was manipulated, the remaining two were controlled. The formulation of the 19 inks were tabulated in Table 2.

Table 2. Formulation of Inks Produced.

Ink Code / Formulation							
Shellac-Based Binder				**Acrylic-Based Binder**			
S1	0.5 g Graphene 5 mL Shellac	S6	0.3 g Graphene 0.3 g Lampblack 9 mL Shellac	A1	0.5 g Graphene 5 g Acrylic binder 7 mL Distilled water	A6	0.3 g Graphene 2 g Charcoal 2 g Acrylic binder 8 mL Ethanol
S2	0.8 g Graphene 7 mL Shellac	S7	3 g Charcoal 0.3 g Lampblack 12 mL Shellac (Filtrated)	A2	0.3 g Graphene 3 g Acrylic binder 6 mL Glycerol	A7	0.3 g Graphene 0.3 g Lampblack 3 Acrylic binder 10 mL Ethanol
S3	0.3 g Graphene 1 g Charcoal 7 mL Shellac	S8	0.3 g Graphene 0.3 g Lampblack 9 mL Shellac (Filtrated)	A3	0.5 g Graphene 3 g Acrylic binder 6 mL Ethanol	A8	0.3 g Lampblack 3 g Charcoal 4 g Acrylic binder 16 mL Ethanol
S4	3 g Charcoal 5 mL Shellac	S9	0.3 g lampblack 13 mL shellac	A4	3 g Charcoal 3 g Acrylic binder 8 mL Ethanol	A9 (Filtered)	0.3 g Graphene 0.3 g Lampblack 3 g Acrylic binder 10 mL Ethanol
S5	3 g Charcoal 0.3g Lampblack 12 mL Shellac		-	A5	0.3 g Lampblack 2 g Acrylic binder 9 mL Ethanol	A10 (Filtered)	0.3 g Lampblack 3 g Charcoal 4 g Acrylic binder 16 mL Ethanol

2.4. *Guideline for Performance Test*

As the inks has been produced, performance test were conducted as based on few characteristics; waterproofness, drying time, feathering, darkness, soft handling, suspended particles. The benchmark and standard had been approved by a calligraphy expert which is explained as in Table 3.

Table 3. Benchmark and standard while conducting performance test.

Subjects to Measure	Scale	Explanation on Scale
Waterproofness	Yes No	The sample inks will be tested with water to determine whether the ink will be taken out from the surface of the paper. The sample inks are not preferable if there is any elements taken out from the surface of the paper.
Drying Time	10 seconds	The sample inks will be written on a paper. At 10 seconds and 1 minute respectively, the written inks on paper will be stroke gently with a ply of tissue. The sample inks are not preferable if the inks do

	1 minute	not dry within those period of time. Smudged (Scale 1)  Not smudged (Scale 5) 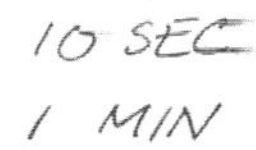
Darkness	1 - 5	1 represent least resulting colour black while 5 shows the best result of black. Light (Scale 1)  Dark (Scale 5) 
Soft Handling	1 - 5	1 represent the least smooth of an ink as it is being written on a paper while 5 shows the smoothest result of an ink on a paper. For scale 1, the pen might get stuck in the middle of writing due to large particle size of pigment or having clumps of the ink on the nib or whether insufficient amount of solvent used. For scale 5, the pen goes smoothly when writing without having any obstacles as stated in scale 1.
Feathering	1 – 5	1 represent the most feather of an ink while 5 shows the least feather of an ink. The least feathering of an ink is the most desired. Feathering (Scale 1)  Non-feathering (Scale 5) 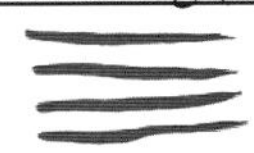
Lifting	1 - 5	1 represent the least suspended particles (pigment) on a paper while 5 has the most suspended particles on a paper. Lifting is when an ink becomes visibly lighter due to erasing. When half of the ink has come off, this shows a bad ink produced. Lifted (Scale 1) 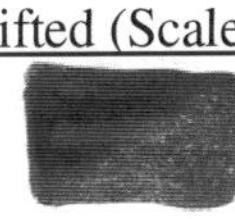 Not lifted (Scale 5) 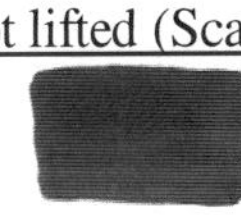

All nineteen produced inks were tested and evaluated by the calligraphy expert as according to the guideline mentioned in Table 3. The method of writing as well as the paper used were as same as usual calligraphy work.

3. Result and discussions

3.1. Performance of Waterproof on Prepared Ink

From the performance test, all ink showed an excellent result on waterproofness from both shellac-based and acrylic emulsion-based binder. Waterproofness of an ink depends largely on solubility of binder used [9]. If the ink is not waterproof, this means that the ink will not be able to bind the pigment on the surface of a paper. An ink is said to be waterproof because it is not soluble in water.

Shellac and acrylic emulsion base binder were not soluble in water. Shellac is soluble in alcohol solvent as well as acetone solution [10] while acrylic emulsion is soluble in ammonia solution, isopropyl alcohol, denatured alcohol, acetone an lacquer thinner [11].

3.2. Comparison on Different Formulated Ink Based on Drying Time

From the performance test for an acrylic-based ink, the solvent used were distilled water (A1), glycerol (A2) and ethanol (A3 until A10). All three solvents showed a discouraging result on drying time; dried in more than 5 minutes, especially glycerol which took more than 1 hour to dry.

Drying time of an ink depends on the boiling point of solvent used whereby a fast drying rate has low boiling point. This proved that solvent with high boiling point takes time to dry. While for a shellac-based ink, there were none of formulations that use element of solvent because the binder itself acts as the binder and solvent at the same time. Shellac binds on the paper easily with pigment and since it already being used in liquid form thus it flowed well on paper. Shellac is approved to have an excellent outcome in quick drying and is reported in many articles [12], [13] and [14].

3.3. Comparison of Ink in terms of Darkness Based on Binder Type
Based on Figure 1, charcoal-based and lampblack-based pigment showed a promising result in terms of darkness as compared to graphene. The difference was because of the nature of pigment colour. Charcoal has the colour of black with the shades of dark grey, lampblack is a fine black pigment with a bluish tint and slightly greys while graphene has a plain dark grey solid colour.

Darkness of an ink is contributed by the source of pigment as well as its particle size. The smaller the particle size of a pigment, the blacker it would be. For example, a mars black pigment of 50 μm particle size is blacker than a 100μm.

However, colour black may change when it is blended with different binders; taking the example of S9 and A5. Lampblack and shellac scored 5 scale while lampblack with acrylic scored only 1. During preparation, when lampblack was mixed with shellac, the black did not change its colour. The case was not the same when it mixed with acrylic emulsion binder. As lampblack came into contact with the white emulsion of acrylic, it changed colour from black into slightly grey. This may due to the chromophore that is reactive to a change in energy; either heat or light. The energy causes a physical change in the chromophore, which changes the way it absorbs light. Ink may also change due to pigment that has chromophore groups which sensitive to the polarity of the solvents [15]. Ink code S4, S5 and S9 appeared to have the most promising result scaled of 5 in terms of darkness.

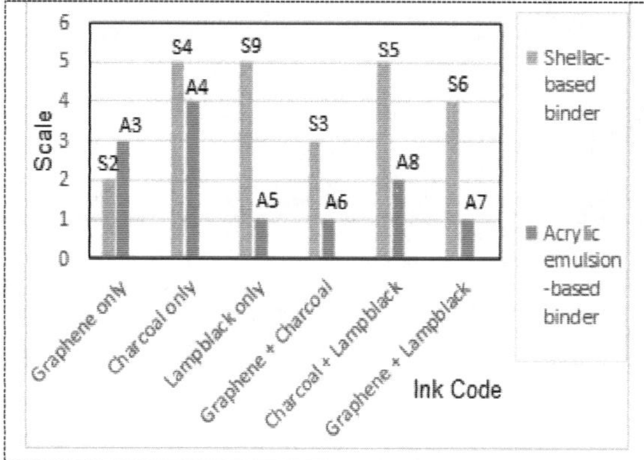

Figure 1. Comparison on Darkness Based on Ink Formulation Using Graphene, Charcoal and Lampblack as Pigment while Shellac and Acrylic Emulsion as the Binder.

3.4. Comparison of Prepared Ink on Soft Handling when Written on Paper
By referring to Figure 2, shellac-based binder showed better result as compared to acrylic-based binder. Ink code S9 scored the highest mark which is 4 while A3, S4, S3, S5 and S6 scored 3 on scale. This shows that binders play an important role in the flow of an ink on paper. The acrylic-based binder inks appeared to have unsatisfactory result because the binder provided a sticky-like texture during preparation thus making the ink to be viscous and thickened that it did not flow well during writing.

According to an expert, inks with acrylic-based binder cannot be stroke more than one letter when written and had a sticky-like texture upon using it which showed a poor ink produced. This refers to the wetting ability of an ink which is the spreading of the liquid ink onto a surface.

Particle size of pigment will affect the flow of an ink on paper. In an article states that flocculation of pigment particles surrounded by the binder can cause the accessibility of ink flow. It is because the attraction between the pigment particles are greater than the adhesive force of the binder [13]. Most inks from both type of binder had a rough texture due to large pigment particles. However, graphene-based pigment had a smooth texture due to small particle size.

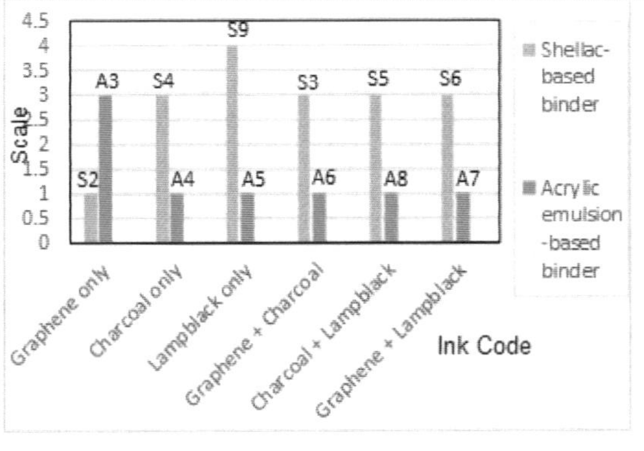

Figure 2. Comparison on Soft Handling Based on Ink Formulation Using Graphene, Charcoal and Lampblack as Pigment while Shellac and Acrylic Emulsion as the Binder.

3.5. Comparison on Selected Inks Before and After Filtration in terms of Darkness
Filtration was made for ink code S5, S6, A7 and A8 to observe the improvement on darkness after large pigment had been filtrate. By referring to Figure 3, the darkness of all four inks remains the same as before filtration.

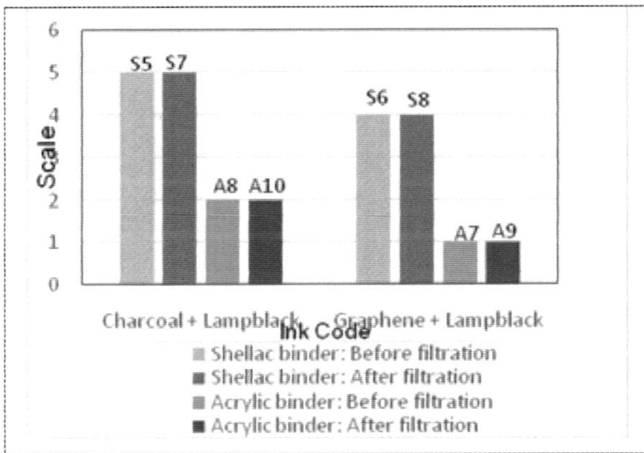

Figure 3. Comparison on Selected Ink (S5, A8, S6 and A7), Before and After Filtration (S7, A10, S8 and A9) by Using Graphene, Charcoal and Lampblack as Pigment while Shellac and Acrylic Emulsion as the Binder in the Determination on Darkness.

The graph shows that by removing larger particles of pigment will not improve the blackness of an ink. This is because there were no actions of force taken on the ink in order for the pigment to enhance its colour capability. Instead, dispersion of pigment should be done which can be achieved by grinding, whereby it is normally done in a mill; e.g. ball mill or pearl mill [16].

3.6. Comparison on Selected Inks Before and After Filtration in terms of Soft Handling
Filtration was also made for ink code S5, S6, A7 and A8 to observe the improvement on soft handling after large pigment had been filtrate. By referring to Figure 4, the soft handling criteria of all four inks remains the same as before filtration.

It is important that the ink is to be free from large pigment agglomerates in order to provide a smooth and clear surface [16], hence the purpose of filtration. This may due to the undue methods of filtration made. During preparations of filtration, material used was just a cloth of gauze. On the other hand in industry, a multiple stage filtration is typically used for pigmented inks [17] . Therefore, the filtration approach made in this work did not show any improvement because it did not fulfill the specification of filtration criteria for pigmented inks.

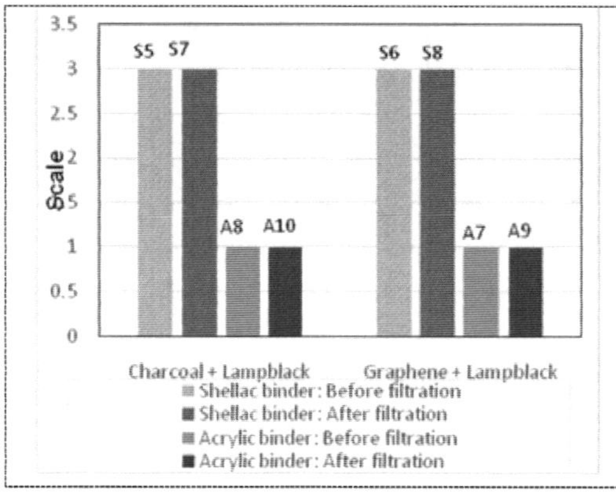

Figure 4. Comparison on Selected Ink (S5, A8, S6 and A7), Before and After Filtration (S7, A10, S8 and A9) by Using Graphene, Charcoal and Lampblack as Pigment while Shellac and Acrylic Emulsion as the Binder, in the Determination on Soft handling When Written.

3.7. Comparison on Ink Lifting Based on Different Binder Used

As based on Figure 5, shellac-based ink showed better performance in ink lifting as compared to acrylic-based binder, whereby scale 5 represent a non-lifting criteria. When half of the ink has come off, this shows a bad ink produced. This was due to the shellac-based binder binds well with pigment but not with acrylic-based binder. The combination of pigment and binder plays an important role in determining ink rheology [18]. Amount of binder must meet the amount of pigment so as to avoid rub-off of pigment substance [19]. Since shellac-based ink showed an excellent performance on ink lifting, thus only ink with shellac-based will be chosen for further selection.

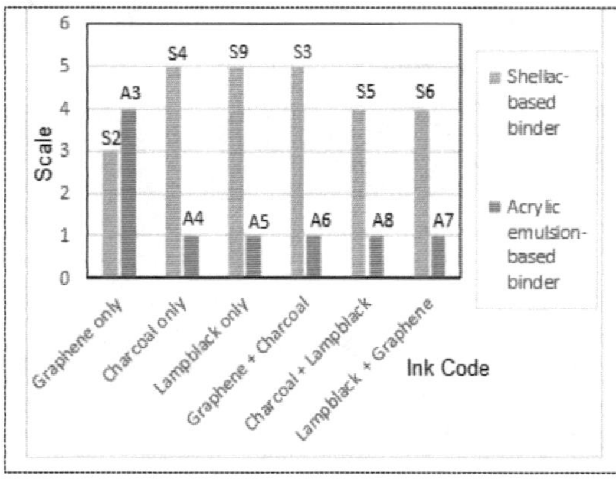

Figure 5. Comparison on Ink Lifting when Erased, Using Shellac and Acrylic Emulsion as the Binder.

3.8. Comparison on Feathering from Selected Inks

From few tests above, ink code S4, S5 and S9 were selected for further selection for it had achieve the highest score for each tests; darkness, soft handling, waterproofness, fast drying time and lifting. Figure 6 shows the performance test on feathering.

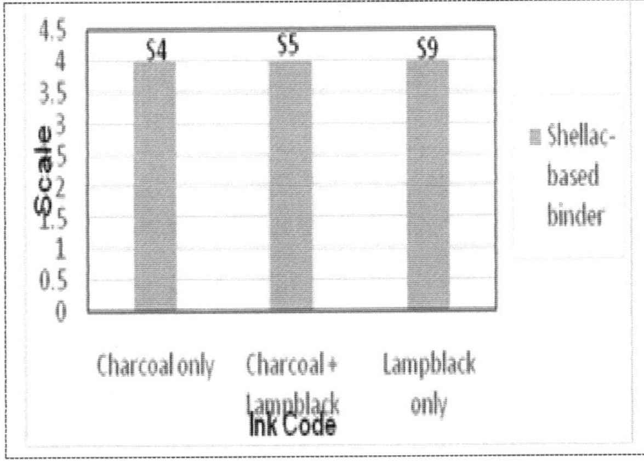

Figure 6. Performance on Feathering when Charcoal and Lampblack were Used as the Pigment with the Assess of Binder Shellac (Selected Inks; S4, S5, S9).

Feathering can be caused by the paper, pen nib, and the nature of the ink. For the case of ink nature, the solvent plays an important role. If the amount of solvent is excessive, blurry lines appears and can be seen on the surface of paper. Since all three inks scored 4 scale, it shows that the inks could be considered to have a good response on feathering.

3.9. Selection of Best Ink Prepared
As according to the result of all inks, tested and evaluated by calligraphy expert, the best ink showing a promising result was S9 with formulation of 0.3 g lampblack and 13 mL shellac. The scores of each properties for ink code S9 is tabulated in Table 4.

Table 4. Performance test of S9.

Waterproof	Drying Time	Darkness	Soft Handling	Feathering	Lifting
Yes	10 seconds: Yes ; 1 minute: Yes	5	4	4	5

4. Conclusion
In conclusion, ink code S9 are able to fulfill the requirement of non-lifting, rich solid black ink, soft handling when written, non-feathering, fast drying time and waterproof. The formulation of S9 of using lampblack and shellac are considered as natural resources thus can be used to write on Al-Quran purposes.

Acknowledgements
I hereby thankful to my supervisor, Prof Madya Dr. Junaidah Jai, Dr. Fazlena Hamzah and Dr. Noor Fitrah Abu Bakar, for the guidance, advice and suggestions leading to better understanding for my research. Not to forget to Encik Mohd Zulfikri Mohd Sobri, an expert calligrapher from Yayasan Restu Malaysia, for contributing his expertise by evaluating the prepared inks. Last but not least, the assistance from the Faculty of Chemical Engineering, University Technology MARA (UiTM) Malaysia and financial support from 600-RMI/DANA 5/3/REI (21/2015) are gratefully acknowledged.

References

[1] R. B. A. Razak, "Traditional Malay Black Ink: Formulation of the Recipe and Analysis on their Characteristics," no. August, 2016.
[2] I. C. C. D.- Copenhagen, "ICCOSS Halal System," 2013.
[3] B. R. Annemarie Schimmel, *Islamic Calligraphy*. Brill Archive, 1992.
[4] A. M. Gueli, S. P. Guido Bonfiglio, and S. O. Troja, "Effect of particle size on pigments colour," *Color Res. Appl.*, vol. 42, no. 2, pp. 236–243, 2016.
[5] C. J. Ridgway, V. Kukkamo, and P. A. C. Gane, "Effects of Binders and Additives on Inkjet

IOP Conf. Series: Materials Science and Engineering **334** (2018) 012050 doi:10.1088/1757-899X/334/1/012050

Coating Pigment Pore Structures," pp. 1305–1319, 2011.
[6] R. Leach, *The Printing Ink Manual*. Springer Science & Business Media, 2012.
[7] J. Felsberg, "Matching solvent-ink parameters to high-speed flexo printing needs," *Sun Chem. Corp*, 2013.
[8] D. Conn, "Protection from Light Damage," *Northeast Doc. Conserv. Cent.*, 2012.
[9] R. J. Varghese, "What makes ink waterproof?," 2015.
[10] Y. Farag and C. S. Leopold, "Physicochemical Properties of Various Shellac Types," no. May, pp. 33–39, 2009.
[11] I. B. Marx, *Professional Painted Finishes*. Whitney Library of Design, 1991.
[12] B. Gibson, *Gourds with Southwestern Motifs: Rainsticks, Masks, Vessels and More*. Sterling Publishing Company, Inc., 2009.
[13] N. Board, *Modern Technology of Paints, Varnishes & Lacquers*, 2nd Editio. Asia Pacific Business Press Inc., 2007.
[14] E. A. Baldwin, R. Hagenmaier, and J. Bai, *Edible Coatings and Films to Improve Food Quality*. CRC Press, 2001.
[15] M. Ferrara and M. Bengisu, *Materials that Change Color: Smart Materials, Intelligent Design*. Springer Science & Business Media, 2013.
[16] C. Sörensson, "An investigation into the Aspects of Grinding Organic Pigments to produce Water Based Dispersions for use in Printing Inks," vol. 1, no. February, 2007.
[17] M. Mehler, "Inkjet ink filtration – Process overview and filtration in the lab," 2016.
[18] M. S. I Hutchinson and E. D. W Grierson, "Pigments and OLigomers for Inks - Moving Towards the Best Combination," pp. 118–125, 1999.
[19] N. C. of Ministers, *Food Contact Materials and Articles, Printing Inks: Check Lists for Compliance in Industry and Trade and Control by Food Inspection*. Business & Economics, 2012.

IOP Conf. Series: Materials Science and Engineering **334** (2018) 012051 doi:10.1088/1757-899X/334/1/012051

Preparation and characterization of a high surface area of activated carbon from *Bambusa vulgaris*—Effect of NaOH activation and pyrolysis temperature

E M Mistar[1], S Ahmad[1], A Muslim[2], T Alfatah[3] and M D Supardan[2,4]

[1] Chemical Engineering Department, Serambi Mekkah University, Indonesia
[2] Chemical Engineering Department, Syiah Kuala University, Indonesia
[3] Environment and Forestry Offices of Aceh Province, Indonesia

E-mail: m.dani.supardan@che.unsyiah.ac.id

Abstract. In this study, *Bambusa vulgaris* was used as raw material of activated carbon. The activated carbon was generated using NaOH as an activating agent. The activated carbon produced with the weight ratio of carbon to NaOH 1:1, 1:2, 1:3 and pyrolysis temperature 600, 700 and 800 °C. The experimental result showed that the best characteristic of activated carbon was obtained at weight ratio of carbon to NaOH 1:3 and pyrolysis temperature 800 °C.

1. Introduction
Activated carbon is an adsorbent produced from materials containing carbon [1]. Activated carbon as charcoal was processed using CO_2, water vapor and chemical materials at high temperatures. It can be used as an absorbent [2]. Activated carbon is a porous substance with a high surface area that can absorb and remove pollutants such as heavy metals, dyes, gases and pesticides [3].

In industry, activated carbon is used in purification units, catalysis, recuperation of solvents and hydrocarbon, natural gas and hydrogen storage and energy storage [4]. Furthermore, activated carbon is used in deodorization processes, filtration, discoloration, modification of liquid, gas concentration, petroleum, pharmaceutical, and food and beverage. This absorption capability is caused by its large number of micro pores, which results in capillary symptoms that can absorb pollutants.

Many materials have been used to produce a good quality activated carbon. The effect of chemical substance addition in the preparation of activated carbon was also studied. Chemical substances used in the activation process give many advantages, including single low activation temperatures, activation time, step activation and better porous structure [5]. Activated carbon consists of 87-97% carbon and the rest of the components are oxygen, hydrogen, sulfur, nitrogen, and other compounds formed in the process of producing activated carbon.

Meanwhile, a recent study reported the potency of *Bambusa vulgaris* as ab activated carbon source. The activated carbon from yellow bamboo was activated by H_2SO_4 to absorb heavy metals of Cd, Hg and Zn [6]. Yellow bamboo has particularly spread and flourished in tropical Asian regions. The chemical compositions of yellow bamboo were 42.4-53.6% of cellulose, 19.8-26.6% of lignin, 17.5-21.5% of pentose and 1.24-3.77% of ash. Adsorption is the process of transferring adsorbate mass

[4] To whom any correspondence should be addressed.

from the moving phase into the absorbent surface, which is influenced by the attractive force between adsorbate molecules with active pores on the absorbent surface [2].

In this study, the *Bambusa vulgaris* was used as a source of activated carbon. The activated carbon was chemically activated using NaOH. The characteristics of activated carbon produced will be investigated.

2. Materials and Methods

Bambusa vulgaris was obtained from local farmers in Aceh Besar District, Aceh, Indonesia. It is locally known as *Treng Gadeng* by Acehnese society. The chemicals used were $HgCl_2$, HNO_3, NaOH, HCl, Aquadest, N_2, $KMnO_4$, $K_2S_2O_8$, NaCl and $SnCl_2$. All chemicals were provided by Merck and used as received.

The carbonization process of *Bambusa vulgaris* was performed with a furnace at temperature of 500 °C and nitrogen gas flows of 200 mL/min for 2 hours. Then the charcoal was cooled and crushed by milling and sieving until -60 +120 mesh of size. After that, *Bambusa vulgaris* charcoal was mixed with NaOH as an activating agent. The weight ratios of carbon to NaOH used were 1:1, 1:2 and 1:3. Then pyrolysis was conducted at temperatures of 600, 700 and 800 °C. Table 1 shows the experimental design. Thereafter, the activated carbon was washed using HCl and stirred for 30 minutes at 85 °C. Finally, residual acid was removed from the product by washing with aquadest.

Table 1. Experimental design.

Sample code	Weight ratio of carbon to NaOH [g/g]	Pyrolysis temperature [°C]
Bambusa vulgaris	-	-
AC-1	without activation	500 °C
AC-2	1:1	600 °C
AC-3	1:2	600 °C
AC-4	1:3	600 °C
AC-5	1:1	700 °C
AC-6	1:2	700 °C
AC-7	1:3	700 °C
AC-8	1:1	800 °C
AC-9	1:2	800 °C
AC-10	1:3	800 °C

The characteristics of the activated carbon include surface area, total pore volume, average pore diameter as determined by the Brunauer, Emmett and Teller (BET) method. It was obtained using Surface Area Analyzer Nova Win Version 11.0 (Quantachrome). The morphology structure of activated carbon was analyzed by means of a Scanning Electron Microscope (TM 3000 Hitachi). All measurements were conducted in the laboratory at ambient temperature; the measurements were collected in duplicate, and the averaged data are reported.

3. Results and Discussion

Figure 1 presents the surface area of activated carbon. The larger the surface area and pore diameter, the better the quality of the activated carbon produced. Figure 1 shows that *Bambusa vulgaris* has 53,5 m^2/g of surface area. The pyrolysis temperature has a significant effect for increasing of the surface area of activated carbon. The activated carbon had a surface 308 m^2/g when prepared at weight ratio of carbon to NaOH 1:1 and pyrolysis temperature 700 °C. At a similar weight ratio of carbon to NaOH, increasing the pyrolysis temperature up to 800 °C caused a much wider surface area, i.e., 560 m^2/g. In addition, the addition of NaOH as the activating agent caused an increase of surface area of activated carbon. The higher the weight ratio of NaOH activator used, the larger the surface area obtained was. At pyrolysis temperature 800 °C, the surface areas of activated carbon at weight ratio of carbon to NaOH 1:1, 1:2 and 1:3 were 560, 635 and 1041,716 m^2/g, respectively.

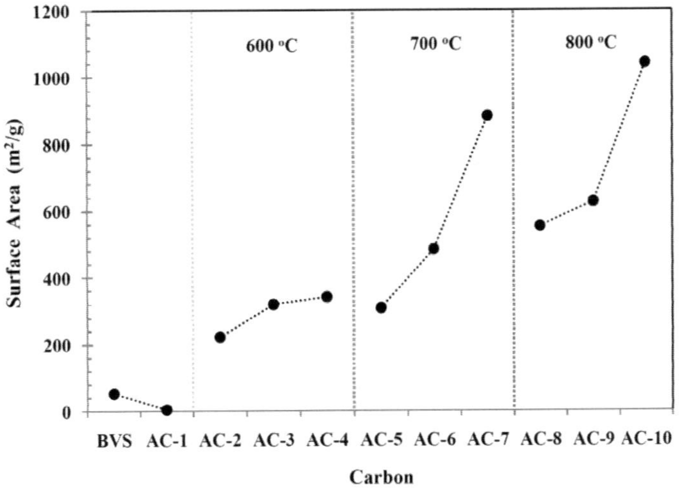

Figure 1. Surface area of the raw material, char, and activated carbon.

Figure 2 presents the total pore volume of activated carbon. There was a similar trend result. The pyrolysis temperature and weight ratio of carbon to NaOH influenced the change in surface area. The total pore volume increased with increasing of pyrolysis temperature and weight ratio of carbon to NaOH.

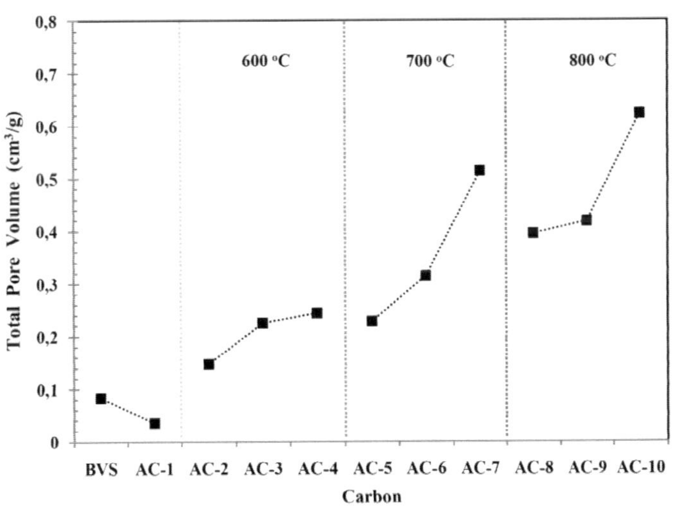

Figure 2. Total pore volume of raw material, char, and activated carbon.

The pyrolysis temperature and weight ratio of carbon to NaOH caused some changes in the physical properties of activated carbon. The increasing pyrolysis temperature and weight ratio of carbon to NaOH potentially increased the residual loss of hydrocarbons occupying the pore space [7]. The additional activator oxidized and eroded the carbon wall so that it formed more pores and increased the surface area of the carbon, and increasing the heat of chemical activity in the activated carbon also contributed to the production of larger pore diameter [8].

The pore size distribution of activated carbon is a significant characterization in determining the quality product. The activated carbon from rice bran has pore volume with the range of 0.025 to 0.14

cm^3/g [9], compared to the total pore volume with the range of 0.15 to 0.622 cm^3/g of activated carbon generated from *Bambusa vulgaris* obtained in this study.

Figure 3 shows the average pore diameter of activated carbon, which was in the range of 23.2 to 30.5 Å. The pyrolysis temperature and weight ratio of carbon to NaOH provided specific pore volume. The decline in physical properties might be caused by carbon damage owing to the heating processes during activation. Thus, it can be concluded that the high pyrolysis temperature can influence average pore diameter. The increasing size of pore diameter occurred in several steps, starting with the opening of pores that were previously closed, formation of new pores, and enlargement of the formed pores [3]. The increased pore diameter could also potentially decrease the surface area.

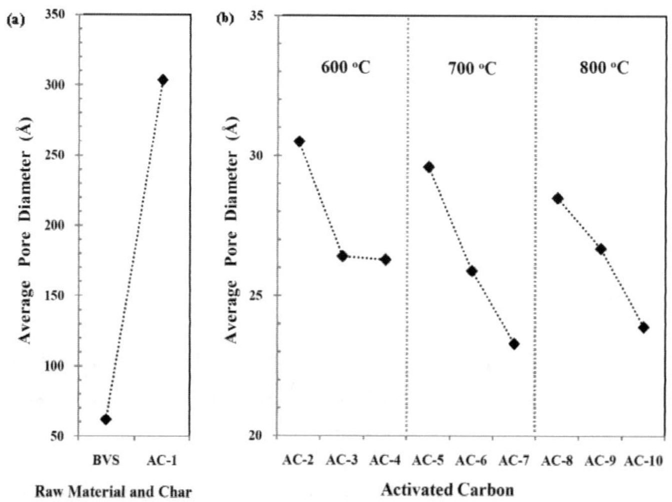

Figure 3. Average pore diameter of the raw material and char (a), activated carbon (b).

The activated carbon produced in this study was then observed with a Scanning Electron Microscope (SEM) to identify the morphological structure, which is shown in Figure 4. As can be seen in Figure 4, *Bambusa vulgaris* and activated carbon had some pores on their surfaces. The walls of macropores contained many subpores that connected the macrospores in the interior of the material.

It can also clearly be seen that a formation of pores and sub-pores rose because of the addition of NaOH as the activating agent. This might be because the activation process with NaOH and the rising pyrolysis temperature increased the reaction between carbon and the activating agent to produced CO_2. The reaction is shown Equations (1)-(3). The CO_2 produced will diffuse to corrode the carbon wall to form new pores and make pores to be more open [2].

$$4\,NaOH + C \leftrightarrow 4\,Na + CO_2 + 2\,H_2O \quad \text{..................} (1)$$
$$6\,NaOH + 2\,C \leftrightarrow 2\,Na + 3\,H_2 + 2\,Na_2CO_3 \quad \text{..................} (2)$$
$$4\,NaOH + 2\,CO_2 \leftrightarrow 2\,Na_2CO_3 + 2\,H_2O \quad \text{..................} (3)$$

4. Conclusion

The effect of NaOH activation and pyrolysis temperature on characteristics of activated carbon produced from *Bambusa vulgaris* was investigated. The experimental result showed that the best characteristic of activated carbon was obtained at weight ratio of carbon to NaOH 1:3 and pyrolysis temperature 800 °C. At this condition, the surface area, total pore volume and average pore diameter obtained were 1042 m^2/g, 0.62 cm^2/g and 24 Å, respectively.

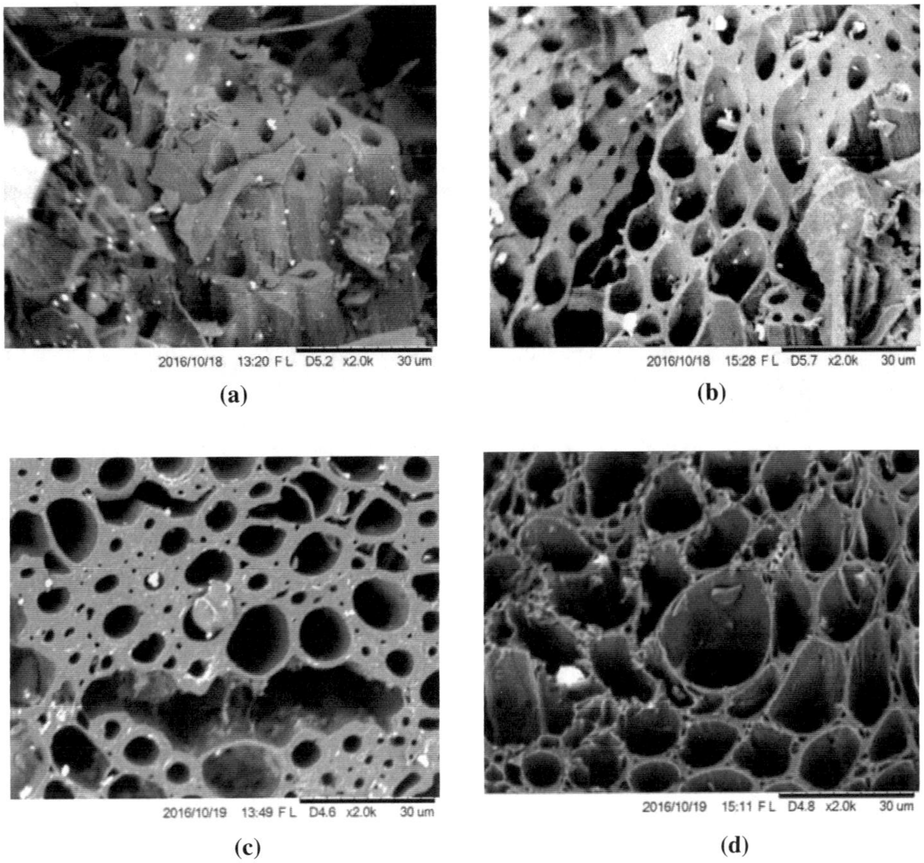

Figure 4. Morphological structure of the activated carbon: (a) AC-2; (b) AC-4; (c) AC-7; (d) AC-10.

Aknowledgement
The authors are grateful to Serambi Mekkah University and Ministry of Research, Technology and Higher Education of the Republic of Indonesia for providing financial support (contract No. 617/K.13.2.2/UM/2016)

References

[1] Dolas, H., Sahin, O., Saka, C., Demir, H., 2011. A new method on producing high surface area activated carbon: The effect of salt on the surface area and the pore size distribution of activated carbon prepared from pistachio shell, *Chemical Engineering Journal* **166**, 191–197

[2] Alfatah, T, Husin, H., Arahman, N., Marya, E., 2015. NaOH - Preparation and characterization activated carbon from coconut shell, *Environment Technology*, **8** 1-8.

[3] Cazetta, A.L., Vargas, A.M.M., Nogami, E.M., Kunita, M.H., Guilherme, M.R., Martins, A.C., Silva, T.L., Moraes, J.C.G, Almeida, V.C., 2011. NaOH-activated carbon of high surface area produced from coconut shell:Kinetics and equilibrium studies from the methylene blue adsorption, *Chemical Engineering Journal* **174** 117– 125.

[4] Hadoun, H., Sadaoui, Z., Souami, N., Sahel, D., Toumert, I., 2013. Characterization of mesoporous carbon prepared from date stems by H_3PO_4 chemical activation, *Applied Surface Science*, **280**, 1– 7.

[5] Yang, K., Peng. J., Srinivasakannan, Zhang. L., Xia, H., Duan, X., 2010. Preparation of high surface area activated carbon from coconut shells using microwave heating, *Bioresource Technology* 101, 6163–6169
[6] Gonzalez, P.G., and Pliego, Y. B., 2015. Adsorption of Cd (II), Hg (II) and Zn (II) from aqueous solution using mesoporous activated carbon produced from *Bambusa vulgaris schrad, Chemical Engineering Research and Design* **92**, 2715–2724.
[7] Setyaningsih, T., Hasanah, U., Darjito. 2008. Study of NaOH – Activation Temperature Influence Toward Character Of Mesoporus Carbon Based on Textile Sludge Waste, *Indo. J. Chem*, **8** (3), 348 – 352.
[8] Shofa, 2012. *Activated carbon from bagasse with potassium hydroxide activation*, Chemical Engineering, Indonesia University.
[9] Suzuki, R.M., Andrade, A. D., Sousa, J. C., Rollemberg, M. C., 2007. Preparation and characterization of activated carbon from rice bran, *Bioresource Technology* **98**, 1985–1991.
[10] Prasetyo, Y and Nasrudin, H., 2013. Determination of $ZnCl_2$ concentration in the process of making corn starch activated carbon and decreasing linear alkylbenzene sulphonate (LAS), *UNESA Journal Of Chemistry* **2**, 3.

Extraction of rare earth elements from low-grade Bauxite *via* precipitation reaction

E Kusrini[1,2], Y Nurani[1] and ZJ Bahari[1]
[1] Department of Chemical Engineering, Faculty of Engineering, Universitas Indonesia, Kampus Baru UI Depok, 16424, Indonesia

E-mail: ekusrini@che.ui.ac.id

Abstract. The aim of this research was to determine the optimum hydrometallurgical parameters to extract the rare earth elements (REE) from low-grade bauxite through acid leaching and precipitation reaction. REE or lanthanide recovery by a precipitation method with sodium sulphate and sodium phosphate as precipitation agents is reported where the effect of pH and recovery of REE are described. The metal composition of REE in low-grade bauxite after treatment were analyzed by ICP-OES. The total recovery values of REE elements at the first precipitation reaction using sodium sulphate as the precipitation agent at pH 3.5 showed ~68.2% of lanthanum, ~18.9% cerium, and ~7.8% yttrium. Lanthanum was the rare-earth element present at the highest concentration in the low-grade bauxite after the series treatments. An optimum pH of 3.5 for precipitation of rare-earth elements using sodium sulphate was demonstrated where this method is recommended for the extraction of REE elements from low-grade bauxite.

1. Introduction

Indonesia is rich mineral resources such as coal, copper, nickel, gold, tin, silver, bauxite and petroleum [1]. In 2017, an estimated total export volume of 6636 tonnes of bauxite occurs from Indonesia. High demand of bauxite by the market has consequently produced a substantial amount of waste without any further use. Low-grade bauxite with alumina content is less than 50% from washing bauxite ore from abundant bauxite sources in Bintan Island, Indonesia. Low-grade bauxite is under-utilized and may pose a potential problem since bauxite mining occurs on open land that leaves unproductive and potential land hazards due to non-reclamation of tailings and disposal.

Demand for lanthanides has been increasing due to the unique luminescent properties and long-lived emission of these elements [2]. They have a very high economic value since they possess very good physical and chemical properties suitable for applications in advanced materials [3]. Therefore, some countries without lanthanides resources must develop the new material resources such as low-grade lanthanide sources. Previous research indicates that low-grade bauxite contains lanthanides, namely lanthanum oxide (La_2O_3) at 0.0052 wt% and yttrium oxide (Y_2O_3) at 0.0041 wt% [4]. Separation of lanthanides from low-grade bauxite is beneficial for both the environment and economy because of waste reduction from mining. Separation of lanthanides using solid-liquid extraction methods by acid, precipitation, crystallization, or ion exchange have been reported [5,6] The alkaline cracking and acid baking methods for rare earth elements concentrate have been reported by Sadri and

[2] To whom any correspondence should be addressed.

partners [6]. A study on the acid leaching of low-grade bauxite after mechanical-roasting treatment was reported to be useful for separation and further purification. A previous study found that substantial reserves of lanthanides exist in low-grade bauxite [4,7], where low-grade bauxite contains La ions with a concentration up to 0.0052% (w/w) [4].

Recently, extraction of lanthanides from Indonesian low-grade bauxite using acid leaching with oxalic acid has been reported [8]. Indonesian low-grade bauxite was pre-treated with mechanical-roasting processes, then followed by oxalic acid leaching at 40°C and concentration of 1 M for 2 hours. This approach has served as an argument for the need to develop a more effective and economical method to separate lanthanides from low-grade bauxite. Therefore, this study evaluates the extraction of lanthanide using a precipitation method with sodium sulphate and sodium phosphate. Precipitation occurs because the substance in compound has a sufficiently low solubility product (K_{sp}) to allow for phase separation from solution in the saturation-state. The lanthanide deposited was carry out by two stages of precipitation reaction by controlling pH and temperature, where the concentration of the lanthanide metal species exceeds the equilibrium condition according to the solubility product.

The filtrate from a one-stage precipitation process still contains lanthanides so a second stage of precipitation is required [9]. The purpose of this work is to determine the optimum pH for lanthanides recovery using oxalic acid leaching [8], along with precipitation using sodium sulphate and sodium phosphate as precipitation agents.

2. Experimental

2.1. Materials

Low-grade bauxite was collected from Madong area, Bintan Island, Indonesia. Oxalic acid dihydrate ($C_2H_2O_4.2H_2O$) was purchased from Merck (Darmstadt, Germany). Other chemicals and materials were used without purification, namely 25% ammonia, 96% sulfuric acid, aquadest, 99% sodium hydroxide, tri-sodium phosphate dodecahydrate, and sodium phosphate.

2.2. Characterization

The metal composition of REE, Al and Fe were analyzed using inductively coupled plasma optical emission spectrometry (ICP-OES) Agilent technology 7000. Samples were dissolved in aqua regia (HCl: HNO_3 = 3:1) solution.

2.3. Preparation of concentrated solution

Precursors for the precipitation were prepared according to the optimum condition as described by Kusrini et al. [8]. In this study, we used the most effective leaching condition at 40°C and a concentration of oxalic acid solution at 1 M. Low-grade bauxite was mechanically milled until it reached a particle size of 200 mesh. The material was roasted at 500°C in order to decompose the remaining organic fraction. Low-grade bauxite was disolved in aquadest with ratio of solid: liquid at 1: 4. The ratio of low-grade bauxite and oxalic acid solution of 1:15 was added into a 500 mL beaker where the mixture was stirred for 2 h on a hot plate at 40°C with 1 M oxalic acid solution.

2.4. Precipitation method

The pH of the leaching solution varies and results in challenges for the control of the acid levels after the roasting-leaching processes. The sample was washed with aquadest and then vacuum pumped until neutralization (pH 7). The lanthanide salt mixture was prepared using NaOH and the added sodium sulphate and sodium phosphate but it was unable to undergo complete precipitate formation. The pH of concentrate solution was adjusted using sodium hydroxide (NaOH) and ammonia solution (25 v/v %).

At pH 1.5, the concentrate was added sodium sulphate to avoid ferric hydroxide precipitation but no precipitate formed. The precipitation reaction continued and the pH was adjusted to pH 3.5 and 9 using NaOH. Both the yield of the lanthanide precipitate with sodium sulphate and/or sodium phosphate was below 1 g. The metal REE element content in a low-grade bauxite was determined by ICP-OES.

In this study, Indonesian low-grade bauxite (200 mesh) after a magnetic separation process was used as raw material (1.5 kg). Based on the ICP-OES analyses, low-grade bauxite is composed of various lanthanide oxide, silicate, alumina and iron oxide. REE existing in low-grade bauxite is La of 0.00157% (0.02355 g), Ce of 0.00176% (0.0264 g) and Y of 0.00018% (0.00277 g). The metal REE element composition from the precipitation reaction using each sodium sulphate and sodium phosphate at each pH condition (pH 3.5 and 9) is shown in Table 1.

In Table 1, the recovery of mixed lanthanides (REE) from Indonesian low-grade bauxite after the precipitation reaction was calculated using equation 1 as follows:

$$\% \text{ Recovery of REE} = \text{m concentrate (REE)/m REE x } 100\%$$

with m is mass (g) and REE is the metal rare earth elements.

Table 1. Elemental composition of REE after precipitation reactions using sodium sulphate and sodium phosphate at each pH 3.5 and 9.

Sodium sulphate	pH 3.5			pH 9		
	1.0368 g			47.322 g		
	La	Ce	Y	La	Ce	Y
	1.55108% (0.01606 g)	0.48129% (0.00498 g)	0.02098% (0.00021 g)	0.00858 % (0.00405 g)	0.00315 % (0.00148 g)	0.00039 % (0.00018 g)
Recovery of REE (%)	68.23	18.88	7.84	17.23	5.64	6.6
Sodium phosphate	pH 3.5			pH 9		
	0.9034 g			52.46 g		
	La	Ce	Y	La	Ce	Y
	0.13839 % (0.00125 g)	0.17297 % (0.00156 g)	0.01841% (0.00016 g)	0.00440 % (0.0023 1g)	0.00694 % (0.00364 g)	0.00017 % (0.000089 g)
Recovery of REE (%)	5.3	5.91	6	9.81	13.78	3.2

3. Results and Discussion

As shown in Table 1, the weight content (%) of Fe element in Indonesian low-grade bauxite before treatment and after magnetic separator process are shown [9] with minor changes observed. In this research, three types of lanthanides were studied; La, Ce and Y. The metal composition (%) of REE elements in low-grade bauxite before treatment is shown in Figure 2. Both La (18.4%) and Ce (19.3%) elements are the major components in low-grade bauxite before treatment. Cerium was the rare-earth element present at the highest concentration in low-grade bauxite before the magnetic separator process.

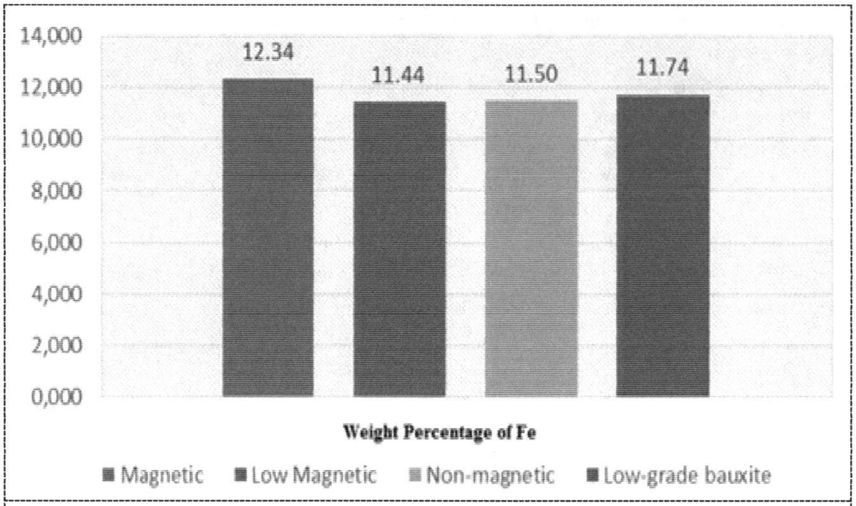

Figure 1. Fe content (%) in Indonesian low-grade bauxite and after the magnetic separator process.

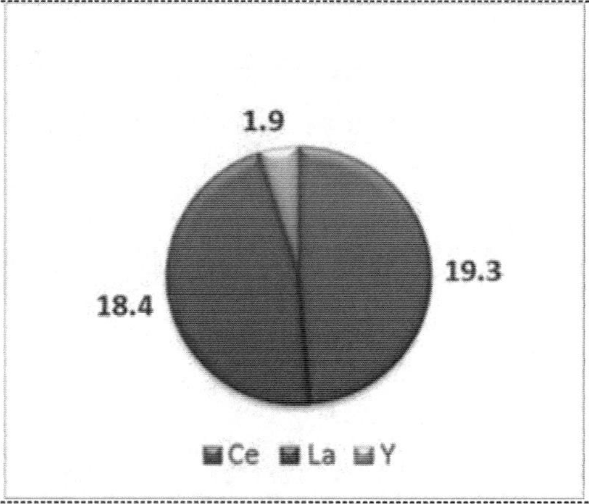

Figure 2. Composition of rare earth element (%) in Indonesian low-grade bauxite before treatment.

To examine the effect of pH on the two step precipitation reactions, the recovery of REE (La, Ce and Y) was calculated from weight content of REE (%) in Table 1. Figure 3 presents the lanthanide recovery for La, Ce and Y from the precipitation reaction. As shown in Figure 3, recovery of lanthanum (~68.2%) was achieved at pH 3.5 using sodium sulphate. On the basis of the total amount, the greatest recovery of lanthanides (~95%) from the low-grade bauxite was achieved using sodium sulphate as the precipitation agent at pH 3.5. The optimal result using sodium sulphate is in accordance with the previous acid leaching process using 1M oxalic acid at 40°C and pH 3.5 [8]. The metal composition of lanthanide species obtained from the second precipitation with sodium phosphate is low when compared with the first stage of precipitation using sodium sulphate. A similar effect was noted by acid leaching with sulfuric acid [9]. In this study, La was the rare-earth element present at the highest level in the low-grade bauxite and similar results were obtained for the leaching of REE slag from Hongcheongun, Korea [2].

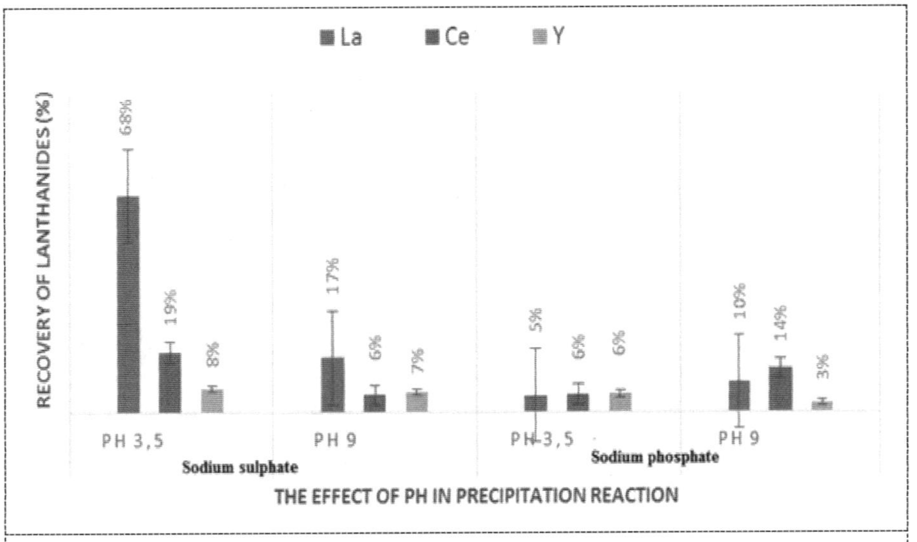

Figure 3. The effect of pH on the recovery of lanthanides.

4. Conclusion

The leaching of lanthanum, cerium and yttrium in oxalic acid and precipitation reactions with sodium sulphate and sodium phosphate at pH 3.5 and 9 were studied. The optimum condition for recovery of La, Ce and Y was pH 3.5 using sodium sulphate as the precipitation agent. The recovery metal composition was analyzed using ICP-OES. The recovery of lanthanum (~68.2%), cerium (~18.9%), and yttrium (~7.8%) were achieved by the first of precipitation step of the reaction using sodium sulphate as precipitation agent.

Acknowledgments

The authors greatly acknowledge the Universitas Indonesia for providing financial support through a grant, Hibah PITTA, No. 773/UN2.R3.1/HKP.05.00/2017.

References

[1] Kuo C S, "*The Mineral Industry of Indonesia*", U.S. GEOLOGICAL SURVEY MINERALS YEARBOOK, 2010, pp. 1-9.

[2] Kim C J, Yoon H S, Chung K W, Lee J Y, Kim S D, Shin S M, Lee S J, Joe A R, Lee S I, Yoo S J, Kim S H 2014, *Hydrometallurgy* **146** 133–137

[3] Gupta C K, Krishnamurthy N 2005, " *Extractive metallurgy of rare earths*", CRC Press, pp 20-250.

[4] Kusrini E, Usman A, Wicaksono W, Rahman A, Gunawan C 2017(a), *J Mater Environ Sci*, Under Review

[5] Binnemans K, Jones P T, Blanpain B, Gerven T V, Pontikes Y 2015, *J Cleaner Product* **99** 17-38.

[6] Sadri F, Nazari A M, Ghahreman A 2017, *J Rare Earths* **35** 739-752.

[7] Herdino F 2016, "Upgrading of rare earth on bauxite residue with mechanical processing and roasting treatments", Final project, Faculty of Engineering, Universitas Indonesia.

[8] Kusrini E, Prasetyanto E A, Bahari Z J, Harjanto S, Usman A 2017(b), Mater Sci Lett, submitted.

[9] Kusrini E, Usman A, Prasetyanto E A, Harjanto S, Trisko N, Rahman A 2017(c), Draft preparation

IOP Conf. Series: Materials Science and Engineering **334** (2018) 012053 doi:10.1088/1757-899X/334/1/012053

Utilization of *aloe vera* extract as electrolyte for an accumulator

F Azmi[1], D Sispriatna[1], K Ikhsan[1], M Masrura[1], S S Azzahra[1], Mahidin[1] and M D Supardan[1,2]
[1] Chemical Engineering Department, Syiah Kuala University

E-mail: m.dani.supardan@che.unsyiah.ac.id

Abstract. Aloe vera contains acid, which has the potential to generate electric current. The objective of this research is to study the potency of aloe vera extract as electrolyte for an accumulator. Experimental results showed that aloe vera extract has no a stable value of voltage and currency. The voltage and currency of aloe vera extract were reduced more than 50% for 60 minutes. Then, aloe vera extract was mixed with accu zuur to produce electrolyte solution. The mixture composition of aloe vera extract to accu zuur of 50:50 (v/v) generated stable voltage and currency. The experimental results showed the potential use of aloe vera extract to reduce the chemicals used in a conventional electrolyte solution.

1. Introduction

Aloe vera (*Aloe vera* L.) is native to Northern Africa, but it is now widely distributed throughout the world. It has thick leaves and thorns on its edges, is green colored, and is characterized by slimy leaves. It grows wild in subtropical parts of the world, and it is considered to be intolerant of low temperatures. Aloe vera has been used for thousands of years in medicine, cosmetics, and as an ornamental plant [1]. The leaf of aloe vera contains latex and gel. The juice or sap of aloe vera is the latex present beneath the epidermis layer, which occupies approximately 20–30% of the weight of the whole leaf. Meanwhile, the inner part of the leaf bears tasteless and colourless gel in the form of pulp or mucilage from the parenchyma cells [2].

Aloe vera is a well-known medicinal plant. It has many biological activities, including anti-viral, anti-bacterial, laxative, protection against radiation, anti-oxidant, anti-inflammation, anti-cancer, anti-diabetic, anti-allergic, and immuno-stimulation [3]. It also can be used as prebiotic [4].

Recently, a new form of energy source based on plants has been investigated by many researchers due to a shortage of energy sources, especially conventional fossil-based energy. Electrical phenomena in plants have attracted researchers for a long time. Some plants have been proven to have the potential to serve as a renewable energy source by embedding pairs of electrodes into them to allow the flow of ions and hence generate electricity. The cells of many biological organs can generate electric potentials that result in the flow of electric currents. Also, electrical impulses may arise as a result of stimulation. Once initiated, these impulses can propagate to adjacent excitable cells. The change in transmembrane potential creates a wave of depolarization that affects the adjoining resting membrane. Electrical signals can propagate along the plasma membrane over short distances in plasmodesmata and over long distances in conductive bundles [1]. The plant tissue of aloe vera has

[2] To whom any correspondence should be addressed.

biologically closed electrical circuits and electric fields that regulate its physiology. The electric circuits spread over aloe vera tissues [5]. Meanwhile, Ningsih et al. [6] also reported the potency of biologic material as an electrolyte for an accumulator. They found that *Manihot Glaziovii* has cyanide acid that potentially reduces the usage of chemicals as electrolyte solution for an accumulator.

The purpose of this study is to evaluate the electrical potency of aloe vera extract. There is no such comprehensive study previously reported on this crop. The information gained from this study can be used to produce an electrolyte that has less chemicals (such as organic acids) compared with electrolyte solution used in conventional accumulators.

2. Materials and Methods

Materials used in this research are aloe vera plant, aquadest, sulfuric acid (H_2SO_4 97%), and accu zuur (conventional electrolyte used in accumulators). The sample used in this research is aloe vera extract from the aloe vera plant (*Aloe Vera* L.), which was obtained from a local farmer in the Aceh Besar District, Aceh, Indonesia.

Firstly, the aloe vera was washed, and then its outer leaf was peeled off. The sample was then sliced to 3 cm of length, then, placed on filter cloths and pressed to produce slime (aloe vera extract). Then, the voltage and electric current measurements were conducted to aloe vera extract. The voltage and electric current measurements of aquadest, sulfuric acid, and accu zuur were also collected for comparison. Subsequently, aloe vera extract and accu zuur were mixed with a specified composition (volume ratio of aloe vera extract to accu zuur 70:30, 60:40, and 50:50). Sample solutions were then tested on an accumulator (12 volt, 5 ampere).

The electric current and voltage of the accumulator were measured with a multitester every 60 minutes for a total period of 300 minutes. Data were obtained by measuring the circuit with the multitester. The obtained data contained voltage and electric current in specified time interval and different concentrations. All measurements were conducted in the laboratory at ambient temperature; the measurements were collected in duplicate, and the averaged data are reported.

3. Results and Discussion

Figure 1 shows the time dependencies of voltage and currency for aquadest, aloe vera, and accu zuur. Experimental results show that accu zuur has higher voltage and currency and is more stable than aquadest and aloe vera. Accu zuur is a solution with H_2SO_4 concentration of 30 %-v. It shows high voltage and currency of about 13.5 volts and 5.6 ampere, respectively. This result has been known previously, although accu zuur has been used as a conventional electrolyte solution in accumulators in which each of 6 cells has2 volts. Aquadest has lower values of voltage and currency. It generates no electricity in the first of 60 minutes. Meanwhile, aloe vera extract has similar results as aquadest. Aloe vera extract also has no stable value of voltage and currency. The voltage and currency of aloe vera extract reduced more than 50% for 60 minutes. Volkov et al. [1] previously reported a similar trend for aloe vera leaf. However, there is the possibility to make an alternative electrolyte solution by mixing of aloe vera extract and accu zuur to reduce the use of chemicals in an electrolyte solution.

An electrolyte is a compound that can conduct electric current. Electrolyte solution can produce electric current because atoms move freely in the solution. Atoms that contain an electric charge are called ions. Ions with positive charge are called cations, and ones with negative charge are called anions. Electrolytes can be acids, bases, and salt solutions. Electrolyte solution has an important role in metal corrosion because this solution can be a medium for contact between anode and cathode. Anodes can be corroded by releasing electrons from neutral metal atoms to create ions. These ions might remain in solution or react and then form dissolved corrosive ions. As observed in the experiments (data not shown), sulfuric acid with concentration of 97% has a higher electricity compared with all electrolyte solutions used. Sulfuric acid with concentration of 97% has voltage and currency of approximately 14 volts and 5.8 ampere, respectively. However, it corrodes and quickly damages the electrodes in the accumulator. Thus, sulfuric acid with concentration of 97% could not be used as an electrolyte solution in accumulator. Accordingly, a proper choice of electrolyte solution is one important aspect in order to achieve good performance of an accumulator.

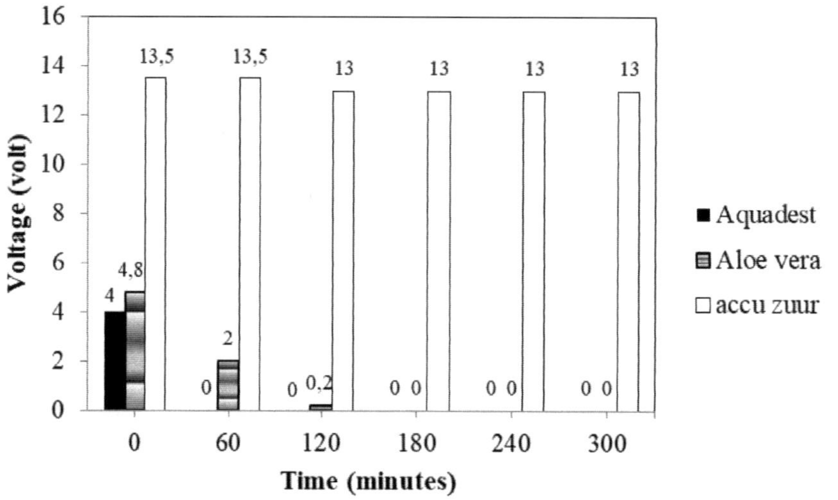

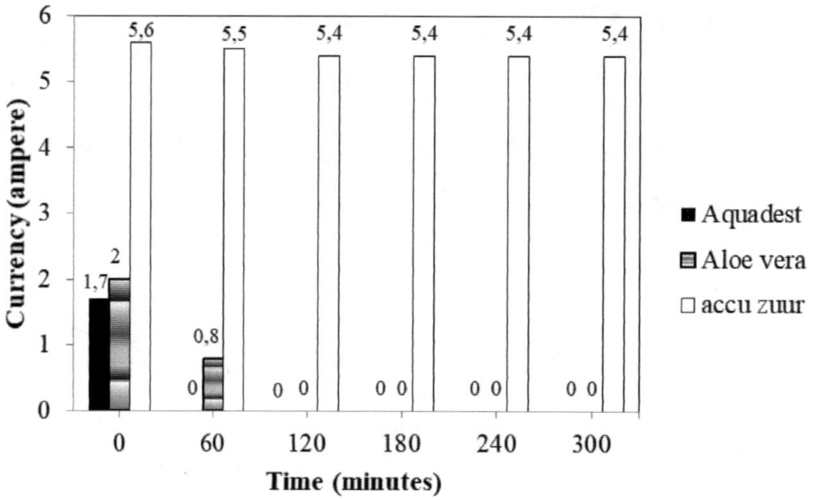

Figure 1. Voltage (V) and currency (I) measurements for aquadest, aloe vera and accu zuur.

Figure 2 shows the time dependencies of voltage and currency for mixtures of aloe vera and accu zuur. The experimental results showed the mixture composition of aloe vera extract to accu zuur of 50:50 as the best composition for reducing the usage of chemicals in the accumulator. This mixture produces voltage that has the same stability as a conventional electrolyte. It also lights the lamp on a simple electrical circuit. The mixture composition of aloe vera extract to accu zuur of 50:50 generates voltage and currency of approximately 12.5 volts and 5.2 ampere, respectively. These values were constant over the entire period of 300 minutes. Ningsih [6] previously reported that another biological material of *Manihot glaziovii* was mixed well with accu zuur (60:40 mixture composition), but only produced 4.8 volts and 1 ampere. It can be concluded that aloe vera has some effects to support the role of accu zuur as an electrolyte solution for accumulators. It was also observed that the mixture of aloe vera and sulfuric acid with concentration of 97% could not be used as an electrolyte on the accumulator, as it will corrode and quickly damage the electrodes in the accumulator.

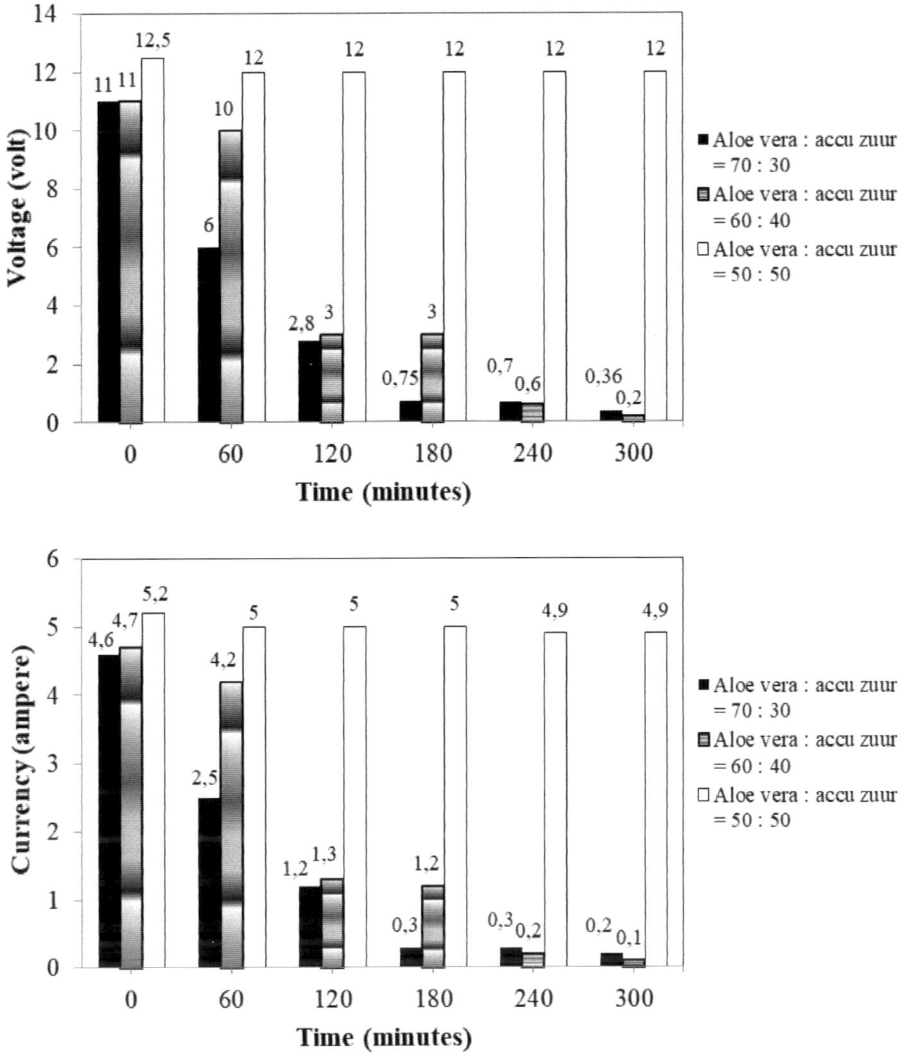

Figure 2. Voltage (V) and currency (I) measurements for mixture of aloe vera and accu zuur.

4. Conclusion

Investigations were carried out to study the electrical behavior of aloe vera extract. Accu zuur is demonstrated as an electrolyte that can be used in many batteries for vehicles through the data that shows high and stable electricity. In order to reduce the use of chemical or organic acid, aloe vera extract can be used as an electrolyte. The mixture composition of aloe vera extract to accu zuur of 50:50 (v/v) generates stable voltage and current.

Aknowledgement

Authors are grateful to University of Syiah Kuala and Ministry of Research, Technology and Higher Education of the Republic of Indonesia for providing financial support during the course of this investigation.

IOP Conf. Series: Materials Science and Engineering **334** (2018) 012053 doi:10.1088/1757-899X/334/1/012053

References

[1] Volkov, AG., Foster, JC., Jovanov, E., Markin, VS. 2011. Anisotropy and nonlinear properties of electrochemical circuits in leaves of *Aloe* vera L., *Bioelectrochemistry*, **81** 4-9

[2] Baruah, A., Bordoloi, M., Baruah, HPD. 2016, *Aloe vera*: A multipurpose industrial crop, *Industrial Crops and Products*, **94** 951-963

[3] Ray A, Ghosh S., Ray A., Aswatha SM. 2015. An Analysis of The Influence of Growth Periods on Potential Functional and Biochemical Properties and Thermal Analysis of Freeze-dried Aloe vera L. Gel. *Industrial Crops and Products* **76** 298-305.

[4] Hussain SA., Patil GR., Yadav V., Singh RRB., Singh AK. 2015, Ingredient Formulation Effects on Physico-chemical, Sensory, Textural Properties and Probiotic Count of Aloe Vera Probiotic Dahi. *LWT – J. Food Science and Technology* **65** 371-380

[5] Volkov AG., Nyasani EK., Tuckett C., Scott JM., Jackson MMZ., Greeman EA., Greenidge AS., Cohen DO., Volkova MI., dan Shtessel YB. 2017. Electronic Potensials in Aloe vera L.: Effects of Intercellular and External Electrodes Arrangement. *Bioelectrochemistry* **113** 60-68

[6] Ningsih YS., Gusnedi, dan Darvina Y. 2014. Effect of aquadest and accu zuur addition of to electric current and voltage of accumulator using Air Singkong Karet (*Manihot Glaziovii* M.A) (in Indonesian). *J. Pillar Physics* **1** 105-112.

IOP Conf. Series: Materials Science and Engineering **334** (2018) 012054 doi:10.1088/1757-899X/334/1/012054

Determination of CEC value (*Cation Exchange Capacity*) of Bentonites from North Aceh and Bener Meriah, Aceh Province, Indonesia using three methods

T Rihayat[1,4], S Salim[2], A Arlina[1], Z Fona[1], R Jalal[1], P N Alam[2], Zaimahwati[1], M Sami[1], J Syarif[3] and N Juhan[3]

[1] Departement of Chemical Engineering, Politeknik negeri Lhokseumawe, 24301, Aceh Indonesia
[2] Departement of Chemical Engineering, Faculty of Engineering, University of Syiah Kuala, Darussalam, Banda Aceh
[3] Department of Mechanical Engineering, Politeknik negeri Lhokseumawe, 24301, Aceh Indonesia

E-mail: teukurihayat@yahoo.com

Abstract. Research on determination of value CEC (Cation Exchange Capacity) Bentonite North Aceh and Bener Meriah with three methods has been studied. The purpose of this study was to determine the value of CEC bentonite North Aceh and Bener Meriah. The methods used in this research were pH equilibrium, $BaCl_2/MgSO_4$ and the adsorption of methylene blue. These three methods used to determine, compare, and calculation of the CEC value and determine the effect of particle size of bentonite on the value of the CEC. Bentonite North Aceh and Bener Meriah sieved with particle sizes of 80, 100, 150, 200, 250 mesh. The results showed that determination of the value of CEC bentonite North Aceh using $BaCl_2/MgSO_4$ with a particle size of 250 mesh is the value of the highest, reaching 79.09 meq/ 100 g.

1. Introduction

Application of bentonite has been widely used for filler, pellet, drilling mud, a catalyst and as adsorbent. Currently it has also been made into several high value products such as nanomaterial materials, medicines and chemicals. Bentonite is widely used for the treatment of esophagitis, gastritis, colitis and so on. Bentonite is also used as a raw material for cosmetics because of its decontamination, detoxification, as a moisturizer. In addition, with its chemical adsorption and stability, bentonite has been widely used in food processing industries that are closely related to health and human life [8].

The use of bentonite as an inorganic compound adsorbent for example such as heavy metal ions because bentonite has a cation exchange capacity and has a hydrophilic property on the surface. The chemical nature of the bentonite pore structure generally determines its adsorptive capacity. Due to the ineffective hydrophilic nature of bentonite to absorb organic compounds, to increase its capacity to

[4] To whom any correspondence should be addressed.

organic compounds, bentonite is intercalated with surfactants which can interact with the negative charge present on the interlayer surface.Bentonite has a surface chemical complex and arises from its ability to form a water gel thixotropic, high water absorption, high surface area, the layered structure and the ability of the cation exchange high, and contains aquartz crystal, cristobalite, feldspar, or other compounds [10]. Mineralogical composition will have an effect on the value of CEC obtained. The analytical problem appear from the CEC determination method result of specific interaction between the bentonite component and cation exchange mode which may cause an increase or decrease in the concentration of cation index [4].

Indonesia is one of the exporting countries of natural materials in form of bentonite which is quite calculated in the world. Indonesia's bentonite reserves are abundant, amounting to approximately 380 million tons spread over several islands, mainly Java and Sumatra. Bentonite contained in Sumatra, especially in North Aceh and Bener Meriah abundance and most of its kind that is Na-bentonite.

Smectite is a group of naturally-coated silica minerals formed naturally. These minerals are included in secondary minerals that can be formed from a hydrothermal alteration process. This mineral group is a very distinctive group, which will expand in wet and shrink at the time of water loss. This is due to the crystal lattice properties that can expand because cations and water molecules easily enter in the cavity between mineral crystal units. An important member of the smectite group is montmorillonite and mineral composition consisting of dust quartz, calcite, accessories, feldspar, mica and illite. The presence of such minerals can reduce the value of CEC bentonite. So most of the low grade bentonite is not suitable for industrial applications, such as water purification [6] commodities [2] and food industry [9].

The nature of smectite is to absorb kinds of cationic from solution by exchange between the balance of cation charged negative layers and cations in solution is referred to as capacity cation exchange (CEC), and the values of CEC can be expressed in centimole of positive charge per kilogram of mineral clay to dry the same with miliequivalent per 100 g of clay (meq/100 g) [1].

There are various methods for determining CEC values. Most of the methods involve natural cation exchange with the clay minerals include cationic species such as ammonium, $K^{+,}$ $N^{+,}$ methylene blue, Co (III) hexamine complex, Ba, Mg, Ag thiourea complex, Cu (II) complex of ethylenediamine and so on others, followed by the removal of their excess species [7].

The determination of CEC values depends on the method used. However, the determination of CEC values is sometimes difficult and time consuming in the process to achieve a perfect cation exchange. researchers determined the value of the CEC Bentonite North Aceh and Bener Meriah by three different methods, namely: pH balance method, the method $BaCl_2$/ $MgSO_4$ and methylene blue adsorption method.

Bentonite were then characterized using X-Ray Difraction (XRD) to determine the content of element/compound (qualitative analysis) and determination of the composition (quantitative analysis), a phase change as a function of temperature and time on bentonite by using Thermogravimetric analysis (TGA) (Rihayat, 2010)

2. Materials and Method

2.1 Materials

Bentonite is collected from nature, North Aceh and Bener Meriah, sodium hexametaphospate $(NaPO_3)_6$, glacial acetic acid (poison, Germany) distilled water, triethanolamine, barium chloride dihydrate ($BaCl_2.2H_2O$) , magnesium sulphate heptahydrate ($MgSO_4.7H_2O$), ammonium chloride, ammonia, erichrome black, ethanol, hydrochloric acid (HCL), Di-sodium EDTA, methlylene blue (Merck, Germany).

2.2 Determination of CEC

Bentonite was modified by preapring and purifying before determination with three methods. Each 100 grams of bentonite is weighed and then pounded with pestle. Performed sizing of 80, 100, 150,

200, 250 mesh. The sieved bentonite was introduced into 1,250 ml of aquades and added 1.24 g of CTAB surfactant then dispersed with a magnetic stirrer for 2 hours. The dispersion result is then senstrifugated for 2 minutes at a rate of 700 rpm to be separated from the liquid. Bentonite is dried in an oven with a temperature of ± 105 °C to dry. The proccess can be seen in Figure 1.

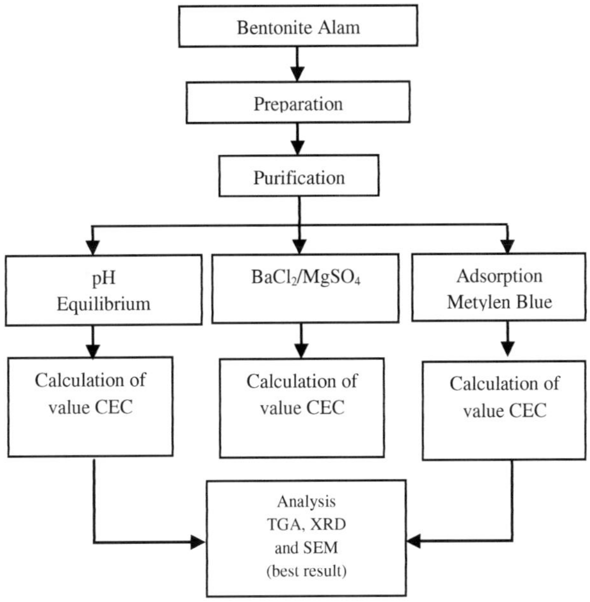

Figure 1. Research procedure.

For the determination of CEC value has done with three methods of analysis were: equilibrium pH, $BaCl_2/MgSO_4$ and methylene blue adsorption. Equilibrium pH method described in the literature, the CEC values can be determined by measuring pH change with pH meters. Measurements and calculations of CEC values were performed in accordance with procedures from the Laboratory of Soil Chemical Analysis. The material used was 1 M acetic acid solution, bentonite sample was inserted into beaker glass and 25 ml of acetic acid added into beaker glass containing sample. The solution stirred for 1 hour using a magnetic stirrer. Then suspension is fall down during sedimentation. Then, supernatant pH is checked. In order to accurate data , three repetitions performed. Measurement of CEC value using the following equation (1)

$$\text{CEC (meq per 100 g of clay)} = (\text{pH observed} - \text{pH acetic acid}) \times 22 \quad (1)$$

Where, 22 is a derived conversion factor of the curve of Brown (1943).

$BaCl_2/MgSO_4$ method involves the substitution of interlayer cations by Ba, which in turn was replaced by Mg, followed by titration with EDTA in the presence of ammonia [9]. The substitution of the interlayer cation with Ba is accomplished by adding a barium buffer chloride solution made by mixing the same volume of barium chloride. Further substitution with Mg is done by adding a solution of $MgSO_4$ with centrifugation and discarding the supernatant. The result of the end of the supernatant centrifugation is used for titration. Then the aliquot pipette of supernatant solution is titrated with standard EDTA by adding buffer ammonia and EBT indicator. The titration was repeated three times. The CEC values are calculated by the equation (20)

$$\text{CEC (meq per 100 g of clay)} = 8\,(B - A2) \quad (2)$$

Where, A2 is the end point is obtained from the equation [A1 x (100 + M2-M1)] / 100 ml, B is the endpoint solution of $MgSO_4$ (ml), A1 is the endpoint in the sample, M1 and M2 mass of bottle plus

IOP Conf. Series: Materials Science and Engineering **334** (2018) 012054 doi:10.1088/1757-899X/334/1/012054

Dry content, and bottle mass plus wet content (g). The detailed procedure based on the procedures of the Industrial Mineral Laboratory Guidelines: Bentonite [5].

Methylene blue adsorption method. Basically this procedure is performed by Pejon (1992), and gives a determination of the CEC value by determining the specific surface on the analyzed bentonite. This method is also easy and fast in analyzing a sample. Before the first analysis was diluted the solution of methylene blue with the need of 125 ml and stirred using a magnetic stirrer for all mixtures dispersed, for 1 minute. Please note the normality of methylene blue is 0.028 N. After the preparation of the material then the next sample is analyzed. Samples were inserted into beaker glass by adding 3 ml of methylene blue solution. Then stirred using a magnetic stirrer for 1 minute. After 1 minute titration with methylene blue solution. Any change at the time of the titration is followed by a dripped on filter paper, the goal is to see a light blue *halo* around dark blue and this is that the substitution between bentonite and methylene blue exists and continues to absorb this is called the principle of methylene blue method. This analysis is repeated three times to keep it accurate. The CEC values are calculated by the equation (3)

$$\text{CEC (meq per 100 g of clay)} = 100 / F \times V \times NMB \quad (3)$$

Where, F is the sample weight, V is the required methylene blue volume at the time of titration, NMB 0.028 N.

3. Results and Discussion

The determination of CEC (Cation Exchange Capacity) value of bentonite North Aceh and Bener Meriah using three methods to get different values.

The following results comparative data of CEC values with particle size of the pH equilibrium method.

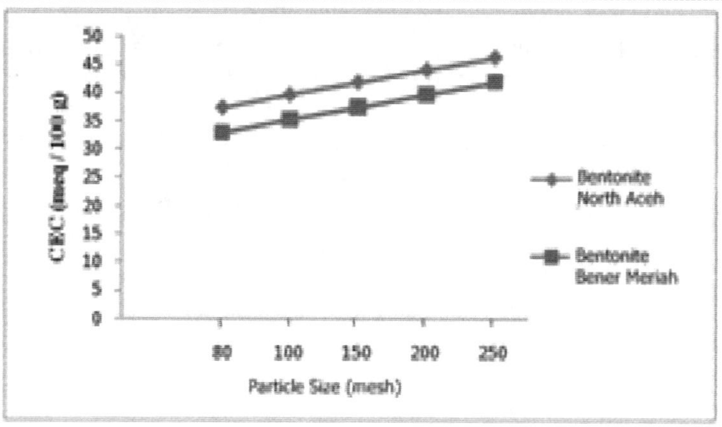

Figure 2. Result CEC Value vs Particle Size by Using pH Equilibrium Method.

Results of the data comparison of the CEC with a particle size of $BaCl_2/MgSO_4$ method.

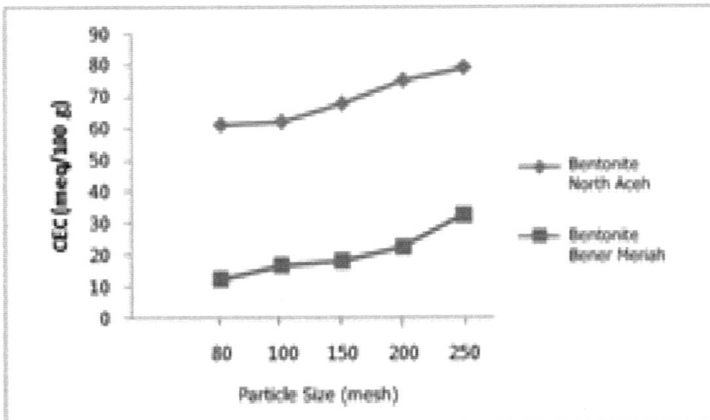

Figure 3. Result CEC Value vs Particle Size by Using $BaCl_2/MgSO_4$ Method.

Results of data comparing the CEC values with particle size of methylene blue adsorption method.

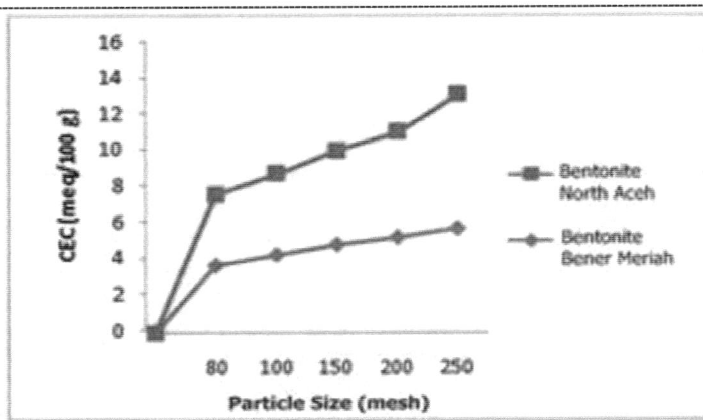

Figure 4. Result CEC Value vs Particle Size by Using Methylene Blue Adsorbtion Method.

From the results of the three methods found that a higher CEC value by using methods $BaCl_2$/ $MgSO_4$ at 250 mesh particle size reached 79.09 meq /100 g, this method is also the most efficient method in the analysis of determining the value of CEC in terms of ease of process and shorter time consumption rather than the pH balance method and the methylene blue adsorption method. This CEC value is greater than the value of CEC bentonite [3] which is 76.4 meq/100 g. Because with the smaller particle size/nano can easily exchange cations present in bentonite. After calculation of the CEC value, then characterized using XRD were done and report in difractogram in Figure 5.

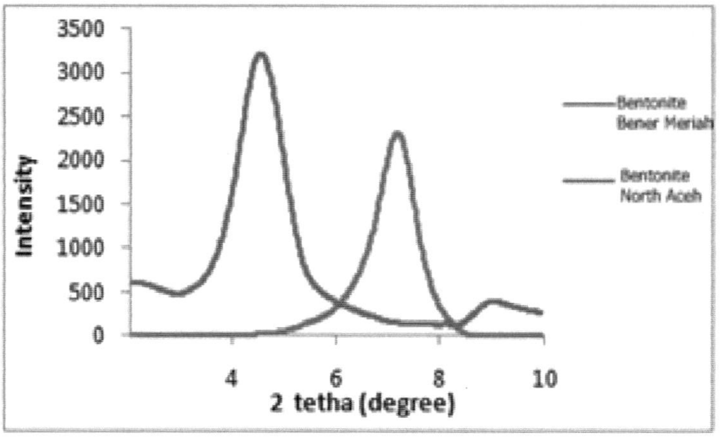

Figure 5. Difractogram Bentonite North Aceh and Bener Meriah Using $BaCl_2$/MgSOmethod.

Figure 5 shows that the XRD peak of bentonite at highest CEC value results using a $BaCl_2/MgSO_4$ method. The X-RD graph plots data between the angle of reflection on the x-axis and the intensity on the y-axis to produce a bentonite d-spacing layer automatically. Using the peak 2θ setting which yields data at the reflection angle from 2° to 10° indicates that the bentonite origin of Bentonite North Aceh has a maximum reflection angle peak at 4.2° with a d-spacing value of 3.281 nm, while bentonite Bener Meriah, it has found that the reflection angle peak at 6.7° with a d-spacing value of 2.415 nm. It seen there is a distinct bentonite diffraction of bentonite North Aceh is higher than Bener Meriah. The difference is caused by the condition where the bentonite located so that the Na^+ ion composition that affects the expanding properties may vary. Thus, from the results of this study can be said that the best swelling power that effect the CEC value shown by bentonite North Aceh. This also proves that the use of surfactant has succeeded in opening the d-spacing layer to make exchangeable higher in bentonite in the presence of organic compounds exchanging between the two cations (ammonium) of surfactants with ions in bentonite (Na^+). .

4. Conclusion

The most efficient method from the results of determining CEC value of bentonite North Aceh and Bener Meriah by three methods is $BaCl_2/MgSO_4$ method. The particle size can affect the value of the CEC, and the maximum value of CEC found in North Aceh Bentonite obtained of 79.09 meq /100 g with particle size of 250 mesh. The size of interlayer bentonite found at reflection angle peak at 4.2° with a d-spacing value of 3.281 nm by XRD graph.

Acknowledgement

The author expresses gratude and thanks to the Ministry of Research, Technology and Higher Education for Financial Support through Stranas Grant, 2017 and Directorate General for Research and Development Reinforcement, Kemenristekdikti.

References

[1] Brigatti, MF, Galan, E., Theng, BKG, 2006. Structures and mineralogy of clay minerals.Chapter 2 in Handbook of Clay Science Developments of Clay Science 1. *Elsevier*, Amsterdam.

[2] Bulut, G., Chimeddorj, M., Esenli, F., Çelik, MS, 2009.Production of desiccants from Turkish Bentonites. *Appl. Clay Sci.* **46**, 141-147.

[3] Choo, KY, Bai, K. 2016. The Effects of The Mineralogy Composition Of Various Bentonites On CEC Value Determined By Three Different Analytical methodes. *Appl. Clay Sci.* **126**, 153-159. Korea.

[4] Dohrmann, R., 2006. The cation exchange capacity methodology II: a modified silver-thiourea method. *Appl. Clay Sci.* **34**, 38-46.

[5] Inglethorpe, SDJ, Morgan, DJ, Highley, DE, Bloodworth, AJ, 1993. Industrial minerals laboratory manual: bentonite. BGS Technical Report WG / 93/20. *British Geological Survey,* Keyworth, Nottingham.

[6] Li, ZH, Chang, PH, Jean, JS, Jiang, WT, Wang, CJ, 2010. Interaction between tetracycline and smectite in aqueous solution. *J. Colloid Interface Sci.* **341**, 311 319.

[7] Meier, LP, Kahr, G., 1999. Determination of the cation exchange capacity (CEC) of the clay minerals using the complexes of copper (II) ions with triethylenetetramine and tetraethylenepentamine. *Clay Clay Miner.* **47**, 386-388.

[8] Mihaljevic, M., Ettler, V., Hradil, D., Sebek, O., Strnad, L., 2006. Dissolution of bentonite and release of rare earth elements at different solid / liquid ratios in a simulated wine purification process. *Appl. Clay Sci.* **31**, 36-46.

[9] Thoruwa, TFN, Johnstone, CM, Grant, AD, Smith, JE, 2001. The novel, low cost CaCl2-based desiccants for solar crop drying applications. *Renew. Energy* **19**, 513-520.

[10] Uddin, F., 2008. Clays, nanoclays and montmorillonite minerals. *Metall. Mater.* Trans. A 39A, 2804-2814.

IOP Conf. Series: Materials Science and Engineering **334** (2018) 012055 doi:10.1088/1757-899X/334/1/012055

Absorption of Carbon Dioxide in the aqueous solution of Diethanolamine (DEA) blended with 1-Butyl-1-Methylpyrrolidinium Trifluoromethanesulfonate [BmPyrr][OTf] at high pressure

S N Jamaludin[1,2] and R M Salleh[1]
[1] Chemical Engineering Department, Faculty of Chemical Engineering, Universiti Teknologi MARA, 40450 Shah Alam, Selangor Malaysia

E-mail: ctnabihahjamaludin@gmail.com, ruzitah@salam.uitm.edu.my

Abstract. Solubility data of carbon dioxide (CO_2) in aqueous Diethanolamine (DEA) blended with 1-Butyl-1-Methylpyrrolidinium Trifluoromethanesulfonate [Bmpyrr][OTf] were measured at temperature 313.15K, 323.15K, 333.15K and pressure from 500psi up to 700 psi. The experiments covered over the concentration range of 0-10wt% for [Bmpyrr][OTf] and 30-40wt% for DEA. The solubility of CO_2 was evaluated by measuring the pressure drop in high pressure stirred absorption cell reactor. The experimental results showed that CO_2 loading in all DEA-[BmPyrr][OTf] mixtures studied increases with increasing of CO_2 partial pressure and temperature. It was also found that the CO_2 loading capacity increase significantly as the concentration of [Bmpyrr][OTf] increases. Jou and Mather model was used to predict the solubility of CO_2 in the mixtures where the experimental data were correlated as a function of temperature and CO_2 partial pressure. It was found that the model was successful in predicting the solubility behavior of the aqueous DEA-[Bmpyrr][OTf] systems considered in this study.

1. Introduction

The growth in worldwide carbon dioxide emission has become a big threat to our environment. Fossil fuels consumption has been found to increase significantly nowadays caused by rapid development of industry as well as population growth. International Energy Agency (IEA) predicts that 40.2GT of fossils fuels utilization by the year 2030 [1]. As a result, it has led to global warming and drastic climate change. Excessive of CO_2 emission consequently will be disastrous to humankind and also a great threat to disrupt. Thus, effective method to capture CO_2 is urgently needed to mitigate this environmental problem [2][3].

Presently, several methods have been implemented for CO_2 removal purposes including absorption, adsorption, membrane separation and cryogenic process [2][3][4][5]. Among these methods, absorption using chemical solvents such as aqueous alkanolamine is one of the most matured and effective methods since it has significant experience for CO_2 capture process besides cost effective for large-scale separations. Diethanolamine (DEA) is one of the common alkanolamine used in the industries due to their high reactivity and low cost however, low loading of this amine developed

[2] To whom any correspondence should be addressed.

researcher's interest to explore new amines/solvents which may improve loading capacity of CO_2 gas [6].

Recently, Ionic Liquids (ILs) has attracted considerable attention due to their significant advantages over basic aqueous alkanolamine [7][8]. ILs possess low vapour pressure, low regeneration energy, tuneable nature and thermally stable over wide range of temperature. These advantages make ILs useful medium for capturing CO_2 [7]. However, most of ILs have revealed that it is not always cost effective, thus developing the most efficient hybrid solvent for CO_2 capture technologies is urgently needed.

In this work, the idea of aqueous amines blended with ionic liquid was comprehensively investigated. The equilibrium CO_2 solubility into aqueous Diethanolamine (DEA) blended with 1-Butyl-1-Methylpyrrolidinium Trifluoromethanesulfonate [BMPyrr][OTf] mixtures experiments were performed at temperature range of 313.15K-333.15K over CO_2 partial pressure range from 500 psi up to 700 psi using different weight percent of [BmPyrr][OTf] (0–10 wt%) and the DEA 30–40 wt%. Jou and Mather model was employed to predict the CO_2 solubility in the mixtures.

2. Experimental

2.1. Material

Purified CO_2 (purity $\geq$ 99.95%) was obtained from Linde Malaysia Sdn Bhd. 1-Butyl-Methylpyrrolidinium Trifluoromethanesulfonate [Bmpyrr][OTf] (purity $\geq$ 98%) and Diethanolamine (DEA) with minimum purity of 99.5% used in this study were purchased from Merck and used as received without further purification. Figure 1 shows the chemical structure for these two types of solvents. Distilled water was used as diluter solvents for all experiments. For this study, the mixtures were prepared by adding [BmPyrr][OTf] at different concentration from 0 to 10wt % to aqueous DEA solutions.

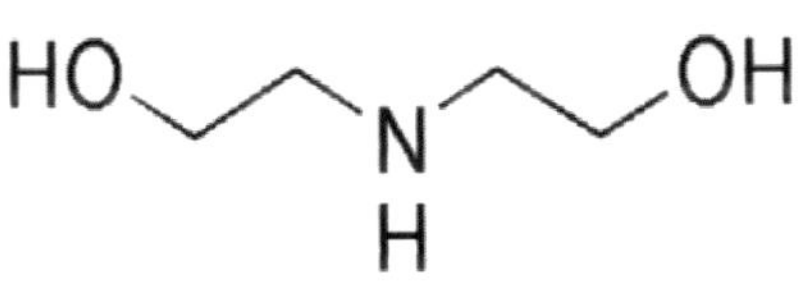

a)

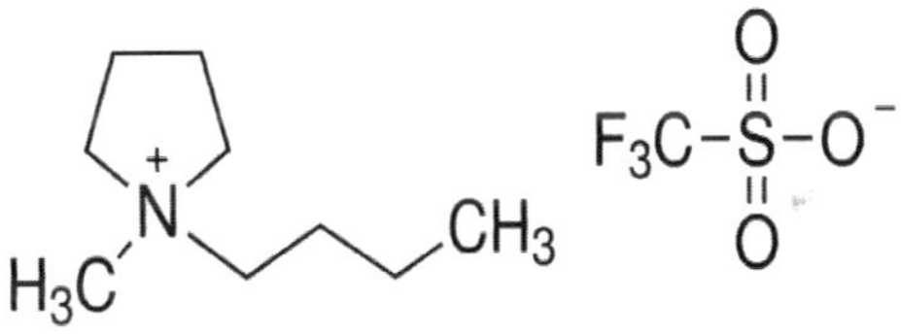

b)

Figure 1. Chemical structure of a) Diethanolamine (DEA), b) 1-Butyl-1-Methylpyrrolidinium Trifluoromethanesulfonates [BmPyrr][OTf].

2.2. Methods

The aqueous solutions were prepared by adding distilled water to the solvent mixtures at the desired concentration in all of the experiments. The CO_2 solubility was measured using high-pressure jacketed reactor, which equipped with magnetic stirrer, glass liner and thermocouple. For this work, the temperatures were set in the ranges of 313.15K-333.15K while pressure up to 700 psi. In each run, 20ml of solvent mixtures was placed into the reactor. The CO_2 gas reservoir and the reactor were heated to the desired temperature. Once temperature was stable, CO_2 gas was released into the reactor and constantly stirred throughout the experiments. Equilibrium was achieved indicated by a constant pressure. Finally, the CO_2 loading for equilibrium cell was calculated based on pressure drop using equation (1).

$$\alpha = \frac{\frac{(PTi-Pf)\times Vvessel}{zRT} - \frac{(Peq-Pvi)\times Vreactor}{zRT}}{n_{total}} \quad (1)$$

IOP Conf. Series: Materials Science and Engineering **334** (2018) 012055 doi:10.1088/1757-899X/334/1/012055

Where α is mol CO_2 loading/ total amine, P_{Ti} is initial total pressure on the vessel (atm), P_f is final pressure in the vessel (atm), V_{vessel} is 1 litre, $V_{reactor}$ is gas volume in reactor (litre), Z is gas compressibility factor, R is gas constant, P_{eq} is equilibrium pressure in reactor (atm), P_{vi} is vapor pressure (atm) n_{total} is no of moles of amine and T is temperature (K)

3. Result

In this work, to validate the experimental data, the solubility of CO_2 in aqueous solution of 1M concentration of 2-Amino-2-Methyl-1-Propanol (AMP) was measured at 333.15K and compared with the literature [9]. Figure 2 demonstrated that the measured data exhibit good agreement with the literature where the average absolute deviation is only 4.28%.

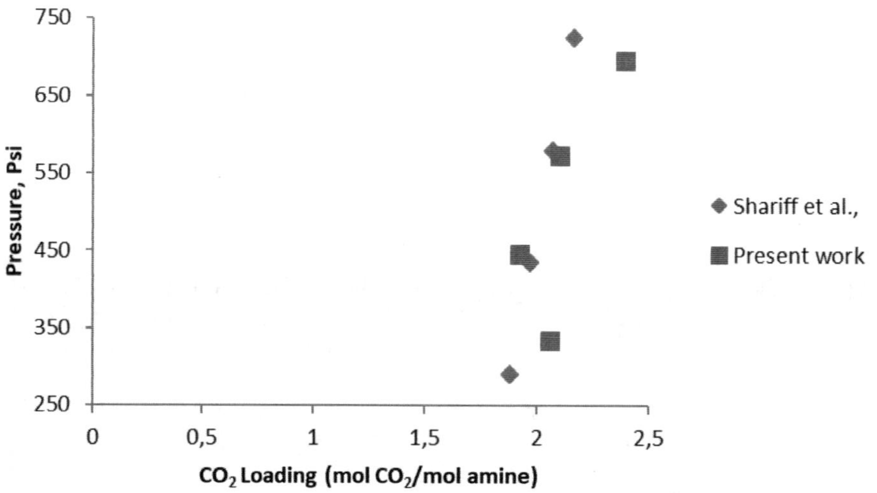

Figure 2. Solubility comparison of this work with literature values at 333.15K for 1M AMP.

3.1. Effect of [BmPyrr][OTf] concentration on CO_2 loading

The experimental focusing on the effect of ILs concentration on CO_2 loading was performed. The results indicated that the addition of [BmPyrr][OTf] into aqueous DEA enhanced the absorption capacity of CO_2. This can be seen from Figure 3 where the uptake of CO_2 into aqueous mixtures of [DEA][BmPyrr] was greatly increased compared to aqueous DEA alone. Addition of 5wt% of [BmPyrr][OTf] to the aqueous DEA increased the CO_2 loading by 56.5% at temperature 333.15K compared to the CO_2 loading in aqueous DEA. In the three aqueous mixtures of [DEA][BmPyrr][OTf] solution studied, the most CO_2 absorption capacity uptake was recorded for 30wt% DEA + 10wt% ILs aqueous solution which is 1.82 mol CO_2/mol amine at pressure 700psi. Therefore, this observation demonstrates that an increase in the concentration of ILs in the solution leads to an increase in CO_2 loading.

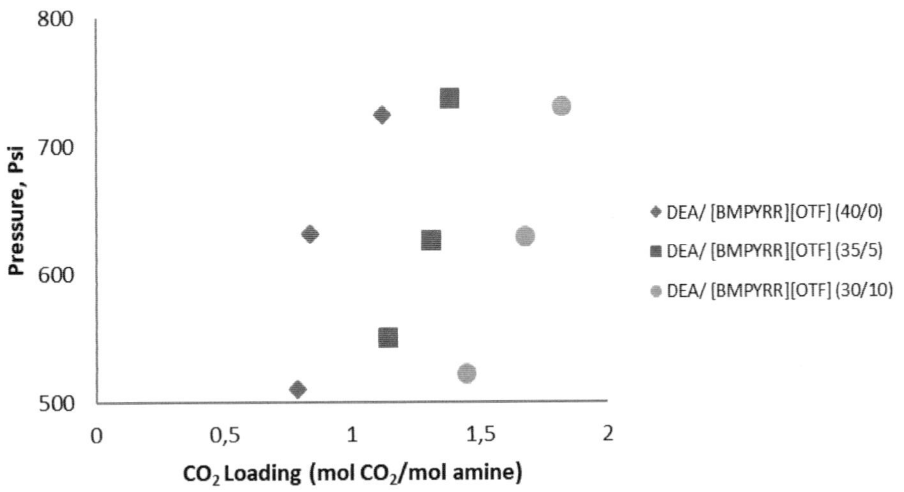

Figure 3. CO_2 loading for aqueous mixtures of DEA + [BmPyrr][OTf] with different composition at 333.15K.

3.2. Effect of temperature and pressure

The experimental results of temperature and pressure effect on CO_2 loading in the aqueous DEA is shown in Figure 4. As expected similar to other single amines behaviour, the result shows that CO_2 solubility decreases with increasing temperature in aqueous DEA solution. There is a clear trend of decreasing CO_2 absorption performance where the curve shifted to the left when the temperature increases indicating that less CO_2 absorbed by the solutions. This CO_2 solubility pattern is in agreement with the findings of other researchers [10][11][12][13]. This behaviour demonstrated that more gas is present in the solution with lower temperature, besides that the increasing of kinetic energy at high temperature has led to vigorous dynamic motion of CO_2 molecules causing the breaking of the intermolecular bond between CO_2 and the solution [14]. As a result, less CO_2 will be absorbed since more CO_2 was escaped from the solutions. This trend can be explained by the understanding of the vapor pressure concept, which increase with temperature [9]

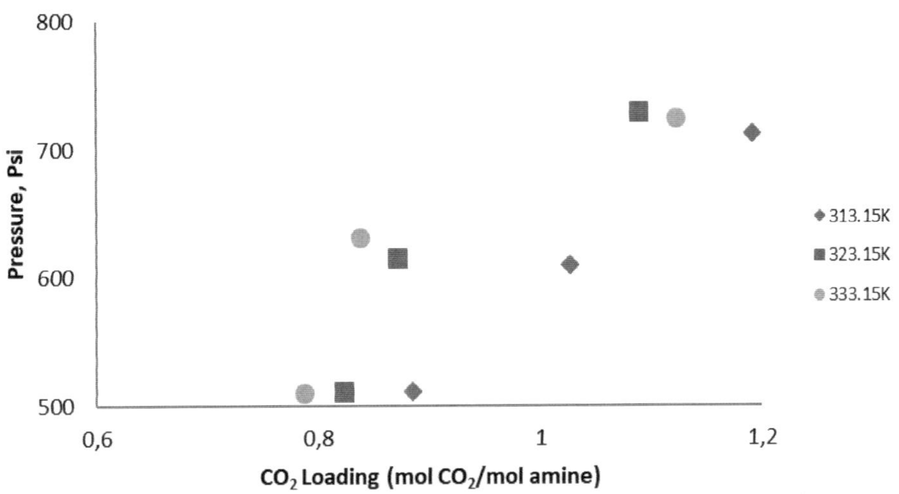

Figure 4. CO_2 loading of 40wt% of aqueous of DEA at different temperature.

However, contrast trends was found for aqueous DEA-[BmPyrr][OTf] mixtures where the CO_2 solubility increases as temperature increases. Figure 5 shows that the curve shifted to the right side when the temperature increases from 313.15K to 333.15K indicating that more CO_2 absorbed by the solutions. This behaviour showing that there is an interaction between water and [BmPyrr][OTf] that consequently causing the absorption process increase with temperature .

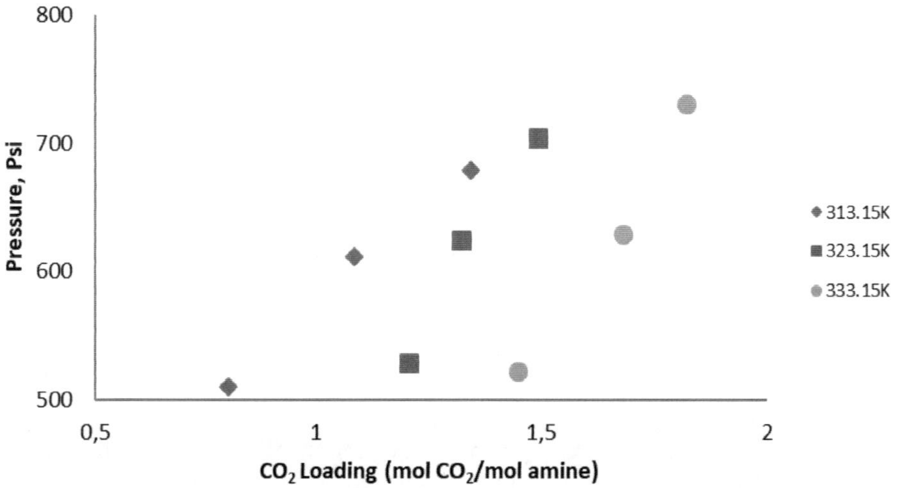

Figure 5. CO_2 loading of 30wt% DEA + 10wt% [BmPyrr][OTf] at different temperature.

In terms of pressure effect toward CO_2 absorption performance, positive impact was observed for aqueous mixtures of DEA + [BmPyrr][OTf] where CO_2 solubility increases with the increasing of the CO_2 partial pressure. This is because the gas pressure increases the number of collision between the gas molecules and the solution surface, therefore more gas is dissolved as the pressure increased [12][15]. This system obeys the Henry's Law concept where the solubility of gas in a liquid is directly proportional to the pressure of the gas above the surface of the solution.

3.3. Correlations of solubility as a function of pressure and temperature

The experimental data of CO_2 loading suggested by Jou and Mather [16] model were correlated linearly fitted as function of CO_2 partial pressure and temperature according to Equation 2 as follow:

$$\ln P = A \ln \alpha + B \tag{2}$$

where P is the pressure in Psi, α is mol CO_2 loading/ mol total amine and A and B were determined by plotting ln P against ln α at temperature of 313.15K, 323.15K and 333.15K. Once the values of A and B were obtained these values were plotted against the temperature to get the coefficients of A and B. The coefficients were listed in Table 1.

Table 1. Jou and Mather coefficients.

System	Coefficients	
	A	B
40 wt% DEA	-0.0137T + 5.4557	0.0083T + 4.5871
35 wt% DEA + 5 wt% [BmPyrr][OTf]	-0.0208T + 8.0386	-0.0617T + 26.558
30 wt% DEA + 10 wt% [BmPyrr][OTf]	0.0438T – 13.052	-0.0312T + 16.41

The solubility correlations as a function of temperature at various pressures were conducted to determine the CO_2 loading measurement. The model was further quantified by determining the average absolute deviation (AAD) using Equation (3) as follows:

$$AAD = \frac{\sum_{i-1}^{N} \left| (\alpha_{exp} - \alpha_{cal}) / \alpha_{exp} \right|}{N} \quad (3)$$

Comparisons between the predicted and experimental values of CO_2 solubility are shown in Figure 6.

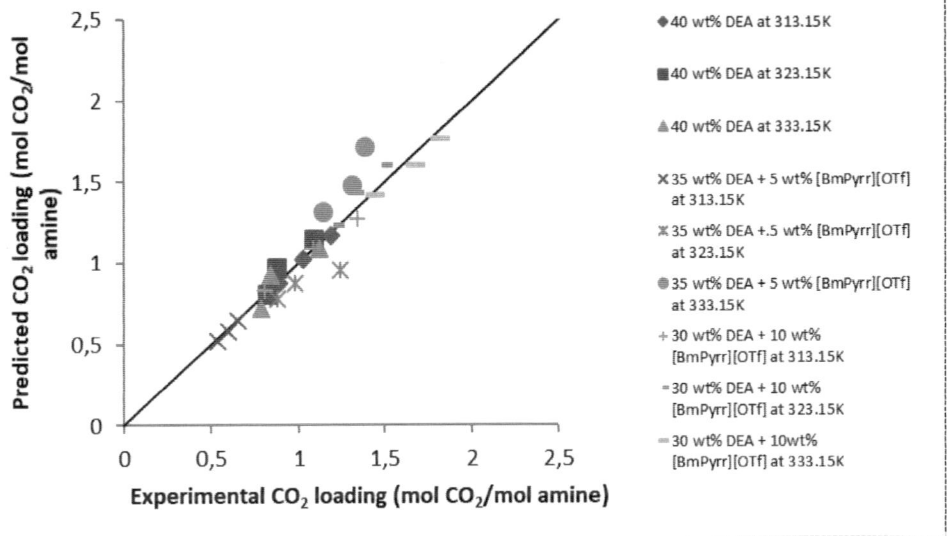

Figure 6. Comparison between correlated and experimental data for CO_2 loading.

It was found that the correlation of CO_2 solubility has AAD percentages of 6.936%. Thus, it shows good agreement between the experimental and predicted data.

4. Conclusion

In this work, the absorption of CO_2 into aqueous DEA blended with [BmPyrr][OTf] was measured. It has been found that addition of small amount of [BmPyrr][OTf] to the aqueous DEA can significantly enhance the CO2bsorption capacity compared to the conventional aqueous single amine. Aqueous solution with 30wt% of DEA + 10wt% of [BmPyrr][OTf] has the highest CO_2 uptake capacity compared with other aqueous mixtures solutions that have been investigated. It was found that increasing the pressure has led to increasing of CO_2 loading and increasing of temperature increases the CO_2 solubility performances for aqueous DEA-[BmPyrr][OTf] mixture however absorption performances reduces for aqueous DEA when temperature rises. Predicted CO_2 solubility using Jou and Mather model was found to be in good agreement with the experimental data.

Acknowledgement

The authors would like to thank the Ministry of Higher Education (MOHE) and Universiti Teknologi MARA (UiTM) for financial support given through the Fundamental Research Grant Scheme (FRGS) (600-RMI/FRGS 5/3 (91/2014)) for completing this research project.

References

[1] I. E. Agency, "World Energy Outlook," 2009.

[2] H. Yang, Z. Xu, M. Fan, R. B. Slimane, A. E. Bland, and I. Wright, "Progress in carbon dioxide separation and capture: A review," *J. Environ. Sci.*, vol. 20, pp. 14–27, 2008.

[3] D. Y. C. Leung, G. Caramanna, and M. M. Maroto-Valer, "An overview of current status of carbon dioxide capture and storage technologies," *Renew. Sustain. Energy Rev.*, vol. 39, pp. 426–443, 2014.

[4] J. D. Figueroa, T. Fout, S. Plasynski, H. McIlvried, and R. D. Srivastava, "Advances in CO2 capture technology-The U.S. Department of Energy's Carbon Sequestration Program," *Int. J. Greenh. Gas Control*, vol. 2, no. 1, pp. 9–20, 2008.

[5] A. S. Bhown and B. C. Freeman, "Analysis and status of post-combustion carbon dioxide capture technologies," *Environ. Sci. Technol.*, vol. 45, no. 20, pp. 8624–8632, 2011.

[6] M. R. M. Abu-Zahra, Z. Abbas, P. Singh, and P. Feron, "Carbon Dioxide Post-Combustion Capture : Solvent Technologies Overview, Status and Future Directions," *Mater. Process. energy Commun. Curr. Res. Technol. Dev.*, pp. 923–934, 2013.

[7] M. Althuluth, M. T. Mota-Martinez, M. C. Kroon, and C. J. Peters, "Solubility of Carbon Dioxide in the Ionic Liquid 1-Ethyl-3-methylimidazolium Tris(pentafluoroethyl)trifluorophosphate," *Fluid Phase Equilib.*, vol. 332, pp. 35–39, 2012.

[8] N. M. Yunus, M. I. A. Mutalib, Z. Man, M. A. Bustam, and T. Murugesan, "Solubility of CO2 in pyridinium based ionic liquids," *Chem. Eng. J.*, vol. 189–190, pp. 94–100, 2012.

[9] A. M. Shariff, G. Murshid, K. K. Lau, M. A. Bustam, and F. Ahamd, "Solubility of CO2 in Aqueous Solutions of 2- Amino-2-Methyl-1-Propanol at High Pressure," *Int. J. Environ. Chem. Ecol. Geol. Geophys. Eng.*, vol. 5, no. 12, pp. 825–828, 2011.

[10] Z. Feng, F. Cheng-Gang, W. You-Ting, W. Yuan-Tao, L. Ai-Min, and Z. Zhi-Bing, "Absorption of CO2 in the aqueous solutions of functionalized ionic liquids and MDEA," *Chem. Eng. J.*, vol. 160, no. 2, pp. 691–697, 2010.

[11] F. Xu *et al.*, "Solubility of CO2 in aqueous mixtures of monoethanolamine and dicyanamide-based ionic liquids," *Fluid Phase Equilib.*, vol. 365, pp. 80–87, 2014.

[12] M. K. Wong, M. a. Bustam, and a. M. Shariff, "Chemical speciation of CO2 absorption in aqueous monoethanolamine investigated by in situ Raman spectroscopy," *Int. J. Greenh. Gas Control*, vol. 39, pp. 139–147, 2015.

[13] S. Nordin, R. M. Salleh, and N. Ismail, "Separation of CO2 by Using Aqueous Mixtures of AMP and [B3MPYR][BF4] at High Pressure," *J. Teknol.*, vol. 67, no. 4, pp. 1–3, 2014.

[14] B. H. Lu, J. J. Jin, L. Zhang, and W. Li, "Absorption of carbon dioxide into aqueous blend of monoethanolamine and 1-butyl-3-methylimidazolium tetrafluoroborate," *Int. J. Greenh. Gas Control*, vol. 11, pp. 152–157, 2012.

[15] A. M. Shariff, M. S. Shaikh, M. A. Bustam, S. Garg, N. Faiqa, and A. Aftab, "High-pressure Solubility of Carbon Dioxide in Aqueous Sodium L- Prolinate Solution," *Procedia Eng.*, vol. 148, pp. 580–587, 2016.

[16] F. Y. Jou and A. E. Mather, "Solubility of carbon dioxide in an aqueous mixture of methyldiethanolamine and N-methylpyrrolidone at elevated pressures," *Fluid Phase Equilib.*, vol. 228–229, pp. 465–469, 2005.

IOP Conf. Series: Materials Science and Engineering **334** (2018) 012056 doi:10.1088/1757-899X/334/1/012056

Investigation on the effect of THF on Nitrogen Hydrate formation under isobaric condition

N Jamil[1], H Husin[2,5], Z Aman[3] and Z Hassan[4]

[1]Faculty of Chemical Engineering, Universiti Teknologi MARA, Shah Alam, Malaysia.
[2]Department of Petroleum Engineering, Universiti Teknologi PETRONAS, 32610 Seri Iskandar, Perak, Malaysia.
[3]School of Mechanical and Chemical Engineering, The University of Western Australia, 35 Stirling Highway, Crawley WA 6009, Australia.
[4]Faculty of Chemical and Natural Resources Engineering, Universiti Malaysia Pahang, Gambang, Pahang, Malaysia.

E-mail: husinhazlina@gmail.com

Abstract. In this paper, we studied nitrogen (N_2) hydrate formation in the presence of tetrahydrofuran (THF) under 3 different conditions; different concentration of THF (0, 3 and 30 %(v/v), different temperature setting (room temperature and induced temperature) and different water content (15, 35 and 55 mL) in an isobaric condition. We found that in the presence of THF which acting as an enhancer, hydrate formation kinetic is highly influenced by these parameters. We observed a striking contrast in hydrate formation behaviour observed at room temperature (RT) and induced temperature (IT) with and without the presence of THF under similar operating conditions. At the presence of 30 %(v/v) of THF in 15 mL water, it can be seen that, hydrate tend to form faster than other samples. Visual observation of N_2 hydrates are also conducted at 30 %(v/v) of THF in 15 mL water.

1. Introduction

Gas hydrate is a crystalline solid that compose mostly of compacted water and gas molecules, which formed at low temperature and high pressure conditions [1]. The formation of hydrate comprises of certain light gases such as methane, ethane, propane, carbon dioxide and nitrogen. The size and shape of hydrate formed depend on the atomic size of gas used and the formation condition that is applied. The most commonly encountered structures are the cubic structures I (sI) and II (sII) [2]. Gas hydrate formation was one of the flow assurance main problems in oil and gas industries [3]. Thus, the gas hydrate particles can agglomerate and progressively build up into a large lump which had the potential to block the flowlines, chokes, valves and also other equipment along the pipelines [3-4]. Hydrate is useful in many industrial applications like storage and transport of natural gas, gas separation, seawater desalination, CO_2 capture, CO_2 sequestration and seawater desalination [5-9]. In recent time, storage of gas in the form of hydrate gains interests due to high gas contents per volume of hydrate, environmentally friendly and as well as its non-explosive nature that possesses no safety risk. Despite

[5] To whom any correspondence should be addressed.

all the applications of hydrate, the hydrate phase equilibria of the carbon dioxide, nitrogen, and water are scarcely studied. Especially in nitrogen hydrates, most of the researchers focus on their application for the dissociation of hydrates in sediment [10]. Nitrogen hydrates were among the hydrate that requires an extreme amount of pressure (up to 1000 atm) to form. Reducing hydrate formation pressure is a hindrance issue for the wide application of hydrate technology.

Tetrahydrofuran (THF) is known as one of a thermodynamic promoter that functions to reduce hydrate formation pressure for methane, ethane, carbon dioxide and nitrogen[11]. It had been proven that THF can form stable sII hydrates with water at atmospheric pressure when the temperature is below 278 K [12]. Therefore, THF is added as guest molecules for the formation of nitrogen hydrates to help stabilize the hydrate cages thereby reducing the formation pressures of nitrogen hydrates. THF is frequently used as a model system for the study of hydrate formation because of the elimination of the mass transfer effect [13]. The use of THF as thermodynamic promoters had been used in previous research, where they use multiple mixed systems, among these, are THF+hydrogen+water [14], [15], THF+carbon dioxide+water [16], THF+methane+water [16], [17], THF+methyl fluoride+water [18] and THF+carbon dioxide+nitrogen+water [19]. Recently, Veluswamy et al. [20] was able to form the hydrate in the laboratory scale in an isobaric system in the present of THF as promoter but at a very high pressures of 30 bar and 283.2 K using methane gas. In this study, the objective was to achieve the enhancement in hydrate formation kinetics for N_2 hydrates in the presence of THF in an isobaric condition at lower pressure and temperatures. The effect of promoter, temperature and volume upon the hydrate formation was also investigated.

2. Experimental Section

2.1 Preparation of Material

High Purity Nitrogen (99.9%) supplied by Pure Dimension Enterprise were used in the formation of hydrates. The promoter used was the analytical reagent grade tetrahydrofuran (THF), supplied by the Fisher Chemical Sdn. Bhd. In this work, the solution used is the mixture of distilled water + THF. All of the solutions were prepared at the room temperature and atmospheric pressure. The preparation of THF solution was done by the following steps: A certain amount of water and THF were weighted successively. Next the water was poured into THF as quickly as one can to avoid volatilization of THF, and subsequently the solution was introduced into the pressure cell after being stirred with a glass rod. Water and THF were weighed on an electronic balance with an accuracy of (0.1 mg).

2.2 Experimental Apparatus

A schematic diagram of the experimental apparatus used to determine the hydrate formation is shown in figure 1. The main part of the apparatus is the high pressure cell with an effective volume of 110 cm^3. The pressure of the cell is controlled by rotating the control knob on to the desired pressure. The pressure cell was secured inside a customized box which acts as the cooling jacket to maintain the experimental temperature inside the pressure cell. The pressure and temperature measurement were made by reading on the pressure that was shown on the pressure gauge attached to the cell as well as from the P-T controller. All of the processes require collection of data manually (table A.1 and table A.2) as the reactor does not connect to any data acquisition (DAQ) system. All experiments performed in a quiescent (unstirred) pressure cell.

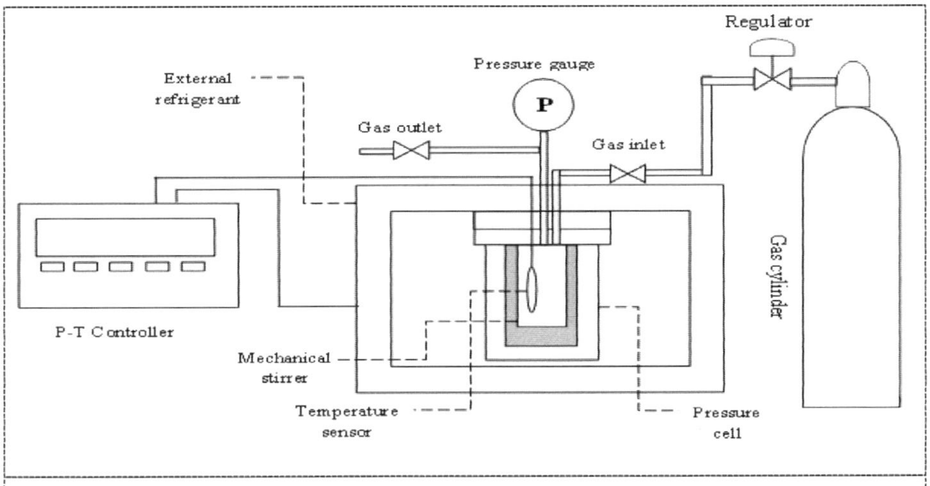

Figure 1. Experimental apparatus for the formation of hydrates.

2.3 Experimental Method

Schematic of the experimental procedure is shown in **Figure 1**. Aqueous solutions (55 cm^3) comprise of different concentration of THF (0 %(v/v), 3 %(v/v) and 30 %(v/v) of THF solution) were prepared and introduced into the cell. The cell and tubing were purged thrice using the nitrogen gas to remove the air remaining and the pressure cell is vacuumed to experimental pressure. The temperature was set at room temperature. The experiment is done in an isobaric system, where the pressure was kept constant throughout the experiment. Once the experimental pressure in the vessel reached desired values, the outlet valves of the cell then closed to concentrate the gas inside the cell. The aqueous solution inside the pressure cell remained quiescent. Temperatures were recorded every 1 hour for six (6) hours. Hydrate formation was detected by a sudden increased in temperature. The time before temperature reading remains stabled is recorded as hydrate induction time.

3. Result and Discussion

3.1 Effect of temperatures and different concentration of THF on hydrate formation.

The pressure required for the formation of hydrate using (N_2) gas is high for a commercial industrial process [21]. This implies significant compression would be required. Therefore, the use of THF has been proposed by Linga et al. [22]. The used of THF is an efficient method for stabilizing the nitrogen hydrates [23]. Usually, the hydrates will form at a pressure of 50-100 bar, but with THF it can be reduced until 5-10 bar at the temperature range between 0 - 10°C [24]. All the experimental results are summarised in table A.1 and table A.2. Figure 2 and figure 3 shows the graph of the effect of different temperatures, one is at room temperature (RT) and at induced temperature (IT) on different concentration of THF. As seen from the graph, there is a huge difference between the 0 %(v/v), 3%(v/v) and 30 %(v/v) RT and IT. During the hydrate formation at room temperature (RT), there is no trace of nitrogen hydrate formed in any of the THF concentration. There is no significant peak is seen and the graph is aligned almost similar to a straight line. The absence of significant peak deduces that there is no hydrate is formed as hydrate formation is an exothermic process [25]. This may also can be linked to less N_2 gas is trapped at high temperature during the nucleation process [26]. Ross et al. [27], in their studies also mentioned that thermal conductivity of THF in hydrate is proportional to temperature and not pressure. There is also a possibility that with the addition of THF in water allow the shift of the promotion effect of THF to inhibition effect at a certain concentration. In all three (3) THF concentrations, none of them performed at room temperature. However, the nucleation process still occurred as the end product (aqueous solution), is in a fizzy and cloudy state. Devarakonda et al. [28] mentioned that the point when the solution turns cloudy is considered the end of nucleation. Perhaps in the next experiment, the time to form N_2 gas hydrates could be lengthened to see whether hydrate will be formed at room temperature. It has been stated in the previous study, in the study of hydrate formation by using THF in order to shorten the induction time the only suitable medium is in an induced system [29]. Nonetheless, hydrate may form if the right temperature and pressure condition are designed.

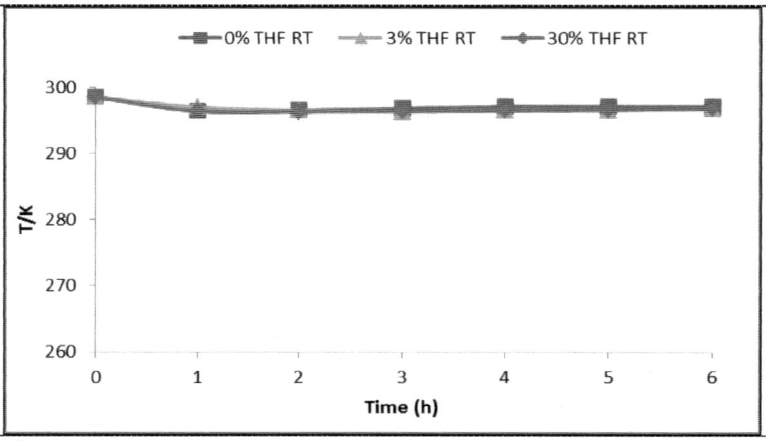

Figure 2. Hydrate formation for the nitrogen+water+THF system at room temperature: 30 %(v/v)THF, 3 %(v/v) THF and 0 %(v/v) THF.

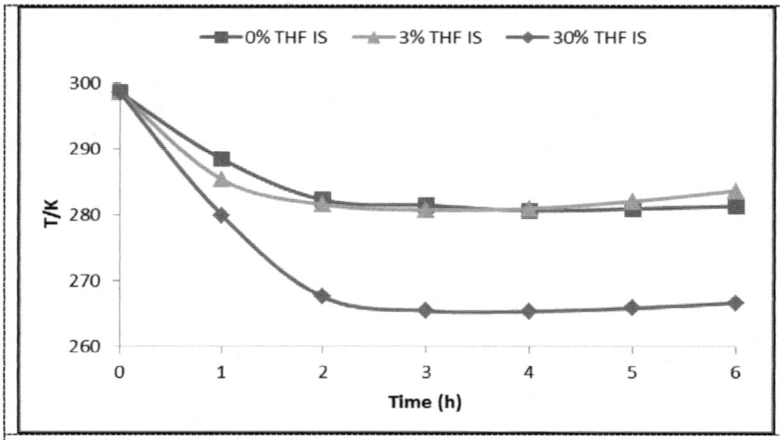

Figure 3: Hydrate formation for an induced temperature nitrogen+water+THF system: 30 %(v/v) THF, 3 %(v/v) THF and 0 %(v/v) THF.

Figure 3 shows the result of hydrate formation at three (3) different concentration of THF (0 %(v/v), 3 %(v/v) and 30 %(v/v)). From the figure, it can be seen that for all 3 concentration in the same induced system, the trend in temperature decrease is different for each concentration. The higher the concentration of THF , the lower the hydrate stability temperature. The temperature continuously decreases until the 2nd hours, at 282.25 K, 281.55 K and 275.75 K for 0 %(v/v), 3 %(v/v) and 30 %(v/v). It can be assumed that the high cooling rates create supercooled water at a temperature of 273.15 K – 278.15 K. Reducing temperature may increase the probability of thermodynamically stable nucleation [30] until the temperature reaches a point that the nucleation is not a random process anymore and stable nuclei (hydrate) become available and grow spontaneously. Similar trend were found by Kang et al. [19], which uses an isocharic method, they observe a remarkable decrease of dissociation pressure for the entire temperature range. It could be the point before 2nd hours is called the induction time.

After the 2nd hours, it can be seen from figure 3, for 0 %(v/v) from 3rd to 5th hours, there is no significant increment in temperature. While for 3 %(v/v), from 3rd to 5th , the increment in temperature started at the 5th hours. Only at 30 %(v/v) THF concentration the temperature keep rising from the 3rd to 5th hours of running. At this stage, it could be the growth of supercritical nuclei which usually controlled by mass and heat transfer [31]. It is obvious that at 30% concentration of THF is the fastest time (3rd hours) stable hydrate was formed compared to the other concentrations (0 %(v/v) at 6th hour, 3 %(v/v) at 5th hour). The increasing of THF concentration from 0 %(v/v) – 30 %(v/v) used in our work caused a substantial decrease in the observed induction time.

3.2 Effect of volumes of aqueous solution.

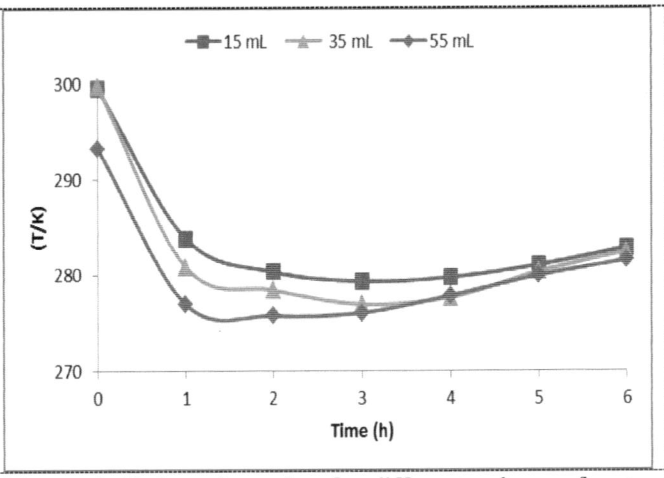

Figure 4: Hydrate formation for different volume of water use in the nitrogen+water+THF system: 15 mL of aqueous solution, 35 mL of aqueous solution and 55 mL of aqueous solution.

Figure 5. N_2 hydrate formation at 30 %(v/v) of THF in 15 mL volume of aqueous solution.

In this section, the highest THF concentration at 30 %(v/v) is used, as in section 3.1, at 30% of THF, hydrate induction time are reduced. From figure 4 and figure 5, it can be seen that at 15 mL, hydrates formed earlier compared to the 35 and 55 mL of (water+THF) at the first 4th hours. All the bonds in hydrates are hydrogen bonds. Water's readiness to make these bonds is responsible for the stability of gas hydrates [32]. It is discovered that when a volume of water to nitrogen ratio are diminished, hydrates can form faster. Hence, the lower the volume, there will be more space available for water molecules to form cages, connected by hydrogen bonding and for N_2 gas molecules to fill in these cavities. The concentration of THF also helps to reduce the equilibrium formation conditions enabling hydrate to form at much lower pressures. THF was found to reduce the induction time and increase the rate of hydrate growth [33]. The effect of volume with the addition of THF as promoter, gives significant effect on the rate of hydrate formation. It is also shown that, if a lower volume of water is used, it will influence the rate of hydrate formation. Moreover, for high volume of water, the time needed for hydrate to form is longer. From a comparison of the variation amounts of volume, the gas consumption due to the hydrate formation at high volume is significantly higher than that at low volume. As the gas solubility at high volume is even lower than at low volume and the decrease of the system temperature is largely derived from the formation of hydrate [34].

4. Conclusion

The effect of THF on N_2 hydrate formation under isobaric condition was studied in three (3) different parameters at different temperature setting, different THF concentration (0 %(v/v), 3 %(v/v) and 30

%(v/v)) and on different volume of water +THF mixtures. There was no hydrate formed at room temperature in all three (3) THF concentrations. However, hydrate may form if the running time is lengthened. This is something that could be done in the future research. In an induce temperature, the induction time is lowered in increasing THF concentration from 0 %(v/v) – 30%(v/v). 30 %(v/v) was used to study the effect of volume (15, 35 and 55 mL) on N_2 hydrate formation and it can be seen that at 15 mL, hydrate tend to form faster. This is because more space is available for water molecules to form cages and for small N_2 gas molecules with the help of THF as promoter to fill in these cavities and form sII hydrate structure. This study in general provides motivation and useful information to develop more effective way to form N_2 hydrate at low pressure and various temperatures setting in the laboratory scale, which may help in natural gas storage application for domestic and industrial purposes.

Acknowledgment
The authors gratefully acknowledge the support from Universiti Teknologi MARA (UiTM) for the fund given through the Institute of Graduate Studies (IPSis). The authors wish to thank the referees for making this a better paper.

Appendices

Table A.1. Room temperature system with different THF concentration.

	Temperature (°C)		
Time (hr)	0 %(v/v) THF	3 %(v/v) THF	30 %(v/v) THF
0	25	24.8	24
1	24.5	23.4	23.3
2	24.1	23.2	23.2
3	23.9	23.1	23.3
4	23.8	23.5	23.4
5	23.8	23.6	23.5
6	23.9	23.7	23.7

Table A.2. Induced temperature system with different THF concentration.

	Temperature (°C)		
Time (hr)	0 %(v/v) THF	3 %(v/v) THF	30 %(v/v) THF
0	25	24.3	20.1
1	15.3	12.2	3.8
2	9.1	8.4	2.6
3	8.6	7.6	2.9
4	7.5	7.8	4.7
5	7.8	8.9	6.8
6	8.2	10.5	8.4

References

[1] Keshavarz Moraveji M, Ghaffarkhah A, Sadeghi A 2017 Effect of three representative surfactants on methane hydrate formation rate and induction time *Egypt J. Pet.* 26(2) 331-9.

[2] Sloan ED and Koh C 2007 Chapter 2: Molecular Structures and Similarities to Ice. *In: Clathrate Hydrates of Natural Gases* CRC Press 3 111-45.

[3] Najibi H, Chapoy A, Haghighi H and Tohidi B 2009 Experimental determination and prediction of methane hydrate stability in alcohols and electrolyte solutions *Fluid Phase Equilibria* 275 127-31.

[4] Ballard A, Creek J, Eaton M, Koh C, Lachance J, McMullen N, et al. 2011 Chapter Seven - Industrial Operating Procedures for Hydrate Control BT - Natural Gas Hydrates in Flow Assurance. *In: Natural gas hydrates in flow assurance* Gulf Professional Publishing 1 145-62.

[5] Sun Z, Wang R, Ma R, Guo K and Fan S 2003 Natural gas storage with the presence of promoters *Energy Convers. Manag.***44** 2733-42.

[6] Kang SP and Lee H 2000 Recovery of CO_2 from flue gas using gas hydrate: Thermodynamic verification through equilibrium measurement *Environ. Sci. Technol.* 34 4397-400.

[7] Babu P, Ho CY, Kumar R and Linga P 2014 Enhanced kinetic for the clathrate process in a fixed bed reactor in the presence of liquid promoters for pre-combustion carbon dioxide capture *Energy* **70** 664-73.

[8] Lee S, Liang L, Riestenberg D, West OR, Tsouris C and Adam E 2003 CO_2 hydrate composite for ocean carbon sequestration *Environ. Sci. Technol* **37** 3701-8.

[9] Park K, Hong SY, Lee JW, Kang KC, Lee YC, Ha MG et al 2011 A new apparatus for seawater desalination by gas hydrate process and removal characteristics of dissolved minarels (Na+, Mg2+, Ca2+, K+ and B3+) **274** 91-6.

[10] Panter JL, Ballard AL, Sum AK, Sloan ED, and Koh CA Hydrate Plug Dissociation via Nitrogen Purge: Experiments and Modeling *Energy & Fuels* **2** 2572–78.

[11] Zeng H, Wilson LD, Walker VK and Ripmeester JA 2003 The inhibition of tetrahydrofuran clathrate-hydrate formation with antifreeze protein *Can. J. Phys.* **81** 17–24.

[12] Anderson R, Chapoy A and Tohidi B 2007 Phase relations and binary clathrate hydrate formation in the system H_2-THF-H_2O *Langmuir* **23** 3440-4.

[13] Sun S, Peng X, Zhang Y, Zhao J, and Kong Y 2017 Stochastic nature of nucleation and growth kinetics of THF hydrate *J. Chem. Thermodyn.* **107** 141–52.

[14] Duarte ARC, Shariati A, Rovetto LJ and Peters CJ 2008 Water cavities of sH clathrate hydrate stabilized by molecular hydrogen: Phase equilibrium measurements *J Phys Chem B* **112** 1888-9.

[15] Komatsu H, Yoshioka H, Ota M, Sato Y, Watanabe M and Smith RL et al. 2010 Phase equilibrium measurements of hydrogen-tetrahydrofuran and hydrogen cyclopentane binary clathrate hydrate systems *J Chem. Eng. Data* **55** 2214-8.

[16] Mohammadi A, Martinez-lopez J and Richon D 2010 Determining phase diagrams of tetrahydrofuran plus methane, carbon dioxide or nitrogen clathrate hydrates using an artificial neural network algorithm *Chem. Eng. Sci.* **65** 6059-63.

[17] Seo YT, Kang SP and Lee H 2001 Experimental determination and thermodynamic modeling of methane and nitrogen hydrates in the presence of THF, propylene oxide, 1,4-dioxane, and acetone *Fluid Phase Equilib.* **189** 99-110.

[18] Mohammadi AH and Richon D 2009 Phase equilibria of clathrates hydrates of tetrahydrofuran + hydrogen sulphide and tetrahydrofuran + methane *Ind. Eng. Chem. Res.* **48** 7838-41.

[19] Kang SP, Lee H, Lee CS and Sung WM 2001 Hydrate phase equilibria of the guest mixtures containing CO_2, N_2, and tetrahydrofuran *Fluid Phase Equilib.* **185** 101-9.

[20] Lee JD and Englezos P 2005 Enhancement of the performance of gas hydrate kinetic inhibitors with polyethylene oxide *Chem. Eng. Sci.* **60** 5323-30.

[21] Linga P, Kumar R and Englezos P 2006 Capture of carbon dioxide from conventional power plants or from integrated gasification plants through gas hydrate formation/dissociation *J. Energy Clim.* **2** 75-82.

[22] Moudrakovski IL, Maclaurin GE, Ratcliffe CI and Ripmeester JA 2004 Methane and carbon dioxide hydrate formation in water droplets: spatially resolved measurements from magnetic resonance microimaging *J. Phys. Chem. B.* **108** 17591-5.

[23] Park J, Seo YT, Lee J and Lee H 2006 Spectroscopic analysis of carbon dioxide and nitrogen mixed gas hydrates in silica gel for CO_2 separation. *Catal Today* **115** 279-82.

[24] Seo YT, Moudrakovski IL, Ripmeester JA, Lee J and Lee H 2005 Efficient recovery of CO_2 from flue gas by clathrate hydrate formation in porous silica gels *Environ. Sci. Technol.* **39** 2315-9.

[25] Park J, Seo YT, Lee J, and Lee H 2006 Spectroscopic analysis of carbon dioxide and nitrogen mixed gas hydrates in silica gel for CO2 separation *Catal. Today* **115** 279–82.

[26] Yang M, Song Y, Jiang L, Jing W, Liu W, and Dou B 2014 The effects of porous medium and temperature on exothermic tetrahydrofuran hydrate formation *J. Chem. Thermodyn.* **78** 167–74.

[27] Devarakonda S, Groysman A, and Myerson AS 1999 THF–water hydrate crystallization: an experimental investigation *J. Cryst. Growth* **204** 525–38.

[28] Sun Q and Kang YT 2015 Experimental correlation for the formation rate of CO2 hydrate with THF (tetrahydrofuran) for cooling application *Energy* **91** 712–19.

[29] Smith JM, Vanness HC and Abbott MM 2005 Introduction to chemical engineering thermodynamics *McGraw-Hill.*

[30] Nesterov AN, Reshetnikov AM, Manakov AY, and Adamova TP 2017 Synergistic effect of combination of surfactant and oxide powder on enhancement of gas hydrates nucleation *J. Energy Chem.* **26** 808–14.

[31] Susilo R, Moudrakovski IL, Ripmeester JA and Englezos P 2006 Hydrate kinetics study in the presence of non-aqueous liquid by nuclear magnetic resonance spectroscopy and imaging *J. Phys. Chem. B.* **110** 25803-9.

[32] Svandal A, Kuznetsova T and Kvamme B 2006 Thermodynamic properties and phase transitions in the $H_2O/CO_2/CH_4$ system *Fluid Phase Equilib.* **246** 177-84.

[33] Xiang CS, Peng BZ, Liu H, Sun CY, Chen GJ and Sun BJ 2013 Hydrate formation/dissociation (natural gas + water + diesel oil) emulsion systems *Energies* 1009-22.

IOP Conf. Series: Materials Science and Engineering **334** (2018) 012057 doi:10.1088/1757-899X/334/1/012057

Effects of process parameters on peanut skins extract and CO_2 diffusivity by supercritical fluid extraction

N R Putra[1], L N Yian[2] , H M Nasir[2], Z binti Idham[2] and M A C Yunus[1,2]

[1] Faculty of Chemical and Energy Engineering, Universiti Teknologi Malaysia, 81310 Johor Bahru, Malaysia
[2] Centre of Lipid Engineering and Advanced Research (CLEAR), Ibnu Sina Institute for Scientific and Industrial Research, Universiti Teknologi Malaysia, 81310, UTM Johor Bahru, Malaysia

E-mail: azizi@cheme.utm.my

Abstract. Peanut skins (*Arachis hypogea*) are an agricultural waste product which has received much attention because they contain high nutritional values and can be potentially utilized in difference industries. At present, only a few studies have been conducted to study the effects of parameters on the peanut skins oil extraction. Therefore, this study aimed to determine the best extraction condition in order to obtain the highest extract yield using supercritical carbon dioxide (SC-CO_2) with co-solvent Ethanol as compared to Soxhlet extraction method. Diffusivity of carbon dioxide in supercritical fluid extraction was determined using Crank model. The mean particle size used in this study was 425 μm. The supercritical carbon dioxide was performed at temperature (40 – 70 °C), flow rate of co-solvent ethanol (0 - 7.5% Vethanol/Vtotal), and extraction pressure (10 – 30 MPa) were used in this studies. The results showed that the percentage of oil yields and effective diffusivity increase as the pressure, rate of co-solvent, and temperature increased.

1. Introduction

Red peanut skins refer to the outer thin layer of the peanut endosperm and it may cause problems in the peanut butter industries. It can also reduce the quality of butter because the skins have astringent taste and it is the primary reason that the skins should be removed. There are 0.74 million metric tons of peanut skins being annually produced worldwide as waste from peanut butter industries[1]. It is commonly known that red peanut skins are rich in antioxidant and may contribute benefits toward the economy and human health. The extract of peanut skins contains 12% protein, 16% fat, and 72% of carbohydrate, and also rich in polyphenols [2]. The total phenolics (TPs) of defatted peanut skin was reported to be 140–150 mg/g dry skins [2]depending on the solvent used. The exploitation of peanut skins as renewable raw material for antioxidant compounds provides protection against human health and enhancement of the sustainable environment [3]. Most of these antioxidant activities on peanut skins rich in catechin and procyanidins have been associated with reduced risk of cardiovascular diseases and cancers [4].

Supercritical fluid extraction (SFE) is a green technology built to extract oil, antioxidant, and bioactive compounds. This method has been conducted by many researchers to effectively extract pharmaceuticals and cosmeceutical products from plants and herbs. The supercritical fluid extraction has many advantages, namely, high purity on solute content, free of organic solvent, easy to separate, SFE has high selectivity, and the solvent used CO_2 which is nontoxic. Generally, carbon dioxide (CO_2) is used as the solvent in food processing as it has low critical temperature and pressure, non-toxic, not expensive as a solvent. The key factor for using supercritical carbon dioxide extraction was the solubility of solvent as it can be manipulated by pressure and temperature [5].

Diffusivity or diffusion coefficient, D_e is defined as a measure of the capability of a substance or energy to be diffused or to allow something to pass by diffusion. The higher the diffusivity of one substance with respect to another, the faster they diffuse into each other. Temperature, pressure, and particle size are the common factors affecting diffusivity in SFE. Diffusion model proposed Crank is more suitable to be used in order to determine diffusivity of carbon dioxide (CO_2) as it is more effective for particle size factor [6].

Although CO_2 supercritical extraction is advantageous, it can only extract nonpolar compound. Supercritical CO_2 only extract oil inside of peanut skins, but it cannot extract bioactive compound (procyanidin, catechin, epicatechin, gallocatechin, and epigallocatechin) [4]. The presence of ethanol is needed for this extraction process to encourage the extraction due to its polarity. Since ethanol is a bipolar compound based on its polarity, ethanol can increase the polarity of CO2 and can also extract polar bioactive compound. The objectives of the study are to determine the best parameter in order to obtain the highest extract yield and to study the diffusivity of carbon dioxide in supercritical fluid extraction using Crank model.

2. Methodology

2.1 Chemicals

Liquid carbon dioxide (99% purity) was used in the supercritical extraction apparatus purchased from Kras Instrument, Johor Bahru, Malaysia. Technical grade of ethanol (99.86%) was obtained from Permula Sdn Bhd, Johor Bahru Malaysia.

2.2 Plant Material

Peanut skins were obtained from G-Tachfood Industries Sdn Bhd, Johor Bahru Malaysia which is the waste of peanut industries. Peanut skins were dried in an oven at 60 °C for 4 hours. The obtained dried peanut skins were blended into powder and classified into different particle sizes which are then stored in air tight plastic and placed in a freezer at -20 °C.

2.3 Supercritical Fluid Extraction

A set of laboratory scale supercritical extraction was used for this experiment. The extraction process was performed at the temperature of 40, 55, and 60 °C, with the pressure of 10, 20, and 30 MPa, the rate of co solvent 0, 5, and 7.5% (Vethanol/Vtotal), the CO_2 flow rate of 3.0 mL/minute, and particle size range of 425 μm. The extraction was carried out for 120 minutes. Chiller temperature was set at 6 °C while the heater on the back pressure regulator was set at 50 °C. Next, 5 ±0.005 g of peanut skins were placed into the extraction vessel. Then, liquid CO_2 was pumped from the CO_2 tank into the system continuously with a supercritical pump at a flow rate of 3 mL/min. The co-solvent ethanol 98.86% was also pumped continuously when the desired temperature was achieved. It fed into the extractor through high pressure. The extracted oil was collected by vial for every 30 minutes of the extraction process. After each extraction process, the extract obtained was sealed and stored at 4°C to prevent any possible degradation.

Table 1. Calculated Parameters for mass transfer using single sphere model for peanut skins.

Pressure (MPa)	Temperature (°C)	Rate of co-solvent (%)	Yield (%)	$D_e \times 10^{12}$ (m^2/s^{-1})	%AARD (%)
20	40	7.5	13.85	6.250	4.39
30	40	7.5	14.77	7.181	2.74
10	40	7.5	13.42	5.159	3.84
30	70	0	6.63	4.989	5.52
20	55	5	14.23	4.731	2.97
30	70	5	15.39	6.793	3.53
20	55	5	14.17	4.398	3.48
20	40	7.5	15.26	8.867	0.72
20	40	5	7.49	3.880	12.85
20	70	5	15.786	6.982	1.627
30	70	7.5	17.55	7.133	2.76

2.4 Percentage of Oil Yield Measurement

Percentage of oil yield can be defined as the percentage weight of oil divided by the weight of sample at a given temperature and pressure at 180 minutes of time extraction. The equation of percentage of oil yield is:

$$Percentage\ of\ Oil\ Yield\ (\%) = \frac{W_a - W_f}{W_o}$$

Where: W_a = weight of vial and weight of oil, g W_f = weight of vial, g, and W_o = weight of raw material,g

2.5 Mathematical Model

Single sphere model which was introduced by Crank considers the relationship between the size of particle and diffusion coefficient [7]. High diffusion coefficient indicates that good extraction phenomena have been obtained. The assumptions made are [8, 9]: Intraparticle mass transfer is the controlling and important factor in the extraction process, mass transfer resistance of the fluid is zero, solute is extracted from particulate bed composed of porous inert sphere, size of sphere is similar, entire particles in the bed are at the same stage of extraction, and components to be extracted move through the particles by process 'similar to diffusion'.

The basic equation for the rate of solute that transports radially across an internal surface within spherical and isotropic particle takes the form of Fick's law as:

$$Rate = -4\pi r^2 D_e \left(\frac{\partial q}{\partial r}\right) \quad (1)$$

Where q is solute concentration inside the particle at radius r (kg solute/ m3 of particle) and D_e is Effective diffusivity. The equation given by Crank to solve the diffusion coefficient is [7]:

$$Y' = \frac{M_t}{M_\infty} = 1 - \frac{6}{\pi^2} \sum_{n-1}^{\infty} \frac{1}{n^2} \exp \frac{D_e t n^2 \pi^2}{R^2} \quad (2)$$

Where M_t is the total amount of solute diffused from the sphere at time t, M_∞ is the total amount of solute, n is number of iterations, D_e is diffusion coefficient, R is particle radius, and t is time.

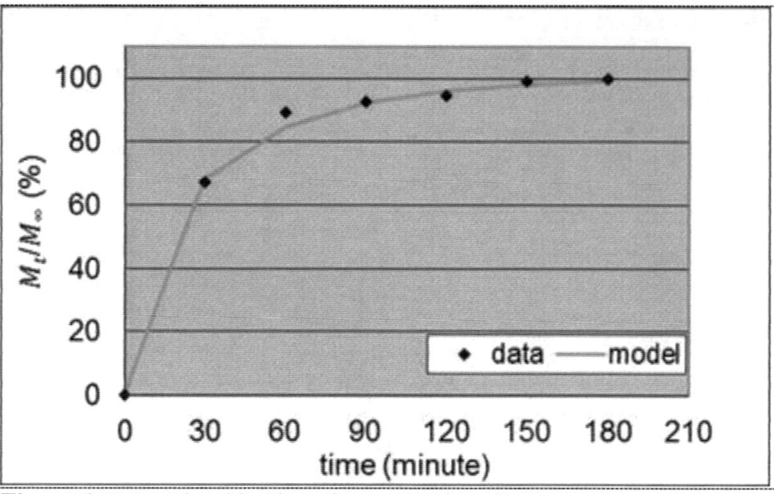

Figure 1. Extraction kinetics of peanut skins oil: comparison of experimental data and model output.

The general equation of single sphere model (SSM) is applicable for high velocity but fails to represent the extraction behavior at low velocity [10]. In this study, the good fitting and correlation have a range from 7.133×10^{-12} to 8.867×10^{-12} m^2/s. For the low velocity, the single sphere model is difficult to represent the extraction behavior due to high %AARD. Based on the result, single sphere model is mainly focused on the solute extraction inside the particles, assuming that the easily accessible oil has been removed from the galls.

Based on the literature, various approaches have been successfully performed which can be adapted to model the supercritical fluid extraction. In this study, in conjunction with SSM, some equations were used based on linear adsorption equilibrium model in order to determine the mass transfer coefficients of the extraction process.

2.6 Soxhlet Extraction

Soxhlet extraction was performed using 10 g of powdered peanut skins with particle size 425 μm and was placed in a thimble holder. Soxhlet extraction was carried out using 100 mL of solvents (n-Hexane, Ethanol, and Water). The heating power was used slightly above the boiling point of the solvent. The arrangement is such that vapors of the solvent are generated from the round-bottomed flask, get condensed into the water as a condenser, and pass through to the thimble that was contained in the sample. The condensed solvent comes into contact with the sample where the extraction occurs. When the liquid reaches the overflow level in the thimble, the liquid moves through the siphon back into the round-bottomed flask, carrying extracted solutes into the bulk liquid. The solvent was removed by rotary vacuum evaporator at the temperature above the boiling point of the solvent to avoid degradation of bioactive compound inside extract.

3. Result and discussion

3.1 Effect of pressure on the percentage of oil yield and diffusivity of extraction.

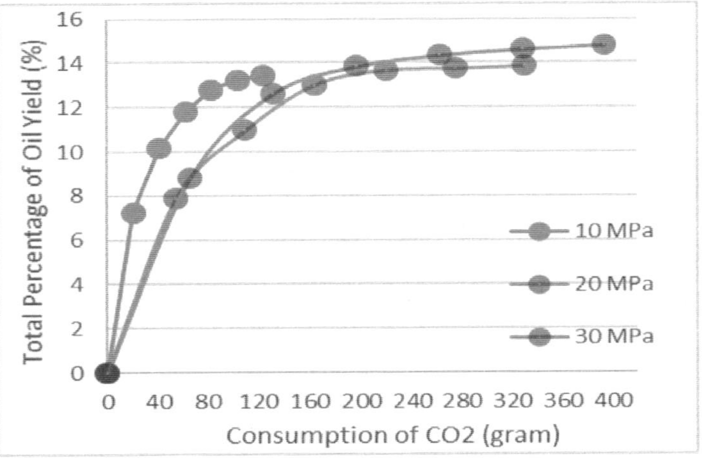

Figure 2. Effect of Pressure on the Percentage of Oil Yield.

Figure 2 shows that the percentage of oil yield using SC-CO_2 extraction with co-solvent of ethanol at different pressure of 10, 20, and 30 MPa, with constant temperature 70 °C, constant flow rate of CO_2 3 mL/min, and 7.5% rate of co-solvent ($V_{ethanol}/V_{total}$). The highest extracted oil yield was 14.7 % at the highest pressure of 30 MPa. Meanwhile, the lowest oil yield was 13.42% at the lowest pressure of 10 MPa. Fig. 2 shows the effect of different pressure at the highest temperature of the process (70 °C). It shows that the curve of all pressure conditions become uniform to each other. In this study, the oil extracted reached the asymptotic yield at the end of the extraction process after 120 minutes of extraction time. Asymptotic yield refers to the maximum amount of oil obtained at the maximum condition of extraction[11]. The asymptotic yield obtained was 14.7% (74.54 mg of oil per 5.0 g sample used). Furthermore, by reaching an asymptotic value, it can be assumed that the extraction was complete where the oil from the seeds was considered completely removed. The results of this study have shown that the effect of pressure is noticeable. This is because the highest pressure can obtain the high percentage of oil yield. Increasing pressure will lead to an increase of density of the solvent in the extraction process. The high density of solvent can create a big opportunity to contact and interact with oil in the inside of particle which is known as the solvating power increased[12].

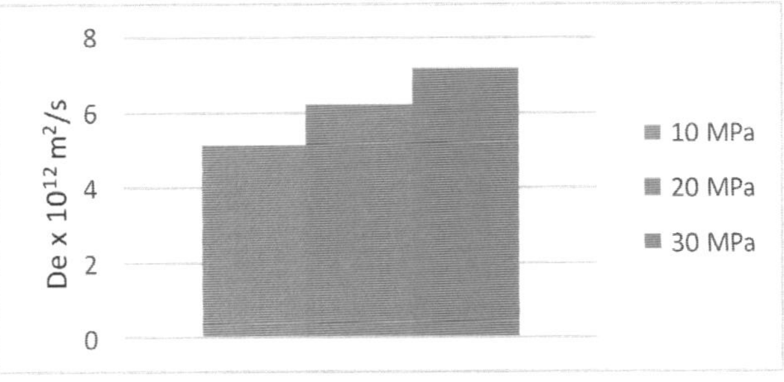

Figure 3. Effect of Pressure on the Effective diffusivity.

Figure 3 shows the effective diffusivity of SC-CO_2 extraction with co-solvent ethanol at a different pressure condition (10-30 MPa), and constant temperature 70 °C, flowrate of CO_2 3 mL/min, and rate of co-solvent 7.5% (Vethanol/Vtotal). The highest effective diffusivity was obtained 7.18×10^{-12} m^2/s at the highest pressure 30 MPa. Meanwhile, the lowest effective diffusivity 5.159×10^{-12} m^2/s was at the lowest pressure of 10 MPa.

From figure 3 and table 1, the single sphere model gives a useful fit to the experimentally determined yield versus data at the high effective diffusivity those in the range (5.1-7.2) x 10^{-12} m^2/s.

The overall mean deviation is below 5%. In this condition, the effect of pressure is noticeable due to effective diffusivity. Increasing pressure enhances the solubility of extraction and also the density. The increasing of solubility and the density in solvent will increase the driving force and extraction rate of the galls [10].

3.2 Effect of Temperature On The percentage of oil yield and diffusivity of extraction

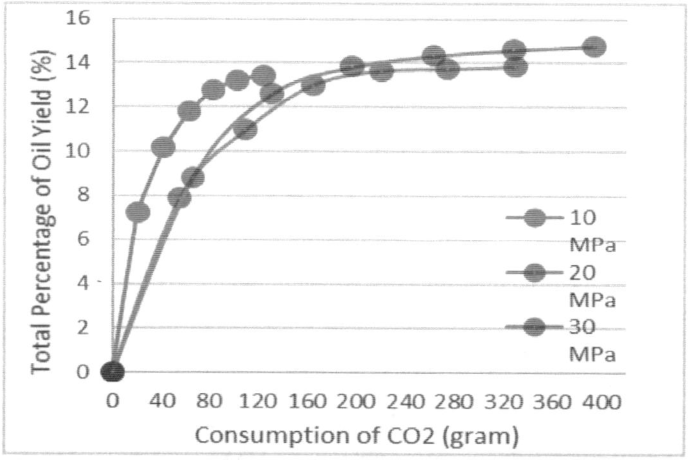

Figure 4. Effect of Pressure on the Percentage of Oil Yield.

Figure 4 shows that the percentage of oil yield by using SC-CO_2 extraction with co-solvent ethanol at different temperature conditions at 40, 55, and 70 °C, constant pressure 20 MPa, constant flow rate of CO_2 3 mL/min, and 5% rate of co-solvent (Vethanol/Vtotal) versus time. In this study, the highest percentage extracted oil yield was 15.78% at the highest temperature of 70 °C. Meanwhile, the lowest oil yield was 7.49% at the lowest temperature of 40 °C. In a different temperature condition, the effect of temperature on the supercritical fluid extraction is dominant. This is because increasing temperature enhances the total percentage of oil yield. In this study, temperature effect has similar behaviors with a previous study that examines the extraction of Pithecellobium Jiringan (Jack) Prain Seeds that the increase of temperature affects the increase of oil yield [13].

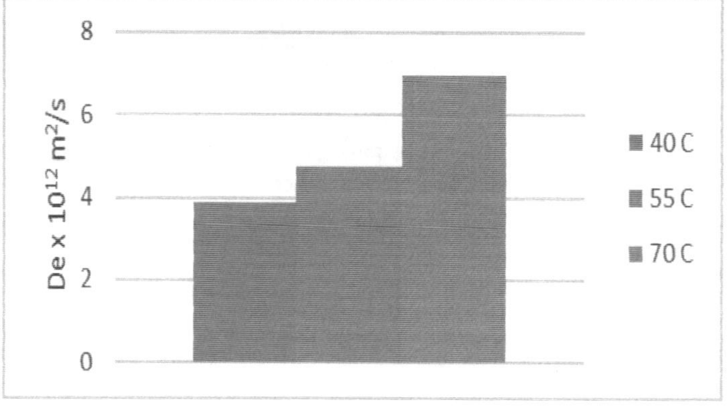

Figure 5. Effect of Temperature on the Effective diffusivity.

Figure 5 illustrates the effective diffusivity of SC-CO_2 extraction with co-solvent ethanol in different temperature conditions (40, 55, and 70 °C). The highest effective diffusivity was obtained 6.98×10^{-12} m^2/s at the highest temperature 70 °C. Meanwhile, the lowest effective diffusivity 3.88 ×

10^{-12} m^2/s at the lowest temperature condition 40 °C. This is because the increase in temperature has resulted in the increase of the solute vapor pressure which contributed to the increase of mass transfer rate of oil into the solvent [13]. It is a different explanation with previous studies since the increase of temperature decreases the solubility and density of solvent that may affect the mass transfer of extract to solute [14].

3.3 Effect of Co-solvent Ethanol

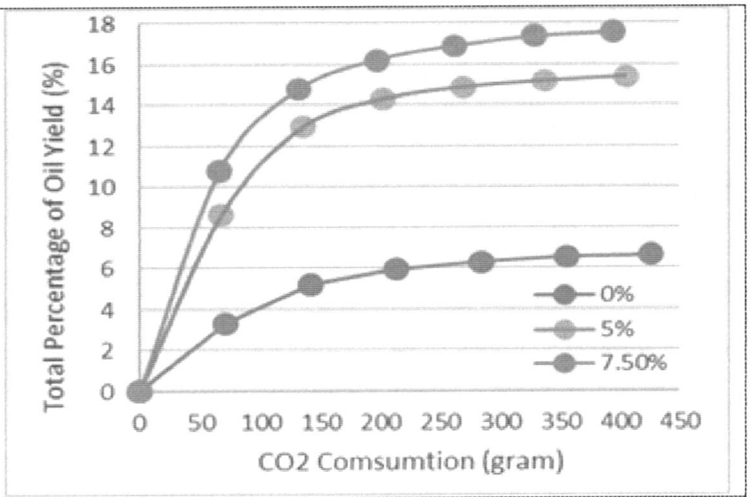

Figure 6. Effect of rate co-solvent ethanol on the Percentage of Oil Yield.

Figure 6 shows that the percentage of oil yield by using SC-CO_2 extraction with co-solvent ethanol at different rates of co-solvent condition 0, 5, and 7.5% ($V_{ethanol}/V_{total}$), constant pressure 30 MPa, constant temperature 70 °C, and constant flowrate of CO_2 3 mL/min. The highest extracted oil yield was 15.39% at the highest rate of co-solvent ethanol. Meanwhile, without co-solvent ethanol, the lowest oil yield was 6.63%. Figure 6 shows the effect of co-solvent is not significant on the 5% and 7.5% ($V_{ethanol}/V_{total}$). This is because the percentage of oil yield is almost the same 14.77% and 15.39%. The rate of co-solvent ethanol enhances the percentage of oil yield because the polarity of the supercritical fluid increases. Therefore, supercritical CO_2 and ethanol can interact and make a bond with nonpolar compound and polar compound in the solute[15].

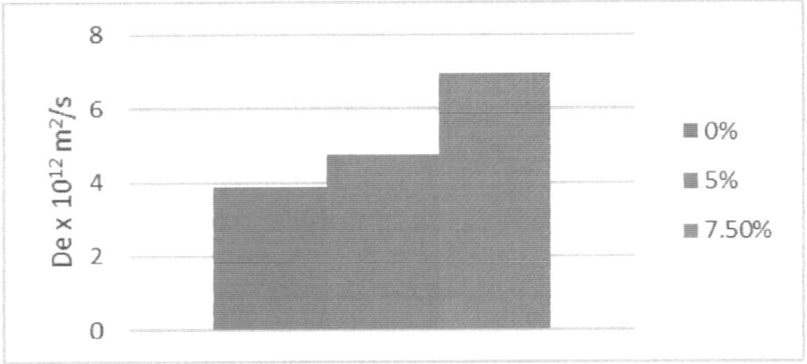

Figure 7. Effect of rate co-solvent ethanol on the Effective diffusivity.

Figure 7 shows that the effective diffusivity of SC-CO_2 extraction with co-solvent and without co-solvent ethanol. The highest effective diffusivity was obtained 7.13×10^{-12} m^2/s at the highest rate of co-solvent 7.5% ($V_{ethanol}/V_{total}$). Meanwhile, the lowest effective diffusivity 4.98×10^{-12} m^2/s at without of co-solvent ethanol condition. The increase in D_e indicated the decrease in the mass transfer resistance in the fluid phase. However, extraction rate in the fast extraction period increased due to the increase in driving force and decrease in mass transfer resistance in the fluid phase [16]. Increase in the oil yield with ethanol addition into SC-CO_2 was reported for fungal lipid [17] and removal caffeine from green tea [18]

4. Conclusion

The extraction of peanut skins using supercritical carbon dioxide extraction and co-solvent ethanol was performed at temperatures ranging from 40 °C to 70 °C, rate of co-solvent ethanol ranging from 0% to 7.5%, and pressures ranging from 10 MPa to 30 MPa. In this study, the results showed that percentage of total oil yield increased as the pressure increased at a constant temperature, flow rate of CO_2, and rate of co-solvent ethanol. Similarly, the percentage of total oil yield has also increased as the temperature increased at constant pressure, flow rate of CO_2, and rate of co-solvent ethanol. The maximum percentage of oil yield obtained was 15.78% mg at the highest supercritical extraction condition of pressure at 30 MPa, the temperature at 70°C, and rate of co-solvent ethanol 5%, respectively. The effective diffusivity, D_e was ranging from 3.881×10^{-11} to 9.181×10^{-12} m^2/s. The value of all adjustable parameters increases with increase in pressure, temperature, and rate of co-solvent.

Acknowledgment

The authors acknowledge the Centre of Lipid Engineering and Applied Research (CLEAR) for the use of equipment. The financial support from Universiti Teknologi Malaysia for professional development research university grant (QJ130000.2509.14H23) is gratefully acknowledged.

References

[1] Ballard, T.S., et al., 2009, *Optimizing the extraction of phenolic antioxidants from peanut skins using response surface methodology.* Journal of Agricultural and Food Chemistry,. **57**(8): p. 3064-3072.
[2] Nepote, V., N.R. Grosso, and C. Guzman, 2002, *Extraction of antioxidant components from peanut skins.* Grasas y aceites, **53**(4): p. 391-395.
[3] Van Ha HOANG, P.A., et al., 2008, *Antioxidant activity of peanut skin extracts from conventional and high-oleic peanuts.* Czech J. Food Sci. Vol, **26**(6): p. 447-457.
[4] Yu, J., et al., 2006, *Peanut skin procyanidins: Composition and antioxidant activities as affected by processing.* Journal of Food Composition and Analysis, **19**(4): p. 364-371.
[5] Mustapa, A., et al., 2011, *Extraction of β-carotenes from palm oil mesocarp using sub-critical R134a.* Food Chemistry, **125**(1): p. 262-267.
[6] Huang, Z., et al., 2016, *Supercritical CO 2 extraction of Chinese lantern: Experimental and OEC modeling.* Separation and Purification Technology,. **159**: p. 23-34.
[7] Crank, J., 1979, *The mathematics of diffusion.*: Oxford university press.
[8] Bartle, K.D., et al., 1990, *A model for dynamic extraction using a supercritical fluid.* The Journal of Supercritical Fluids,. **3**(3): p. 143-149.
[9] Reverchon, E., G. Donsi, and L. Sesti Osseo, 1993, *Modeling of supercritical fluid extraction from herbaceous matrices.* Industrial & engineering chemistry research,. **32**(11): p. 2721-2726.
[10] Esquıvel, M., M. Bernardo-Gil, and M. King, 1999, *Mathematical models for supercritical extraction of olive husk oil.* The Journal of Supercritical Fluids,. **16**(1): p. 43-58.

[11] Arsad, N.H., et al., 2016, *Effect of Operating Conditions of Supercritical Carbon Dioxide on Piper Betle Leave Oil Yield and Antioxidant Activity.* International Journal of Applied Chemistry,. **12**(4): p. 741-751.

[12] Mohd-Setapar, S.H., et al., 2013, *Extraction of rubber (Hevea brasiliensis) seeds oil using supercritical carbon dioxide.* Journal of Biobased Materials and Bioenergy,. **7**(2): p. 213-218.

[13] Yunusa, M.A.C., et al., 2012, *Effect of Supercritical Carbon Dioxide Condition on Oil Yield and Solubility of Pithecellobium Jiringan (Jack) Prain Seeds.*.

[14] Allada, S.R., 1984, *Solubility parameters of supercritical fluids.* Industrial & Engineering Chemistry Process Design and Development,. **23**(2): p. 344-348.

[15] Casas, L., et al., 2007, *Effect of the addition of cosolvent on the supercritical fluid extraction of bioactive compounds from Helianthus annuus L.* The Journal of Supercritical Fluids,. **41**(1): p. 43-49.

[16] Özkal, S., M. Yener, and L. Bayındırlı, 2005, *Mass transfer modeling of apricot kernel oil extraction with supercritical carbon dioxide.* The Journal of supercritical fluids,. **35**(2): p. 119-127.

[17] Cygnarowicz-Provost, M., et al., 1995, *Supercritical-fluid extraction of fungal lipids: Effect of cosolvent on mass-transfer rates and process design and economics.* The Journal of Supercritical Fluids,. **8**(1): p. 51-59.

[18] Park, H.S., et al., 2007, *Effect of mass transfer on the removal of caffeine from green tea by supercritical carbon dioxide.* The Journal of supercritical fluids,. **42**(2): p. 205-211.

IOP Conf. Series: Materials Science and Engineering **334** (2018) 012058 doi:10.1088/1757-899X/334/1/012058

Optimization of Lactide synthesis from Lactic Acid in biorefinery of palm oil waste using Response Surface Methodology

M Gozan[1,4], F Kamilah[1], Y Whulanza[2] and Rahmayetty[3]

[1] Department of Chemical Engineering, Universitas Indonesia, Kampus UI Depok 16424, Indonesia
[2] Department of Mechanical Engineering. Universitas Indonesia. Kampus UI Depok 16424, Indonesia
[3] Department of Chemical Engineering. Faculty of Industrial Technology. Universitas Sultan Ageng Tirtayasa, Banten, Indonesia

E-mail: mgozan@che.ui.ac.id

Abstract. Ring open polymerization is one of the production polylactic acid by formation of monomer before. Lactic acid is converted into lactide in two stages, polycondensation and depolymerization. Yield lactide will determine the molecular weight which produced. This study is to optimize the lactide production from lactic acid 90% by the variations of temperature (190-220°C), vacuum pressure (5-15 cmHg), and zinc acetate catalyst (0,3-0,6% w/w). As the temperature, vacuum pressure, and catalyst is increased, lactide that is produced also increases. Optimum condition of lactide production is obtained by Response surface methodology at the temperature 220°C, catalyst 0,45%w/w, and 10 cmHg in vacuum pressure. Equation or model from this study by using RSM is *yield* lactide = -258,75 + 7,79A + 2,90B + 3,51C + 0,48AB - 0,06AC – 3,97x10^{-3}BC – 105,42A^2 – 7,17B^2 – 0,10C^2 (A:catalyst; B:temperature; C:pressure).

1. Introduction

Utilization of conventional plastics are used as packaging materials not only get a lot of profit, but also causes serious environmental problems such as accumulation of garbage. Furthermore, plastic raw materials requires many natural resources that has declined and difficult to renewable (non-renewable). Polymer demand in 2020 is predicted to reach 7149.8 kilo ton increased by 5.6% from 2014 to 2020 [1]. Fossil resources has been reduced and the increasing concentration of carbon dioxide in the atmosphere has focused attention on the development of bio-based plastics [2].

Polylactic acid (PLA) has great potential to be developed as a replacement for conventional plastic. PLA is a multifunctional polymer biodegradable, and derived from renewable resources [3]. PLA has been widely used for various applications, especially in the fields of medicine, packaging, and textiles. In the medical field, PLA is used in the sewing thread for operating purposes (surgical implants) and wrapping materials of capsules for drug delivery systems (drug delivery) and also for repair of body tissue (mooring new cell growth). In the field of packaging, PLA is developed for the manufacture of

[4] To whom any correspondence should be addressed.

plastic bags (retail bag), containers, and edible films for vegetables and fruit. In the field of textiles, PLA is used in the manufacture of shirts and bags [4, 5].

Poly lactic acid is a plastic material that often used as a food packaging because it has the characteristics of transparency, maintaining, and biodegradability. Ester bond in the PLA can cause either naturally degraded by heat, light, or bacteria (Fusarium moniliform and Penicillium roquefort) [6].

In general, PLA is used to replace a transparent material with high density and prices. Plastic materials are replaced into the type of PET (1.4 g / cc, 1.4 usd / kg), flexible PVC (1.3 g / cc, 1 usd / kg). Polylactic acid needs in Indonesia each year is quite large, visible from export and import data in Figure 1. The increasing needs of PLA in Indonesia, make we should develop technology to produce PLA.

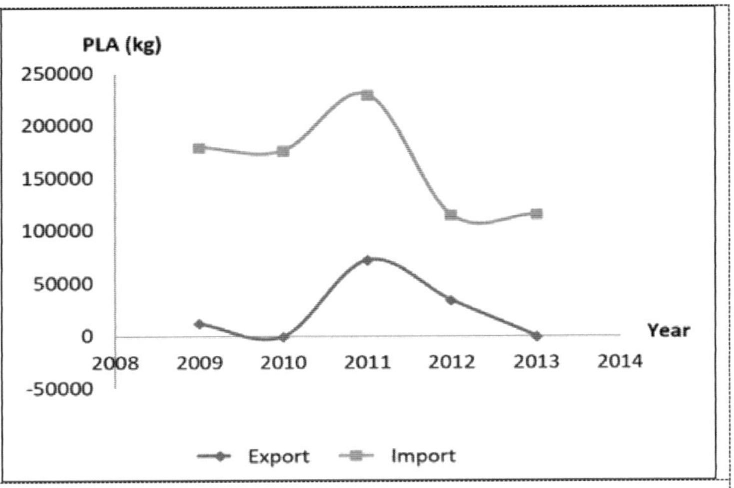

Figure 1. Export and import of PLA.
Source: www.bps.go.id, 2013

Poly lactic acid is synthesized in two stages, with fermentation process to produce lactic acid and ring opening polymerization of a cyclic dimer (lactide) of lactic acid-based polycondensation [7]. In the ring-opening polymerization process, there are three steps are being taken, namely polycondensation of lactic acid, depolymerization thus forming a cyclic dimer (lactide) and ring-opening polymerization to obtain poly lactic acid with a high molecular weight. The third process is crucial to the molecular weight of poly lactic acid produced.

Depolymerization reactions which produce compounds cyclic dimer (lactide), resulting from molecular weight between 100-5000 [8]. To produce poly lactic acid with a high molecular weight lactide monomer required a high yield. The monomer forming reaction is an equilibrium reaction which depends on operating conditions and the presence or absence of a catalyst. We have had a lot of research done on the production of lactide using a metal catalyst to obtain a high yield of lactide. But no one has done to optimize lactide formation process. Zinc acetate is one of the metal catalysts that has not been much research done on the results obtained lactide yield. This catalyst is more secure than other zinc-based catalysts. Zinc acetate as a catalyst in the thermal degradation of poly lactic acid [6]. Zinc acetate is able to accelerate the OH groups in attacking the C = O group on the oligomer chains so as to obtain lactide yield 65-70% [9].

Optimization point will be done by using statistical analysis method, Response Surface Methodology (RSM). To obtain an optimum process is influenced by a number of variables and necessary experimental data in large quantities and take a long time, automatically also will require a great cost. By using the RSM through Expert design software will get high accuracy, reducing the time and cost of research and minimize the occurrence of errors than using conventional methods.

IOP Conf. Series: Materials Science and Engineering **334** (2018) 012058 doi:10.1088/1757-899X/334/1/012058

2. Methods

In this research, we have stages such as preparation, polycondensation, and depolymerization.

2.1 Design with response surface methodology

To determine the optimize point using the matrix, first, decide temperature range, the amount of catalyst, and vacuum pressure of forming lactide to proceed to the stage of optimization. That researcher use experimental design is the design matrix Central Composite Design (CCD).

Table 1. Matrix Design of CCD.

	Satuan	-alpha	-1	+1	+ alpha
Katalis	%	0,198	0,3	0,6	0,702
Suhu	°C	179,773	190	220	220,227
Tekanan	cmHg	1,591	5	15	18,409

After these parameters are determined, the experimental matrix by the Design-Expert software is turned up.

2.2 Poly condensation of Lactic Acid

Merck lactic acid 90% by volume of 50 ml put in a four neck flask and heated gradually at 120-180° C for 5 hours, and nitrogen flowed.

2.3 Depolymerization of Oligomer PLA

Poly lactic acid oligomers which have been formed through polycondensation stage, then catalyst added 0.3 to 0.6% w/w to weight oligomers. The catalyst variations according to RSM design optimization. Oligomers have been added catalyst then heated at 190-220° C and vacuum pressure by 5-15 cmHg until the oligomer doesn't produce steam again. After the depolymerization process is complete, the vacuum valve is closed slowly. Then lactide is collected on the sample container.

2.4 Analysis of Lactide Concentration

High-Performance Liquid Chromatography (HPLC) analysis is used to determine the concentration of lactide. The column is used Agilent eclipse XDB C18 Two Agilent 1200 series (Agilent, USA). Mobile phase or eluent is water/acetonitrile. Water mobile phase starting with 98% vol, after 2 minutes streamed acetonitrile concentration increased linearly up to 100% vol. Solvents for diluting the sample in the form of ethanol. A total of 0.05-0.1 grams of lactide dilute into 25 mL ethanol. H_3PO_4 buffer used is 0.03 M. The ratio of Buffer-to-eluent of 88:12. Lactide concentration can be calculated using by equation (1)

$$\text{Concentration lactide} = \frac{\text{Sample area}}{\text{Standard area}} \text{ X C} \ldots (1)$$

where,

C : concentration lactide standard (ppm)

3. Results and Discussion

3.1. Bonding test with FTIR

Bonding test with Fourier Transform Infrared (FTIR) on the first sample to confirm that the depolymerization process produce lactide or not.

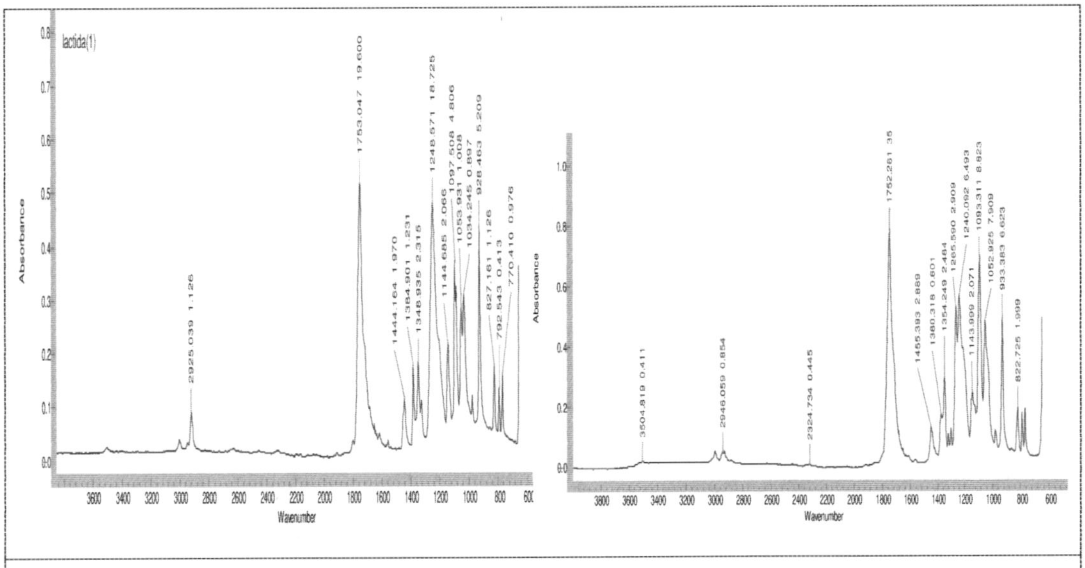

Figure 2. Chromatogram of Lactide, Standard Chromatogram (left); Sample Chromatogram (right)

On the spectrum can be seen that the OH bond in water in the area of 3504.819 cm-^{1}. Ester structure, a cyclic -C = O very strong look at the area 1752.261 cm-^{1}. Then four substituents -CO appear on the area 1265.590; 1240.092; 1093.311; 1052.925 cm-^{1}. Three substituents -CH appear on 1455.393; 1380.318; 1354.249 cm-^{1}. Spectrums with the bond -C = O, -CO, and -CH indicate that there is a ring-shaped compound (lactide).

3.2. Optimization of temperature, zinc acetate catalyst and pressure using Response Surface Methodology (RSM)

Research with response surface methodology to get an optimum condition of lactide formation is getting results with optimum operating conditions at 220° C temperature, amount of catalyst is 0.45% w/w and vacuum pressure is 10 cmHg. In this condition lactide is produced which is 78.469%. At the contours of response, there is a central point experiment conducted 3 points. While the optimum point at the high response. This point is the alpha value +1 of the draft model of central composite design (CCD). This value is not different from the predicted values which use the equation.

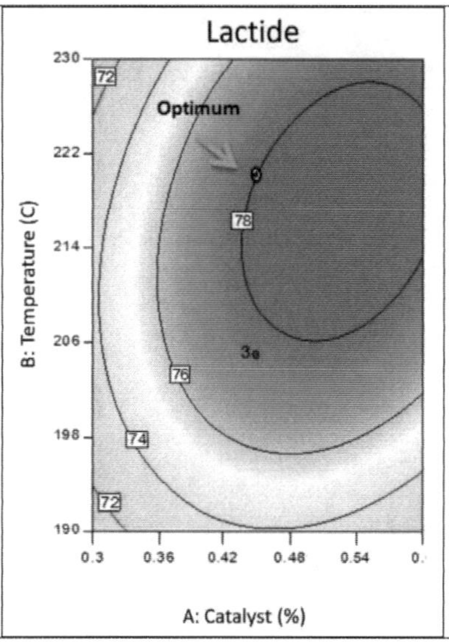

Figure 3. Response 2D of lactide yield

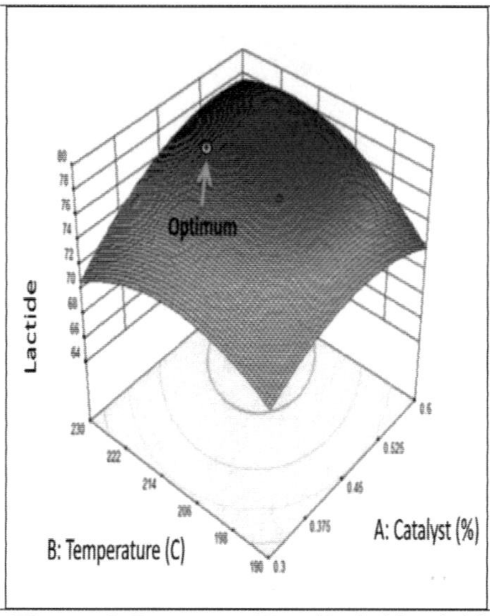

Figure 4. Response 3D of Laktide yield

The highest response (red area) is the highest yield of lactide. But the optimum point is in the region shown in arrows and aren't on the highest response (red). This is because the lowest point in the response seen imprecise. At low catalyst percent, response decreases drastically than at high catalyst. So that the optimum point shifted toward with smaller catalyst and not be in the highest response (red entirely).

When the pressure vacuum is increased and the catalyst is added more, the production of lactide is increased. According to the reference auras, 2010, that the increase of catalyst concentration and vacuum pressure will increase the production of lactide. However, increasing catalyst is not always linear to increase production of lactide, it is linked to the rate of mass transfer between the catalyst with the substrate [8]. Moreover, when the temperature is increased, the reaction rate will be increased so that the reaction goes to the right. The amount of catalyst will also accelerate the rate of reaction this is due to the increased surface area of contact between the PLA oligomers with a catalyst.

3.3 Lactide Analysis by HPLC

Lactide analysis by HPLC method used to determine the concentration of lactide.

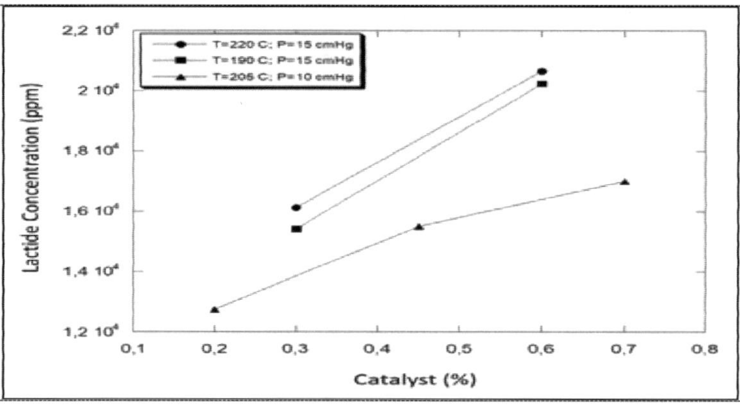

Figure 5. Effect of catalyst zinc acetate on lactide concentration.

IOP Conf. Series: Materials Science and Engineering **334** (2018) 012058 doi:10.1088/1757-899X/334/1/012058

From figure 5, show that when the catalyst is increased, lactide concentration is also increased. From that figure also seen that the curve with conditions of 190° C and vacuum pressure 15 cmHg and the curve with conditions of 220°C and vacuum pressure 15 cmHg, have the same precision visually. Both of these curves differ only in the temperature conditions where the temperatures are higher at 220° C was resulted in higher concentrations as well. The relationship between the amount of catalyst added to the linear lactide concentration also occurs on other research [10], which found the addition of catalyst will increase yield lactide and increase conversion of depolymerization reactions.

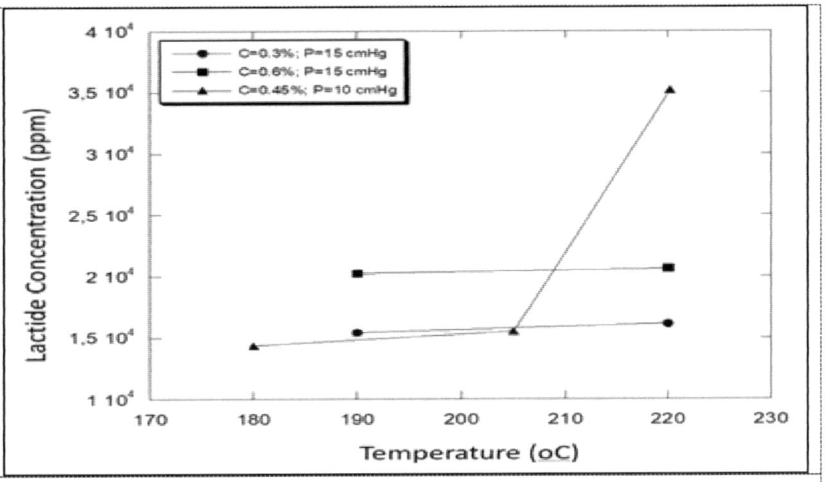

Figure 6. Effect of temperature depolymerization on lactide concentration.

From the Figure 6 show that the increasing temperature accompanied by increasing concentration of lactide. Same as in Figure 5, there are two curves which have the same precision visually. The curve a condition of 0.3% catalyst and a pressure of 15 cmHg with the curve a condition of 0.6% catalyst and the vacuum pressure of 15 cmHg. These curves have the same pressure, but have the difference in the amount of catalyst is used. The catalyst that is increased will generate a lot of lactide concentration. However, the significant increase in concentration occurs 220° C to 205° C temperature, with a concentration of 0.45% catalyst and 10 cmHg vacuum pressure. So, the increasing of temperature causes increasing concentration of lactide [10]. Increasing temperature cause deprotonation process of PLA oligomers faster. Deprotonation is the process of releasing a proton of oxygen atom to a carbonyl carbon groups due to the addition of a catalyst.

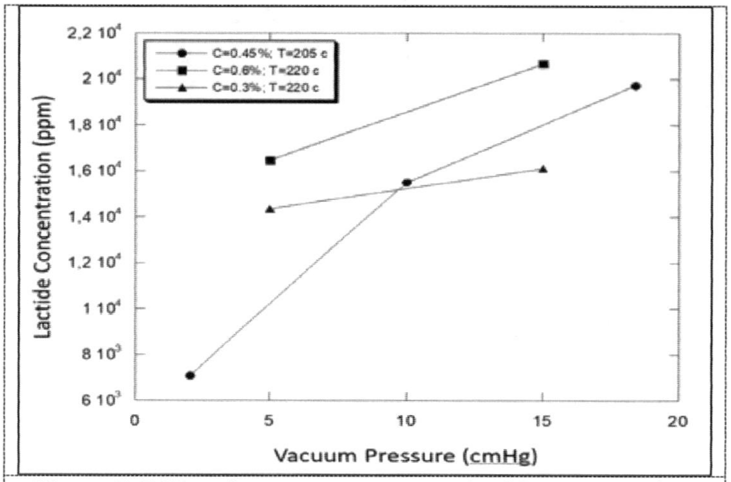

Figure 7. Effect of vacuum pressure on lactide concentration.

In Figure 7, can be seen that the higher vacuum pressure was given or more vacuum pressure in the system, due to higher concentration of lactide. However, 0.45% of catalyst, and 205° C of temperature, and 2 cmHg vacuum pressure, a concentration of lactide is decreased. This is because the vacuum pressure hasn't been able to attract steam in the flask to flow to the condenser.

It also happens in research conducted by Yoo et al, 2006. The concentration of lactide is increased when the pressure is lowered (preferably vacuum). This is due to the lower pressure, monomer and dimer are more volatile than long-chain [11]. In addition to lower pressure, the concentration of impurities such as water also decreases [11].

4. Conclusion

Optimal conditions for Lactide production from lactic acid was obtained at temperature 220°C; 0.45%w/w zinc acetate catalyst; and a vacuum pressure is 10 cmHg. The model from RSM is,

$$\text{Yield lactide } (\%) = -258{,}75 + 7{,}79A + 2{,}90B + 3{,}50C + 0{,}48AB - 0{,}06AC - 3{,}97.10^{-3}BC - 105{,}42A^2 - 7{,}17B^2 - 0{,}10C^2$$

with A: %zinc acetate catalyst; B: temperature (°C); C: vacuum pressure (cmHg)

Acknowledgement

The author acknowledge the funding from Ministry of RISTEKDIKTI through grant of *Publikasi Internasional terindeks Tugas Akhir Mahasiswa* (PITTA nr. 855/UN2.R3.1/HKP.05.00/2017). Special gratituted to grant from Centre for Development of Sustainable Region (CDSR) SHERA by USAID-RISTEKDIKTI in 2017.

References

[1] Pusat Penelitian Karet. 2014. Perspektif Pasar Polimer untuk Perangkat Medis. Balai Penelitian Sungai Putih. Retrieved from http://balitsp.com/perspektif-pasar-polimer-untuk-perangkat-medis/.

[2] Matsumoto, Ken'ichiro, & Taguchi, Seiichi. 2010. Enzymatic and whole-cell synthesis of lactate-containing polyesters: toward the complete biological production of polylactate. Applied microbiology and biotechnology, 85(4), 921-932.

[3] Huang, Wei, Qi, Yunbiao, Cheng, Na, Zong, Xupeng, Zhang, Tianrong, Jiang, Wei, Hong, Li, Zhang, Quanxing., Green synthesis of enantiomerically pure l-lactide and d-lactide using biogenic creatinine catalyst. Polymer Degradation and Stability.2014, 101, 18-23.

[4] Madhavan Nampoothiri, K, Nair, Nimisha Rajendran, & John, Rojan Pappy., An overview of the recent developments in polylactide (PLA) research. Bioresource Technology. 2010, 101(22), 8493-8501.

[5] Kamel, Gihan, Bordi, Federico, Chronopoulou, Laura, Lupi, Stefano, Palocci, Cleofe, Sennato, Simona, & Verdes, Pedro V., Adsorption of Candida rugosa lipase at water-polymer interface: The case of poly (DL) lactide. Surface Science. 2011, 605(23), 2017-2024.

[6] Vichaibun V., Chulavatnatol M. 2003. A new assay for tne enzymatic degradation of polylactic acid. ScienceAsia. (29): Hal 297-300.

[7] Lasprilla, Astrid JR, Martinez, Guillermo AR, Lunelli, Betânia H, & Jardini, André L. 2012. Poly-lactic acid synthesis for application in biomedical devices—A review. Biotechnology advances, 30(1), 321-328.

[8] Auras, R., L.T. Lim., S. E. M. Selke., dan H. Tsuji, 2010. Poly(lactic acid): Synthesis, Structures, Properties, Processing, and Applications. John Wiley & Sons, Inc.

[9] Abe H, Takahashi N, Kim KJ, Mochizuki M, Doi Y. 2004. Biomacromolecules. *Polymer Journal*. 5, 1606.

[10] Yoo, D.K., D. Kim, D.S. Lee,. 2006. Synthesis of lactide from oligomeric PLA: Effects of temperature, pressure, and catalyst. *Macromolecular Research*, 14(5): Hal. 510-516.

[11] Xavier, Ana Marta. 2010. Study of Lactic Acid Polycondensation and Lactide Production. University of Porto. Hal 9-15

IOP Conf. Series: Materials Science and Engineering **334** (2018) 012059 doi:10.1088/1757-899X/334/1/012059

Discrimination of several Indonesian specialty coffees using Fluorescence Spectroscopy combined with SIMCA method

D Suhandy[1,3] and M Yulia[2]

[1] Department of Agricultural Engineering, Faculty of Agriculture, The University of Lampung, Jl. Prof. Dr. Soemantri Brojonegoro No.1, Bandar Lampung, 35145, Indonesia
[2] Department of Agricultural Technology, Lampung State Polytechnic, Jl. Soekarno Hatta No. 10, Rajabasa Bandar Lampung, Indonesia

E-mail: diding.sughandy@fp.unila.ac.id

Abstract. Indonesia is one of the important producers of several specialty coffees, which have a particularly high economic value, including Civet coffee ('kopi luwak' in Indonesian language) and Peaberry coffee ('kopi lanang' in Indonesian language). The production of Civet and Peaberry coffee is very limited. In order to provide authentication of Civet and Peaberry coffee and protect consumers from adulteration, a robust and easy method for evaluating ground Civet and Peaberry coffee and detection of its adulteration is needed. In this study, we investigate the use of fluorescence spectroscopy combined with SIMCA (soft independent modelling of class analogies) method to discriminate three Indonesian specialty coffee: ground Peaberry, Civet and Pagar Alam coffee. Total 90 samples were used (30 samples for Civet, Peaberry and Pagar Alam coffee, respectively). All coffee samples were ground using a home-coffee- grinder. Since particle size in coffee powder has a significant influence on the spectra obtained, we sieved all coffee samples through a nest of U. S. standard sieves (mesh number of 40) on a Meinzer II sieve shaker for 10 minutes to obtain a particle size of 420 μm. The experiments were performed at room temperature (around 27-29°C). All samples were extracted with distilled water and then filtered. For each samples, 3 mL of extracted sample then was pipetted into 10 mm cuvettes for spectral data acquisition. The EEM (excitation-emission matrix) spectral data of coffee samples were acquired using JASCO FP-8300 Fluorescence Spectrometer. The principal component analysis (PCA) result shows that it is possible to discriminate types of coffee based on information from EEM (excitation-emission matrix) spectral data. Using SIMCA method, the discrimination model of Indonesian specialty coffee was successfully developed and resulted in high performance of discrimination with 100% of sensitivity and specificity for Peaberry, Civet and Pagar Alam coffee. This research has opened the possibility to develop a promising method to detect and evaluate authentication of Indonesian specialty coffees using fluorescence spectroscopy.

1. Introduction

Indonesia is well known as one of the main exporting coffee country especially for robusta type with average production is around 540 million kg in 2014 [1]. Indonesia is also rich with some specialty

[3] To whom any correspondence should be addressed.

coffee with unique taste and high price. Some specialty coffee is associated with specific origin of coffee plantation like Gayo coffee in Aceh (arabica), Pagar Alam coffee in South Sumatera (robusta) and others. Some specialty coffee is very delicious and very limited production due to special fermentation like Civet coffee. Peaberry coffee is also one of the specialty coffee in Indonesia with very limited production and unique taste.

Food authentication is the process that verifies that a food is in compliance with its label description [2]. The declaration of specific quality attributes in high-value products is of particular interest since these products are often target of fraudulent labeling. Proof of provenance is an important topic for food safety, food quality and consumer protection, as well as the compliance with national legislation, international standards, and guidelines [3]. In coffee trading, authentication is becoming important issue. Specialty coffee is limited in production, with unique taste and high price compare to normal coffee. Unfair trading is usually happen by mixing normal coffee with some specialty coffee. In this step, the discrimination between authentic specialty coffee and mixed or fake specialty coffees is highly required. The discrimination and authentication process is becoming very difficult and highly complex in roasted and ground coffee considering that after roasting and grinding, the addition of others materials cannot be detected visually [4].

The use of several conventional methods to identify the adulteration of roasted and ground coffee have been reported including the use of optical and electron microscopy [4]. Recently, other techniques have been studied in order to provide new analysis method that are more reliable, reproducible, and able to identify the many different types of possible adulteration include chromatographic analysis and infrared spectroscopy. Several spectroscopic based method include NIR spectroscopy and mid infrared have been reported for successful application in adulteration detection in coffee [5], [6], [7]. The use of spectroscopic method for authentication in coffee is also frequently combined with the power of pattern recognition methods such as soft independent modeling of class analogy (SIMCA), partial least square discriminant analysis (PLS-DA), back propagation neural network (BPNN) and least squares-support vector machine (LS-SVM) [8]. Recently, ultraviolet-visible (UV-Vis) spectroscopy also has been used for promising application of authentication in ground coffee [9]. Suhandy et al. [10] used UV-vis spectroscopy combined with SIMCA for authentication of Civet coffee. The discrimination between Peaberry and normal coffee was studied and successfully separated using UV-vis spectroscopy combined with SIMCA and PLS-DA [11].

Fluorescence spectroscopy has been extensively used for the classification of food [12]. The advantages of fluorescence spectroscopy as analytical method are fast, selective, and sensitive. However, the report for application of fluorescence spectroscopy for coffee discrimination is very limited. Fluorescence spectroscopy was recently reported has been used as tool for the geographical discrimination of coffees in Brazil [13]. As far as the authors are aware, no record has been found in the literature on the discrimination of Indonesian specialty coffee using fluorescence spectroscopy techniques. In this study, the feasibility of the technique in discrimination of Indonesian specialty coffee was evaluated, and models were established using SIMCA.

2. Methodology/ Experimental

2.1. Coffee samples

In this study, we prepare three types of Indonesian specialty coffee: Peaberry (robusta), Pagar Alam (robusta) and Civet coffee (robusta). Total 90 samples were used (30 samples for Civet, Peaberry and Pagar Alam coffee, respectively). The coffee samples were collected from several coffee traders in Lampung province. The preparation of coffee powder and coffee extraction were conducted based on the previous reported studies [9], [10], [14]. 3 mL of each liquid samples was pipetted into 10 mm cuvette for fluorescence spectral acquisition.

IOP Conf. Series: Materials Science and Engineering **334** (2018) 012059 doi:10.1088/1757-899X/334/1/012059

2.2. Fluorescence spectral data acquisition

All the EEM (excitation-emission matrix) spectral data of coffee samples were acquired using a spectrofluorometer (JASCO model FP-8300, Tokyo, Japan) in the excitation range from 200 nm to 500 nm (5 nm steps) and the emission range from 210 nm to 750 nm (1 nm steps). The excitation and emission bandwidth were both 5.0 nm. The software of Spectra Manager (JASCO Co., Tokyo, Japan) was used to control EEM spectral data acquisition. This software was also used for spectral data pre-treatment such as fluorescence correction and finding peak for high intensity of EEM spectral data.

2.3. Data analysis

First, the fluorescence spectra were analysed using principal component analysis (PCA) [15]. In PCA, the main systematic variation in data was extracted by breaking down the information into principal components (PCs) based on equation (1).

$$\mathbf{X} = 1\bar{x} + \mathbf{TP}' + \mathbf{E} \tag{1}$$

The emission or fluorescence spectra data are contained in a matrix, X, and decomposed into a score matrix, T, and a loading matrix, P. The scores contain information about the amount of each PC in every sample and the loadings hold information about the contribution of the variables to each PC. The residuals, which are the variations that are not described by the scores and loadings, are collected in a residual matrix, E.

SIMCA is a supervised pattern recognition technique in which an individual PCA sub-model is created for each class and its boundary is calculated. The established PCA sub-models are then used to assess to which classes the new samples belong. Moreover, the results of SIMCA analysis allow us to evaluate the discrimination power of the individual wavelength variables. In SIMCA, the samples were separated into two groups: training and prediction sample set. Training sample set consist of 20 samples for each type of coffee and was used to develop SIMCA calibration model for each type of coffee. Prediction sample set consist of 10 samples for each type of coffee and was used to perform prediction step. To evaluate the result of SIMCA discrimination in prediction step, two parameters were used: sensitivity (samples belonging to the class and classified correctly in this class) and specificity (samples not belonging to the modeled class and correctly classified as not belonging) [16]. PCA and SIMCA analysis were performed with a statistical software package of 'The Unscrambler' V9.7 (CAMO AS, Trondheim, Norway). SIMCA was calculated at 10% of significance level in the F test.

3. Results and Discussion

3.1. Excitation and fluorescence spectra of coffee samples

Figure 1 shows the EEM spectra of Peaberry, Pagar Alam and Civet coffee with emission spectra in horizontal axis and excitation spectra in vertical axis. It can be seen that there is no data recorded in the area that wavelength of excitation is higher than that of emission. Using peak finder option in Spectra Manager (JASCO Co., Tokyo, Japan) software, we can identify several peaks with high intensity of emission or fluorescence spectra. Based on Figure 1, it can be inferred that the peak at excitation wavelength of 370 nm resulted in high intensity of fluorescence at 453 nm.

Figure 2 shows emission or fluorescence spectral data of all coffee samples (Peaberry, Pagar Alam, and Civet coffee) at 370 nm of excitation wavelength. The peak of fluorescence spectra was located at 453 nm which is based on previous report may be related to the typical emission spectra of flavonoid content in coffee.

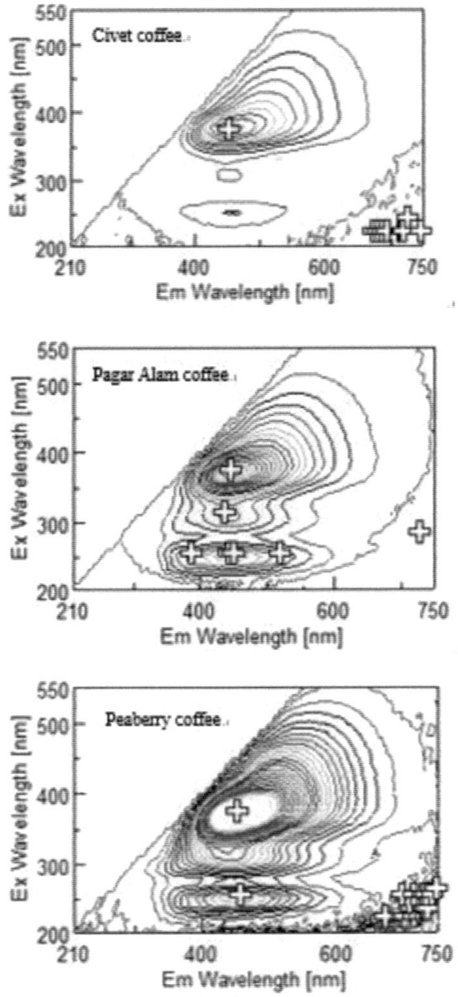

Figure 1. The contour of excitation emission spectra of three types of coffee samples.

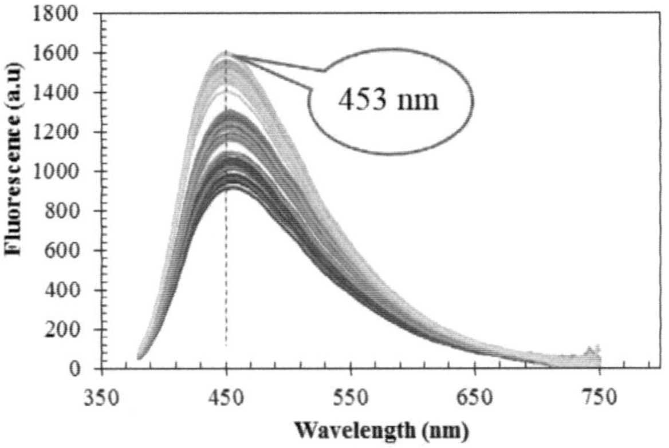

Figure 2. Emission/fluorescence spectra of coffee samples at 370 nm excitation.

3.2. Principal component analysis

In order to evaluate the discriminating ability of fluorescence spectra regarding the type of coffee (Peaberry, Pagar Alam and Civet), a principal component analysis was performed. Figure 3 shows the score plots (PC1×PC2) resulting from the application of PCA to the raw spectra. Figure 3 shows a clear of separation between Peaberry, Pagar Alam and Civet coffee using the raw spectra. Thus, it can be verified that the data pre-processing of obtained fluorescence spectra is not necessary.

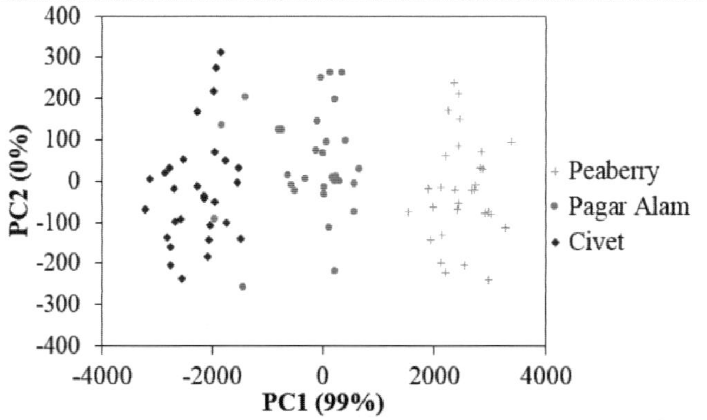

Figure 3. PC1×PC2 score plots for the 90 studied coffee samples using the raw spectra. Peaberry coffee (green), Pagar Alam coffee (red circle), Civet coffee (blue square). The explained variance in each PC is indicated in parentheses.

3.3. Soft independent modelling of class analogy (SIMCA)

SIMCA, a superior pattern recognition method, applied in fluorescence spectroscopy to develop classification models. SIMCA describes each class separately in a PC space. When building SIMCA model, the three classification models (Peaberry, Pagar Alam and Civet) were developed by PCA method, respectively. The samples of the calibration and prediction sets were used to calibrate SIMCA model and evaluate its performance. The determination of the appropriate number of PCs is a critical step for developing SIMCA model. The number of PCs was determined by the full-cross validation method. In our case, two PCs were used for Civet coffee, Pagar Alam and Peaberry coffee, respectively.

Table 1. Discrimination results between Peaberry, Pagar Alam and Civet samples in training and prediction samples using the soft independent modeling of class analogy (SIMCA) method using original fluorescence spectra at 370 nm of excitation wavelength.

Types of samples	Samples	Classified correctly to proper class	Classified to none	Classified to both	Sensitivity (%)	Specificity (%)
	Peaberry	20	0	0	100%	100%
Training	Pagar Alam	20	0	0	100%	100%
	Civet	20	0	0	100%	100%
	Peaberry	10	0	0	100%	100%
Prediction	Pagar Alam	9	1	0	100%	100%
	Civet	10	0	0	100%	100%

The results for sensitivity and specificity among categories are presented in Table 1. As can be seen in Table 1, most samples in training and prediction were correctly classified as falling inside the class model. However, there is one sample from Pagar Alam coffee in prediction sample set was not correctly classified as Pagar Alam (classified as none). This can be seen in Figure 4 and Figure 5, where the Coomans plots of the SIMCA models used are shown. It can be seen that none of the models built admitted samples from the other category, and thus the specificity is 100% in all cases. The sensitivity of the three models is also 100% for all coffee samples.

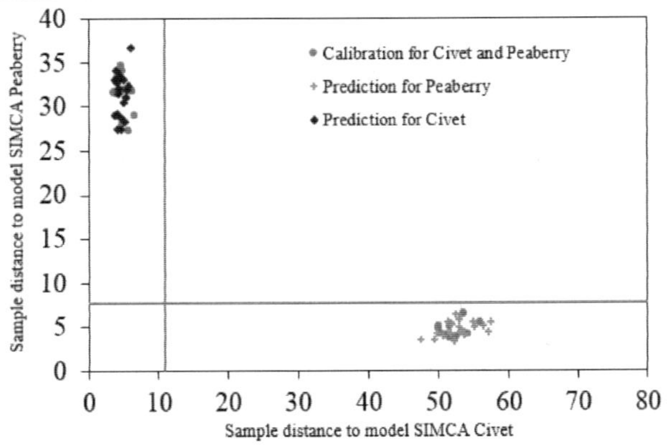

Figure 4. Coomans plot with the squared distance SIMCA for Peaberry and Civet coffee samples. Continuous red lines are the critical SIMCA distance for each class.

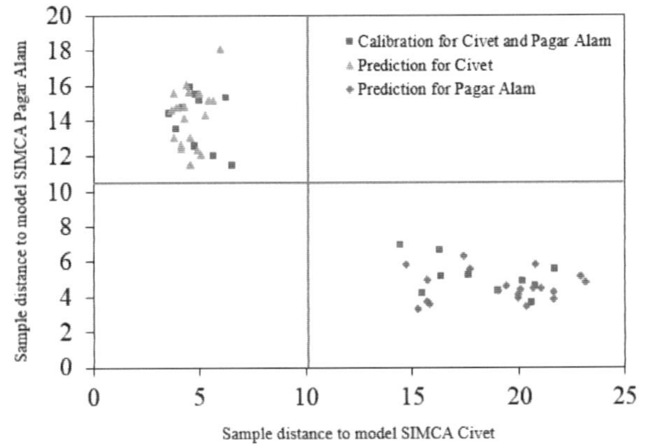

Figure 5. Coomans plot with the squared distance SIMCA for Pagar Alam and Civet coffee samples. Continuous red lines are the critical SIMCA distance for each class.

4. Conclusion

In this study, the discrimination of Indonesian specialty coffee using fluorescence spectroscopy and SIMCA method has been successfully conducted. Analysis of excitation emission spectral data showed that the peak of fluorescence spectra was located at 453 nm, which is based on previous report, may be related to the typical emission spectra of flavonoid content in coffee. The PCA on fluorescence spectral data at 370 nm of excitation wavelength resulted in a clear separation between the three types of Indonesian specialty coffees. Classification using SIMCA method showed a high degree of sensitivity and specificity. This result has shown a great potential of using fluorescence spectroscopy and SIMCA for accurate discrimination of Indonesian specialty coffees.

Acknowledgement

The authors would like to thank Professor Naoshi Kondo and Associate Professor Yuichi Ogawa of Laboratory of Bio-Sensing Engineering of Kyoto University for giving the author (Diding Suhandy) permission to use fluorescence spectrometer and its supporting devices (through JASSO Follow up

Research Program). Some part of this research is also funded by Kemenristekdikti Indonesia (Hibah Strategis Nasional/STRANAS 2016–2017 No. 585/UN26.21/KU/2017).

References

[1] ICO 2014 World coffee trade 1963–2013 http://www.ico.org
[2] Danezis G P, Tsagkaris A P, Camin F, Brusic V and Georgiou C A 2016 *Trends Analyt. Chem.* **85** 123–32.
[3] Aung M M and Chang Y S 2014 *Food Control*. 39 172–84.
[4] Toci A T, Farah A, Pezza H R and Pezza L 2016, *Crit. Rev. Anal. Chem.* **46** 83–92.
[5] Pizarro C, Esteban-Diez I and Gonzalez-Saiz J M 2007 *Anal. Chim. Acta.* **585** 266–76.
[6] Tavares K M, Pereira R G F A, Nunes C A, Pinheiro A C M, Rodarte M P and Guerreiro M C 2012 *Quim. Nova.* **35** 1164–68.
[7] Reis N, Franca A S and Oliveira L S 2013 LWT - *Food Sci. Technol.* **50** 715–22.
[8] Li X, Yi S, He S, Lv Q, Xie R, Zhen Y and Deng L 2016 *Precis. Agric.* **17** 365–74.
[9] Souto U T C P, Barbosa M F, Dantas H V, Pontes A S, Lyra W S, Diniz P H G D, Araújo M C U and Silva E C 2015 LWT - *Food Sci. Technol.* **63** 1037–41.
[10] Suhandy D, Yulia M, Waluyo S, Sugianti C, Iriani R, Handayani F N and Apratiwi N 2016 *Proc. of the USR International Seminar on Food Security* (Bandar Lampung: Universitas Lampung) pp 245–53.
[11] Suhandy D and Yulia M 2017 *Int. J. Food Prop.* In press.
[12] Sádecká J and Tóthová J 2007 *Czech J. Food Sci.* **25** 159–73.
[13] Botelho B G, Oliveira L S and Franca A S 2017 *Food Control*. **77** 25–31.
[14] Yulia M, Suhandy D, Waluyo S and Sugianti C 2016 *Proc. of the USR International Seminar on Food Security* (Bandar Lampung: Universitas Lampung) pp 254–61.
[15] Wold S, Esbensen K and Geladi P 1987 *Chemometr. Intell. Lab. Syst.* **2** 37–52.
[16] Grasel F S and Ferrao M F 2016 *Anal. Methods.* **8** 644–49.

The influence of addition of papain enzyme and Carboxyl Methyl Cellulose on the textural properties of Tofu

Faridah[1,3], Fachraniah[1], Arifien[2] and C M Sari[1]

[1] Department of Chemical Engineering , Lhokseumaawe State Polytechnic, Jl. Banda Aceh-Medan Km 280, Buketrata, Lhokseumawe, Indonesia
[2] Department of Mechanical Engineering, Lhokseumaawe State Polytechnic, Jl. Banda Aceh-Medan Km 280, Buketrata, Lhokseumawe, Indonesia

E-mail: faridahtki@pnl.ac.id

Abstract. Papain enzyme and carboxyl methyl cellulosa was used in tofu production as coagulant and thickener. Papain enzyme is a protease enzyme that can break proteins. Papain enzymeuseful as coagulant to replace acid and base coagulant. The goal of this study is to observe papain enzyme as coagulant and carboxyl methyl cellulose as thickener to increase characteristic of tofu. Tofu is from soy milk has been pasteurized at 70 °C with the addition of papain enzyme and carboxyl methly cellulose. The concenration of papain enzyme is varied such as 200, 400, 800, and 1000 ppm. After Temperature reachs at 90 °C, carboxyl methyl cellulosa is added in soy milk to produce tofu. This study focuses on introducing papain enzyme as coagulant as well as investigating its potential in improving tofu making process productivity. Further the present work attempts to determine the synergistic effect of combining–CMC/enzyme in tofu characteristic. This research was conducted with soy milk pasteurized at 70 °C with increasing papain enzyme. Protein from tofu was determined by using Spectrophotometer UV-VIS Shimadzu UV-1800 type. The highest protein concentration of the papain enzyme was found in 1000 ppm with CMC concentration of 0.6% w/v and based on organoleptic tests that the adding CMC and enzyme papain does not effect the taste, smell, texture and color of tofu. The taste of tofu produced is similar to the taste of tofu in the market.

1. Introduction

Tofu production is usually made by using a coagulant such as base, acid and enzyme coagulants. Many coagulants were used in tofu production. The use of coagulant ingredients in the process of tofu production will affect the high protein level and also influence the acidic, slightly bitter, taste and the crumbly texture of tofu [1,2,3].

Papain enzyme is a protease (proteolytic) that can break down proteins. Papain enzyme is able to agglomerate protein, resistant to high temperatures and able to function properly at low concentrations, it is also free of chemicals, accessible, easily own-made, non-toxic, and no side reactions. Papain enzyme can agglomerate milk protein in the manufacture of cottage cheese [4]. The papain enzyme obtained from the latex of the papaya fruit. In order to observe the activity of the enzyme papain as a coagulant in the process of tofu production, a research is need to be conducted to

[3] To whom any correspondence should be addressed.

find the optimum conditions of the enzyme. The use of coagulant in tofu production will affect the level of protein held out, which resulted different characteristic such as sense of knowing that acid, a little bitter and easy destroyed [4].

This research aimed to observe the use of papaya latex as papain enzyme source, which used as a coagulant in the process of tofu production. In addition, this study also examined the addition of carboxyl methyl cellulose in tofu production. Carboxyl methyl cellulose is a derivative of cellulose and is often used in the food industry or used in foodstuffs to prevent retrograde or inverse of the gelatinization process, where starch crystals converge to form a particular form which can affect the texture.

2. Material and Method

The latex of papaya fruit, which is used as an ingredient in the process of enzyme papain. The enzyme papain obtained with the addition of 0.7% $NaHSO_3$. Before papain enzyme was used in the processing of tofu, the MCU of the enzyme was calculated. MCU used for to determinee proteolytic activity of the papain enzyme. Here, papain enzyme derived from the latex of papaya fruit tested to observe proteolytic enzyme. Next step is the process of making soy milk from soybeans. Soybeans used as a source of protein, weighed 500 grams. The soybeans then were soaked for one night to tenderize and enlarge the size. After that, the soybeans husk were cleaned, and then the soaked and cleaned ssoybeans were weighed and added to the hot water with a ratio 1: 3 of soy beans and water. The soybeans were blended until smooth and then filtered with a white cloth or gauze to separate the residue. After getting soymilk, the process of making papain enzyme derived from the sap of the papaya fruit performed by adding the carboxyl methyl cellulose. Soymilk fed into a 500-ml beaker glass. The milk was pasteurized up to 70^0 C –boiling point- ,pipette enzyme papain with 200, 400, 600, 800, 1000 ppm repeated treatments with soy milk. Following conducted clots at temperatures of 90 ^{0}C with the addition of carboxyl methyl cellulose with concentrations of 0; 0.2; 0.3; 0.4; 0.5% w/v. The tofu molded on a silk mold for thirty minutes. Then the protein and organoleptic test analysis were tested and using statistic to determine on the organoleptic parameter.

Statistical test used by Q Cochran test method, this test used three or more sets of scores (proportions or frequencies) paired significantly. The scale is used the ordinal scale. If the value of Q chochran $\geq$ the value of Chie squared at a significant level of 5% then Ho rejected and Ha accepted. Conversely, if the value of Q Cochran $\leq$ the value of Chie squared on 5% significant level then Ho accepted and Ha rejected. The formula is

$$Q = \frac{\left[(k-1)k\sum_{j=1}^{k} ci^2 - \left(\sum_{j=1}^{k} ci\right)\right]}{k\sum_{i=1}^{n} Li - \sum_{i=1}^{N} Li^2} \qquad \textbf{(1)}$$

Where :
Q approaches X^2 with db = k-1
Test rule: Reject H0 if $Q \geq X^2$ -tables with db = k-1
K = number of samples (treatment)
N = number of replications
Ci = number of successes in each treatment (1 to k)
Li = the number of successes in each repetition (1 to n)

Hypothesis:
Ho: There is no effect of adding papain enzyme and carboxyl methyl cellulose for tofu production to organoleptic quality and favorability

Ha: There is an effect of adding papain enzyme and carboxyl methyl cellulose for tofu production in oorganoleptic quality and favorite level

3. Result and Discusion

3.1 Test of Enzyme Activity Papain from the latex of Papaya Fruit

This research has been conducted to observe the use of crude extract papain from papaya latex as a coagulant in the tofu production by looking at the carboxyl methyl cellulose as a thickening agent. The papain enzyme which is a rough extract latex papain from papaya fruit is used as a coagulant the activity of the enzyme was tested. The activity of papain enzyme has conducted by using the MCU method. The table 1. shown the activity of the papain enzyme.

Table 1. Test of papain enzyme activity

Temperature °C	Agglomeration Time Minute	Enzyme Activities MCU
40	2,28	439,28
45	2,61	385,92
50	1,75	574,47
55	2,97	354,66
60	3,54	283,80
65	4,63	216,95
70	3,29	305,69
75	3,60	280,10

Papain enzyme produced from the latex of the papaya fruit. Before this enzyme used as the ingredient in Tofu production, its activity tested against the temperature. Before papain enzyme used as coagulant in tofu production, the papain enzyme tested its activity. The activity enzyme was used MCU method. With the variation of temperatures, it reviewed that the coagulant formed in the process of enzyme activity by using milk. The selected papaya fruits were unripe papaya fruits aged 2-3 months as they have a lot of sap, The papain enzyme used in this study were taken from the plants grown in marshland. The papaya sap tapped for ± 2 hours. The reason for using 2 hour taping was because papaya latex will be contaminated with air and the color will be brown if it was left for too long and it can affect the outcome of papain flour made. Papain from papaya latex is a substance which easily damaged by air oxidation occurring either during manufactureing or storage, for this reason, the process of making the papain enzyme used sodium bisulfite which functions as a preservative. The best concentration of sodium bisulfite is 0.7%. This definition includes microbial inhibitors, antioxidants, marinade ingredients and binder. The solution mixed homogeneously dried in an oven using temperature of 70 °C because principally papain enzyme will work well at low temperatures, 55-70 °C. The drying step was done for 3 hours until the papain enzyme was completely dry, the previous research used period of 6 hours to dry the enzyme papain. The reason for using 3 hours drying time because the papain solution that has been pipetted is homogeneous 10 ml per petri dish while the previous research used 6 hours with 30 ml. Once the flour papain processing was finished, the next step was to see the level of enzyme activity. The activity of papain enzyme is using MCU (Milk clotting Units) method against the influence of several factors; time, temperature of drying, and the addition of sodium bisulfite ($NaHSO_3$) to determine the quality of the enzyme papain generated, MCU were best contained in the leaves and the rod was 200 MCU while the fruit was 400 MCU.

3.2 Test of functional groups for the enzyme papain

Papain enzyme produced from the latex of the fruit papaya. In this study , it was compared functional groups of enzyme papain between enzyme commercial with enzyme papain from latex papaya. To compare functional group was used FTIR. FTIR analysis performed to measure the functional groups in papain enzymes.

This analysis aims to identify organic compounds based on a reading group functions held in the form of the spectrum. Here, FTIR analysis was conducted on a commercial papain enzyme and papain enzyme derived from the sap of papaya fruit. FTIR analysis conducted for both of these enzymes; the results of the test obtained at Figure 5.1 and 5.2. From the pictures shown in the figure 5.1 and 5.2, it shows that the wave spectrum of the papain enzyme and papain enzyme commercial were same, it addressed that the papain enzyme derived from the sap of the papaya has the same structure spectrum as enzymes commercial papain. Wave spectrum at the peak was seen in the range of 2300 cm-1 showed a C≡C group, in the range of 3750 cm-1 it showed the appearance of the wave peaks that showed the presence of O-H groups which are pure cellulose main functional groups. C-H group is the group-forming cellulose detected at wave number 300 cm-1. Carbonyl group C = O as cellulose binding was detected at wave number 1650 cm-1.

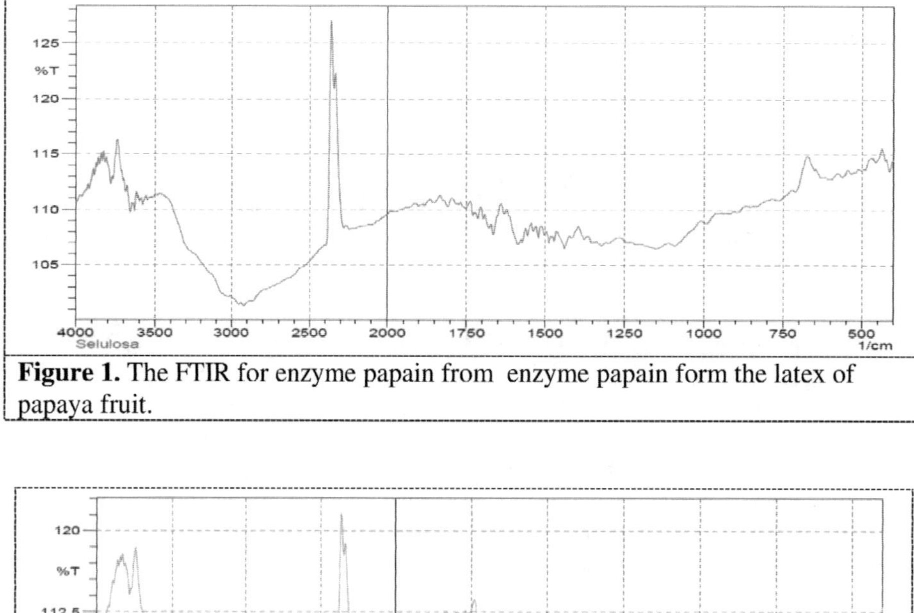

Figure 1. The FTIR for enzyme papain from enzyme papain form the latex of papaya fruit.

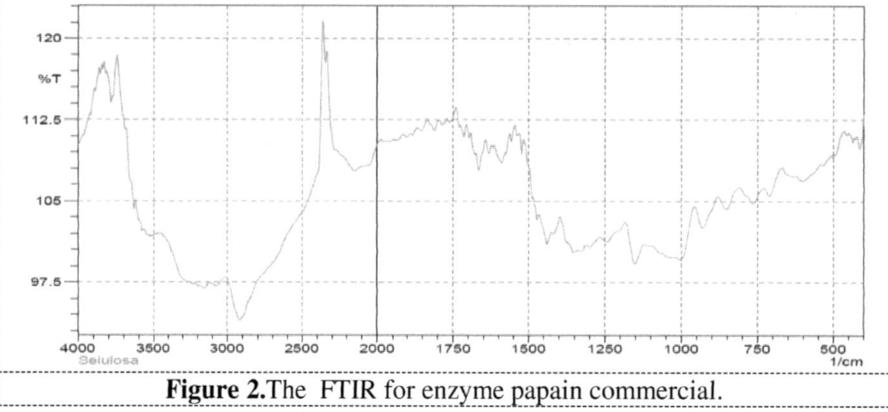

Figure 2. The FTIR for enzyme papain commercial.

3.3 Effect of Enzyme Papain enzyme, CMC and their combination on Protein production

Papain enzyme derived from the papaya fruit sap used in the manufacturing process. From the research that has been done previously, it can be concluded that the enzyme papain can be used as a coagulant. Clotting that occured in the manufacturing process carried out at temperature of 70 °C. At this temperature, the enzyme papain added in the process of tofu production as in this temperature the papain enzyme is still active. The coagulation of tofu that carried out the papain enzyme is very influential in forming protein, where the papain enzyme as coagulant material served as a catalyst that accelerates the reaction, in addition to the papain enzyme used to sever the bond of peptide soymilk into acid amino acids.

Spectrophotometry was used to analyzee the protein. The method used is Lowry method to examine the levels of soluble protein found in tofu. From Figure 5.3, it shown that the higher the concentration of carboxyl methyl cellulose the higher protein produced. From the graph in Figure 5.3, it also can be seen that the higher the concentration of enzyme protein produced also the highest protein content in tofu, here the highest protein concentration of the enzyme papain was found in 1000 ppm with carboxyl methyl cellulose concentration of 0.6% w/v . This range showed the influence of the addition of the enzyme papain that increase the protein content. While the lLowest solubility protein, soy milk without the addition of carboxyl methyl cellulose was found in the addition of a concentration of 200 ppm.

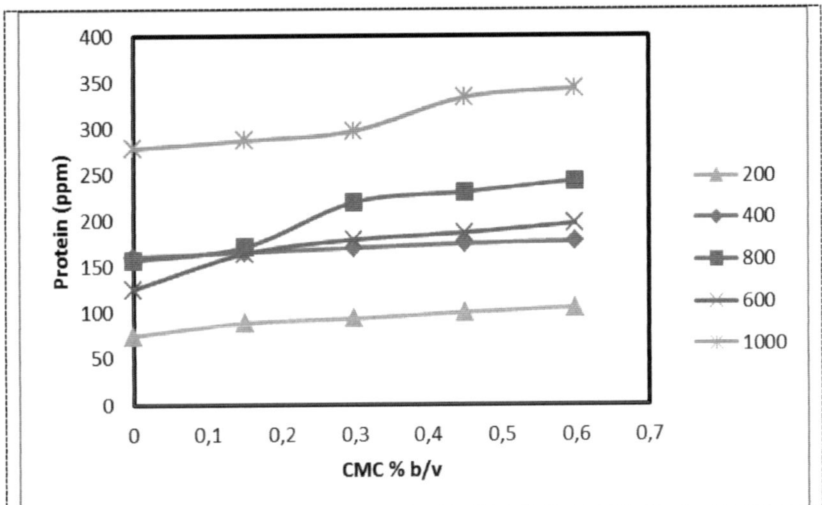

Figure 3. The effect of CMC addition and enzyme papain toward protein in tofu.

3.4 Organoleptic test

In this research, organoleptic test did to taste, color, texture, aroma and favorite level. The addition of papain enzyme and carboxyl methyl cellulose were made in various concentrations and tested on a rather well trained by panelist. Panelists argue that the concentration of papain enzyme 200 ppm with the increment of carboxyl methyl cellulose produces tofu which with a very soft texture.

Table 2. Analysis of organoleptic test based on the formula Q Cochran and chi squared.

Papain Enzyme Concentration (ppm)	Q cohcran	Chi square
200	9,7	10,8
400	0,7	5,7
600	1,8	2,4
800	2,05	18
1000	4,2	8,9

According to the data collected know that the resulting soy-scented soybean like tofu market, and the addition of carboxyl methyl cellulose with a concentration of 0.15% b/v, 0.30% b/ 0.45% w / v and 0.60% b / v the resulting flavor is still the same as soy-scented with a very soft texture. Tthe panelists like tofu very much with the addition of papain enzyme and carboxyl methyl cellulose, due to the bitter taste contained in papaya sap is not felt in the tofu, the texture of tofu is very soft and chewy also become a separate reference in the panelist assessment. This table show that the statistic data which use chochran equatio to determine the organoleptic level from panelists.

Based on taste levels of tofu was determined with using panelist. The results from panelist is calculated by Cochran equation and Chi square (table 2). Table 2 shows that panelists like tofu by adding papain enzyme and carboxyl methyl cellulose. Papain enzyme and carboxyl methyl cellulose addition does not effect for taste, aroma, texture, color and favorite level. It can be describe from the value of Q Cochran and Chi Square. This calculation to know the level of the organoleptic parameter. It can be shown that The value of Q cochran is lower than Chi square for organolepticparameter. It means that organoleptic parameter from tofu was illustrating that panelists like tofu with adding papain enzyme and carboxyl methyl cellulose. For example, 400 ppm of papain enzyme addition has Q chochran around 0,75 and Chi square around 15,2. The value of Q chochran was lower than Chi square. It meat that Ho is accepted and Ha is rejected. There is no effect of adding papain enzyme and carboxyl methyl cellulose forTofu production to organoleptic quality and favorability. The hypotesis explans that the papain enzyme and cmc addition do not effect for the quality and favorability of organoleptic test for variuos concentration of papain enzyme and carboxyl methyl cellulose.

4. Conclusions

From this research, it concluded that the enzyme papain from papaya latex used as a material for coagulant in tofu production. It found that the best concentration was 1000 ppm and 0.6% carboxyl methyl cellulose and the result of organoleptic test show that it does not affect for organoleptic test for taste, aroma, texture and favorite level from tofu with adding papain enzyme and carboxyl methyl cellulose.

Acknowledgment

The author would like to thank KEMENRISTEK DIKTI who has provided the financial support for this research through a competitive grant program in 2016. The author would like also thank the students and laboratory staff that have been involved in this study.

References

[1] Oboh, G., M.M. Ekperigin and M.I. Kareem, 2004. Inhibitors and chymotrypsin inhibitors can interact with Nutritional and haemolytic property of eggplant the protein and thereby altering the digestibility of tofu. (*Solanum macrocarpon*) leaves. *Journal of Food*. The higher apparent digestibility values could be because Composition and Analysis. In Press.

[2] Hou, H. J., Chang, K. C., & Shih, M. C. 1997, Yield and textural properties of soft tofu as affected by coagulation method. *Journal of Food Science*, 62(4), 824–827. doi:10.1111/j.1365-2621.1997. tb15464.x

[3] Kao, F.-J., Su, N.-W., & Lee, M.-H.. 2003, Effect of calcium sulfate concentration in soymilk on the microstructure of firm tofu and the protein constitutions in tofu whey. *Journal of Agricultural and Food Chemistry*, 51(21), 6211–6216. doi:10.1021/jf0342021

[4] Lampert, L.M., 1970. Modern Dairy Products, Chemical Publising New York, Company. Inc

IOP Conf. Series: Materials Science and Engineering **334** (2018) 012061 doi:10.1088/1757-899X/334/1/012061

Isolation of fish skin and bone gelatin from tilapia (*Oreochromis niloticus*): Response surface approach

N Arpi[1], Fahrizal[1] and M Novita[1]

[1] Agricultural Product Technology Department, Faculty of Agriculture, Syiah Kuala University, Jl. Krueng Hasan Kalee 3, Kopelma Darussalam, Banda Aceh, 23111 Indonesia

E-mail: normalina.arpi@unsyiah.ac.id

Abstract. In this study, gelatin from fish collagen, as one of halal sources, was extracted from tilapia (Oreochromis niloticus) skin and bone, by using Response Surface Methodology to optimize gelatin extraction conditions. Concentrations of alkaline NaOH and acid HCl, in the pretreatment process, and temperatures in extraction process were chosen as independent variables, while dependent variables were yield, gel strength, and emulsion activity index (EAI). The result of investigation showed that lower NaOH pretreatment concentrations provided proper pH extraction conditions which combine with higher extraction temperatures resulted in high gelatin yield. However, gelatin emulsion activity index increased proportionally to the decreased in NaOH concentrations and extraction temperatures. No significant effect of the three independent variables on the gelatin gel strength. RSM optimization process resulted in optimum gelatin extraction process conditions using alkaline NaOH concentration of 0.77 N, acid HCl of 0.59 N, and extraction temperature of 66.80 °C. The optimal solution formula had optimization targets of 94.38%.

1. Introduction

Gelatin is a translucent, colorless, brittle (when dry), flavorless, edible, multifunctional natural macromolecule protein polymer which is extracted by thermal hydrolysis of collagen tissue found in bones, skin, and connective tissue of various animal and fish [1][2][3]. Industries utilize gelatin extensively as a gelling agent, emulsifier, stabilizer, adhesives, viscosity agent and binder agent. Worldwide gelatin consumption is to reach 395.84 thousand metric tons by the year 2017. The vast majority of commercial gelatin is derived from porcine skin (46%), bovine hide (29.4%), bones (23.1%) and other sources (1.5%) [4]. Demand for alternative gelatin source is increasing due to religious reasons. Porcine gelatin cannot be used in kosher (Jewish) and halal (Muslims) foods, while Hindus do not consume bovine gelatin. The use of fish byproducts for alternative gelatin production is growing. Fish byproducts such as skin, bones, fins and scales account for ± 21% [5], even ± 50% for certain fish. Tilapia (*Oreochromis niloticus*) consumption in Indonesia is rapidly increasing and utilization of skin and bone are expected to reduce waste and increase the value of the waste. Fish gelatin properties influenced by many factors, such as species, water temperature fish habitats, and the

[1] To whom any correspondence should be addressed.

gelatin extraction process, which may depend on pH, temperature, and time during pretreatment and extraction process [6]. The objectives of this research were to study the effects of pretreatments using alkaline and acid on gelatin extraction, to characterize tilapia gelatin, and to determine optimum conditions for gelatin extraction from tilapia fish skin and bone.

2. Method

In order to obtain the optimum conditions of gelatin extraction, a Response Surface Methodology (RSM) approach was used. RSM using design expert software (9.0.5 version) was used to obtain several optimum gelatin extraction conditions. Mathematical techniques combine with statistical techniques used in the RSM method to create and analyze Y response (e.g. yield, gel strength, etc.) which are influenced by independent variables or X factor. Three independent variables (X factors) used in this study, i.e. X_1= alkaline (NaOH) concentration, and X_2= acid (HCl) concentration in the pretreatment process, and X_3= temperature of the extraction process which were tested at 5 different levels using 3 factors and 5 levels central composite rotatable design (Table 1). Levels of independent variables were obtained after the central composite rotatable design was run (Table 2). Response factors (=Y response) observed were yield [7], emulsion activity index [8], and gel strength [9].

Table 1. The independent variables and the levels of independent variables in 3 factors and 5 levels central composite rotatable design.

Independent Variables	Symbol	Code Level				
		-α	-1	0	+1	+α
The concentration (N) of alkaline NaOH pretreatment	X_1	0.58	0.75	1	1.25	1.42
The concentration (N) of acid HCl pretreatment	X_2	0.08	0.25	0.5	0.75	0.92
Extraction temperature (°C)	X_3	33.2	40	50	60	66.8

The gelatin extraction process based on [10] method with adaptations. The raw materials (200 g) were soaked in a solution of NaOH (1: 5 w/v) with various concentrations (factor X_1, N) for 1 hour. Then, the samples were washed with water (1: 5 w/v) three times, filtered and then squeezed. Subsequently, the samples were immersed in HCl (1: 5 w / v) with various concentrations (factor X_2, N) for 1 hours, then the samples were washed with water (1: 5 w/v) three times, filtered and then squeezed. After that, the samples were put into glass beakers, added with distilled water (1: 4 w/v) and covered with aluminum foil. Subsequently, the samples were extracted using a water bath with various extraction temperatures (factor X_3, °C) for 3 hours. Next, the filtrate was obtained by filtering using cheesecloth and dried in a glass container at a temperature of 60°C for 72 hours.

Table 2. Central composite rotatable design with 3 factors (X) and 5 levels of independent variables, along with dependent variables (Y).

Run	X_1[a] (N)	X_2[b] (N)	X_3[c] (°C)	Y_1 = Yield (%)		Y_2 =Emulsion activity index (m²/g)		Y_3 =Gel strength (g)
				Exp.	Pred.	Exp.	Pred.	Exp.
1	0	0	0	6	5,81	32,08	38,95	236,7
2	-α	0	0	8,87	7,51	19,36	22,48	189,3
3	0	-α	0	5,45	5,13	40,48	34,6	53,83
4	0	0	0	5,44	5,81	19,84	28,52	127,7
5	0	0	-α	6,14	8,32	17,65	18,39	218,2

6	-1	1	-1	5,09	5,73	19,58	21,91	138,8
7	0	0	0	6,13	5,81	27,19	28,52	174
8	0	0	0	5,8	5,81	22,39	28,52	297,7
9	α	0	0	4,3	4,11	23,88	14,5	269,7
10	-1	-1	1	9,06	7,91	31,51	32,89	119
11	1	1	-1	3,5	3,71	31,45	34,49	48,33
12	0	0	-α	3,83	3,3	31,71	24,71	198
13	0	α	0	6,17	6,49	39,34	30,81	133,3
14	-1	-1	-1	3,58	4,92	25,5	25,6	10,5
15	-1	1	1	9,5	8,72	36,28	30,81	9,5
16	1	1	1	8,25	6,7	15,35	19,68	3,833
17	1	-1	-1	3,76	2,9	17,34	27,23	66,5
18	0	0	0	4,92	5,81	34,42	28,52	140
19	0	0	0	4,83	5,81	34,12	28,52	109,7
20	1	-1	1	5,62	5,88	8,715	10,81	2,833

[a]The concentration of pretreatment alkaline NaOH (N)
[b]The concentration of acid HCl pretreatment (N)
[c]The extraction temperature
[d] Experiment
[e] Prediction

3. Results and Discussion

3.1. Response Surface Model Building of Gelatin Extraction

The response surface models for every response investigated showed in Table 3. Based on the results, in the optimization of the gelatin extraction, the response variables optimized were yield, emulsion activity index, and gel strength.

Table 3. Response surface models of tilapia gelatin extraction.

Response	Model	P- Value		Lack of fit		R^2
Yield	Linear	<0.0001	Significant	0.0544	Non significant	0.7375
Emulsion activity index	Quadratic	0.1723	Non significant	0.2655	Non significant	0.6269
Gel strength	Mean	-	-	0.2814	Non significant	-

3.2. Yield

The bone and skin tilapia gelatin yield ranged from 3.49% to 9.5%. Generally, the average of fish gelatin yield is 6-19% [11]. Response surface model of tilapia gelatin yield was linear (Table 3). The RSM equation of the gelatin extraction optimization towards the yield response is: Yield = +5,81-1,01A+0,41B+1,49C, where A= the concentration of alkaline pretreatment, B= the concentration of acid pretreatment, and C= the extraction temperature. The yield response will increase proportionally with the decrease of alkaline pretreatment concentration, and the increase of acid pretreatment also extraction temperature which are indicated by negative and positive value, respectively. This presumably because acid produced a pH that was suitable for the extraction of gelatin. Acid and temperature increase also caused some cross-linkages easily to break with little damage to the polypeptide chain [12]. Therefore, a higher yield could be generated.

Figure 1 and 2 show surface shape on three-dimensional graph of interaction among components (X_1, X_2, X_3) interplay yield response values. Different colors on the graph indicate different yield values. Blue indicates the lowest yield response value and red indicates the highest yield response. The lines which composed of dots on the graph represent the combination of three components with different values which produced similar yield response. Figure 1 and 2 showed that yield increase at combination of lower alkaline concentration, and higher acid concentration and extraction temperature. Kittiphattanabawon et al., 2016 [13] was using only 0.1-3% HCl in the pretreatment process with 40-80°C extraction temperature stated that gelatin extracted at higher temperature for longer time (>12 hours) had higher yield but poorer gel strength. However, Niu et al., 2013 [14] indicated that adjusting acids concentrations to appropriate levels resulting in optimal protein yield and gelatin physicochemical properties.

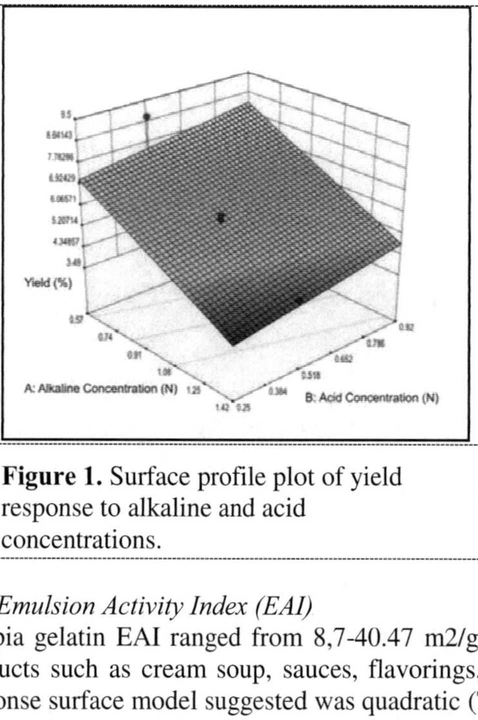

Figure 1. Surface profile plot of yield response to alkaline and acid concentrations.

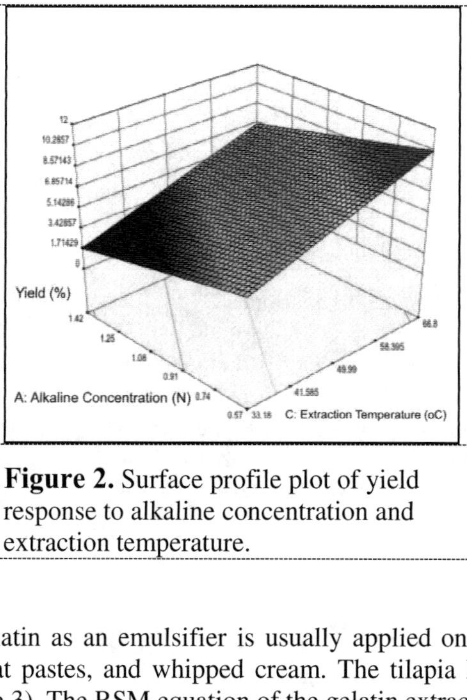

Figure 2. Surface profile plot of yield response to alkaline concentration and extraction temperature.

3.3. Emulsion Activity Index (EAI)

Tilapia gelatin EAI ranged from 8,7-40.47 m2/g. Gelatin as an emulsifier is usually applied on the products such as cream soup, sauces, flavorings, meat pastes, and whipped cream. The tilapia EAI response surface model suggested was quadratic (Table 3). The RSM equation of the gelatin extraction optimization towards the EAI response is: $EAI= 28,52-2,37A+1,29B- 1,88C+2,74AB-5,93AC+0,4BC-3,55A^2+ 2,92B^2- 2,46C^2$. Figure 3 shows that the EAI increase as the alkaline concentration decrease and the acid concentration increase. On the other hand, the EAI also increase as the extraction temperature decrease when the alkaline concentration decrease or the acid concentration increase (Figure 4 and 5).

3.4. Gel Strength

The gel strength of commercial gelatin range between 50-300 g [9]. The skin and bone tilapia gelatin have gel strength between 2.83-297.67 g, which suitable for product such as wafers, dairy products, desserts, and marshmallow. Depend on fish habitats, gelatin of other fish such as cold water salmon has a gel strength of 195 g [15], while warmer water cat fish at 252 g [16]. Furthermore, da Silva et al., 2017 [17] was using NaOH, acetic acid and kumakuma (Brachyplatystoma filamentosum) fish skin from the Amazonian industries produced gelatin with high yield and gel strength, 19.7% and 244.3g. Response surface model of tilapia gelatin gel strength was mean (Table 3 and Figure 6).

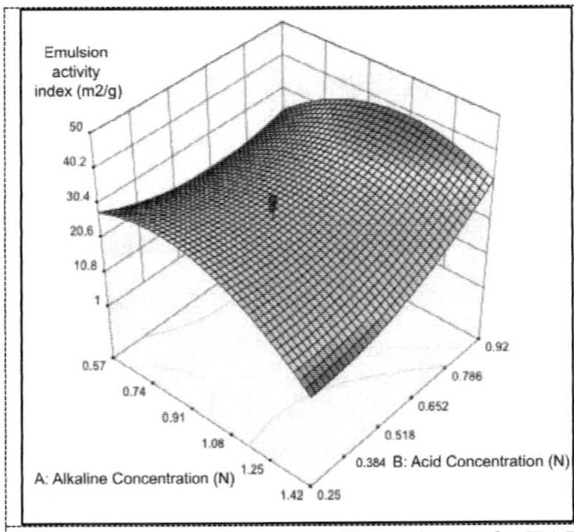

Figure 3. Surface profile plot of EAI response to alkaline and acid concentrations.

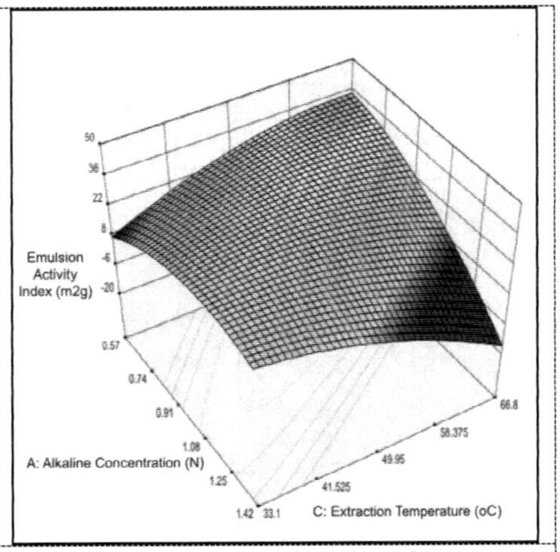

Figure 4. Surface profile plot of EAI response to alkaline concentration and extraction temperature.

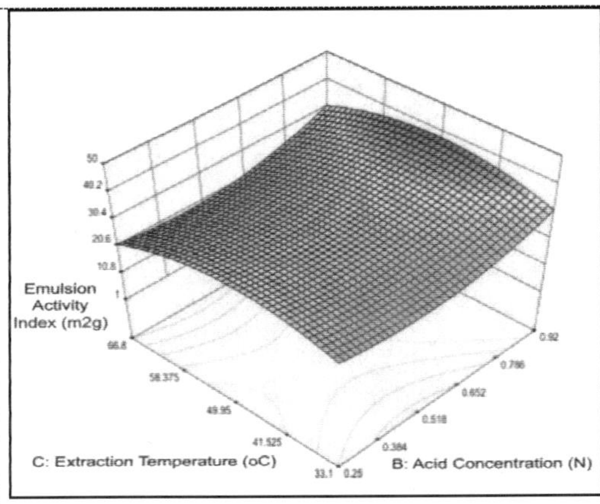

Figure 5. Surface profile plot of EAI response to acid concentration and extraction temperature.

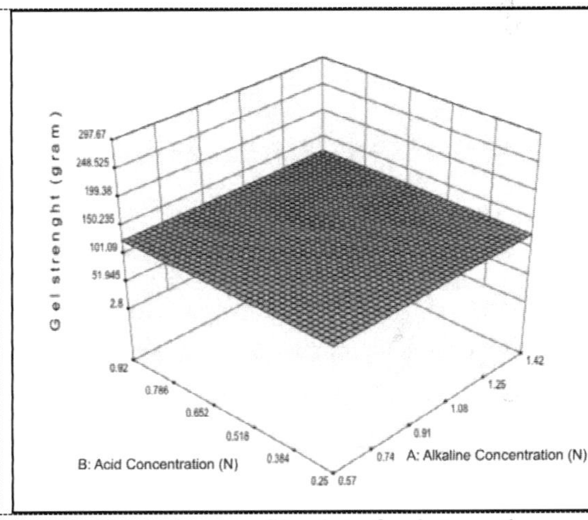

Figure 6. Surface profile plot of gel strength response to alkaline and acid concentration (N).

3.4. Gelatin Extraction Optimization

The purpose of optimization was to minimize the effort required and to maximize the desired results. Through this approach, optimization of multi responses could be obtained using desirability functions. Table 4 shows the components which were optimized, targeted, limits, as well as the level of importance at the stage of formulating the optimization. The concentration of alkaline and acid, and extraction temperature will affect the quality of the resulting gelatin. Since yield, emulsion activity index (EAI), and gel strength responses have the highest level of importance (5=+++++), therefore only these responses that would be optimized.

Table 4. Optimization parameters in the response optimizer.

Component/ Response	Target	Lower Limit	Upper Limit	Importance[a]
Alkaline Concentration	In range	0.75	1.25	3(+++)
Acid Concentration	In range	0.25	0.75	3(+++)
Extraction Temperature	In range	40	60	3(+++)
Yield	Maximum	3.50	3.50	5(+++++)
Emulsion Activity Index (EAI)	In Target	8.72	40.48	5(+++++)
Gel Strength	In Target	2.83	297.70	5(+++++)

[a]Importance level range from 1 to 5 with 1 as the lowest importance and 5 as the highest importance

There were 33 optimization solution formulas obtained from the optimization process which had the desirability values all over 0.91, except one with the lowest value of 0,89. Gelatin extraction process conditions recommended was using alkaline (NaOH) pretreatment concentration of 0.77 N, acid (HCl) pretreatment concentration of 0.59 N, and gelatin extraction temperature of 66.79°C. The optimal solution formula had optimization targets of 94.38% and was predicted to produce gelatin with yield of 9.4%, emulsion activity index of 26.97 m^2/g, and gel strength of 120.96 g.

4. Conclusion

The decrease in alkaline pretreatment concentration combine with the increase of extraction temperature cause an increase in yield of the extracted tilapia gelatin. On the other hand, the decrease of alkaline concentration and extraction temperature resulted in high emulsion activity index (EAI). There is no significant effect of alkaline and acid pretreatment concentration, and gelatin extraction temperature on gel strength of the extracted tilapia gelatin. Gelatin yield were 3.49-9.5%, while EAI and gel strength were 8.71-40.47 m2/g and 2.83-297.67 g, respectively. Optimization using expert design program 9.0.5 via response surface methodology approach generates optimal formula for gelatin extraction with alkaline concentration (NaOH) of 0.77 N, acid (HCl) of 0.59 N and temperature of 66.80°C with desirability value of 94.38% and was predicted to produce gelatin with 9.40% of yield, gel strength of 127.37 g, and emulsion activity index of 26.97 m2/g.

Acknowledgment

We thank Syiah Kuala University, Banda Aceh and Kemenristekdikti Indonesia "Fundamental Research Project", for financial support provided.

Reference

[1] Ahmad T, Ismail A, Ahmad SA, Khalil KA, Kumar Y, Adeyemi KD, and Sazili AQ 2017 Recent advances on the role of process variables affecting gelatin yield and characteristics with special reference to enzymatic extraction: A review *Food Hydrocolloids* **63** 85-96

[2] Vergauwen B, Stevens P, Prawitt J, Olijve J, Brouns E, Babel W, Schulz D, Giesen-Wiese M, Seybold U, Gareis H, and Dick E 2016 *Gelatin* Ullmann's Encyclopedia of Industrial Chemistry

[3] Alfaro AT, Balbinot E, Weber CI, Tonial IB, Machado-Lunkes A 2015 Fish gelatin: Characteristics, functional properties, applications and future potentials *Food Engineering Reviews* 733-44

[4] Chuaynukul K, Prodpran T and Benjakul S 2015 Properties of thermo-compression molded bovine and fish gelatin films asinfluenced by resin preparation condition *International Food Research Journal* **22** 1095-1102

[5] Vannuccini S 2004*Overview of Fish Production, Utilization, Consumption and Trade* FAO, Fishery Information, Data and Statistics Unit

[6] Gomez-Guillen MC and Montero P 2001 Extraction of gelatin from megrim (*Lepidorhombus boscii*) skins with several organic acids*Journal of Food Science* **66** 213-216

[7] AOAC 1995 *Official Methods of Analysis of The Association of Official Analytical Chemist* Inc. Washington DC

[8] Aewsiri TS, Benjakul W, Visessanguanb BJ, Eunc AP, Wierenga H and Gruppene 2009 Antioxidative activity and emulsifying properties of cuttlefish skin gelatin modified by oxidised phenolic compounds*Food Chemistry* **117** 160–168

[9] GMIA (Gelatin Manufacturers Institute of America) 2013*Standard Testing Methods for Edible Gelatin* Official Prosedure of The Gelatin Manufactures Institute of America, Inc. Available from :http://www.gelatine-gmia.com/

[10] Arpi N, Fahrizal, Satriana and Edward S 2016 Extraction and properties of gelatin from spotted oceanic triggerfish (*Canthidermis maculata*) skin and bone *International Journal on Advance Science, Engineering and Information Technology (IJASEIT)* **6** 561-569

[11] Karim AA and Bhat R 2009 Review Fish gelatin: properties, challenges, and rospects as an alternative to mammalian gelatins *Food Hydrocolloids* **23** 563-576

[12] Zhou P and Regenstein JM 2005 Effects of alkaline and acid pretreatment on Alaska Pollock skin gelatin extraction *Journal of Food Science* **70** C392-C396

[13] Kittiphattanabawon P, Benjakul S, Sinthusamran S and Kishimura H 2016 Gelatin from clown featherback skin: Extraction conditions *LWT – Food Science and Technology* **66** 186-192

[14] Niu L, Zhou X, Yuan C, Bai Y, Lai K, Yang F and Huang Y 2013 Characterization of tilapia (*Oreochromis niloticus*) skin gelatin extracted with alkaline and different acid pretreatments *Food Hydrocolloids* **33** 336-341

[15] Arnesen AA and Gildberg A 2007 Extraxtion and characterisation of gelatine from Atlantic salmon (*Salmo salar*) skin *Bioresource Tehcnology* **98** 53-57

[16] Yang H, Wang Y, Jiang M, Oh JH, Herring J and Zhou P 2007 2-step optimization of the extraction and subsequent physical properties of channel catfish (*Ictalurus punctatus*) skin gelatin*Journal of Food Science* **72** 188-195

[17] da Silva EVC, Lourenço LFH and Pena RS 2017 Optimization and characterization of gelatin from kumakuma (*Brachyplatystoma filamentosum*) skin *CyTA – Journal of Food* http://dx.doi.org/10.1080/19476337.2016.1266391

The feasibility of using explicit method for linear correction of the particle size variation using NIR Spectroscopy combined with PLS_2 regression method

M Yulia[1,3] and D Suhandy[2]

[1] Department of Agricultural Technology, Lampung State Polytechnic, Jl. Soekarno Hatta No. 10, Rajabasa Bandar Lampung, 35144, Indonesia
[2] Department of Agricultural Engineering, Faculty of Agriculture, The University of Lampung, Jl. Prof. Dr. Soemantri Brojonegoro No.1, Bandar Lampung, 35145, Indonesia

E-mail: meinilwitayulia@polinela.ac.id

Abstract. NIR spectra obtained from spectral data acquisition system contains both chemical information of samples as well as physical information of the samples, such as particle size and bulk density. Several methods have been established for developing calibration models that can compensate for sample physical information variations. One common approach is to include physical information variation in the calibration model both explicitly and implicitly. The objective of this study was to evaluate the feasibility of using explicit method to compensate the influence of different particle size of coffee powder in NIR calibration model performance. A number of 220 coffee powder samples with two different types of coffee (civet and non-civet) and two different particle sizes (212 and 500 μm) were prepared. Spectral data was acquired using NIR spectrometer equipped with an integrating sphere for diffuse reflectance measurement. A discrimination method based on PLS-DA was conducted and the influence of different particle size on the performance of PLS-DA was investigated. In explicit method, we add directly the particle size as predicted variable results in an X block containing only the NIR spectra and a Y block containing the particle size and type of coffee. The explicit inclusion of the particle size into the calibration model is expected to improve the accuracy of type of coffee determination. The result shows that using explicit method the quality of the developed calibration model for type of coffee determination is a little bit superior with coefficient of determination (R^2) = 0.99 and root mean square error of cross-validation (RMSECV) = 0.041. The performance of the PLS2 calibration model for type of coffee determination with particle size compensation was quite good and able to predict the type of coffee in two different particle sizes with relatively high R^2_{pred} values. The prediction also resulted in low bias and RMSEP values.

1. Introduction

Near infrared (NIR) spectroscopy is widely used as a rapid, low-cost and non-destructive method of analysis, which can discriminate between varieties and measure internal properties. For coffee, NIR spectroscopy has been applied for evaluation of geographic and genotypic origin of arabica coffee [1].

[3] To whom any correspondence should be addressed.

NIR spectroscopy also is often used to discriminate blends of arabica and robusta grains, obtaining satisfactory results [2], [3], [4] or to detect defects or adulteration in coffees [5], [6]. However, no studies reported NIR spectroscopy for discrimination of civet and non-civet coffee.

The discrimination between civet and non-civet coffee is important due to massive unfair trading of civet coffee. In practice, it is very difficult to discriminate between authentic civet coffee and adulterated civet coffee. The discrimination is more challenging in ground coffee samples. After roasting and grinding, authentic and adulterated civet coffee is visually impossible to differentiate. Therefore, an attempt to develop NIR spectroscopy based method for discrimination between civet and non-civet coffee is important.

The quality of NIR calibration is highly affected by spectral data itself. During spectral data acquisition, both chemical information of samples as well as physical information of the samples, such as particle size and bulk density are included [7], [8]. Before developing any NIR calibration model, it is suggested to remove physical information of the samples since it can be highly affected the ability of the developed model to predict the coming new samples. Physical properties that have been correlated to NIR diffuse reflectance spectra include sample temperature, particle size, particle size distribution and compact density [9], [10], [11].

There are several ways to remove those physical properties from spectral data. First, by using various spectral transformations to suppress the physical information in NIR spectra because it obscures the chemical information [7]. Several spectral data pre-processing have been used to remove the influence of physical properties and to improve the quality of the calibration model such as SNV-DT (standard normal variate detrending) [7], and multiplicative scatter correction (MSC), also known as multiplicative signal correction [12].

Several methods have been also established for developing calibration models that can compensate for sample temperature variations [13]. One common approach is to include temperature variation in the calibration model both explicitly and implicitly. In implicit approach, a global calibration model is developed in combination with calibration samples from a broad range of temperatures [10]. Using same approach, recently, Suhandy and Yulia [14] reported the successful application of NIR spectroscopy and implicit method to compensate the influence of particle size variation of ground coffee samples. In this present study, a feasibility of using NIR spectroscopy and explicit method using PLS2 regression to compensate the influence of particle size variation on NIR spectroscopy measurement of ground coffee samples is evaluated. In PLS2 model, responses are multivariate. In this study, the response variables Y are particle size of coffee powder (mesh 212 and 500 micrometer) and type of coffee (civet and non-civet coffee). Using PLS2 regression, a discrimination between civet and non-civet coffee was evaluated based on PLS-DA analysis.

2. Methodology/ Experimental

2.1. Samples

Civet and non-civet coffee samples were collected directly from the same coffee farmers at Liwa, Lampung, Indonesia. All samples were roasted and then grinded using home coffee grinder (Sayota). To investigate the influence of particle sizes on NIR spectra, coffee powder from ground roasted civet and non-civet coffee was separated into two different particle sizes (212 and 500 μm) by sieving through a nest of U. S. standard sieves on a Meinzer II sieve shaker (CSC Scientific Company, Inc. USA) for 10 minutes. The experiments were performed at room temperature (around 27-29°C). In this study, a number of 220 coffee powder samples with two different types of coffee (civet and non-civet) and two different particle sizes (212 and 500 μm) were prepared for samples.

2.2. NIR Spectral data acquisition

All the NIR spectra were collected using a V-670 UV-Visible-NIR Spectrophotometer (JASCO Co., Japan). The spectrometer was equipped with an integrating sphere beside detector (ISN-723, JASCO Co.) which is connected to the spectrometer by an optical fibre. The NIR instrument was controlled by a compatible Windows XP system, and Spectra Manager (V-670, JASCO Co.) was used to acquire spectra data. A 4 cm by 4 cm square sample holder was used; with care taken to ensure every sample (weight of approx. 0.2 g) was placed into the sample holder and the sample surface kept flat. Each spectrum was measured from 1300 to 2500 nm, with a bandwidth of 20.0 nm, and a scan speed of 1000 nm/min. All experiments were conducted at room temperature (around 20°C).

2.3. Multivariate data analysis
Multivariate data analysis based on projection methods was applied for statistical data analysis. Specifically, a projection to latent structures (PLS-DA)-based method was applied to discriminate the samples. PLS2-based approach is a robust regression technique used to investigate the relationships existing between two blocks of data, usually called X- and Y-block and has predictive applications. Here, X block contain NIR spectral data and Y block contain two variables (PLS2), particle sizes (212 and 500 μm) and type of coffee (civet = 1 and non-civet = 2). In the present study, discriminant classification was carried out using partial least squares-discriminant analysis (PLS-DA), a method based on the PLS regression algorithm. PLS-DA was performed by using a statistical software package of 'The Unscrambler' V9.7 (CAMO AS, Trondheim, Norway).

3. Results and Discussion

3.1. NIR absorbance spectra of coffee samples with different particle size
Figure 1 shows the NIR spectra of civet coffee samples with two different particle size (212 and 500 μm). Based on Figure 1, it can be seen that the spectral data of coffee samples is highly influenced by particle size difference. The absorbance of coffee samples with larger particle size is higher than that of smaller one. It can be explained that the coarser the particles, the greater the penetration of light and the greater the absorbance [15].

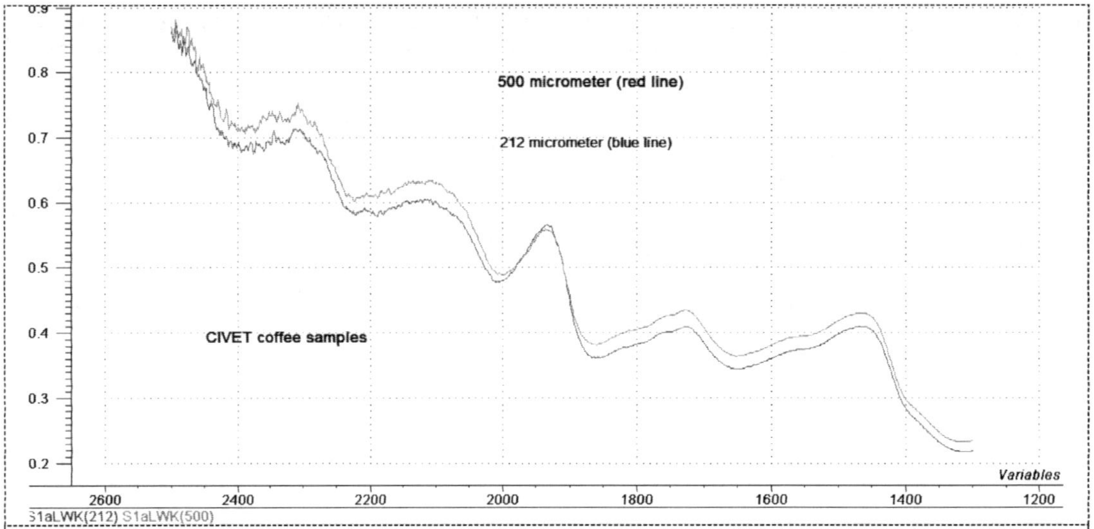

Figure 1. NIR spectra data of civet coffee samples with different particle size (212 and 500 μm).

3.2. Developing and predicting PLS-DA model using PLS1 (without compensation)
Figure 2 demonstrated the calibration model for discrimination between civet and non-civet coffee using PLS-DA with PLS1 regression (without particle size compensation). In this model, the samples have same size particle (212 μm). From Figure 2 we can see that the correlation between measured and predicted y (type of coffee) is very high with coefficient of determination (R^2) = 0.99. The cross-

validation of the model is also resulted in high coefficient of determination (R^2) = 0.99. The error of calibration model is 0.042 and error in validation is 0.044.

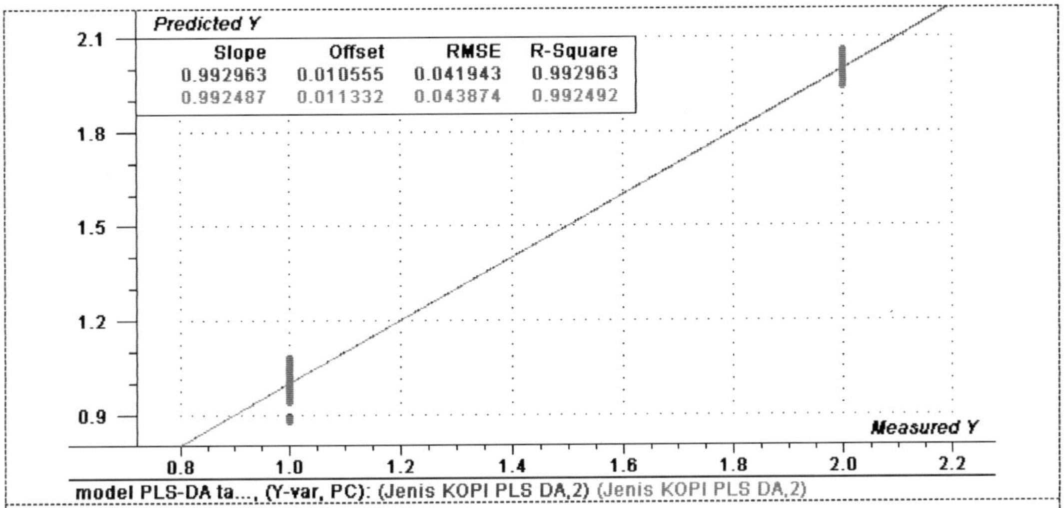

Figure 2. Calibration model of PLS-DA for civet and non-civet discrimination using PLS1 regression (without compensation).

In order to evaluate the performance of the developed calibration model, a prediction was performed using samples having two different particle sizes (212 and 500 μm). The result was presented in Figure 3. The prediction of samples having same particle size (212 μm) resulted in good prediction performance with all samples are located close to the target line (left side of Figure 3). However, when the calibration model was applied to predict the samples having different particle size (500 μm), the result is poor with almost all samples are located far away from the target line. It can be inferred that the performance of NIR calibration model in discriminating between civet and non-civet coffee is highly affected by the particle size of ground coffee samples.

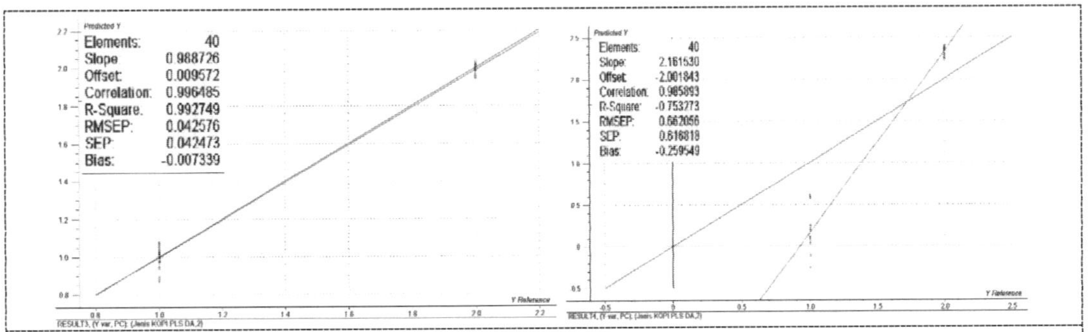

Figure 3. The result of prediction using calibration model without compensation for discrimination between civet and non-civet coffee using PLS-DA for particle size 212 μm (left) and 500 μm (right).

3.3. Developing and predicting PLS-DA model using PLS2 (with compensation)

To compensate the influence of particle size in NIR calibration model, in this study we propose to use PLS2 regression. Here, explicitly, the variable of particle size was added into y-variables (response variable). For this, we now have two y-variables: type of coffee (civet and non-civet coffee) and particle size (212 and 500 μm). The calibration model of PLS-DA for discrimination between civet and non-civet coffee using PLS2 regression was presented in Figure 4. Quality of the model is high with R^2 = 99 both for calibration and validation. The error is also low with RMSE = 0.0386 for calibration and RMSE = 0.0413 for validation, respectively.

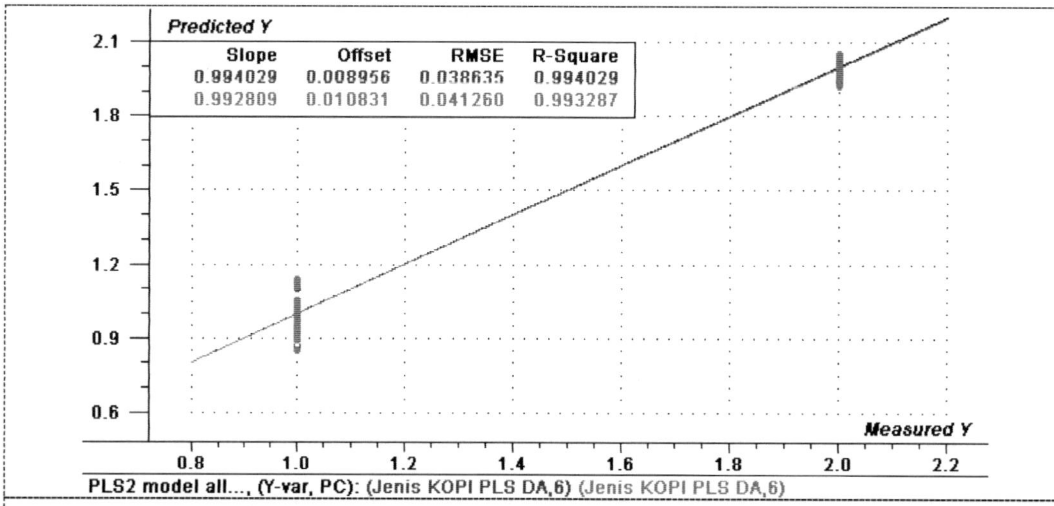

Figure 4. Calibration model of PLS-DA for civet and non-civet discrimination using PLS2 regression (with compensation).

The performance of calibration model with PLS2 for discrimination civet and non-civet coffee was evaluated with particle size 212 and 500 μm in the prediction sample set. The result was depicted in Figure 5. It can be seen that the influence of particle size was completely removed with good prediction result for all samples (both 212 and 500 μm). All samples were close to the target line. It can be said that PLS2 regression with explicit inclusion of particle size information on modelling the calibration model was acceptable for compensation the influence of particle size on NIR calibration model.

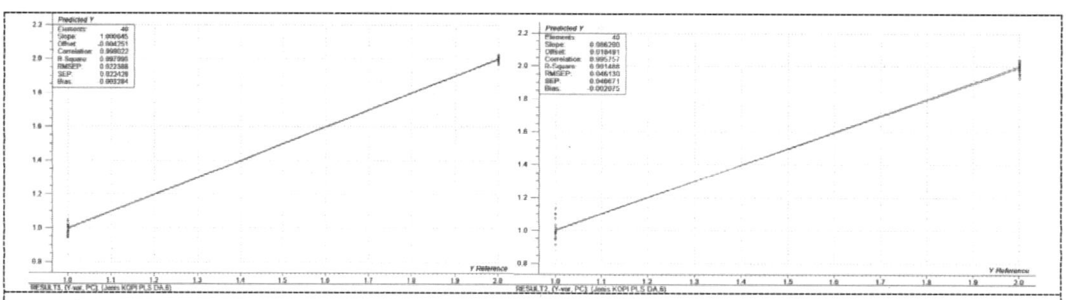

Figure 5. The result of prediction using calibration model with compensation for discrimination between civet and non-civet coffee using PLS-DA for particle size 212 μm (left) and 500 μm (right).

4. Conclusion

In the present research, an investigation on the use of PLS2 regression and PLS-DA for discrimination between civet and non-civet coffee with two different particle size of samples (212 and 500 μm) was conducted. Both calibration models (with and without particle size compensation) for discriminating the type of coffee were very good with coefficient of determination 0.99. However, the calibration model without compensation is highly influenced by the different of particle size. The prediction was poor when the particle size of samples for developing calibration model was different with that for prediction. The use of calibration model with particle size compensation using PLS2 regression was promising. All samples with particle size 212 and 500 μm in prediction sample set could be predicted well.

Acknowledgement
The authors would like to thank Professor Naoshi Kondo and Associate Professor Yuichi Ogawa of Laboratory of Bio-Sensing Engineering of Kyoto University for giving the author (Diding Suhandy) permission to use UV-Visible-NIR spectrometer (JASSO Follow up Research Program). Some part of this research is also funded by Kemenristekdikti Indonesia (Hibah Strategis Nasional/STRANAS 2016–2017 No. 585/UN26.21/KU/2017).

References

[1] Marquetti I, Link J V, Lemes A L G, Scholz M B dos S, Valderrama P and Bona E 2016, *Comput. Electron. Agr.* **121** 313–319.
[2] Esteban-Díez I, Gonzáliz-Sáiz J M, Sáenz-González C and Pizarro C 2007 *Talanta* **71** 221– 9.
[3] Santos J R, Sarraguça M C, Rangel A O S S and Lopes J A 2012 *Food Chem.* **135** 1828–35.
[4] Bertone E, Venturello A, Giraudo A, Pellegrino G and Geobaldo F 2016 *Food Control* **56** 683–9.
[5] Craig A P, Franca A S, Oliveira L S, Irudayaraj J and Ileleji K 2015 *Talanta* **134** 379–86.
[6] Winkler-Moser J K, Singh M, Rennick K A, Bakota E L, Jham G, Liu S X and Vaughn S F 2015 *J. Agric. Food Chem.* **63** 10662–8.
[7] Barnes R J, Dhanoa M S and Lister S J 1989 *Appl. Spectrosc.* **43** 772–7.
[8] Pasikatan M C, Steele J L, Spillman C K and Haque E 2001 *J Near Infrared Spec.* **9** 153–164.
[9] Gupta A, Peck, G E, Miller R W and Morris K R 2005 *J. Pharm. Sci.* **94** 1589–97.
[10] Yulia M, Suhandy D, Ogawa Y and Kondo N 2014 *Eng. Agric. Environ. Food* **7** 148–54.
[11] Barajas M J, Cassiani A R, Vargas W, Conde C, Ropero J, Figueroa J and Romanach R J 2007 *Appl. Spectrosc.* **61** 490–6.
[12] Gómez A H, He Y and Pereira A G 2006 *J. Food Eng.* **77** 313–19.
[13] Swierenga H, Wulfert F, de Nord O E, de Weijer A P, Smilde A K and Buydens L M C 2000 *Anal. Chim. Acta* **411** 121–35.
[14] Suhandy D and Yulia M 2016 *Proc. of the 1st International Conference "The Role of Agricultural Engineering for Sustainable Agriculture Production* (Bogor: Institut Pertanian Bogor) pp 6–10.
[15] Shan J, Suzuki T, Suhandy D, Ogawa Y and Kondo N 2014 *Eng. Agric. Environ. Food* **7** 139–42.

Performance evaluation of extractor cutting blade configuration in Inulin extraction process from *Dahlia sp. L* tuber

E Sundari[1,2], E Praputri[1], W Marthiana[1] and M Jaya[1]
[1] Chemical Engineering Department, Bung Hatta University, Jl. Gajah Mada No. 19 Olo Nanggalo Padang, Sumatera Barat, Indonesia

E-mail: elmisundari@yahoo.com

Abstract. Inulin, a polysaccharide plant-based nutrient, can be isolated from dahlia flower tubers by liquid-solid extraction processes and is generally carried out in an extractor tank equipped with an agitator. To accelerate the diffusion rate of solute from the solid phase (bulk phase) to the external surface (boundary layer) in order to increase yield of inulin, the size reduction of material is required. The purpose of this research was to design the cutting blade needed for dahlia tuber size reduction and investigate the effect of blade types, agitator speed (350, 700, 1050, and 1400 rpm), and configuration of cutting blade to material fineness at 90 minutes of contacting time. The results showed that higher cutting blade speed results in higher cut material fineness rate. The best conditions was achieved by the configuration of two four-blade turbine combined with one three-blade turbine with fineness rate more than 90% in 30 minutes of contacting time at every variation of agitator speed. The cutting blade designed in this study can be used for size reduction purpose of tubers other than dahlia tubers.

1. Introduction

The dahlia flower (*Dahlia coccinea Cav.*) is an ornamental plants that have not been widely used in Indonesia. The tubers of dahlia contain about 12.5% of inulin as a storage polysaccharide [1]. Inulin, a plant-based nutrient, is polysaccharide chains that serves as an effective prebiotic additive in beverages, yogurts, biscuits, spreads, and also dietary supplements. Inulin can be isolated from dahlia flower tubers by solid-liquid extraction processes and generally carried out in an extractor tank equipped with an agitator.

The mechanisms of solid-liquid extraction involves three major steps; diffusion of solvent into solid phase, diffusion of solute from the solid phase (bulk phase) to the external surface (boundary layer) and mass transfer of solute from the solid phase to the liquid phase (solvent) [2]. The diffusion rate of solute from the solid phase (bulk phase) to the surface is dependent on molecular size as larger molecules diffuse slower than smaller molecules. The sizes of the particles involved in the diffusion are important because they closely relate to the concepts of heat and energy in the context of diffusion. If there is not sufficient energy to move a larger molecule, it resists the effects of diffusion and is unable to move from one area to the next [3].

[2] To whom any correspondence should be addressed.

IOP Conf. Series: Materials Science and Engineering **334** (2018) 012063 doi:10.1088/1757-899X/334/1/012063

To achieve the fastest and most complete solid-liquid extraction possible, the solvent must be provided with large exchange surfaces. This can be done by reducing the size of solids to be extracted. On a large scale, this can be done by designing an agitator that functions as cutting blade. The blade type was designed to creates enough shear to reduce the size of materials through dispersion process. The process of dispersion is a two possible mechanism. First, the materials will hit the cutting blade and are broken apart. And second, in the turbulence surrounding the cutting blade, the particles will hit one another and hit the cutting blade at high speeds, further reducing the material size [4].

The purpose of this research was to design the cutting blade needed for dahlia tuber size reduction and investigate the effect of blade types, agitator speed (350, 700, 1050, and 1400 rpm), and configuration of cutting blade to material fineness at 90 minutes of contacting time.

2. Materials and Methods

The research was conducted in two steps, designing and evaluating the performance of cutting blade. The material used as cutting blade was 2 mm stainless steel in types of four-blade turbine, three-blade propeller and twelve-blade turbine. The dahlia flower tubers sample was taken from Biaro, West Sumatera. The performance of cutting blade was evaluated against contacting time (15, 30, 45, 60, 75, and 90 minutes), agitator speed (350, 700, 1050, and 1400 rpm), agitator configuration and cut material fineness. Other variables, including samples and solvent ratio, were conducted according to previous study in laboratory scale [5]. The specification of cutting blade designed is described in Table 1.

Table 1. Cutting blade specifications.

Cutting Blade Types	Number of Blades	Diameter (cm)	Weight (gr)
 Turbine	4	19.5	222
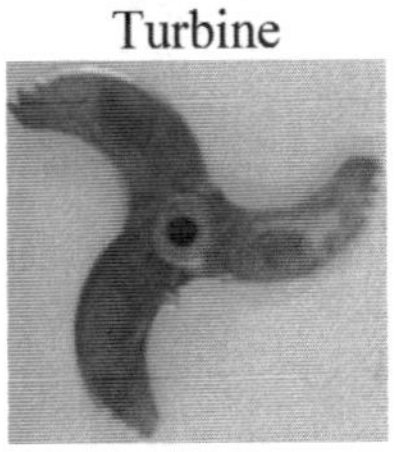 Propeller	3	18	195
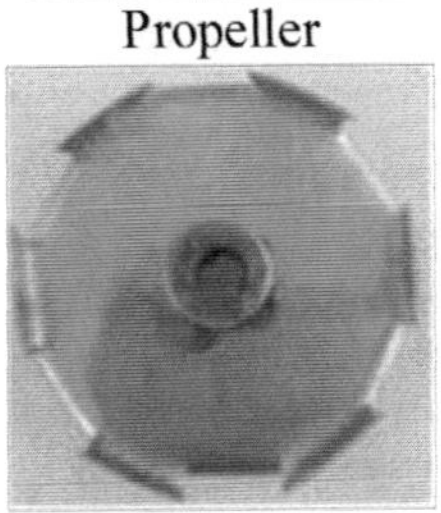 Turbine	12	10.5	250

IOP Conf. Series: Materials Science and Engineering **334** (2018) 012063 doi:10.1088/1757-899X/334/1/012063

3. Results and Discussions

3.1. Effect of agitator speed, types, and configuration of cutting blade on cut materials fineness

This research was conducted to investigate the performance of cutting blade based on mass flow produced by the turbine and propeller types. The results showed that the types, configuration of the blades and agitator speed affected the cut materials fineness. The cut materials fineness is increasing as the contacting time increase as shown in Figure 1. Fineness rate higher than 90% was achieved at 60 minutes of contacting time with an agitator speed of 350 rpm, and a rate of 100% was achieved at 90 minutes. At speeds of 700, 1050 and 1400 rpm, fineness rate higher than 90% were achieved at 45, 30 and 30 minutes of contacting time, respectively. Fineness rate of 100% was achieved at 90 minutes of contacting time. The four-blade turbine has an excellent feature in creating the radial flow and axial flow in the reactor vessel. This will result in the mass flow in every direction that will lead to higher cut material fineness rate [6,7]. The cut material fineness rate is shown in Figure 2.

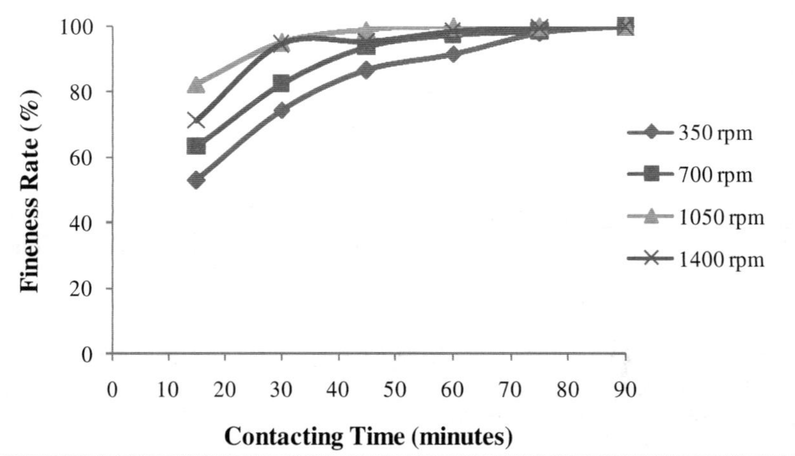

Figure 1. Effect of four-blade turbine configuration on fineness rate at agitator speed variation.

Furthermore, agitator speed showing the positive tendency on the fineness rate. Higher speed results in higher cut material fineness rate. This result was due to the turbulence established by the cutting blade speed that creates enough shear force to reduce the size of materials. However, the fineness rate showing a negative tendency for cutting blade speed of 1400 rpm. At vigorous cutting blade speed a vortex regime appeared, thus reducing the blade performance.

Three-blade propeller gave 94-98% of the fineness rate at 45-90 minutes of contacting time for both agitator speed of 350 and 700 rpm. At agitator speeds of 1050 and 1400 rpm, the cut material fineness rate of 96.5-100% were achieved at 30-90 minutes of contacting time as described in Figure 3. Moreover, the agitator speed also showing the positive tendency on the fineness rate. Higher speed results in higher cut material fineness rate due to turbulence flow established during the process.

Meanwhile, twelve-blade turbine types were only able to achieve maximum of 90% cut material fineness rate at 90 minutes of contacting time as shown in Figure 5. The compact design of blade limits the performance of blade in creating the mass flow, which is only axial flow, that leads to limited cutting performance [8,9].

Likewise, the contacting time also showing positive correlation with the cut materials fineness rate for all types of cutting blade and agitator speed variation. The lengthened time will encouraged the contact between the cutting blade and materials and promoted better size reduction performance [10].

Additionally, variations of four-blade turbine and three-blade propeller types were conducted to evaluate the performance of configuration combination on cut material fineness rate. The best configuration was shown by a configuration of two four-blade turbine and one three-blade propeller that gave fineness rate higher than 90% at 30 minutes of contacting time as shown in Figure 7.

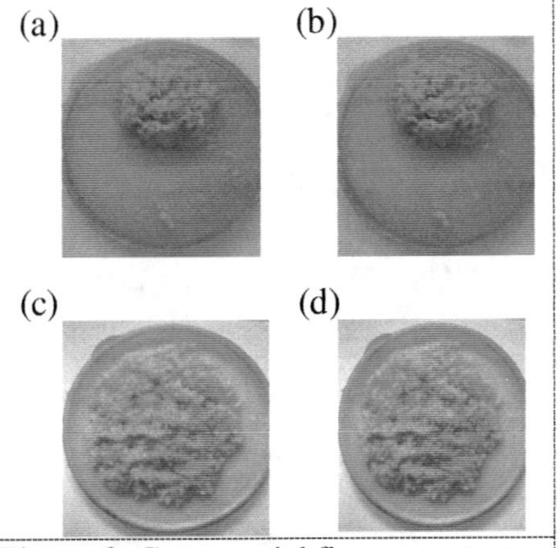

Figure 2. Cut material fineness rate at 90 minutes, four-blade turbine type. (a) 350 rpm (b) 700 rpm (c) 1050 rpm (d) 1400 rpm.

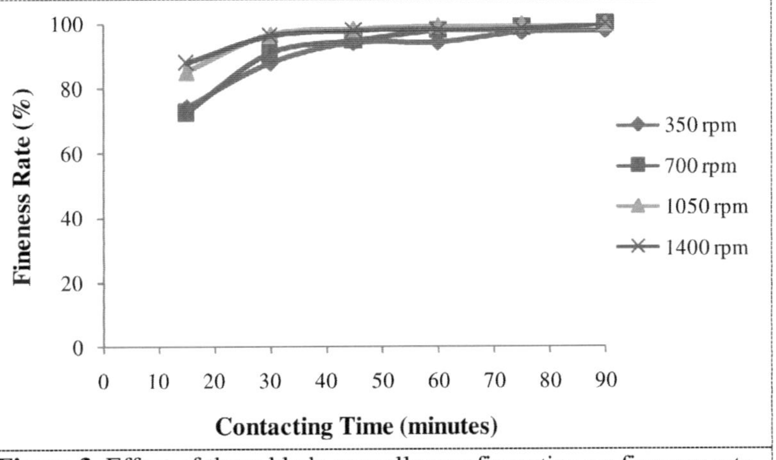

Figure 3. Effect of three-blade propeller configuration on fineness rate at agitator speed variation.

IOP Conf. Series: Materials Science and Engineering **334** (2018) 012063 doi:10.1088/1757-899X/334/1/012063

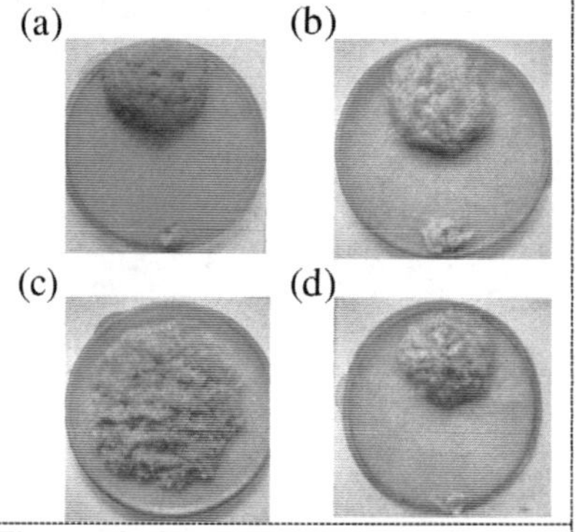

Figure 4. Cut material fineness rate at 90 minutes, three-blade propeller type. (a) 350 rpm (b) 700 rpm (c) 1050 rpm (d) 1400 rpm.

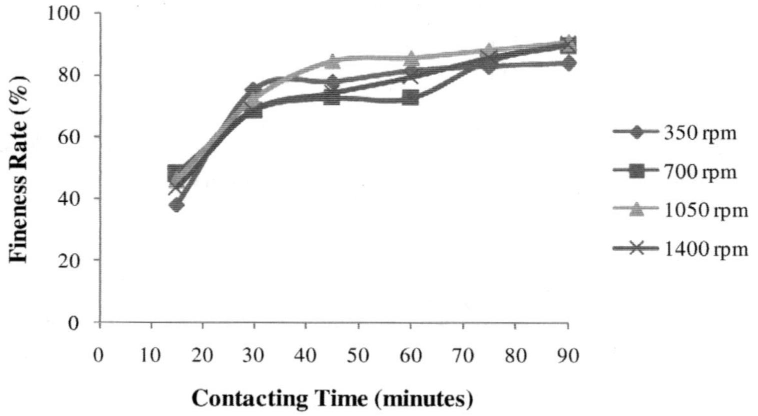

Figure 5. Effect of twelve-blade turbine configuration on fineness rate at agitator speed variation.

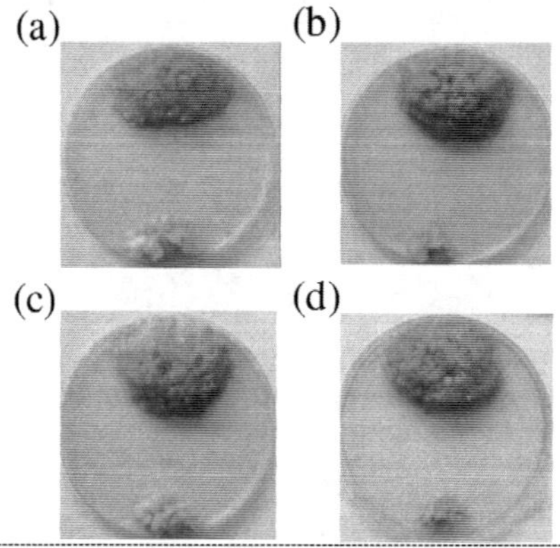

Figure 6. Cut material fineness rate at 90 minutes, twelve-blade turbine type. (a) 350 rpm (b) 700 rpm (c) 1050 rpm (d) 1400 rpm.

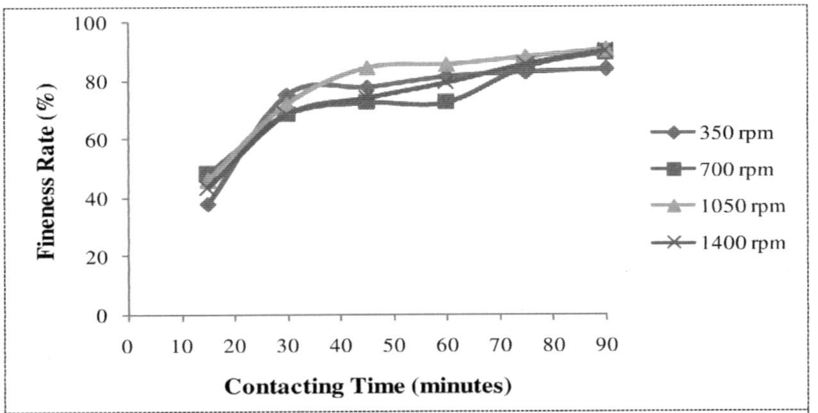

Figure 7. Effect of two four-blade turbine and one three-blade propeller configuration on fineness rate at agitator speed variation.

3.2. Effect of cutting-blade types on electrical power consumption.

Electrical power consumption is equal to the power required for material size reduction process [11,12]. The power needed to rotate cutting blade with drive motor efficiency of 90% and design margin of 1.15 can be calculated using Equation (1). The results are shown in Table 2.

$$P = N_p \rho N^3 D^5 \tag{1}$$

From Table 2, the results showed that, with constant mass of the cut material, every increase of the cutting blade shaft speed by two times led to statistically significant increase in electrical power consumption. The highest energy consumption was demonstrated by four-blade turbine with the cutting blade shaft speed of 1400 rpm. The configuration was found to be inefficient from the energy consumption perspective. The average power consumption per unit volume for industrial reactors ranges from 10 kW m^{-3} for small vessels (approximately 0.1 m^3) to 1-2 kW m-3 for large vessels (approximately 100 m3) [13].

Table 2. Electrical power consumption for variation of agitator speed and cutting blade types.

Cutting Blade Types	Density (kg/m^3)	Agitator Speed (rpm)	Blade Diameter (cm)	Power Number (N_P)	P (kW)
Twelve-blade turbine	0.00104	350	10.5	1.94433E-08	0.1
		700		2.43042E-09	1.8
		1050		7.20124E-10	9.1
		1400		3.03802E-10	28.7
Four-blade turbine		350	19.5	3.03553E-09	2.5
		700		3.79442E-10	39.6
		1050		1.12427E-10	200.3

	1400		4.74302E-11	633.1
Three-blade propeller	350	18	3.85941E-09	1.7
	700		4.82427E-10	26.5
	1050		1.42941E-10	134.3
	1400		6.03033E-11	424.3

4. Conclusion

This research showed that cutting blade configuration, agitator speed and contacting time affected the cut material fineness rate. Agitator speed variations showing positive tendency on the materials fineness rate. Higher agitator speed resulted in higher cut materials fineness rate. The lengthened contacting time also showing positive correlation with the fineness rate. The best operation condition was achieved by the configuration of two four-blade turbine combined with one three-blade turbine with fineness rate more than 90% in 30 minutes of contacting time at every variation of agitator speed (350, 700, 1050 and 1400 rpm). However, the electrical power consumption was found to be high for agitator speed of 1400 rpm and this speed is not favourable. The cutting blade designed in this study can be used for size reduction purpose of tubers other than dahlia tubers.

References

[1] Cruz V A, Belote J G, Belline M Z, Cruz R 1998 *Rev. Microbiol.* **29**, 301-306.
[2] Ahangari H T, Noeparvar P and Moghaddas J Sadegh 2016 *An Indian J* **11** **Issue 4** 123-132 ISSN: 0974-7443.
[3] Mardina P, Gunawan A dan Nugraha M I 2012 *Konversi* **1** 1.
[4] McDowell R I 2006 (University of Louisville: Electronic Theses and Dissertations).
[5] Sundari E, Desfitri E R, Martynis M and Praputri E\ 2014 *Prosiding SNSTL* **I** ISSN 2356-4938, OP-022 174-179.
[6] Geankoplis C J 1993 (New Jersey: Prentice Hall).
[7] Driss Z, Karray S, Chturou W, Kchaou H dan Abid M Salah 2012 *The Arch. of Mec. Eng.* **LIX** No.1.
[8] Jirout T and Rieger F 2011 *Chem. Eng. Res. Des.* **89** 7 1144-1151.
[9] Kurniawan R, Juhanda S, Syamsudin R dan Lukman M 2011 (Bandung: Itenas).
[10] Widjanarko S B, Suwasito T S 2014 Jurnal Pangan dan Agroindustri **2** 1 79-85.
[11] Issa H M 2016 *J Eng.* **2016**
[12] John A 1998 (UK: University of Birmingham).
[13] Doran P M 2013 (UK: Elsevier Ltd.)

IOP Conf. Series: Materials Science and Engineering **334** (2018) 012064 doi:10.1088/1757-899X/334/1/012064

Tumeric oil as the antioxidation agent in edible coating film

N A Ahmad[1], Z I M Sharif[1], J Jai[1,2], N M Yusof[1] and F A Mustapha[1]
[1] Faculty of Chemical Engineering, Universiti Teknologi MARA (UiTM), 40450 Shah Alam, Selangor, Malaysia.

E-mail: junejai@salam.uitm.edu.my

Abstract. Turmeric oil (TO) has been studied for its potential as an antioxidation agent in starch edible coating for fresh cut apples and its degree of oxidation was analysed. TO incorporate with starch edible coating was examined using FT-IR Spectroscopy to determine the presence of secondary metabolites. The presence of alcohol and aromatic ring in the edible coating film proved that the secondary metabolites from TO were existed. The fresh cut apples were underwent the sensory test and six out of ten panellist concluded that coated fresh cut apples have good appearance and surface colour. Fresh cut apples were coated with edible coating incorporated with different concentrations of TO (uncoated, 0μL, 5μL, 10μL, 15μL. Percentage weight loss for 15μL were the least which were 1.98% (day 6) and 3.95% (day 12). Colour measurement were done for few days and it shows that the total colour difference (ΔE) for 15μL were the lowest. Thus, the oxidation activities for 15μL is the slowest compared to the others. These can be proved through the degree of oxidation analysis using UV-Vis spectroscopy. Uncoated fresh cut apples have the highest degree of oxidation while those with 15μL have the lowest. This study can be illustrated that the oxidation activities of fresh cut apples could be postponed using edible film incorporated with TO.

1. Introduction

Fresh cut fruits are basically ones that have been trimmed or reshaped from its original state to become another usable products but still remain in its fresh state. However, it is hard to maintain its fresh state mainly due to the removal of their pure vigilant skin and cutting process [1]. The tissue of the fruits will become wounded and it will lead to the enzymatic browning process on the product. Nowadays, consumers are aware about the importance of healthy eating habits. Therefore, the basis appearance and the freshness of the fresh cut fruits during the time of the purchase will be judged by the consumer. Enzymatic browning process occurred is due to presence of polyphenol oxide (PPO), which is a copper-containing oxidoreductase basically driven the browning reactions to occur where it catalyses two different reactions involving phenolic compounds and oxygen. Through this reaction, O-quinones will be developed from the oxidation of O-diphenols and will lead to the build-up of melanin and the development of brown compounds [2]. Accordingly, there are ways to improve their quality and postponing the enzymatic browning of the fresh cut fruits by coated them with edible coatings incorporated with antioxidation agents [3]. Antioxidation agent is needed to prevent the enzymatic browning towards the fresh cut fruits by prohibiting the initiation of browning by reacting with the oxygen such as turmeric oil. Turmeric oil has been showing antibacterial and antioxidant properties.

[2] To whom any correspondence should be addressed.

IOP Conf. Series: Materials Science and Engineering **334** (2018) 012064 doi:10.1088/1757-899X/334/1/012064

Most of the turmeric oil components are bisabolane type sesquiterpenoids [3]. The application of edible coatings from the turmeric oil and the use of turmeric oil as antioxidant, anti-browning and texture enhancers to improve the quality and functionality of fresh cut fruits will be further reviewed.

2. Materials and Methodology

2.1. Coating materials
Materials used in this experiment were cassava starch (Cap Kapal ABC.Co), glycerol (Merck, United State) as plasticizers and turmeric oil (Soul, Malaysia) as antioxidation agent. Apple was purchased from the local market located in Shah Alam prior to start this experiment. Undamaged apple in uniform size and similar maturity stage was selected.

2.2. Coating Preparation
The coating formulation were made by dissolving starch (2 % and 3 % w/v) in 100 ml of distilled water with constant agitation at temperature of 70 °C until it reached homogeneity. Ratio of starch: glycerol ratio were at 1:0.25, and glycerol was added after starch's gelatinization. Afterwards, turmeric oil (0.05 % to 0.30 % v/v) was added to the formulation and stirred for 5 minutes at 1600 rpm [4].

2.3. Fruits selection and preparation
The apples that were bought from the local market at Shah Alam were clear up by cleaning it and sorted to remove damaged apples and foreign materials. The apples were sliced in equal slices of 1.5 cm in length to minimize browning [5]. Apples were divided into two parts, apples coated with emulsion and apples without coating for blank. The fresh cut apples were dipped in the emulsion for 30 s and then dried at room temperature before stored in the chiller (4 °C) [6].

2.4. Weight loss determination
The weight losses of non-coated and coated apples during storage were compared. Weight loss in this matter happened because of the water loss during storage. Weight loss were disclosed as percentage decrease in fruit weight, using the following formula

$$Weight\ Loss = \frac{W_i - W_t}{W_i}\ x\ 100\% \quad (1)$$

where, W_i is initial weight (g) and W_t is fruit weight after a designated period of storage (g) [7].

2.5. Colorimetric determination
Colour was measured using a colorimeter HUNTERLAB-D25-9000 to obtained CIE values (L*,a*,b*). The subscript 0 in the equation stand for the colour parameters of fresh apple slices. The greater the total of ΔE, it represents greater colour transition from the fresh apple [8]

$$\Delta E^* = [\,\Delta L^{*2} + \Delta a^{*2} + \Delta b^{*2}]^{1/2} \quad (2)$$

2.6. Sensory Evaluation
For the sensory evaluation, coated and uncoated apples were analysed. Ten sensory panellists were picked and for each sensory test panel, in coded dishes, samples containing coated and uncoated apples were given. The panellists were given a hedonic questionnaire to test appearance and surface colour of coded samples of apples. Below table 1are the sample for the hedonic questionnaire and the scoring scale [6].

Table 1. Scorecard of Hedonic Rating Scale [9].

Scorecard- Hedonic Rating Scale	
Dish Number:	**Name:**
In front of you are the samples of coated and uncoated apple slices inside a numbered dish. Please rate the apple inside the dish according to the scale given below:	
Food Characteristics	
Appearance	**Surface colour**

Table 2. Scale and description for the Scorecard of Hedonic Rating Scale [6].

Scale	Description
5	Excellent
4	Very Good
3	Good
2	Fair
1	Poor

2.7. Methods to Monitor Enzymatic Browning Oxidation (Assay for PPO Activity)
Enzymatic activity was assayed by checking the rate of increment in absorbance at a range of 420 nm and at 25 °C in a double beam model UV-1601 UV/VIS spectrophotometer. The reference cuvette has only the substrate solution. Meanwhile, the straight-line section of the activity curve as a function of time was used to figure out the enzyme activity (Units/g of fruit/min). A unit of enzyme activity was defined as the change of 0.001 in the absorbance value under the conditions of the assay [10]. When a lag phase happened, the reaction rate was measured after the lag phase [11].

3. Results and Discussion

3.1. FTIR Analysis
Turmeric consists more than 500 bioactive components which major of them gives benefits to medical value. Turmeric rhizomes can be extracted as essential oils and has being discovered that turmeric oil has the potential to be used as antimicrobial and antioxidant agents due to the presence of curcumin [12].

As it can be seen in Figure 1, the functional groups that can be found in curcumin are alkane, alkene, alcohol, aromatic ring, and ketone. Thus, an analysis of coating mixture contains 100 mL distilled water, 5 g of starch, 1.25 mL of glycerol and 15 μL of turmeric oil was done. The result can be seen on Figure 2 and Table 3 below. From the result obtained all functional groups that contain inside the curcumin except for ketone were presence. However, even the presence of the aromatic ring is enough for the turmeric oil to show its antioxidant properties.

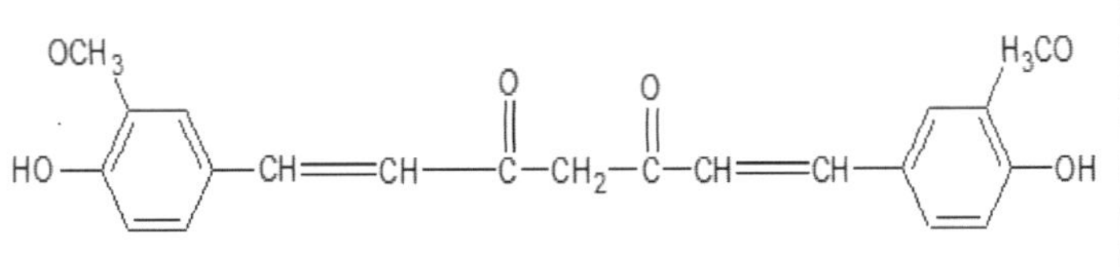

Curcumin-I

Figure 1. Antioxidant compound in curcumin.

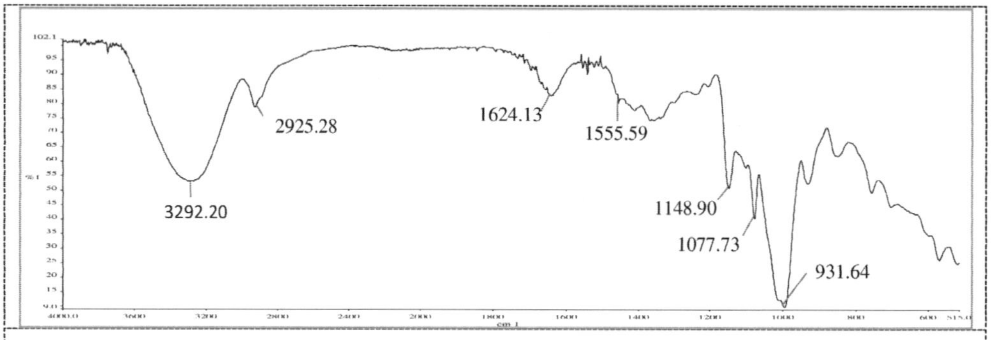

Figure 2. FTIR results for coating solution contained 15μL of turmeric oil.

Table 3. Functional groups in edible coating film incorporated with turmeric oil.

Functional Group	Type of Vibration	Characteristics Absorptions (cm^{-1})
Alcohol		
O-H	Stretch, H-bonded	3292.20
C-O	Stretch	1148.90
Alkane		
C-H	Stretch	2925.28
Alkene		
C=C	Stretch	1642.13
=C-H	Bending	931.64
Aromatic		
C=C	Stretch	1555.59
Ether		
C-O	Stretch	1077.73

3.2. Sensory Evaluation

Sensory evaluation has been done for the coated and uncoated apple cubes that were stored in a chiller for 12 days at 4°C. For the coated apple cubes, apple cubes that were coated with coating solution that contain 15μL turmeric oil were presented as a sample. A total of ten sensory panellists were gathered during the 4th of May 2016 to evaluate the sample of apple cubes that were being presented in a coded dish labelled as A and B. The panellists were given a hedonic questionnaire to test the appearance and surface colour of coded samples of apple cubes.

Based on Figure 3 and Figure 4 above, there are variety of ratings given from the sensory panellist. However, the ratings for the coated apple cubes were quite high both for the appearance and the surface colour of the apple cubes. Six panellist rate the appearance to be in very good appearance and one panellist rate the appearance of the apple cubes as excellent. It can be concluded that even from the sensory evaluation, the appearance of the apple cubes that were coated with the turmeric oil seems to be having a great appearance towards the panellist. This proved that the turmeric oil can preserve the apple cubes from the enzymatic browning by using its antioxidant characteristics.

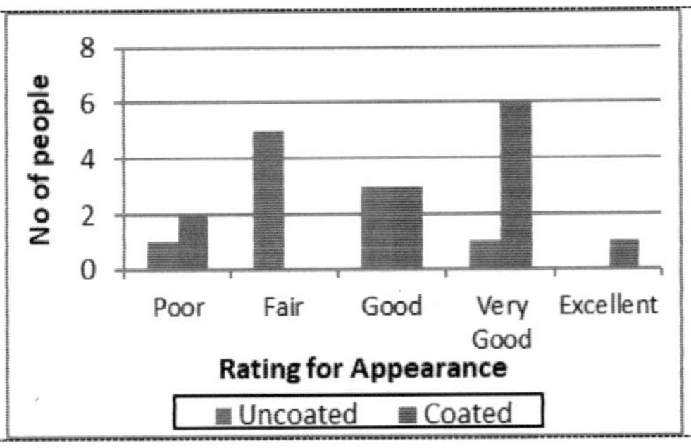

Figure 3. Rating for appearance of uncoated and coated apple cubes.

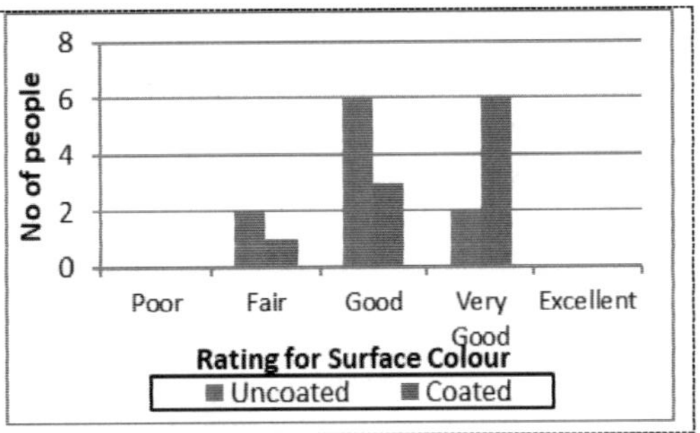

Figure 4. Rating for surface colour of uncoated and coated apple cubes

3.3. Weight loss

The percentage of weight loss for apple cubes with different types of coating at their sixth day and twelfth day of storage were shown in Figure 5. All apple cubes with different types of coating were being stored inside a chiller at 4 °C. Analysis of divergence showed that apple cubes being coated with 15 μL turmeric oil presented notably the lowest weight loss at their sixth and twelfth day with 1.98 % (sixth day) and 3.95 % (twelfth day) compared to other apple cubes, whereas, apple cubes that were uncoated had no significant effect on the weight loss since the percentage of their weight loss on their twelfth day (11.85 %) is quite high. Moreover, the effect of using the turmeric oil can be seen through the weight loss where apple cubes that were being coated with turmeric oil, even the smallest concentration, showed lesser weight loss compared to the apple cubes that were being uncoated and coated without turmeric oil. At the same time, it can be seen that the higher the concentration of turmeric oil being used in a coating solution, the lower the weight loss occurred to the apple cubes. This shows that coatings that contained even the slightest amount of turmeric oil played an important role in avoiding weight loss of apple cubes, meanwhile coating that does not contain turmeric oil (only contain starch and glycerol) showed poor moisture barrier.

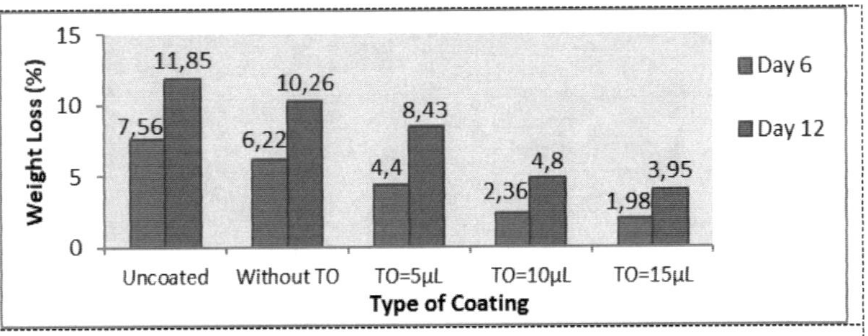

Figure 5: Weight loss of different type of coating

3.4. Colour Measurement

The apple cubes were stored in a chiller with a mean temperature of 4°C for a few days. Since the apples were dipped into difference concentrations of turmeric oil, the apple cubes were kept in different containers. As the apple cubes were being observed for a few days, the apple cubes underwent the color changes as there were increased in a* and b* values and decrease in L* value. According to Perez-Gago *et. al.* (2006), an increase in colorimetric a* and b* values and decrease in L* value (represent lightness), indicate that the enzymatic browning in the apple cubes were also increased. The lower the value of L*, the higher the loss of the lightness of the apple cubes [13].

Table 4 shows the result obtained for the value of L*, a* and b* for apple cubes with different turmeric oil concentrations on their sixth day and twelfth day of storage. Meanwhile, Figure 6, 7 and 8 shows the result being plotted into the graph. From Figure 6, it can be seen that the value of a* seems to be increasing from the control value for all apple cubes in their sixth and twelfth day. However, after being observed, the value appeared to start slowing down on their increment with the apple cubes being coated with a solution that contain 5 μL of turmeric oil. The value of a* further slowing down with higher concentration of turmeric oil. The same trend can also be implied with b* values as shown in Figure 7. An increase in a* and b* values indicates an increase in overall pigmentation of the apple cubes. However, the pigmentation starts to slow down with the usage of turmeric oil.

Table 4: Value of L*a*b* for different turmeric oil concentration on their 6th and 12th day.

	Control	Uncoated		Without TO		TO = 5μL		TO = 10μL		TO = 15μL	
Day		6	12	6	12	6	12	6	12	6	12
L*	77.38	66.01	54.70	69.24	54.78	72.83	67.91	74.13	70.92	75.53	72.01
a*	-1.80	2.17	4.72	3.64	4.95	2.28	3.17	2.07	2.63	1.05	1.33
b*	19.15	26.11	33.94	36.45	34.25	31.91	34.92	29.03	32.93	30.28	32.30
ΔE		13.91	27.85	19.88	28.01	14.15	19.05	11.10	15.85	11.64	14.54

At the same time, the value of L* also further proving that the usage of turmeric oil can slow down the enzymatic browning of the apple cubes and also act as an antioxidant agent. In Figure 8, the value of L* seems to be decreasing from its control value. L* value in the CIE L*a*b* coordinated represent the lightness of an item and decrease in L* value indicate the loss of lightness. In this study, the lower the L* value of the apple cubes means the higher the changes in colour happened towards the apple cubes. Hence, the higher the rate of enzymatic browning of the apple cubes. In Figure 8, apple cubes that were uncoated and coated without the turmeric oil are showing the most drastic decrease in their L* value. Meanwhile, the apple cubes that are being coated with the turmeric oil but with different concentrations are only showing a little decrease in their L* value. This also proved that the use of turmeric oil in coating the fresh cut fruits can help them decrease the rate of the enzymatic browning.

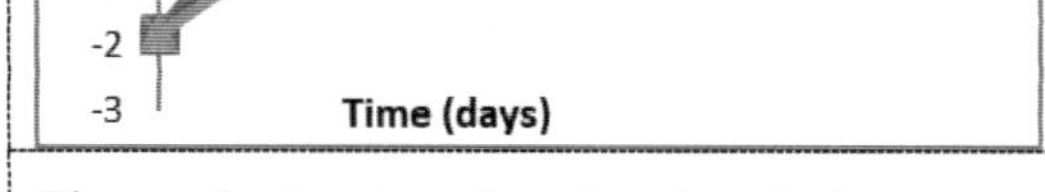

Figure 6: a* value of apple cubes during storage at 4 °C in the chiller.

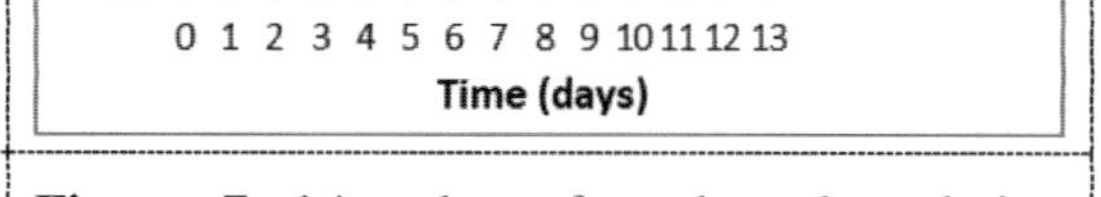

Figure 7: b* value of apple cubes during storage at 4 °C in the chiller.

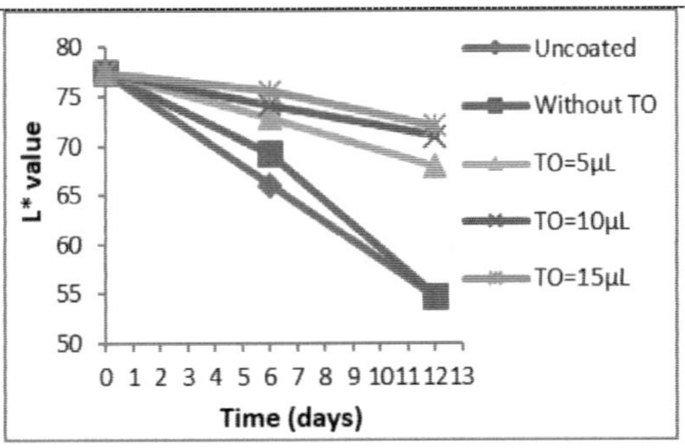

Figure 8: L* value of apple cubes during storage at 4 °C in the chiller.

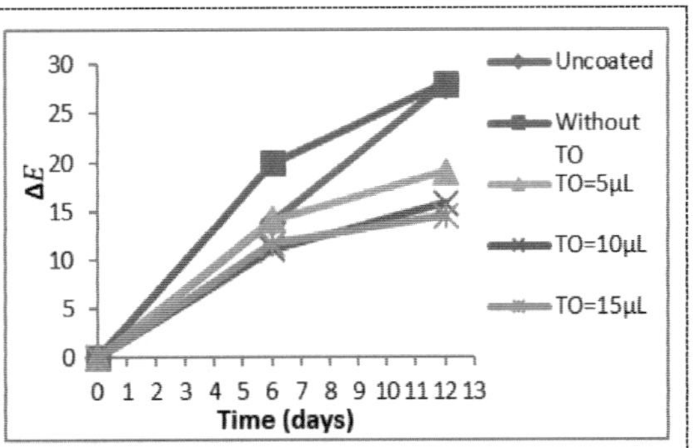

Figure 9: Total colour difference (ΔE) value of apple cubes with different turmeric oil concentration at 4 °C in the chiller.

To further verifying that turmeric oil can act as an antioxidation agent by slowing down the rate of the enzymatic browning, the total colour difference (ΔE) were calculated. Figure 9 shows the result being plotted to a graph. As being shown in Figure 9, the total colour difference for the apple cubes increased significantly between day 0 and 6 of storage. However, for the apple cubes that were coated with the turmeric oil, the colour difference seems to be remained approximately constant thereafter. But, apple cubes that were uncoated and coated without the turmeric oil continue to increase their colour difference on their twelfth day of storage. The increases in the total colour difference suggest an increase in the proportion of oxidized phenols during the storage. This further proving the characteristics of turmeric oil as antioxidation agent by slowing down the enzymatic browning reaction of the apple cubes.

3.5. Degree of Oxidation

The changes in absorbance at 420 nm were studied with the time of treatment and also with the type of different coating for apple cubes. Significant differences for the degree of oxidation were observed among the apple cubes as shown in Table 5. Results showed that increasing processing time also increase the absorbance at 420 nm. However, that was not the only factor since it can be seen from Figure 10 that different types of coating also affected the absorbance rate at 420 nm. Apple cubes that were uncoated and coated without the turmeric oil were leading the degree of oxidation followed by coating that contains 5 μL, 10 μL and 15 μL of turmeric oil respectively.

Table 5: Value of absorbance at 420 nm during storage at 4 °C in the chiller.

Day	Uncoated	Without TO	TO = 5μL	TO = 10μL	TO = 15μL
4	2.9606	2.9339	2.6242	1.9789	1.4300
8	3.2251	3.3048	2.7694	2.1552	1.5069
12	3.6001	3.5984	2.9537	2.3511	1.7082

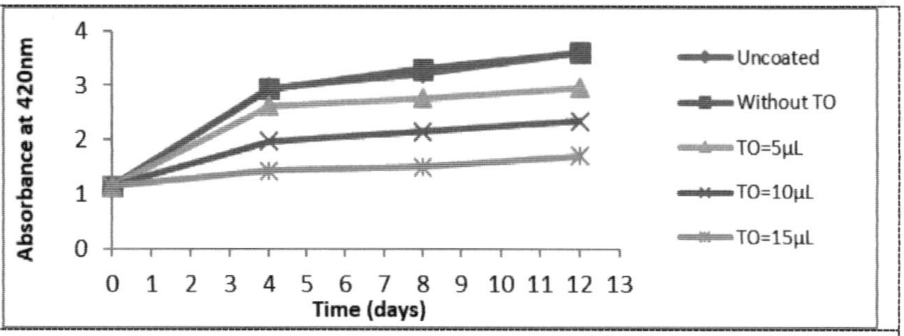

Figure 10: Change in the degree of oxidation at absorbance at 420 nm during storage at 4°C chiller

As stated before, the degree of oxidation was monitored during cold storage in order to observe relationships with total phenolic or PPO activity. At zero time measurement of the change in absorbance at 420 nm, the PPO activity already registered an initial reading at 1.1413. Similar to what was reported, the result for PPO activity of the apple cubes varied for the next fourth, eight and twelfth days of storage [14]. Apple cubes that were coated with coating contained turmeric oil showed very little change during the storage especially apple cubes that coated with high concentration of turmeric oil. As for the apple cubes that were uncoated and coated without the turmeric oil, there were an enormous increased in their degree of oxidation on the fourth day of storage and it keeps on increasing till the twelfth days of storage.

After comparing the degree of oxidation of the apple cubes with five different types of coating, there was a correlation between the degree of oxidation and PPO activity with the type of the coating used. Apple cubes that were uncoated showed a very high degree of oxidation and rate of enzymatic browning were also very high. However, for the apple cubes that were coated with coating contained 15 µL turmeric oil, it shows a slight increase from the initial reading and it proves that turmeric oil can preserve the apple cubes from the enzymatic browning. Also, the higher the concentration of turmeric oil used in a coating solution, the lower the rate of the enzymatic browning.

4. Conclusion

Based on the findings, it can be concluded that turmeric oil shown its antioxidant characteristics when it incorporated with starch edible coating film for the fresh cut apples. All the apple cubes were stored inside a plastic container at a chiller with a temperature of 4 °C. Apple cubes were divided into categories which were uncoated, coated without turmeric oil and coated with turmeric oil with three different concentrations which were 5 µL, 10 µL and 15 µL. Sensory evaluation has been done and within ten sensory panellist, 6 panellist rate the appearance and the surface color to at a very good rate. Meanwhile for the weight loss, the uncoated apple cubes showed the highest percentage of weight loss during the twelfth day of storage which is 11.85 % whereas apple cubes that were coated with coating that contain 15 µL of turmeric oil showed the lowest percentage of weight loss which is 3.95 %. This proved that the edible coating incorporated with turmeric oil able to preserve the apple cube. Same goes for the colour measurement where the apple cubes that were coated with coating that contain 15µL of turmeric oil showed the lowest total color difference which is $\Delta E = 14.54$ at the twelfth day of storage. As for the degree of oxidation, the oxidation activity was measured at absorbance at 420 nm the results showed that increasing usage of turmeric oil decrease the absorbance at 420 nm. Thus, it can be said that the turmeric oil can postponed the enzymatic browning from happening. In conclusion, turmeric oil has been showing its antioxidant activities for the fresh cut apple throughout this entire study and therefore the objectives for this research have been achieved.

Acknowledgement
Assists from the Faculty of Chemical Engineering, Universiti Teknologi MARA (UiTM) Malaysia and financial support from LESTARI (UiTM) (600-RMI/DANA 5/3/LESTARI (30/2015)) are gratefully acknowledged

References

[1] A. E. Watada and L. Qi, "Quality of fresh cut produce," *Postharvest Biol. Technol.*, vol. 1, no. 1, pp. 201–205, 1999.
[2] C. Queiroza, M. L. M. Lopes, E. Fialbo, and V. L. Valente-Mesquita, "Characteristics and mechanisms of browning control," *Food Rev. Int.*, vol. 1, no. 1, pp. 361–375, 2008.
[3] A. R. Maria, S. Robert, and M. Olga, "Edible coatings to incorporate active ingredients to fresh-cut fruits: a review," *Trends Food Sci. Technol.*, vol. 1, no. 1, pp. 438–447, 2009.
[4] V. B. Oriani, G. Molina, M. Chiumarelli, G. M. Pastore, and M. D. Hubinge, "Properties of cassava starch based edible coating containing essential oils," *J. Food Sci.*, vol. 1, no. 1, pp. 189–194, 2014.
[5] M. Shahnawaz, A. Wali, and S. Abbass, "A study on the shelf life of minimally processed apple with edible coating," *Chem. J.*, vol. 1, no. 1, pp. 20–22, 2013.
[6] M. Shariatifar and E. Jafarpour, "Edible coating effects on storage life and quality of apples," *J. basic Appl. Sci. Res.*, vol. 1, no. 1, pp. 24–27, 2013.
[7] Z. H. Hassan, S. Lesmavati, R. Oomariah, and A. Hasbianto, "Effect of wax coating applications and storage temperature on the quality of tangerines citrus (Citrus reticulata)," *Int. Food Res. J.*, vol. 1, no. 1, pp. 641–648, 2014.
[8] M. Radi, E. Firouzi, H. Akhavan, and S. Amiri, "Effect of Gelatin-Based Edible Coatings Incorporated with Aloe vera and Black and Green Tea Extracts on the Shelf Life of Fresh-Cut Oranges," vol. 2017, 2017.
[9] D. S. Ackharali and R. Maharaj, "Sensory evaluation as a tool in determining acceptability of innovative products developed by undergraduate students in food science and technology at the University of Trinidad and Tobago," *J. Curric. Teach.*, vol. 1, no. 1, pp. 10–27, 2014.
[10] M. A. M. Galeazzi and V. C. J. Sgarbieri, "Substrate specificity and inhibition of polyphenoloxidase from a dwarf variety of banana," *J. Food Sci.*, vol. 1, no. 1, pp. 1401–1406, 1981.
[11] A. M. C. N. Rocha and A. M. M. B. Morais, "Characterization of polyphenoloxidase (PPO) extracted from Ô Jonagored Õ apple," *J. food Control*, vol. 1, no. 1, pp. 1–6, 1981.
[12] M. T. Sultan, M. S. Butt, S. Akhtar, A. N. Ahmad, M. Rauf, M. S. Saddique, and A. Naz, "Antioxidant and antimicrobial potential of dried cumin (Cuminum cyminum L.), caraway (Carum carvi L.) and turmeric powder (Curcuma longa L.)," *J. Food, Agric. Environ.*, vol. 12, no. 3–4, pp. 71–76, 2014.
[13] M. B. Perez-gago, M. Serra, M. Alonso, M. Mateos, and M. A. R, "Effect of whey protein- and hydroxypropyl methylcellulose-based edible composite coatings on color change of fresh-cut apples," *Post*, vol. 36, no. 1, pp. 77–85, 2005.
[14] A. Podsedek, J. Wilska-Jexska, B. Anders, and J. Markowiski, "Compositional characterisation of some apple varieties," *Eur. Food Res. Technol.*, vol. 2, no. 1, pp. 268–272, 2010.

Influence sample sizing of citrus hystrix essential oil from hydrodistillation extraction

A Yahya[1,2], I Amadi[1], S A Hashib[1] and F A Mustapha[1]
[1] Faculty of Chemical Engineering, Universiti Teknologi MARA, 45050 Shah Alam, Selangor, Malaysia.

E-mail: asda1205@gmail.com

Abstract. Essential oil extracted from kaffir lime leaves through hydrodistillation. The objective of this study is to quantify the oil production rate by identify the significant influence of particle size on kaffir lime leaves. Kaffir lime leaves were ground and separated by using siever into 90, 150, 300 μm and other kaffir lime leaves. The mean essential oil yield of 0.87, 0.52, 0.41 and 0.3% was obtained. 90 μm of ground gives the highest yield compared to other sizes. Thus, it can be concluded that in quantifying oil production rate, the relevance of different size of particle is clearly affects the amount of oil yield. In analysing the composition of kaffir lime essential oil using GC-MS, there were 38 compounds found in the essential oil. Some of the major compounds of the kaffir lime leave oils were detected while some are not, may due to oil experience thermal degradation which consequently losing some significant compounds in controlled temperature.

1. Introduction

Kaffir lime (Citrus hystrix DC) is a sort of plant from Southeast Asian with exceptionally solid scent. It is otherwise called makrut lime, which is a typical tropical herb in the family Rutaceae start all over in Southeast Asia [1]. Kaffir lime has a prickly shrubbery with fragrant leaves and the organic product is dull green with unpredictable uneven surface as appeared in Figure 1. The leaves and organic product peel is the crucial parts of kaffir lime and it is a key fixing in numerous Thai, Cambodian, Indonesian, Laotian, Malaysian, and Philippine cooking styles [2].

Figure 1. Kaffir lime (Ranjit *et al.*, 2013)

[2] To whom any correspondence should be addressed.

By and large, fundamental oil is one of a rich source of biological active compounds. This is on the grounds that key oils have exceptionally complex regular blends which can contain 20 to 60 components. Some of the major component of kaffir lime oil are mostly monoterpene hydrocarbon, with limonene (30.73%) and β-pinene (18.76%).For the minor components are terpinene-4-ol (10.63%), α-terpineol (8.35%), γ-terpinene (6.18%), α-terpinene (5.09%) and terpinolene (4.33%)[3]. In this way, they have been utilized broadly as bactericides, virucides, fungicides, against parasites, and bug sprays in different applications, particularly in the pharmaceutical, sterile, restorative, sustenance and farming commercial enterprises [1].

There are many studies reported on extraction of essential oil from peel of citrus fruits using different methods [4,5,6]. Sud and Badyal [4] extracted peel oil from the rind at various stages of fruit maturity by steam distillation and reported maximum yield in terms of peel oil as 2.05ml/100 g peels in orange. Brat [5] reported essential oil extraction by flash vacuum expansion of peel of lemon, orange, mandarin and grapefruit with maximum yield of 2.41 kg/tonne of fruits. Bousbia [6] extracted peel essential oil from citrus varieties such as Eureka, lime, Marsh Seedless, Torroco, Washington Naval were extracted by Clevenger apparatus using microwaves hydro-diffusion and gravity reported a significant reduction in extraction time (15 min).

Therefore, the major research efforts have been on extraction of essential oil from various citrus species but some people are still not received attention. If production of essential oil is related to the contact of water with particle size, then exposing to larger surface area of kaffir lime leaves will result more oil yield[7]. The objective of present research work was to evaluate the effect of size on yield of the essential oil from kaffir lime leaves by using hydrodistillation. The size reduction of leaves was aimed to increase the yield of essential oil. The chemical composition of kaffir lime oil produced was also analyzed.

2. Methodology/ Experimental

2.1. Materials

Kaffir lime leaves are obtained from night market at Seksyen 7, Shah Alam, Selangor. n-hexane used in the experimental work were analytical grade.

2.2. Sample preparation

Fresh kaffir lime leaves were washed and dried for two days in a Universal MEMMERT Oven (model 100-800) at 40°C until reach the constant humidity. The final humidity of the leaves is 60.33%. Finally, the air-dried samples were grind by using Waring commercial laboratory blender. The ground kaffir lime leaves were separated according to different sizes by using siever shaker (Octagon Digital Siever) with the range size of 300μm, 150μm and 90μm.

The humidity of kaffir leaves was calculated using Equation 1 as follows:

$$Humidity\,(\%) = \frac{fresh\ leaves - dried\ leaves}{fresh\ leaves} \times 100\% \tag{1}$$

2.3. Hydrodistillation extraction

Once the ground kaffir lime leaves were sieve and collected according to the sizes, the experiment was performed by setting up the hydrodistillation equipment. 25g of kaffir lime leaves mix with distilled water was taken into hydrodistillation (Clevenger type) apparatus using ground leaves. The amount water in the ratio of 1:16 for 6 hours. As the boiling of water progressed when reach 100^0 C, the water and volatile oil in the form of vapor started rising which got condensed in the condenser tube. The condensed water and oil were collected into the graduated cylinder.

2.4. Percentage yield
On completion of extraction process, essential oil that was collected. The oil yield was calculated using Equation 2 as follows:

$$Yield\ (\%) = \frac{weight\ of\ essential\ oil\, recovered}{weight\ of\ fresh\ leaves} \times 100\% \tag{2}$$

2.5. Gas Chromatography Mass Spectrometer (GC-MS) analysis
When the essential oil samples collected were ready to be analyzed, n-hexane solvent was added into all oil samples. The mixture of the essential oil and solvent was shake and sent to the Gas Chromatography (GC-MS) GC-MS analysis was performed using Varian 450-GC model, mass spectrometry Varian 240-MS inert, while the operation temperature was programmed at a rate of 3°Cmin-1 to a maximum temperature of 230°C and held for 10 minutes. The carrier gas for GC was helium with flow rate 1.0 ml/min. The spectrometers were operated in electron-impact (EI) mode, while the scan range was 45-450 amu, and the ionization energy was 70 eV. The inlet, ionization source temperatures were 250°C respectively. The components of the oil were identified by comparing their mass spectra with the recorded spectra in the MS library [8].

3. Results and Discussion

3.1. Effect of size of grind leaves on essential oil yield
The essential oil yield on dry basis varied from 0.1 to 1.0% in kaffir lime leaves. The percentage yields of the essential oil were based on different sizes of kaffir lime leaves. The essential oil was collected as shown in Figure 2. The color of kaffir lime essential oil was colorless. The most oil extracted from hydrodistillation was 90 μm with 0.25 ml, followed by 150 μm and 300 μm with oil extracted of 0.15 and 0.12 ml. The least amount oil extracted to be 0.1 ml from the other size of ground kaffir lime.

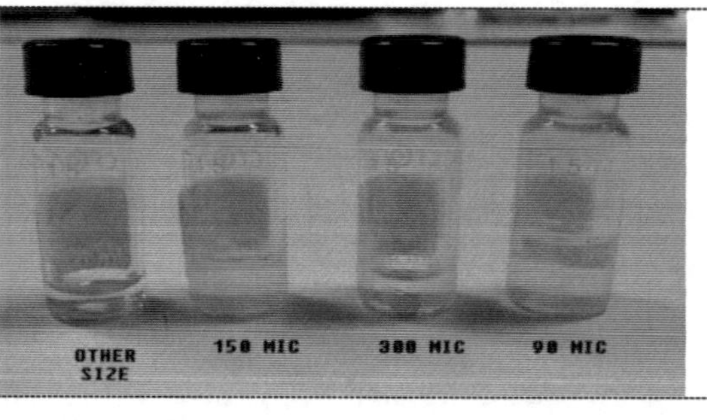

Figure 2. Oil extracted with different particle size.

Figure 3 shows the relationship between percentage of essential oil yield with the size of kaffir lime leaves was plotted and observed. The percentage of kafir lime essential oil from 90 μm leaves was 0.81% while for150 μm and 300 μm leaves, the percentage yields were 0.5226% and 0.418%. However, for other size of grounds kaffir lime leaves, the percentage yield was 0.3484%.

Essential oil yield increased with decrease in particle size (i.e increase in grinding time) of the leaves and reached to almost maximum at 10 min of grinding time. In the mundane grinding process, heat is engendered when energy is utilized to fracture a particle into a more minute size, that engender heat causes temperature elevate in the grinder which is responsible for a loss of volatile oil content[8].

It is evident that oil yield varied widely among particle size. The oil recovery less when extraction was carried out from ground kaffir lime leaves. This is due to small surface area of the leaves and thereby having less contact with water during extraction [9]. The mean yield increased to 0.418% on the grinding kaffir lime leaves at 300 μm. With further grinding to 150 μm and 90 μm, the oil yield increased further to 0.522% and 0.87%, respectively.

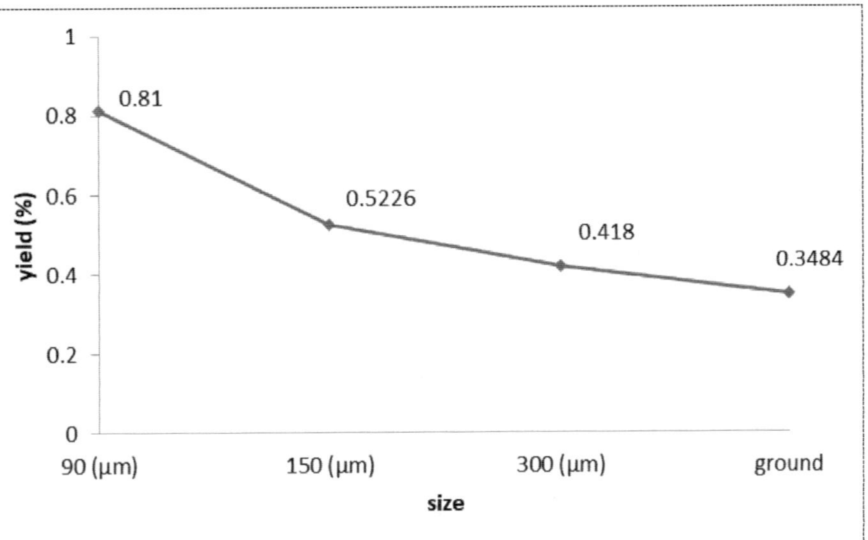

Figure 3. Percentage of oil yield at different sizes of ground kaffir lime leaves.

Thus, a very significant increase in oil yield was found on reducing particle size of kaffir lime leaves. This increase in oil yield might be due to increased surface area of kaffir lime leaves and thereby having close contact with water during extraction as well as having reduced resistance to oil water mixture from oil sac to the solution.

3.2. Composition of Volatile Compounds in Essential Oil

There were 38 compounds was found and identified in kaffir lime leaves extracted using hydrodistillation shown in Table 1. There were different numbers of compounds extracted in different sizes of kaffir lime leaves where a total of 31 compounds, 27 compounds, 29 compounds and 30 compounds were identified in size of 90 μm, 150 μm, 300 μm and ground kaffir lime leaves. From here, it can be said that 90 μm size of ground kaffir lime leaves can extract more chemical compounds.

Table 1. Compound of essential oil obtained from the Kaffir lime leaves.

No	Chemical compounds	Amount (%)			
		90μm	**150 μm**	**300 μm**	**Ground**
1	Arginine	0.02	-	0.02	-
2	Ecgonine	0.01	0.02	0.03	0.04
3	α-Pinene	0.01	0.06	0.01	0.00
4	6-Octen-1-ol	0.00	0.02	0.05	0.10
5	Naphthalene	0.06	0.09	0.08	0.04
6	6-octenal	1.30	0.31	16.99	0.42
7	1-Penten-3-one	0.00	0.00	0.01	-
8	3-Carene	0.02	0.01	0.00	0.00
9	1H-Imidazole	0.00	0.00	0.05	0.03
10	Limonene oxide	0.20	0.12	0.02	0.08
11	Piperidine	0.20	0.00	0.00	0.01
12	α-Phellandrene	0.01	0.01	0.00	0.01
13	7-octen-1-ol	1.61	0.14	0.24	0.02

14	Mequinol	0.01	0.00	0.00	0.01
15	Squalene	0.75	0.01	0.01	4.67
16	Cis-α-Terpineol	0.00	0.01	0.00	0.01
17	Camphene	0.04	0.36	1.64	0.01
18	α-Cubebene	0.00	0.01	0.00	0.01
19	Citronellyl Isobutyrate	0.00	0.00	-	-
20	3-Cyclohexen-1-ol	0.00	0.02	0.17	0.04
21	α-Farnesene	0.03	0.12	0.13	0.67
22	Alloaromadendrene oxide	0.00	-	0.18	0.04
23	Spathulenol	0.01	0.09	0.02	0.08
24	Humulane	0.00	-	0.02	-
25	Aristolene	0.00	0.03	-	0.02
26	Azulene	0.15	0.25	0.18	0.12
27	α-Gualene	0.01	-	-	-
28	Caryophyllene	0.00	0.03	0.06	0.01
29	α-Vatirrnene	0.00	-	-	-
30	Phenanthrene	0.01	0.00	0.00	0.00
31	Eicosene	0.01	-	0.00	0.00
32	ς-Elemene	-	0.64	-	-
33	ς-Himachalene	-	0.00	-	-
34	α- Panasinsene	-	-	0.02	0.00
35	Ferrocene	-	-	0.00	0.01
36	Camphor	-	-	-	0.00
37	ς-Gurjunenepoxide	-	-	-	0.00
38	1-9-Dihydropyrene	-	-	-	0.00
39	others	95.54	97.64	80.08	93.53

Some of the major compounds of the kaffir lime leave oils such as *α-pinene, limonene* and *citronellal* were detected while some of the major compounds such as *sabinene* and *β-pinene were* not detected. This is due to the high temperature while grinding of kaffir lime leaves. This condition may due to oil experience thermal degradation which consequently losing some significant compounds in uncontrolled temperature [7,8,9,10,11]. Throughout the hydrodistillation extraction, the temperature somehow could accidently exceed 100°C and cause some of the volatile component like *sabinene* and *β-pinene* lost during extraction process. The are many beneficial chemical compounds contains in kaffir lime essential oil. For 90 µm sample, the highest chemical contains is *7-octen-1-ol* where is suitable not used for fragrances or flavors while the highest chemical compounds for 150 µm is *ς-Elemene*. It can be use as flavor and fragrance agents. 300 µm shows 6-octenal is the highest compond and for ground sample indicate that *Squalene* is the major compounds contains in kaffir lime essential oil. *6-octenal* and *sequalene* can be use as natural substances and extractives [12].

4. Conclusion

In conclusion, by the result obtain from the test and analysis, hydrodistillation is still a very good method to extract kaffir lime essential oil although there are many other and modern methods such as automated steam distillation, flash vacuum expansion and many more. The analysis on qualitative and quantitative of Kaffir lime leaves oil extracted using hydrodistillation had been accomplished successfully. In quantifying essential oil production, the relevance of different size of particle is clearly affects the amount of oil yield. Increasing the surface area will cause the contact of water and the kaffir lime leaves particle during the hydrodistillation and reduce resistance to oil water mixture from oil sac in the solution. In analyzing the composition of kaffir lime essential oil using GC-MS, there were 38 compounds found in the essential oil. Some of the major compounds of the kaffir lime leave oils were detected while some are not, may due to oil experience thermal degradation which consequently losing some significant compounds in uncontrolled temperature.

Acknowledgement
Deepest gratitude to Malaysia Ministry of Education and Universiti Teknologi MARA(UiTM) grant 600-IRMI/MYRA 5/3/LESTARI (0047/2016) for the financial support for this project.

References

[1] Doreen SH, Rose LC, Suhaimi H, Mohamad H, Rozaini MZH, Tai M (2011). Preliminary evaluation on the antibacterial activities of Citrus hystrix oil emulsions stabilized by tween 80 and span 80. Int J Pharm Pharm Sci 3 (Suppl 2), 209–11.

[2] Vimol Srisukh, Chanwit Tribuddharat, Veena Nukoolkarn, Nuntavan Bunyapraphatsara, Kulkanya Chokephaibulkit, Siwimol Phoomniyom, Sirirat Chuanphun and Somporn Srifuengfung (2012). Antibacterial activity of essential oils from Citrus hystrix (makrut lime) against respiratory tract pathogens. ScienceAsia 38 (2012): 212–217.

[3] Hongratanaworakit T, Buchbauer G (2007). Chemical compositions and stimulating effects of *Citrus hystrix* oil on humans. Flavour & Fragrance J., 22:443- 449.

[4] Sud R G and Badyal J M. (1993). Utilization of citrus rind for peel oil: stage of fruit and evaluation of methods. Indian Food Packer 45-7.

[5] Brat P, Olle D, Grance A L and Brillouet M (2001). Essential oils obtained by flash vacuum-expansion of peels from lemon, sweet orange, mandarin and grapefruit. Fruits 56 : 395-402.

[6] Bousbia N, Vian M A, Ferhat M A, Meklati B Y and Chermat F. (2009). A new process for extraction of essential oil from Citrus Peel: Microwave hydro diffusion and gravity. Journal of Food Engineering, 90: 409-13.

[7] Y. Asdarina & M.Y.Rosli (2010) Study the effect of sample preparation during extracting patchouli Oil.Natural Product R&D: Leads from Nature. Publisher Universiti Malaysia Pahang. ISBN 978-967-0120-04-1: 269-274.

[8] Rohit Saxena1 , Aditi Soni (2013) Cryogenic Grinding: A Physical Technique To Retain Volatile Content In Natural Products. International Journal of Modern Physics: Conference Series Vol. 22:589–592.

[9] Kirti R Jalgaonkar, S K Jha, R. K Pal and D V K Samuel (2013) Effect of species and particle size on essential oil yield of citrus peels (Citrus spp). Indian Agricultural Research Institute, 83 (12): 1285-88

[10] Lijun Wang and Curtis L. Weller (2006). Recent advances in extraction of nutraceuticals from plants. Trends in Food Science & Technology, 17, 300–312.

[11] AsdarinaYahya & Rosli MohdYunus (2013) Influence of Sample Preparation and Extraction Time on Chemical Composition of Steam Distillation Derived Patchouli Oil. Procedia Engineering .Volume 53:1-6

[12] Bill (2017,August 21). Chemical Compounds. Retrieved from http://www.thegoodscentscompany.com.

The processing of used cooking oil (yellow grease) using combination of adsorption and ultrafiltration membrane processes

C M Rosnelly[1,3], Sofyana[1], D Amalia[2] and S Sarah[2]

[1] Faculty of Engineering, Syiah Kuala University
[2] Department of Chemical Engineering, Faculty of Engineering, Syiah Kuala University

E-mail: cutnelly@ymail.com

Abstract. Yellow grease is used cooking oil whose quality has degraded due to the oxidation, polymerization, or hydrolysis process. In previous studies, yellow grease refining had been conducted either by adsorption or by using membrane. In this study, adsorption process using adsorbent from bagasse activated with H_3PO_4 12.5%, and ultrafiltration using Polyethersulfone (PES) membrane were combined. In adsorption stage, several variation of bagasse mass was fed into 200 ml of yellow grease and stirred for 60 minutes at 60 rpm. Yellow grease produced from adsorption with best condition was then processed using ultrafiltration membran that is PES membran with concentration by 15 wt % with transmembrane pressure variation by 0.5; 1; 1.5; 2; and 2.5 Bar. Analysis of yellow grease characteristics before refined showed its acid number, peroxide number, iodine number, and water content respectively by 2.68 mgKOH/Kg; 5.97 Meq/Kg; 51,48; and 1.29%. Characteristics of yellow grease after adsorption at its best condition on the parameters of acid number, peroxide number, iodine number, and water content are respectively by 2.55 mgKOH/Kg; 4.19 Meq/Kg; 40,02; and 0.27%. Characteristics of yellow grease after ultrafiltration at its best condition on the parameters of acid number, peroxide number, iodine number, and water content are respectively by 1.12 mgKOH/Kg; 1.8 Meq/Kg; 41,36; and 0.02%. Combination of adsorption and ultrafiltration processes for yellow grease processing showed decreasing value on the parameters of acid number, peroxide number, and water content that conforms to the SNI quality standard, but has not been able to increase the iodine number.

1. Introduction

Cooking oil is a medium for frying foods. The purpose of using cooking oil is as a medium for conducting heat in order to change the physical appearance and texture of the foodstuffs that will give savory taste, and also to add nutritional value and calorie in the foodstuffs [1]. Cooking oil which has repeatedly been used and is damaged is called as yellow grease (used cooking oil). The oil damage is caused by oxidation, polymerization, or hydrolysis process during the frying which causes the formation of unwanted carcinogenic compounds (peroxide and aldehyde) that harmful for health.

[3] To whom any correspondence should be addressed.

Research regarding yellow grease refining has been widely conducted and the methods that are commonly used in between are adsorption and membrane process. Yellow grease refining process is purposed to reduce compounds that were performed due to the oil damaging process [2], and then the refined yellow grease can be utilized as a benefit things such as material for soap and biodiesel. However, the anvanced studies is needed to innovate method in refining yellow grease.

Adsorption is a method that is commonly used in refining yellow grease. Adsorption is chosen because its process is easy to perform, and usually is effective in reducing free fatty acid in yellow grease. The previous studies of refining yellow grease using adsorption process were conducted by Muallifah (2009) using activated carbon from *moringa oliefera* seed; Ramdja et.al (2010) using adsorbent from bagasse; Yustinah and Hartini (2011) using activated carbon from coconut fiber; and Triyanto (2013) using activated carbon from bagasse. Membrane technology has been extensively used for refining of oil [3]. Membrane separation is based on a physical retention of compounds present in treated stream and depend mainly on molecular weight of a separated substance [4]. Membrane process for increasing the oil quality is easy to perform and has many advantages compared to other process, because of its low energy consumption, ability to operate in room temperature, and because the addition of chemicals is not necessary [5]. The previous studies of refining yellow grease using membrane were conducted by Widyaningsih and Purwati (2013) using *nata de coco* membrane; Miyagi et.al (2003) using nonporous composite membrane; and Subramanian et.al (2000) using various of polymer membranes. In this research was conducted a process combination of adsorption and ultrafiltration, where the adsorption process is on the pretreatment stage that is purposed to reduce membrane workload in ultrafiltration process, so that it can increase the quality of refined yellow grease. The aim of this research was to study the effect of process combination of adsorption and ultrafiltration on characteristics of refined yellow grease. The adsorbent used is bagasse activated with H_3PO_4 12.5%. The high content of carbon in bagasse after activation become the reason why using bagasse activated carbon in yellow grease refining. The membrane used is Polyethersulfone membrane which is hydrophobic that it suits for processing yellow grease. Characteristics of oil before and after refined are including acid number, peroxide number, iodine number, and water content.

2. Materials and Methods

This study was conducted a process refining of yellow grease that consists of two stages that are adsorption and ultrafiltration processes. The using combination processes of adsorption and ultrafiltration is purposed to produce the better quality of refined yellow grease than refined yellow grease from adsorption or ultrafiltration process only.

2.1. Materials

Materials used in this study were yellow grease (used cooking oil), bagasse, technical H_3PO_4, distillate water, Polyethersulfone (PES), N-Methyl-2-pyrrolidone (NMP), ethanol 95%, KI, KOH, IBr_2, Na_2SO_3, glacial acetic acid, chloroform, amylum 1%.

2.2. Methods

The processes of yellow grease refining are schematically arranged as shown on figure 1 and figure 2.

IOP Conf. Series: Materials Science and Engineering **334** (2018) 012066 doi:10.1088/1757-899X/334/1/012066

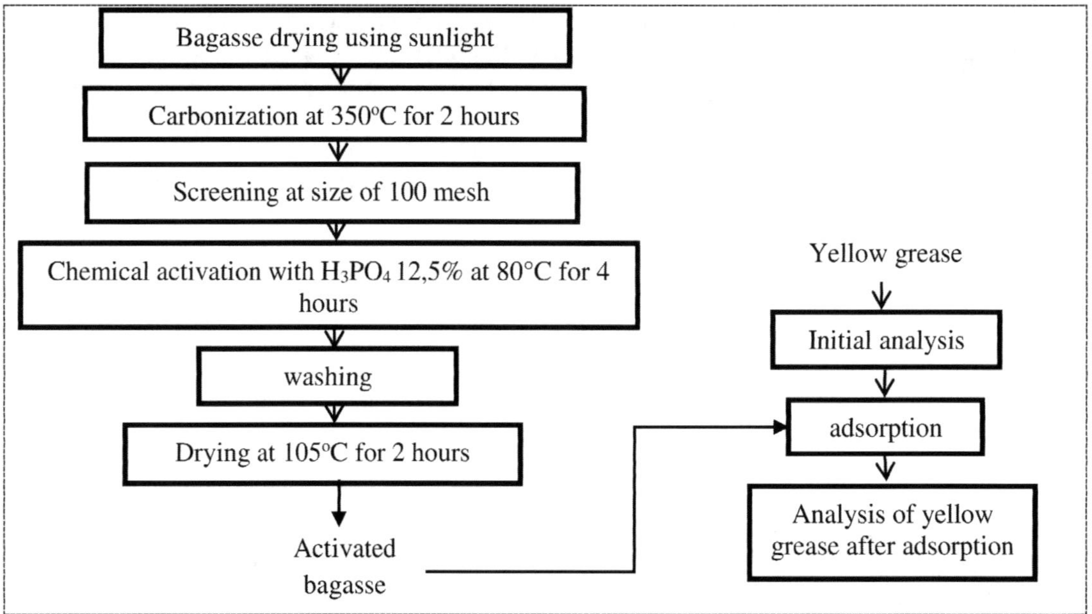

Figure 1. Flow chart of adsorbent manufacture and adsorption process.

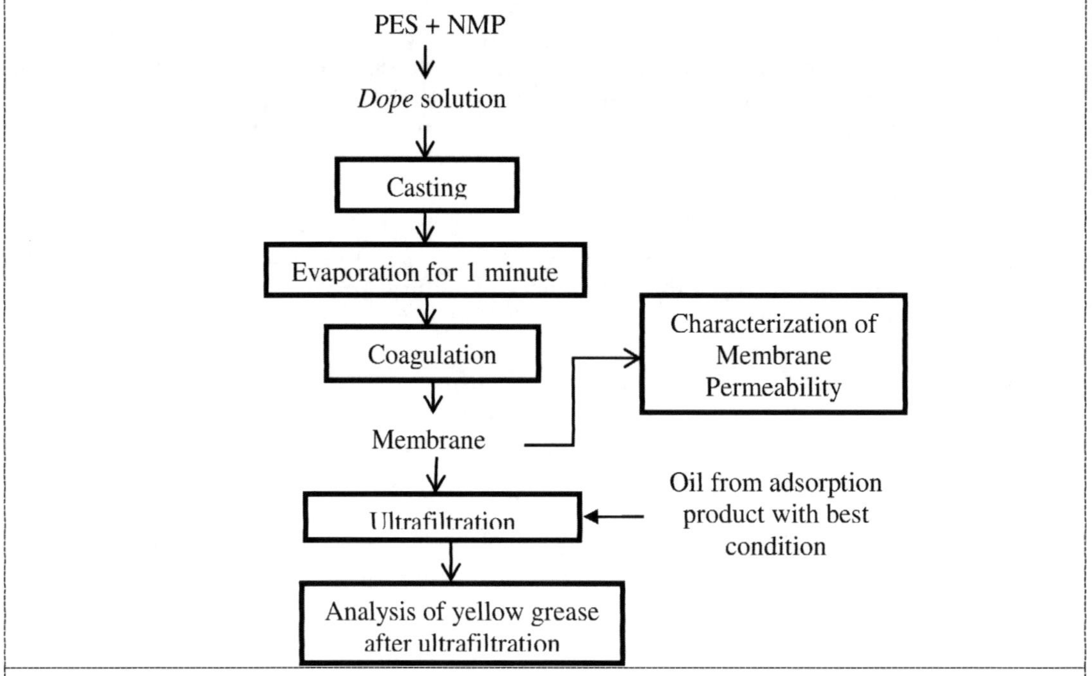

Figure 2. Flow chart of membrane manufacture and ultrafiltration process.

2.2.1 Manufacture of adsorbent from activated bagasse. Bagasse was dried in advanced with sunlight. Then, bagasse was cut in 5 mm size and was then carbonized using tubular furnace for 2 hours at 350°C. After that, bagasse carbon was finely grounded and screened using screener until it sized 100 mesh. The bagasse carbon was then chemically activated with H_3PO_4 12.5%, and stirred for

IOP Conf. Series: Materials Science and Engineering **334** (2018) 012066 doi:10.1088/1757-899X/334/1/012066

4 hours at 60 rpm and 80°C. Next, bagasse was filtrated using a 5 micron filter paper and washed until its pH is neutral. Finally, the activated carbon was dried in oven at 105°C for 2 hours [6].

2.2.2 Processing yellow grease with adsorption process using adsorbent from bagasse. In adsorption stage, several variation of bagasse mass was fed into 200 ml of yellow grease and stirred for 60 minutes at 60 rpm. Sample was then filtrated using a 5 micron filter paper to separate yellow grease from adsorbent, and then the sample's acid number, peroxide number, iodine number, and water content were analyzed. Adsorption result with best condition will be further processed using membrane.

2.2.3 Manufacture of Polyethersulfone (PES) membrane with phase inversion by immersion precipitation method. Dope solution was made by solving Polyethersulfone (PES) and N-Methyl-2-pyrrolidone (NMP) with polymer concentration by 15 wt %, 20 wt %, and 25 wt % of weight. Dope solution was casted on glass plate with 200 μm of thickness, and then it went through solvent evaporation process for 1 minute before coagulated in coagulation bath filled with non-solvent (distilled water). After the membrane was prepared, the permeability test was done using deionized distilled water at various transmembrane pressures (TMP) which are 0.5; 1; 1.5; and 2; 2.5 Bar. While the membrane used in yellow grease refining is classed as ultrafiltration membrane.

2.2.4 Processing of yellow grease resulted from adsorption process by using ultrafiltration membrane. Yellow grease produced from adsorption with best condition was then processed with ultrafiltration membrane process. Feed of yellow grease was flown through dead-end system membrane module with transmembrane pressure variation by 0.5; 1; 1.5; 2; and 2.5 Bar. Permeate resulted was then analyzed to know its acid number, peroxide number, iodine number, and water content.

3. Result and Discussion

3.1. Analysis of Initial Characteristics of Yellow Grease

In this study, the initial characteristics of yellow grease were determined as a comparison between the qualities of yellow grease before and after refined. Characteristics of yellow grease were described by several numbers that had been determined from calculation result with several analyses conducted. Comparison of initial characteristic of yellow grease with predefined SNI standard can be seen on Table 1.

Table 1. Comparison of initial characteristic of yellow grease with predefined standard: SNI 3741 2002- SNI 3741 1995.

Characteristic	SNI	Analysis results
Acid Number	Max. 2 Mg KOH / g	2.68 Mg KOH / g
Peroxide Number	Max. 2 Meq / kg	5.97 Meq / kg
Iodine Number	45-46	51.48
Water content	Max 0.3 %	1.29 %

From table 1 It can be seen that yellow grease quality was degraded because its acid number, peroxide number, iodine number and water content were not up to or above value interval that has been predefined in SNI regulation. It was caused by the repetitive use of oil that the oil quality was degraded.

3.2. Analysis of Yellow Grease Characteristic in Adsorption Stage

The adsorption step is a preliminary treatment of yellow grease purifications. This step was conducted by varying of amount of mass of adsorbent, then the effect of it on yellow grease characteristics were investigated. The results of characteristic analyses of yellow grease purified are shown in Figure 3.

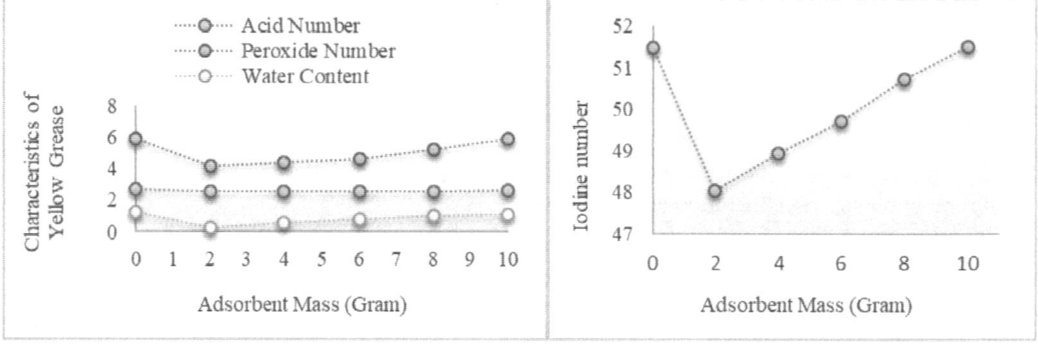

(a) **(b)**

Figure 3. Characteristic of Yellow Grease from Adsorption Process **(a)** Acid Number, Peroxide Number, Water Content; and **(b)** Iodine Number.

Figure 3 describe that adsorption process could increase the quality of yellow grease, compared to its quality before the adsorption process was performed. The values of acid number, peroxide number, iodine number, and water content on the best condition, with 2 grams of adsorbent and 200 mL oil, were 2.55 MgKOH/g; 4.19 Meq/kg; 48.02; and 0.27% respectively.

3.3. Membrane Characterization

Membrane characterization can be done by determining membrane permeability coefficient. Permeability coefficient of membrane (Lp) is carried out through flux measurement, which is by flowing pure water as feed at various transmembrane pressures (TMP). By plotting transmembrane pressure to the flux values, a linear regression curve will be obtained, where the slope is the permeability coefficient.

Table 2. Permeability Coefficient of PES Membranes.

PES membrane	Permeability Coefficient (L/m^2.h)
PES 15 wt %	10.03
PES 20 wt %	5.414
PES 25 wt %	1.592

According to permeability coefficient of membranes, PES membrane with concentration by 15 wt% is classified as ultrafiltration membrane, while PES membranes with 20 wt % and 25 wt % concentration are classified as nanofiltration membrane. Classification of separation process types of these membranes is based on the range of permeability coefficient for each separation process type. Permeability coefficient of ultrafiltration membrane is ranged between 10-20 L/m^2.h at operating pressure range between 1 to 5 bar, while the coefficient of Nano filtration membrane is ranged from 1.4 to 12 L/m^2.h at operating pressure range between 5 to 20 bar [7].

3.4. Analysis of Yellow Grease Characteristic in Ultrafiltration Stage

Ultrafiltration is advanced step of yellow grease processing after finishing adsorption stage. Yellow grease sample used in this stage was oil with the best characteristic resulted from adsorption process. Analysis of yellow grease permeate characteristic was also performed for the acid number, peroxide number, iodine number, and water content. The following is the chart of yellow grease characteristic resulted from ultrafiltration process at various transmembrane pressures.

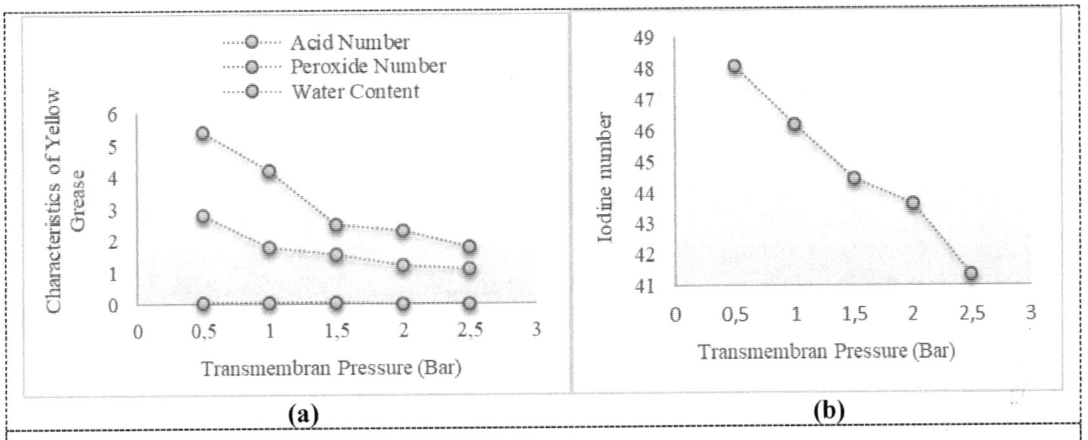

Figure 4. Characteristics of Yellow Grease (**a**) Acid Number, Peroxide Number, Water Content; and (**b**) Iodine number Resulted from Ultrafiltration.

After finishing ultrafiltration process using PES membrane 15 wt %, characteristics of yellow grease including acid number, peroxide number, iodine number, and water content were decreasing along with the increase of transmembrane pressure.

Free Fatty Acid (FFA). Based on Figure 4 (a), acid number is decreasing as the increase of transmembrane pressure. The highest acid number was obtained at 0.5 bar pressure, which is 2.8 Mg KOH/gr, and the lowest was obtained at 2.5 bar pressure, which is by 1.12 Mg KOH/gr. Compared to the one resulted from adsorption process, the acid number was decreased at ultrafiltration process. It is because the free fatty acids from yellow grease in the form of dimer and trimer are retained by the membrane whose pore size is smaller than fatty acid molecule size. In addition, the higher transmembrane pressure, the more acid number in yellow grease to decrease. This may be due to the increasing driving force (pressure) that cause an increase in the density of membrane material (compaction) so as to reduce the flux and diffusion of dissolved material through membrane [8]. Flux value is always inversely proportional to the membrane rejection, so the decrease of acid number in yellow grease is because the membrane rejection towards free fatty acid molecules is increasing.

Peroxide Value. According to Figure 4 (a), it is seen that peroxide number is also decreasing along with the increase of transmembrane pressure. The decrease of peroxide number in ultrafiltration stage is because peroxide has size that relatively bigger that membrane pore size, so peroxide molecule cannot pass through membrane pores [2]. As well as acid number, the decrease of peroxide number along with the increase of transmembrane pressure is also caused by compaction phenomenon in membrane due to the increase of driving force (transmembrane pressure) that cause the membrane material density to increase and decrease the flux and diffusion of dissolved material through membrane [8].

Iodine Value. Iodine number shows unsaturation level of oil and relates to the amount of double bond contained in fatty acid or oil. The more double bond contained in oil, the higher iodine number in it and vice versa [2]. Yellow grease has very low iodine number. This is because to the amount of double bond in used cooking oil is getting smaller due to the high-temperature heating or oxidation reaction that generate free fatty acid, alcohol, aldehyde, free radicals, and single bond [9]. According to Figure 4 (b), iodine number is decreasing as the increase of transmembrane pressure. This is caused by the more short-chain free fatty acids that can pass through membrane pores due to the increase of given driving force (pressure) that the saturated fatty acid content in yellow grease is still high. Thus, one can say that ultrafiltration of yellow grease using membrane has not been able to increase iodine number [2].

Water Content. Water is an undissolved impurity in oil and can be separated through filtration. According to Figure 4 (a), water content in yellow grease resulted from ultrafiltration tends to be stable against transmembrane pressure, where the water content is ranged between 0.02% - 0.05%. Water content in yellow grease resulted from ultrafiltration is significantly reduced, compared to water content in initial sample of yellow grease. Ultrafiltration of yell grease using PES membrane can reduce water contained in yellow grease. It is because separation process by using membrane PES - which has hydrophobic nature - is based on the difference of membrane wettability by fluids. When water particles are in contact with membrane matrix, they will be trapped or bonded on the membrane surface that they cannot escape through membrane pores [10]. Hydrophobic membrane is easily wetted by oil, so the use of hydrophobic membrane in separating water contained in yellow grease can effectively increase the efficiency of separation process [8].

Recapitulation of the characteristics of yellow grease refined from adsorption and ultrafiltration can be seen on the following Table 3.

Table 3. Recapitulation of the characteristics of yellow grease refined from adsorption and ultrafiltration.

Parameters	SNI	Initial sample	Refining Stages		Refining Efficiency (%)	
			Adsorption [a]	Ultrafiltration [b]	Adsorption	Process Combination [c]
Acid Number (MgKOH/g)	2	2,68	2,55	1,12	4,85	58,21
Peroxide Number (meq/kg)	2	5,97	4,19	1,8	29,82	69,85
Iodine Number	45-46	51,48	48,02	41,36	6,72	19,66
Water Content (%)	0,3	1,29	0,27	0,02	79,07	98,45

[a] Adsorption with 2 grams adsorbent (bagasse) mass.
[b] Ultrafiltration with PES 15 % and 2,5 bar transmembrane pressure.
[c] Process combination of adsorption and ultrafiltration.

According to the recapitulation result, it can be seen that the characteristics of yellow grease resulted from ultrafiltration has reduction in acid number, peroxide number, iodine number, and water content, compared to the result obtained in adsorption stage. Characteristics of oil that meet the SNI standards are acid number, peroxide number, and water content, while the iodine number has not meet the range set by SNI. The change of colour of yellow grease before and after processing can be seen on Figure 5.

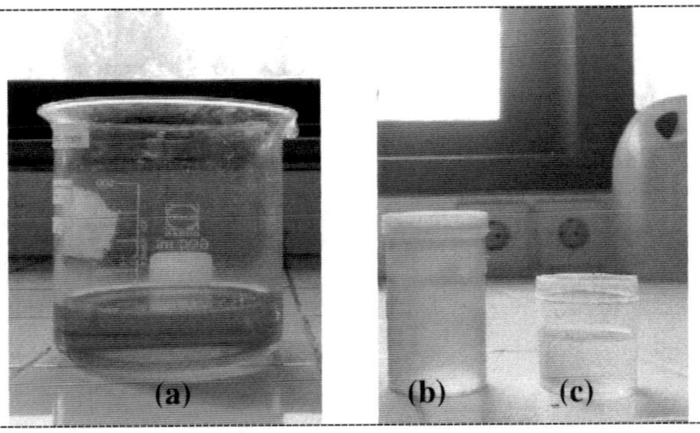

Figure 5. **(a)** Yellow grease; **(b)** Yellow grease resulted from adsorption process **(c)** Yellow grease resulted from combination of adsorption and ultrafiltration membrane process.

4. Conclusion

Variation of adsorbent mass in adsorption stage had effect on yellow grease characteristic. Adsorbent mass that was efficient to use for yellow grease preliminary treatment was by 2 grams, with acid number, peroxide number, iodine number, and water content decreased to 2.55 Mg KOH/gr; 4.19 Meq/kg; 48.02; and 0.27% respectively. Transmembrane pressure that was efficient in increasing yellow grease characteristics in ultrafiltration membrane refining process was at 2.5 bar, which could reduce the value of acid number, peroxide number, and water content respectively to 1.12 Mg KOH/gr; 1.8 Meq/kg; and 0.02%. The yellow grease's acid number, peroxide number, and water content from the processes combination has meet the range set by SNI, while the iodine number has not.

References

[1] Ketaren S 1986 *Introduction to Technology and Fatty Food* (Jakarta: University of Indonesia Press)
[2] Wulyoadi S and Kaseno 2004 *Proc. Nat. Conf. on Chemical and Process Engineering* Vol 11 (Semarang: University of Diponegoro) p 1-7
[3] Niazmand R, Farhoosh R, Razavi S, Mousavi S M and Noghabi M S 2011 Investigated of Quality and Stability of Canola Oil Refined by adding Chemical Agents and Membrane Processing *J. Procedia Food Sci.* **1** 90-94
[4] Kwiecińska A, Iluk T, Kochel M and Szul M 2016 The Application of Ultrafiltration for Preliminary Treatment of Liquid Waste streams Generated in Gasification Process *J. Civil Environ Eng.* **6** 1-4
[5] Miyagi A, Subramanian R, and Nakajima M 2003 Membrane and Additional Adsorption Processes for Quality Improvement of Used Frying Oil *J. Am. Oil Chem. Soc.* **80** 927-932
[6] Farihah T and Mawarani L J 2013 Effect of Activated Carbon Size from Bagasse as Biomaterials Pretreatment on Characteristics of Biodiesel from Used Cooking Oil *J. POMITS Technical.* **2** 268-272
[7] Mulder 1996 *Basic Principles of Membrane Technology* (Netherland: Kluwer Academic Publishers)
[8] Wenten I G, Himma N, Annisa S, and Prasetya N 2015 *Dictat of Preparation Superhidrofobic Membrane, Its Characterization and Applications* (Bandung: Bandung Institute of Technology)
[9] Triyanto A 2013 *Improving the Quality of Cooking Oil Using Activated Bagasse and Neutralization with $NaHSO_3$* (Semarang: Semarang State University)

[10] Widyaningsih S and Purwati 2013 Utilization of Nata de Coco Membrane as a Filtration Media For The Recovery of Used Cooking Oil *J. Molecular*. **8** 20-30

IOP Conf. Series: Materials Science and Engineering **334** (2018) 012067 doi:10.1088/1757-899X/334/1/012067

The utilization of *Eschericia coli* and *Shewanella oneidensis* for microbial fuel cell

S R Juliastuti[1,2], R Darmawan[1], A Ayuningtyas[1] and N Ellyza[1]

[1] Department of Chemical Engineering, Institut Teknologi Sepuluh Nopember, Kompleks ITS, Keputih Sukolilo, Surabaya, Indonesia, 60111

E-mail: julizainul@gmail.com

Abstract. Microbial Fuel Cell (MFC) is a technology that convert chemical energy into electrical energy with catalytic reaction from microorganism. The research method using bacteria in organic waste on anode compartment and ferricyanide solution on cathode compartment. Wastewater from sugar factory was used as organic waste with bacterial concentration of 10%, 12.5%, 15%, 17.5% (v/v) and with bacteria mixture ratio 1:1, 1:2, 2:1. The result of the research showed that the best voltage of bacteria concentration was 12.5% for *Eschericia coli* and *Shewanella oneidensis* bacteria, which were 847 mV and 988 mV, and for the mixed bacteria variable was 1:2 ratio with the voltage was 1261 mV. For 12 days, the largest percentage of the decrease of BOD_5 was 12.5% *Eschericia coli* bacteria concentration variable reached 84.531% and 17.5% *Shewanella oneidensis* was 73.779%. The best Fe^{3+} reduction was 53.52% for *Escherichia coli* at 10% concentration (v/v), and for *Shewanella oneidensis* bacteria reached out of 62.22% at 15% concentration (v/v). In the variable with mixed bacteria was obtained the best reduction result on the ratio of *Eschericia coli* : *Shewanella oneidensis* 1:2 was 77,44%.

1. Introduction

MFC is one of the ways to produce energy continuously in the form of electricity from degradable materials. MFC is a tool for converting chemical energy into electrical energy with the aid of catalytic reactions of microorganisms [1]. MFC generate electricity by oxidizing anaerobic organic matter through the aid of bacteria. The catalytic activity and proton transfer are carried out using enzymes or additional mediators [11]. The working principle of MFC is the use of microbes that perform metabolism of the medium at the anode to catalyze the converter of organic matter into electrical energy by transferring electrons from the anode through a cable and generating current to the cathode [13]. The electron transfer of the anode is received by the complex ion at the cathode which has free electrons. Protons were also delivered to the cathodic through cation exchange membrane. The protons are transported through the salt bridge. In the cathodic electrons were received by Fe(III). And the power was generated by the movement of electrons from the anodic to the cathodic. Fe(III) reduction and current production occurred because of the movement of these protons and electrons [6]. MFC has several advantages that can generate electricity from organic waste and renewable biomass [7]. Bacteria are able to catalyze and adapt well to different organic materials present in environmental

[2] To whom any correspondence should be addressed.

waste to produce electrons. In the cathode compartment, there is a conductive electrolyte solution. Potassium ferricyanide ($K_3Fe(CN)_6$) is very well known as an electron acceptor in MFC systems [6]. $K_3Fe(CN)_6$ is an electroactive species capable of capturing electrons well with a standard reduction potential value of +0.36V. The greatest advantage in the use of potassium ferricyanide is the resulting low overpotential when using a carbon electrode [11].

2. Materials And Methods

2.1 MFC construction

Dual chamber microbial fuel cell was constructed, with a tubular junction of 2 cm diameter, connecting the compartments contained the salt bridge [9] at a height of 4 cm above the base. The salt bridge was formed by dissolving 4% KCl salt in 10% agar [12]. The working volumes for the anode and cathode chambers were 800 mL. The graphite size 22.844 mm were used as electrodes for both the chambers. The electrodes in the two chambers were connected with a copper wire for electron (e^-) transfer. The exposed metal surfaces were sealed with non-conductive epoxy tapes. Dry electrodes were introduced into anode and cathode chambers. The whole anaerobic set up was autoclaved before assembling [2].

2.2 Cathodic and anodic reagent preparation

The cathodic chamber was filled with the solution of $K_3Fe(CN)_6$ made by mixing 400 ml a solution of 0.1 M and 400 ml of buffer phospate 0.1 M with pH 7.0 [10]. The anodic chamber was inoculated with the bacteria (*Escherichia coli* and *Shewanella oneidensis* and binary culture of both bacteria) and sugar waste with 5% concentration [14].

2.3 Electrode preparation

Carbon electrodes soaked into a solution of HCl of 1 M for 1 day later flushed by using aquades. After that, the electrodes soaked again into a solution NaOH of 1 M for 1 day with aquades. Electrodes soaked in solution aquades until the electrodes will be used.

2.4 Procedure

A digital multimeter was connected on both the electrodes. The postive end at a cathodic chamber and the negative end at a anodic chamber. The voltage was recorded every 24 h. Then, the power was calculated by using Eq 1

$$P = (V\,I)/A \qquad (1)$$

Where P is the power density in milli watts (mW), V is the voltage in milli volts (mV), I is the current density in milli ampere (mA) and A is the surface area of electrode. Voltage and current measurement is carried out by using a multimeter with a system OCV (Open Circuit Voltage) [8]. By getting the voltage and the current, then it can be calculated that the resulting power density.
The variable of bacteria concentration of this experiment is 10%, 12.5%, 15% and 17.5% (v/v) with volume of chamber is 800 ml. For example, to obtain the variable of bacteria concentration 10% is measured 10 ml for the volume of bacteria and 790 ml for the wastewater. In anodic chamber was record the voltage, pH and BOD, otherwise in cathodic chamber was measured the voltage, pH and the Fe reduction. The BOD value is obtain with measured the liquid of the anodic chamber with DO meter, and then the result value of DO meter was calculated by using Eq 2

$$BOD5 = DO_s^o - DO_d^f + \text{volume of bottle/volume of sample}\ (DO_d^f - DO_s^f) \qquad (2)$$

Where,
DO_s^o = initial dissolved oxygen concentration in sample
DO_d^f = final dissolved oxygen concentration in water
DO_s^f = final dissolved oxygen concentration in sample

3. Result And Discussion

3.1 Voltage and Power Density

Analysis of voltage and current are performed every day. The result of this experiment, maximum voltage of *Eschericia coli* was generated at a concentration of bacteria 12.5% (v/v) was 847 mV and maximum voltage of *Shewanella oneidensis* was generated at a concentration of bacteria 12.5% (v/v) was 988 mV. For the mixed bacterial variables, the voltage was generated in ratio 1:2 was 1261 mV.

Power density is affected by voltage, electric current and the surface area of electrode. In this experiment, an electric current recorded was 0.01 mA, and for the surface area of the electrodes was 22,844 cm^2. Maximum power density of *Eschericia coli* at a 12.5% (v/v) was 0.371 mW/m^2 and maximum power density of *Shewanella oneidensis* at a 12.5% was 0.432 to mW/m^2. For the mixed bacterial variables, the maximum power density at a 1:2 was 0.552 mW/m^2.

The number of microorganisms can affect in the voltage, because of the more microorganisms it will be more electrons were released. The substrate on the anode chamber become food for the microorganism and also energy to release electrons. In this experiment, there are a many bacteria but the voltage generated is small. This is due to the presence of a biofilm layer, formed by bacterial metabolic activity, where the layers can also be barriers to bacteria to release electrons and degrade organic materials in the substrate.

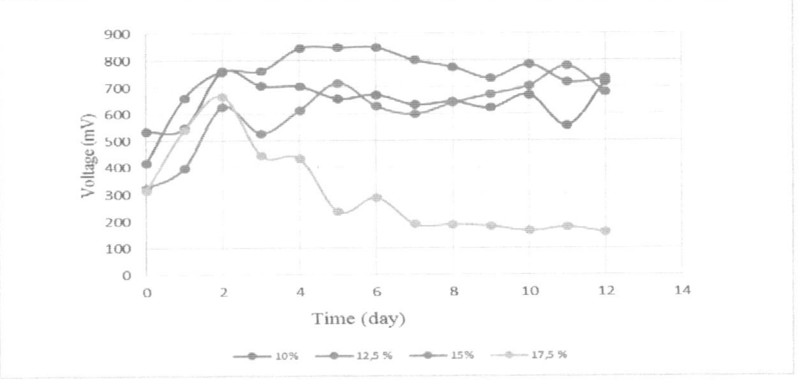

Figure 1. Voltage of *Eschericia coli* for 12 days

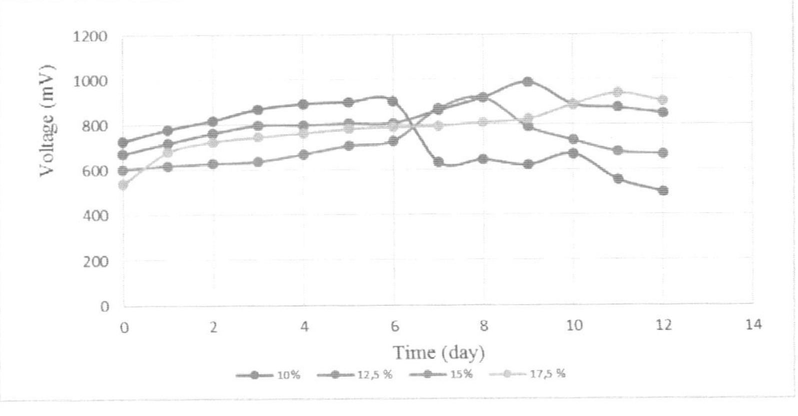

Figure 2. Voltage of *Shewanella oneidensis* for 12 days

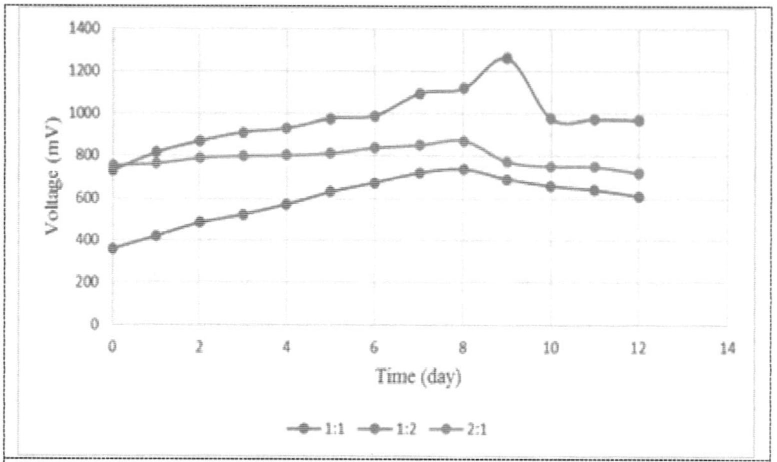

Figure 3. Voltage of mixture bacteria *Eschericia coli:Shewanella oneidensis* for 12 days

3.2 Reduction of Ferricyanide

In the cathode chamber of the MFC reactors, used 400 mL ferricyanide solution 0.1 M mixed with 400 mL phospate buffer solution. In the cathode chamber there is a reduction reaction of Fe(III) to Fe(II). The results of this experiment indicated that the Fe(III) level was decrease. In variable of *Eschericia coli*, Fe(III) was reduced 53.52 %, 12.56 %, 48.86% and 47.16 % for bacteria concentrations of 10%; 12.5%; 15% and 17.5%. In variable of *Shewanella oneidensis*, Fe(III) was reduced 37.14 %, 44.20 %, 62.22% and 55.78 % at bacteria concentrations of 10%; 12.5%; 15% and 17.5%. And then in variable of ratio of *Eschericia coli* and *Shewanella oneidensis*, Fe(III) was reduced 76.54 %, 77.45 %, and 66.74 % at bacteria ratios of 1:1 ; 1:2 and 2:1. In the cathode, reduction reaction occurs, where Fe^{3+} will turn into Fe^{2+} with the help of the electrons coming from the anode. Fe^{2+} is oxidized to Fe^{3+} by releasing the electrons and will reacted with H^+ coming from the anode chamber through the salt bridge to form water molecules. The Fe(III) reacted with these electrons and reduced Fe(III) to Fe(II) form as in equation below :

$$Fe^{3+} + 8H^+ + 8e^- + 8O_2 \rightarrow Fe^{2+} + 8H_2O \qquad E^o = +0.791 \text{ V}$$

The half-cell reaction has a redox potential (E^o) of 0.791 V [5]. The positive potential (E^o= + 0.791 V) indicates the greater affinity of Fe(III) for electrons and tendency to be reduced [4].

IOP Conf. Series: Materials Science and Engineering **334** (2018) 012067 doi:10.1088/1757-899X/334/1/012067

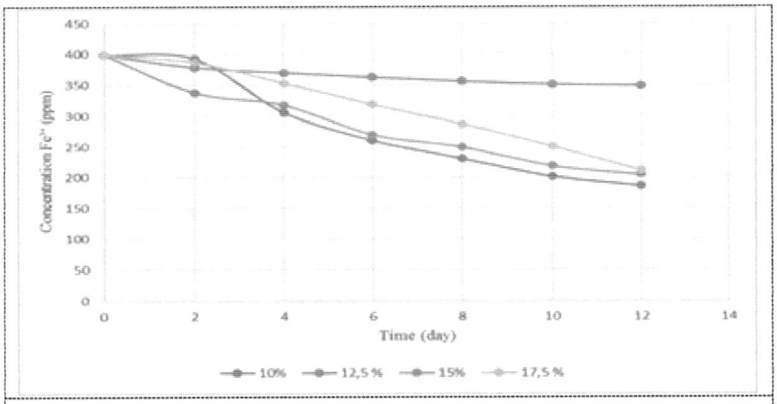

Figure 4. Reduction of Ferricyanide concentration of *Eschericia coli* for 12 days

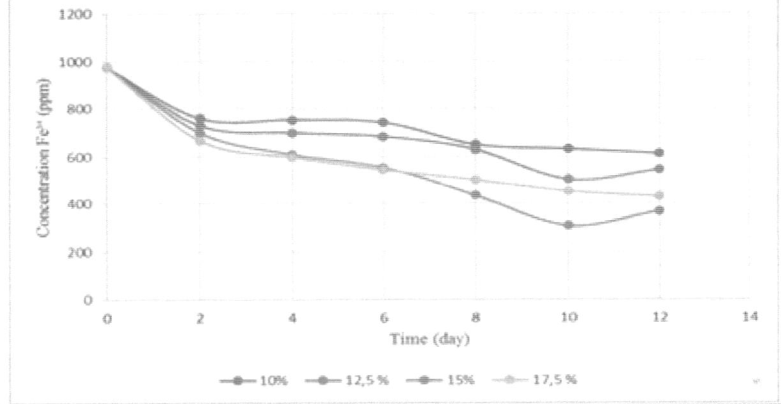

Figure 5. Reduction of Ferricyanide concentration of *Shewanella oneidensis* for 12 days

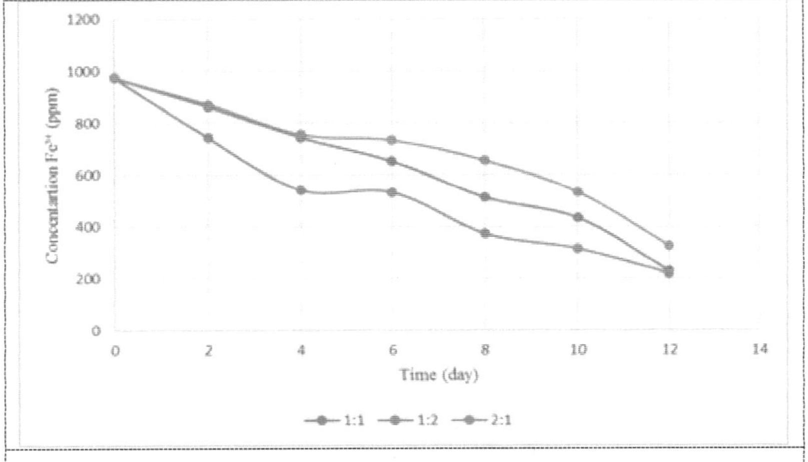

Figure 6. Reduction of Ferricyanide concentration of mixture bacteria *Eschericia coli:Shewanella oneidensis* for 12 days

4. Conclusion

In this experiment, it can be concluded that the higher density power generated at *Eschericia coli* concentration of 12.5% was 0.371 mW/m^2, whereas *Shewanella oneidensis* concentration of 12.5% was 0.432 mW/m^2, and for variable ratio of *Eschericia coli* and *Shewanella oneidensis* is in ratio 1:2 was 0.552 mW/m^2. The largest decrease in the reduction of Fe of *Eschericia coli* are at concentrations of 10% was 53.52%, *Shewanella oneidensis* at concentrations of 15% was 62.22%, while for a variable mixture of bacteria on the ratio 1:2 was 77.45%.

Acknowledgments

We would thank to Wastewater Treatment Laboratory, Chemical Engineering Department, Institut Teknologi Sepuluh Nopember, for the facilities provided to accomplish the research.

References

[1] Allen RM and Bennetto HP. 1993. *Microbial fuel cells: electricity production from carbohydrates.* J. Appl. Biochem. Biotechnol. 39: 27-40.

[2] A. Carmalin Sophia and Saikanth Sai. 2016. *Modified microbial fuel cell for Cr(VI) reduction and simultaneous bio-electricity production*, Journal of Environmental chemical engineering 4 2402–2409

[3] Gralnick A. Jeffrey dan Heidi H. Hau. 2007. *Ecology and Biotechnology of the Genus Shewanella.* Annual Review of Microbiology. 61:237-58

[4] Karthikeyan. 2016. *Effect of composites based nickel foam anode in microbial fuel cell using Acetobacter aceti and Gluconobacter roseus as a biocatalysts*, J. Bioresource Technology.

[5] Liu H and Logan BE. 2004. *Electricity Geneterion Using An Air Chatode Singel-Chamber Microbial Fuel Cell In The Presence and Absence of Proton Exchange Membrane,* J. Environmental Science Technology 38: 4040

[6] Logan E. Bruce, et al. 2006. *Microbial Fuel Cells : Methodology and Technology.* Environmental Science & Technology. Vol. 40, No. 17.

[7] Lovley, Derek R. 2008. *The microbe electric: conversion of organic matter to electricity*, Current Opinion in Biotechnology, 19:564-571

[8] Miroliaei M.R, Abdolreza Samimi, Davod Mohebbi-Kalhori. 2015. *Competition Between E. Cli and Shewanella s. for Electricity in Air Cathode MFC in Presence of Methylene Blue as Artificial Mediator.* Environmental Progres & Suistaiable Energy. Vol. 34 No. 4.

[9] Miroliaei M.R, Abdolreza Samimi, Davod Mohebbi-Kalhori. 2014. *Kinetics Investigation of Diversity Cultures of E.coli and Shewanella sp., and their Combined Effect with Mediator on MFC Performance.* Journal of Industrial and Engineering Chemistry. Vol 36 No.4

[10] Mohan S.V, G Mohanakrishna, B Purushotham Reddy, R Saravanan, P.N Sarma. 2007. *Bioelectricity generation from chemical wastewater treatment in mediatorless (anode) microbial fuel cell (MFC) using selectively enriched hydrogen producing mixed culture under acidophilic microenvironment.* Biochemical Engineering Journal 39 121-130

[11] Rabaey K, Boon N, Siciliano SD, Verhaege M, dan Verstaete W. 2004. *Biofuel cell select for microbial consortia that self-mediate electron transfer*. J. Applied Environmental Microbiology 70: 5373-5382.

[12] Ramya N, Renganathan K, S Barathi, Venkatraman. K. 2013. *Performance of Salt-Bridge Microbial Fuel Cell at Various Agarose Concentrations Using Hostel Sewage as Substrate.* International Journal of Advancements in Research & Technology, Volume 2, Issue 5, May

[13] Rozendal, R. A., Hamelers, H. V. M. & Buisman, C. J. N. 2006. *Effects of Membrane Cation Transport on pH and Microbial Fuel Cell Performance*. Journal Environmental Science Technology, Vol. 40, No.17, 5206–5211.

[14] Scott, Keith dan Murano, Cassandro. 2007. *Microbial Fuel Cells Utilising Carbohydrates.* Journal of Chemical Technology and Biotechnoogy (82):92-100

IOP Conf. Series: Materials Science and Engineering **334** (2018) 012068 doi:10.1088/1757-899X/334/1/012068

Tofu wastewater treatment by sediment microbial fuel cells

W Rinaldi[1,2], Abubakar[1], R F Rahmi[1] and Silmina[1]
[1] Teknik Kimia Universitas Syiah Kuala, Banda Aceh, Aceh, Indonesia

E-mail: wahyu.rinaldi@che.unsyiah.ac.id

Abstract. This research aimed to measure power density generated by sediment microbial fuel cells (SMFCs) by varying anode position and wastewater concentration. Anode position was varied at 2 cm and 4 cm under the surface of sediment, while wastewater concentration varied into 25%, 50%, 75% and 100%. The electrodes employed was stainless steel mesh, while the organic subtrate source was taken from wastewater of soybean washing and boiling process. The sediment was taken from the Lamnyong River around the outlet of tofu industry wastewater. SMFCs was run until the power density was relatively small. The produced electricity represented in power density. The results of this research showed that power density was decreased over time. Generated power density by varying 2 cm and 4 cm position of anode under the sediment surface was not significantly different, while the lowest wastewater concentration, 25%, gave the highest power density.

1. Introduction

The volume of wastewater from 8400 tofu industries across Indonesia in 2010 reached the number of 20 million m^3[2]. Most tofu industries in Indonesia are home scale industries which have minimum wastewater treatment installation. They usually let the wastewater flow into the river.

Table 1. Tofu industry wastewater parameters and quality standards.

Parameters	Concentration (mg/L)	
	Tofu industry in Banda Aceh[2]	Quality standards[3]
BOD	4415.5	150
COD	8500	300
TSS	841	200

With the BOD, COD and TSS levels above the quality standard as be shown in Table 1, the tofu wastewater can potentially pollute the environment and harm people's health. Tofu wastewater comes from boiling, filtering, pressing, and washing process of soybean. The wastewater contains organic compounds such as protein, carbohydrate and fat[4,5].

The most appropriate wastewater treatment installation to most tofu industries in Indonesia is the most convenient and cost effective method. The conventional methods of tofu wastewater treatment utilized are aerobic[6] and anaerobic[1] processes. Both of those processes are giving side effects;

[2] To whom any correspondence should be addressed.

aerobic process requires wide pond area and energy to run aeration pump, while anaerobic process produces methane that will pollute environment. By those considerations, the previous methods are ineffective. Microbial fuel cells (MFCs) is a new technology that can provide solution for wastewater treatment problem of tofu industry in Indonesia, as this process requires small area, no energy input, no by side product, and another advantage is MFCs can produces electrical energy.

SMFCs is one type of MFCs that takes advantage of metabolic activity of bacteria present in sediment to develop energy while degrading organic compound. According to current SMFCs researches, the primary focus of the study is to obtain higher energy, but not many of them take SMFCs application to treat wastewater as center of interest[14]. By advocating SMFCs method to treat tofu wastewater, not only it will reduce the pollution, but also it will produce electricity as well.

Several factors affecting SMFCs performance are materials and arrangement of electrodes[8]. Carbon and stainless steel are the most common materials used, whether use in one or both electrodes[9]. Stainless steel that is more robust and relatively cheaper than carbon is more applicative to be used in SMFCs to treat wastewater of home scale tofu industry. Stainless steel mesh (SSM) size chosen is 40 mesh type 304 which is the alloy of 18% chromium and 8% nickel. This choice is based on the result of research done by Zhang et al.[9], where the highest power density obtained is from SSM size of 30 mesh, small difference from 50 mesh, and the power density is linearly decreasing as the SSM size is smaller (70 mesh, 90 mesh and 120 mesh).

Based on Huang et al.[11], the most suitable internal resistor will give the highest power. Thus, in this research, the selection of the optimum internal resistance is completed using potentiometric test, where it will give the highest electric power.

High substrate concentration is also adequate to increase electric power in certain condition, that is because higher concentration will facilitate proton transfer to cathode surface[12]. High concentration also means more organic material to be degraded by microbial activity, thus more electric potential can be obtained[13].

Therefore, the objective of this study is to learn the effectivity of SMFCs in treating wastewater of tofu industry, SMFCs performance in decreasing COD level of wastewater, and the effect of anode position and wastewater concentration in developing power density obtained.

2. Materials and method

2.1. Preparation of SMFCs reactor

SMFCs reactor consisted of easily obtained materials, glass jar with volume of 0.5 L and 5 cm x 5 cm stainless steel wire 40 mesh as the electrodes. Electrodes consisted of anode and cathode. While anode was put in the sediment, cathode was put in the wastewater. Both electrodes were connected by electrical wire to facilitate electron transfer. The sediment was collected near the wastewater outlet on Lamnyong River where tofu industries flow their wastewater into. The first pretreatment was sieving the sediment using 5mm-opening wire to separate gravel and rubbish. It was then left for 24 hours to minimize water content. Then, the treated sediment was put into the jar.

Wastewater was collected from soybean boiling water and washing process in tofu industry around Lamnyong River. Collected wastewater was then varied into several concentrations and poured into each jar.

2.2. Tofu wastewater treatment using SMFCs

In this study, the thickness of sediment was set to 4 cm from the bottom of the reactor, while the heigth of wastewater was 4 cm from top of the sediment. To learn the effect of anode position in degrading COD and producing electric power, the anode position was varied on 2 cm and 4 cm from top of sediment. On the other side, cathode position was fixed on 2 cm from wastewater surface. This point will give the maximum power, as the dissolved oxygen level is stable to accommodate reaction at cathode surface. This is because when cathode too close to wastewater surface or too deep the

SMFCs performance will be affected by dissolved oxygen deficiency due to evaporation and depth factor[14].

The effect of wastewater concentration was studied by varying the concentration into 25%, 50%, 75% and 100%. After electrodes, sediment and wastewater were set, cables from anode and cathode were attached with 50kΩ resistor and datalogger DATAQ DI-710 to record electricity generated over time.

Efficiency of SMFCs is determined from COD removal in wastewater. COD was analyzed using titrimetric method. The COD removal efficiency is calculated using Equation 1.

$$\text{COD removal}=\frac{\text{COD}_{\text{initial}}-\text{COD}_{\text{final}}}{\text{COD}_{\text{initial}}}\times 100\% \quad (1)$$

To compare SMFCs performance, control reactor was used for each variation. The control reactor use the same equipment but electrodes, thus in terms of reducing wastewater, the process was conducted aerobically and anaerobically. Data recorded by data logger are voltage (E_{cell}) produced. The electric power (P) is then calculated using equation 2.

$$P = IE_{cell} \quad (2)$$

where I is electric current. As the Ohm law stated $E_{cell} = IR_{ext}$, then

$$P = \frac{E_{cell}^{2}}{R_{ext}} \quad (3)$$

with R_{ext} is the external resistance used during this research.

Power density indicated electric power obtained per electrode surface area, but it is often expressed per anode surface area (A_{anode}). The calculation is using Equation 4[8]:

$$P = \frac{E_{cell}^{2}}{A_{anode}R_{ext}} \quad (4)$$

3. Result and disscussion

3.1 Effect of anode depth to power density

With relatively small depth variation, Figure 1 shows that power densities was not significantly affected by anode position. In initial wastewater concentration of 25%, reactor with 4 cm anode depth produced power density of 215.954 mW/m^2, while 2 cm-deep anode yielded 217.142 mW/m^2 after one hour. The amount of power generated decreased with time due to the formation of thin biofilm on wastewater surface which inhibit oxygen transfer into cathode chamber.

These results were different from research conducted by Lehman et al.[16], who varying spacing between anode and cathode, ~3cm (reactor 1) and ~10.5cm (reactor 2). Lehman's research showed that highest power density developed in reactor after 60 days operation. It is due to sediment which is closer to the surface contains more dissolved oxygen[17], therefore major bacteria, heterotrophic bacteria, existed in that position will consume more substrate compared to electron donor bacteria (anodophile and electrodophile) presented in deeper sediment. This deviation is caused by variable difference of anode spacing is too small.

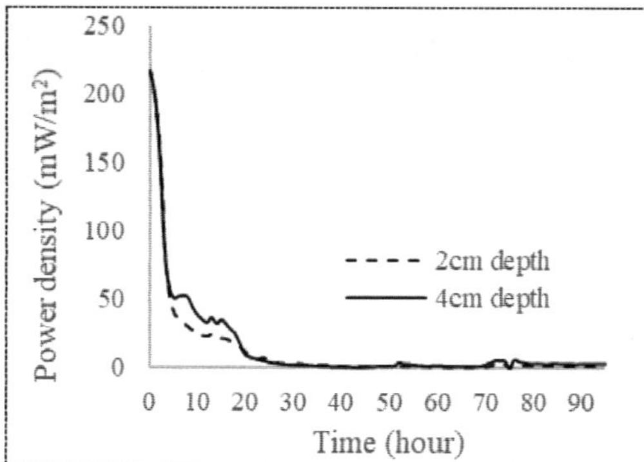

(a) Wastewater concentration 25%

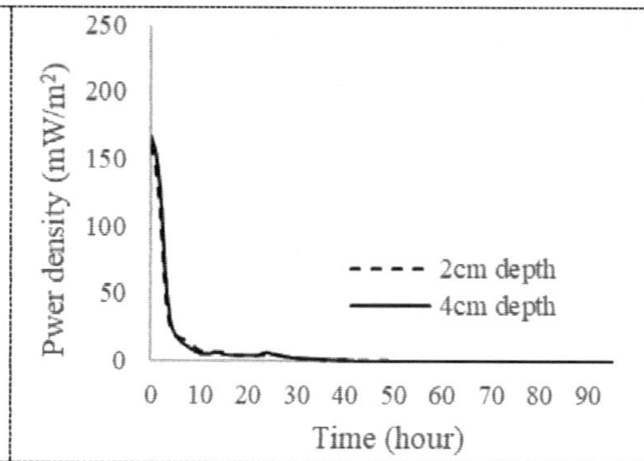

(b) Wastewater concentration 50%

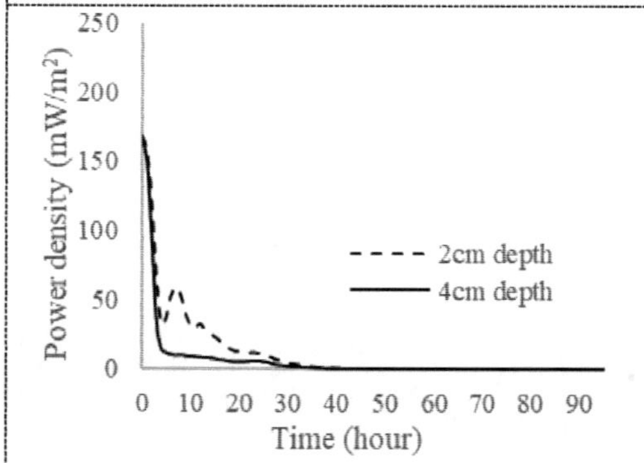

(c) Wastewater concentration 75%

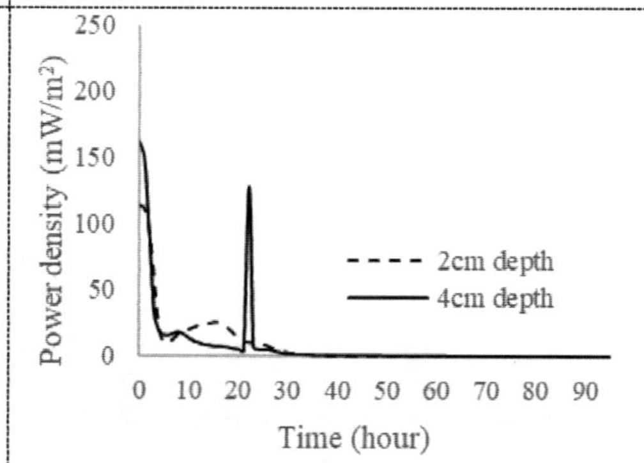

(d) Wastewater concentration 100%

Figure 1. Power densities from anode depth of 2 cm and 4 cm in various initial wastewater concentration.

3.2 Effect of wastewater concentration to power density

Wastewater of tofu industry used in this study contains complex mixture of organic substrates, thus this kind of wastewater is tend to form biofilm on its surface.

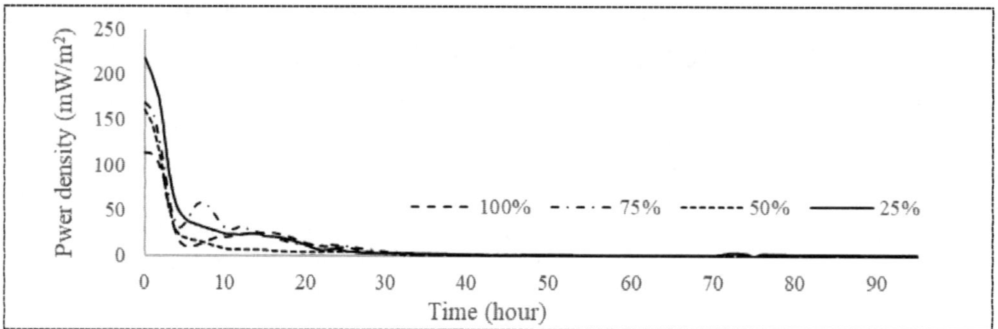

(a) Depth of 2 cm

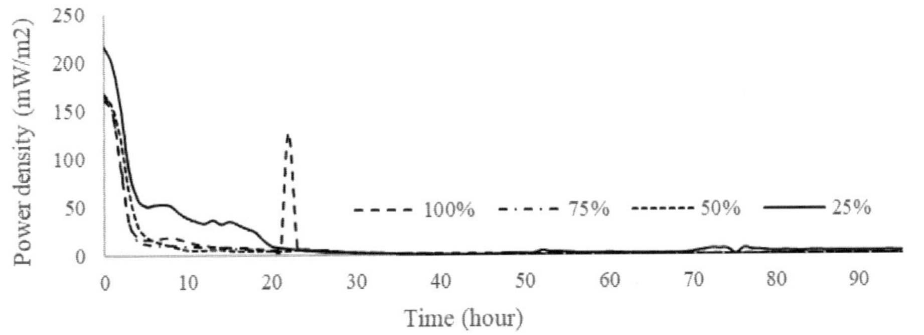

(b) Depth of 4 cm

Figure 2. Correlation between power densities and initial wastewater concentration.

Figure 2 showed that electricity generated by SMFCs is affected by wastewater concentration. The highest power density is obtained from wastewater concentration of 25%; 217.142 mW/m^2 and 215.954 mW/m^2 for each anode depth of 2 cm and 4 cm. This is due to wastewater with higher concentration tend to form thicker biofilm layer on the wastewater and cathode surface, therefore it will deplete electric potential generated[19,20].

3.3 COD Removal using SMFC

After 95 hour operation, COD removal in control reactor, SMFCs reactor with anode position of 2 cm and SMFCs reactor with anode position of 4 cm respectively are 60%, 40% and 30%. As mentioned result, it can be seen that highest COD removal is achieved by control reactor. Similar result is also obtained by Xu et al.[7] and Liu et al.[21], where control reactor showed best performance in COD removal.

This occured since COD removal is affected by the growth of microbes that is able to degrade organic substrates[22]. The growth process is concerned by dissolved oxygen level in substrates, while in SMFCs reactor, oxygen can also fixate electron (electron acceptor), thus dissolved oxygen content in SMFCs tend to be lower than that of control reactor. On the other side, even though the COD removal rate is low, SMFCs will give advantages as it can perform better removal of several specific compound, such as nitrate removal and decolorization[7].

4. Conclusion

Based on the research of tofu wastewater treatment using SMFCs done, it can be known that effect of anode position of 2 cm and 4 cm from sediment surface is not significant to electric power densities. This is because of small differences between the two variables. Highest power density is generated from wastewater concentration of 25%. This occurred as in lower concentration, biofilm formation that will inhibit oxygen transfer to electrode is thinner than in the higher concentration. Power densities in SMFCs is decreasing with time because of oxygen transfer to electrode decreased which mainly affected by biofilm layer formation. COD removal by SMFCs is relatively smaller than of control reactor due to lesser dissolved oxygen in SMFCs which will affect growth of degrading bacteria in wastewater. To raise power densities of SMFCs, one requires further study in resolving biofilm or thick layer formed on tofu wastewater surface or on cathode surface, optimum anode position, and other reactor operating condition.

IOP Conf. Series: Materials Science and Engineering **334** (2018) 012068 doi:10.1088/1757-899X/334/1/012068

References

[1] Sutisna M and Pratama Y 2014 Batch anaerobic tofu wastewater treatment *J. Institut Teknologi Nasional* **2**(1) 1-10 (in Indonesian)

[2] Faisal M, Maulana F, Gani A and Daimon H 2015 Physical and chemical properties of wastewater discharged from tofu industries in Banda Aceh city, Indonesia *Research J. Pharmaceutical. Biological. Chemical Sciences* **6**(4) 1053-58

[3] Environment Minister Regulation No. 5 2014 about Wastewater Standard (in Indonesian)

[4] Belen F, Sanchez J, Hernandez E, Auleda J M and Raventos M 2012 One option for the management of wastewater from tofu production: freeze concentration in a falling-film system *J. Food Engineering* **110** 364-73

[5] Kim D, Lee D and Kim M 2011 Enhanced biohydrogen production from tofu residue by acid/base pretreatment and sewage sludge addition *International J. Hydrogen Energy* **36** 13922-7

[6] Mulyana Y, Purnaini R and Berlian 2013 Domestic wastewater treatment for water reuse *J. Mahasiswa Teknik Lingkungan UNTAN* **1**(1) 1-10. (In Indonesian)

[7] Xu B, Ge Z and He Z 2015 Sediment microbial fuel cells for wastewater treatment: challenges and opportunities *Water Resource Technol.* **1** 279-84

[8] Liu H, Cheng S and Logan B E 2005 Production of electricity from acetate or butyrate in a single chamber microbial fuel cell *Environmental Science & Technology* **39** 658-62

[9] Zhang X, Merril M D, Tokash J C, Saito T, Cheng S, Hickner M A and Logan B E 2010 Mesh optimization for microbial fuel cell cathodes constructed around stainless steel mesh current collectors *J. Power Sources* **196** 1097-102

[10] Kristin E 2012 *Electric production from tempe industry waste through microbial fuel cell* (Depok: Skripsi Jurusan Teknologi Bioproses Fakultas Teknik Universitas Indonesia Depok) (In Indonesian)

[11] Huang Y, He Z, Kan J, Manohar A K, Nealson K H and Mansfeld F 2012 Electricity generation from a floating microbial fuel cell *Current Microbiology* **62** (2) 3029-35

[12] Najafgholi Z, Rahimnejad M and Najafpour G 2015 Effect of electrolyte conductivity and aeration performance of sediment microbial fuel cell *J. Renewable Energy and Environment* (1) 49-55

[13] Xu X, Zhao Q L and Wu M S 2014 Removal and changes of sediment microbial fuel cells and amorphous ferric hydroxide *Chemical and Biochemical Engineering* **28**(4) 561-6

[14] Muler A I C 2015 *Effect of experimental parameters on the voltage output of a sediment microbial fuel cell* (Islandia: Tesis Master of Science in Sustainable Energy Engineering Reykjavik University Islandia)

[15] Campo AG, Canizares P, Lobato J, Rodrigo M and Morales F J F 2014 *Effects of external resistance on microbial fuel cell's performance* (Switzerland: Springer International Publishing)

[16] Lehman J, Lin H and Hu B 2015 Optimization of sediment microbial fuel cell for power generation *Scientific J. UROP*

[17] An J, Kim B, Nam J, Ng H Y and Chang I S 2013 Comparison in performance of sediment microbial fuel cells according to depth of embedded anode *Bioresource Technology* **127** 138-42

[18] Mardanpour M M, Esfahany M N, Behzad T, Sedaqatvand R, Sharifi F and Naderi F 2013 Factors affecting the performance of single chamber microbial fuel cell using a novel configuration *Iranica* J. *Energy & Environment* **4** (4) 343-7

[19] Hunter W J and Manter D K 2011 Increased electrical output when a bacterial ABTS oxidizer is used in microbial fuel cell *Current Microbiology* **62** (2) 633-8

[20] Izadi P and Rahimnejad M 2014 Simultaneous electricity generation and sulfide removal via a dual chamber microbial fuel cell *J . Biofuel Research* **1** 34-8

IOP Conf. Series: Materials Science and Engineering **334** (2018) 012068 doi:10.1088/1757-899X/334/1/012068

[21] Liu H, Ramnarayanan R and Logan B E 2004 Production of electricity during wastewater treatment using a single chamber microbial fuel cell *Environ. Sci. Technol.* **38** 2281-5
[22] Zhang X, He W, Ren L, Stager J, Evans P J and Logan B E 2015 COD removal characteristics in air-cathode microbial fuel cells *Bioresource technol.* **176** 23-31

IOP Conf. Series: Materials Science and Engineering **334** (2018) 012069 doi:10.1088/1757-899X/334/1/012069

Formulation of oleozon with *Phaleria macrocarpa* and *Cinnamomum burmanii* extract for diabetic wound treatment

N Salsabila[1], F Moulydia[1] and S Bismo[1,2]
[1] Department of Chemical Engineering, Faculty of Engineering, Universitas Indonesia, Kampus Baru UI Depok 16424, Indonesia

E-mail: setijo.bismo@ui.ac.id

Abstract. In this work, the effect of ozonation on coconut oil and mixture of coconut oil and olive oil was studied. The properties of ozonated oils (oleozon) were analytically tested by the method of iodine number, acid number, peroxide number, and FT-IR as general chemical substances. Ozonation may increase the peroxide and acid number for both oils but decrease the iodine number. The best ozonation condition has been seen from an increase of 277.52% acid number, peroxide number about 114.77 meq O_2^{2-}/kg oil, and decrease of iodine number up to 22%. Furthermore, ozonated oils were mixed with herbal extract and be tested the diabetic wound healing ability through antibacterial activity test. A mixture of 160 mL coconut oil that ozonated for 72 hours and 0.18 gram herbal extracts with *n*-hexane solvent showed the highest inhibition zone of 18.3 mm in *Staphylococcus aureus* bacteria.

1. Introduction

One of the most potentially dangerous complications in people with diabetes is diabetic wounds. High sugar levels in the blood induce to the growth of anaerobic bacteria, resulting an infection in the diabetic wound so that it becomes difficult to recover and get worse called gangrene or diabetic ulcers.

The ozonation of vegetable oils may produce a substance having the function of antiseptic against bacteria which cause skin diseases [4]. Olive oil is the most widely used oil as an ozonated vegetable oil, this is due to the high content of unsaturated fatty acids. In addition, studies on olive oil show the ability to heal skin wounds [8]. However, the price of olive oil is expensive and olive oil does not come from Indonesia, so research on the ability of other ozonated vegetable oil derived from Indonesia was carried out. Coconut oil is chosen because of its relatively cheap price and its huge availability in Indonesia. In addition, the content of lauric acid and capric acid in coconut oil is able to accelerate wound healing and reduce infection [7].

The reaction of ozone with vegetable oils produces *Criegee ozonide*, hydroperoxide, and aldehyde products [3]. In addition, mixing vegetable oil with natural ingredients can also increase the effectiveness of the oil. Saponins in *Phaleria macrocarpa* and sinamaldehid in *Cinnamomum burmannii* are also known to have antibacterial effects [2,5].

There are three goals to be achieved from this research. The first goal is to get the best ozonation condition on vegetable oil based on its physical and chemical properties. Second, to get the oleozon composition with *Phaleria macrocarpa* and *Cinnamomum burmannii* extract which is effective in healing diabetic wound based on their antibacterial effect. Third, to

[2] To whom any correspondence should be addressed.

compare the ability of coconut oil and olive oil in healing diabetic wound based on their bacterial disinfection capabilities.

2. Experimental section

2.1. Materials
Materials used in this research are coconut oil (Barco), olive oil (D'VIRGIN), *Phaleria macrocarpa*, *Cinnamomum burmannii*, isopropyl alcohol and *n*-hexane solvents, material for the determination of iodine number, material for the determination of acid number, material for the determination of peroxide number, and material for the determination of antibacterial activity.

2.2. Instrumentation
Equipment used in this research are ozonator, reactor, stirrer, chiller, laboratory glassware for the extraction and determination of iodine, acid, peroxide number, analytical balance, FT-IR spectrometer, LC-MS spectrometer, and equipment for the determination of antibacterial activity.

2.3. Procedure

2.3.1. Oil ozonation and herbal extraction. This study uses the main component of ozonator as an ozone-producing tool to apply in the ozonation process of vegetable oil mixture. The ozonation reaction of the oil will occur in a batch reactor equipped with a 4 L/min air flow rate and a coolant to keep the reaction temperature at 25 °C. At this stage, the ozonation of vegetable oil has been done continuously for 72 hours to see the relationship between the length of time ozonation with the resulting antibacterial effects. The vegetable oil mixture will be ozonated with variation of oil concentrations.

After oil ozonation, herbal extraction then performed. This stage aims to produce herbal plant extracts from *Phaleria macrocarpa* and *Cinnamomum burmannii*. These two herbs will be macerated using isopropyl alcohol (IPA) and *n*-hexane solvents for 5 days and stirring every 4 hours. After obtaining the extract by evaporating, the extract will be mixed with the ozonated vegetable oil to add antibacterial properties.

2.3.2. Characterization of oleozon. The test method used iodine number, acid number, peroxide number, and FT-IR spectrum. Iodine number is used to know the amount of double bond in oil. Acid number to indicate the amount of free fatty acids present in the oil. Peroxide number to show the level of oxidation of ozone to fatty acids. Then, FT-IR spectrum is used to find out the differences in functional groups that increase or decrease before and after the ozonation.

2.3.3. Characterization of compound in herbal extract. The test method used Liquid Chromatography–Mass Spectrometry (LC-MS). This test was performed to identify the compounds contained in the *Phaleria macrocarpa* and *Cinnamomum burmannii* extract.

2.3.4. Antibacterial activity determination. In this stage, ozonated vegetable oil with herbal extracts is put into the culture of *Staphylococcus aureus* bacteria that have been prepared before. The ability of bacterial disinfection qualitatively seen in the form of visual observations by looking at the diameter of the inhibit zone formed. The test was conducted by using the paper disc method.

3. Result and Discussion

3.1. Iodine number
Iodine number can be used to determine the content of double bonds in vegetable oils. The success of ozonation is marked by a decrease of the iodine number obtained. Based on Table 1 it can be seen that

on all variations of oil mixture the oil ozonation for 72 hours has decreased the iodine number. It can be inferred that the ozonator succeeds in attacking and lowering the double bond on pure coconut oil and also a mixture of coconut and olive oil.

The decrease of iodine number occurs because ozone reacts with double bonds in unsaturated fatty acids in vegetable oils. This reaction is called the Criegee reaction that produces ozonide, hydroperoxide, aldehydes, peroxides, dioxide, and polyperoxide [3]. The highest decrease in iodine number is in the ozonation process of pure coconut oil, which changes up to 22%. This is due to coconut oil is MCFA (Medium Chain Fatty Acid) which more responsive or faster to receive ozone than olive oil. Thus, coconut oil requires a faster ozonation time than olive oil.

Table 1. Iodine numbers for pure and oils ozonated after 72 h.

	Iodine number (g iod/100 g oil)		
Oil	Untreated oil	72 h ozonated oil	% Change
Coconut oil	2.06	1.61	21.84
Mix of coconut oil and olive oil (85:15)	3.45	3.40	1.45
Mix of coconut oil and olive oil (75:25)	4.85	4.42	8.86

3.2. Acid number

The acid number indicates the degree to which the triglycerides in the oil have broken down to release free fatty acids. The success of ozonation is marked by an increase of the acid number obtained. Based on the available data in Table 2, it can be seen that the acid numbers of oil after the ozonation have increased. It means that the ozonation in this research was successful.

The increase in acid number is triggered by increased peroxide decomposition and oxidation of aldehydes to carboxylic acids during the ozonation reaction. The highest increase in acid number is in the ozonation process of pure coconut oil, which changes up to 277.52%. This is proportional to the results obtained in the iodine number, where the highest iodine number decline belongs to pure coconut oil. The content of polyphenols and tocopherols in olive oil that serves as an antioxidant slows or prevents other molecular oxidation processes so that the ozonation process in pure coconut oil runs better than a mixture of coconut oil and olive oil.

Table 2. Acid numbers for pure and oils ozonated after 72 h.

	Acid number (mg KOH/g oil)		
Oil	Untreated oil	72 h ozonated oil	% Change
Coconut oil	0.2154	0.8131	277.52
Mix of coconut oil and olive oil (85:15)	0.8968	2.9160	225.15
Mix of coconut oil and olive oil (75:25)	0.7139	1.4203	98.94

3.3. Peroxide number

Peroxide number shows the level of oxidation of ozone to fatty acids, how much ozone binds with fatty acids to produce peroxide. The success of ozonation is marked by an increase of peroxide number obtained. The higher peroxide number, the more ozones that react with the carbon-carbon double bond and indicates the more ozones content in the oil. The following Table 3 presents the results of peroxide numbers. The highest numbers were obtained for a mix of ozonated coconut oil and olive oil with a concentration ratio of 85:15.

High peroxide number is connected with the reaction of ozone with carbon-carbon double bonds in unsaturated fatty acids. It produces ozonides, hydroperoxides, aldehydes, peroxides, diperoxides and polyperoxides according to well-known mechanism [3]. The resulting peroxide acts as an oxidizing agent in lowering the growth factor of *Staphylococcus aureus* bacteria.

Table 3. Peroxide numbers for pure and oils ozonated after 72 h.

	Peroxide number (meq O_2^{2-}/kg oil)	
Oil	Untreated oil	72 h ozonated oil
Coconut oil	0	102.91
Mix of coconut oil and olive oil (85:15)	0	114.77
Mix of coconut oil and olive oil (75:25)	0	110.64

3.4. FT-IR spectroscopy

FT-IR spectroscopy is used to highlight differences in the functional groups during the oil ozonation, in particular, the decrease of the bands corresponding to both C=C and =C–H stretches, and the increase of the band corresponding to ozonide CO stretch. Based on the spectrum obtained, the percent value of transmittance can be seen in Table 4.

The result shows that the concentration decreases in the bonds, C=C, and =C-H when an increase in transmittance occurs. Increased transmittance percent indicates a reduced number of moles of some group or bond in oleozon (percent transmittance is inversely proportional to the absorbance value). It shows also that the concentration increases in the C=O and C-O bonds when a decrease in transmittance occurs. It is because the reaction of ozone with carbon-carbon double bonds in unsaturated fatty acids produces some compounds include ozonides and aldehydes [3]. From the result of FT-IR spectra obtained for each bond, there are some percentage of transmittance which are not constant, therefore it can be said that FT-IR analysis is not as sensitive as other oleozon properties test.

Table 4. FT-IR Analysis results.

	Aromatic bond (C=C) (1400-1600 cm^{-1})		Alkene bond (=C-H) (675-1000 cm^{-1})		Carbonyl bond (C=O) (1650-1820 cm^{-1})		Ozonide bond (C-O) (1050-1260 cm^{-1})	
Oil	Wave length (cm^{-1})	%T	Wave length (cm^{-1})	%T	Wave length (cm^{-1})	%T	Wave length (cm^{-1})	%T
Pure coconut oil	1462.94	86.29	721.62	88.79	1742.85	57.24	1110.07	76.16
Ozonated coconut oil	1463.08	89.36	721.59	90.79	1743.02	66.96	1109.36	81.85
Pure coconut oil and olive oil (85:15)	1462.65	86.14	721.61	88.03	1743.01	57.97	1110.58	77.18
Ozonated coconut oil and olive oil (85:15)	1462.42	85.99	721.65	88.09	1742.93	58.61	1110.07	76.58
Pure coconut oil and olive oil (75:25)	1462.36	86.04	721.65	87.58	1743.15	58.46	1110.98	77.91
Ozonated coconut oil and olive oil (75:25)	1462.43	85.99	721.03	86.86	1743.00	58.82	1110.19	77.03

3.5. LC-MS spectroscopy

The compounds contained in herbal extract were analyzed by LC-MS. The observed results are mass arrays which containing information about the mass to charge ratio (m/z) of the detected peak and peak intensity (% area). The alleged composition contained in the extract was carried out by matching the m/z value of the spectra mass array to the accurate mass of the literature compound using a database in ChemSpider or PubChem applications. The analysis results of the compound based on the LC-MS method for each extract sample were summarized in Table 5.

Table 5. LC-MS Analysis results.

Compound	m/z (theoretical)	100 g herbal with *n*-hexane solvent	200 g herbal with *n*-hexane solvent	100 g herbal with IPA solvent	200 g herbal with IPA solvent
Cinnamaldehyde	132	-	✓	✓	✓
Cinnamyl acetate	176.215	-	-	-	-
Linalool	154.253	-	✓	✓	-
Eugenol	164.204	-	-	-	-
Caryophyllene	204.357	✓	✓	✓	✓
Diosgenin (Saponin)	414.63	-	-	✓	-
Bacoside A3 (Saponin)	929.107	-	-	-	✓
Asparagoside A (Saponin)	578.787	✓	✓	-	✓
Asparagoside B (Saponin)	596.802	-	✓	-	-
Elatin (Flavonoid)	594.522	✓	-	-	-
Eupatin (Flavonoid)	360.318	✓	✓	-	-
Hispidone (Flavonoid)	346.335	-	✓	-	-
Laurifolin (Flavonoid)	356.374	✓	-	-	-

The type of extraction performed in this study is liquid-solid extraction which there are several factors that must be considered for the extraction to run properly, such as particle size, solvent type, operating temperature, and stirring. It can be seen in table above that the compounds detected on extraction using *n*-hexane solvent are more than using isopropyl alcohol solvent. In the extraction process, the compound will be more easily attracted or dissolved with the solvent having the same polarity level. In this case, herbal extract is an organic compound that tends to have nonpolar properties, making it more soluble in nonpolar solvents (*n*-hexane) as well.

When analyzed from the differences in feed mass used, the extraction results of 100 g feed mass should be better than 200 g feed mass. Although the operating temperature used was same, the 200 g feed mass has a longer time to reach the operating temperature than the 100 g feed mass. Thus, the diffusion process at 100 g feed mass runs faster than the 200 g feed mass. In addition, stirring can make the diffusion rate increase and the transfer of material from the surface of the particles into the solution increases rapidly. Stirring also prevents precipitation. Stirring on 100 g feed mass is easier than on 200 g feed mass. As a result, the diffusion process at 100 g feed mass runs faster than the 200 g feed mass.

3.6. Antibacterial activity

The antibacterial activity test is necessary to know how effective the mixed sample of the ozonated vegetable oil with herbal extract to inhibit the growth of *Staphylococcus aureus* bacteria. As we know, *Staphylococcus* is a group of bacteria that can cause various diseases from the infection of some tissues in the body. Diseases caused by *Staphylococcus* are varying from no treatment to severe (such as diabetic ulcer) and potentially cause death. Table 6 and Figure 1 present the results of antibacterial activity test.

Based on the results, it was found that ozonated oil (oleozon) is a good oil in disinfecting bacteria, seen from the inhibition zone that formed. Moreover, after adding *Phaleria macrocarpa* and *Cinnamomum burmannii* extract the antibacterial effect will increase. A mixture of 160 mL coconut oil that ozonated for 72 hours and 0.18 gram herbal extracts with *n*-hexane solvent showed the highest inhibition zone of 18.3 mm in *Staphylococcus aureus* bacteria. This result is much better when compared with a study conducted by Nisa (2014) i.e. the 42-hour ozonated olive oil sample showed a 2 mm inhibition zone in *Staphylococcus aureus* bacteria. The highest inhibition zone produced by a mixture of coconut oil and herbal extracts with *n*-hexane solvent was also comparable with the test results of LC-MS extracts, where the compounds on the extraction using the *n*-hexane solvent were more detected than using the isopropyl alcohol (IPA) solvent. 72 hours of ozonation time is a good time in the ozonolysis reaction of oleozon and the addition of herbal extracts proven to increase disinfection properties of oleozon.

Table 6. Antibacterial activity results.

Sample	Symbol	Diameter of inhibition (mm)
160 mL Ozonated coconut oil (CO)	Blank	12.2
160 mL Ozonated CO+0.16 g extract, IPA solvent	N	13.7
160 mL Ozonated CO+0.17 g extract, IPA solvent	M	10.1
160 mL Ozonated CO+0.18 g extract, *n*-hexane solvent	J	18.3
160 mL Ozonated CO+0.2 g extract, *n*-hexane solvent	K	12.7
160 mL Ozonated coconut oil (CO) & olive oil (OO) (85:15)	L	15.1
160 mL Ozonated CO&OO (85:15)+0.16 g extract, IPA solvent	B	12.6
160 mL Ozonated CO&OO (85:15)+0.17 g extract, IPA solvent	I	16.1
160 mL Ozonated CO&OO (85:15)+0.18 g extract, *n*-hexane solvent	D	14.0
160 mL Ozonated CO&OO (85:15)+0.2 g extract, *n*-hexane solvent	C	12.6
160 mL Ozonated coconut oil (CO) & olive oil (OO) (75:25)	E	11.3
160 mL Ozonated CO&OO (75:25)+0.16 g extract, IPA solvent	H	12.5
160 mL Ozonated CO&OO (75:25)+0.17 g extract, IPA solvent	G	11.7
160 mL Ozonated CO&OO (75:25)+0.18 g extract, *n*-hexane solvent	F	12.6
160 mL Ozonated CO&OO (75:25)+0.2 g extract, *n*-hexane solvent	A	14.6

Figure 1. Inhibition of bacteria *Staphylococcus aureus* on medium with ozonated oil.

4. Conclusion

The results presented here show that oil ozonation was successful, either for pure coconut oil or a mixture of coconut oil and olive oil. Ozonation may increase the peroxide and acid numbers for both oils and also decrease the iodine numbers. Evaluating from the analysis of chemical and physical properties results, the best ozonation condition is picked up from an increase of 277.52% acid number, peroxide number about 114.77 meq O_2^{2-}/kg oil, and decrease of iodine number up to 22%.

The addition of herbal extracts proven to increase disinfection properties of oleozon. A mixture of 160 mL coconut oil that ozonated for 72 hours and 0.18 gram herbal extracts with *n*-hexane solvent showed the highest inhibition zone of 18.3 mm in *Staphylococcus aureus* bacteria.

The highest increase in acid number and highest decrease in iodine number is owned by coconut oil, which means the acceptance of ozonation in coconut oil is better than mixture of coconut oil and olive oil. These results are also proportional to the antibacterial activity test result where the highest inhibitory zone is generated by coconut oil. It means that the acid number and iodine number determining for antibacterial test, but not with peroxide number. Meanwhile, high peroxide number is also effective to inhibit bacteria. From the properties of ozonated oils and antibacterial activity test results, it can be concluded that coconut oil is potential to become Olozon® as an alternative to olive oil which is more expensive and less availability in Indonesia.

Acknowledgments

The authors acknowledge Universitas Indonesia and PITTA program for the financial support. We are also thankful to Laboratorium Intensifikasi Proses Teknik Kimia Universitas Indonesia, Laboratorium AKA Bogor, Laboratorium Kesehatan Daerah Provinsi DKI Jakarta and Fakultas Farmasi Universitas Pancasila for the instruments to characterize the oleozon.

References

[1] Bismo S, Ikramina D, Susilarini R E and Nisa H I 2014 *Proc. Seminar Nasional Integrasi Proses 2014* 169–77
[2] Dewanti T, Wulan S N and Nur I 2005 *J. Agric. Technol.* **6** 29–36
[3] Diaz M F, Nunez N, Quincose D, Diaz W and Hernandez F 2005 *J. Ozone Sci. Eng.* **27** 153–7
[4] Guzel-Seydim Z B, Greene A K and Seydim A C 2004 Use of ozone in the food industry *J. Lebensm.-Wiss. u.-Technol.* **37** 453–60
[5] Hayati L 1990 [Bachelor's thesis] Depok: Universitas Indonesia
[6] Nisa H I 2014 [Bachelor's thesis] Depok: Universitas Indonesia
[7] Nurdiana, Hariyanto T and Musfirah 2008 [Bachelor's thesis] Malang: Universitas Brawijaya
[8] Schulz S 1981 *Dtsch Tierärztl Wochenschr* **88** 60–64

IOP Conf. Series: Materials Science and Engineering **334** (2018) 012070 doi:10.1088/1757-899X/334/1/012070

Evaluation of power density on the bioethanol production using mesoscale oscillatory baffled reactor and stirred tank reactor

H W Yussof[1], S S Bahri[1,2] and N A Mazlan[1]
[1] Faculty of Chemical Engineering & Natural Resources, Universiti Malaysia Pahang, 26300 Gambang, Pahang, Malaysia

E-mail: syamsutajri@gmail.com

Abstract. A recent development in oscillatory baffled reactor technology is down-scaling the reactor, so that it can be used for production of small-scale bioproduct. In the present study, a mesoscale oscillatory baffled reactor (MOBR) with central baffle system was developed. The reactor performance of the MOBR was compared with conventional stirred tank reactor (STR) to evaluate the performance of bioethanol fermentation using *Saccharomyces cerevisiae*. Evaluation was made at similar power density of 24.21, 57.38, 112.35 and 193.67 Wm-3 by varying frequency (f), amplitude (xo) and agitation speed (rpm). It was found that the MOBR improved the mixing intensity resulted in lower glucose concentration (0.988 gL^{-1}) and higher bioethanol concentration (38.98 gL^{-1}) after 12 hours fermentation at power density of 193.67 Wm^{-3}. Based on the results, the bioethanol yield obtained using MOBR was 39% higher than the maximum achieved in STR. Bioethanol production using MOBR proved to be feasible as it is not only able to compete with conventional STR but also offers advantages of straight-forward scale-up, whereas it is complicated and difficult in STR. Overall, MOBR offers great prospective over the conventional STR.

1. Introduction

The development of high efficiency bioreactors has been an important research objective in the field of bioprocesses. Appropriate selection and design could greatly improve the efficiency of the overall process. The oscillating reactors were firstly used in separation processes in order to enhance the contact between the phases and, consequently, to improve mass transfer rates. Since then, they have been applied to a number of systems, either chemical or biochemical, under several configurations [1]. Oscillatory baffled reactor (OBR) consists of a cylindrical column or tube as illustrated in Figure 1. It is similar to the continuous plug flow reactor. The reactor tubes fitted with equally-spaced, low constriction orifice plate baffles that have an oscillatory motion ranging from 0.5 to 10 Hz superimposed upon the net flow of the process fluid. The oscillatory flow reactor creates an oscillatory motion when the flow pumped back and forth through the orifice. Each baffle behaved like a uniformly mixed stirred tank, leads to an excellent mixing and suspension by creating vortices between orifice baffles and oscillatory fluid [2]. Moreover, eddies are generated when fluid flow passes through the baffles that enable radial motions.

[2] To whom any correspondence should be addressed.

The production of biofuel using OBR was first introduced using NaOH as a homogeneous catalyst [3]. The most important application of OBR is it can provide long residence time processes in continuous mode. Many such processes are currently run in batch because conventional designs of continuous reactor are impractical in term of cost, control and size [4]. In addition, the OBR offers the prospect of a compact plug flow reactor with uniform and controllable mixing. The mesoscale oscillatory baffles reactor (MOBR) was first been developed for laboratory-scale processes [5]. MOBRs are millilitre scale OBRs designed for small scale screening of reactions. These mesoreactors combine the basic capabilities of the conventional OBRs such as enhanced heat and mass transport, and product uniformity with small scale necessary for screening and production of specialist chemicals. MOBRs have received considerable attention due to their small volume and ability to operate at low flow rates, reducing reagent requirements and waste [6].

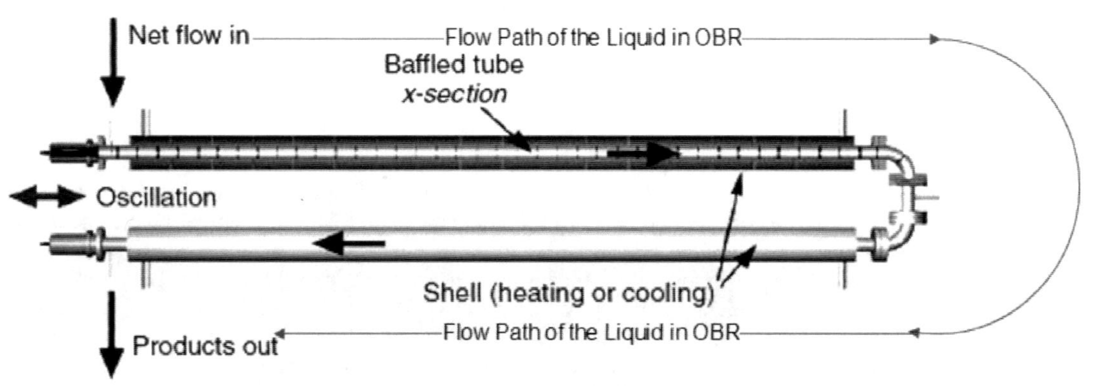

Figure 1. The oscillatory baffled reactor design [4].

MOBR are also suitable for continuous high throughput screening, and rapid determination of reaction order and rate constant as plug flow behaviour can be easily achieved [7-8]. There have been several studies in the literature reporting that STR is complicated and difficult to scale up, as the degree of mixing and heat transfer diminish with increasing scale. This leads to substantial reductions in rate. Whereas, a straight forward scale up processes can be done by maintaining geometrical and dynamic similarity. This allows mixing and flow conditions produced at laboratory-scale to be easily replicated for pilot-scale and industrial-scale processes [9-10].

Geometric and dynamic parameters play an important role in the reactor design and operation. Scale-up can be performed directly by maintaining geometric and dynamic similarity. Geometric similarity is achieved by maintaining the baffle-spacing-to-diameter ratio and the fractional open area of the baffle while dynamic similarity is achieved by maintaining the net and oscillatory flow Reynolds numbers [11]. Reynolds number (R_e) is a dimensionless number that describes the flow conditions either laminar or turbulent. It measures the ratio of inertial forces to viscous forces for given flow conditions. The oscillatory Reynolds number (R_{eo}) measures the intensity of mixing inside a column or baffled reactor [6]. It is similar to the net flow Reynolds number, R_{en}, in steady flow, except that the superficial velocity (u) has been replaced by maximum oscillatory velocity ($2\pi f x_o$). A velocity ratio is introduced in order to describe the interaction between oscillatory and net flows. It is the ratio of oscillatory R_{eo} to net flow R_{en}, and simplifies to the ratio of maximum oscillatory velocity to superficial velocity. In addition, the ratio of the two Reynolds numbers is a measure of the degree of plug flow [4]. The summary of the geometric and dynamic parameters are expressed as in Eq. (1) – Eq. (5) as below [12-14].

Net flow Reynold number:
$$Re_n = \frac{\rho du}{\mu} \quad (1)$$

Oscillatory Reynold number:
$$Re_o = \frac{2\pi f x_o \rho d}{\mu} \quad (2)$$

Velocity ratio:
$$\psi = \frac{Re_o}{Re_n} = \frac{2\pi f x_o}{u} \quad (3)$$

Baffle spacing:
$$L = 1.5d \quad (4)$$

Open cross sectional area:
$$S = \left(\frac{d_o}{d}\right)^2 \quad (5)$$

where f is the frequency of oscillation (Hz); x_o is the centre-to-peak amplitude of oscillation (m); u is the superficial velocity of the liquid (m/s); ρ is density (kg.m^{-3}.s^{-1}); μ is viscosity (Pa.s); d is tube diameter (m); d_o is orifice diameter (m) and L is baffle spacing (m). The oscillation frequency (f) and amplitude (x_o) are the most important operational parameters in OFR. At a given L and d_o, changing the combination of f and x_o allows control the generation of eddies and produces a range of fluid mechanical conditions as broad as required.

This work aims to evaluate the bioethanol production using MOBR and compare their performance with conventional stirred tank reactor (STR) at the same power density of 57.38, 112.35, and 193.67 Wm^{-3}. The fermentation was carried out using *Saccharomyces cerevisiae* along with glucose as medium in both MOBR and STR at 30°C.

2. Materials and methods

2.1 Experimental set-up

The set-up of MOBR consists of glass tube, one central baffle, and two syringe pumps as in Figure 2. Two syringe pumps (Eurodyne Ltd., UK) were provided to inject the inoculum inside the reactor and to oscillate the fluid inside the reactor. The MOBR was run at 8 mL working volume including inoculum. The reactor was sterilized using sodium hypochlorite (NaOCl), 5.25% prior to fermentation. Fermentations were performed at 30°C with power density value of 57.38, 112.35, 193.67 Wm^{-3} in order to compare with STR by changing amplitude and frequency. Samples were taken twice at 6 and 12 hours and stored at 4°C for further analysis.

The STR was performed using a 2-L fermenter as in Figure 3. The fermenter was autoclaved at 121°C, for 15 minutes prior to fermentation with the pH and the dissolved oxygen (DO) electrodes installed. The pH and DO were always monitored but not controlled. Agitation was varied at 190, 260, 320 and 384 rpm in order to produce a power density of 57.38, 112.35, 193.67 Wm^{-3} to compare with MOBR. The fermentation was maintained constantly at pH 5.0 by adding 2M sodium hydroxide solution and 2M hydrochloric acid. The working volume was approximately 1.5 L. The temperature was kept constant at 30°C. An initial sample was taken to measure their initial concentration, and then the inoculum was added to the fermenter. Samples were taken periodically every two hour for the first 12 hours and every 6 hours subsequently up to 48 hours and stored at 4°C for further analysis.

IOP Conf. Series: Materials Science and Engineering **334** (2018) 012070 doi:10.1088/1757-899X/334/1/012070

2.2 Sample analysis

The glucose, biomass cell concentration, and bioethanol concentrations were measured according to 3, 5-dinitrosalicyclic (DNS), cell dry weight (CDW) and high performance liquid chromatograph (HPLC) methods, respectively. Glucose concentration in the fermentation broth was analysed by 3, 5-dinitrosalicylic acid (DNS) using 1% DNS reagent based on method applied in previous study [15]. The fermentation broth was centrifuged at 13,000 rpm for 10 min. One mL of the supernatant was mixed with 3 mL of 1% DNS reagent and incubated at 90°C for 10 min. After the mixture cooled to room temperature, the absorbance was measured at 540 nm, using a UV-VIS spectrophotometer (Hitachi, Japan). Standard solutions of anhydrous D-glucose containing 0.5–3 g glucose/L deionized water were prepared and 1 mL of each standard solution were mixed with 3 mL 1% DNS reagent and incubated 90°C for 10 min. A blank (deionized water) was incubated with the reagent and was used for zero adjustment of the spectrophotometer. The glucose concentration in the sample was compute by least squares linear regression, using a standard curve [16]. Biomass cell concentration was determined by the cell dry weight (CDW) at optical density of 660 nm using UV-VIS spectrophotometer. A correlation between the CDW and optical density needs to be established beforehand, that is CDW (gL^{-1}) = (0.407 × OD600) + 0.0014 [16]. Bioethanol was analysed using HPLC (Shimadzu Prominence® LC-20A). The HPLC system was equipped with an auto-sampler, a Rezex ROA-Organic Acid column (7.8 x 300 mm; Rhenomenex, Torrance, CA, USA) as an analytical column, and a refractive index detector. The temperature was maintained at 60°C in a column oven and sulphuric acid (0.005 N) was used as a mobile phase at 0.6 mL/min under isocratic condition. All the samples were clarified by filtration with a 0.45 μm filter before injected to the HPLC column at 10 μL.

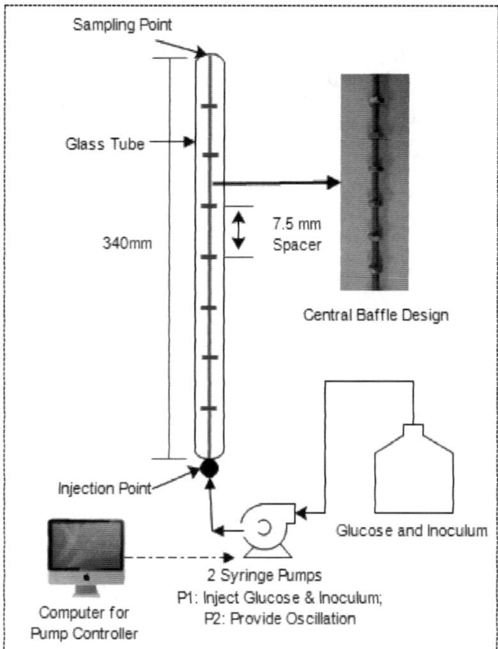

Figure 2. MOBR set-up for bioethanol production.

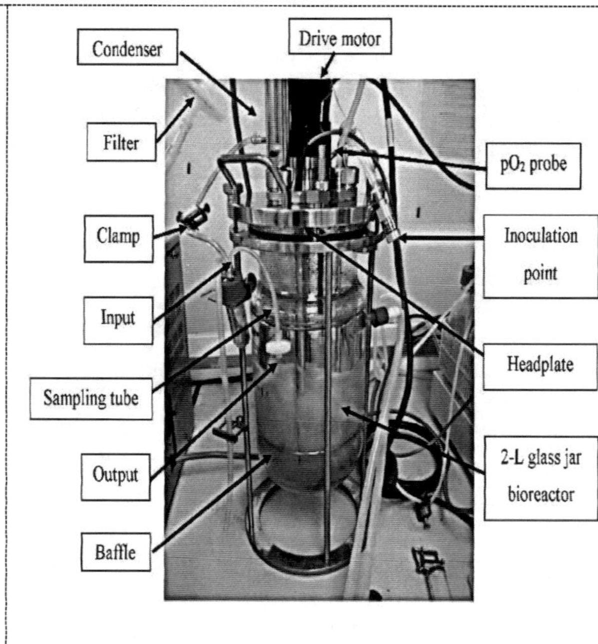

Figure 3. A 2-L STR setup for bioethanol production.

3. Results and discussion

A series of batch fermentations using *Saccharomyces cerevisiae* were conducted in mesoscale oscillatory baffled reactor (MOBR) with comparison to the conventional stirred tank reactor (STR). The fermentation was conducted in order to evaluate the MOBR reactor performance and its ability to perform anaerobic fermentations in batch culture. The comparison was made on the basis of power consumed per unit volume (power density). The fermentation was carried out at the same power density of 57.38, 112.35, and 193.67 Wm^{-3} in both MOBR and STR using glucose medium at 30°C. In terms of Reynolds numbers (R_e), the fermentation in STR was performed at R_e = 8,538, 10,672 and 12,787 while the fermentation in MOBR was carried out at oscillatory Reynolds numbers, R_{eo} of 247.59, 309.49, 372.39.

The effect of initial glucose concentration at 10 gL^{-1} and cell growth (biomass cell concentration) were summarized in Figure 4 and Figure 5. It can be seen from Figure 4 that the MOBR give better glucose conversion when compared to the STR at the same power density. The lowest glucose concentration of 0.988 gL^{-1} was obtained by MOBR compared to STR of 2.339 gL^{-1} at 193.67 Wm^{-3}. The bioethanol yield was recorded at 39.4% when using MOBR as STR only produced the highest yield of 15.4%. Nonetheless, the bioethanol yield obtained using MOBR was 39% higher than the maximum achieved in STR. This result is similar to those findings by Masngut et al. [16] where 0.22 g/L/h of biobutanol was produced from oscillatory baffled bioreactor which is 38% higher than the maximum achieved in the STR. Higher bioethanol productivity in MOBR is likely to be due to early initiation of solventogenic phase, eventually accumulating more solvent earlier than the STR [16]. In terms of Reynolds number, it is thought that both fermentation in MOBR (Reo = 373) and STR (Re = 12,787) were performed at turbulent condition. It is suggested that the STR system was fully turbulent when operated at values of R_e above 10,000 [17]. Even at the lowest power density of 112.35 Wm^{-3} and laminar condition, the difference in glucose conversion was nearly 1 gL^{-1} between the MOBR (R_{eo} = 247.59) and STR (R_e = 8,538). It appears from Figure 4 that the advantages of using MOBR is not always prominent. Taking data at 6 hours, for 193.67 Wm^{-3} and 112.35 Wm^{-3}, the advantage of MOBR is not obviously. However for 57.38 Wm^{-3}, the advantage of MOBR is clear.

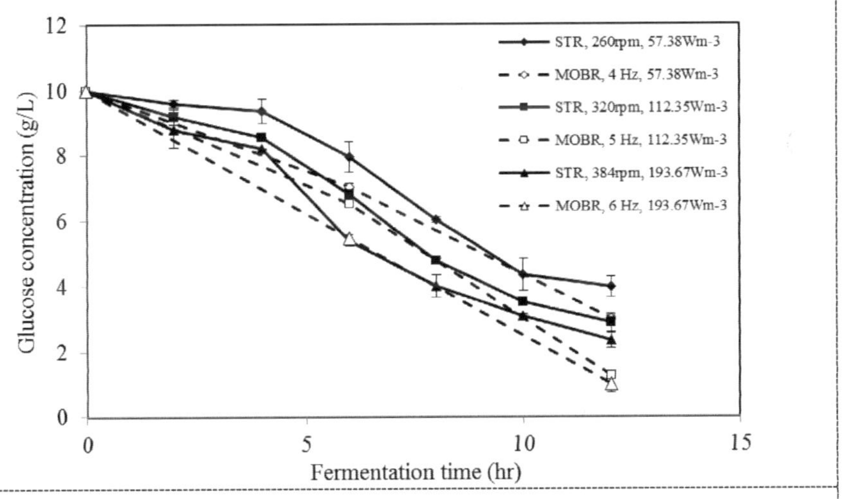

Figure 4. Glucose consumption during fermentation.

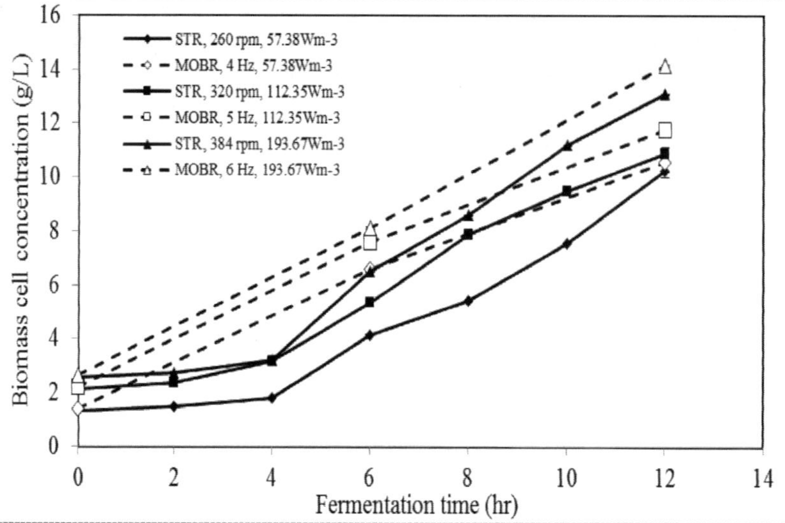

Figure 5. Biomass cell concentration during fermentation.

This result was supported by the highest biomass cell concentration of 14.14 gL^{-1} obtained after 12 hours using MOBR at 193.67 Wm^{-3} as in Figure 5. Increase the MOBR mixing rate has increase the bioethanol concentration as in Table 1. The highest bioethanol concentration of 38.98 gL^{-1} was obtained using the MOBR at 193.67 Wm^{-3} and R_{eo} of 371.39. In contrast, the highest bioethanol concentration of 36.11 gL^{-1} was obtained using the STR at the same power density of 193.67 Wm^{-3}. The finding is consistent with findings of past studies by Abbott et al. [10] in the enzymatic saccharification of pure α-cellulose using cellulase conducted in oscillatory baffled (OBR) and stirred tank (STR) reactors. Enzymatic saccharification requires a cellulase enzyme system to convert cellulose to glucose. Their finding highlights that the highest glucose conversion of 23 gL^{-1} was obtained in the OBR at a relatively low power density (2.36 Wm^{-3}) and highly turbulent with R_{eo} = 600 compared to the STR at an impeller speed of 350 rpm (250 Wm^{-3}). Their work also demonstrated a 94 – 99% decrease in the required power density to achieve maximum conversion rates and showed a 12% increase in glucose consumption after 24 h at 2.36 Wm^{-3}.

In a different study, a series of pullulan fermentation experiments were carried out in OBR using *Aureobasidium pullulan*, which produces the polymer pullulan [18]. The growth rate and pattern were compared with the traditional STR, including the influence of aeration on pullulan and biomass cell concentration. The result show that the OBR gives a much higher rate of growth as well as higher yield of pullalan as compared to the STR. For example, 11.3 gL^{-1} and 12.1 gL^{-1} pullulan were produced in the STR after 96 and 144 hours fermentation, respectively compared to 37 and 39 hours were achieved when using OBF. Overall, most of these studies give an indication that the MOBR (including OBR) performance significantly better than the conventional STR when compared at the same power density.

4. Conclusion

A series of batch fermentations using *Saccharomyces cerevisiae* were performed in MOBR and compared with the conventional STR in term of power consumed per unit volume (power density). This study has shown that the MOBR give better glucose conversion when compared to the STR at the same power density. The lowest glucose concentration of 0.988 gL^{-1} was obtained by MOBR compared to STR of 2.339 gL^{-1} at 193.67 Wm^{-3}, when performed at turbulent condition. This result was supported by the highest biomass cell concentration of 14.14 gL^{-1} and the highest bioethanol concentration of 38.98 gL^{-1} obtained after 12 hours using MOBR at 193.67 Wm^{-3} and R_{eo} of 371.39. In contrast, the highest bioethanol concentration of 36.11 gL-1 was obtained using the STR at the same power density of 193.67 Wm^{-3}. Overall, findings from this study suggests that the MOBR is suitable for performing fermentation in a power efficient manner compared to conventional STR. Results from this study provided essential data on power input to allow any economic assessment on the bioethanol production processes based on both MOBR and STR later. In addition, MOBR may offer potential for more biological application i.e. animal cell where expensive chemical usage can be significantly reduced. However, more research needs to be done to prove this in practice, but this possible benefit offers great prospective over the conventional STR.

Acknowledgement

The authors wish to express their thanks to Ministry of Higher Education and Universiti Malaysia Pahang through the financial aid from grant RDU160901 and Universiti Malaysia Pahang's postgraduate research grants scheme (GRS170000).

References

[1] Reis N M F 2006 *Novel oscillatory flow reactors for biotechnological applications*. Doctoral Thesis. Universidade do Minho,.
[2] Zheng M, Skelton R L, and Mackley M R 2007 Biodiesel reaction screening using oscillatory flow mesoreactors. *Trans Icheme, Part B, Process Safety and Environmental Protection*. **85**(B5): 365-371.
[3] Harvey A P, Mackley M R, Seliger T 2003 Process intensification of biodiesel production using a continuous oscillatory flow reactor. *Journal of Chemical Technology and Biotechnology*. **78**(2-3): 338-341.
[4] Reay D, Ramshaw C, and Harvey A P 2013 A process intensification. engineering for efficiency, sustainability and flexibility. IChemE. UK.
[5] Reis N, Harvey A P, Mackley M R, Vicente M R, and Teixeira J A 2005 Fluid mechanics and design aspects of a novel oscillatory flow screening mesoreactor. *Chemical Engineering Research and Design*. **83**(A4): 357-371.
[6] Phan A N, Harvey A P, and Rawcliffe M 2011 Continuous screening of base-catalysed biodiesel production using new designs of mesoscale oscillatory baffled reactors. *Fuel Processing Technology*. **92**: 1560-1567.
[7] Rasidi F R M, Phan A N, and Harvey A P 2013 Rapid determination of reaction order and rate constants of an imine synthesis reaction using a mesoscale oscillatory baffled reactor. *Chemical Engineering Journal*, **222**: 282-291.
[8] Reis N, Vicente A A, Teixeira J A, and Mackley M R 2004 Residence times and mixing of a novel continuous oscillatory flow screening reactor. *Chemical Engineering Science*. **59**: 4967-4974.
[9] Masngut N, Harvey A P, and Ikwebe J 2010 Potential uses of oscillatory baffled rectors for biofuel production. *Biofuels*. **1**(4): 605-619.
[10] Abbott M S R, Perez G V, Harvey A P, Theodorou M K 2014 Reduced power consumption compared to a traditional stirred tank reactor (STR) for enzymatic saccharification of alpha-cellulose using oscillatory baffled reactor (OBR) technology. *Chemical Engineering Research Design* **92**(10): 1969-1975.
[11] Stonestreet P and Harvey A P 2002 A mixing-based design methodology for continuous oscillatory flow reactors. *Trans IChemE*. **80**: 31-44.
[12] Phan A N, and Harvey A P 2010 Development and evaluation of novel designs of continuous mesoscale oscillatory baffled reactors. *Chemical Engineering Journal*. **159**: 212-219.
[13] Phan A N and Harvey A P 2011 Effect of geometrical parameters on fluid mixing in novel mesoscale oscillatory helical baffled designs. *Chemical Engineering Journal*. **169**; 339-347.
[14] Phan A N and Harvey A P 2012 Characterization of mesoscale oscillator helical baffled reactor – Experimental approach. *Chemical Engineering Journal*. **180**; 229-236.

[15] Miller R 1958 Enzyme test for glucose. *The Lancet*, **271**(7014): 263.
[16] Masngut N and Harvey A P 2012 Intensification of biobutanol production in batch oscillatory baffled reactor. *Procedia Engineering*. ***42***: 1079-1087.
[17] Sinnott R K and Towler G 2009 *Chemical engineering design*. 5th Edition, Butterworth-Heinemann.
[18] Gaidhani H K, McNaeil B, and Ni X 2005 Fermentation of pullulan using an oscillatory baffled fermenter. *Chemical Engineering Research Design.* **83**(A6): 640–645.

IOP Conf. Series: Materials Science and Engineering **334** (2018) 012071 doi:10.1088/1757-899X/334/1/012071

Effect of solvent polarity levels on separation of xanthone and coumarin from *Calophyllum inophyllum* leaves extract

D F Susanto[1], S Hapsari[1], Z Trilutfiani[1], A Borhet[1], H W Aparamarta[1], A Widjaja[1] and S Gunawan[1,2]

[1] Department of Chemical Engineering, Faculty of Industrial Technology Institut Teknologi Sepuluh Nopember, Surabaya 60111, Indonesia

E-mail: gunawan@chem-eng.its.ac.id

Abstract. *Calophyllum inophyllum* has various benefits that can be utilized from root, stem, leaf, until seed. *C. inophyllum* leaves contain many bioactive compounds, such as xanthone and coumarin which are useful as antioxidant, and inhibitors of enzyme activity from HIV virus. The aim of this research was to investigate the effect of solvent polarity levels on the separation of xanthone and coumarin compounds contained in the crude extract of *C. inophyllum* leaves. Crude leaves extract was obtained by percolation method. Moreover, Liquid Liquid Extraction (LLE) was used for separating xanthone and coumarin compounds. It was performed by methanol (polar solvent) and hexane (non-polar solvent) with solvent ratio of 1. Methanol concentration in water used were 20%, 50%, 80%, and 100%. Each fraction obtained was tested qualitatively using Thin Layer Chromatography (TLC) and quantitatively using Gas Chromatography (GC) to analyze xanthone and coumarin. The best separation result was obtained by using 50% methanol. In this results, coumarin and xanthones were separated in methanol fraction (81.18% recovery) and in hexane fraction (81.91% recovery), respectively.

1. Introduction

Indonesia has many varieties of flora and fauna. The most common plant species in coastal areas in Indonesia are mangroves. According to FAO (2007), in 2005, mangrove forests in Indonesia reaced 3.062.300 ha (19% of the world's mangrove forest). The largest areas of mangroves in Southeast Asia are found in Indonesia (almost 60 percent of Southeast Asia total), Malaysia (11.7%), Myanmar (8.8%), Papua New Guinea (8.7%) and Thailand (5.0%) [1]. Mangroves contains a number of minerals, vitamins, amino acids that are essential for the growth of the world used for health [2].

One type of mangrove plant that has high economic value is *Calophyllum inophyllum* because almost all parts of this plant (stems, leaves, flowers, seeds, and sap) can produce various benefit products. *C. inophyllum* is a mangrove species from the family Clusiaceae. *C. inophyllum* is an evergreen tree with supports a dense canopy of glossy, elliptical leaves, fragrant white flowers and large round nuts. This habitat is primarily coastal and adjacent to lowland forests. This tree grows to height 8-20m (25-26 ft), sometimes reaching up to 35 m (115 ft) [3]. It has milky white sap. Leaves are opposite, deep glossy green, glabrous, simple, coriaceous with broadly elliptic or obovate-elliptic

[2] To whom any correspondence should be addressed.

lamina 10-20 cm long by 6-9 cm wide, rounded or emarginate apex, rounded or cuneate base, entire margin and distinct parallel lateral veins, perpendicular to mid rib [4].

Higher plants as sources of bioactive compounds play a dominant role in the maintenance of human health. The various parts of *C. inophyllum* plant contain many bioactive compounds, including among others: xanthones, coumarins, chromanones (flavonoids, biflavonoids), triterpenes and steroids [5]. The phytochemical compounds of *C. inophyllum* leaves had been studied previously [6]. Meanwhile, this plant contains phytochemical compounds that can be used as a cure of various diseases [7]. These compounds include inophynone, canophyllol, canophyllic acid, calophyllolide, inophyllolide, jacareubin, calanolide A, calophynone, and others. *C. inophyllum* contains xanthones have chemopreventive cancer agents and coumarins have antimicrobial activity [8]. Coumarin is one of the secondary metabolites in plants. Coumarin compounds exhibit anti-HIV drugs belonging to the *Non-nucleoside reverse transcriptase inhibitor* (NNRTI) [9]. NNRTI is a group of compounds that inhibits the activity of the reverse transcriptase enzyme from HIV-1 [10]. Coumarin compounds and their derivatives have many biological activities such as anticoagulant blood, antibiotics, and inhibiting carcinogenic activity [11]. Some of xanthone compound such as caloxanthone A, inoxanthone, macluraxanthone, and caloxanthone B have antimicrobial activity [12]. In order to increase the utilization of *C. inophyllum* leaves in medical function, bioactive compounds must be extracted, concentrated, separated and purified.

Xanthones are polyphenolic compounds that have molecular formula of $C_{13}H_{18}O_2$. Xanthones consist of a unique backbone with two benzene rings bridged across a carbonyl group and an oxygen. The most bioactive property of xanthones is their antioxidant ability. The derivatives of xanthones which isolated from *Garcinia mangostana* was found to act as the free radical scavenger and hence prevented the the oxidative damage of low density lipoprotein [13]. Coumarins are one of the member of the benzopyrone compounds. Coumarins have a benzene ring linked to a pyrone [14]. Coumarins are attractive to human for their physiological, bacteriostatic, and anti-tumour activity. 4-phenylcoumarins might be valuable as potential cancer chemopreventive agents (anti-tumour-promotors) [15].

Therefore, the objective of this work was to separate coumarin and xanthone contained in *C. inophyllum* leaves by liquid-liquid extraction method. This separation is based on the polarity index solvent. The separation technology of xanthone and coumarin is needed due to they have different benefits. The effect of solvent polarity levels on the separation of xanthone and coumarin compounds contained in the crude extract of *C. inophyllum* leaves were investigated.

2. Materials and methods

2.1. Materials

Dried *C. inophyllum* leaves were obtained from the "Koperasi Jarak Lestari", Cilacap, Central Java, Indonesia. Thin-layer chromatograph (TLC) aluminum plates (20 x 20cm x 250 m) were purchased from Merck (Darmstadt, Germany), Advantec filter papers were obtained from Toyo Roshi Kaisha Ltd. (Tokyo, Japan). Standard of xanthone and coumarin were obtained from Sigma Aldrich (St. Louis, MO). Methanol, hexane, aquadest, ethyl acetate, acetic acid were obtained from commercial sources.

2.2. Preparation of the crude extract

C. inophyllum leaves were dried in the sun for 3 days to reduce their water content. They was chopped to a homogeneous size by a mill and soaked in 3 L of methanol for two times (72 h each time). Then, the solutions were filtered through filter paper. The filtrate were combined and then evaporated to dryness by distillation at 80 °C.

2.3. Separation of coumarin and xanthone from Calophyllum inophyllum leaves
In this study, Liquid-Liquid Extraction was used to separate the mixture. This method was carried out to remove impurities simultaneously, purify and separate the polar and nonpolar fraction in the crude extract. Crude extract (3 g, 12 g) was put into beaker glass to be mixed with methanol (150 g) with variable methanol concentration of 20%, 50%, 80%, 100%. Then, it was stirred for 30 minutes with magnetic stirrer. The mixture was added with hexane (150 g), then it was stirred again for 30 minutes. The whole mixture was separated in separation funnel, the mixture split into layers. Each layer's content was analyzed.

2.4. Analysis by Thin Layer Chromatography (TLC)
TLC was employed to qualitatively analyze the sample, using authentic standards, as described research [16]. TLC plates that has been stained by the sample was immersed in a mobile phase of hexane : ethyl acetate : acetic acid at 90:10:1 (v/v/v). Then, TLC plates were analyzed by UV light at 254 nm.

2.5. Analysis by Gas Chromatography (GC)
The contents of bioactive compounds in each fraction were determined by GC [17] with some modification. External standard calibration curves were obtained by using 0,2-20 mg pure standard. Gas chromatographic analysis was performed on a Shimadzu GC-2010 (Kyoto, Japan) gas chromatography equipped with a Flame Ionization Detector. Separations were carried on a DB-5HT (5%-phenyl)-methylpolysiloxane non-polar column (15m x 0.32mm i.d.; Agilent Tech. Palo Alto, California). Temperatures of the injector and the detector were both set at 310°C. The temperature of the column was started at 80°C, increased to 300°C at rate of 15°C/min, and maintained at 300°C for 8 min. The split ratio was 1:50 using nitrogen as carrier gas with a linear velocity of 30 cm/s at 80°C. A twenty-milligram sample was dissolved in 1 mL ethyl acetate, and a 1 μL sample was taken and injected into the GC instrument.

3. Results and discussion

3.1. Identification of xanthone and coumarin in C. inophyllum leaves crude extract
C, inophyllum leaves contain many bioactive compounds, such as xanthone and coumarin which are useful for medicinal value. Figure 1 shows the TLC of crude extract of *C. inophyllum* leaves, standard xanthone, and coumarin. It can be known that coumarin is more polar than xanthone. The coumarin spot (Rf = 0.28) was located below the xanthone spot in TLC plate (Rf 0.56). Component having a smaller Rf value, have more polar properties.

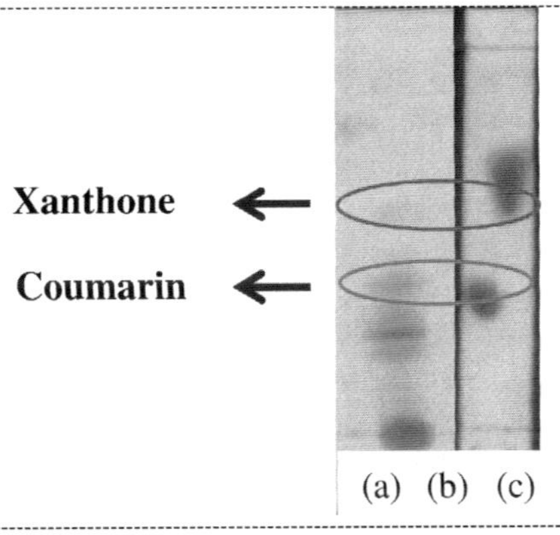

Figure 1. TLC analysis of *C. inophyllum* leaves crude extract (a); standard coumarin (b) and standard xanthone (c).

The aim of this research was to know effect of solvent polarity level on the separation of xanthone and coumarin compounds contained in the crude extract of *C. inophyllum* leaves. It was expected that the largest recovery of coumarin was in polar fraction, and the largest recovery of xanthone was in non-polar fraction.

3.2. Separation of xanthone and coumarin from C. inophyllum leaves crude extract

The separation process was done by Liquid - Liquid Extraction method were used polar solvent (methanol / water combination) and nonpolar solvent (hexane). Selection of solvent based on polarity index. Water is a polar solvent with a polar solvent with polarity index of 9, while methanol is semipolar solvent with a polarity index of 5.1 [18]. Hexane is a non-polar solvent with polarity index of 0. In this research, first, there were formed three layers. The lowest layer was insoluble fraction. The middle layer was the methanol fraction. The top layer was the hexane fraction. The insoluble fraction was separated by filter paper. Then, methanol and hexane fractions were inserted in separating funnel as shown in figure 2. Top layer (transparent yellow color) was hexane fraction which contained non-polar compounds, while the bottom layer (light brown color) was methanol fraction which contained polar compounds.

Figure 2. The immiscible layer in LLE process.

Table 1 shows the effect of methanol concentration to separation of coumarin and xanthone. The combination of methanol and water is expected to dissolve polar compounds contained in crude extract, while hexane is expected to dissolve non-polar compounds in crude extract. From **Table 1**, it appears that methanol 50% was excellent solvent for the separation of xanthones and coumarin. This is evident from the mass of coumarin in the methanol fraction is greater than the mass of coumarin in hexane fraction, while the mass of xanthones in the methanol fraction is smaller than the mass of xanthones in hexane fraction. Methanol 20% was not the good solvent for separation process of coumarin and xanthone, because the mass of coumarin was more on hexane fraction. No separation between xanthone and coumarin on the use of methanol 80% and 100%. This is because coumarin is more polar than xanthone (figure 1b). Water and its combination with organic solvent commonly used to extract bioactive components from plants. The largest extraction yield of flavonoid and phenolic content of the *Macademia tetraphylla* skin was obtained on the use of solvents with 50% methanol, 50% ethanol, 50% acetonitrile, and 50% acetone concentration (v/v) in water, respectively [19].

From table 1, it shows that the best separation of coumarin and xanthone was at methanol 50%, whereas %recovery coumarin in methanol fraction (3.35%) larger than in hexane fraction (0%). The different of solvent affects the recovery of bioactive compound such as total phenolic compound and total flavonoid compound, where the % recovery is best for bioactive compounds separation on the combination of water and organic solvent (methanol, ethanol, acetonitrile) by 30-60% concentration [20]. This is caused by differences in the dielectric constant and polarity index of solvents. The dielectric constant of the solvent is the magnitude of the force acting between two payloads in the solvent. These constant determines the extent to which the level of ability of the solvent to dissolve components. Solvent that has a low dielectric constant is a good solvent for dissolve non-polar compounds and vice versa [21].

Table 1. Effect of methanol concentration and solvent to crude extract ratio (mL/g) on separation of xanthone and coumarin.

Methanol concentration	Compound	Solvent / crude extract = 50			Solvent / crude extract = 12.5		
		Methanol Fraction	Hexane Fraction	Insoluble Fraction	Methanol Fraction	Hexane Fraction	Insoluble Fraction
20%	Coumarin	0,011[a]	0.04	0.06	0.05	0.10	0.09
		(0.96)[b]	(3.13)	(95.90)	(1.04)	(10.05)	(88.90)
	Xanthone	0.001	0.13	0.19	0.04	0.28	0.29
		(0.04)	(3.70)	(96.26)	(0.26)	(9.65)	(90.09)
50%	Coumarin	0.026	0.29	0.04	0.03	0	0.08
		(34.56)	(4.30)	(61.13)	(3.35)	(0)	(96.65)
	Xanthone	0.13	0.69	0.13	0.02	12.12	0.25
		(7.55)	(27.97)	(64.48)	(1.02)	(0.22)	(92.5)
80%	Coumarin	0.06	0.09	0.03	0.05	0.12	0.09
		(50.74)	(5.46)	(43.79)	(15.69)	(8.74)	(75.56)
	Xanthone	0.08	0.27±0.14	0.10	0.02	0.22	0.25
		(24.38)	(23.96)	(51.66)	(1.74)	(5.72)	(92.5)
100%	Coumarin	0.04	0.05	NA	0.09	0.09	NA
		(51)	(48.99)		(68.52)	(31.48)	
	Xanthone	0.07	0.05	NA	0.11	0.45	NA
		(63.29)	(36.71)		(36.20)	(63.79)	

[a] Content (%wt)
[b] Recovery (%)
NA : Not available

In this research, besides there methanol fraction and hexane fraction, there are also insoluble solid fraction. From table 1, % recovery xanthone and coumarin in solid fraction is still very large, so we need to get back extraction of solid fraction to get greater of % recovery xanthone and coumarin in methanol and hexane fraction. The insoluble fraction from variable 50 (mL/g) solvent / crude extract ratio was chosen to re extracted again. The best separation between xanthone and coumarin occurs in the solvent / crude extract variable ratio of 50 mL/g. This can be seen from the differences of %recovery xanthone and coumarin far enough between in methanol fraction and hexane fraction. from table 1, it appears that for the variable methanol 50%, %recovery of coumarin in insoluble fraction of variable solvent / crude extract ratio 50 mL/g (61.13%) smaller than variable solvent / crude extract ratio 12.5 mL/g (96.65%), and %recovery of xanthone in insoluble fraction of variable solvent / crude extract ratio 50 mL/g (64.48%) smaller than variable solvent / crude extract ratio 12.5 mL/g (92.5%). The insoluble solid fraction of variable solvent / crude extract ratio 50 mL/g was extracted with the same method and solvent (methanol 50% and hexane) repeatedly. The insoluble fraction of variable solvent / crude extract ratio 50 mL/g (2.424 g) extracted with methanol 50% as polar solvent and hexane as non-polar solvent with the same amount with first LLE. Extraction was done in a glass beaker with stirring with magnetic stirrer for 30 minutes. Once extraction was complete will be formed of three layers, namely the lower layer insoluble hereinafter referred to solid fraction, the middle layer is methanol fraction, and the top layer is hexane fraction. But, extraction is done with new solvent repeatedly until the insoluble fraction depleted. Repeated extraction of this insoluble fraction has done as much as 6 stages (table 2).

Table 2. Result of each stage of repeated extraction of insoluble fraction.

Stage	Compound	Fraction		
		Methanol Fraction	Hexane Fraction	Solid Fraction
1	Coumarin	0,24[a] (39.75)[b]	0.03% (3.60%)	0.05% (61.13%)
	Xanthone	0.13 (7.55)	0.70% (27.97%)	0.16% (64.48%)
2	Coumarin	0.04 (5.37)	0.01% (0.91%)	0.06% (54.85%)
	Xanthone	0.03 (1.20)	0.62% (22.34%)	0.17% (58.15%)
3	Coumarin	0.04 (4.91)	0.01% (3.26%)	0.09% (46.68%)
	Xanthone	0.03 (1.06)	0.15% (16.81%)	0.22% (40.28%)
4	Coumarin	0.04 (4.53)	0.09% (7.48%)	0.1% (34.66%)
	Xanthone	0.08 (2.72)	0.29% (8.58%)	0.24% (28.98%)
5	Coumarin	0.02 (1.47)	0.02% (2.50%)	0.18% (30.69%)
	Xanthone	0.06 (1.55)	0.22% (7.83%)	0.33% (19.60%)
6	Coumarin	0.001 (0.24)	0.25% (10.50%)	NA
	Xanthone	0.03 (0.48)	0.46% (19.12%)	NA

[a] Content (%wt)
[b] Recovery (%)
NA : not available

Based on table 2, it can be seen that the recovery of xanthone and coumarin on the solid fraction will decrease. So, enrichment coumarin in methanol fraction and xanthone in hexane fraction can be done. table 3 shows the total distribution of xanthone and coumarin in the methanol and hexane fraction.

Table 3. Total xanthone and coumarin in methanol and hexane fraction.

Compound	Methanol Fraction	Hexane Fraction
Coumarin	0,089[a] (81.18)[b]	0.057 (18.82)
Xanthone	0.067 (18.09)	0.351 (81.91)

[a] Content (%wt)
[b] Recovery (%)

Based on table 3, it appears that there is a separation between xanthones and coumarin between hexane fraction and methanol fraction, where the % total recovery coumarin in the methanol fraction is equal to 81.18%, while the % total recovery xanthone in hexane fraction is equal to 81.91%. So, it can be indicated that this multiple extraction method with methanol 50% and hexane solvent suitable for separation of xanthones and coumarin from methanolic *C. inophyllum* leaves extract.

4. Conclusion

Repeated Liquid – liquid Extraction (LLE) with methanol and hexane can be used for separating xanthone and coumarin from *C. inophyllum* leaves crude extract. Solvent polarity levels have effect for separating xanthone and coumarin. Methanol 50% gives the best performance for separating xanthone and coumarin from *C. inophyllum* leaves crude extract. Coumarin was succeeded separately in the methanol fraction with recovery of 81.18%, while xanthone was succeeded separately in the hexane fraction with recovery of 81.91%.

Acknowledgement

The authors would like to convey a great appreciation for Directorate General of Resources for Science, Technology and Higher Education, Ministry of Research, Technology and Higher Education of Republic Indonesia which funding the current project under a scheme (700/PKS/ITS/2017) called "Penelitian Pascasarjana", by Institute of Research and Public Services (LPPM-ITS).

References

[1] Giesen W, Wulffraat S, Zieren M, Scholten L 2006 *Mangrove Guidebook for Southeast Asia* (FAO and Wetlands International)
[2] Bandarnayake W M 2002 Bioactivities, bioactive compounds and chemical constituents of mangrove plants *Wetlands Ecology and Management.* **10** 421-452
[3] Prabakaran K, John Britto S 2012 Biology, Agroforestry and medicinal value of Calophyllum inophyllum L. (Clusiaceae) : A review *International Journal of Natural Products Research.* **1** 24-33
[4] Lim T K 2012 *Edible Medicinal and Non-Medicinal Plants Vol 2, Fruits* (New York: Springer Dordrecht Hdelberg)
[5] Su X H, Zhang M L, Li L G, Huo C H, Gu Y C, Shi Q W 2008 Chemical constituents of the plants of the genus Calophyllum *Chem Biodivers.* **5** 2579-2608
[6] Susanto, D F, Aparamarta H W, Widjaja A, Gunawan S 2017 Identification of phytochemical compounds in *Calophyllum inophyllum* leaves *Asian Pac J Trop Biomed* **7** 773-81
[7] Ling K H, Kian C T, and Hoon T C 2009 *A Guide To Medicinal Plant* (Singapore: World scientific)
[8] Yimdjo M C, Azebaze A G, Nkengfack A E, Meyer A M, Bodo B, and Fomum Z T, 2004. Ántimicrobial and Cytotoxic Agents from *Calophyllum inophyllum. Phytochemistry*, **65** 2789-95
[9] Laure F, Raharivelomanana P, Butaud F F, Bianchini J P, Gaydou E M 2008 Screening of Anti-HIV Inophyllums by HPLC-DAD of Calophyllum Inophyllum Leaf Extracts from French Polynesia Islands *Analytica Chimica Acta.* **624** 147-153
[10] Pawar K D, Swati P J, Sunil R B, and Shubhada R T 2007 Pattern of Anti-HIV Dypyranocoumarin Expression in Callus Cultures of Calophyllum inophyllum Linn *Journal of Biotechnology*. **130** 346-353
[11] Murray R D H 1982 *The natural Coumarins* (New York: John Wiley and Sons)
[12] Noldin V F, Isaias D B, and Filho V C 2006 *Calophyllum* Genus: Chemical and Pharmacological Importance *Quimica Nova* **29** 549-554
[13] Williams P, Ongsakul M, Proudfoot J, Croft K, and Beilin L 1995 Mangostin inhibits the oxidative modification of human low density lipoprotein *Free Radics Res* **23** 175-84
[14] Bezwada R 2008 *Chemistry of couamrins* (Indofine Chemical Company)

[15] Itoigawa M, Ito C, Tan H T, Kuchide M, Tokuda H, Nishino H, and Furukawa H 2001 Cancer chemopreventive agents, 4-phenylcoumarins from Calophyllum inophyllum *Cancer Lett.* **169** 15-9

[16] Gunawan S, Kasim N S, and Ju Y H 2008 Separation and Purification of squalene from soybean oil deodorizer distillate *Separ Purif Tech.* **60** 128-135

[17] Gunawan S, Fabian C, Ju Y H 2008 Isolation and Purification of Fatty Acid Steryl esters from Soybean Oil Deodorizer Distillate *Ind. Eng. Chem. Res.* **47** 7013-7018.

[18] Sadek P 2002 *The HPLC Solvent Guide* (United States of America: Wiley of Interscience)

[19] Dailey A and Quan V V 2015 Effect of extraction solvents on recovery of bioactive compounds and antioxidant properties from macadamia (*Macadamia tetraphylla*) skin waste *Coagent Food & Agriculture.* **1** 1115646

[20] Wang H, Provan G J, Helliwell K 2004 Determination of rosmarinic acid and caffeic acid in aromatic herbs by HPLC *Food Chem.* **87** 301-311

[21] Effendy 2006 *Teori VSEPR, Kepolaran dan Gaya Antar Molekul* Bayumedia Publishing pp. 122-140

Lactic acid fermentation of dahlia tuber starch and waste using *Lactobacillus bulgaricus*: A comparative study

E Praputri[1,2], E Sundari[1], M Martynis[1] and P Agenta[1]
[1] Chemical Engineering Department, Bung Hatta University, Jl. Gajah Mada No. 19 Olo Nanggalo Padang, Sumatera Barat, IDN

E-mail: ertipraputri@yahoo.co.id

Abstract. Lactic acid fermentation of dahlia tuber starch and waste was performed by means of *Lactobacillus bulgaricus* through enzymatic hydrolysis followed by fermentation process. The effect of pH condition on lactic acid production was investigated during the process. The selected bacteria produced lactic acid after 24 hours of fermentation and the productivity was increase after 24 hours of fermentation. After 120 hours of fermentation, it was found that dahlia tuber starch can produce up to 16.18% of lactic acid, whereas lactic acid produced from dahlia tuber waste was only 0.40% at pH of 4. The lactic acid production increase significantly for pH 3.5 and 4 until 96 hours of fermentation, then slowed down. On the other hand, for pH 4.5 the lactic acid production increase until 48 hours of fermentation and then slowed down. The identification of fermentation product indicated that the lactic acid produced in this study was 16.20%, acidic, yellow and cloudy with pH 3.4 – 4.2. The density of lactic acid produced ranged between 1.21 to 1.25 gr/ml.

1. Introduction

Utilization of excess biomass or waste concept from agricultural and agroindustrial residues to produce energy, feeds or foods, and other useful products is not necessarily new. Though they are rich in carbohydrate, their utilization is very less due to the low content of protein and poor digestibility [1]. Recently, fermentation of biomass has gained considerable attention because of it was found to be able to solve environmental problems associated with their disposal.

Dahlia is an annual ornamental plant and tuber that have not been widely used in Indonesia. Dahlia tubers contain polyfructans, and especially inulin, that can be chemically or enzymatically hydrolyzed to fructose and sucrose. These are fermentable sugars that can be transformed into lactic acid or other substances by fermentation. Lactic acid, a 2-hydroxypropionic acid or 2-hydroxypropanoic acid compound, has many applications in food, pharmaceutical, textile, leather, and other chemical industries as mild acid flavour, pH regulator, or as a preservative. It has attained a high demand as it is the precursor for polylactic acid (PLA), a biodegradable plastic [2]. Biotechnologically, lactic acid can be produced by SHF - separate hydrolysis and fermentation, by SSF – simultaneous saccharification and fermentation, and by CHF – combined hydrolysis and fermentation, using raw materials contain glucose, sucrose, maltose, and lactose [3,4]. Numerous studies on lactic acid fermentation from agricultural residues have been reported, but there have been few studies on lactic acid production

[2] To whom any correspondence should be addressed.

from dahlia tuber starch and waste. Previous study reported that 1 kg of dahlia tuber would produce 200 grams of dahlia tuber waste (20% of weight) [5]. The technique of lactic acid fermentation is important because it can reduce the time of fermentation or increase the yield of lactic acid. To avoid any contaminants, all processes in fermentation phase are performed aseptically [6]. The present work reports on the investigation results of fermentation time and pH on reduction sugar content and lactic acid yields of dahlia tuber starch and waste.

2. Materials and methods

2.1. Materials
The dahlia tubers used in this study were purchased from West Sumatera area. The starch and waste were taken from post-production of inulin process in Bung Hatta University laboratory.

2.2. Hydrolysis and fermentation
The dahlia tubers starch and waste were subjected to enzymatic hydrolysis using α-amylase and glucoamylase for 2 hours. The resultant hydrolysate contained the reducing sugar necessary for the fermentation process. The hydrolysate pH was adjusted to 3.5, 4 and 4.5, then sterilized and cooled off at 25°C. The hydrolysates was then inoculated with *Lactobacillus bulgaricus* bacteria and immediately incubated at 30°C with agitation at 250 rpm for 120 hours [7]. For pH stabilization, $CaCO_3$ was used as a buffering agent. The analysis of reduction sugar content was performed every 24 hours.

2.3. Analysis
The total reduction sugar concentration in hydrolysates was estimated by Luff-Schoorl. The deviation between initial reduction sugar (after hydrolysis) and residual reduction sugar (after fermentation) was counted as lactic acid content. The identification of lactic acid is performed through measurement of density, pH, odor, color, and esterification process.

3. Results and discussions

3.1. Effect of fermentation time on reducing sugar content from dahlia tuber starch and waste at pH 4
The fermentation of lactic acid from reduction sugar of dahlia tuber starch and waste were performed for 120 hours and the reduction sugar content are presented in Figure 1.

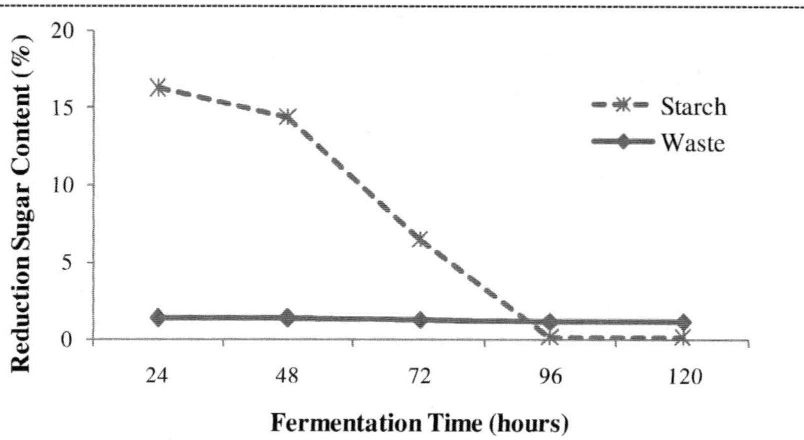

Figure 1. Effect of fermentation time on reducing sugar content from dahlia tuber starch and waste at pH 4.

After 24 hours, the dahlia tuber starch produced 16.27% of reduction sugar, approximately 12 times higher than those of dahlia tuber waste. The difference in carbohydrates content will affects the reducing sugar content obtained [8]. The residual reducing sugar can be used as an indicator of fermentation process. During the fermentation process, the decrease of reduction sugar was attributed to microbes activity that use reduction sugar as a source of food [9]. After 48 hours of fermentation, dahlia tuber starch show faster decrease of reduction sugar content than dahlia tuber waste and at the end of the fermentation period both of samples showed that the reduction sugar content were approximately reached 0%.

3.2. Effect of fermentation time on lactic acid yield from dahlia tuber starch and waste at pH 4
Lactic acid yields from dahlia tuber starch and waste hydrolysates are presented in Figure 2.

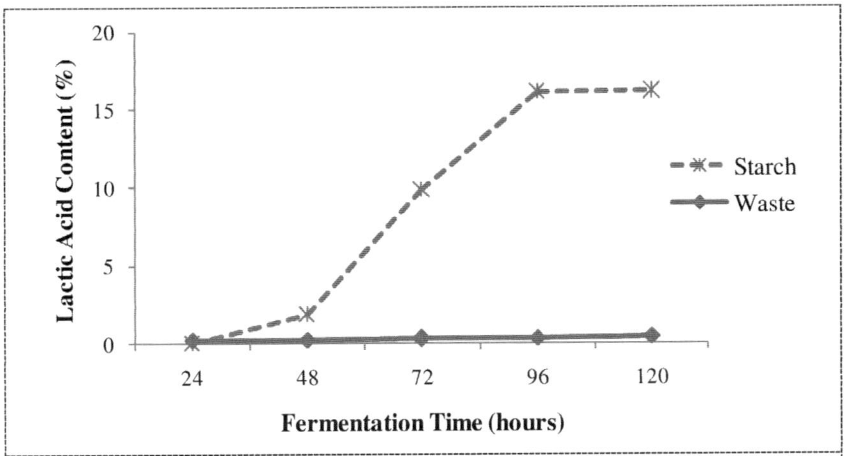

Figure 2. Effect of fermentation time on lactic acid yield from dahlia tuber starch and waste at pH 4.

After 48 hours of fermentation, the lactic acid produced from dahlia tuber starch was increasing significantly. The maximum 16.18% of lactic acid content was achieved after 120 hours of fermentation. This result is 15 times higher than those produced of dahlia tuber waste. The fermentation proceeded very rapidly at the early stage of fermentation, but slowed down at the later stage producing 16.14% at 96 hours because less reduction sugar is available, hence slow down the fermentation process. At the same time, lactic acid accumulation at the early stage of fermentation process and low pH values have led to inhibition of lactic acid production by the selected bacteria.

3.3. Effect of pH on lactic acid content of dahlia tuber starch fermentation
The effect of pH on lactic acid content of dahlia tuber starch fermentation is shown in Figure 3. After 24 hours of fermentation, the lactic acid started to produced within the three conditions of pH. The lactic acid production increase significantly for pH 3.5 and 4 until 96 hours of fermentation, then slowed down. On the other hand, for pH 4.5 the lactic acid production increase until 48 hours of fermentation and then slowed down. The lactic acid produced indicated that *Lactobacillus bulgaricus* bacteria can tolerate pH values as low as 3.5. However, low lactic acid yields suggested that the fermentation process were not optimized as *Lactobacillus bulgaricus* bacteria optimum pH condition range between 5.5-5.8 [10].

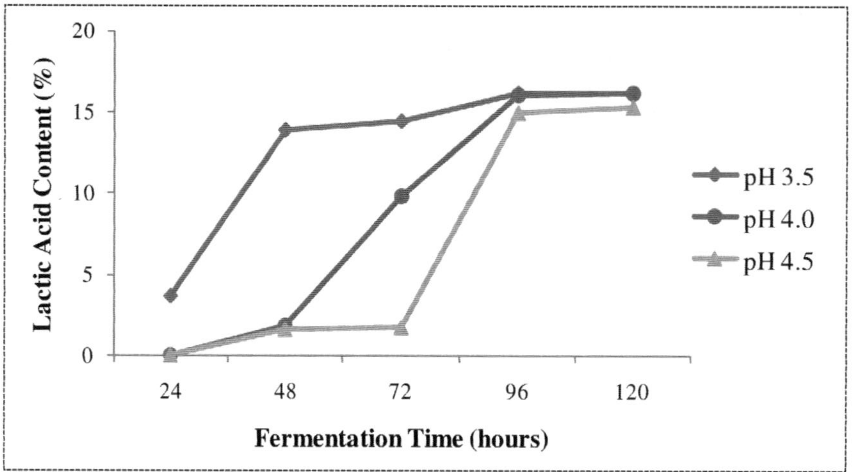

Figure 3. Effect of pH on lactic acid content of dahlia tuber starch fermentation.

The identification of fermentation product indicated that the lactic acid produced in this study was 16.20%, acidic, yellow and cloudy with pH 3.4 – 4.2. The density of lactic acid ranged between 1.21 to 1.25 gr/ml.

4. Conclusion

The carbohydrates derived from dahlia tuber starch and waste could be converted to lactic acid by *Lactobacillus bulgaricus* bacteria through enzymatic hydrolysis followed by fermentation process. The identification through measurement of density, pH, odor, color, and esterification process showed that the combined hydrolysis and fermentation performed on dahlia tuber starch and waste will produce lactic acid that matched the specification. After 120 hours of fermentation, the dahlia tuber starch can produce up to 16.18% of lactic acid content, whereas lactic acid obtained from dahlia tuber waste was only 0.40%, thus suggesting that lactic acid production from dahlia tuber waste is not economically preferable.

References

[1] Nassar N M A, Elkholy H and Eltantawy A 2000 The world cassava production: an overview *J Root Crops* **26** 2 1-5.

[2] Rojan P J, Nampoothiri K M, Nair A S and Pandey A 2005 L(+)-lactic acid production using Lactobacillus casei in solid-sate fermentation *Biotehnol. Lett.* **27** 1685-1688.

[3] Baston O and Constantin O E 2012 Selection of lactic acid bacteria able to ferment inulin hydrolysates *Food Tech.* **36(2)** 31-40.

[4] Purnavita S 2014 Rekayasa produksi asam laktat dari limbah ampas pati aren sebagai bahan baku poli asam laktat (Semarang: Akademi Kimia Industri Santo Paulus).

[5] Sundari, Elmi 2014 Identifikasi dan kondisi ekstraksi inulin dari umbi dahlia di sumatera barat *Prosiding: Seminar Nasional Sains dan Teknologi Lingkungan (SNSTL)* **1**.

[6] Stanbury P F, Whitaker A and Hall S J 1995 Principles of fermentation technology (New York: Butterworth–Heinemann)

[7] Hamamci H and Ryu D D Y 1994 Production of L(+)- lactic acid using immobilized *Rhizopus oryzae*: reactor performance based on kinetic model and simulation *Appl. Biochem. Biotech.* **44** 125-133.

[8] Rahmawati I S 2015 Evaluasi pertumbuhan isolat prebiotik (L. casei dan L. plantarum) dalam medium fermentasi berbasis ubi jalar (Ipomea batatas L.) selama proses fermentasi (kajian jenis isolat dan jenis tepung ubi jalar) *Jurnal aplikasi teknologi pangan* **4** (4).
[9] Andarti I Y and Wardani A K 2015 The influence of fermentation time to chemical, microbiological, and organoleptic characteristic of black soybeans (glycine max (L)) miso *J Pangan dan Agroindustri* **3** 3 889-898.
[10] Hutkins R W and Nannen N L 1993 pH homeostatis in lactic acid bacteria (Lincoln: University of Nebraska).

IOP Conf. Series: Materials Science and Engineering **334** (2018) 012073 doi:10.1088/1757-899X/334/1/012073

Characterization of bioplastic based from cassava crisp home industrial waste incorporated with chitosan and liquid smoke

U Fathanah[1,2], M R Lubis[1], F Nasution[1] and M S Masyawi[1]
[1] Department of Chemical Engineering, Faculty of Engineering, Syiah Kuala University, Darussalam, Banda Aceh 23111, Indonesia

E-mail: umi_fathur@yahoo.com

Abstract. Cassava peel (*Manihot utilissima*) is waste of agricultural result that is much potential as raw material of bioplastic making. This research focuses on bioplastic making from cassava peel. It aims to characterize the resulted bioplastic (mechanical and physical properties, SEM analysis, FTIR analysis and time test of bioplastic degradation). The bioplastic preparation takes place by mixing starch of cassava peel and chitosan (20, 30, 40 and 50% w/w), glycerol 30% w/w as plasticizer, and liquid smoke (0, 1 and 2 mL) as antimicrobial agent. The research result shows the highest value of tensile strength is 96.04 MPa, the highest elongation at break is 52.27%, and the value of water-resistant test is 22.68%. Morphology analysis by using SEM shows uneven surface and there is fracture in its cross-section. The analysis of functional group by FTIR shows the presence of functional groups of O–H (hydroxyl), N–H (amine), dan CH_3–O (ether). The fastest complete degradation of bioplastic occurs in 45 days, and the longest occurs in 57 days.

1. Introduction

The usage of plastic as packaging keeps increasing, so that it could cause plastic waste accumulation. The waste has impact on environmental pollution because it is hard to degraded by microorganism [1]. Various industrial sectors produces synthetic packaging plastic about 100 million tons/year. They produce plastic waste about as much as that amount every year. Meanwhile the need of packaging plastic achieves 2.3 million tons/year [2]. Therefore, development of biodegradable plastic technology is necessary, because biodegradable plastic can decompose faster than commercial plastic in the soil that can reduce a plastic waste.

Biodegradable plastic is plastic that could be degraded by degrading-microorganism activity. Biodegradable plastic has benefit that is the same as synthetic plastic or conventional plastic. It is bioplastic, namely plastic that all or almost all of its components are from renewable raw materials. They are environmentally friendly because of their properties that could be back to nature. Biodegradable packaging is packaging film that could be recycled and destroyed naturally [3]. Raw material often used in biodegradable plastic is tuber that has starch such as sweet potato and cassava [1].

Chitosan is another polymer that is the second most natural polymer occurring after fiber. It has interesting biological activities and biodegradable, non-toxic, excellent strength and elongation properties. Chitosan combined with cellulose increases the mechanical strength of paper board [4]. Its

[2] To whom any correspondence should be addressed.

IOP Conf. Series: Materials Science and Engineering **334** (2018) 012073 doi:10.1088/1757-899X/334/1/012073

advantageous properties cover hydrophilicity, bio-compatibility, degradability, and anti-bacteria. It is hydrophobic, water-resistant in its structure, forms gel spontaneously and membrane or film easily [5].

Liquid smoke is liquid that has anti-microbe property. Liquid smoke so far is useful as preservative because acid content contained inside is anti-microbe. Compounds of phenol and acetic acid could inhibit bacterial growth. Liquid smoke with such property could be additive in making biodegradable plastic [6].

Waste in the form of cassava peel from cassava-chip industry could be raw material of bioplastic. According to Badan Pusat Statistik (2016), cassava production in Indonesia in 2015 is 21.8 million tons/year [7]. Cassava peel has starch content of 4 – 59% [8]. Up to present, not many industries benefits cassava peel maximally. It is just thrown away so that it becomes waste that pollutes environment. Therefore, improvement of bioplastic from solid waste of cassava-chip industry needs to be further investigated.

2. Materials and Preparation

2.1. Materials

Materials used were cassava peel, distilled water, chitosan (20, 30, 40, and 50% w/w), glycerol PA (30% w/w) as plasticizer, acetic acid as chitosan solvent, sodium metabisulphyte, and liquid smoke (0, 1, and 2 mL) as anti-microbial agent. Equipments used were blender, oven, hot plate, magnetic stirrer, thermometer, desiccator, digital balance, and mirror plate.

2.2. Procedure

2.2.1. Extraction of Cassava Peel, Cassava peel of 100 grams was washed and soaked in sodium metabisulphyte 1% for 1 hour. The next process was separation process between cassava peel and sodium metabisulphyte. It was continued by crushing of cassava peel by blender, added 100 ml water to ease crushing. Slurry resulted was mixed by water with ratio of 1 : 1 to extract starch in cassava peel. After that it was filtered by using gauze and the result was called filtrate I. The extract was added with water at the same ratio and filtered so that filtrate II was obtained. The rest of the waste was thrown away. Filter results (filtrate of I and II) were precipitated for 3 hours. The starch precipitation was taken and the filtrate was thrown away. The starch precipitation was dried in oven at temperature of 70°C for 30 minutes [9]. Dried precipitate was refined by grinding and screened by gauze. The screening results in cassava starch that was ready used.

2.2.2. Bioplastic Making, Cassava-peel starch of 5 grams was put into Erlenmeyer I, and added by aquadest until 50 grams. It was stirred by magnetic stirrer for 25 minutes. Chitosan was put into Erlenmeyer II and dissolved with glacial acetic acid 1% (%v). Chitosans of 20, 30, 40, and 50% of starch solution mass was put into Erlenmeyer I. They were put in hot plate and heated at temperature of 70°C while stirred for 15 minutes. Glycerol was added 30% mass of starch solution, and various volumes of liquid smoke of 0, 1, and 2 ml. Furthermore, it was stirred back until homogeneous. Casting was carried out on glass plate of 0.02 mm thickness [10] to form thin sheet of bioplastic. Thin sheet (bioplastic) was removed from glass plate, kept in desiccator and ready for analysis.

2.3. Sample Characterization

2.3.1. Mechanical Properties, Tensile strength and elongation at break were determined with ASTM D 882-10 (Standard Test Method for Tensile Properties of Thin Plastic Sheeting). Tensile strength test measured the highest tensile stress attained by a sample during test. It also determined elongation of the sample at rupture, and expressed as percent of the gage length. Tensile strength and elongation at break used Equations of (1) and (2) [4].

$$Tensile\ strength\ (\sigma) = \frac{F}{A} \quad (1)$$

$$Elongation\ (\varepsilon) = \frac{L - L_0}{L_0} \quad (2)$$

F is the highest tensile; A is the cross-section of the gage section; L_0 is the initial gage length; L is the final length.

2.3.2. Physical Properties (Water Resistance). Test of water resistance was carried out by cutting plastic film of 1 cm × 1 cm. Initial mass was weighed and put into measuring cup that contained distilled water in 10 seconds. Sample was lifted from measuring cup, removed from distilled water by tissue, after that just weighed. Sample was put back into measuring cup that contained distilled water for 10 seconds. Then sample was lifted from measuring cup and weighed back. Procedure of soaking and weighing was carried out back until final sample mass was constant. Test of water-resistant was calculated as the following Eq. (3) [11].

$$Water\ Resistance\ (W') = \frac{W - W_0}{W_0} \times 100\% \quad (3)$$

W_0 is the initial sample mass; W is the final sample mass.

2.3.3. Morphology Analysis by Using SEM (Scanning Electron Microscope). Analysis of sample morphology was in Geology Laboratory of Institut Teknologi Bandung. Sample was cut in size of 2 × 3 cm then it was adhered in holder. Sample was put into chamber of SEM equipment for setting place and recording picture. Photos of SEM taken were top and cross sections of film with magnitude of 5000 × [11].

2.3.4. Analysis of Functional Group by Using FTIR (Fourier Transform Infrared Spectroscopy). Functional group was analyzed in Chemistry Laboratory of Universitas Negeri Padang. It was done by Fourier Transform Infrared Spectroscopy PerkinElmer® (FTIR). Sample of film was placed into set holder, and then appropriate spectrum was searched. The result was in the form of diffractogram of relation between wave number and intensity. Spectrum of FTIR was recorded by using spectrometer at room temperature [12].

2.3.5. Test of Biodegradability on Environment. Biodegradability was tested by method of soil burial test by cutting plastic film in size of 3 × 3 cm. After that plastic was put in desiccator for 24 hours then it was weighed. Plastic burial used soil with soil depth of 5-cm. Sample checking took place once in every three days, followed by weighing and putting back in soil. Observation was carried out until sample was degraded completely in soil [13].

3. Result and Discussion

3.1. Mechanical Properties of Biodegradable Film

Tensile strength is the highest stress needed by a specimen to suppress force which given [5]. The highest tensile strength obtained is at chitosan of 30% without liquid smoke namely 96.04 MPa. The lowest strength is at chitosan of 40% without liquid smoke namely 35.28 MPa as in Figure 1. The high value of tensile strength indicates that film has strong structure and not easy to break. It is because chitosan could form inter-molecular hydrogen bond that makes bioplastic denser, strong, and hard to break [11].

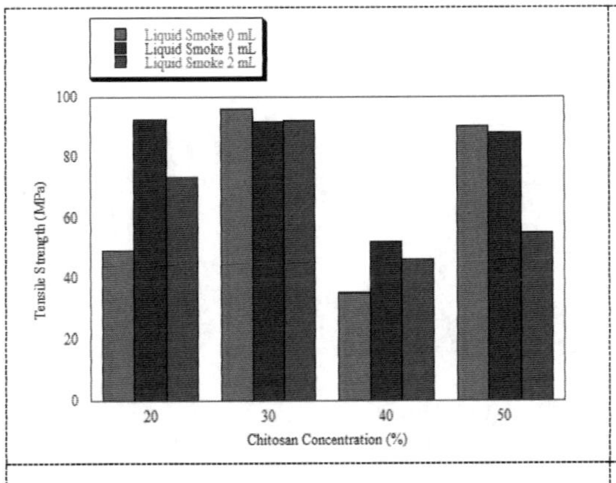

Figure 1. The effect of chitosan on tensile strength of bioplastic at various liquid smoke.

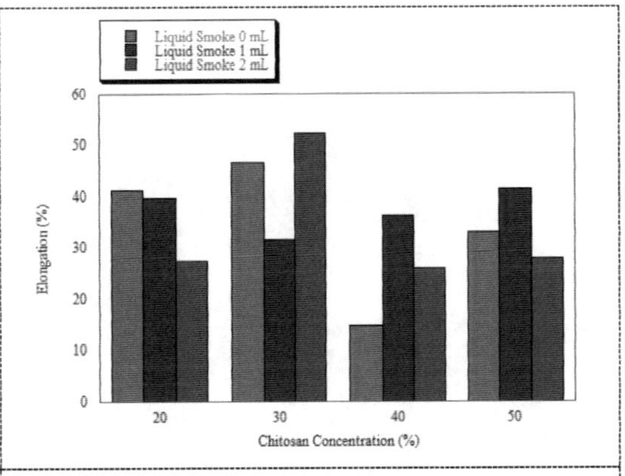

Figure 2. The effect of chitosan on elongation of bioplastic at various liquid smoke.

Elongation means the residual relative change in length compared with the starting length after the break of the test piece in a tensile test [14]. The highest elongation obtained is at chitosan of 30% with liquid smoke volume of 2 mL namely 52.27%. The lowest is 14.87% at chitosan of 40% without liquid smoke as indicated in Figure 2. The lowest elongation value indicates that film's structure is not plastic (stiff) and hard to stretch. Strong interaction between starch and chitosan makes molecule bond denser and strong. Therefore, film is hard to stretch or elongate [11].

3.2. Physical Properties (Water Resistance of Biodegradable Film)

Water resistance of the cassava starch-based film increases as chitosan concentration increases (Fig. 3). Hydrophobic chitosan property and the film that could not be in water cause it [3].

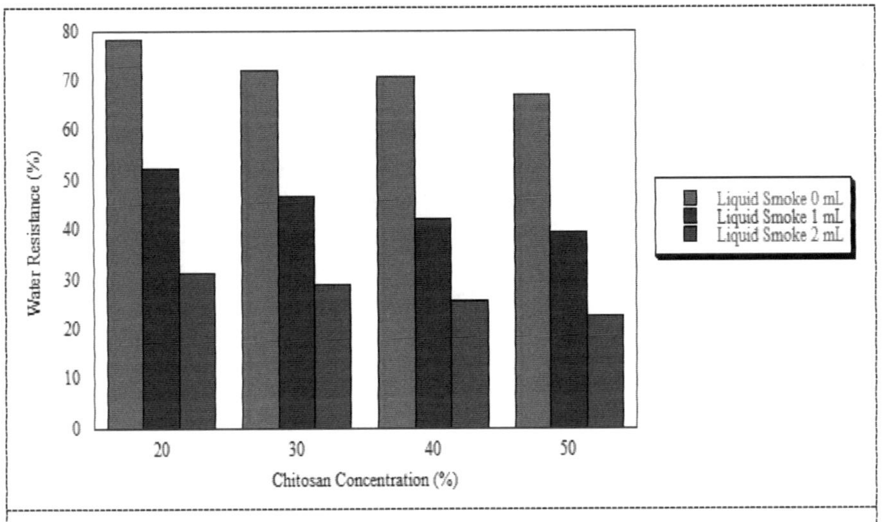

Figure 3. The effect of chitosan on swelling at various liquid smoke volumes.

The highest water resistance obtained is at chitosan of 20% without liquid smoke namely 78.40%. The lowest water resistance is 22.68% at chitosan of 50% and liquid smoke of 2 mL.

3.3. Analysis by SEM

Analysis of bioplastic morphology carried out is to explain morphology of bioplastic formed. The part of sample analyzed is top and cross sections as shown in Figures of 4 and 5. Micro-structure of bioplastic with chitosan of 30% and liquid smoke of 1 mL is in Figure 4. It is not distributed completely (uneven) as shown by agglomerating part (laminar).

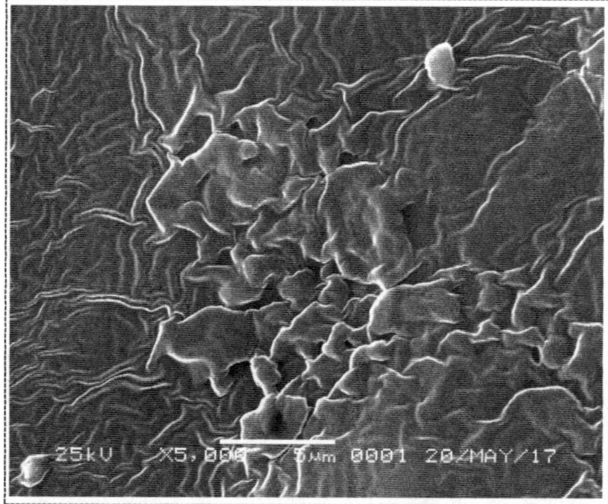

Figure 4. The surface of bioplastic with magnitude of 5000 × at chitosan of 30% and liquid smoke of 1 mL.

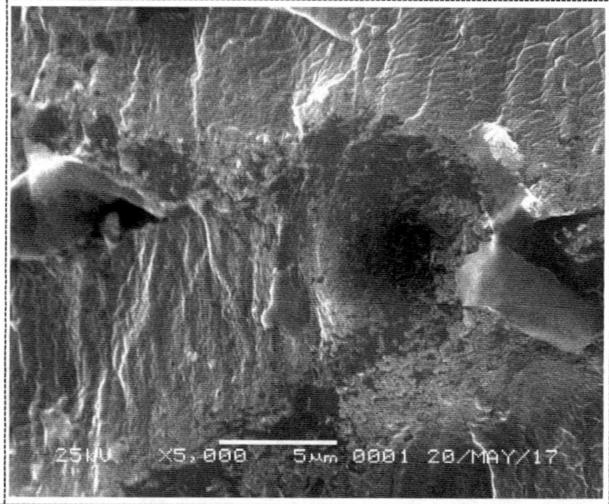

Figure 5. Cross section of bioplastic with magnitude of 5000 × at chitosan of 30% and liquid smoke of 1 mL.

Disorder while casting causes uneven sample so that its morphology structure is not homogeneous. It is also caused by the presence of insoluble chitosan in organic solvent [15]. Film layer is formed based on chemical bond of polymer and another compound added. Chemical bond that occurs is interaction of H–H or hydrogen bond [16]. In Figure 5, it could be observed that bioplastic resulted is non-porous. It seems fracture (wrinkle) caused by the presence of irregularity at the time of casting.

3.4. Analysis by FTIR

Bioplastic functional group is analyzed to identify its structure (organic and inorganic compounds). Figure 6 shows the result of functional group analysis. It also shiws that product resulted is bioplastic (based on standard IR spectrum). It indicates strong absorption at wave numbers of 3294.31 and 3291.54 cm^{-1} from O–H vibration. At the same wave number, it seems N–H group that shows the N-acetyl presence. Sharp absorption at wave number of 1645.45 and 1645.86 cm^{-1} indicates vibration of C=O. Sharp absorption at wave number of 1018.84 and 1020.66 cm^{-1} is from vibration of C–O.

Based on result of FTIR, it seems specific functional group contained in its constituent part. Groups of O–H show the material presence of bioplastic composing i.e. glycerol [5]. Meanwhile, the N–H group identification shows the presence of chitosan. The new functional group is not found (no chemical reaction occurs). For sample without liquid smoke addition, CH_3–O group does not appear. It is because at the chitosan concentration, the sample uses liquid smoke whose part is ether.

3.5. Degradation Time of Bioplastic

Degradation time of the cassava-starch- based film increases as liquid smoke volume increases (Fig. 7). A decrease of film mass marks it until the film completely degraded. It occurs because in liquid smoke there is active substance namely phenol. In liquid smoke, it could inhibit bacterial growth, so that the more liquid smoke, the longer degradation time needed.

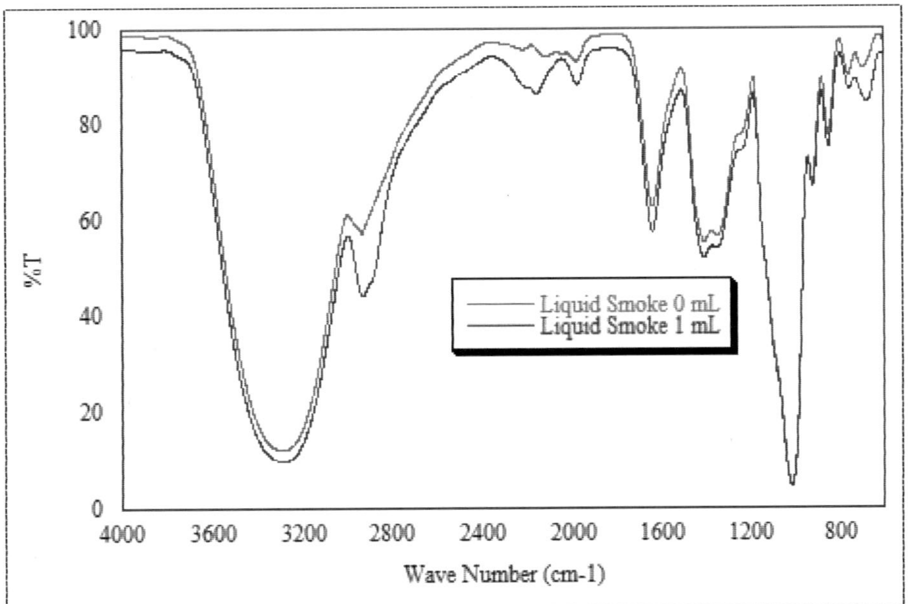

Figure 6. FTIR spectrum of bioplastic with chitosan concentration of 30% and liquid smoke of 1 mL.

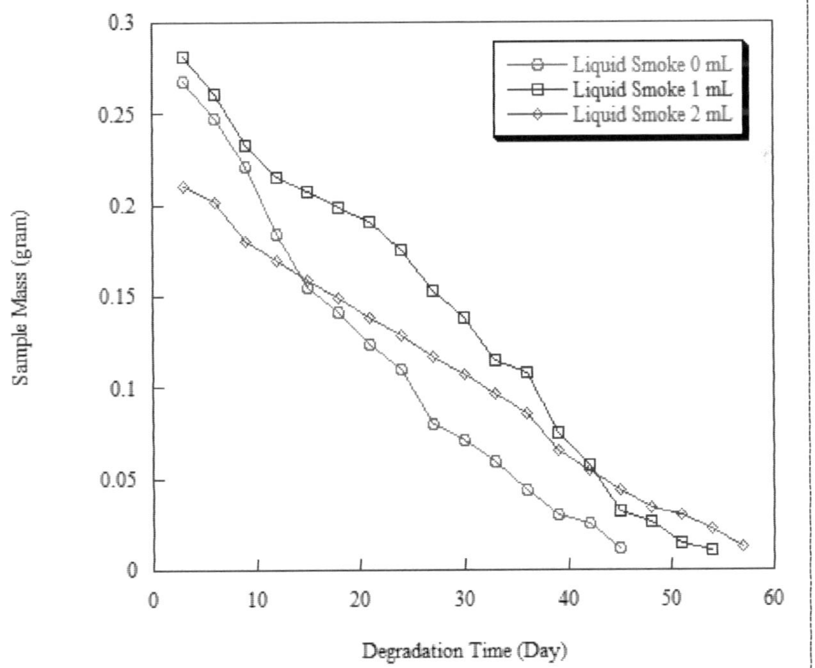

Figure 7. The correlation of bioplastic mass reduction and time at chitosan of 30% and various volume of liquid smoke.

In Figure 7 the fastest degradation is in 45th day for film with chitosan of 30% and without liquid smoke. The longest degradation is at film with chitosan of 30% and liquid smoke of 2 mL in 57 days.

4. Conclusion

This research focuses on characteristic of cassava-peel-starch bioplastic with chitosan and various volumes of liquid smoke. Bioplastic with chitosan of 30% without liquid smoke has the highest tensile strength of 96.04 MPa. Bioplastic with chitosan of 30% and liquid smoke of 2 mL has the highest elongation of 52.27%. Bioplastic with chitosan of 50% and liquid smoke of 2 mL has the lowest water absorbtion of 22.68%. SEM analysis of bioplastic with chitosan of 30% and liquid smoke of 1 mL indicates uneven surface. There is fracture in its cross-section. FTIR analysis of bioplastic with chitosan of 30% and liquid smoke 1 mL indicates O–H, N–H, and CH_3–O groups. Such groups show glycerol, chitosan, and liquid smoke presence in film. Degradation test shows that liquid smoke could prolong degradation time. For bioplastic with chitosan of 30% without liquid smoke, the fastest degradation time is 45 days. For bioplastic with chitosan of 30% and liquid smoke of 2 mL, degradation is longer namely in 57 days.

References

[1] Utami R M, Latifah L, and Widiarti N 2014 Bioplastic synthesize from banana peel incorporate with chitosan and glycerol as plasticizer *Indonesian Journal of Chemical Science* **3** 163 – 167
[2] Utomo W A, Argo B D, and Hermanto M B 2013 Effect of Temperature and drying duration toward psychochemical characteristic of biodegradable plastic from starch composite of aloevera-chitosan *Jurnal Bioproses Komoditas Tropis* **1** 77 – 78
[3] Coniwanti P, Laila Land Alfira M R 2014 Preparation of bioplastic from corn starch incorporate with chitosan and glycerol as plasticizer *Jurnal Teknik Kimia* **4** 118 – 123
[4] Kaisangsri N, Kerdchoechuen O, and Lauhakunjit N 2012 Biodegradable foam tray from cassava starch blended with natural fiber and chitosan *Industrial Crops and Products* **37** 542 – 546
[5] Lazuardi G P, and Cahyaningrum S E 2013 Preparation and characterization based bioplastic chitosan and cassava starch with glycerol as plasticizer *UNESA Journal of Chemistry* **3** 161 – 166
[6] Morelli C L, Mohamed N B, Márcia C B, Rosario E S, and Julien B 2015 Natural copaiba oil as antibacterial agent for bio-based active packaging *Journal of Industry and Corps.* **70** 134 – 141
[7] BPS (Badan Pusat Statistik) 2016 *Cassava production in 1993-2015* (Jakarta: Badan Pusat Statistik)
[8] Richana N 2013 *Discover a potential of sweet potato and cassava* (Bandung: Nuansa Cendekia)
[9] Akbar F and Anita Zand Harahap H 2013 Effect of storing time of biodegradable film based cassava peel starch toward its mechanical properties *Jurnal Teknik Kimia* **2** 11 – 15
[10] Nurviantika L, Sedyawati S M, and Mahatmanti W F 2015 Comparison of bioplastic properties based rice waste and cassava peel with limonene as aditive *Indonesian Journal of Chemical Science* **4** 52 – 55
[11] Setiani W, Darmadji P, Setiaji B, and Pranoto Y 2014 Preparation and characterization of edible film from breadfruit starch *Valensi* **2** 100 – 109
[12] Darni Y and Herti U 2010 Preparation and characterization study of bioplastic mechanical properties and hydrophobicity from sorghum starch *Jurnal Rekayasa Kimia dan Lingkungan* **7** 88 – 93
[13] Anggarini F, Latifah, and Miswadi S S 2013 Application of plasticizer glycerol at preparation of bioplastic from jackfruit seed *Indonesian Journal of Chemical Science* **2** 173 – 178
[14] Lenze F J, Sikora S and Banik J 2017 Method for producing a component having improved elongation at break properties *U.S. Patent Application* **15/461,783** 1 – 4
[15] Kurniasih M and Dwi K 2011 Synthesis and psychochemical characterization of chitosan *Jurnal Inovasi* **5** 42 – 48

[16] Susilawati, Irfan M, and Desy M 2011 Biodegradable plastics from a mixture of low-density polyethylene (LDPE) and cassava starch with the addition of acrylic acid *Jurnal Natural* **11** 12 – 19

Acidity, oxophilicity and hydrogen sticking probability of supported metal catalysts for hydrodeoxygenation process

A Ng K Lup[1], F Abnisa[1,2,4], W M A W Daud[1] and M K Aroua[3]

[1] Department of Chemical Engineering, Faculty of Engineering, University of Malaya, 50603 Kuala Lumpur, Malaysia
[2] Department of Chemical Engineering, Faculty of Engineering, University of Malikussaleh, Lhokseumawe 24351, Indonesia
[3] School of Science and Technology, Sunway University, No.5 Jalan Universiti, 47500 Selangor Darul Ehsan, Malaysia

E-mail: faisal.abnisa@gmail.com

Abstract. Hydrodeoxygenation is an oxygen removal process that occurs in the presence of hydrogen and catalysts. This study has shown the importance of acidity, oxophilicity and hydrogen sticking probability of supported metal catalysts in having high hydrodeoxygenation activity and selectivity. These properties are required to ensure the catalyst has high affinity for C-O or C=O bonds and the capability for the adsorption and activation of H_2 and O-containing compounds. A theoretical framework of temperature programmed desorption technique was also discussed for the quantitative understanding of these properties. By using NH_3-TPD, the nature and abundance of acid sites of catalyst can be determined. By using H_2-TPD, the nature and abundance of metallic sites can also be determined. The desorption activation energy could also be determined based on the Redhead analysis of TPD spectra with different heating rates.

1. Introduction

Bio-oil has been extensively researched on for its application as one of the sources for liquid fuel. Bio-oil is produced through the pyrolysis of lignocellulosic biomass. However, bio-oil contains significant amount of oxygen-containing compounds due to depolymerization and defragmentation of cellulose, hemicellulose and lignin within biomass feedstock during pyrolysis[1]. The high oxygen content of bio-oil would cause issues such as low heating value, thermal instability and corrosivity that might affect its performance as liquid fuel [2, 3]. Thus, hydrodeoxygenation is often employed as an upgrading treatment for bio-oil in producing liquid fuel of low oxygen content and for the valorization of oxygen-containing compounds into high-added value products [4].

Hydrodeoxygenation (HDO) is a deoxygenation reaction which involves oxygen abstraction from the compound in the presence of hydrogen. The reaction is commonly catalyzed by zeolites and supported metal catalysts of various forms such as carbide, sulfide, nitride, phosphide and oxide. Numerous studies on HDO process using various oxygen-containing compounds and catalysts were done by researchers. The HDO studies using various oxygenated model compounds are instrumental in the understanding of the reaction mechanisms with respect to the functionalities of model

[4] To whom any correspondence should be addressed.

compounds such as phenolics, furans, acids, carbonyls, alcohols, ethers and esters. This is because a typical bio-oil sample has more than 400 different oxygenated compounds of phenolics, furans, acids, carbonyls, alcohols, ethers and esters [5]. With the proper elucidation of the respective HDO reaction pathways of model compounds, HDO study of real feed can then be accurately studied with the consideration of synergistic, inhibitive and bulk effects owing to the chemical interactions of the complex mixtures [6].

The studies on HDO process using various catalysts have also shown the various effects of chemical compositions, physicochemical properties and electronic interactions of catalysts on the HDO performance. Thus, in this paper, several major properties of HDO catalysts such as hydrogen activation, acidity and oxophilicity were analyzed for the prediction of excellent HDO catalysts with high deoxygenation activity. Temperature programmed desorption for HDO catalysts were also discussed as the theoretical frameworks for the quantitative interpretation of such properties.

2. Hydrodeoxygenation mechanism

The mechanism of HDO generally involves the scission of C-O or C=O bonds for the oxygen removal from model compounds. HDO process can proceed through three major reaction pathways: direct deoxygenation route (DDO), hydrogenation-dehydration route (HYD) or tautomerization-deoxygenation route [7]. In DDO route, a direct cleavage of C-O or C=O bonds via hydrogenolysis occurs without the hydrogen saturation of aromatic or furanic rings. In HYD route, hydrogen saturation of aromatic or furanic rings would occur first before the cleavage of C-O bonds via dehydration. The selectivity of DDO or HYD pathways is of greater interest when it comes to HDO of model compounds containing aromatic or furanic rings and unsaturated bonds. DDO route may be more advantageous than HYD route as it does not require saturation of bonds by H_2, resulting in a lower H_2 consumption. However, it was also noted that DDO requires higher activation energy than HYD since the bond dissociation energy of $C_{Aromatic}$-OH bond (468 kJ/mol) is higher than that of C_{Cyclic}-OH bond (385 kJ/mol) [7, 8].

Tautomerization-deoxygenation route is not as common as the former two pathways as tautomerization is only characteristic to model compounds that are of enol form. The tautomerization of oxygen-containing compound in its enol form will result in the formation of keto form to form a keto-enol equilibrium mixture. For instance, phenol HDO could undergo tautomerization to form a highly unstable 3,5-cyclohexadienone which would readily undergo C=O hydrogenation to form 3,5-cyclohexadienol. The 3,5-cyclohexadienol would then dehydrate to form benzene. This pathway is kinetically more favorable than DDO as tautomerization brings about an alteration of $C_{Aromatic}(sp^3)$-O bond into $C_{Aromatic}(sp^2)$=O bond which has lower bond energy, without resulting in the saturation of aromatic or furanic rings [9, 10]. However, this pathway requires metal catalysts of high oxophilicity such as C, Ti, Zr, V, W, Co, Fe, Ru and Sn [7].

3. Properties of supported metal catalysts

3.1. Hydrogen sticking probability

The adsorption and activation of hydrogen at catalyst surface are integral studies for HDO process which utilizes hydrogen species for the hydrogenolysis of C-O and C=O bonds. For HDO process, supported metal catalysts should have metallic sites that are active for H_2 adsorption and activation [5]. With the saturation of metal sites by activated hydrogen adsorbates, a facile hydrogen spillover to the metal support interface site will then be induced for the hydrogenolysis of the activated O-containing compounds [11].

Supported transition metal catalysts were commonly used in HDO because transition metals were noted to exhibit high H_2 sticking coefficient and hydrogenation activity [3, 12]. For instance, transition metals such as Cu, Co, Ir, Ni, Pd, Pt, Rh, Ru were noted to have H_2 sticking probability increasing in the order of Cu, Ni, Co, Ir, Pd, Pt, Rh and Ru at 150°C below and at 1 bar [12]. However, it was also noted that H_2 sticking probability and surface coverage on transition metals are also thermodynamically dependent on the reaction temperature. Since H_2 sorption is a reversible exothermic process, an increase in reaction temperature would result in lower H_2 surface coverage, which then result in lower HDO activity [13, 14]. Thus, an optimized reaction temperature is important to ensure sufficient H_2 surface coverage and sufficient energy for bond dissociation during HDO process.

3.2. Acidity

Acidity of supported metal catalysts is another important parameter in affecting HDO activity and selectivity. Acidity of catalyst is generally dependent on the nature and abundance of acid sites within the catalyst. The nature of acid sites can be classified as Lewis and Brønsted acid sites. Lewis acid sites are commonly the unsaturated metal sites which are the good electron acceptors while Brønsted acid sites are –OH or –SH groups of supports which are the good proton donors [3]. The function of acid sites in HDO process is to act as the adsorption and activation sites for the oxygen from the O-containing compounds. Subsequently, the activated O-containing compounds will then be hydrogenolyzed by activated hydrogen species from metallic sites.

Likewise, catalyst with higher abundance of acid sites for O-containing compound activation would also result in higher turnover rate. Zeolites and catalysts supported on C, TiO_2, ZrO_2, Al_2O_3, CeO_2 and SiO_2 were some of the catalysts with good HDO activity that exhibit acidic characters. Zeolites were known for their high acidity for HDO while metal oxide supports were noted for their mild acidity. Nevertheless, acidity of catalysts must also be optimal since catalysts of high acidity were also noted to catalyze side reaction such as condensation to produce carbonaceous species that would deactivate the catalyst through coking [15].

3.3. Oxophilicity

Oxophilicity of metal is associated with the tendency of metal being attracted to oxygen and its abstraction. Supported metal catalysts with high oxophilicity would show higher affinities to C-O and C=O bonds, resulting in effective oxygen removal from these bonds. Likewise, several studies have also reported that the use of oxophilic metal catalysts such as Fe, Ti and Zr have resulted in the preference of tautomerization-deoxygenation pathway to produce deoxygenated aromatics as major products [16, 17]. Kepp [18] has analyzed the quantitative scale of the oxophilicity of metals as shown in table 1. Based on table 1, it can be inferred that HDO supports such as C, TiO_2, ZrO_2 and metals such as Ti, V, W, Zr, Co, Fe, Ru and Sn are highly oxophilic for the facile oxygen abstraction.

Table 1. Normalized oxophilicity of several elements.

Element	C	Ti	V	W	Zr	Co	Fe	Ru	Sn
Normalized oxophilicity	1.4	1.0	0.8	0.8	0.8	0.4	0.4	0.4	0.4

4. Recommendations for effective HDO process

From the reaction chemistry perspective, an effective hydrodeoxygenation reaction quintessentially entails high catalytic deoxygenation activity, high conversion of O-containing compounds and high yield and selectivity of deoxygenated products. As for the process design of such reaction, optimum reaction conditions must also be determined to optimize the production and design costs while maximizing the yield of desired products. Based on the desired properties of HDO catalysts as examined, we can see that supported metal catalysts with transition metals such as Cu, Co, Ir, Ni, Pd, Pt, Rh, Ru being supported on oxophilic and mild acidic supports such as C, TiO_2 and ZrO_2 would be recommended for an effective HDO process [19]. Zeolites were also noted to be excellent HDO catalysts; however, their high acidity may pose severe coking issues during catalytic regeneration. Likewise, HDO via DDO route is often preferred due to formation of deoxygenation aromatics and lower H_2 consumption. For DDO route to be the dominant pathway, a reaction temperature of more than 573 K is often recommended [14]. For reaction pressure, HDO at atmospheric pressure was also noted to be successful. For instance, de Souza et al. [16] reported that the vapor phase HDO of phenol to benzene over Pd/ZrO_2 at 573 K and 1.01 bar was able to achieve 75% phenol conversion and 54% benzene selectivity. Likewise, Boullosa-Eiras et al. [20] also reported that the liquid phase HDO of phenol to benzene over Mo_2C/TiO_2 at 673 K and 25 bar was able to achieve 90% phenol conversion and 90% benzene selectivity. A higher H_2 pressure would certainly enhance the adsorption of H_2 for HDO but the handling of highly pressurized process may require additional design costs for reactor designing and process safety.

5. Temperature programmed desorption for HDO catalysts

Temperature programmed desorption (TPD) is a technique used to study chemisorption at catalyst surface. By analyzing the chemisorption behavior, the physicochemical properties of catalyst surface can then be determined. TPD uses different probe molecules such as NH_3, H_2, CO_2, CO, etc. for different analyses of chemisorption properties. For hydrodeoxygenation process, NH_3-TPD and H_2-TPD are needed for the analysis of the physicochemical properties of HDO catalysts. NH_3-TPD involves the chemisorption of basic NH_3 molecules on the acid sites of catalyst which enables the quantitative study of the acid sites. Likewise, H_2-TPD involves the chemisorption of H_2 molecules on the metallic sites of catalyst which enables the quantitative study of the metallic sites. These quantitative studies rely on the characteristics of desorption peaks as shown in figure 1.

Every desorption peak has the following characteristics: desorption phenomenon, peak area, peak temperature and peak width. The physical significance of a desorption peak is that it tells of a unique desorption phenomenon which occurs at that particular desorption condition. The unique desorption phenomenon is often attributed to a particular chemisorption state or interaction of the probe molecules with the different chemisorption sites of the samples. Thus, based on figure 1, the three distinct desorption peaks may represent three unique chemisorption states of probe molecules due to the presence of three different chemisorption sites of the sample.

The physical interpretation of peak area would be the amount of desorbed probe molecules. By knowing the amount of desorbed probe molecules and the chemisorption mode, the number of chemisorption sites of samples can then be calculated. Likewise, peak temperature is the desorption temperature in which desorption rate is at maximum. A desorption phenomenon which occurs at higher desorption temperature would indicate the stronger chemisorption interaction of the probe molecule with the chemisorption site of sample. As for peak width, it represents the range of temperature at which desorption occurs.

The physical interpretation of NH_3-TPD for HDO metal catalyst is that it gives a quantitative understanding of the acid sites while H_2-TPD gives a quantitative understanding of the hydrogen interaction with metallic sites. For NH_3-TPD spectra, the number of desorption peak represents the type of acid sites while the peak area correlates to the amount of acid sites. Peak temperature would represent the acidity strength of the acid sites. Based on these information, the nature and abundance of acidic sites of HDO metal catalysts can then be determined based on NH_3-TPD. Likewise, the identification of Lewis and Brønsted acid sites can also be further determined using FTIR spectroscopy. For H_2-TPD spectra, the number of desorption peak represents the chemisorption mode of hydrogen species while the peak area correlates to the amount of metallic sites. Peak temperature would represent the chemisorption bond between hydrogen and metallic sites. Based on this information, H_2 sticking probability and its chemisorption mode for HDO metal catalysts can then be determined based on H_2-TPD.

IOP Conf. Series: Materials Science and Engineering **334** (2018) 012074 doi:10.1088/1757-899X/334/1/012074

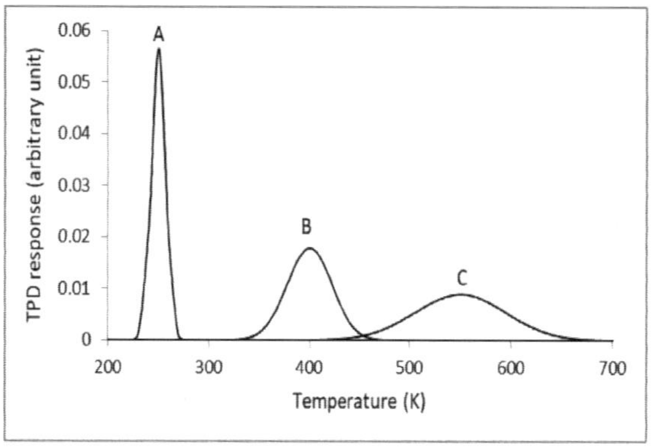

Figure 1. Simulated TPD spectra with three desorption peaks.

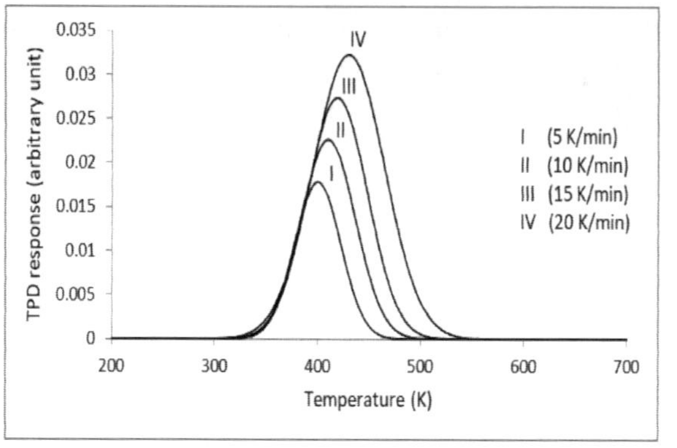

Figure 2. Simulated TPD spectra with different heating rates.

The desorption activation energy can also be determined using Redhead analysis which utilizes TPD spectra of different heating rates [21]. Firstly, Polanyi-Wigner equation was used to describe desorption rate equation that is equation (1). Redhead equation was obtained by differentiating equation (1) with respect to temperature and was then equated to zero. Equation (2) was then obtained by substituting the coverage and temperature at maxima.

$$r_d = -\beta \frac{\mathrm{d}\theta}{\mathrm{d}T} = A\theta^m \exp\left(\frac{-E_d}{RT}\right) \quad (1)$$

$$\frac{E_d}{RT_p^2} = \frac{A}{\beta} m\theta_p^{\ m-1} \exp\left(\frac{-E_d}{RT_p}\right) \quad (2)$$

where r_d is desorption rate, β is heating rate, θ is surface coverage of probe molecules, θ_p is surface coverage at maxima, θ_0 is initial surface coverage which is 2 times of θ_p, T is absolute temperature, A is pre-exponential factor, m is order of desorption, E_d is desorption activation energy, R is ideal gas constant and T_p is peak temperature. By expressing equation (2) in its linear form, we obtained the following:

$$\ln\left(\frac{\beta}{T_p^2}\right) = -\frac{E_d}{R}\left(\frac{1}{T_p}\right) + \ln\left(\frac{mAR\theta_p^{\ m-1}}{E_d}\right) \quad (3)$$

$$\ln\left(\frac{\beta}{T_p^2}\right) = -\frac{E_d}{R}\left(\frac{1}{T_p}\right) + \ln\left(\frac{AR}{E_d}\right) \quad \text{where } m \text{ is } 1 \quad (4)$$

$$\ln\left(\frac{\beta}{T_p^2}\right) = -\frac{E_d}{R}\left(\frac{1}{T_p}\right) + \ln\left(\frac{AR\theta_0}{E_d}\right) \quad \text{where } m \text{ is } 2 \quad (5)$$

Equation (4) is used when desorption is of first order while equation (5) is used when desorption is of second order. Based on equation (4) and (5), desorption activation energy can then be calculated from the gradient of the plot of $\ln\left(\frac{\beta}{T_p^2}\right)$ against $\left(\frac{1}{T_p}\right)$. H_2 desorption activation energy is one of the parameters required for the kinetic studies of HDO. By conducting Redhead analysis for H_2-TPD spectra, H_2 desorption activation energy can then be calculated.

6. Conclusion

Hydrodeoxygenation of O-containing compound involves the scission of C-O or C=O bonds which can be accomplished through DDO, HYD or tautomerization-deoxygenation routes. The acidity, oxophilicity and hydrogen sticking probability are the important properties of supported metal catalysts to ensure a facile hydrodeoxygenation. These properties would result in the catalyst having high affinity for C-O or C=O bonds, facile adsorption and activation of hydrogen and O-containing compounds. The use of temperature programmed desorption technique has been noted to give a theoretical understanding of such properties. By using TPD such as NH_3-TPD and H_2-TPD, the nature and the abundance of acid sites and metallic sites can be elucidated.

Acknowledgement

The authors would like to thank University of Malaya for fully funding this work through the project number "MO008-2015" and Postgraduate Research Grant (PPP) "PG081-2016A".

References

[1] Abnisa F, Wan Daud W M A, Arami-Niya A, Ali B S and Sahu J N 2014 *Energy Fuels* **28** 3074-85
[2] Mortensen P M, Grunwaldt J-D, Jensen P A, Knudsen K G and Jensen A D 2011 *Appl. Catal., A* **407** 1-19
[3] He Z and Wang X 2012 *Catal. Sustainable Energy* **1** 28-52
[4] Jacobson K, Maheria K C and Dalai A K 2013 *Renewable Sustainable Energy Rev.* **23** 91-106
[5] Bu Q, Lei H, Zacher A H, Wang L, Ren S, Liang J, Wei Y, Liu Y, Tang J, Zhang Q and Ruan R 2012 *Bioresour. Technol.* **124** 470-477
[6] Choudhary T V and Phillips C B 2011 *Appl. Catal., A* **397** 1-12
[7] Kay Lup A N, Abnisa F, Wan Daud W M A and Aroua M K 2017 *Appl. Catal., A* **541** 87-106
[8] Furimsky E 2000 *Appl. Catal., A* **199** 147-190
[9] Hensley A J R, Wang Y and McEwen J-S 2015 *ACS Catal.* **5** 523-536
[10] Robinson A, Ferguson G A, Gallagher J R, Cheah S, Beckham G T, Schaidle J A, Hensley J E and Medlin J W 2016 *ACS Catal.* **6** 4356-68
[11] Moberg D R, Thibodeau T J, Amar F G and Frederick B G 2010 *J. Phys. Chem. C* **114** 13782-95
[12] Johansson M, Lytken O and Chorkendorff I 2008 *J. Chem. Phys.* **128** 1-11
[13] Derrouiche S and Bianchi D 2004 *Langmuir* **20** 4489-97
[14] Gutierrez A, Kaila R K, Honkela M L, Slioor R and Krause A O I 2009 *Catal. Today* **147** 239-246
[15] To A T and Resasco D E 2015 *J. Catal.* **329** 57-68
[16] de Souza P M, Rabelo-Neto R C, Borges L E P, Jacobs G, Davis B H, Sooknoi T, Resasco D E and Noronha F B 2015 *ACS Catal.* **5** 1318-29
[17] Nie L, de Souza P M, Noronha F B, An W, Sooknoi T and Resasco D E 2014 *J. Mol. Catal. A: Chem.* **388-389** 47-55
[18] Kepp K P 2016 *Inorg. Chem.* **55** 9461-70
[19] Kay Lup A N, Abnisa F, Wan Daud W M A and Aroua M K 2017 *J. Ind. Eng. Chem.* http://dx.doi.org/10.1016/j.jiec.2017.06.049
[20] Boullosa-Eiras S, Lødeng R, Bergem H, Stöcker M, Hannevold L and Blekkan E A 2014 *Catal. Today* **223** 44-53
[21] Redhead P A 1962 *Vacuum* **12** 203-211

Comparative study of activated carbon, natural zeolite, and green sand supports for CuO_X and ZnO sites as ozone decomposition catalyst

A S Azhariyah[1], A Pradyasti[1], A G Dianty[1] and S Bismo[1,2]

[1] Chemical Engineering Department, Universitas Indonesia, Kampus Baru UI Depok, 16426, Indonesia

E-mail: setijo.bismo@ui.ac.id

Abstract. This research was based on ozone decomposition in industrial environment. Ozone is harmful to human. Therefore, catalysts were made as a mask filter to decompose ozone. Comparison studies of catalyst supports were done using Granular Activated Carbon (GAC), Natural Zeolite (NZ), and Green Sand (GS). GAC showed the highest catalytic activity compared to other supports with conversion of 98%. Meanwhile, the conversion using NZ was only 77% and GS had been just 27%. GAC had the highest catalytic activity because it had the largest pore volume, which is 0.478 cm^3/g. So GAC was used as catalyst supports. To have a higher conversion in ozone decomposition, GAC was impregnated with metal oxide as the active site of the catalyst. Active site comparison was made using CuO_X and ZnO as the active site. Morphology, composition, and crystal phase were analyzed using SEM-EDX, XRF, and XRD methods. Mask filter, which contained catalysts for ozone decomposition, was tested using a fixed bed reactor at room temperature and atmospheric pressure. The result of conversion was analyzed using iodometric method. CuO_X/GAC and ZnO/GAC 2%-w showed the highest catalytic activity and conversion reached 100%. From the durability test, CuO_X/GAC 2%-w was better than ZnO/GAC 2%-w because the conversion of ozone to oxygen reached 100% with the lowest conversion was 70% for over eight hours.

1. Introduction

Ozone is commonly used in many industries. Ozone is a strong oxidizing agent in the disinfection process, which not only removes bacteria, viruses, and germs, but also improves the quality of water and preserves food [1]. However, some industries do not have an ozone destructor unit and the excess of ozone is released into the air and have reached 0.1 – 0.4 ppm, which exceeds the allowed threshold [2], [3]. The allowed threshold for ozone is 0.1 ppm in 8 hours (REL, Recommended Exposure Limits). High concentration ozone is harmful to human, causing respiratory diseases, and leading to death.

Catalytic ozone decomposition is preferred because it can be operated at room temperature [4]. Active sites of the catalysts have either metal or metal oxide. CuO_X and ZnO were chosen as the variation of the active sites for this research because those have good catalytic activity for ozone decomposition and have been used commercially [5]. Catalyst support used to increase the conversion

[2] To whom any correspondence should be addressed.

of ozone decomposition. The presence of a catalyst support with a high surface area can improve the catalytic activity of catalysts such as Granular Activated Carbon (GAC), Natural Zeolite (NZ), and Green Sand (GS). All of them have high adsorption ability and a wide surface area. With a wider surface area, active sites of metal oxide are easier to be dispersed on the catalyst support.

2. Methodology/ Experimental

2.1. Material

All chemicals were analytical grade reagents and used as received without further purification. GAC and NZ were obtained from CV Ady Water. GS was obtained from CV Mitra Water. Deionized water, HCl, NaOH, H_2SO_4, KI, and $Na_2S_2O_4$ and were obtained from CV Sinar Kimia. $CuCO_3$ was obtained from PD Cipta Bangun Nauli. $ZnCO_3$ was obtained from Inti Periodik Laboratory. Tetra cloth for mask filter was supplied from CV Cahaya Textile. Mask was supplied from Asemka.

2.2. Instrumentation

Instruments that used to investigate the catalytic activity were fixed bed reactor and commercial ozone generator. X-Troy Ozonator was used to produce ozone. ASAP 2020 by Micromeritics was used for BET analysis. JEOL JSM 6510L A was used for SEM-EDX analysis. TORONTECH TT-EDXPRT was used for XRF analysis. Empyrean XRD was used for XRD analysis.

2.3. Procedure

2.3.1. Preparation and characterization of catalyst support GAC, NZ, and GS with size of 18 – 100 mesh were used for investigation. First, GAC, NZ, and GS were pre-treated with 0.05 mol/L HCl and 0.05 mol/L NaOH both for an hour to remove the impurities or ash content on the surface.

2.3.2. Impregnation of catalyst The best catalyst support which has the highest catalytic activity and widest surface area was impregnated to $CuCO_3$ and $ZnCO_3$ solutions for 12 hours with loading percentage of 0-2%-w. After impregnated onto the catalyst support, catalysts were calcined at 300 °C for 1 hour to release CO_2.

2.3.3. Catalytic activity test Catalytic activity of the samples was investigated using a fixed bed reactor with a diameter of 10 cm and height of 1 cm. Catalyst with mass of 10 grams was used for investigation. Operating conditions of the reactor were room temperature and atmospheric pressure. Commercial ozone generator was used to produce ozone from air with concentration of 0.388 mg/minute. Excess of ozone from the reaction was analyzed using iodometric titration method. The amount of ozone was determined by reaction of I^- ion with ozone, which produces iodine (I_2) at acidic condition. Acidic condition was fulfilled by the addition of H_2SO_4. Then, I_2 was titrated with sodium thiosulfate ($Na_2S_2O_4$) at pH 2 (± 10%) and the volume of $Na_2S_2O_4$ could be used to calculate the volume of ozone which had not yet been decomposed to oxygen.

2.3.4. Characterization of catalyst Treated catalyst supports were observed by BET method to determine pore volume and specific surface area. Before measurements, samples were degassed at 300 °C. Surface morphology of catalysts was observed by scanning electron microscopy (SEM) and surface composition analysis was carried out by energy dispersive X-ray spectroscopy (EDX) and X-Ray Fluorescence (XRF). The magnitude for SEM was x1.000 and the voltage was 20 kV. The accelerating voltage for EDX was 20 kV and the energy range for EDX was between 0 – 20 keV. XRF used soil method. Crystal phase was studied by X-ray diffraction (XRD) with diffraction interval 2θ of 5 – 80°.

2.3.5. Mask preparation Two pieces of tetra cloth with a size of 20 x 12 cm were filled with catalyst and stitched to form 1x1 matrices. Tetra clothes which were used as a mask filter was inserted into a mask which already contained a zipper.

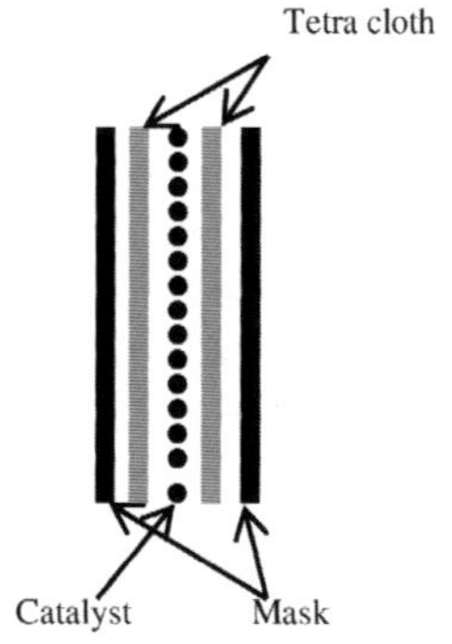

Figure 1. Mask configuration.

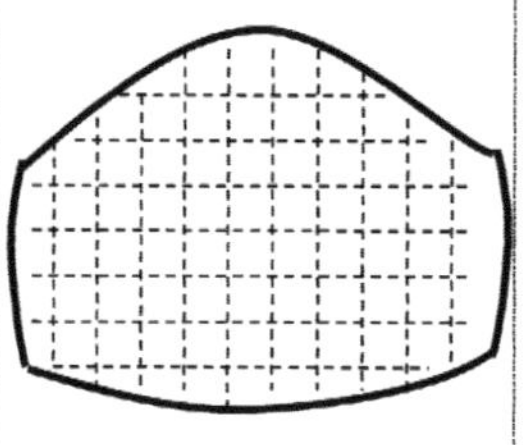

Figure 2. Matrices of tetra clothes.

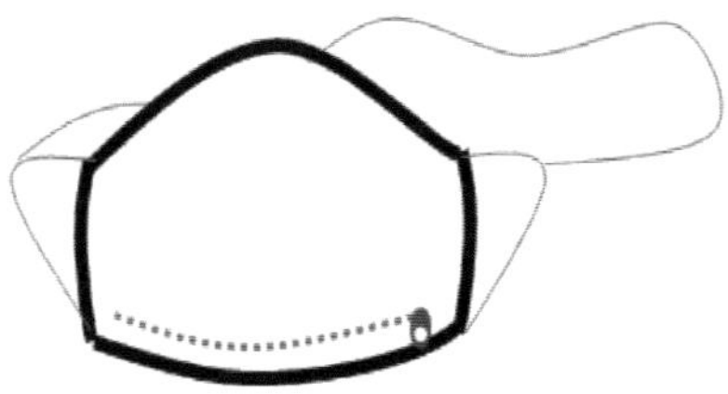

Figure 3. Ozone decomposition mask.

3. Results and Discussion

3.1. Catalytic activity analysis of catalyst support

Figure 4 showed that GAC with a size of 60 – 100 mesh had the highest conversion compared to other supports and sizes. Catalyst supports with smaller size (60 – 100 mesh) had the largest surface area for ozone to be decomposed to oxygen. GAC also had oxygen functional groups that can be used to decompose ozone [6].

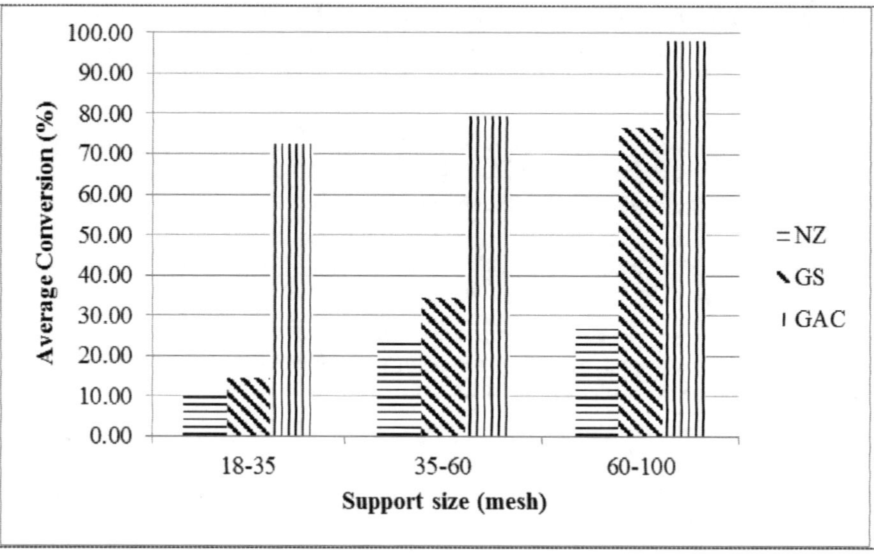

Figure 4. Average conversion of ozone decomposition of catalyst supports.

3.2. BET analysis of catalyst support

Specific surface area and pore volume of GAC, NZ, and GS with a size of 60-100 mesh could be seen in Table 1. It showed that there was an increase in surface area and pore volume for GAC. But, contrary to GAC, surface area and pore volume for NZ and GS were decreased. These were occurred because of desilication. Desilication happened when silica was dissolved during NaOH treatment and the outer layer of the structure shrunk. Partial dealumination also happened because of HCl treatment. When desilication and dealumination occurred at a serious level, the structure of the catalyst would be deformed and decrease the surface area [7].

Meanwhile, surface area and pore volume increased for GAC because impurities on the surface were removed. Aqueous HCl solution was dissolved and Cl^- ion was bonded with any positive charged impurities and H^+ ion bonded with GAC. With the addition of NaOH, H^+ ion, which was previously bonded with GAC, then bonded with OH^- ion and formed water, while Na^+ ion was bonded with negative charge impurities. Excess of Na^+ was bonded to GAC. Because GAC had the highest surface area and pore volume, it was used for the catalyst support of CuOx and ZnO.

Table 1. BET characterization of catalyst support.

Parameter	GAC		NZ		GS	
	Before pre-treatment	After pre-treatment	Before pre-treatment	After pre-treatment	Before pre-treatment	After pre-treatment
Pore volume (cm^3/g)	0.184	0.317	0.092	0.093	0.049	0.015
Surface area (m^2/g)	333.545	558.754	53.334	40.931	55.563	5.574

3.3. Catalytic activity analysis for CuOx/GAC and ZnO/GAC

Comparison of catalyst activity between CuOx/GAC and ZnO/GAC catalyst at different loading percentage could be seen in Figure 5. It turned out that both catalysts have the same average conversion value in the first 30 minutes. The catalyst with loading 1%-w and 2%-w had a perfect ozone conversion which was 100%.

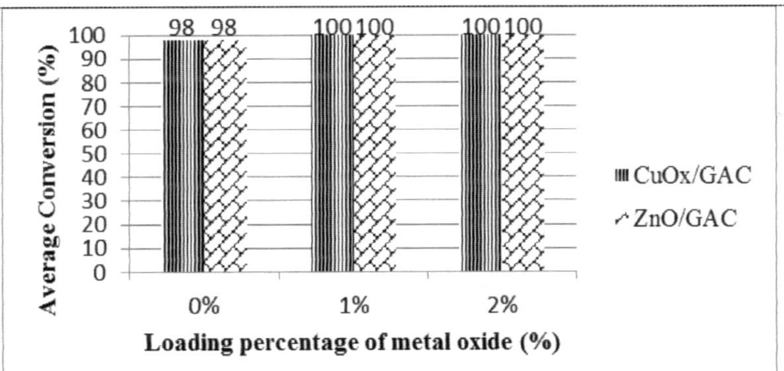

Figure 5. Average conversion of ozone decomposition of CuOx/GAC and ZnO/GAC catalysts.

Based on the literature, CuOx had a conversion of ozone decomposition, which was 21 – 22% [8]. Therefore, doping of CuOx to catalyst support increased catalytic activity significantly. Another literature explained that when the content of CuOx increased from 1%-w to 5%-w, the catalytic activity was increased. However, further increasing the CuOx content to 7%-w and 9%-w resulted in a decrease of the catalytic activity. From literature, CuOx with 5%-w loading exhibited the highest catalytic activity [9]. Meanwhile, for the previous research of ZnO catalyst for ozone decomposition, conversion only reached 30% at room temperature [10]. Nevertheless, by increasing the loading of ZnO in catalysts, would increase the defect sites in catalysts. Defect sites such as oxygen vacancies were important parts of ZnO structure because oxygen vacancies were the place of ozone decomposition occurred [11].

IOP Conf. Series: Materials Science and Engineering **334** (2018) 012075 doi:10.1088/1757-899X/334/1/012075

3.4. Characterization of CuOx/GAC and ZnO/GAC

3.4.1. SEM-EDX analysis Surface morphology and composition of GAC before and after pre-treatment were shown in Figure 6, Figure 7 and Table 2. From Table 2, it could be concluded that many impurities existed on the surface of GAC before pre-treatment. The amount of impurities such as Mg, Fe, Cl, Ca, and K decreased after pre-treatment. Amount of Na was increased because NaOH had been used in pre-treatment and bonded to GAC surface.

Figure 6. SEM analysis of GAC before pre-treatment.

Figure 7. SEM analysis of GAC after pre-treatment.

With a wider surface area, CuOx would be easier to be dispersed into the GAC. Figure 8-11 displayed the surface morphology of CuOx/GAC and ZnO/GAC 1-2%-w.

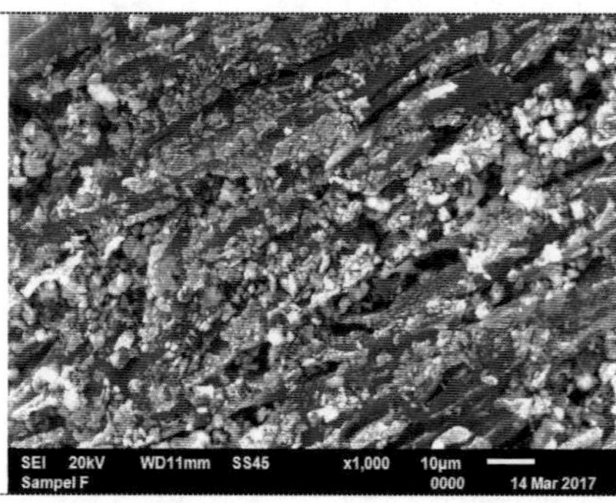

Figure 8. SEM analysis of CuOx/GAC 1%-w.

Figure 9. SEM analysis of CuOx/GAC 2%-w.

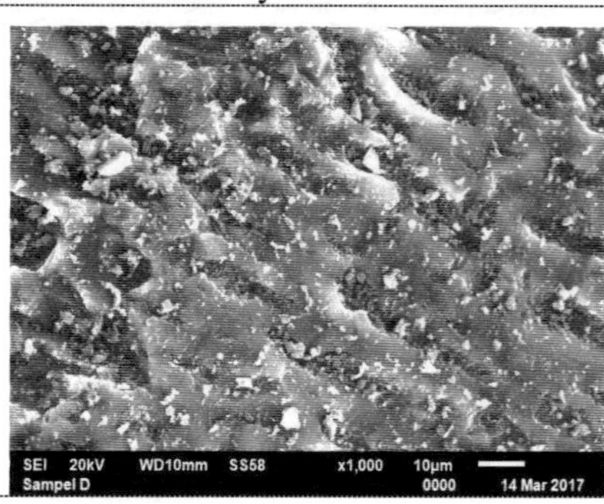

Figure 10. SEM analysis of ZnO/GAC 1%-w.

Figure 11. SEM analysis of ZnO/GAC 2%-w.

Table 2 also gave data of Cu and Zn on GAC that have been impregnated. It seems that the composition of Cu and Zn in any samples did not match the loading percentage. EDX only analyzed some chemical elements that were several microns from the surface of the catalyst. With higher amount of Cu and Zn, metal oxides were not absorbed into the pores of the GAC and just attached to the outer surface of the GAC. Because GAC possessed a very small micropore size, metal oxide particles were hardly accessed the micropore and only attached to the outer surface, not the deepest pore. Moreover, EDX methods only shot at one sample point, not analyzed all the surface of the catalyst.

Table 2. EDX characterization of GAC before and after impregnation.

Composition	Before pre-treatment (%)	0%-w	CuO_x/GAC		ZnO/GAC	
			1%-w	2%-w	1%-w	2%-w
C	81.34	78.89	70.89	73.09	77.91	72.74
O	9.42	17.53	20.63	14.07	20.23	21.46
Na	0.75	2.02	1.78	0.00	0.00	0.00
Mg	0.10	0.00	0.00	0.00	0.00	0.00
Al	0.00	0.44	0.00	0.00	0.00	0.29
Si	0.13	0.00	0.00	0.00	0.00	0.00
P	0.50	0.00	0.00	0.00	0.00	0.00
Zn	**0.00**	**0.00**	**0.34**	**0.56**	**0.98**	**3.50**
Cl	1.50	0.00	0.00	0.00	0.00	0.23
K	3.39	0.00	0.00	0.00	0.00	1.76
Cu	**0.00**	**0.00**	**6.19**	**12.16**	**0.00**	**0.00**
Ca	0.00	0.63	0.17	0.00	0.88	0.00
Fe	2.56	0.00	0.00	0.00	0.00	0.00
Mo	0.31	0.00	0.00	0.00	0.00	0.00

3.4.2. XRD analysis Figure 12 and Figure 13 consecutively showed the XRD analysis of CuO_x/GAC before and after calcined and Table 3 showed the intensity. JCPDS 76-660, JCPDS 65-2309, JCPDS 05-0667 were used as references for $CuCO_3$, CuO, and Cu_2O. It showed that there was a decrease in peak position after calcined process. Calcined process decomposed CO_2 from $CuCO_3$ to create CuO_x. From XRD analysis, after calcined, only CuO_x (CuO and Cu_2O) remained in GAC.

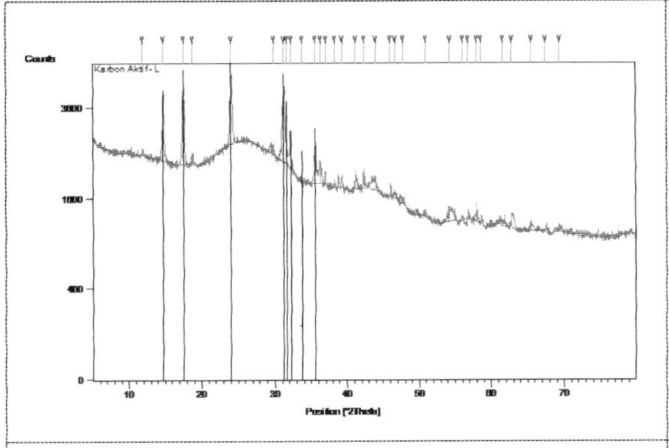

Figure 12. XRD analysis of CuO_x/GAC 1%-w.

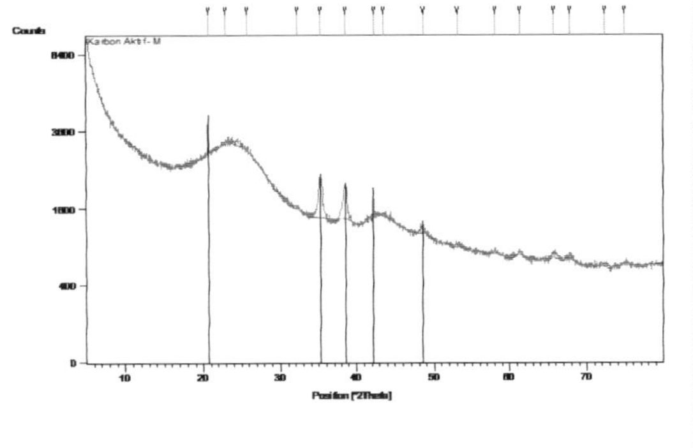

Figure 13. XRD analysis of CuO_x/GAC 2%-w.

Table 3. XRD peak list of CuO_X/GAC.

Before calcined			After calcined		
Compound	2θ (°)	Rel. Intensity (%)	Compound	2θ (°)	Rel. Intensity (%)
$CuCO_3$	14.7756	62.29	CuO	35.2256	100.00
	17.5614	86.46		38.6532	79.66
	24.0875	97.95	Cu_2O	42.3340	43.34
	31.2850	100.00			
	31.6315	69.99			
	32.1952	38.91			
	35.5834	49.00			

Figure 14 and Figure 15 displayed the XRD analysis of ZnO/GAC before and after calcined and Table 4 showed the intensity. JCPDS 8-0449 and JCPDS 36-1451 were used consecutively as references for $ZnCO_3$ and ZnO. It seems that there was a decrease in peak position after calcined process. This process also decomposed CO_2 from $ZnCO_3$ to create ZnO. From XRD analysis, after calcined, only ZnO remained in GAC.

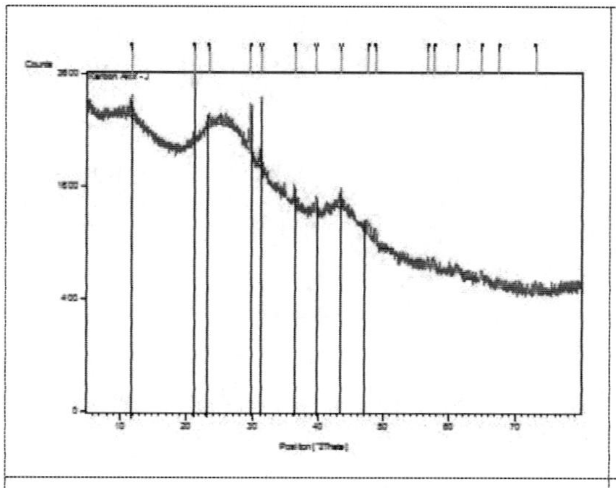

Figure 14. XRD analysis of ZnO/GAC 1%-w.

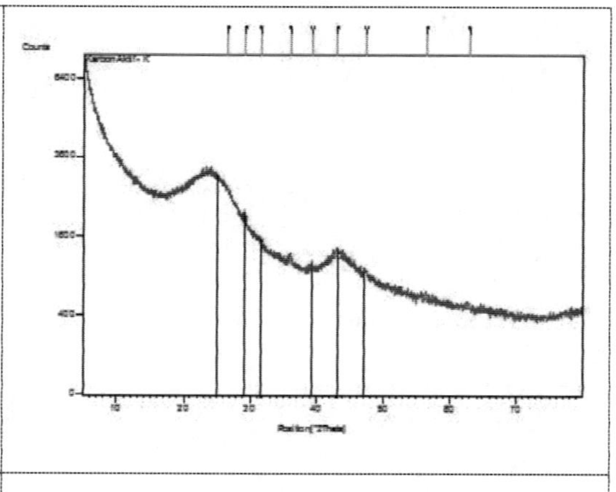

Figure 15. XRD analysis of ZnO/GAC 2%-w.

Table 4. XRD peak list of ZnO/GAC.

Before calcined			After calcined		
Compound	2θ (°)	Rel. Intensity (%)	Compound	2θ (°)	Rel. Intensity (%)
$ZnCO_3$	11.6863	44.39	ZnO	26.5884	23.03
	21.1760	77.81		29.2305	100.00
	23.3473	15.40		31.5557	17.78
	29.5453	100.00		39.2556	34.16
	31.1068	64.92		42.9412	24.26
	36.1921	17.73		47.2526	37.45
	39.5114	21.77			
	43.3993	22.96			
	47.6005	19.23			

3.4.3. XRF analysis XRF method was used to see the metal oxide composition in GAC after impregnation. Table 5 showed the loading percentage of CuOx and ZnO. CuOx 2%-w had a near-real value, while CuOx 1%-w was greater than the desired loading percentage. This could happen because the impregnation process was uneven, so the loading on the GAC was not homogeneous. Additionally, ZnO 1%-w and 2%-w have less value than the desired loading percentage. This might be due to agitation process that was not perfect, so ZnO did not attach to the whole GAC.

Table 5. XRD peak list of ZnO/GAC.

Catalysts	Loading percentage	
	1%-w	2%-w
CuOx/GAC	1.66	1.90
ZnO/GAC	0.47	1.58

3.5. Catalyst durability
CuOx/GAC and ZnO/GAC catalysts were tested for 8 hours. The results could be seen in Figure 16.

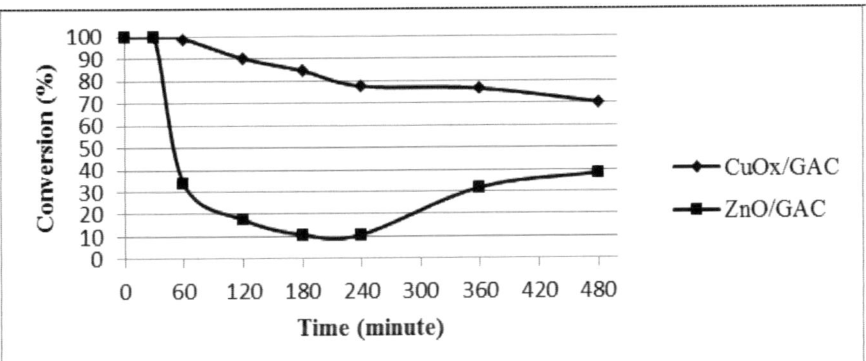

Figure 16. Average conversion of ozone decomposition of GAC.

CuOx/GAC catalyst performance was better than ZnO/GAC because it produced the highest ozone conversion. ZnO/GAC conversion decreased significantly after 30 minutes while CuOx/GAC catalyst was after 60 minutes. There was a slight decrease in the conversion of CuOx/GAC catalyst from 240 minutes to 360 minutes. It happened because the data were taken discretely, which was tested for 2 days. There was a 15-hour break after 240 minutes before the catalyst was tested again. It means that the catalyst was allowed to regenerate for 15 hours. This break gave time to O2 and CO2 to desorb from the catalyst. So, for the next day, the surface of the catalyst would be wider to decompose ozone. While for ZnO/GAC catalyst, after 240 minutes, conversion was increased. It also happened because of a 15-hour break. O_2 gas near the catalyst was bonded with the catalyst to make other oxygen functional group that could be used to decompose ozone. The use of mask to decompose ozone is important to our community because it decomposed ozone at least 75.45 ppm for CuOx/GAC catalyst and 11.57 ppm for ZnO/GAC catalysts. Therefore, mask that contained CuOx/GAC and ZnO/GAC can be used to reduce the ozone concentration to be below the allowed threshold.

4. Conclusion

The best catalyst support was GAC with a size of 60-100 mesh because it had the largest surface area and pore volume, which was 558.754 m^2/g and 0.317 cm^3/g. CuOx/GAC and ZnO/GAC catalysts with loading 1%-w and 2%-w resulted in the highest conversion of 100% within the first 30 minutes.

However, CuO_X/GAC catalysts with loading percentage of 2%-w had the best catalyst activity because it could convert ozone to oxygen with a minimum of 70% within 8 hours.

Acknowledgement
The authors acknowledge Universitas Indonesia and PITTA program for the financial support. We are also thankful to Laboratorium Intensifikasi Proses Teknik Kimia Universitas Indonesia, Badan Pengkajian dan Penerapan Teknologi, Badan Tenaga Nuklir Nasional, Laboratorium Pusat Penelitian Fisika-LIPI, and FMIPA Universitas Negeri Jakarta for the instruments to characterize the catalysts.

References

[1] O'Donnel C, Tiwari B K, Cullen P J, Rice G 2012 *Ozone in Food Processing* (UK: John Wiley & Sons, Ltd.).
[2] Subrahmanyam C, Bulushev D A, Kiwi-Minsker L 2005 *Applied Catalysis B: Environmental* **61** 98-106.
[3] McKee D 1993 *Tropospheric Ozone: Human Health and Agricultural Impacts* (USA: CRC Press, Inc.).
[4] Sigmund J 2001 *Catalytic Destruction of Ozone: A Cost Effective Approach to Controlling Off-Gas Ozone Emissions* (USA: EIJ Company LLC).
[5] Dhandapani B, Oyama S T 1997 *Applied Chemistry B: Environmental* **11** 129-166.
[6] Wang M, Zhang P, Li J, Jiang C 2014 *Chinese Journal of Catalysis* **35** 335-341.
[7] Ates A, Akgul G 2016 *Powder Technology* **287** 285-291.
[8] Dan C, Popovici E J, Imre-Lucacia F, Popovici N, Marginean P, Silaghi-Dumitrescu I 2008 *Journal of Optoelectronics and Advanced Materials* **10** 2234-2236.
[9] Zhao F, Gong M, Zhang G, Li J 2015 *Journal of Rare Earths* **33** 604-610.
[10] Milenova K I, Nikolov P M, Kasabova N A, Avramova I A 2014 *Polish Journal of Chemical Technology* **16** 55-59
[11] Zhu G, Zhu J, Jiang W, Zhang Z, Wang J, Zhu Y, Zhang Q 2017 *Applied Catalysis B: Environmental* **209** 729-737.
[12] Gurrath M, Kuretzky T, Boehm H P, Okhlopkova L B, Lisitsyn A S, Likholobov V A 2000 Carbon **38** 1241-1255.

IOP Conf. Series: Materials Science and Engineering **334** (2018) 012076 doi:10.1088/1757-899X/334/1/012076

Catalytic effect of different reactor materials under subcritical water conditions: decarboxylation of cysteic acid into taurine

M Faisal[1]
[1] Department of Chemical Engineering, Syiah Kuala University, Jl. Tgk. Syech Abdul Rauf No.7 Darussalam, Banda Aceh 23111, INDONESIA

E-mail: mfaisal@unsyiah.ac.id

Abstract. In order to understand the influence of reactor materials on the catalytic effect for a particular reaction, the decomposition of cysteic acid from Ni/Fe-based alloy reactors under subcritical water conditions was examined. Experiments were carried out in three batch reactors made of Inconel 625, Hastelloy C-22 and SUS 316 over temperatures of 200 to 300 °C. The highest amount of eluted metals was found for SUS 316. The results demonstrated that reactor materials contribute to the resulting product. Under the tested conditions, cysteic acid decomposes readily with SUS 316. However, the Ni-based materials (Inconel 625 and Hastelloy C-22) show better resistance to metal elution. It was found that among the materials used in this work, SUS 316 gave the highest reaction rate constant of 0.1934 s−1. The same results were obtained at temperatures of 260 and 300 °C. Investigation of the Arrhenius activation energy revealed that the highest activation energy was for Hastelloy C-22 (109 kJ/mol), followed by Inconel 625 (90 kJ/mol) and SUS 316 (70 kJ/mol). The decomposition rate of cysteic acid was found to follow the results for the trend of the eluted metals. Therefore, it can be concluded that the decomposition of cysteic acid was catalyzed by the elution of heavy metals from the surface of the reactor. The highest amount of taurine from the decarboxylation of cysteic acid was obtained from SUS 316.

1. Introduction

Supercritical fluid technology (SFT) is of significant scientific and technological importance, and has been extensively demonstrated to be one of the most effective media for a large number of processes. Water at subcritical conditions has been used as a solvent in natural extraction [1–3], the recovery of useful materials from organic wastes [4,5], depolymerization [6,7], chemical reactions [8,9] and organic synthesis [10–12] due to its high ion product. Supercritical water conditions can be applied for the complete oxidation of organic waste [13,14]. Alternatively, supercritical CO2 has been used for various purposes, such as the recycling of precious metals [15], leaf oil extraction and quinone analysis, among others [16]. As SFT simultaneously uses extreme temperatures and pressures, consequently, chemical corrosion increases [17,18]. The corrosion of various types of materials under high temperature and pressure conditions has been investigated intensively [19–22]. These types of corrosion could be hazardous and their presence in the products is not desirable. Conversely, the

[1] To whom any correspondence should be addressed.

corrosion may act as a catalyst for some reactions. Therefore, selection of the reactor material and reaction conditions is crucial and special attention should be paid when applying SFT.

Under subcritical water conditions and different reactor materials, a small amount of metals (i.e., Fe, Ni, Cr, Mn and Mo) was found to be eluted during the reaction [17]. Although their presence is still within the acceptable environmental limit, these metals, as well as reactor materials, might influence the reaction rate of some reactions. Studies on metal elution from Fe/Ni-based reactor materials using water and cysteic acid have been conducted at short reaction times [17,23]. However, detailed investigations of the effects of metal elution on the reaction media are also important. The objective of this research is to investigate the possible catalytic effect on a particular reaction under subcritical conditions using different reactor materials.

In this research, the chemical reaction of interest for studying the catalytic effect under subcritical water conditions is the preparation of taurine from the decarboxylation of cysteic acid. Taurine is one of the most abundant free amino acids in the body and is essential in human metabolism [24]. From the viewpoint of the complete degradation of organic waste, the decarboxylation of cysteic acid is not a rate limiting step compared to acetic acid. However, since cysteic acid is the intermediate product of taurine production, this decarboxylation information is highly useful, especially when considering resource recovery and organic reactions under supercritical fluids (SF) conditions. Reactor materials and the presence of heavy metals under subcritical water conditions during the decarboxylation of cysteic acid are significant factors that might influence taurine production.

2. Methods

Experiments were performed in three batch reactor materials made of SUS 316 (consisting of Fe: 65%, Cr: 18%, Ni: 12%, Mn: 1% and Mo: 2%), Hastelloy C-22 (consisting of Ni: 60%, Cr: 20%, Mo: 12.5% and W: 3%) and Inconel 625 (consisting of Ni: 60%, Cr: 20%, Mo: 9% and Fe: 5%), all with the same size of 1 cm ID and 12 cm in height (a volume of 11 ml). The schematic diagram of the batch reactor apparatus is shown in Figure 1. The apparatus (TSC-006, Taiatsu Glass Corp. consists of a stirrer, pressure gauge, molten salt bath (mixture of KNO3 and NaNO3) and the reactor.

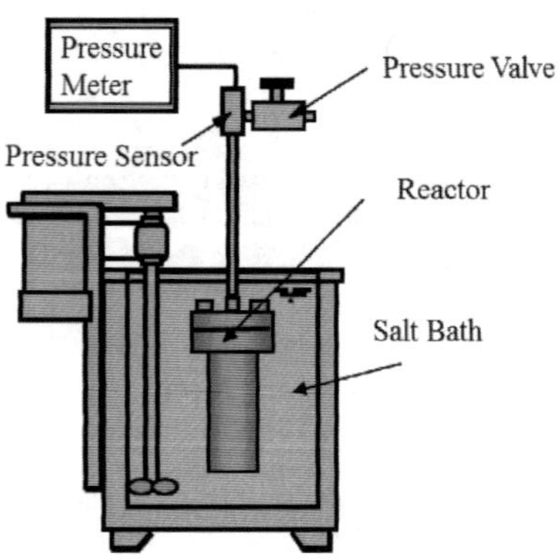

Figure 1. Schematic diagram of batch reactor apparatus.

The experiments were conducted over a temperature range of 200 to 300 °C with the corresponding saturated vapor pressures (2-9 MPa) and a time range from 2 to 10 min. During a typical run, about 9 ml of 10 mmol/L cysteic acid was charged into the reactor. The reactor was sealed and the air inside was removed by purging with argon gas. The reactor was then immersed into the preheated molten salt at a desired time and temperature. The reaction was stopped by removing the reactor from the salt batch and immersing the reactor in a cold water bath to bring it quickly to room temperature. The experimental procedure was similar to that described by Faisal et al. [6]. The products were analyzed for taurine and other derived products using an amino acid analyzer (LC-10AD, Shimadzu Corp.).

Metal elution was measured using an inductively coupled plasma atomic emission spectrometer (SPS 7700R, Seiko Instrument Inc.).

3. Results and discussions

3.1. Effect of reaction time on eluted metals

Figures 2–4 show the effect of reaction time on the amount of metal elution from the various reactor materials (batch, SUS 316, Hastelloy C-22 and Inconel 625) at 260 °C. As can be seen from Figure 2, the eluted metal from SUS 316 was mainly Fe. The amount of eluted Fe is significantly increased with increasing reaction time. Mo was not detected at this temperature; however, a small amount of Cr was detected that was almost stable with increasing reaction time. Ni was detected at an even shorter reaction time and again increased with reaction time. Although Mn was also observed at a short reaction time, a small amount of Mn was detectable after 6 min and increased with reaction time.
Figure 3 shows the effect of reaction time on the amount of metal elution from Inconel 625 at 260°C. The main eluted metal was Ni, followed by Cr and Mo. The total amount of metal elution was much lower than that for SUS 316. The amount of elution from Inconel 625 follows the trend of SUS 316 except for Fe, in which the trend of eluted Fe was unstable with reaction time. Interestingly, the elution of heavy metals from Inconel 625 was not dominated only by Ni, as a significant amount of Fe was also observed at this condition.

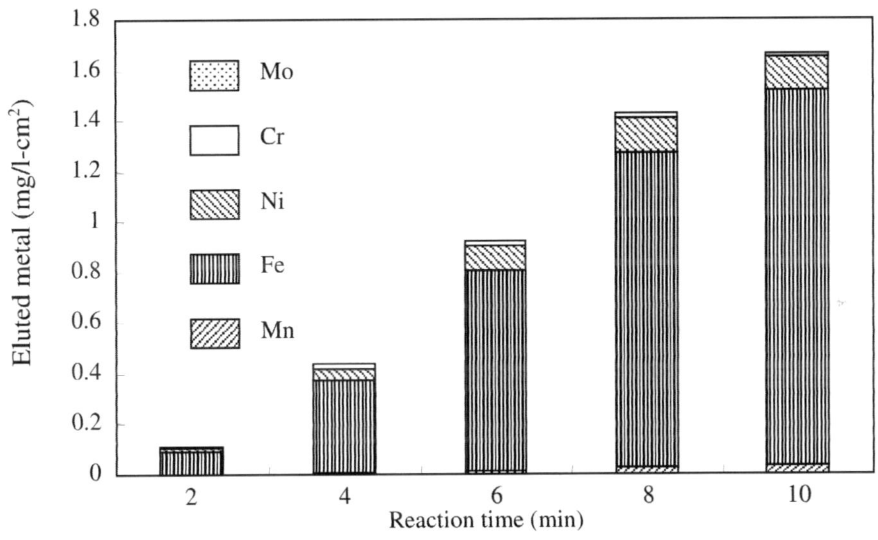

Figure 2. Effect of reaction time on the amount of metal elution from SUS 316 at 260 °C.

The amount of metal elution from Hastelloy C-22 at 260 °C at various reaction times is shown in Figure 4. A significant amount of Ni was observed from Hastelloy C-22 over the reaction time of 4–10 min. Cr and Mo were detectable and almost stable with reaction time, while W was not detected even after 10 min of reaction time. This observation allows us to conclude that the total eluted metals from SUS 316 were higher than Hastelloy C-22 or Inconel 625. It was also found that the total amount of elution in cysteic acid solution was higher than that in water [23] Moreover, the erratic behavior of metal elution under SF conditions in water and cysteic acid was also different.

3.2. Effect of reactor materials on decomposition rate of cysteic acid

Figure 5 shows the time course for the decomposition of cysteic acid at 280 °C. The results show that the decomposition of cysteic acid was higher for SUS 316 compared with Hastelloy C-22 and Inconel 625. Among the materials used in this work, SUS 316 gives the highest reaction rate constant (k) of

0.1934 s^{-1}. The same results were obtained at temperatures of 280 and 300 °C. It was found that the amount of eluted metals follows the decomposition rate of cysteic acid. The presence of these metals

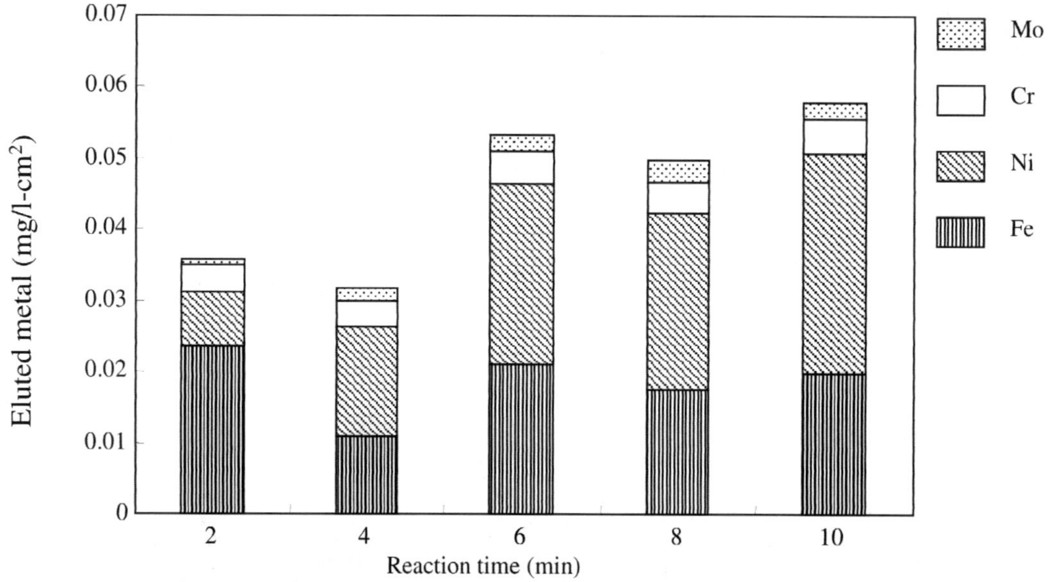

Figure 3. Effect of reaction time on the amount of metal elution from Inconel 625 at 260 °C.

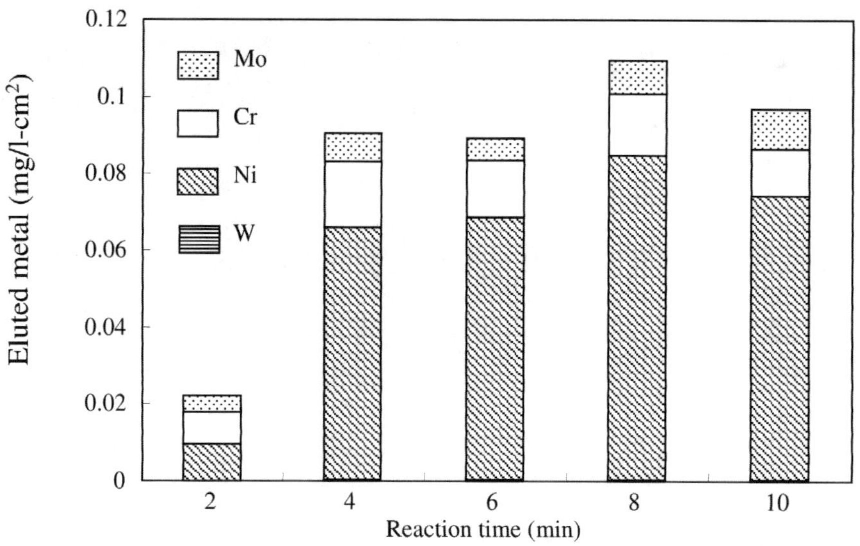

Figure 4. Effect of reaction time on the amount of metal elution from Hastelloy C-22 at 260 °C.

might affect the acceleration of chain reactions in transforming organic compounds into other chemical compounds. Therefore, it can be concluded that the decomposition of cysteic acid was catalyzed by the metals eluted from the reactor surface. The chemical reaction in SUS 316 was faster than in Inconel 625 or Hastelloy C-22. In addition, to prove that the reactor materials affect the reaction, one piece of SUS 316 metal was added into the SUS 316 reactor (20 cm^2). As can be seen from Figure 5, the increased surface area (SUS + metal) strongly affects the decomposition rate of cysteic acid. The reaction rate constant of SUS + metal was 0.2609 s−1. However, the current results indicated that increased surface area was not favorable for taurine production. This finding agreed well with the observation of Belsky et al. [25], who reported that the rate of reaction at sub-supercritical fluid conditions was found to depend significantly on the materials of the reactor. Houser and Liu [10] also examined the role of different reactor materials in high-temperature reactions. They reported that chloride attacked the metal walls of the reactor to produce metal chlorides and these compounds may act as catalysts for the decomposition reactions. In fact, both the metal wall and the eluted metals had a catalytic effect on the reaction. In addition, preliminary experiments also revealed that the use of new reactors produced different results from those obtained from reactors that have been used at least once. It was observed that the new reactors, especially on SUS 316, led to a higher decomposition of cysteic acid and a higher amount of eluted heavy metals. High Fe elution, which produced yellowish decomposition products, was observed for the first run. Consequently, all the reactors used in this study were seasoned in order to eliminate these effects.

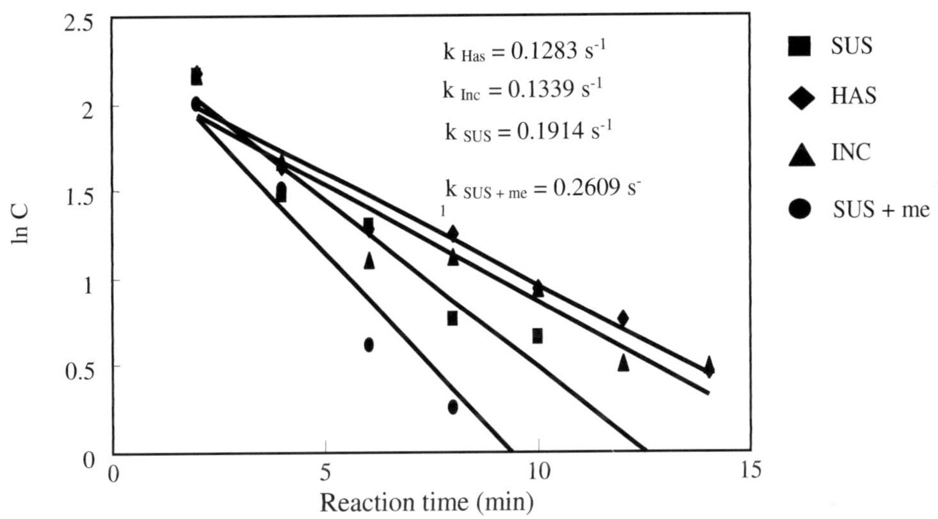

Figure 5. Time course for decomposition of cysteic acid at 280 °C.

The effect of reactor materials on the reduction of the Arrhenius activation energy (Ea) for the first-order decomposition of cysteic acid is summarized in Table 1. These results are obtained from the plot of the three reaction rate constants (k) at temperatures of 260, 280 and 300 °C.

Table 1. Effect of reactor materials on the reduction of activation energy.

Reactor materials	k = A exp (-Ea/RT)			Activation Energy Ea (kJ/mol)
	260 °C	280 °C	300 °C	
SUS 316	0.1078	0.1914	0.0757	70
Inconel 625	0.0640	0.1339	0.0332	90
Hastelloy C-22	0.0531	0.1283	0.0658	109

The results show that the highest activation energy was for Hastelloy C-22 (109 kJ/mol), followed by Inconel 625 (90 kJ/mol) and SUS 316 (70 kJ/mol). Based on the assumption that the highest Ea controls the reaction rate, the elution from the reactor surface might increase the reaction rate by helping to break the chemical bonds in reactant molecules. The results suggest that the decomposition reaction in SUS 316 was catalyzed by the SUS 316 material. Compared to Hastelloy C-22, the activation energy was reduced by around 40 kJ/mol. These results agree with the findings of Mingotaud et al. [26]. Their experiments on the effect of the reactor wall of depolymerization of styrene in supercritical fluid CO_2/ethanol suggests that a teflon wall improved the yield and the metals itself can interact with the reactants.

3.3. Formation of taurine
The effect of reaction time on the formation of taurine from decarboxylation of cysteic acid at 280 °C is illustrated in Figure 6. Cysteic acid started to decompose from 2 min to produce taurine. For all reactor materials, the trend of decarboxylation is almost the same at <4 min of reaction time, after which (>4 min) the trend differs slightly. As can be seen, the formation of taurine increased with increasing reaction time up to 10 min. The highest amount of taurine obtained was from SUS 316, which has the highest metal elution, suggesting that decarboxylation of cysteic acid was catalyzed by the elution, as well as by the reactor wall itself. At 4 min of reaction time, the addition of SUS (Tau SUS+me) results in faster decomposition of cysteic acid, followed by a significant increase in the amount of taurine. However, at these conditions, the formation of taurine decreased after 10 min, while cysteic acid almost completely decomposed at 14 min, indicating that taurine might also decompose into other compounds.

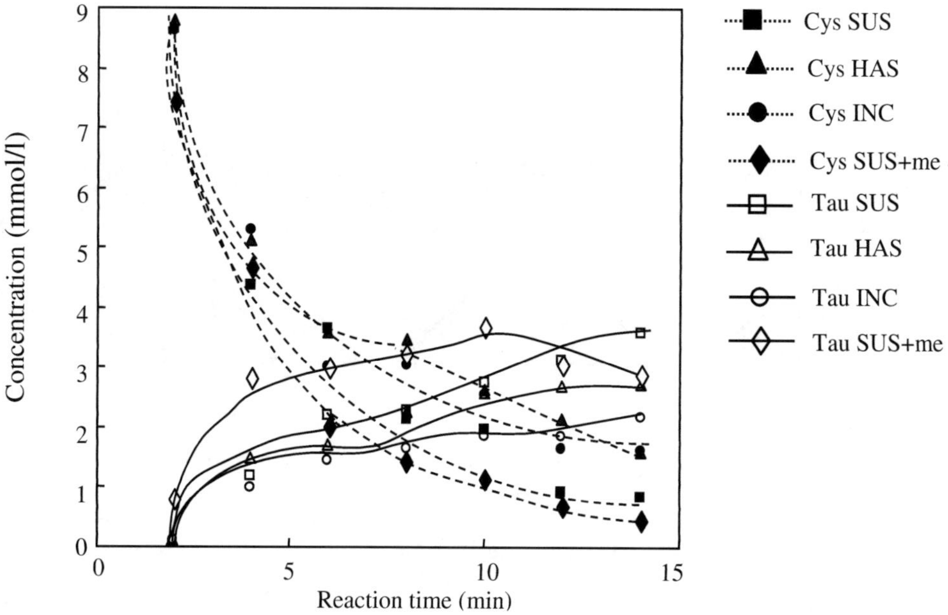

Figure 6. Decarboxylation of cysteic acid into taurine at 280 °C with reaction time.

4. Conclusions
We examined the influence of reactor materials on the catalytic effect for decarboxylation of cysteic acid into taurine. We demonstrated that reactor materials highly affect the resulting product. Under the tested conditions, cysteic acid decomposes readily in SUS 316, where the highest elution was obtained. The results for the amount of eluted metals show that the elution rate tends to follow the decomposition rate of cysteic acid. Therefore, based on these results, it can be concluded that the decomposition of cysteic acid was catalyzed by the reactor wall and the metals eluted from the reactor surface. The highest amount of taurine from decarboxylation of cysteic acid was obtained from SUS 316. An increase in surface area was also favorable for the decomposition rate of cysteic acid, but does not promote taurine production.

IOP Conf. Series: Materials Science and Engineering **334** (2018) 012076 doi:10.1088/1757-899X/334/1/012076

Acknowledgments
The author would like to thank the Syiah Kuala University and the Ministry of Research, Technology and Higher Education of Indonesia for their financial support.

References

[1] Švarc-Gajić J, Cvetanović A, Segura-Carretero A, Linares I B and Mašković P 2017 *J. Supercrit. Fluids* **123** 92
[2] Koyu H, Kazan A, Ozturk T K, Yesil-Celiktas O and Haznedaroglu M Z 2017 *J. Supercrit. Fluids* **127** 15
[3] Naffati A, Vladić J, Pavlić B, Radosavljević R, Gavarić A and Vidović S 2017 *J. Supercrit. Fluids* **121** 1
[4] Kang K, Quitain A T, Daimon H, Noda R, Goto N, Hu H and Fujie K 2001 *Can. J. Chem. Eng.* **79** 65
[5] Shanableh A 2000 *Wat. Res.* **34** 945
[6] Faisal M, Saeki T, Tsuji H, Daimon H and Fujie K 2007 *Asian J Chem.* **19** 1714
[7] Faisal M, Saeki T, Tsuji H, Daimon H and Fujie K 2006 *WIT Transactions on Ecology and the Environment* 92
[8] Faisal M, Sato N, Quitain A T, Daimon H and Fujie K 2007 *Int. J. Chem. Kinet.* **39** 175
[9] Faisal M, Saeki T, Daimon H and Fujie K 2006 *Asian J. Chem.* **18** 248
[10] Houser T J and Liu Xu 1996 *J. Supercrit. Fluids* **9** 167
[11] Savage P E 1999 *Chem. Rev.* **99** 603
[12] Holliday RL, Jong B Y M and Kolis J W 1998 *J. Supercrit. Fluids* **12** 255
[13] Goto M, Nada T, Kawajari S, Kodama A and Hirose T 1997 *J. Chem. Eng. Japan* **30** 813
[14] Mizuno T, Goto M, Kodama A and Hirose T 2000 *Ind. Eng. Chem. Res.* **39** 2807
[15] Faisal M, Atsuta Y, Daimon H and Fujie K. 2008 *Asia-Pac. J. Chem. Eng.* **3** 364
[16] Hasanudin U, Faisal M Daimon H and Fujie K 2006. *Appl. Microbiol. Biotechnol.* **69** 506
[17] Faisal M, Quitain A T, Urano S Y, Daimon H and Fujie K 2004 *J. Hazard. Mater.* **108** 213
[18] Kriksunov L B and Macdonald D D 1995 *J. Electrochem. Soc.* **142** 4069
[19] Kritzer P, Boukis N and Dinjus E 1999 *J. Supercrit. Fluids* **15** 205
[20] Gao W, Guo X, Shen Z and Zhang L 2017 *J. Nucl. Mater.* **486** 1
[21] Kim S Y, Mitton D B and latanision R M 2000 *Korean J. Chem. Eng.* **17** 58
[22] Carranza R M and Alvarez M G 1995 *Corrosion Sci.* **38** 909
[23] Faisal M, Daimon H and Fujie K 2012 *REAKTOR* **14** 9
[24] Gaull G E 1986 *J. Am. Coll. Nutr.* **5** 121
[25] Belsky A J, Maiella P G and Brill T B 1999 *J. Phys. Chem. A* **103** 4253
[26] Mingotaud AF, Bègue G, Cansell F and Gnanou Y 2001 *Macromol. Chem. Phys.* **202** 2857

Preparation of CaO/Fly ash as a catalyst inhibitor for transesterification process off palm oil in biodiesel production

Z Helwani[1,3], W Fatra[2], E Saputra[1] and R Maulana[1]

[1] Departement of Chemical Engineering, Riau University, Pekanbaru, 28293, Indonesia
[2] Departement of Mechanical Engineering, Riau University, Pekanbaru, 28293, Indonesia

E-mail: zuchra.helwani@lecturer.unri.ac.id

Abstract. A palm fly ash supported calcium oxide (CaO) catalyst was prepared and used in transesterification from off-grade palm oil for biodiesel production. The catalyst synthesized by loading CaO of calcium nitrate tetrahydrate ($Ca(NO_3)_2.4H_2O$) into fly ash through impregnation method. The optimum catalyst preparation conditions were determined by influence of calcination temperature and weight ratio of $Ca(NO_3)_2.4H_2O$ and fly ash. Catalyst with highest catalytic activity was achieved when calcined at 800 °C and proportion of $Ca(NO_3)_2.4H_2O$ to fly ash is 80:20. Under the conditions of oil : methanol ratio of 1:6, catalyst dosage of 6 wt% and temperature of 70 °C for 2 h, the biodiesel yield reaches to 71.77%. CaO, SiO_2, $Ca(OH)_2$ and Ca_2SiO_4 were found in the catalyst through X-ray diffraction (XRD) while the basic strength of the catalyst $H\_$ in the range 9.3 – 11. Surface area of the developed catalyst is 24.342 m^2/g through Brunauer-Emmett-Teller (BET). Characteristics of biodiesel such as density, kinematic viscosity, acid value, flash point has been matched with standard for biodiesel specification of Indonesia.

1. Introduction

Oil-palm off grade is a source of vegetable oil available in large quantities, which is about 7-10 tons produced from palm oil mill with a capacity of 30 tons/hour and sold at prices 30-40% cheaper than good quality CPO [1]. Besides having a great availability and low price, oil-palm off grade has not been utilized to the maximum and potentially be used as raw material for biodiesel. The weakness of oil-palm off grade is to have high levels of FFA. The weakness can be overcome with the use of two-stage reaction due to high FFA levels. FFA can be reduced through esterification reaction before continuing to the transesterification reaction. FFA levels allowed in the transesterification reaction is <2% [2].

The biodiesel production is commonly performed by transesterification of refined vegetable oils with methanol in the presence of homogeneous base catalysts, such as sodium or potassium hydroxides, carbonates or alkoxides. But these catalysts have some problems such as difficulty in removing the catalysts after the reaction, production of large amount of wastewater and emulsification [2]. In order to overcome the drawbacks of homogeneous catalytic process, recently profound

[3] To whom any correspondence should be addressed.

attention has been devoted to the development and application of heterogeneous catalysts for synthesis of biodiesel. The advantage of using heterogeneous catalysts in the transesterification reaction is not corrosive, environmentally friendly, easily separation, high selectivity, reusable, and not a lot of waste [3]. Calcium Oxide (CaO) is one of basic heterogeneous catalyst that most developed, they are CaO, $CaCO_3$ and Ca $(OH)_2$ [4], CaO / Fe_3O_4 [5], and the CaO/*fly ash* [6]. However, the use of CaO as a catalyst has disadvantages, the oxygen ions (O^{2-}) on the surface easily form hydrogen bonds with methanol or glycerol, increasing the viscosity of glycerol and forms suspensoid with CaO, that it is difficult to be separated [5]. To overcome these problems, CaO may be supported by an active metal [3].

CaO catalyst has prepared by supported with palm oil fly ash from the combustion of light fractions of husk and shell at boiler [6]. However, fly ash can be used as a source of silica in the catalyst preparation, whereas it can acts as adsorbent, either as an active sites in the reaction. High silica content in CaO will improve the mechanical, hardness properties and morphology [7]. On the other hand, high mechanical properties of the catalyst will prevent *leaching* [8].

Fly ash, the chief inorganic waste arises through palm oil mill waste boiler ashes, has a typical chemical composition (on dry basis): 55.19% SiO_2, 30.01% Al_2O_3, 4.58% Fe_2O_3, 2.12% Na_2O, 0.77% CaO, 1.91% MgO, 2.74% TiO_2, 1.28% BaO and 1.40% K_2O. The presence of high amount of SiO_2 and Al_2O_3 prompts its potential applicability as a low-cost catalyst support. In a very recent publication, fly ash supported calcium oxide has been employed as a recyclable solid base catalyst for Knoevenagel condensation reaction [7]; however, the CaO was derived from reagent grade $CaCO_3$ which might increase the cost of catalyst preparation.

Calcination and temperature time can affect the formation of CaO [10]. The study by Ho et al. (2014) found that the higher calcination temperature and the longer time of calcination, the *yield* of biodiesel tends to decrease. Calcination temperature depends on the primary material (precursor) for the formation of CaO. The basic strenght explains the active site of catalyst for the transesterification. The basic strength determined using Hammett indicators such as phenolphthalein ($H\_ = 9.8$), Nile blue ($H\_ = 10.1$) and 2,4-Dinitroaniline ($H\_ = 18.4$). Kesic et al. (2012) using CaO:ZnO, with optimum conditions of yield biodiesel reached 99%. The higher basic strenght, the higher catalytic activity of catalyst. Tang et al. (2012) using a catalyst CaO/Fe_3O_4 to see the effect of the molar ratio of n(Ca): n(Fe) at the calcination temperature of 600 °C for 6 h. The optimum conditions are obtained with a molar ratio of Ca:Fe by 5:1, the yield increases with increasing biodiesel content of Ca.

Ho et al. (2012) studied the effect of CaO catalyst using bottom ash, ash boiler, and fly ash palm as its support material. CaO was obtained from calcium carbonate (CaCO3) calcination with temperature of 800, 900 and 1000 °C. The optimum temperature condition of catalyst preparation were obtained at a temperature of 800 ° C for 30 minutes, by addition of 15%-w calcined $CaCO_3$ into fly ash palm. The catalyst was to use for transesterification process of CPO and obtained the conversion result of 94.48%. The use of fly ash as a CaO support material can provide a larger catalyst surface area. The larger surface area of the catalyst would achieve better productivity of catalyst. The extended research was conducted by Ho et al. (2014) reviewing the development and evaluation of the CaO/palm fly ash catalyst. The catalyst was prepared by a wet impregnation method of $CaCO_3$ and fly ash with composition 45% wt% and 55 wt %, respectively. $CaCO_3$ was calcined at 800 ° C for 1.5 h. Catalyst was used in transesterification process of CPO with methanol in the biodiesel production. The optimum condition of the transesterification reaction was reached at 60 °C, with a stirring rate of 700 rpm, and catalyst composition of 6 wt% of oil with molar ratio of methanol: oil = 12 : 1. Hence, these condition was obtained the biodiesel yields approximately 75.73%.

CaO can be synthesized using a thermal decomposition process of $CaCO_3$ with calcination [9]. However, high calcination temperatures require more energy to decompose $CaCO_3$ into CaO. Ho et al. (2014) obtained CaO from the precursor $CaCO_3$ with the highest concentration of 39%-w after calcination at 1000 °C [6]. Nugraha et al. (2016) have been used CaO catalysts from precursors of calcium nitrate tetrahydrate ($Ca(NO_3)_2.4H_2O$) and fly ash as support [10]. The advantage of $Ca(NO_3)_2.4H_2O$ can be directly used without calcination with high temperature to obtain CaO.

Preparation of catalyst using wet impregnation method with composition 50% wt% $Ca(NO_3)_2.4H_2O$ and 50 wt% fly ash. The amount of CaO in the catalyst was 17.07%. Catalysts are used in the transesterification of biodiesel production from off-grade palm oil. The optimum condition was obtained at 60 ° C, CaO/fly ash concentration of 7% -w of oil with mole ratio of methanol: oil 8: 1. The highest biodiesel yield was 61.72%.

Thus, the present work attempts to develop and characterize a novel low cost, highly efficient supported base catalyst through utilization of the two waste materials viz. fly ash and $Ca(NO_3)_2.H_2O$ as a precursor to derived CaO for transesterification of oil-palm off grade to yield biodiesel.

2. Experimental

2.1. Preparation of off grade fruit bunch oil

The first step is the washing of the fruit from the dirt. Furthermore the fruit steamed as much as 3 kg in the stall for 120 minutes. After that, the fruit is pressed using a spindle hydraulic press. The extraction results are inserted into the separating funnel. The oil obtained is then analyzed to determine the level of FFA and moisture content.

2.2. Preparation of CaO/fly ash catalyst

2.3. Biodiesel Production

The extraction oil is weighed as much as 60 grams and fed into the esterification reactor. After the temperature reaches 60 °C, methanol reactants with a mole ratio of methanol: oil = 12: 1 and a 1% -w H_2SO_4 catalyst are added. The reaction lasts for 1 hour and the speed of stirrer 400 rpm. The mixture of H_2SO_4 catalyst and residual methanol is separated by separating funnel. The lower layer is checked for FFA content [11], then proceed to the transesterification reaction stage. The lower layers of esterification product separation were fed into 50 g of transesterification reactor, heated to 70 °C, methanol was added with the mole ratio of methanol: oil = 6: 1, CaO / fly ash catalyst of 6% -w oil and 400 rpm stirring rate. After 3 hours, the mixture is cooled and filtered. The catalyst precipitate is separated from the filtrate. The obtained filtrate proceeds to the process of separation and purification of biodiesel.

2.4. Catalyst Characterization

There are a few characterization method for photocatalyst such as X-Ray *diffraction* (XRD) and scanning electron microscopy (SEM) observation to study composition and morphology which contain important information. XRD measurement was conducted by using Shimadzu XRD 600 X-ray diffractometer (30 kV, 30 mA). The morphology of the prepared sample was recorded by Scanning electron microscope (SEM) using a Philips XL-30 (15 kV, 70 Pa). A prepared sample was filled through gold sputtering in the surface of CaO/Fly ash. The picture was gained through scanning electron microscope gained at 15 kV with 10000 times enlargement.

3. Result and Discussion

3.1. Extraction of CPO Off-grade

The CPO off-grade fruit was extracted using a spindle press tool and produced an off-grade palm oil of 18.53% or about 556 g. High FFA therefore through esterification process before transesterification. FFA levels of CPO off-grade which initially was 6.19% decreased to 0.96%. Characteristics of CPO off-grade are shown in Table 1.

Table 1. Characteristics of Crude Palm Oil Off-grade.

No	Characteristic	Unit	Characterization	Standard CPO SNI 01-2901-2006
1	Density (40 °C)	kg/m^3	892.11	-
2	Viscosity (40 °C)	mm^2/s	29.47	-
3	Moisture	%	3.5	Max 0,5
4	FFA	%	6.19	Max 0,5
5	Color		Orange	Orange

3.2. Basic Strength

Fly ash (FA) and CaO/FA on variation of weight ratio of $Ca(NO_3)_2.4H_2O$:FA 80:20, calcination temperature 800, 850 and 900 °C tested using Hammett indicator i.e phenolphthalein. Based on the result, FA does not show the color change. This indicates that the FA has a strong base of $H\_ < 8.2$. High SiO_2 content causes FA to have no alkaline properties, does not contain oxide ions and does not react with acids. Instead, it is a weak acid and reacts with a strong base. Therefore, the FA is particularly suitable as a support material [12]. However, the 3 variations of the catalyst CaO/FA show the change to purple. This indicates that the catalyst has a basic strength of $H\_ > 9,3$ [13]. According to Ho et al. (2014) a strongly developed base catalyst of $H\_ > 9.3$ is considered a relatively strong base catalyst for the transesterification process although its basic strength is lower than pure CaO ($15 < H\_ < 18.4$) [6].

3.3. X-Ray Diffraction (XRD)

The catalysts used in the test using XRD method were five variations, i.e catalyst with percent weight ratio of $Ca(NO_3)_2.4H_2O$:fly ash of 70:30, 80:20, and 90:10 calcinedat 800 °C. The comparison of XRD pattern three different composition of catalyst is shows in Figure 1.

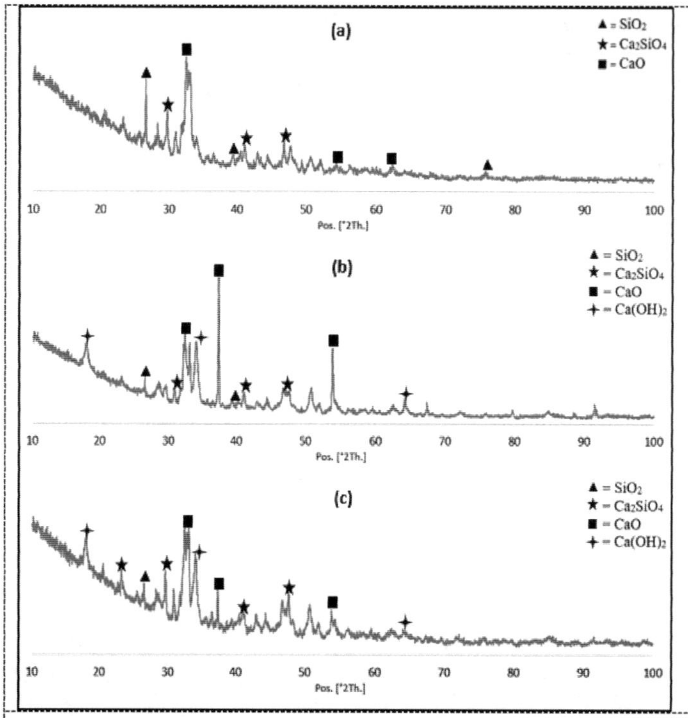

Figure 1. XRD Patterns of 800 °C Calcination Temperature Catalysts on Variations $Ca(NO_3)_2.4H_2O$: FA a) 70:30, b) 80:20 and c) 90:10.

Figure 1 (a) there is a peak of the highest crystals with an intensity of 1157.63 indicating the CaO content of the catalyst at 2θ: 32.29°. Next the CaO peak appears at 2θ: 54.23° and 64.67°. In addition, there is a SiO_2 peak at 2θ: 26.59°, 39.38° and 75.51°, indicating the high content of silica compound from impregnated fly ash. Other compounds are dicalcium silicate (Ca_2SiO_4) at 2θ: 29.62°, 39.38° and 75.51°.

Figure 1 (b) it still shows the highest CaO content at 2θ: 37.35° with an intensity of 1647.77 and subsequently at 2θ: 32.19° and 53.85° (JCPDS File No. 37-1497). There is a SiO_2 peak at 2θ: 26.58° and 39.38° (JCPDS File No. 29-0085) where the silica intensity is reduced because the weight percent or fly ash content is also reduced in the catalyst. Dicalcium silicate peak (Ca_2SiO_4) appears at 2θ: 32.46°, 41.14° and 46.71°. Other compounds are calcium hydroxide ($Ca(OH)_2$) at 2θ: 18.06°, 34.10° and 64.29°.

Figure 1 (c) the XRD pattern shows the SiO_2 peak at 2θ: 26.58° the silica intensity decreases because of the small fly ash content in the catalyst. The $Ca(OH)_2$ peak appears at 2θ: 17.98°, 34.02° and 62.49°. The highest peak is CaO at 2θ: 32.19° with intensity 958.03 and then at 2θ: 37.35° and 53.85° (JCPDS File No. 37-1497). CaO intensity decreases because CaO has reacted with SiO_2 from fly ash in the impregnation process to form dicalcium silicate (Ca_2SiO_4). So the intensity of Ca_2SiO_4 is greater and the peak is increasing at 2θ: 23.26°, 29.62°, 41,18° and 47.56°. The presence of dicalcium silicate indicates that CaO has interacted with the fly ash surface, resulting in a reaction as in the following equation 1:

$$2CaO + SiO_2 + H_2O \rightarrow (CaO)_2(SiO_2)(H_2O) \qquad \text{................................ (3.1)}$$

In the presence of water molecules, CaO and SiO_2 react to form dicalcium silicate hydrate ($Ca_2SiO_4H_2O$). However calcination at 800-900 °C will released the water molecule and produced dicalcium silicate (Ca_2SiO_4) [14]. The more dicalcium silicate (Ca_2SiO_4) formed by the reaction with fly ash shows the more CaO impregnated on the fly ash surface, the active side of the catalyst becomes more numerous. With the increase of the active side of the catalyst, the likelihood of the reaction is higher, the resulting product more and more. Meanwhile, according to Setiawan et al. (2014) CaO component will be impregnated in the SiO_2 pores present in fly ash [15].

Then another catalyst variation is a catalyst with a weight percent ratio of $Ca(NO_3)_2.4H_2O$: fly ash precursor of 80:20 calcined at 850 and 900 °C. The XRD pattern of the catalyst is shown in Figure 2.

Figure 2 (b) shows the highest CaO content at 2θ: 32.22° with an intensity of 1186.53 and then at 2θ: 37.37° and 53.89°. The higher calcination temperature at 850 °C makes SiO_2 more reactive with CaO to form Ca_2SiO_4 at 2θ: 28,32°, 33,10°, 41,14° and 46,71° so that no SiO_2 peaks appear. According to Soleimani et al. (2012) the SiO_2 intensity decreases after calcination because Si-O-Si or Si-O-Al bonds become loose so that there will decrease the amount of crystals and new crystalline formation during the calcination process [16]. Meanwhile, other compounds are $Ca(OH)_2$ at 2θ: 17.95° and 63.93°.

Figure 2 (c), the highest peak is CaO at 2θ: 32.29° with an intensity of 1491.24 and then at 2θ: 37.46°. The higher calcination temperature at 900 °C makes SiO_2 more reactive with CaO to form Ca_2SiO_4 at 2θ: 28.32°, 33.10°, 41.14°, 46.71° and 51.89°. Another compound is Ca $(OH)_2$ at 2θ: 18.08°. The presence of $Ca(OH)_2$ compounds is thought to have formed due to contact between the surface of CaO and water vapor (H_2O) solids from free air. Kouzu et al. (2008) also reported that the highest order of base strength was in CaO, then $Ca(OH)_2$ and the lowest was $CaCO_3$. Thus, the presence of $Ca(OH)_2$ compound on the surface of CaO solids results in a decrease of base strength on the catalyst [4].

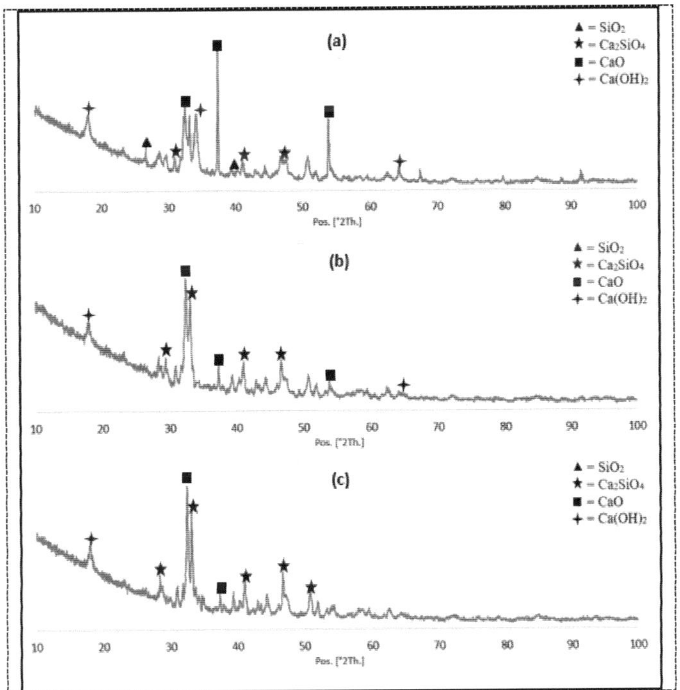

Figure 2. XRD pattern of $Ca(NO_3)_2.4H_2O$: FA 80:20 calcination temperature a) 800 °C, b) 850 °C and c) 900 °C.

From the result of XRD pattern in Figure 2 (a) - (c) shows the intensity and the peak of Ca(OH) 2 decreases. This is due to the high temperature of calcination which causes $Ca(OH)_2$ to decompose into CaO. This corresponds to the purpose of calcination to modify the chemical composition of the catalyst by heating at high temperatures. Thus, the intensity of CaO increases with increasing calcination temperature (Wong et al., 2014). On the other hand, the temperature of calcination is required to increase the mechanical forces preventing leaching [12].

3.4. Brunaeur-Emmet-Teller (BET)
The catalysts were characterized by 2 variations, CaO/FA catalysts with a calcination temperature at 800 and 850 °C and a weight ratio of $Ca(NO_3)_2.4H_2O$: FA of 80:20 are shown in table 2.

Table 2. Catalyst Surface Area of the catalys FA/CaO at Ratio $Ca(NO_3)_2.4H_2O$: FA 80:20.

Catalyst	CalcinationTemperature (°C)	Surface Area (m^2/g)
CaO/fly ash 80:20	800	24.342
	850	909.781

Based on the result of BET in table 2, the large catalyst surface area in the temperature 850 °C is 909.781 m^2/g. This is due to an increase in the temperature of calcination, then the catalyst gradually turns into a stable crystal and increased surface area [17]. Closing of pores from fly ash by CaO crystals causes a decrease in the surface area of the catalyst [13]. Therefore, the magnitude of catalyst surface area obtained due to the possibility of CaO is not evenly distributed across the pores and fly ash surfaces in the impregnation process.

3.5. Transesterification Reaction
The lowest biodiesel yield obtained was 45.19% using a catalyst of calcination temperature of 900 °C and the ratio of $Ca(NO_3)_2.4H_2O$: FA by 90:10, where the CaO content in the catalyst was 68.11%. The highest biodiesel yield was obtained by 71.77% using a catalyst calcined at 800 °C and the ratio of $Ca(NO_3)_2.4H_2O$: FA by 80:20, where CaO content in the catalyst was 48.69%. Catalyst $Ca(NO_3)_2.4H_2O$: FA 80:20 has a high CaO content among other catalysts, thus providing a higher basic strength. The larger the basic strength, the higher the catalytic activity of the catalyst so that the yield of biodiesel produced is also higher [18].

The calcination temperature of 800 °C did not cause the chemical structure of the catalyst to change which caused the deactivation of the catalyst. The surface area of the catalyst at a temperature of 850 °C which is too large causes the distribution of active sites uneven across the pore and the catalyst surface so that it does not react entirely [19]. Meanwhile, according to Liu et al. (2010) the amount of CaO is too small then the active side of the catalyst is less so the resulting yield is small [3]. Thus, the surface area (BET) does not significantly affect the activity of the CaO/FA catalyst.

On the other hand, the high temperature calcination of 900 °C of catalyst with a weight ratio of $Ca(NO_3)_2.4H_2O$: FA 90:10 lowers the catalyst base side. Further increase of the calcination temperature causes the active component parts of the catalyst to be lost and cause catalyst agglomeration. Too high CaO contained in the catalyst of 68.11% cause CaO intensity decreases due to CaO has been much reacted with SiO_2 from fly ash to form Ca_2SiO_4 thereby decreasing its strong base. The excess of CaO in the catalyst may lead to the possibility that the surface of the CaO solid can no longer decompose methanol into a methoxide ion so that the transesterification reaction cannot proceed well [20].

3.6. Characteristics of Biodiesel
Biodiesel results are tested and characterized prior to use to the machine to work optimally and more durable. The specifications or quality standards of biodiesel obtained compared to the characteristics of biodiesel based on SNI 7182: 2015 are shown in Table 3.

Table 3. Characteristics of Biodiesel.

No	Characteristics	Unit	Result	SNI 7182:2015
1	Density	kg/m^3	860.49	850 – 890
2	Kinematic Viscosity	mm^2/s	3.75	2.3 – 6.0
3	Flash point	°C	125	Min. 100
4	Acid Number	mg-KOH/g-biodiesel	0.24	Max. 0.5

Biodiesel with highest yield using catalyst FA/CaO calcined at 800 °C and ratio of $Ca(NO_3)_2.4H_2O$: FA of 80:20 identified its methyl ester using Gas Chromatography - Mass Spectrometry (GC-MS) at UPT Diponegoro University Integrated Laboratory. The results of the sample analysis with GC-MS can be seen in the methyl ester chromatography of the CPO off-grade shown in Figure 3.

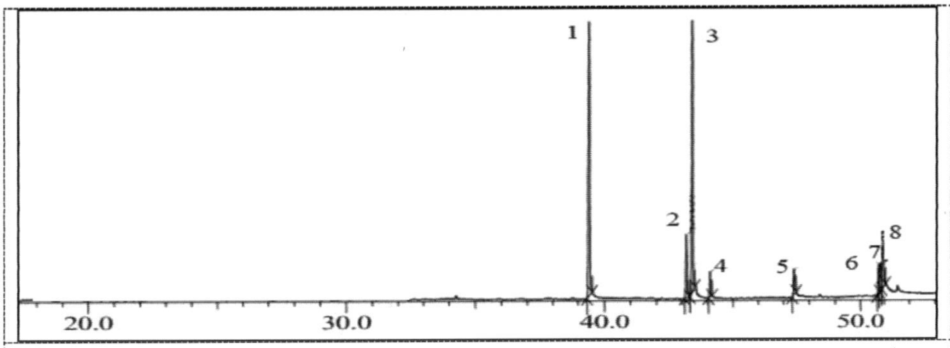

Figure 3. Methyl Ester Chromatogram.

This identification aims to determine the composition of the suspected compounds as methyl esters of the synthesis. The result of GC-MS fragmentation pattern analysis as shown in Figure 3.3 shows that large methyl esters have retention time (tr) (39.407; 43.167; 43.402 and 44.095) minutes. The graph of biodiesel characterization results using GC-MS shows the chemical composition and methyl ester conversion as shown in Table 4.

Table 4. Chemical Composition of Palm Oil and Biodiesel.

Komposisi	Methyl Palmitate (C16:0)	Methyl Stearate (C18:0)	Methyl Oleate (C18:1)	Methyl Linoleate (C18:2)	Reference
Sludge Palm Oil	42.84%	4.21%	39.58%	9.92%	[21]
CPO Off-Grade	33.74%	3.10%	36.04%	7.91%	The Result of Research

It can be seen in table 4 that the largest composition of biodiesel is methyl palmitate of 42.84%. While the least composition on biodiesel is methyl stearate of 3.10%. The conversion of the carboxylic acid group to methyl ester in the biodiesel production process reaches 80.79%. This indicates that the biodiesel product produced is not very pure or still other impurities such as catalyst residue, water, methanol and glycerol. This is evidenced by the presence of glycerol that is converted from GC-MS yield of 7.27%. The residual methanol on the use of the catalyst shows that between oil and methanol does not react completely. Substantial amounts of methanol and glycerol form emulsions with the catalyst thus affecting the yield of biodiesel and the conversion of methyl esters.

IOP Conf. Series: Materials Science and Engineering **334** (2018) 012077 doi:10.1088/1757-899X/334/1/012077

4. Conclusion

Biodiesel can be produced from CPO off-grade using a CaO/Fly ash catalyst at the transesterification. Heterogeneous CaO/fly ash can be synthesized and have a basic strength value of H_> 9.3 using the Hammett phenolphthalein. The highest yield of biodiesel was obtained 71,77% using catalyst at 800 °C calcination temperature and weight ratio of $Ca(NO_3)_2.4H_2O$: FA 80:20, CaO level in catalyst was 48,69%. The higher the calcination temperature, the lower the yield of biodiesel. The biodiesel yield tends to increase with increasing CaO levels in the catalyst. The activity of the CaO/fly ash is not affected by its surface area (BET) but is affected by its basic strength.

References

[1] Arifin, J.K., 2009. Pemanfaatan Buah Sawit Sisa Sortiran sebagai Sumber Bahan Baku Asam Lemak. *Tesis*. Program S2 Teknik Kimia Universitas Sumatra Utara. Medan.

[2] Helwani, Z., M. R. Othman, N. Aziz, J. Kim and W. J. N. Fernando. 2009. Solid Heterogeneus Catalyst for Transesterification of Triglycerides with Methanol : A Review. *Applied Catalysis A : General*. 369: 1 -10.

[3] Liu, X.J., He, H.Y., Y. Wang and S. Zhu. 2008. Transesterfication of Soybean Oil to Biodiesel Using CaO as A Solid Base Catalyst. *Fuel*. 87 :216-221.

[4] Kouzu, M., T. Kasuno, M. Tajika, Y. Sugimoto, S. Yamanaka and J. Hidaka. 2008. Calcium Oxide as Solid Base Catalyst for Transesterification of Soybean Oil and its Application to Biodiesel Production. *Fuel*. 87:2798 -2806.

[5] Liu, C., L.V., Pengmei. Yuan, Z., F. Yan and W. Luo. 2010. The Nanometer Magnetic Solid Base Catalyst for Production of Biodiesel. *Renewable Energy*.15 : 1531-1536.

[6] Ho, W.W.S., H.K. Ng, S.Gan and S.H. Tan. 2014. Evaluation of Palm Oil Mill Fly Ash Supported Calcium Oxide as A Heterogenous Base Catalyst in Biodiesel Synthesis from Crude Palm Oil. *Energy Conversion and Management*. 88 : 1167-1178.

[7] Zahrina, I., Yelmida and F. Akbar. 2012. Sintesis ZSM-5 dari Fly Ash Sawit Sebagai Sumber Silika dengan Variasi Nisbah Molar Si/Al dan Temperatur Sintesis. *Jurnal Rekayasa Kimia dan Lingkungan*. 9 : 94-99.

[8] Nugraha, P. Z., Z. Helwani and E. Saputra. 2016. Penggunaan Katalis Fly Ash yang Diimpregnasi dengan CaO dari $Ca(NO_3)_2$ pada Tahap Transesterifikasi Minyak Sawit Off-grade Menjadi Biodiesel. *Jom FTEKNIK*. 3 (1).

[9] Budiawan, R. Zulfansyah, W. Fatra and Z. Helwani. 2013. Off Grade Palm Oil as A Reneweble Raw Material for Biodiesel Production by Two-Step Processes. *ChESA Conference*. Januari. Banda Aceh. 7 : 40-50.

[10] Marinkovic, D.M., M.V. Stankovic, A.V. Velickovic, J.M. Avramovic, M.R. Miladinovic, O.O. Stamenkovic, V.B. Veljkovic and D.M. Jovanovic. 2016. Calcium Oxide as a promising heterogenous catalyst for biodiesel production: Current state and perspectives. *Renewable and Sustainable Energy Reviews*. 56 : 1387-1408.

[11] Hindryawati, N., G. P. Maniam, Md. R. Karim and K. F. Chong. 2014. Transesterification of used cooking oil over alkali metal (Li, Na, K) supported rice husk silica as potential solid base catalyst. *Engineering Science and Technology, an International Journal*. 1-9.

[12] Helwani, Z., N. Aziz, J. Kim and M. R. Othman. 2016. Improving The Yield of Jatropha Curcas's FAME through Sol-Gel Derived Meso-porous Hydrotalcites. *Renewable Energy*. 86: 68-74.

[13] Ho, W.W.S., H.K. Ng and S. Gan. 2012. Development and characterization of novel heterogeneous palm oil mill boiler ash-based catalyst for biodiesel production. *Bioresource Technology*. 125 : 158-164.

[14] Setiawan A., Widiyastuti, S. Winardi and A. Nugroho. 2014. Kinetika Reaksi Sintesis Biomaterial Hydroxyapatite Dengan Prekursor Nitrat dan Asetat. *Reaktor*. 15 (2) : 104-110.

[15] Soleimani, M.A., R. Naghizadeh, A.R. Mirhabibi and F. Golestanifard. 2012. Effect of Calcination Temperature of The Kaolin and Molar Na_2O/SiO_2 Activator Ratio on Physical and Microstructural Properties of Metakaolin Based Geopolymers. *Iranian Journal of Materials Science & Engineering*. Vol. 9. Number 4.

[16] Tang, S., L. Wang, Y. Zhang, S. Li, S. Tian, and B. Wang. 2012. Study on Preparation of $Ca/Al/Fe_3O_4$ Magnetic Composite Solid Catalyst and Its Application in Biodiesel Transesterification. *Fuel Processing Technology*. 95 : 84-89.

[17] Kesic, Z., I. Lukic, M. Zdujic, H. Liu and D. Skala. 2012. Mechanochemically synthesized CaO·ZnO catalyst for biodiesel production. *Procedia Enginering*. 42 : 1278-1287.

[18] Wong, Y.C., Y.P. Tan, Y.H. Taufiq-Yap and I. Ramli. 2014. Effect of Calcination Temperatures of CaO/Nb_2O_5 Mixed Oxides Catalysts on Biodiesel Production. *Sains Malaysiana*. 43 (5) : 783-790.

[19] Samik, R. Ediati and D. Prasetyoko. 2014. Review: Pengaruh Kebasaan dan Luas Permukaan Katalis Terhadap Aktivitas Katalis Basa Heterogen untuk Produksi Biodiesel. *Jurusan Kimia Fakultas MIPA Institut Teknologi Sepuluh Nopember,* Surabaya.

[20] Fanny W. A., Subagjo and T. Prakoso. 2012. Pengembangan Katalis Kalsium Oksida untuk Sintesis Biodiesel. *Jurnal Teknik Kimia Indonesia*. 11 (2) : 66-73.

[21] Hayyan, A., Md. Z. Alam, M.E.S. Mirghani, N.A. Kabbashi, N.I.N.M. Hakimi, Y.M. Siran and S. Tahiruddin. 2010. Sludge Palm Oil as A Reneweble Raw Material for Biodiesel Production by Two Step Processes. *Bioresource Technology.*101 : 7804 – 7811.

Impact-induced fracture mechanisms of immiscible PC/ABS (50/50) blends

M N Machmud[1,5], M Omiya[2], H Inoue[3] and K Kishimoto[4]

[1] Division of Alternative Materials for Components and Structures, Department of Mechanical Engineering, Syiah Kuala University, Darussalam, Banda Aceh 23123, Indonesia
[2] Department of Mechanical Engineering, Keio University, 3-14-1 Hiyoshi, Kohoku-ku, Yokohama, Kanagawa 223-8522, Japan
[3] Department of Mechanical and Control Engineering, Tokyo Institute of Technology, 2-12-1 Ookayama, Meguro-ku, Tokyo 152-8552, Japan
[4] Department of Mechanical Sciences and Engineering, Tokyo Institute of Technology, 2-12-1, O-okayama, Meguro-ku, Tokyo 152-8552, Japan

E-mail: mnizar.machmud@unsyiah.ac.id

Abstract. This paper presents a study on fracture mechanisms of polycarbonate (PC)/ acrylonitrile-butadiene-styrene (ABS) (50/50) blends with different ABS types under a drop weight impact test (DWIT) using a circular sheet specimen. Formation of secondary crack indicated by a stress-whitening layer on the mid-plane of scattered specimens and secondary surface of fracture perpendicular to primary fracture surface were captured under scanning electron microscope (SEM). Although the both blends finally failed in brittle modes, SEM observation showed that their secondary fracture mechanisms were completely different. Observation through the thickness of the etched PC/ABS specimen samples using SEM also clearly showed that PC and ABS were immiscible. The immiscibility between PC and ABS was indicated by presence of their layer structures through the thickness of the blends. It was revealed that layer of ABS structure was influenced by size of rubber particle and this latter parameter then affected microstructure and fracture mechanisms of the blends. Impact-induced fracture mechanisms of the blends due to such microstructures are discussed in this paper. It was also pointed out that the secondary cracking was likely caused by interface delamination between PC and ABS layers in the core due to transverse shear stress generated during the impact test.

1. Introduction

PC/ABS blends have been the subject of many studies and their behaviours have been reported in a numerous published scientific works due to widely application of the blends [1-3]. They have revealed that properties of such ternary blends strongly depend on the structure properties of the components while morphology of the blends depends on blending and injection moulding conditions [4,5]. At an ABS volume fraction of about 50%, for instance, microstructure of the blend shows a co-continuous morphology [4]. Whilst, a moulded material produced from an injection moulding process may

[5] To whom any correspondence should be addressed.

experience the effect of significant normal stresses, especially at the flow front. Because of such effect, the obtained moulding has a structure gradient through the thickness, varying also along the respective flow path [5-11]. Furthermore, characteristics exhibited in the final polymer blend products depend on the properties of the individual components as well as on the microstructure formed mainly in the process [12].

Many research works have reported that fracture mechanisms of PC/ABS blends were influenced by the ABS types [13-15]. In an investigation of correlation between morphology formation and fracture mechanism of injection-moulded PC/ABS (60/40) blends under Izod impact test, Bärwinkel et al. [15], for instance, revealed that with dependent on molecular weight of SAN, there was a correlation between interphase adhesion among PC and ABS layers and fracture mode of the blends. However, study on influence of the microstructure gradient through the thickness produced after moulding due to ABS type on fracture mechanisms of the PC/ABS blends under impact has been very rare.

In a preliminary study using a drop weight impact test, it was pointed out that similarity of failure mode exhibited by the PC/ABS (50/50) blends at a certain speed eliminated the role of ABS type in the blends and performance of the blends under such impact has therefore not clearly been understood. Since microstructure can provide a useful basis for understanding the fracture of materials, this paper is therefore presented to gain an understanding of the influence of ABS type on the microstructure gradient through the thickness of PC/ABS (50/50) blends as well as on their impact-induced fracture mechanisms. The latter type of PC/ABS blends was selected since they have become frequently used engineering materials combining several advantageous properties of both components.

2. Materials and Experiments

2.1. Materials
Two grade specimens of PC/ABS (50/50) blends containing butadiene rubber of about 10 wt% with large (∅270 nm) and smaller rubber (∅150 nm) particle size in ABS system, further respectively called as PA-01 and PA-02, were used. Both of the PC/ABS (50/50) blends used in this study were supplied by Japan Synthetic Rubber Co. Ltd.

2.2. Experiments
2.2.1. Impact test. Most recent reported experimental studies have been carried out using Izod and Charpy impact tests on notched specimens [16]. Due to use of such specimens, they only measure crack propagation energy, not crack initiation energy, and thus, they do not give a 'true' indicating impact resistance of the materials. In this study, impact resistance characteristics of PC/ABS (50/50) blends were hence carried out on sheet specimen by use of a drop weight impact testing (DWIT) apparatus outfitted with instruments. The instrumented DWIT apparatus which was used then to study the fracture mechanism of the materials is similar to that of used by Machmud et al. [17]. Samples of circular sheet specimen with a diameter of 99.3 mm and initial thickness of 2 mm were used. The specimen was simply-supported on its bottom surface at 8.5 mm inside around the outer edge of the sheet. A mass was dropped from a certain height and the striker struck perpendicularly the centre of the circular sheet specimen. Body of the mass attached to guiding rods brought a total weight of 2.4 kg. Response of the materials was then induced by a hemispherical tip-ended cylindrical striker made of stainless steel (SS55) with diameter of 20 mm. Three samples for each material were tested at room temperature (25±2°C) under a rate of 4 m/sec.

2.2.2. Microstructure test. As microstructure plays an important role on characteristics of materials under impact, it is necessary to characterize their through-thickness microstructures using scanning electron microscope (SEM). Before performing this work, the sample of the PC/ABS blends was prepared by cutting the sheet specimen into small pieces (25 mm × 25 mm) with 2 mm thickness using diamond-tip cutter. After polishing, they were etched then in Potassium Hydroxide (KOH) solution with a certain concentration for ten days to remove the PC phase. After coating the etched samples with a thin layer of gold of approximately 2 nm in thickness, observation was then carried out using SEM (JEOL JSM-5300T) at accelerating voltage of 15 kV.

2.2.3. SEM observation on fracture. Secondary surface of fracture perpendicular to primary fracture surface and secondary crack formation captured on the mid-plane were found in the both scattered PA-01 and PA-02 samples. For as an example, Figure 1(a) and Figure 1(b), respectively, show the secondary fracture and the secondary formation of crack on PA-01 captured under SEM with low magnification. The secondary fracture mechanisms exhibited by PA-01 and PA-02, however, were completely different. Since there is a correlation between microstructure and fracture exhibited by both blends, it is therefore necessary to characterize such fracture using SEM. After cutting these latter samples into pieces (25 mm × 25 mm) with thickness of approximately 2 mm using diamond-tip cutter and then coating samples with a thin layer of gold of approximately 2 nm in thickness, investigation on the fracture was also carried out using SEM (JEOL JSM-5300T) at accelerating voltage of 15 kV. In situ observation of the primary and the secondary fractures using SEM on the piece of fractured PC/ABS blend samples is illustrated in Figure 2.

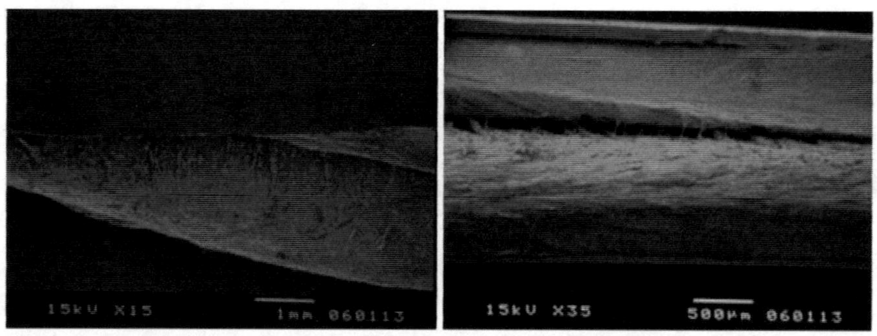

(a) (b)

Figure 1. SEM micrograph of: (a) secondary fracture and (b) secondary crack captured on the mid-plane of a PC/ABS blend specimen.

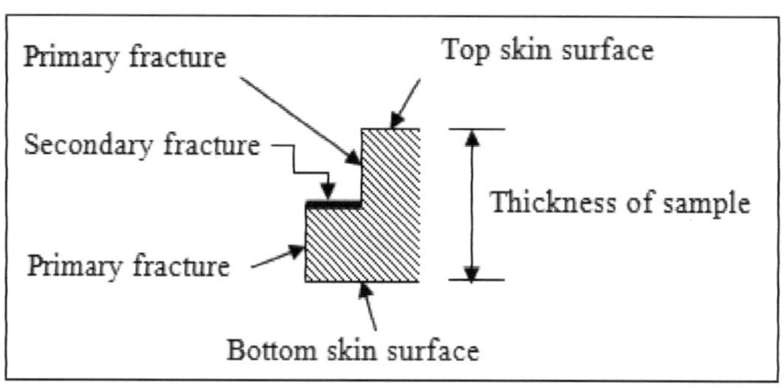

Figure 2. In situ observation of primary and secondary fractures using SEM. The shaded part is the observed sample.

3. Results and discussions

3.1. Through thickness microstructure of PC/ABS (50/50) blends

Through thickness skin-core-skin microstructures of PA-01 and PA-02 are respectively showed in Figure 3(a) and Figure 3(b). Observation of through thickness microstructure of the specimen using SEM captured immiscibility between PC and ABS. Skin of the specimen made of the blends was mostly occupied by ABS. The immiscible PC and ABS could be identified from the combination of PC and ABS layers, especially in the core of the specimen. One also can see co-continuous structure as well as continuous structure formed, respectively, in the skin and the core of both PA-01 and PA-02. Seemingly, such structures in both skin and core did not depend on the size of rubber particles distributed in the ABS systems.

On the core of specimen, structures of PC and ABS layers were also easily distinguished. Although structures on the core formed in the both blends were PC and ABS, composition of PC and ABS represented by both thickness and number of their layers in each blend was not similar. It was therefore revealed that the size of rubber particle influenced the thickness of ABS layer structure formed in the core of the blends. As shown in Figure 3(a), large size of rubber particle led to increase thickness of the ABS layers and hence reduced number of the ABS layers in the core of PA-01. As shown in Figure 3(b), due to its smaller size of rubber particles, PA-02 had a much more number of the ABS layers in the core rather than PA-01. There was thus a contradiction on the composition described by Hamada et al. [7] who indicated that composition of PC was highly distributed in the core.

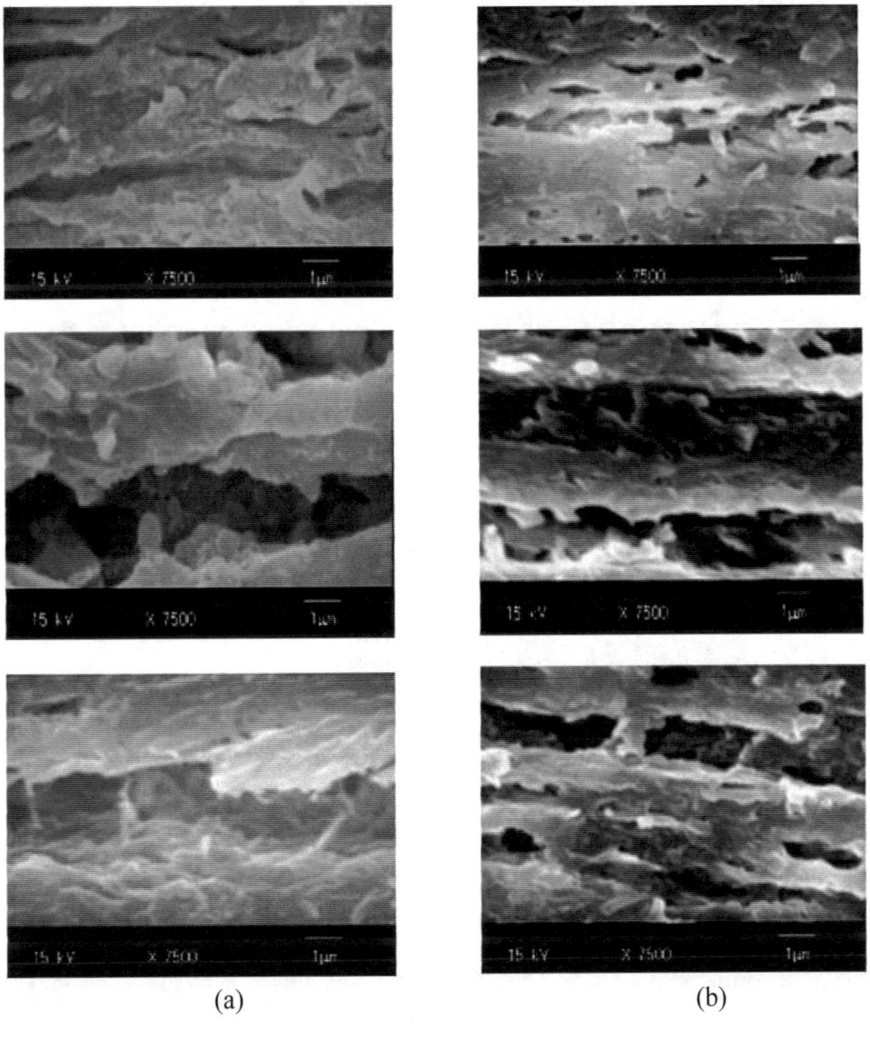

(a) (b)

IOP Conf. Series: Materials Science and Engineering **334** (2018) 012078 doi:10.1088/1757-899X/334/1/012078

Figure 3. From the top to the bottom, SEM micrograph views of through-thickness skin-core-skin structures for: (a) PA-01 and (b) PA-02.

3.2. Fracture mechanisms analysis

3.2.1. Primary fracture. Both PA-01 and PA-02 finally failed in brittle modes under the impact test. Primary fracture surface of PA-01 and PA-02 obtained under SEM observation in low magnification is shown in Figure 4 and Figure 5, respectively. Formation of the secondary crack exhibited by each of the blends was also captured on the surface as seen in the both figures. Since formation of the secondary crack was transverse to the structure gradient through the thickness, such crack was likely caused due to interface delamination between PC and ABS layers in the core induced by transverse shear stresses generated during the impact test.

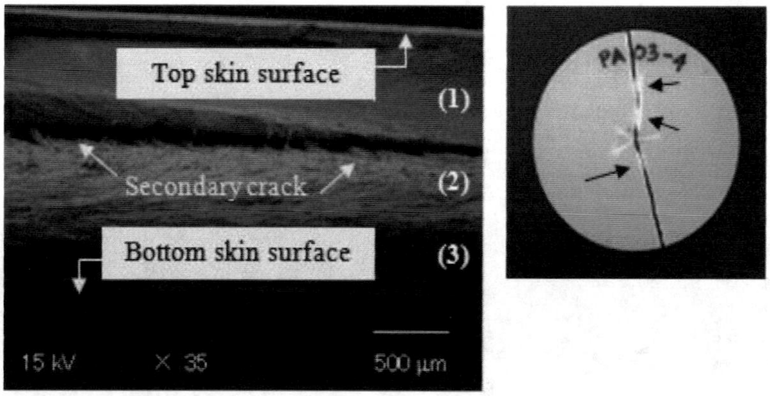

Figure 4. From the left to the right, respectively: secondary crack formation and failure mode with the location of secondary fracture surfaces of the scattered PA-01 as indicated by arrows.

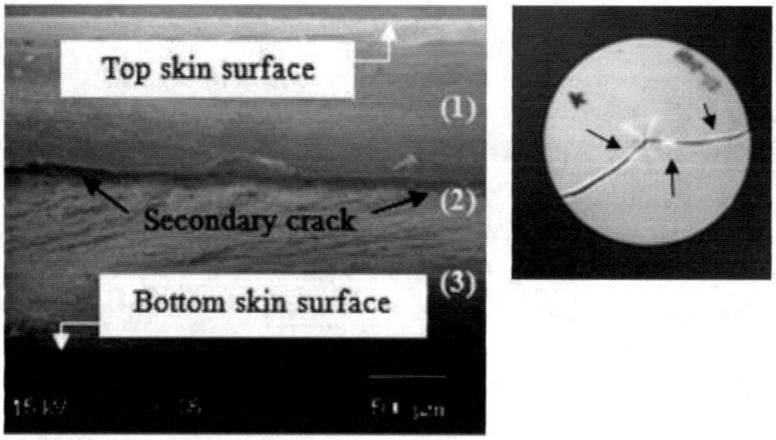

Figure 5. From the left to the right, respectively: secondary crack formation and failure mode with the location of secondary fracture surfaces of the scattered PA-02 as indicated by arrows.

Fracture surface of PA-01 and PA-02 obtained under SEM observation in higher magnification are shown in Figure 6(a) and Figure 6(b), respectively. These latter fracture surfaces were taken from the three locations in the primary fracture surface of the blends as indicated by (1), (2) and (3) in Figure 4 and Figure 5. SEM observation on the primary fracture surface of PA-01 showed that PA-01 actually failed in mixed modes fracture (semi ductile to brittle modes). As shown in Figure 6(a), the skin of PA-01 exhibited a number of pores (usually voids formed due to rubber cavitation) and failed in semi ductile modes while its core failed in brittle mode. It was indicated that large size of rubber particles played a role in generating such formation of pores in the skin of PA-01. With significant concentration of stresses associated with the pores, increase in load induced crack to initiate in the skin of the blend. Further increase in load led then the crack to propagate fast through the core. As exposed earlier in subsection 3.1, blending ABS with large size of rubber particles into PC led to reduce composition of the ABS in the core of PA-01 (Figure 3(a)). Since composition of ABS determines ductility of the core of PA-01, reducing composition of ABS led the core to fail in brittle mode and finally led the PA-01 to fail in brittle mode too. At this point, it could be pointed out that blending ABS with large size of rubber particles into PC improved ductility of the skin but reduced the ductility of the core of the blends.

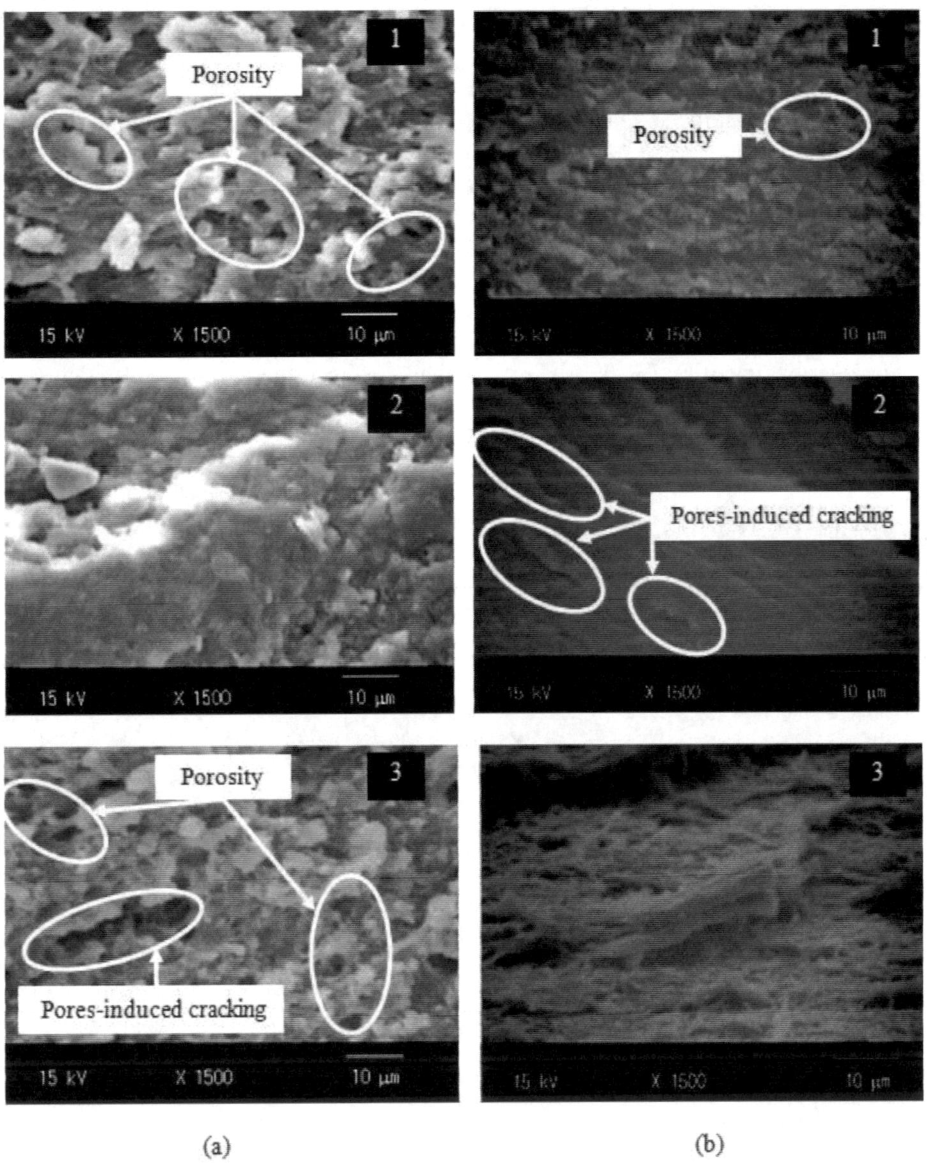

Figure 6. SEM micrographs with higher magnification of primary fracture surface for: (a) PA-01 and (b) PA-02.

SEM observation on the primary fracture surface of PA-02 showed that PA-02 actually failed in mixed modes fracture (brittle to ductile modes). As shown in Figure 6(b), the skin of PA-02 with a bit number of pores exhibited brittle modes of fracture while the core with the pores-induced cracking formations failed in ductile mode. With a bit number of pores in the skin of PA-02, it was indicated that increase in load did not significantly deform the smaller size of rubber particles in the skin. Without significant deformation of the rubber, crack initiated fast in the skin and it definitely led the skin to fail in brittle mode. Further increase in load led then the crack to propagate through the core. With a much more number of the ABS layers in the core rather than PA-01 (Figure 3(b)), the core of PA-02 failed in ductile mode as indicated by formation of the shear bands. Presence of pore-induced cracking formation in the core of PA-02 also indicated that the rubber was ever partly deformed. At this point, it could be pointed out that blending ABS with smaller size of rubber particles into PC improved ductility of the core but reduced ductility of the skin of the blends. Such different fracture mechanisms on the skin-core structure of the blends subsequently generated high stresses to take place along the interface of PC and ABS that would finally lead the blends to fail earlier.

3.2.2. Secondary fracture. Figures 7(a) and Figure 7(b), respectively, show the SEM micrographs of the secondary surface of fracture found in the scattered specimen for each of PA-01 and PA-02. Secondary surface of fracture exhibited by the both blends was perpendicular to the primary fracture surface. Such fracture was hence induced by shear stress working at the interface between PC and ABS layers in each blend. Their fracture mechanisms were completely different, however. In the figures, one can distinguish that fracture exhibited by PA-01 was collateral to propagation of crack. In vice versa, fracture exhibited by PA-02 was perpendicular to propagation of crack. Such mechanisms indicated that PA-02 showed a little bit more resistance to the crack propagation rather than PA-01.

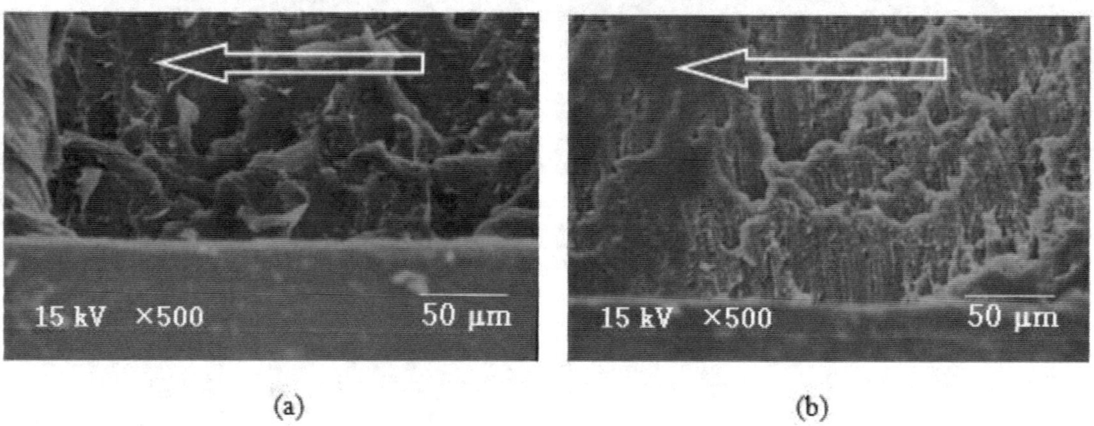

Figure 7. SEM micrograph of secondary fracture mechanism for: (a) PA-01 and (b) PA-02. White arrow in each of the micrographs shows direction of the crack propagation.

3.2.3. Energy-based approach. Energy-based approach for identify initiation and propagation of fracture in polypropylene/polyethylene blends as expressed by Tai et al. [16] and Saifullah et al. [18] could be used to explain the difference of fracture mechanisms exhibited by the PC/ABS (50/50) blends. In such approach, an impact force-displacement curve derived from an impact force-time history of tested material obtained under a certain rate of impact was divided into two distinct regions, a region of crack initiation and a region of crack propagation. The absorbed energy in these latter regions was then characterized by area under the impact force-displacement curve, before and after the maximum value of the impact force, respectively.

The impact force-time history of a PA-01 and a PA-02 tested under 4 m/sec is presented in Figure 8 and Figure 9, respectively. After deriving their impact force-time histories into their impact force-displacement relationships, energy absorbed for both initiation and propagation of crack in PA-01 and PA-02 is then presented in Figure 10. The presented results were the average of three measurements performed on each blend. As shown in Figure 10, crack initiation energy absorbed by PA-01 was 3.86 J, slightly higher than that of absorbed by PA-02 (3.61 J). Whilst, crack propagation energy absorbed by PA-01 was 2.95 J, was slightly lower than that of absorbed by PA-02 (3.07 J). It was therefore indicated that initiation of crack in PA-02 was slightly faster than that of in PA-01 and propagation of crack in PA-01 was slightly faster than that of PA-02. Such results were in correlation with both primary and secondary fracture mechanisms exhibited by both PA-01 and PA-02.

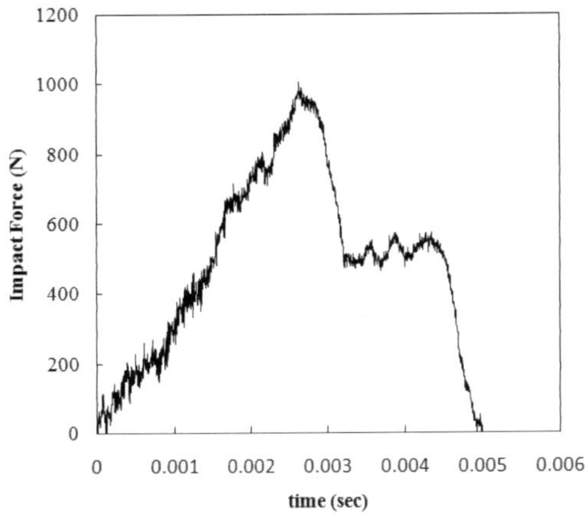

Figure 8. Impact force-time history of a PA-01 under 4 m/sec.

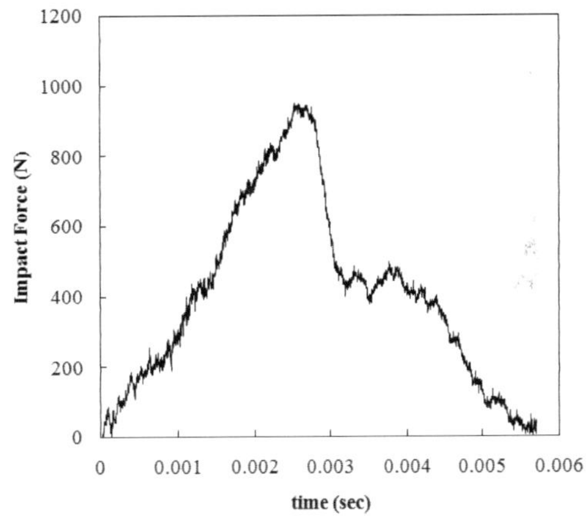

Figure 9. Impact force-time history of a PA-02 under 4 m/sec.

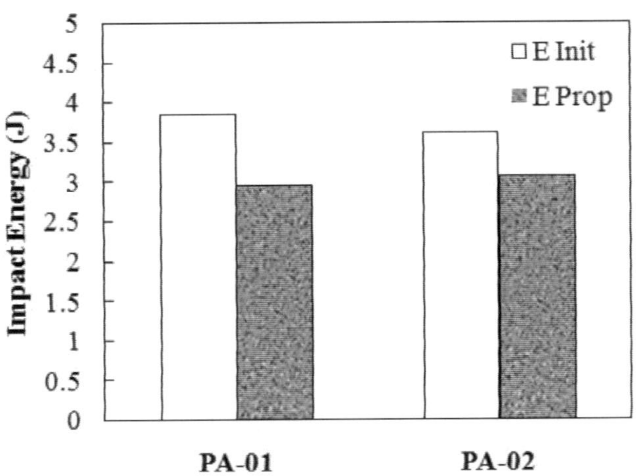

Figure 10. Crack initiation energy (E_{Init}) and crack propagation energy (E_{Prop}) absorbed by PA-01 and PA-02 under impact.

There was a correlation between size of rubber particles in ABS and composition of the ABS layers in the core of the blends. Both of the latter parameters significantly then contributed to determine

IOP Conf. Series: Materials Science and Engineering **334** (2018) 012078 doi:10.1088/1757-899X/334/1/012078

fracture mechanisms and failure modes of the skin-core structures under the impact. However, with such energy absorbed by PA-01 and PA-02, it could be pointed out that at rubber content in ABS of about 10wt%, macroscopically, size of rubber particles in ABS did not significantly affect the toughening in the blends. In overall, PA-01 and PA-02 exhibited a similar mode of failure at their fracture and showed slightly different in toughness since they absorbed 35.5% and 34.8% of the impact energy, respectively, before failure. Such results indicated that fracture energy of both PA-01 and PA-02 did not significantly differ. That is why both PA-01 and PA-02 failed in the similar brittle modes.

4. Conclusions

Fracture mechanisms of immiscible PC/ABS (50/50) blends with different ABS types under a drop weight impact test (DWIT) using a circular sheet specimen have been studied. Observation on through thickness structure of the blends using SEM clearly captured that PC and ABS were immiscible. The immiscibility between PC and ABS could be identified from presence of the PC and the ABS layer structures in skin and core of the specimen. SEM observation also revealed that size of rubber particles in ABS influenced thickness and number of ABS layer structures formed in the core of the blends. Large size of rubber particle led to increase of thickness of the ABS layer and hence reduced number of ABS layer in the core of the blends. In vice versa, smaller size of rubber particle led to reduce of thickness of ABS layers and therefore increase of number of ABS layers in the core of the blends. The number of PC and ABS layers, respectively, represents the composition of PC and ABS in PC/ABS blends. At rubber content of 10wt%, ABS exhibited co-continuous structures in the skin as well as continuous structures in the core of the blends. Such structures formed in both skin and core of the blends did not depend on size of rubber particles in ABS.

Microscopically, there was a correlation between size of rubber particles in ABS and failure mode exhibited by both skin and core of the PC/ABS (50/50) blends. Skin and core of the blends occupied by ABS with large size of rubber particles failed in semi-ductile and brittle modes, respectively. Whilst, skin and core of the blends occupied by ABS with smaller size of rubber particles failed in brittle and ductile modes, respectively. It could be therefore pointed out that blending ABS with large size of rubber particles into PC improved ductility of the skin but reduced ductility of the core of the blends while blending ABS with smaller size of rubber particles into PC reduced ductility of the skin and improved ductility of the core. Such different fracture mechanisms, exhibited by both skin and core of the blends, subsequently generated high stresses to take place along the interface of PC and ABS that would lead the blends to fail earlier. Such early failure of the blends was macroscopically indicated by their brittle fracture under the impact. Size of rubber particle in ABS, therefore, did not significantly play a role in determining failure modes of the blends under impact. At a certain content of rubber, blending ABS with smaller size of rubber particle into PC, even, did not properly work to significantly improve toughness of the blends. In such case, such size of rubber particles in ABS only contributed to influence thickness and number of the ABS layers in the skin-core structures of the PC/ABS (50/50) blends. Composition of ABS represented by both thickness and number of its layers, further, significantly influenced fracture mechanisms of the skin and the core of the blends under impact.

It was revealed that the secondary fracture mechanisms exhibited by the blends were strongly influenced by the type of ABS. Due to different type of ABS, shear stress working at the interface between PC and ABS layers in each blend led the secondary fracture mechanisms of the blends under impact were completely different. Fracture mechanisms of the blend with smaller size of rubber particle showed a little bit more resistance to the propagation of crack rather than that of the blend with large size of rubber particle.

With energy-based approach, it was also revealed that the similar mode of failure at fracture exhibited by both of the blends under impact was due to their similarity in fracture energy.

IOP Conf. Series: Materials Science and Engineering **334** (2018) 012078 doi:10.1088/1757-899X/334/1/012078

References

[1] Ma Q, Zhao W X, Li X R, Li L S and Wang Z L 2013 *International Journal of Adhesion and Adhesives* **44** 243

[2] Zhang J 2016 *Procedia Engineering* **135** 83

[3] Wang H T, Zhang Y, Huang Z G, Gao H, Zhang Y, Gao X P and Zhou H M 2016 *Polymer Testing* **56** 115

[4] Inberg J P F and Gaymans R J 2002 *Polymer* **43** 2425

[5] Inberg J P F and Gaymans R J 2002 *Polymer* **43** 3767

[6] Lee M P, Hiltner A and Baer E 1992 *Polymer* **33** 685

[7] Hamada H, Shonaike G O, Miyaji S and Yamane H 1994 *Journal of Materials Science Letters* **13** 1352

[8] Makata Y, Wei L Y, Hamada H, Tomari K and Yamada K 2009 *17th Proc. Int. Conf. On Composite Material (ICCM-17)* vol 17 *(Edinburgh, UK, 27-31 Juli 2009)*

[9] Sang Z H, Xie X L, Zhou S Y, Li Y, Yan Z, Xu L, Zhong G J and Li Z M 2017 *Ind. Eng. Chem. Res.* **56** 6295

[10] Barbosa C N, Simoes R, Franzen M, Baranowski T and Viana J C 2017 *Polym Eng Sci.* doi: 10.1002/pen.24650

[11] Giboz J, Copponnex T and Mélé P 2009 *Journal of Micromechanics and Microengineering* **19** 1

[12] Zhong G J and Zhong M L 2005 *Polym Eng Sci.* 45 1655

[13] Lee M P, Hiltner A and Baer E 1992 *Polym Eng Sci.* **32** 909

[14] Fang Q Z, Wang T J, Beom H G and Zhao H P 2009 *Polymer* **50** 296

[15] Bärwinkel S, Seidel A, Hobeika S, Hufen R, Mörl M and Altstädt V 2016 *Materials* **9** 659

[16] Tai C M, Li R K Y and Ng C N 2000 *Polymer Testing* **19** 143

[17] Machmud M N, Omiya M, Inoue H and Kishimoto K 2005 *Key Engineering Materials* **297-300** 1297

[18] Saifullah A, Thomas B, Tabeshfar K, Cripps B and Wang L 2017 *Polym Eng Sci.* doi: 10.1002/pen.24531

IOP Conf. Series: Materials Science and Engineering **334** (2018) 012079 doi:10.1088/1757-899X/334/1/012079

Effect of fiber fibrillation on impact and flexural strength of coir fiber reinforced epoxy hybrid composites

I Mawardi[1,2], Jufriadi[1] and Hanif [1]
[1]Department of Mechanical Engineering, Politeknik Negeri Lhokseumawe, Lhokseumawe, 24301, Indonesia

E-mail: ddx_72@yahoo.com

Abstract. This study aims to develop fiber-reinforced epoxy resin composites. This study presents the effect of fiber fibrillation on the impact and flextural strength of the epoxy hybrid composite reinforced by coir fiber. Coir is soaked in 5% NaOH solution for 5 hours. Then fiber is rocessed using a blender of 2000 rpm density fibrillation. The length of time the fibrillation varied for 10, 20 and 30 minutes. Volume fraction of 30% fiber and matrix 70% composited. The composite uses a matrix of epoxy by hand lay up method. The implemented tests are impact and flexural tests. The test results show fiber fibrillation treatment can improve the composite mechanical properties. The highest impact and flexural strength, 24.45 kJ/m^2 and 87.91 MPa were produced with fiber fibrillation for 10 minutes.

1. Introduction

Composites with synthetic fiber reinforcement materials have been used in various aspects of life, starting with household needs, land, air and sea vehicle industries, sports equipment, health and bulletproof vests. However, the use of synthetic fibers as a composite amplifier has a negative impact on the environment since its waste is unable to decompose naturally but able to interfere with up to several generations. The development of composite technology is now begin to experience a shift from composite materials to synthetic fibers to composite materials of natural fibers. This technological trend shift is based on the nature of the composite fiber-making nature that is more environmentally friendly, because natural fibers are able to decompose naturally and reneweble. Although not completely shifting synthetic fibers, the utilization of environmentally friendly natural fibers is a wise move to save the environment. Natural fiber-loaded composites also have a high density strength ratio so that the resulting components are lighter. Many varieties of natural fibers are available such as coconut fiber, hemp fiber, pineapple fiber-nanasan,fibers, and banana fiber.

Coir fiber is one of the natural fiber materials that can be forgiven in the manufacture of composites. This coconut fiber is being used because it is easy to obtain and widely available. Coir fiber as a reinforcing element greatly determines the mechanical properties of the composites as it forwards the loads which is distributed by the matrix. Coir fiber which is combined with polymers as a matrix, will produce alternative composites that are useful for the industrial world. The initial treatment of coco fiber is expected to produce maximum composite mechanical properties to support the utilization of alternative composites.

[2] To whom any correspondence should be addressed.

Some research on the mechanical properties of coconut fiber has been done. Udaykumar, et al [1] in their research work stated that short coir fibers reinforced polymer composite have been developed by hand layup techniques with varying fiber percentages (5%, 10%, 15%, 20%, 25% by weight), natural fiber (coir) reinforced polymer matrix (vinyl ester). The developed coir fiber reinforced composites were then tested for their flextural properties. The result shows that a flexural property increases with increases in fiber percentage; however after a certain fiber weight percentage the properties are decreased. From the data of tested results conclude that 20% of coir reinforcement at 4mm thickness of polymer composite exhibited higher percentage of elongation. The coir fiber in the present study could be used as an effective reinforcement for making composites, which have an added advantage of being light.

Sarocha Charoenvai, et al [2] investigated the mechanical properties of coir based green composites were prepared using coir fiber treated with varying pre-treatment condition. The changes in the proportion of chemical composition and morphological properties of coir fibres with different coir pre-treatment condition were discussed. It is observed that the mechanical properties of coir-based green composites; modulus of rupture and internal bond, increase as a result of chemical composition modification and surface modification. Scanning electron microscopy (SEM) investigations show that surface modifications improve the fiber/matrix adhesion. Bakri [3] has investigated that the mechanical properties of coir rope-glass fibers reinforced polymer hybrid composites. These results mechanical properties of impact energy and strength of coir rope and glass fiber as reinforcement in polymer hybrid composite is 15C15G, it is higher than 10C20G and 20C10G also 2.8 J and 177.1 kJ/m², respectively. The possibility of surface chemical modification of Coir fibers have been extensively used in a wide variety of application, e.g., packaging, furnitures etc. [4]. Sakthivel, et al [5] have conducted a research of the mechanical properties of epoxy composites with natural fiber reinforcement (banana, coir, sisal). His research yielded impact energy for each fiber ie, coir 4 joule,banana 5, sisal 4 joule. Karthikeyan *et al* [6] compared the impact property of sodium lauryl sulphate (SLS) and NaOH treated coir fiber reinforced epoxy composites. Result shows, SLS treated coir fiber reinforced composite shows more impact energy than NaOH treated coir fiber reinforced composite. Haishal *et al* [7] have analysed the mechanical properties of randomly oriented coir fiber/epoxy resin composite. The result shows tensile strength of 31.08 MPa, impact strength of 11.49 kJ/m^2 and suggested for low load applications. According to Akash *et al* [8] hemp fiber/epoxy resin composite exhibits more tensile strength than coir fiber/ epoxy resin composite and coir fiber/epoxy resin composite exhibits more bending strength than hemp fiber/epoxy resin composite.

The strength of natural fiber composites is highly depends on *mechanical bonding* between fiber and matrix. It is supported by Alex.S, et al, [9] which have stated in their works of the strength of hybrid composites depends on the properties of fiber, aspect ratio of fiber in composites, orientation of fiber, intermingling of fiber and fiber matrix interface. The bond interface is improving by the addition of coupling agents and chemical treatments. Chemical treatment is relatively easy to do and more economical. Commonly the chemical element is using NaOH. Narendra *et al* [10] suggested that the presence of nylon fabric and chemically treated pith could contribute to longer durability of the panels in moisture condition. 18% NaOH treated sisal fiber reinforced composite shows higher tensile and flexural strength than 5%, and 10% NaOH treated sisal fiber reinforced composites [11]. The tensile strength, tensile modulus and creep resistance of composite increases with increasing of the sisal fiber content in composite up to a certain limit [12].

In addition, the use of fiber as an amplifier in natural composites is very high by diameter and fiber length. The smaller the fiber diameter will be the more matrix bonds with the fibers so as to improve the composite properties and the properties of the composite structure. Reducing the fiber size is not sufficient by chemical treatment (soaking the fibers in NaOH solution) however; it is required further treatment such as fibrillation process or reducing fiber size. One study to minimize fiber diameter has been done by Gery P [13] examining the manufacture of flax fiber nanofiber. First, the fibers were processed pre-treatment chemically with 5% NaOH solution at 105 ° C for 60 minutes followed by mechanical fibrillation process with blender at 13,000 rpm for 20 minutes. From the above process it is able to change the diameter of the hemp fiber from > 10um to 100-300 nm in size. This study aims to examine the effect of coir fiber fibrillation on the impact and flexural strength of epoxy hybrid composites reinforced by coir fiber.

2. Experimental

2.1. Materials

The coir fibers were collected from local resources, Bireun, Aceh, Indonesia (Figure 1). The epoxy resin is used *Eposchon* type A for epoxy resin se (*bispenola epichlorohydrin*) and type B for epoxy hardener (*polyaminoamide*) with a ratio of 1: 1. PT. Justus Kimia Raya has supplied the epoxy resin.

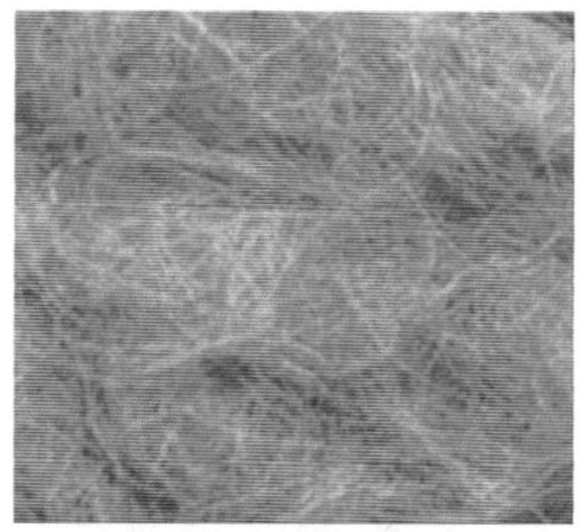

Figure 1. Coir

2.2. Fiber pre-treatment

The Coir fiber is first sorted in uniformity of diameter and length. Fibers cleaned and dried ± 2 days. The fibers are cut with a length of 30-50 mm. Coir fiber is soaked in 5% NaOH solution for 5 hours [14] (Figure 2). The function of this treatment is to reduce soluble cellulose, hemi cellulose, pectin, lignin, etc., from fiber.

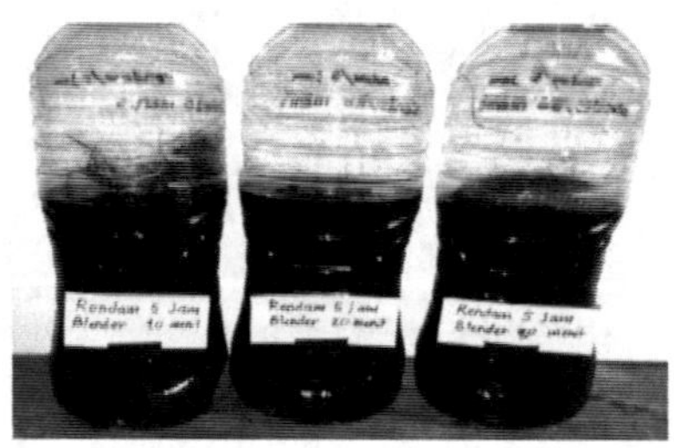

Figure 2. Soaking fiber in 5% NaOH solution

2.3. Fiber fibrillation

Fiber that has been pre-treatment is further process of fibrillation. Fiber fibrillation process is done semi-mechanically; it is by blending fibers at 2000 rpm for 10, 20 and 30 minutes (Figure 3). This process aims to reduce fiber diameter. The fibrillated fibers are then washed with distilled water and dried for 3 days prior to composite fabricating.

IOP Conf. Series: Materials Science and Engineering **334** (2018) 012079 doi:10.1088/1757-899X/334/1/012079

Figure 3. Process fiber blender

2.4. Composite preparation

The formation or printing of composite hybrid panels is carried out using a mould made of steel. Various composite materials were fabricated by hand lay up technique at room temperature (Figure. 4). Epoxy resin and hardener were mixed in a bowl to prepare the matrix materials. Coir fibers were mixed and added in to matrix. Hybrid composite was fabricated with 30% volume fraction of fiber and 70% volume fraction of matrix. A well-mixed matrix and fibers is poured into the mould. Steps involved in fabrication of hybrid composites are shown in Figure.4.

Mould to fabricated composite

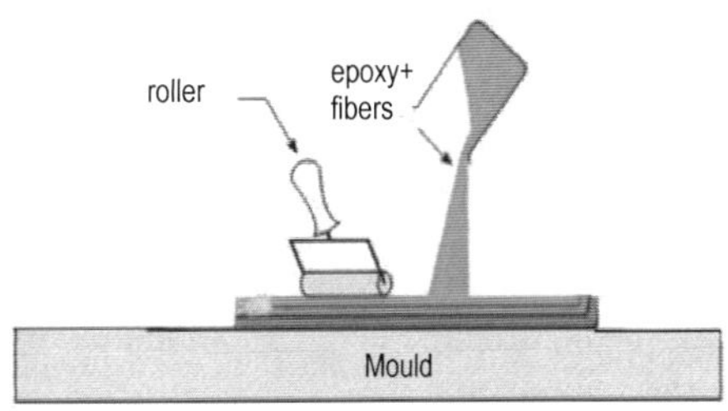

Hand lay up technique

Mixture of matrix and reinforcement filled in mould

Fabricated composite specimen

Figure 4. Various stages of composite fabrication

2.5. Testing of the composites

The fabricated composite is formed into a specimen for mechanical testing; they are impact and flexural test. The flexural test (three- point bend) was carried out according to ASTM D790 standard and impact test with Charpy method used ASTM D6110 standard. The fractured surface morphology of the composite specimens was examined using digital microscope. Each variation was made 5 specimens. Figure 5 shows samples of impact and flexural test specimens that have been formed.

Spesimen impact test Spesimen flextural test

Figure 5. Sample specimen impact and flexural test

3. Results and Discussion

Composites with a ratio of volume fraction of 30% fiber and 70% epoxy matrix with various variations of fiber treatment have been tested impact and bending. From the results of hybrid composite test of epoxy reinforced coconut fiber for various variations of fiber treatment to impact and flexural strength can be tabulated in Table 1.

Table 1. Mechanical properties of epoxy and coir composites

Type of composites	Fibrillation (minute)	Impact strength (kJ/m^2)	Flexural strength (Mpa)
Fibrillation Untreated Coir Fiber (FUCF)	0	9.63	42.81
Fibrillation Treated Coir Fiber (FTCF)	10	24.45	87.91
	20	18.96	72.42
	30	17.46	51.37

3.1. Impact strength

Figure 6 shows coir fiber reinforced composite without fibrillation treatment (FUCF) having a significantly lower impact strength compared with fiber treatment with fibrillation (FTCF), reaching 100%. In the composites with fiber fibrillation treatment there is a decrease in the value of impact energy and impact strength as the fibrillation time increases. The highest value of impact energy measured is 14.06 joule, which is induced by the 10 minutes of fiber fibrillation. Meanwhile, the lowest value of impact energy measured is 10, 04 joule that is induced by the 30 minute of fiber fibrillation.

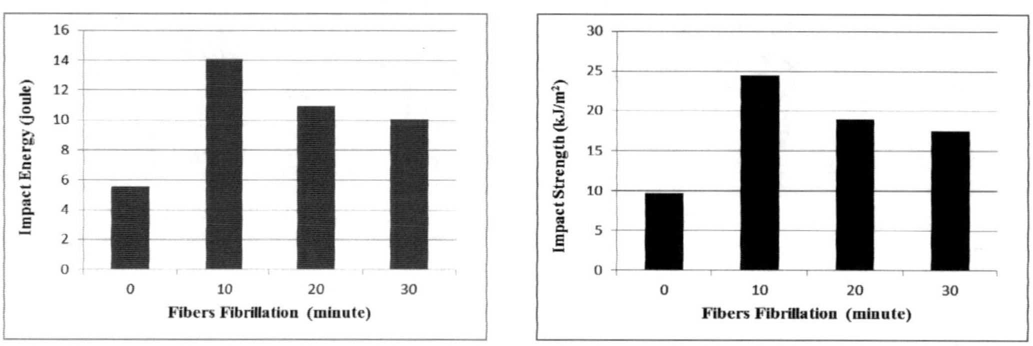

(a) Impact energy (b) Impact strength

Figure 6. The effect of fiber fibrillation on impact energy and strength of composites

The mean impact strength of FTCF composites is increased by 110% against FUCF composites. The highest increase of 153.75% occurred in FTCF composites around 10 minutes of fibrillation, followed by fibrillation of 20 and 30 minutes (96.75% and 81.23%). The highest impact strength was obtained on the composite with FTCF for 10 minutes, i.e. 24.45 kJ/m^2. Decrease in impact strength occurs in FTCF for 20 and 30 minutes. Decline in composite impact strength value with FTCF for 30 minutes as 8% against FTCF for 20 minutes. From the impact strength result, there is a significant effect of fiber treatment with fibrillation using a blender on the impact strength of the epoxy composite hybrid.

Impact strength of a coir fiber reinforced epoxy composites is a measure of the ability of the composites to resist fracture failure at high velocity under a sudden applied force. Impact energy also defined the toughness of the composites. In this case, the increase in impact strength value is unable to be separated from the reduced factor fiber diameter after experiencing the process of fibrillation using blender. The contribution is longer will damage the fiber structure thereby decreasing the value of the impact strength of coir fiber reinforced polymer hybrid composites.

Treatment of fiber with NaOH and followed by fibrillation process using blender. After fibrillation due to the reduced diameter of the fibre, the aspect ratio of the fiber increases and yields rough surface topography, which in turn offers a better fibre-matrix interface. This also results when obtaining the enhanced properties. The effect of fibrillation in modifying surface of fibre can be observed by digital microscopy as shown in the fig 7. On the fracture surface, Visible fiber break and no visible hole marks due to fiber release (pull outs bonding).

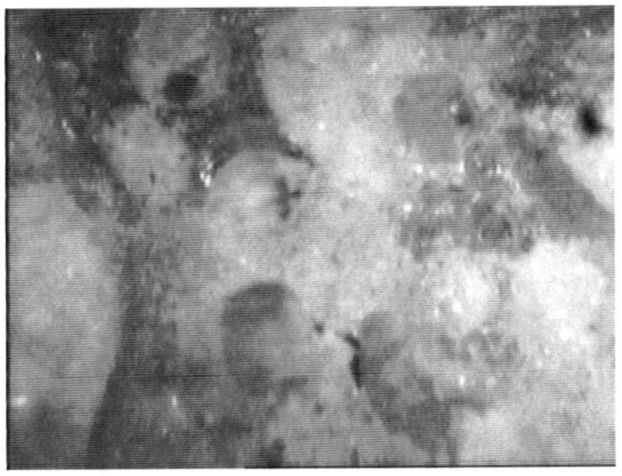

Figure 7. Surface morphology of the hybrid composite epoxy

3.2. Flexural strength

The result of flexural strength testing for various fiber fibrillation treatment variations on composite hybrids has been tabulated in table 1 and shown in Figure 8. In Figure 8, the flexural strength trend is similar to the impact strength. Composites with FTCF have a higher flexural strength of 51% compared to composites with FUCF.

Composite with FTCF for 10 minutes has the highest flexural strength, which is 87.91 MPa and the lowest in FTCF for 30 minutes, i.e. 51.37 MPa. Flexural strength will decrease with increasing time of fibrillation. Flexural strength decreased 17.62% and 29.1% in 20 and 30 minutes of fiber fibrillation. Decreased flexural strength at 20 and 30 minutes of fiber fibrillation due to fiber damage due to blender or fibrillation process. Therefore, to increase flexural strength in composites with FTCF compared to FUCF indicates the presence of fiber fibrillation treatment. The fiber diameter becomes smaller so that it can increase the interface between fiber and matrix.

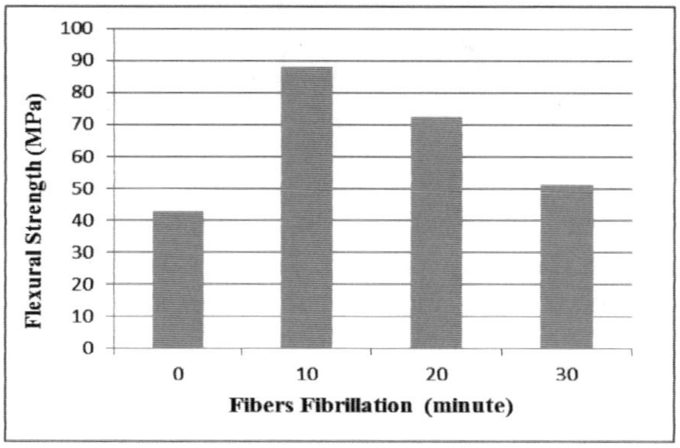

Figure 8. The effect of fiber fibrillation on flexural strength of composites

4. Conclusions

The results of the impact and flexural test by varying the length of fiber fibrillation treatment in coir fibers reinforced epoxy hybrid composites can be concluded that the successful fabrication of coir fibers reinforced hybrid composites can be achieved by hand lay up method. It is also shown that the highest impact and flexural strength was produced by composite with FTCF for 10 minutes, i.e. 24.45 kJ/m^2 and 87.91 MPa. In addition, for 10 minutes fiber fibrillation can produce better impact and flexural strength than 20 and 30 minutes. Treatment of Fiber fibrillation treatment can improve the fiber-matrix interface, so that it is able to raise the composite mechanical properties.

References

[1] P A Udaykumar, Rajanna S, Ramalingaiah 2014 *IJRET* **03** 37
[2] Asasutjarit, Chankan, Charoenvai, Sarocha, Hirunlabh, Jongit, Khendari, Joseph 2009 *ELSEVIER Composite Part B* **7** 633-637
[3] B Bakri, S Chandrabakty, R Soe 2015 *IJJSMM* **2** 132-135
[4] D Verma, P C Gope, A Shandilya, A Gupta, M K Maheshwari 2013 *J. Mater Environ Sci.* **4** 263-276
[5] M Sakthivel, S Ramesh 2013 *SCIENCE PARK* **1** 1-6
[6] A Karthikeyan, K Balamurugan and A Kalpana 2013 *Journal of Scientific &Industrial Research* **72** 132-136.
[7] S Harish, D Peter Michael, A Bensely, D Mohan Lal, A Rajadurai 2009 *Materials Characterization* **60** 44–49.
[8] Akash, Venkatesh Gupta, K V Sreenivas Rao, Prasad C B, Prabilsonkhadka 2015 *International Journal of Applied Engineering Research* **10** 55.
[9] Alex S, Stanly Johns Retnam, M Ramachandran 2015 *International Journal of Applied Engineering Research* **10** 11.
[10] R Narendar, K Priya Dasan, Muraleedharan Nair 2014 *Materials and Design* **54** 644–651.
[11] P Noorunnisa Khanam, H P S Abdul Khalil, G Ramachandra Reddy, S Venkata Naidu 2011 *J Polym Environ* **19** 115-119.
[12] Xuefeng Zhao, Robert K Y Li, Shu-Lin Bai 2014 *Elsevier Composite Part A* 65 169-174.
[13] Gery P, Heru SBR 2014 *Prosiding SNaPP 2014* (Bandung: Unisba)
[14] Indra M 2009 *Prosiding Seminar Nasional* (Medan: UISU)

IOP Conf. Series: Materials Science and Engineering **334** (2018) 012080 doi:10.1088/1757-899X/334/1/012080

Investigation on the feasibility of coffee husk (endocarp) as efficient filler material for enhancing physical and mechanical properties of styrofoam based particleboard

Milawarni[1,2], Nurlaili and Ernayusnianti[1]
[1] Departement of Electrical Engineering, Lhokseumawe State Polytechnic, Buketrata, Lhokseumawe, Aceh, Indonesia

E-mail: wmila12345@gmail.com

Abstract. This research focuses on introducing a coffee husk as viable and efficient filler for enhancing physical and mechanical properties of Styrofoam based particleboard. Heat treatment method was adopted to produce the-particleboard from the mixture of coffee husk (CH) with Styrofoam (PS). Styrofoam is material derived from polystyrene. The aim of this research is to get the appropriate weight composition between coffee husks with PS and to identify the physical and mechanical properties of the produced particleboard. The composition of coffee husk varies between 0-90%wt. The manufacture of particleboard i.e. coffee husk milled with size 20/10 mesh then soak with 10% NaOH for 2 hours, rinsed with clean water and dried and weight according to the composition. The mixture of CH and PS is inserted into mold and put into hot-press. The result shows from physical properties that density, water absorption and thick development test corresponding with SNI 03-2105-2006 standard, the mechanical properties shows MOR test meets the standard on the addition of CH 10-50%, while the MOE test has not meet the standard.

1. Introduction

Central Aceh is the largest Arabica coffee-producer region in Indonesia [1]. The abundance of waste of coffee processing (endocarp) is a huge potential that has only been burned for free. The use of coffee husk (endocarp) as a raw material for making particleboard is considered as one of the best choice. The particleboard is a board product that is produced by compressing the wood particles and simultaneously binding with an adhesive [2].

The development of particleboard technology is now beginning to shift from composite materials to synthetic fibres to natural fibre-making materials [3]. Natural fibre-reinforced particleboard began to be observed by various industries, such as railway industry, ships, automotive, sports, and construction of civil buildings even to household industries. Consideration of fibre selection for composites is strongly influenced by several parameters such as strength and modulus of elasticity, elongation when fracture, thermal stability, bonds between fibres, matrix, dynamic behaviour, density, price, process cost, availability and ease of recycle [4].

Styrofoam made from raw polystyrene (PS) is an inelastic plastic, so it is able to obtained that is not vulnerable essential plastic addition. Styrofoam, which was originally fragile has made it more plastic with the addition of plastic material that is dioctyl phthalate (DOP) [5].

[2] To whom any correspondence should be addressed.

IOP Conf. Series: Materials Science and Engineering **334** (2018) 012080 doi:10.1088/1757-899X/334/1/012080

The fundamental of this study was conducted to examine the extent to which coffee husk and PS can be used as a new material particleboard. Particleboard of fibre reinforced coffee husk and PS as a matrix; it is used as an engineering material, in terms of physical and mechanical properties. The purpose of this research is to make particleboard material with Styrofoam binder. In addition to obtaining a suitable weight volume composition between the husk fibre and Styrofoam fibres which will serve as particleboard that meets JIS A 5908; 2003 standards.

2. Experiment

This section briefly discusses the preparation of specimen and the procedure of the research, which are including the provision of CH, provision of PS and finally the preparation and testing of the samples.

2.1 Speciment Preparation

The used materials in this study are Styrofoam (PS), toluena, coffee husk (*endocarp*), and clean water (distilled water). While the used the equipment are 500 ml beaker glass, sieves, spatula, analytical balance, hot plate, moulds, electronic system universal tensile machine type SC-2DE, and aluminum foil.

2.2. Research Procedure

The Preparation of Material Composition: the percentage of Coffe husks weighed (0,10,20,30,40,50,60,70,80,90) %. The percent of PS weighed (100,90,80,70,60,50,40,30,20,10) %.

2.2.1. *Provision coffee husk (CH)* Coffee husk was left in open air for one week, and then soaked in the 10%NaOH then it was washed with water until its pH is neutral. Additionally, it was dried in 800^0C oven. After that it was milled (20/10 mesh) and sieved so it was aready-made coffee parchment skin. Alkali treatment improves the mechanical bond better [8].

2.2.2. *Provision of Styrofoam (PS)* The used Styrofoam obtained from the packaging waste is washed and dried then cut to size 0.5 x 0.5 cm and weighed according to the composition, then dissolved with toluene and then added 5% MEXPO catalyst. This mixture is stirred using mixer until evenly distributed.

2.2.3. *Preparation of Samples* The mixture is between coffee husks and was ready to use PS. The PS was stirred until it was homogeneous. Then put in mould. The samples were in hot-press at a temperature of 170^0C and pressure 25 kgf/cm^3. Once it was completed, the hot press was turned off and the samples were removed and conditioned at room temperature for 7 days and cut to its standard.

2.2.4. *Testing* Physical testing was performed to test the density, water absorption and the thick development and mechanical testing including the Mechanical of Rapture (MOR) and Modulus of elasticity (MOE) were done by mean of the Electronic System Universal Testing Machine Type SC-2DE MFG No.6079 ASTM D 3039 [9].

3. Results and discussion

The effect of added composition of coffe husk weighted the percentage on physical and mechanichal properties such as density, water absorption, tick development, modulus of raputer (MOR) and modulus of elasticity (MOE), respectively as follow,

3.1 Density

The test results of the density is illustrated in Figure 1.

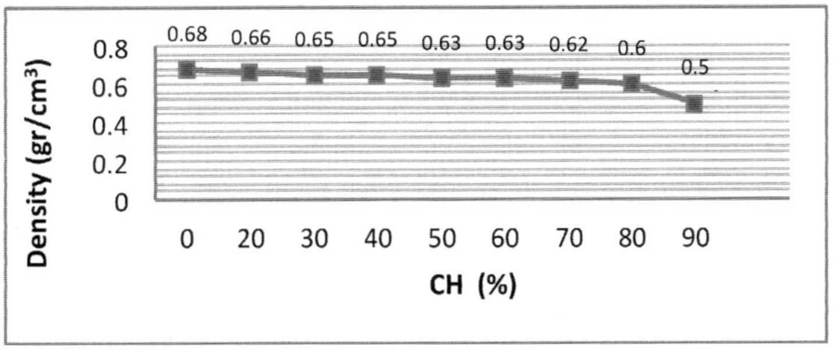

Figure 1. Graph of the density of the weight percent of the coffee husk

Figure 1 shows that the presence coffee husk influenceds its density value, including the value of density was lowest for the 90% CH of 0.51 g/cm^3, while the value of the highest densities obtained in the composition 0% CH by 0,68 g/cm^3. The density value was affected by the volume fraction of the sample, a decrease in density along with weight reduction of pp and additional of husks particles coffee resulted in the addition of a volume fraction, this is due to differences in density PP and shell of coffee beans (endocarp), with the same mass between PP and coffee husks but the second volume was different therefore PP reduction and the additional of coffee husks in the same weight fraction can increase the volume of the resulting particleboard, and this would affect the value of density. The greater volume produced, the density value would be diminished.

Density of the particleboard tends to increase along with the addition of adhesive, this occurs due to physical force between the adhesive interactions with the filler through cavities that filled it. Moreover, the results indicate that the presence of the addition coffee husk (CH) can improve the physical properties of the resulting particleboard, but if the particles used has exceeded the limits of the binding matrix to fiber, composites produced is damaged. All composition have met the SNI 03-2105-2006 standard [10].

Density of the particleboard tends to increase along with the addition of adhesive, this happens due to physical force between the adhesive interactions with the filler through cavities that filled it. These results indicate that the presence of the addition coffee husk (CH) can improve the physical properties of the resulting particleboard, but if the particles used has exceeded the limits of the binding matrix to fiber, composites produced will be damaged. All composition have met the SNI 03-2105-2006 standard [10].

3.2. Water Absorption

The water absorption resulting samples ranged 0.01-7.6 %. Figure 2 illustrates the results of the water absorption weighted the coffee husk percentage.

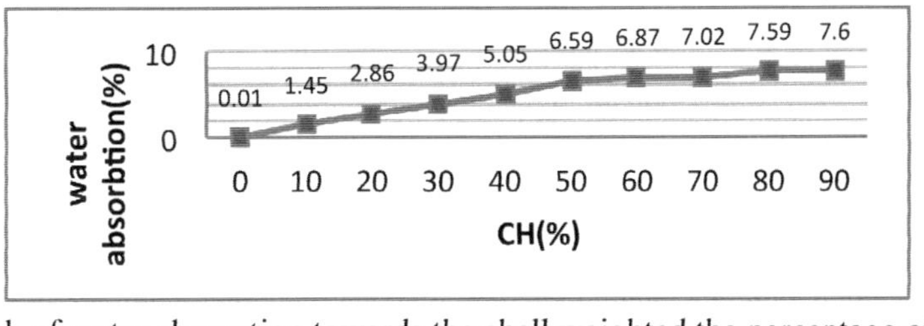

Figure 2. Graph of water absorption towards the shell weighted the percentage of coffee beans (endocarp).

Figures 2 illustrates that the minimum water absorption in 0% CH in this sample it still dominated by the adhesive properties of density Styrofoam and its high density was 0,9 g/cm^3. It was difficult to make it into the water between the particles. Moreover, the effectiveness of the used of Styrofoam as the matrix is suitable, this is due to the hydrophobic properties of Styrofoam. The particleboard is not easy to absorb water from the environment. The slightly increased of water content occured in the 10-90% so that the content of the particles reduced the resistance of PS mixture which caused erosion. So with reduced Styrofoam materials, the percentage of water absorption becomes smaller.

Water absorption is highest in the 90% CH, where coffee husks a material that tends to absorb water, therefore the increase in the percentage of coffee husks lead to the increasing of water in the sample during the manufacturing process. Water absorption increases with the addition of skinweight percentage of the coffee, this is because the skin of coffee is a lightweight aggregate that has many pores, so that the water absorption percentage is greater than the water absorption without fiber. SNI 03-2105-2006, particleboard, requires the water absorption value for all samples <14%. From the test results indicated that of all particleboard produced met the standards. This result is very suitable to use for interior or exterior panels since the water content is very low.

3.3 Thick Development

Development of a minimum thickness of 0.4% was in 0% CH and a maximum of 0.04% was in the sample 5. When it is compared with SNI 03-2105-2006, thick indigo development of the required maximum was 12%, it is thus the particleboard can be said to have fulfilled standard for all compositions on the test thickness swelling as illustrated in Figure 3.

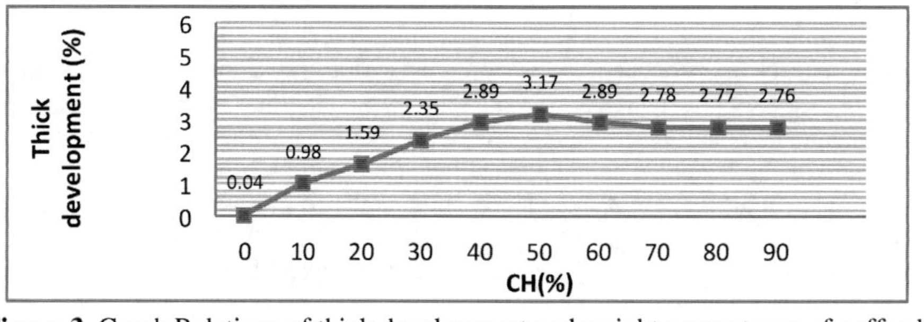

Figure 3. Graph Relations of thick development and weight percentages of coffee husk (CH).

Figure 3 illustrates that the thick development particleboard tends to increase with the increasing weight percentage of coffee husk (endocarp). This is because the nature of coffee husk that absorbs water (hydrophilic) so the amount of water absorbed more and more resulted in the development of particleboards. However, if it is compared to the SNI 03-2105-2006 standard, which requires maximum thickness development value more than 12%, then the rate of progression thickness development of particleboard have met the standard.

3.4. Modulus of Rapture (MOR)

The value of the maximum MOR is 40% CH, which amounted to 99 kgf/cm^2 and the lowest flexural strength values obtained in 90% CH is 90 kgf/cm^2. Figure 4 illustrates that the addition of particle composition of CH tends to increase MOR strength. This suggests that the presence of CH may increase the MOR of the particleboard material. The maximum MOR strength results on samples is 40% CH due to the adhesive force strong enough on Styrofoam, which creates better bonding. In composition of 60-90% CH MOR values begin to decrease; this occurs due to not maximal PS binds the CH. The bond between the particle and matrix PS is easily discharge. Resulting in the emergence of shear stress. The failure is dominated by loose bonding of particles and matrix. This is often called "fibre pull out". Particleboard with 10-50% CH composition meets SNI 03-2105-2006 standard. The particleboard requires minimum MOR value of 82 kgf/cm^2. The probability SEM viewing on the board at its breaking section explains the increase in mechanical properties.

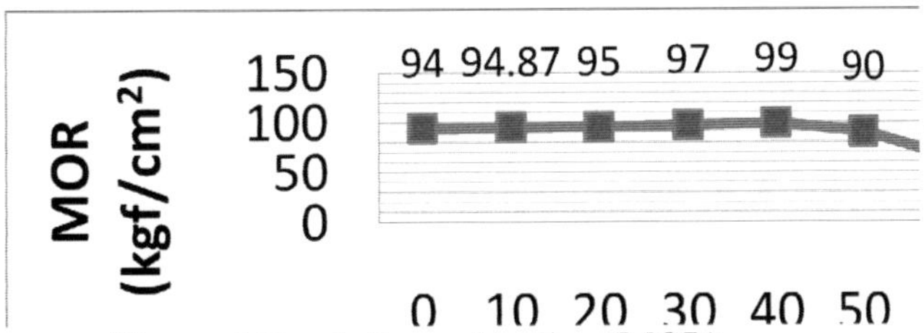

Figure 4 Graph flexural testing (MOR)

3.5. Modulus Of Elasticity (MOE)

MOE maximum possible value currently on the sample 5 with a composition of 70:30, in the amount of 27987.90 kgf/cm2 and the lowest value of MOE obtained on the sample 1 with a composition of 30:70 is 25416.67 kgf/cm^2.

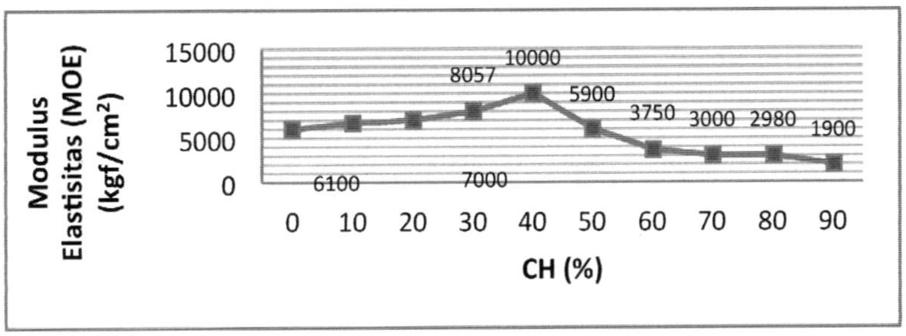

Figure 5 Graph Testing Modulus Of Elasticity (MOE)

Figure 5 illustrates the MOE that tends to increase with the increasing percentage of heavy CH. This is caused by the skin effect of coffee, which is natural fibber that has good elasticity. The increase in MOE value tends to increase until the composition of 40% CH. However, at 50% interval MOE value decline, this is due to the gap on the particles so that when the composite is loaded then the tension moves the void area and reduces the strength of particleboard. All samples have not met the requirement of SNI i.e., minimum 20.400 kgf/cm^2.

4. Conclusion

The results that obtained from present work can be summarized as follow; Physical test results such as density, water absorption and thickness development of coffee husk (CH) and Styrofoam (PS) particleboard have met the SNI 03-2105-2006 standards. Increased coffee husk percentage composition indicated that the addition of particle composition CH tends to increase MOR strength. This suggests that the presence of CH may increase the MOR of the particleboard material. The maximum MOR strength results on samples 40% CH due to the adhesive force strong enough on Styrofoam, which creates better bonding. In composition 60-90% CH MOR values begin to decrease. This occurs due to not maximal PS binds CH. The bond between the particle and matrix PS is easily discharge. Resulting in the emergence of shear stress. The failure is dominated by loose bonding of particles and matrix. This is often called "fibre pull out". Particleboard with 10-50% CH composition has met the SNI 03-2105-2006 standard.

In addition, the increase in water content and thickness development decrease the value of MOR and MOE. Finally, The result of mechanical test of bending strength of CH-PS particleboard fulfilling SNI 03-2105-2006 standard is 10-50% KTBK composition, while 60-90% KTKB has not met the standard.

Reference

[1] Surip, M, A Brief History of Development of the coffee in the Gayo highlands in: Free Cultivation and processing of Coffee Arabica Gayo, (2008) ICRRI.

[2] Raudah and Ernawati. 2012. Utilization of Arabica Coffee Leather Pulping Processes for making bioethanol. Reactions Journal Vol. 10, No. 1, June 2012, ISSN 1693-248x. Chemical Engineering Department of the Polytechnic Lhokseumawe.

[3] Siswati, N. D, Orphans, M. Hidayanto, R, 2011, Bioethanol from Waste Leather Coffee with Fermentation. Faculty of Industrial Technology, National Development University "Veteran" East Java.

[4] Sudarsono, T. R. (2010). The making of particelboard based of coconut husks as the natural bond. *Jurnal Teknologi* , 22-32.

[5] Milawarni 2013. Preparation and Characterization of Polymer Composite Tile Using Long Fiber Coconut Fiber With Polypropylene Resin. Proceeding SNYuBe, ISBN 978-602-2-2. Polytechnic State Lhokseumawe.

[6] Odi Iskandar, 2009. Test Potential Waste Coffee for Particle Board Using UF, PF and Thermoplastics Journal of Technology, 23 (2009), 12-19, july-24-2009

[7] Enisafriana, 2013. Making the Absorbent Material Particle Board As Beep Using Coffee Waste formaldehyde adhesives. (Theses, Yogyakarta, UGM).

[8] George J., Janardhan R., J.S., Bhagawan S.S., Thomas S.1996 Melt Rheological behavior of Short Pineaple Fiber Reinforced low density polyethylene Composite, Journal of polymer, Volume 37, No.24, Get Britain.

[9] ASTM, 1987, Annual Standard Book Composite Material, ASTM D 3039.

[10] Standard National Indonesia (SNI), Indonesia standard of Particle Board SNI 03-2105-2006

IOP Conf. Series: Materials Science and Engineering **334** (2018) 012081 doi:10.1088/1757-899X/334/1/012081

Mechanical properties evaluation of single and hybrid composites polyester reinforced bamboo, PALF and coir fiber

T Rihayat[1,*], S Suryani[1,2] , T Fauzi[1], H Agusnar[2], B Wirjosentono[2], Syafruddin[1], Helmi[1], Zulkifli[1] and P N Alam[3], M Sami[1]

[1] Departement of Chemical Engineering,
Politeknik Negeri Lhokseumawe, 24301, Aceh Indonesia
[2] Department of Chemistry, Faculty of Mathematics and Natural Science,
Universitas Sumatera Utara, 20155, Medan, Indonesia
[3] Department of Chemical Engineering, Faculty of Engineering
Univeritas Syiah Kuala, Darussalam, Banda Aceh

[*]Email : teukurihayat@yahoo.com

Abstract. This study aims to determine the composition fiber natural of bamboo, pineapple leaf and coir in single and hybrid composite to see the best characteristics of tensile strength and flexural test by using a Universal Testing Machine (UTM) and observe the effect on the microstructure of the composite through optical and scanning electron microscopy. Bamboo, Palf and coir have synthesis from natural fiber was used as reinforcement in polyester composite using hand lay up or a hot-compression moulding while filler:matrix was used (45%:55wt.%, 70%:30wt.% and 15%:85wt.%). From the variation of the volume fraction between filler and matrix show that mechanical properties of composites increased with increasing amount of filler in the matrix. This is evidenced by the high mechanical properties A:B:C/Ps in compositions 45%: 55wt.% 136 Mpa while flexural strength 93 N and good structure surface morphology. This research has produced a hybrid composite materials that have high mechanical properties and bending compared with conventional synthetic fibers and other materials.

1. Introduction

Regardless of environmental benefits, the availability of abundant raw material resources and renewable instead of fossil resources, low cost and high specific strength while due low density [1,2] when compared to the classic mineral filler / plastic composite. Plastic reinfored fibers have been widely used in various industries such as the aircraft industry, automobile, recreational, electronic, medical and furniture. Composites industry is always looking for alternative sources of low cost, which can decrease overall manufacturing costs. Natural fiber polymer composite hybrid produces a variety of designs to customize and achieve cost-effective composite properties as compared to a single fiber as filler dispersed in the matrix. merging two or more fibers in a matrix called a hybrid composite [3].

The advantages of natural fibers joined to form hybrid composites have been reported previously. Therefore, the natural fiber is now a new alternative materials that can replace petroleum-based products either alone or combined with other ingredients to produce green composites such as bambooa fast-growing plants, bio abundant resources included in the family bambusae, under the genus Gramineae. Variations of structural, mechanical properties, fiber extraction, chemical modification, and thermal properties that make it a versatile material for composite applications-industries has been investigated [4,5], Pineapple leaf fibers rich cellulose content, are relatively cheap and abundant availability and potential for polymer reinforcement filler. Currently pineapple leaf fiber is a waste product of cultivation of pineapple and therefore relatively inexpensive. Coir fibers lies between the skin and the outer shell of the coconut [6] the use of coconut fiber, including its use as reinforcement in polymer composites have been investigated [7],

Advantages of natural fibers, such as low cost, low density, availability, sustainability, recycling and biodegradability, making it an area of research that is interesting and great effort is made to exploit the potential of the source of reinforcing material polymer in the form of bamboo, pineapple leaf and coconut fiber to be used as filler to development of fiber as a renewable natural materials in polymer composites. Therefore, in this study the synthesis of the fiber of bamboo, pineapple leaf and coconut fiber as filler in fiber-reinforced polymer single and hybrid to be seen also compares the characterization of the mechanical properties of tensile strength, flexural and morphology by scanning electron microscopy for each material composites.

2. Methodelogy/Experimental

A. *Samples*

In this study the bamboo fiber (namely the first fiber = A) and pineapple leaf fiber fiber / PALF is collected from Bener meriah (it was namely the second fiber = B) while coir was called C). Matrix used is polyester resin (Ps) from Sigma Aldrich with etyl methyl ketone peroxide as a hardener. Physical, mechanical, and thermal natural fiber is strongly influenced by the composition of chemical constituents. Cellulose as the major structural component gives strength to the walls of plant and fiber [8]. Preparations of Bamboo, PALF and Coir describes in figure 1.

B. *Fabrication of the composite*

Reinforced polyester composite material of natural bamboo fiber, PALF and coir is fabricated by hand lay-up method and hot compression molding. the mixture of raw material bamboo, PALF and coir as filler in Isothalic polyester resin according with volume fraction and added 1% catalyst Methyl Ethyl Ketone Peroxide (MEKP) from the total weight of the matrix in 25cm × 5cm × 0,32cm.. the rule of mixtures such as bamboo-PLF-coir fiber, each fiber type consisted of (45%: 55 wt.%, 30%: 70wt.% and 15%: 85wt.%) for single fiber in polyester resin, during the hybrid composites while combination three fibers in polyester resin is consisted of (15%: 15%: 15%: 55.% wt, 10%: 10%: 10%: 70wt.%, 5%: 5%: 5%: 85wt .%), the composite is molded by a hot compression with a temperature of 85 ^{0}C [9] for 15 minutes. Specimens have been removed from the mold were cutting According with ASTM D 638 for tensile and flexural strength D 790 standard method.

C. *Tensile and flexural test*

Fig 2 (a) and (c) testing the tensile properties of the hybrid composite material refers to standard ASTM D 638. The tensile strength is measured using by universal testing machine (UTM) Exceed Model E43. Dimensions, gauge length and cross-head speed selected according to standard ASTM D-638. Fig 4 (a) and (b) flexural tests are performed in the same instrument According to ASTM D790 at a cross head speed of 0,8 mm / min and a gauge length of 25 mm.

D. *Scanning electron microscopy*

Scanning electron microscopy (SEM) was performed on a microscope JEOL-T20(Japan), operating at 20 kV accelerating voltage, in high vacuum mode, to study the morphology of the fracture surface of a

IOP Conf. Series: Materials Science and Engineering **334** (2018) 012081 doi:10.1088/1757-899X/334/1/012081

sample. Fractures hybrid composite sample surface was observed using SEM to investigate the adhesion of the interface between the fiber and matrix hybrid composites. For this purpose, the sample is mounted on bronze stubs, using double-sided tape, and coated with a layer of gold (40-50 nm).

Bamboo was cut into a small pieces

Insert bamboo to decortication machine, usually it was called "Ketam"

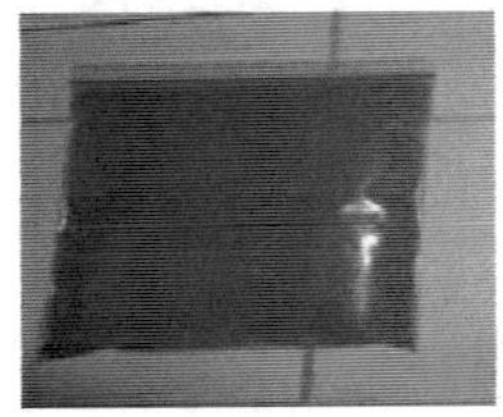
bamboo fiber

PALF is scrapped the bottom layer until the fiber pull out from the layer

Fiber is crushed by the machine to Obtain fiber in powder

PALF fiber

Coir fiber is collected from various local source

Fiber is separated from the shell of coconut insert in crusher machine

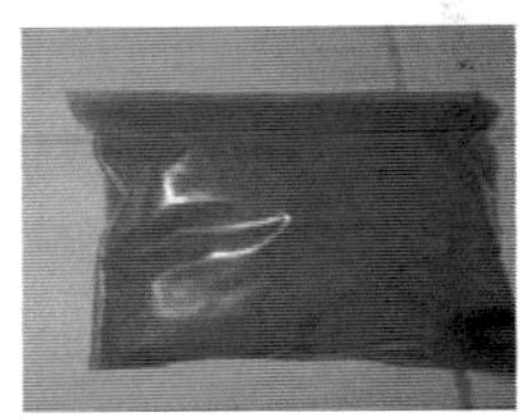
Coir fiber

Figure 1. Preparations of Bamboo, PALF and Coir

3. Results and Discussions

a. Tensile strength

Tensile testing of fiber reinforced composite material nature ployester analyzed using a Universal Testing Machine (UTM) Model E43 with a specimen sample used is ASTM D-638 is shown in Figure 1. Samples to be tested are clamped with a clamp located on the tool and the sample is pulled up to drop out at a rate of time t is 0.8 mm / s. The results are shown in graphical form when the sample experienced a tensile strain or stress vs. strain values at the break.

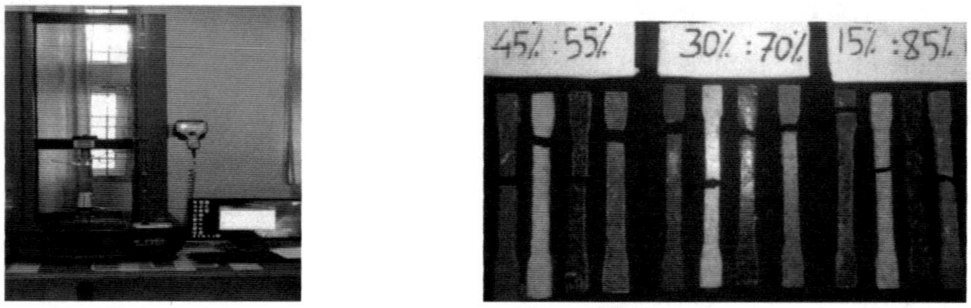

II. (b)

Figure 2. (a) Equipment sample test, (b) sample specimens ASTM D-638

Data from a composite material tensile test are compared with the theoretical data it shown in figure 2. Theoretical calculations for single fiber in matrix can be calculated using by equation (1) [10], during the hybrid composites consisting of a mixture of two or three fibers in the matrix is calculated using equation (2). The result of test composite polyester is presented in Table 1.

$$\sigma c = \sigma m V m + \sigma f V f \quad \text{.....................} \quad (1)$$
$$\sigma H = \sigma a V f a + \sigma b V f b + \sigma c V f c + \sigma m\,(1 - V f a - V f b - V f c) \quad \text{....................} \quad (2).$$

Table 1. Result of Tensile Strength Test

The volume fraction (Filler: Matrix)	Tensile strenght (Mpa)				Flexural strength (KN)			
	A/Ps	B/Ps	C/Ps	A: B: C/Ps	A/Ps	B/Ps	C/Ps	A: B: C/Ps
45%: 55%	255	116	92	136	91	74	47	93
30%: 70%	183	81	58	103	79	62	33	89
15%: 85%	129	61	30	90	68	45	24	83

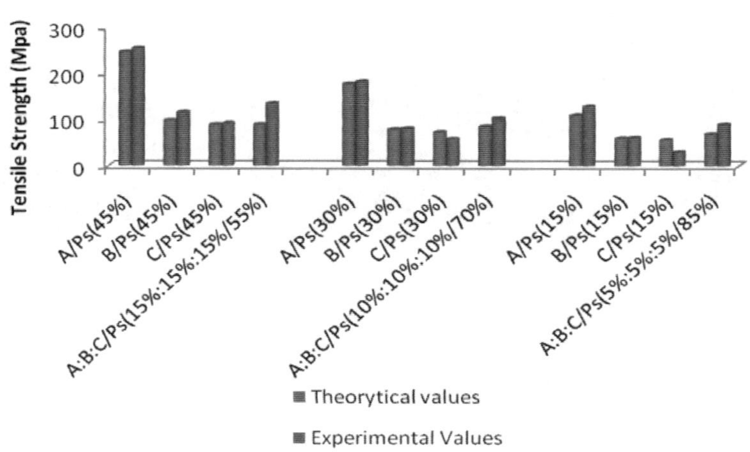

Figure 3. Comparison of experimental and theoretical tensile strength (MPa) versus Natural fiber loading (wt.%).

Based on Figure 3. of testing the tensile strength of the hybrid composite material polyester reinforced bamboo, PALF and coir fiber while the ratio of 45% fiber and 55% matrix has a tensile strength of 136 MPa. The tensile strength of 103 MPa produced from the hybrid composite ratio of 30%: 70% and 90 MPa for a ratio 15%: 85. This shows that hybrid composites are capable of producing the tensile strength values are better than the single fiber mixture of coconut coir (C: Matrix) and PALF fibers (B: Matrix) in the polyester matrix.

Disadvantages of the single fiber composites which is less properties than optimal characteristic can be improved by combining two or more types of fibers in the matrix so that it will produce a composite that has properties better mechanical characteristics. This effect is due to the good mechanical properties of bamboo fiber is associated with the natural fiber composition as shown in Table1[11] its will being hybrid composite polyester has tensile strength with a value of 136 MPa.

However, the tensile properties of hybrid composite polyester under the price of bamboo/ Polyester (A / Ps) and its because of difference volume fraction between mixing the three fibers (A: B: C / Ps) is less than the volume fraction of a single fiber as reinforcement matrix polyester indicating differences between the tensile strength properties of single fiber composites with hybrid composite. Recent studies have produced promising results with hybridization of natural fibers as a reinforcement of the mechanical properties and the impact strength of hybridization bamboo, PALF and coir fiber as reinforcement of polymer composites is very influential as shown in Table 2. Evaluation of the effect of hybridization on the mechanical performance of hybrid bamboo, PALF and coir fiber reinforced polyester composites found the tensile properties of NFCs increased with the addition of bamboo fiber. The mechanical properties of natural fiber is highly dependent on cellulose content results. The different cellulose content affects the mechanical properties of natural fibers [12].

b. flexural strength

Bending is a test to see durability due to loading application of a material is strongly influenced by the physical and mechanical properties of material. The Physical and mechanical properties of a material can be seen when they are done testing. The purpose of doing a mechanical testing is to determine the response material of a construction, component or assembly fabrication when subjected to a load or deformation from the outside [13]. The flexural strength of composite determined from 3-point bend test or with three point bend test on a sample of which is shown in Figure 4. The flexural strength for a variety of single and hybrid fiber composites are presented in Table 2.

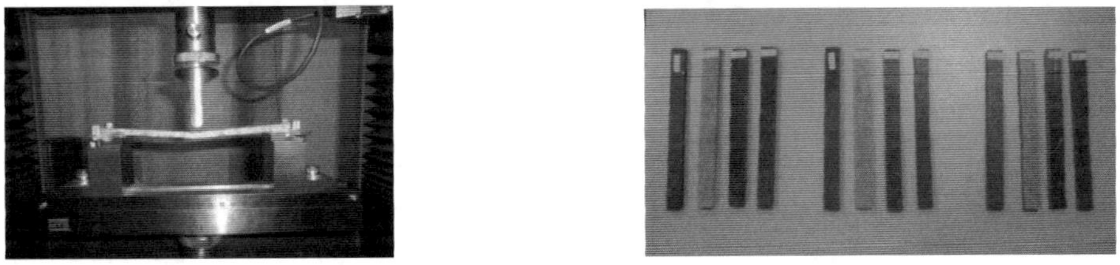

(a) (b)

Figure 4. (a) Example flexural test, (b) flexural test samples ASTM D-790

Flexural properties of fiber reinforced composite samples were single and hybrid obtained optimum results at the composition of the mixture A: B: C/Ps. show directly from testing using Exceed UTM machine Model E43 during the loading of bending to the best samples of hybrid composites.

Comparison of flexural strength of single fiber composite and hybrid polyester are presented in Figure 5. From the figure, confirmed that the bending load hybrid composite is better than single fiber-reinforced polyester composites A/Ps, B/Ps and C/Ps. Increased load bearing capability of composite materials increased significantly with the addition of fiber volume fraction in the mixing to form a composite.

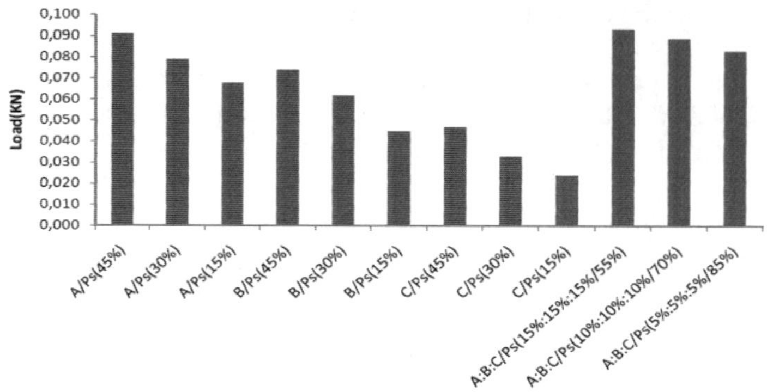

Figure 5. Comparison of Bending Strength of Single Fiber Reinforced Composites

The results showed flexural optimum value for hybrid composite (A: B: C / Ps) is 0.093 KN. As for the single fiber (A/Ps) is 0091 KN, (B/Ps) and (C/Ps) each are having flexural optimum of 74 N and 47 N. Research on hybrid polyester composite bending properties was also investigated by[11], That mixing sisal-jute-GFRP / Ps produces a maximum of 3 KN flexural strength than mixing between the sisal-GFRP / Ps and jute-GFRP / Ps. From these experiment results that the hybrid composite play a role in shaping a composite material that has a value and a better flexural characteristics and this is evidenced by scanning electron microscopy analysis of the results contained in Figure 6. which shows the hybrid composite A:B:C/ Ps have a better surface structure [14].

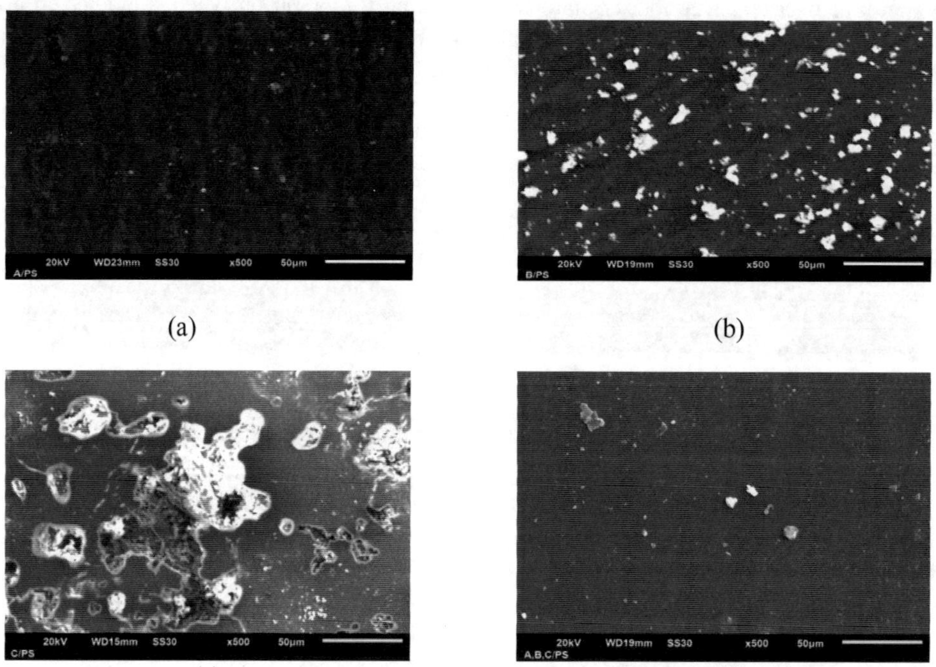

(c) (d)

Figure 6. Analysis of scanning electron microscopy (a) A/Ps 500x, (b) B/Ps 500x, (b) C / Ps 500x and (d) A:B:C/Ps 500x.

From the results of the analysis are shown in Figure 6. (a) that (A/Ps) has a good dispertions fiber in surface structure than fig 6. (b) (B/Ps), which contains a lot of impurities on the surface of the fiber and fig 6. (c) (C/Ps) coir fiber reinforced polyester composites showed some void between fiber and matrix composites that influent interfacial bond and will not be effective when the transfer voltage from the matrix to the fiber, where the maximum usage in the composite fiber strength is achieved and these interactions affect the mechanical properties of composites that have been confirmed by the results of tensile tests in previous discussions [15].

While in Figure 6. (d) hybrid composite (A:B:C/Ps) it is apparent that fibers evenly dispersed into the matrix and provide a flatter surface properties. This shows the interface better adhesion properties, where there is a better interpenetration at the interface with the polyester fiber. This could be due to the high levels of delamination on the outer surface of polyester.

4. Conclusions

Comparison of bamboo fiber, PALF and coir in a matrix of polyester discovered that bamboo / polyester generate the highest characteristic value of 255 MPa compared PALF/ polyester and coir fiber / polyester. The addition of bamboo in the hybrid composite polyester can increases mechanical properties of the composite is 136 MPa.. The structure of the composite morphology indicating the presence of the number of voids and the interlocking matrix interfacial filler thus affecting the mechanical properties of the composite.

5. Acknowledgements

The authors express their gratitude and thanks to the Ministry of Research, Technology and Higher Education for Financial Support through the Stranas Grant, 2017 and the Directorate of Student Affairs and Education, on funding through grants Student Creativity Program (PKM), 2017.

References

[1] Alireza Ashori. 2017. Hybrid Thermoplastic Composites Using Plant Fibers Nonwood. *Hybrid Polymer Composite Materials*, pages: 39-56.

[2] Rihayat, T., Suryani., 2010. Synthesis and properties of biobased polyurethane/clay nanocomposites. *World Academy of Science, Engineering and Technology*. **65**, 918-922

[3] John, MJ, Thomas, S., 2008. Biofibres and Biocomposites. *Carbohydr Polym.*Vol **71**, pages: 343-364.

[4] Okubo, K., Fujii, T., Yamamoto, Y., 2004. Development of Bamboo-Based Polymer Composites and Their Mechanical Properties. *Compos. Part A Appl. Sci. Manuf.* Vol.**35**, pages: 377-383.

[5] Rihayat, T., Suryani, S., Zaimahwati, Z. 2014. Effects of heat treatment on the properties of polyurethane/clay nanocomposites paint. *Applied Mechanics and Materials*. 525, 97-100.

[6] Emad Omrani, Menezes Pradeep L., Pradeep K., Rohatgi. 2016 State Of The Art on the tribological Behavior of Polymer Matrix Composites Reinforced With Natural Fibers in The Green Materials World: Review. *Engineering Science and Technology, an International Journal.* Vol **19**, pages: 717-736.

[7] Yashwanth, MK, Easwara Prasad, GL, and N. K, Akshay. 2016. Comparative Study on the Properties of Coir And Sisal Fiber Reinforced Composites. *Inovative International Journal of Research in Science, Engineering and Technology*. Vol.**5**, Special Issue 9, pages: 992-926.

[8] Ahmed Edhirej., Sweep SM, Jawaid Mohammad., 2017. Cassava / Sugar Palm Cassava Starch Hybrid Fiber Reinforced Composites: Physical, Thermal and Structural Properties. *International Journal of Biological Macromolecules*. Vol.**101**, pages: 75-83.

[9] Akil, HM, Omar, M. F., Mazuki, A. A. M., Safiee. S., Isaac, ZAM, Abu Bakar, A. 2011. Kenaf Fiber Reinforced Composites: A review. *Materials and Design*. Vol.**32**, pages: 4107-4121.

[10] Yusoff, BR, Takagi, H., Nakagaito, AN 2016. Tensile and Flexural Properties of Polylactic Acid-Base Hybrid Green Composites Renforced by Kenaf, Bamboo And Coir Fibers. Industrial Crops and Products. Vol.94, pages: 562-573.

[11] Ramesh. M., Palanikumar. K., Reddy, Hemachandra, K ,. 2013. Evaluation of Mechanical Property-Jute Sisal-Glass Fiber Reinforced Polyester Composites. Composites: Part B. Vol.48, pages: 1-9.

[12] Easwara Prasad, GL, B. S, Gowda Keerthi., Velmurugan, R. 2017. A Study on Impact Strength Characteristics of Coir Polyester Composites. *Procedia Engineering* , Vol.**173**, Pages: 771-777.

[13] Kasim, AN, Congratulations, M. Z, David, MAM, Yaakob, MY, Son, A., Sivakumar, D., 2016. Mechanical Properties of Polypropylene Composites Reinfprced Alkaline Treated with Pineapple Leaf Fiber from Josapine Cultivar. *International Journal of Automative and Mechanical Engineering*. Vol.**13**, pages: 3157-3167.

[14] Umar Nirmal, JH. 2015. A Review on the tribological Performance of Natural Fiber Polymeric Composites. *Tribology International*. Vol 83, pages: 77-104.

[15] Rihayat, T., Suryani. 2013. Morphology properties of polyurethane/clay nanocomposites base on palm oil polyol paint. Advanced Materials Research. 647, 701-704

IOP Conf. Series: Materials Science and Engineering **334** (2018) 012082 doi:10.1088/1757-899X/334/1/012082

Thermal degradation of Shredded Oil Palm Empty Fruit Bunches (SOPEFB) embedded with Cobalt catalyst by Thermogravimetric Analysis (TGA)

R Alias[1,2], N H Hamid[1], J Jaapar[1], M Musa[1], H Alwi[1]and K H Ku Halim[1]

[1] Faculty of Chemical Engineering, Universiti Teknologi MARA, Shah Alam, Malaysia

E-mail: rusmi@salam.uitm.edu.my

Abstract. Thermal behavior and decomposition kinetics of shredded oil palm empty fruit bunches (SOPEFB) were investigated in this study by using thermogravimetric analysis (TGA). The SOPEFB were analyzed under conditions of temperature 30 °C to 900 °C with nitrogen gas flow at 50 ml/min. The SOPEFB were embedded with cobalt (II) nitrate solution with concentration 5%, 10%, 15% and 20%. The TG/DTG curves shows the degradation behavior of SOPEFB following with char production for each heating rate and each concentration of cobalt catalyst. Thermal degradation occurred in three phases, water drying phase, decomposition of hemicellulose and cellulose phase, and lignin decomposition phase. The kinetic equation with relevant parameters described the activation energy required for thermal degradation at the temperature regions of 200 °C to 350 °C. Activation energy (E) for different heating rate with SOPEFB embedded with different concentration of cobalt catalyst showing that the lowest E required was at SOPEFB with 20% concentration of cobalt catalyst..

1. Introduction

Palm oil production in Malaysia listed as second largest exporting country after Indonesia surpassed in 2007, and the Federal Land Development Authority, FELDA is provided 375 hectares for Palm Oil plantations besides the Malaysian Palm Oil Council, MPOC plays the roles as promoter in the vegetable oils and fat market to the international demand [8]. While the production of palm oil in Malaysia rapidly expand, the oil palm empty fruit bunches is considered as solid waste from palm oil industry gives benefits by using process of thermal degradation to converts the biomass to bio-oil.

Before the researcher discovered the specialty of OPEFB in bio-oils, OPEFB was used as organic fertilizer and burned in oil palm mills as a source of heat and energy [2]. OPEFB is considered as fibrous material and has 65% of moisture content. Based on these, the pre-treatment of OPEFB is contains almost zero sand and other contaminants (Al).

There are three components mainly consist in oil palm empty fruit bunches which are hemicellulose, cellulose, and lignin that effects the texture, physical properties and thermal and combustion reactivity. To converts biomass into bio-fuels, these three components must be decomposed properly. The idea to decomposed these components is by pyrolysis which from Thermo-

[2] To whom any correspondence should be addressed.

Gravimetric Analyzer, TGA and through thermal degradation process. The data from the TGA shows that the decomposition of hemicellulose occurs at temperatures 160-360 °C, while volatilization of cellulose at temperature 240 - 400 °C and decomposition of lignin are at temperature 200-700 °C [10].

To upgraded bio-oil into these conditions, the biomass is go through under catalytic fast pyrolysis and considered as most promising way [13]. To enhanced bio-oils, the catalyst used should be selected based on highly active, resistant to deactivation, readily recycled and cheap. The active metals recommended are cobalt, ferum, zinc and etc, is used in this catalytic fast pyrolysis. The purpose of this study is to investigate the thermal degradation rate and activation energy required from variation of heating rate and concentration of cobalt catalyst by analyzing the T_G/D_{TG} curves.

2. Materials and Methods

2.1 Materials

The shredded oil palm empty fruit bunches (SOPEFB) obtained for this experiment was from Kilang Sawit FELDA Sungai Tengi. Sabak Bernam, Selangor. Sample of SOPEFB was weight 10 grams before soaked into cobalt (II) nitrate solution.

2.2 Catalyst Preparation

Cobalt catalyst solutions were prepared by mixing the Cobalt (II) nitrate powder and water into 200ml beaker with four concentrations which were 5, 10, 15, and 20% w/w. The samples were soaked for 24 hours. The SOPEFB then dried in the oven with temperature below than 60oC for 5-6 hours

2.3 Thermogravimetric and differential thermal analyses (T_G/D_{TG})

Thermo-gravimetric and differential thermal analyses (T_G/D_{TG}) were conducted by using nitrogen gas which is flow at 50ml/min and heat from 30 °C to 900 °C at four different heating rates (10, 20, 30, and 40 °C/min) adopted from Pichet Ninduangdee (2015).

2.4 Kinetic Analysis

Kinetic parameters of SOPEFB such as activation energy (E), the pre-exponential factor (A), and the reaction order (n), can be determined by using Arrhenius equation and modified Arrhenius equation.

$$k = A \exp\left(-\frac{E}{RT}\right)$$

Modified Arrhenius equation:

$$k = AT^m \exp\left(-\frac{E}{RT}\right)$$

Where m is mass and at m = 0, the modified equation is become Arrhenius equation.
Hence, for 1st order reaction:

$$\frac{d\alpha}{d\tau} = k(1-\alpha)$$

For n^{th} order reaction:

$$\frac{d\alpha}{d\tau} = k(1-\alpha)^n$$

By combining modified Arrhenius equation and n^{th} order reaction equations,

$$\frac{d\alpha}{d\tau} = AT^m\, exp\left(-\frac{E}{RT}\right)(1-\alpha)^n$$

Where:

A is the pre-exponential factor or frequency factor,
E is activation energy,
R is the universal gas constant,
T is temperature at the current time,

α is the biomass decomposition rate as a time-related parameter in the dimensionless from as below:

$$\alpha = \frac{w_o - w_\tau}{w_o - w_f}$$

Where:

w_o is initial weight,
w_τ is weight at time τ,
w_f is final weight.

At constant heating rate, β:

$$\beta = \frac{dT}{d\tau}$$

Then, rewritten equation is:

$$\frac{d\alpha}{dT} = \frac{A}{\beta} T^m \, exp\left(-\frac{E}{RT}\right)(1-\alpha)^n$$

Consider as m = 0 and n = 1,

$$\ln\frac{d\alpha}{dT} = \ln\frac{A}{\beta} - \frac{E}{RT} + \ln(1-\alpha)$$

Consequently, the final equation can be obtained by simplified the equation above:

$$\ln\frac{\frac{d\alpha}{dT}}{(1-\alpha)} = \ln\frac{A}{\beta} - \frac{E}{RT}$$

Therefore, the activation energy, E and frequency factor, A can be deduced from the slope and intercept of ln [(dα/dT)/(1-α)] vs 1/T.

3. Results and Discussion

3.1 The effects of heating rate on SOPEFB

The data collected from TGA for pyrolysis of SOPEFB, and treated SOPEFB were presented in TG and DTG curves. The TG indicating weight loss (mg) while DTG indicating the derivative mass loss (mg/min) under the presence of nitrogen gas at 50 ml/min and heat from temperature 30 °C until 900° C with four different heating rates (10, 20, 30, and 40°C/min) . From the TG curves, the shaped form was S-shaped temperature dependent weight loss. The DTG curves show some peaks and represented the phases in pyrolysis of SOPEFB. Phase I which occurs at temperature below than 200°C was where the remaining water drying and evaporation light component [7]. Phase II where hemicellulose and cellulose and lignin were devolatilized at temperature 200 to 400°C. At temperature 400 to 600°C was where the remaining lignin was devolatilized and the char degradation started at temperature 600 °C and above [10]

IOP Conf. Series: Materials Science and Engineering **334** (2018) 012082 doi:10.1088/1757-899X/334/1/012082

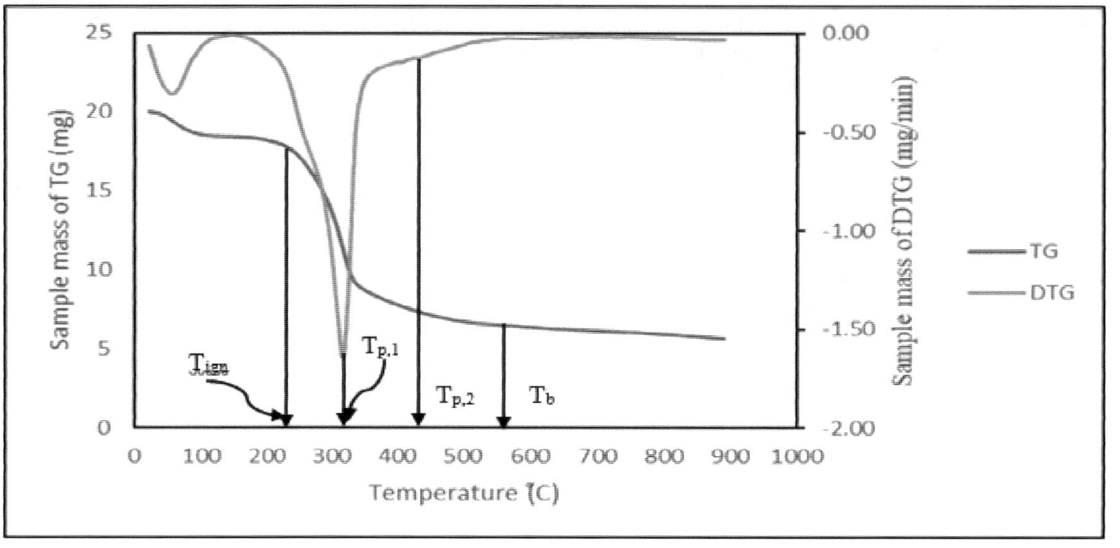

Figure 1. TG/DTG curves of fresh SOPEFB with 10 °C/min.

Figure 1 shows the ignition temperature (Tign), temperature peaks (Tp,1 and Tp,2), and burnout temperature (Tb). The ignition temperature (Tign), is at the certain temperature, the weight of SOPEFB start to loss because of the degradation of hemicellulose and cellulose. The temperature peak 1 (Tp, 1), is at that temperature, the moss of the weight is loss during decomposition of hemicellulose and cellulose. The temperature peak 2 (Tp,2), is at that temperature, most of the remaining lignin is decomposed before char conversion. The burnout temperature (Tb), is where the char production is start.

From table 1 below, the Tp,1 average for fresh SOPEFB with different heating rate was at temperature 329°C and with the average Tign at temperature 225°C. For SOPEFB with 5% concentration of cobalt catalyst, the average Tp,1 was at temperature 348°C and Tign was at 208˚C and average Tb with temperature 530˚C. SOPEFB with the 10% concentration of cobalt catalyst was recorded the average of Tp,1, Tign and Tb were 345˚C, 201˚C and 586˚C respectively. There were no Tp,2 data recorded for fresh SOPEFB and SOPEFB with 5% and 10% concentration of the cobalt catalyst. SOPEFB with 15% concentration of the cobalt catalyst has average temperatures of Tp,1 at 330˚C, Tp,2 at 491˚C, Tign at 200˚C and Tb at 592˚C while the average temperatures for SOPEFB with 20% concentration of cobalt catalyst has Tp,1 at 335˚C, Tp,2 at 485˚C, Tign at 196˚C and Tb at 604°C. The Tp,2 is present at these two concentrations.

The heating rate used in TGA is 10, 20, 30, and 40 °C/min and samples SOPEFB used were fresh SOPEFB, and 5%, 10%, 15% and 20% concentration of cobalt catalyst. From the TG curves, it was observed that the char production is decreased with heating rate increase. However, the most effective with lowest char production was at heating rate 10˚C/min. It was proven that the effective heat transfer of the samples happens when it under lower heating rate.

Table 1. Thermogravimetric parameters of SOPEFB with different heating rate and concentration of cobalt nitrate.

SOPEFB	Heating rate (℃/min)	Temperature(℃)			
		$T_{p,1}$	$T_{p,2}$	T_{ign}	T_b
Fresh	10	317	-	153	555
	20	326	-	194	533
	30	334	-	160	607
	40	338	-	167	629
5% of cobalt catalyst	10	313	-	210	507
	20	354	-	200	509
	30	360	-	203	504
	40	363	-	217	601
10% of cobalt catalyst	10	335	-	186	562
	20	345	-	200	600
	30	348	-	209	620
	40	353	-	207	562
15% of cobalt catalyst	10	329	478	195	552
	20	337	487	201	570
	30	339	494	201	606
	40	314	504	204	639
20% of cobalt catalyst	10	326	476	185	588
	20	333	482	192	595
	30	340	488	206	612
	40	342	493	200	619

Figure 2, 3, 4, 5, and 6 shows the lowest char production was at 10°C/min of heating rate for sample of SOPEFB.

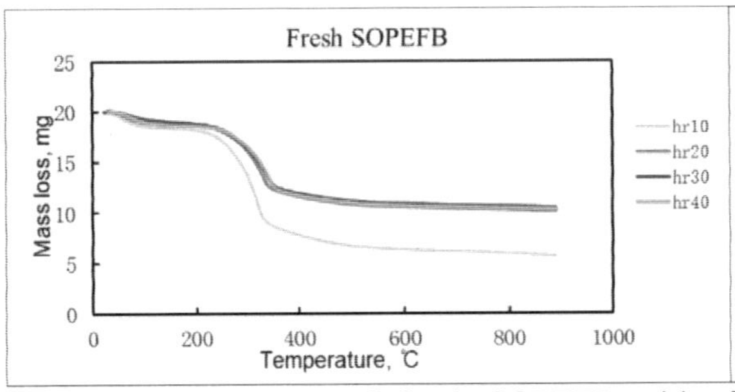

Figure 2. TG curves of fresh SOPEFB with different heating rate.

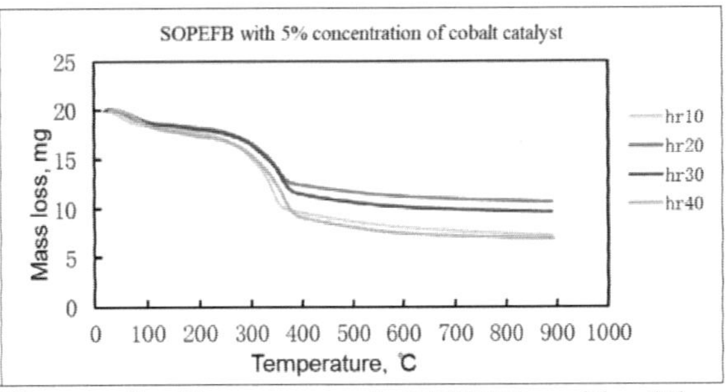

Figure 3. TG curves of SOPEFB with 5% concentration of cobalt catalyst with different heating rate.

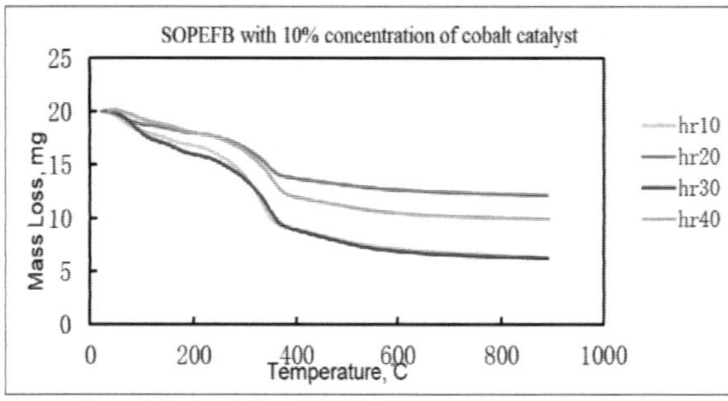

Figure 4. TG curves of SOPEFB with 10% concentration of cobalt catalyst with different heating rate.

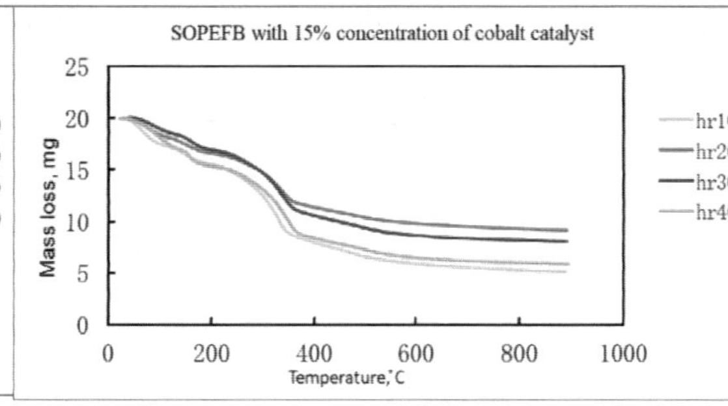

Figure 5. TG curves of SOPEFB with 15% concentration of cobalt catalyst with different heating rate.

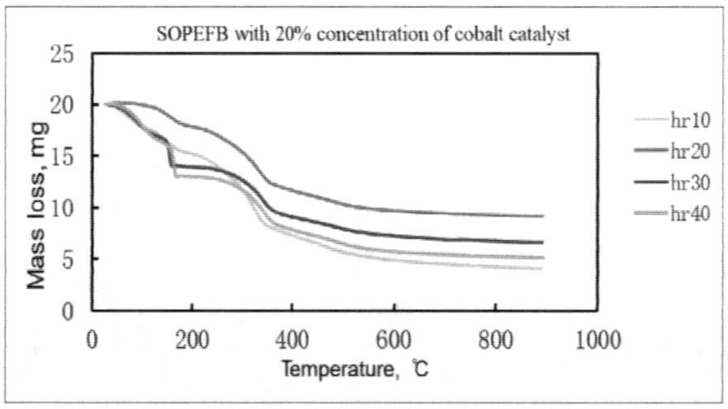

Figure 6. TG curves of SOPEFB with 20% concentration of cobalt catalyst.

From DTG curves, it was observed that the peak of the mass loss at phase II which decomposition of hemicellulose and cellulose were shifted to the right with increased heating rate which means that the highest weigh loss was at higher temperature with higher heating rate. Figure 7, 8, 9, 10, and 11 shows the patterns of the DTG curves with different heating rate.

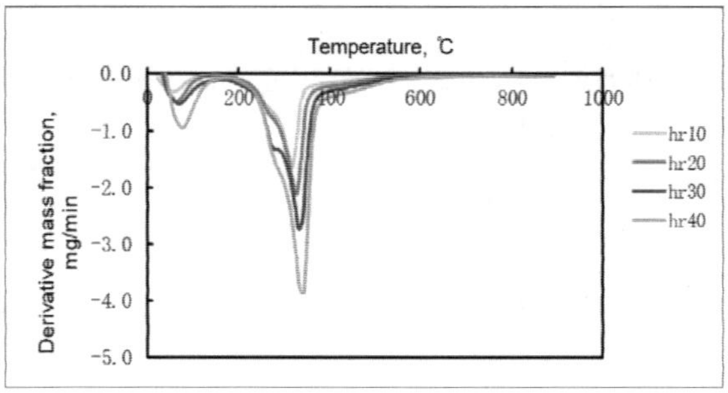

Figure 7. DTG curves of fresh SOPEFB.

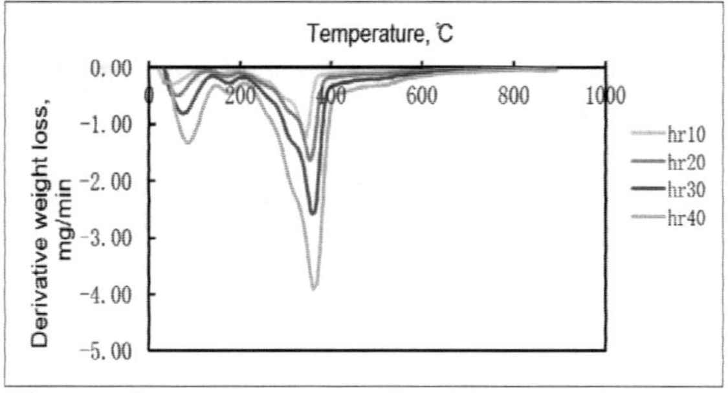

Figure 8. DTG curves of SOPEFB with 5% concentration of cobalt catalyst with different heating rate.

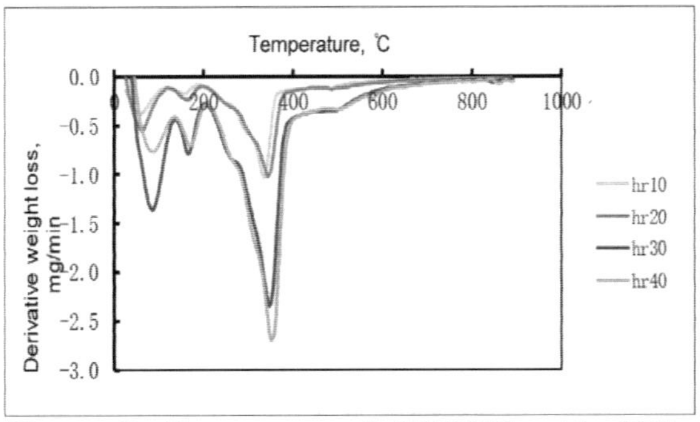

Figure 9. DTG curves of SOPEFB with 10% concentration of cobalt catalyst with different heating rate.

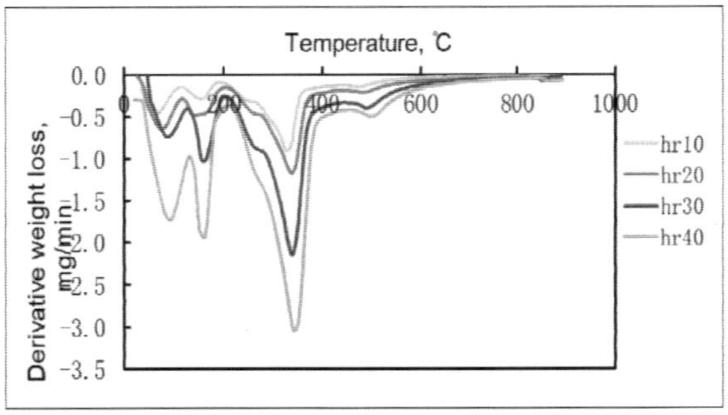

Figure 10. DTG curves of SOPEFB with 15% concentration of cobalt catalyst with different heating rate.

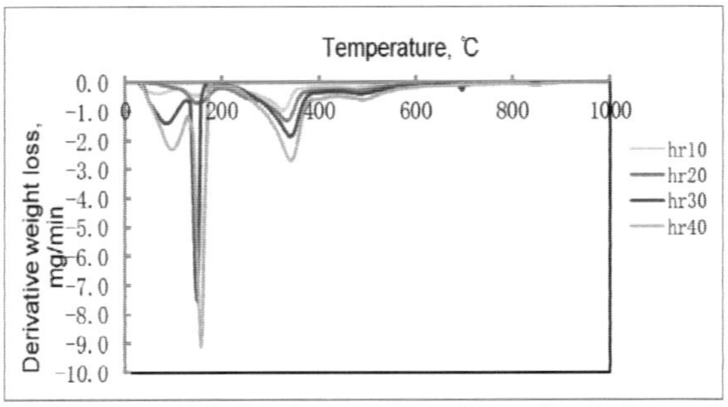

Figure 11. DTG curves of SOPEFB with 20% concentration of cobalt catalyst with different heating rate.

3.2 The effects of different concentration of cobalt catalyst

From TG curves which is in figure 12, 13, 14 and 15, it was observed that the char production was decreased with increased concentration of cobalt catalyst. When heating rate as constant, lowest char production was sample of SOPEFB with 20% concentration of cobalt catalyst.

If higher concentration is embedded with the samples of SOPEFB, the char production is less. However, at 15% and 20% of concentration of cobalt catalyst, at the phase I which at the water evaporation, second peak was appeared because of the coating from the cobalt catalyst like shows in DTG curves.

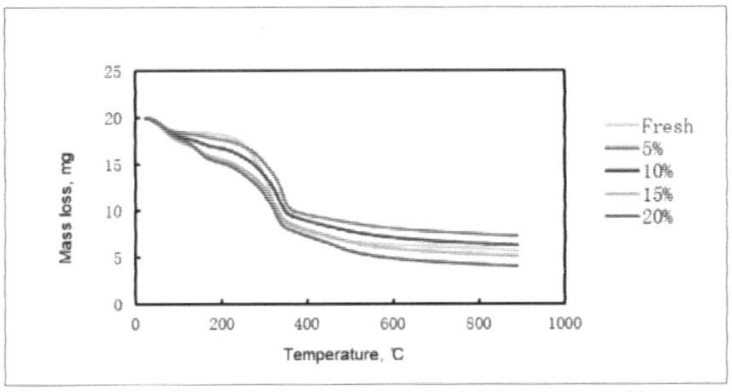

Figure 12. TG curves of SOPEFB with different concentration of cobalt catalyst at heating rate 10 °C/min.

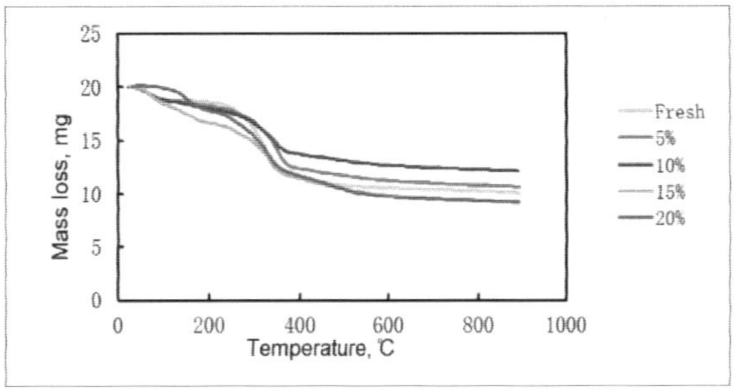

Figure 13. TG curves of SOPEFB with different concentration of cobalt catalyst at heating rate 20 °C/min.

IOP Conf. Series: Materials Science and Engineering **334** (2018) 012082 doi:10.1088/1757-899X/334/1/012082

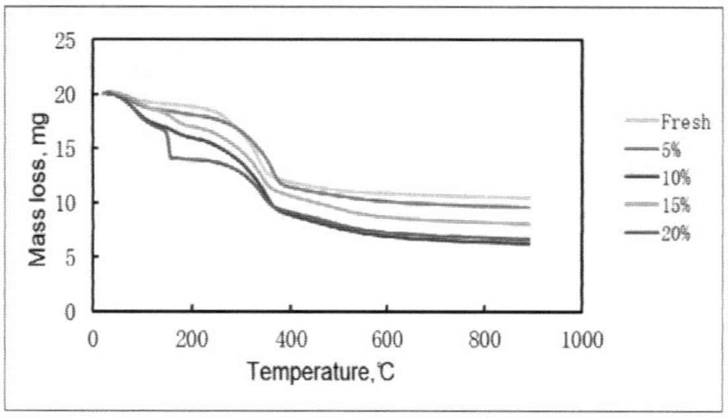

Figure 14. TG curves of SOPEFB with different concentration of cobalt catalyst at heating rate 30 °C/min.

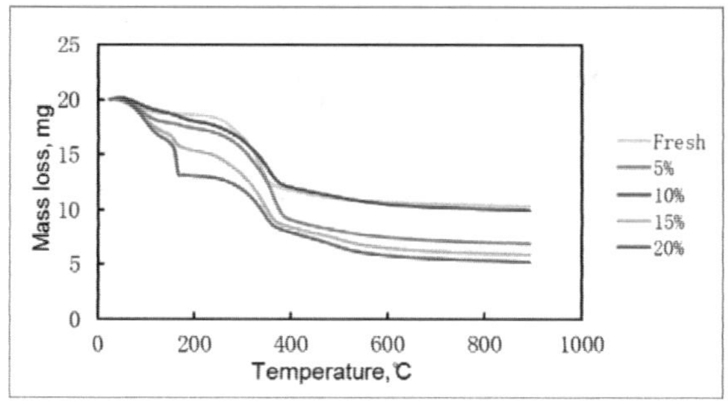

Figure 15. TG curves of SOPEFB with different concentration of cobalt catalyst at heating rate 40 °C/min.

In DTG curves, at the samples of fresh SOPEFB and SOPEFB with concentration 5% and 10% cobalt catalyst, there were no second peak at the third phase which at lignin decomposition. It was because of most of the lignin was decomposed at the second phase with hemicellulose and cellulose. It also does not have second peak at the first phase unlike SOPEFB with concentration 15% and 20% cobalt catalyst because the coating of cobalt nitrate solution was not thick. Figure 16, 17, 18, and 19 shows the DTG curves of different cobalt catalyst concentration with different heating rate.

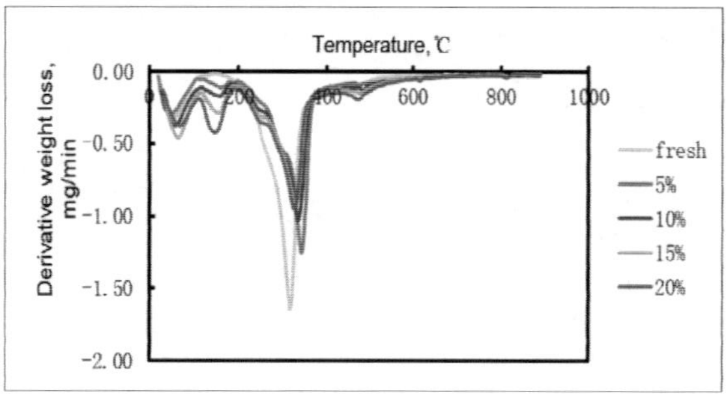

Figure 16. DTG curves of SOPEFB with different concentration of cobalt catalyst at heating rate 10 °C/min.

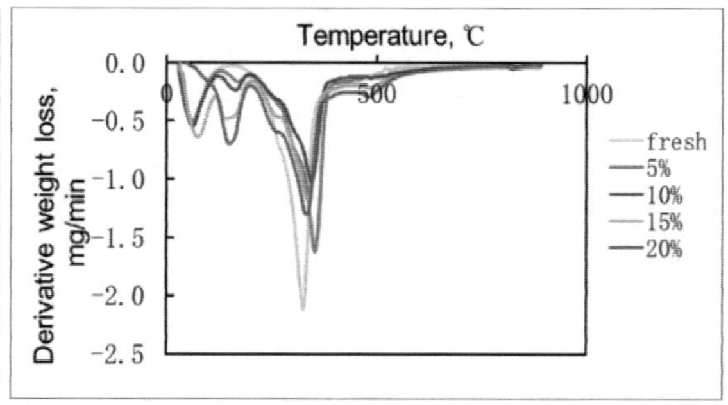

Figure 17. DTG curves of SOPEFB with different concentration of cobalt catalyst at heating rate 20 °C/min.

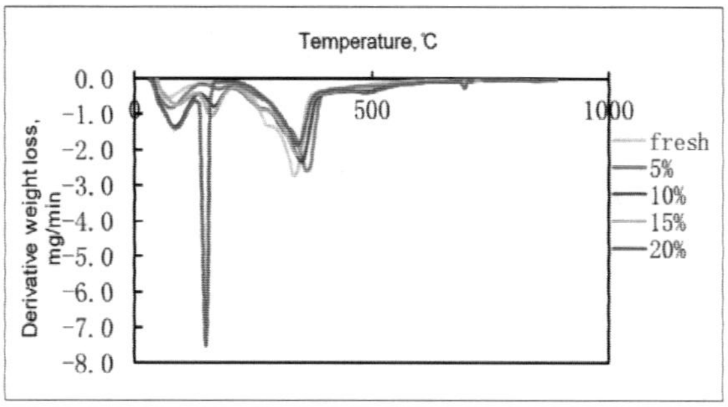

Figure 18. DTG curves of SOPEFB with different concentration of cobalt catalyst at heating rate 30 °C/min.

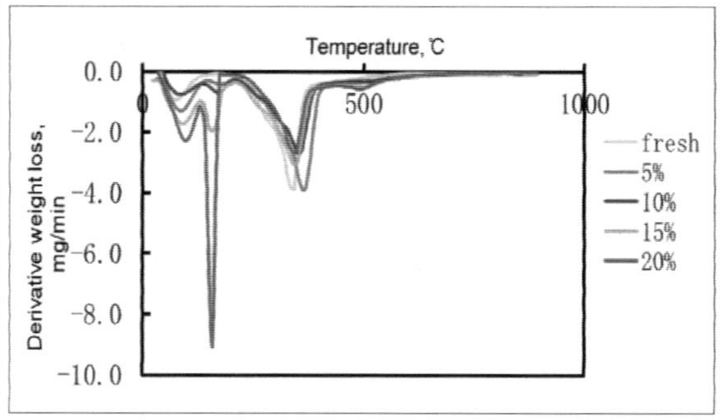

Figure 19. DTG curves of SOPEFB with different concentration of cobalt catalyst at heating rate 40 °C/min.

3.3 Kinetic parameters of SOPEFB

Table 2 shows the kinetic parameters of samples of SOPEFB with different concentration of cobalt catalyst for different heating rate (10, 20, 30 and 40°C). The kinetic parameters were determined for temperature regions of 200 to 350°C to give the accurate fitting of the experimental data. The accuracy of the graph proven with the correlation coefficient (r) adjacent with value 1.0.

From the table 2, shows that the activation energy (E) at heating rate 10 and 20 °C/min were decreased with increased SOPEFB with concentration of cobalt catalyst. However, at heating 30 and 40°C/min the value of activation energy (E) were scattered and SOPEFB with 20% concentration cobalt catalyst was higher than 5%, 10% and 10%. This was because of the thick coated of cobalt nitrate solution required more energy to degraded the SOPEFB

Table 2. Kinetic parameters of SOPEFB.

Heating rate, β	Samples of SOPEFB	Fitting equation	Correlation coefficient, r	Activation energy, E (J/mol)	Pre exponential factor, A(1/min)
10	Fresh	y= -1924.7x-2.5876	0.9706	16002	0.752
	5%	y= -1712.9x-3.807	0.9800	14241	0.222
	10%	y= -1549.8x-4.2169	0.9787	12885	0.147
	15%	y= -1432.2x-4.5745	0.9829	11907	0.103
	20%	y= -1369.2x-4.7891	0.9835	11384	0.0832
20	Fresh	y= -2093.7x-2.1253	0.9894	17407	2.388
	5%	y= -1520.7x-4.63	0.9829	12643	0.195
	10%	y= -1453x-4.7817	0.9853	12080	0.168
	15%	y= -1361x-4.9242	0.9851	11835	0.165
	20%	y= -1341.3x-4.9242	0.9872	11318	0.145
30	Fresh	y= -1939x-2.7277	0.9856	16121	1.961
	5%	y= -1561.5x-4.6681	0.9863	12983	0.282
	10%	y= -1401.4x-4.983	0.9839	11651	0.206
	15%	y= -1478.5x-4.6115	0.9850	12292	0.298
	20%	y= -1995.2x-3.153	0.9880	16588	1.282
40	Fresh	y= -2309.2x-1.5985	0.9907	19199	8.088
	5%	y= -1624.2x-4.4788	0.9901	13504	0.454
	10%	y= -1490.6x-4.7628	0.9872	12393	0.342
	15%	y= -1555.7x-4.4268	0.9842	12934	0.478
	20%	y= -2147.9x-2.707	0.9948	17858	2.669

4. Conclusions

The data collected from TGA were used to construct T_G and D_{TG} curves. From T_G and D_{TG} curves, it shows that, the thermal degradation of the shredded oil palm empty fruit bunches of were divided into three phases which were phase I, phase II and phase III. At phase I, this was the temperature region where water and cobalt nitrate coating were evaporated. The average temperature region of this phase was below 200°C. Second phase was at temperature range of 200 to 400°C where the highest peak of the weight loss of shredded oil palm empty fruit bunches samples because of decomposition of hemicellulose and cellulose. The third phase was for decomposition of lignin at range temperature of 600°C and above. The peak of this phase is low rather than at phase II because the lignin also started to be decomposed at second phase besides hemicellulose and cellulose.

The effect of heating rate and different concentration was decreased in char production. The production of char was decreased with increased heating rate and increased concentration of cobalt catalyst embedded into shredded oil palm empty fruit bunches. The most effective with less char production is at 10°C/min because of the heat was effectively transferred to the samples.

Activation energy, E required for degraded the shredded oil palm empty fruit bunches samples were decreased with the increased concentration of cobalt catalyst. However, at the higher heating rate, the activation energy of 20% concentration of cobalt catalyst higher than others. It can simply conclude that the energy was required more to decomposed the coating of cobalt nitrate before started to decomposed hemicellulose and cellulose. But, at lower heating rate, the heat transfer was more efficient to decomposed the coating of cobalt nitrate.

The best criteria in order to get less char production and heat transfer efficient was at lower heating rate which were at 10 or 20 °C/min and at the higher concentration which are 15% or 20% concentration of cobalt catalyst.

Acknowledgment
This work was supported by the Universiti Teknologi MARA, (UiTM), through the research code number: 600-IRMI/Dana KCM5/3/LESTARI (130/2017) is gratefully acknowledgement.

References

[1] Fernquest, J. (2011). Palm oil price guarantees. Bangkok: Bangkok Post. Retrieved 1 December, 2016, from http://www.bangkokpost.com/learning/ learning-news/220525/palm-oil-price-guarantees
[2] Jesus David Coral Medina, A. L. (2016). Biological activities and thermal behavior of lignin from oil palm empty fruit bunches as potentil source of chemicals of added value. Industrial crops ans products, 630-637.
[3] M. A. Mohd Salleh, N. H. (2010). Gasification of Biochar from Empty Fruit Bunch in a Fluidized Bed Reactor. Energies, 1344-1352.
[4] M. T. Azizan, S. Y. (n.d.). Production of bio-oil from oil palm's empty fruit bunch via pyrolysis. Renewable Energy Sources, 228-233.
[5] N. Abdullah, F. S. (2011). Characterisation of Oil Palm Empty Fruit Bunches for Fuel Application. Journal of Physical Science, 1-24.
[6] N. Abdullah, H. G. (2010). Fast pyrolysis of empty fruit bunches. Fuel, 2166-2169.
[7] N. Alias, N. I. (2014). Pyrolysis of empty fruit bunch by thermogravimetric analysis. Energy Procedia, 2532-2536.
[8] Nickson E. O, X. D. (2016). Palm Oil Production in Malaysia: An Analytical Systems Model for Balancing Economic Prosperity, Forest Conservation and Social Welfare. Agricultural Sciences, 55-69.
[9] Niessen, W. R. (1978). Combustion and Incineration Processes. CRC Press Book.
[10] Pichet Ninduangdee, V. I. (2015). Thermogravimetric studies of oil palm empty fruit bunch and palm kernel shell: TG/DTG analysis and modelling. Energy procedia, 453-458.
[11] R. Alias, K. H. (2011). Co-pyrolysis and Catalytic Co-pyrolysis of Waste Tyres with Oil Palm Empty Fruit Bunches. Journal of Applied Science, 2448-2451.
[12] Rajvanshi, A. (2014). Biomass Gasification. Nimbkar Agricultural Research Institute.
[13] Soria, T. D. (2013). Catalytic Fast Pyrolysis: A Review. Energies, 514-538.
[14] Y. Basiron & C. K. Weng. (June, 2004). The oil palm and its sustainability. Journal of Oil Palm Research, 16, 1-10.

IOP Conf. Series: Materials Science and Engineering **334** (2018) 012083 doi:10.1088/1757-899X/334/1/012083

Composite material making from empty fruit bunches of palm oil (EFB) and Ijuk (*Arengapinnata*) using plastic bottle waste as adhesives

T Rihayat[1,4], S Salim[1], N Audina[1], N S P Khan[1], Zaimahwati[1], M Sami[1], M Yunus[1], Z Salisah[1], P N Alam[2], Saifuddin[3] and I Yusuf[3]

[1] Departement of Chemical Engineering,
Politeknik Negeri Lhokseumawe, 24301, Aceh Indonesia
[2] Department of Chemical Engineering, Faculty of Engineering
University of Syiah Kuala, Darussalam, Banda Aceh
[3] Departement of Mechanical Engineering,
Politeknik Negeri Lhokseumawe, 24301, Aceh Indonesia

E-mail : teukurihayat@yahoo.com

Abstract. Reviewed from the current technological required a new methods to capable offering a high profit value without overriding the quality. The development of composite technology is now beginning to shift from traditional composite materials based petroleum to natural fibers composite. In the present study, aim to made specimens using natural fibers in form of EFB as a composite reinforcedment with Polyethylene Terephtalate (PET) derived from Plastic bottles waste as matrix with mixed composition parameters and time-tolerance in the mixing process to build a biocomposite material. The characterization of mechanical properties includes tensile test (ASTM D638-01) and bending test (ASTM D790-02) followed by thermal analysis using Thermogravimetric Analysis (TGA), and morphological analysis using scanning electron microscope (SEM). The analysis effect of EFB, Ijuk and PET mixtures on the composite matrix is very influential with mechanical properties characterization, including tensile test and bending strength. The results demonstrated that from the sample named : 50 : 25: 25, hybrid composites showed improved properties such as tensile strength of 167 MPa while the 90:05:05 based composites exhibited tensile strength values of 30 MPa, respectively. In term the flexural test the best result of composition on the properties with 10 minutes duration time its load value 7,5 Mpa for 80:10:10.

1. Introduction

Indonesian s a major producer of palm oil. Indonesian oil palm plantations are found in Sumatra, West Java, Kalimantan, Sulawesi, Bangka Belitung and Papua. Indonesia has the number of oil palm plantations is about 13.5 million hectares [1]. The main product of fruit bunches palm is palm oil, after the processing, the remaining empty bunches of oil palm (EFB) as a by-product. Physically EFB has fibers that can be utilized for organic fertilizer, paper-making raw materials, briquettes, and also used as a filler medium such as filler cavity and car seat cushions. Therefore, awareness appears to

[4] To whom any correspondence should be addressed.

examine further on the potential of EFB as a natural fiber material that can be processed into composite matrix material.

In addition to EFB, palm fiber is also widely available in Indonesia, which can be used as a composite matrix. The fibers are obtained from palm trees (*Arenga Pinnata*), which are traditionally often used as a base material for the construction of wooden structures planted in the soil to prevent termites. The usefulness is supported by the elastic nature of the elastic, hard, waterproof, and difficulty to digest by destructive organisms.

Currently, the bunch of empty Palm Oil Bunches (EFB), and Ijuk (*Arenga pinnata*) is easy to obtain. EFB tend not to be utilized and thrown away and there is still little information about the utilization of EFB, so it tends to become waste. Potential fibers almost 73% fiber is very potential to be utilized as research materials, especially low cost composite matrix materials. Therefore, due to the availability of abundant EFB and Ijuk and cost effectiveness, TKSS and Ijuk fibers can be excellent for composite matrix productio[2].

Plastic bottles are chemically polymeric materials widely used in human life. The waste of plastic bottles from the drink is wasted and underused, so it can pollute the environment. Plastic waste is generally difficult to decipher by bacteria and tends to stay inside and cause one of the degradation of environmental quality. Therefore, this study attempted to utilize plastic bottle waste combined with EFB fiber for the manufacture of composite matrix [3].

The development of the composites leads to a more natural composite as has been observed by some researchers. Ferdy Arif (2014) and Rihayat (2015), who has developed composite technology is now beginning to experience a shift from composite materials made synthetic fibers into composite materials fiber natural, in this research the specimens form is used natural fiber as a composite reinforcement to mixing polypropylene as a matrix with temperature and time-resistant parameters in the injection molding process and its not only easly to reused but also renewable too. This research has given a good effect with reducing the consumption of difficult chemicals in recycling that can disrupt the environment. Composites with natural fibers have other advantages when compared to synthetic composites. Natural composites are more environmentally friendly because they are able to degrade naturally and the price of natural fiber is cheaper than synthetic fibers. Have superior characteristics such as resistant to attack of destructive organisms, weather and moisture, thus relatively resistant to fire.

Therefore, the research aims to make a composite matrix by combining fiber, in this case empty palm bunches (EFB), palm fiber (*Arenga pinnata*), with polyethylene terephthalate (PET) derived from waste plastic bottles as matrix.

2. Materials and Methodes

2.1 Material

2.1.1 ***Fiber treatment*** In this study the materials used are PET (Polyethylene Terephtalat) as a matrix derived from plastic bottle waste around Lhokseumawe City North Aceh Indonesia, oil palm empty bunches (EFB) will be obtained from PTP II operating in Cot Girek North Aceh Indonesia, while Ijuk (Arengapinnata) is obtained from Sawang North Aceh Indonesia plantation, both fibers are used as filler. Sodium Hydroxide (NaOH) merck, and Aquadest. Furthermore, fibers and matrix is mixed by the Extruder and fabricated with Hot Press for pressing process. while for the analysis phase using UTM (universal Testing Machine) to analyze mechanical properties include tensile and flexural test according with (ASTM D638-01) and (ASTM D790-02) respectively. Thermal analysis using the Termogravimetric Analysis (TGA) while morphological analysis using a scanning electron microscopy (SEM).

In this research the first step procedure is EFB and Ijuk fiber in crusserh then soaked in 2% NaOH solution at room temperature for 30 minutes, then fiber is filtered and washed several times with aquades to reducing sodium hydroxide and alkalinity. Furthermore, the waste of plastic bottles made from PET (Polyethylene Terephtalat) material that has been collected and then cut into small pieces, then PET, EFB and is mixed by Extruder between matrix and filler volume fraction 90:05:05, 80:10:10, 50:25:25 and fabricated with Hot Press machine while retention time variations is 5, 10, 15, 20 and 25 minutes at temperature 140°C.

2.1.2 Scanning electronmicroscopy Scanning electron microscope (SEM) LEO 1455 VPSEM equipped with EDX is used to identify the morphology of commercial samples. The surface of the composite specimen is examined directly by scanning an electron microscope. The EFB bench is mounted on an aluminum specimen stamp using a double-sided carbon adhesive tab. The sample is washed, cleaned properly, dried with air and coated with thick gold in JEOL ion sputter coil and SEM observed at 20 kV. To improve the conductivity of the sample, a thin layer of platinum vaporously evaporates forward before the photomicrograph is taken.

2.1.3 Tensile properties The tensile test is performed with according to a standard (ASTM D 638) used universal testing machine (UTM) with 10 kN load to determine tensile strength. Tensile test is performed to determine the tensile strength of the composite material.

Composite materials have better compression properties than their tensile properties. Tensile strength is influenced by the molecular bonds of the constituent material and Bending test specimens are made according to ASTM D790 standard with three point bend to ivestigations the ability of the material to a given load.

2.1.4 Thermogravimetric analysis(TGA) The study of thermal stability, determined by thogogravimetric analysis (TGA) was performed to understand the degradation characteristics of specimens using Tho-gravimetric analyzer TGA Q500 (Perkin Elmer thermal analyzer). Thermal analysis of the material provides some basic information about the thermal stability of the material. All measurements were made under nitrogen flow (30 ml / min), keeping the heating rate constant 10°C/min and using alumina container with pinhole. Symmetric larvae analysis (TGA) was used to characterize the decomposition and thermal stability of the material under various conditions. In principle TGA analysis changes the thermal stability are examined in terms of percentage weight loss as a function of temperature. Samples with weight between 7 and 20 mg were placed on ceramic crucibles, while the tests were carried out in nitrogen atmosphere.

3. Result and Discussions

3.1 Mechanical properties

Mechanical properties are the most important for composite matrices. The mechanical properties of composite matrix, it can be seen by trought tensile strength and flexibility (Bending). This aim was done to investigation materials composite reinforced natural fibers.

3.1.1 Tensile strength The tensile test is performed to determine the maximum tensile load to material test. Tensile strength is one of the most important properties that often is used to investigated the characteristics of a polymeric material. Therefore the tensile strength of a material is defined as the maximum load (Fmax) used to decide the specimen. The fracture samples test for tensile is shown in figure 1, while the average result of tensile test is present by figure 2. Its explained that highest tensile strength was produced at 167 Mpa for volume fraction 50:25:25 and hold time 20 minutes while graph nominal stress vs strain is presented by Figure 3.

Enhancement amount of fiber as fillers in the matrix composites has shown increasing the meachanical properties its depend on the holding time when the matrix is pressed and the optimal temperature, because many fiber breaks. Fiber break being by a strong bond between the matrix and filler. The composite bonding power affects the composite strength of the load while the hotest compress it will be effect which breaks of bonds on Polyethylene Terefthalat (PET) if the durability is too long.

The relationship between tensile strength with addition of EFB fibers and variation of rate time has given a different tensile strength, compared with a pure matrix, whereas in mixing of 90:05:05 at 5 minutes duration it shows the lowest tensile strength value. The bonding between the matrix and filler at the time of the breaking specimen and get some void can reduces the ability to receive the tensile load [5].

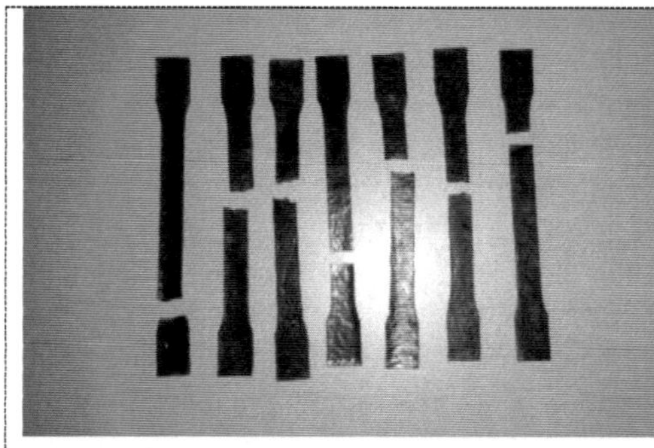

Figure 1. Fracture specimen.

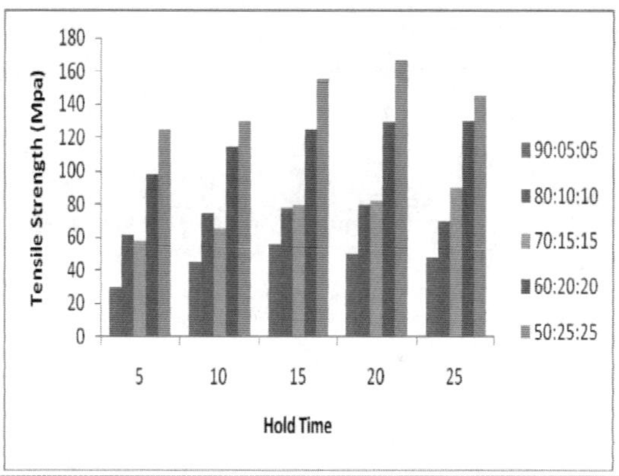

Figure. 2 Pull Test Grap.

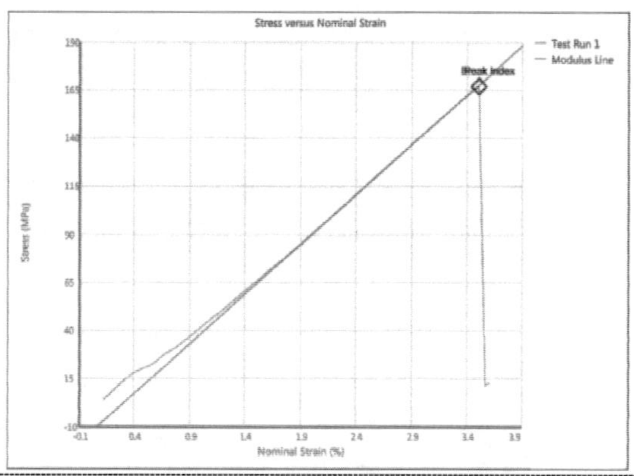

Figure. 3 Graph Nominal stress vs strain for sample 50:25:25 with hold time 20 minutes.

3.1.2 Flexural strength The bending test specimen is made according to ASTM D790 standard. It can be seen from the result of bending test analysis in Figure 4. After calculated using by the formula, its can be seen the results in the above table shows that the highest flexural strength value is got at the mixture of 80:10:10 with hold time 10 minutes and its load value 93,2 N during the flexural strength of 5,1 Mpa.

Figure. 4 shows the 50:25:25 composition variation with 20 minutes rate time indicating the lowest flexural strength value of 3.4 Mpa with the load value of 61.5 N. Then the bending strength shows an increase in the mixing composite of 70:15:15 while rate time 25 minute shows an increased flexural strength value of the mixing composite of 50:25:25 is 5 Mpa with a load value of 90.5 N. The best result of bending test was get from the average data during the test.

It was concluded that composites made from PET wastes had lower grades of flexibility compared to composites made of polyester (resins), indicating that polyesters were synthesized by means of petroleum-based acids and alcohol would be placed in high-temperature vacuum to create a process called "Condensation polymerization". Long chemical bands will be generated from condensed polymers and when drawn up to five times longer than their original size so that polyester elasticity is higher than the PET value [6].

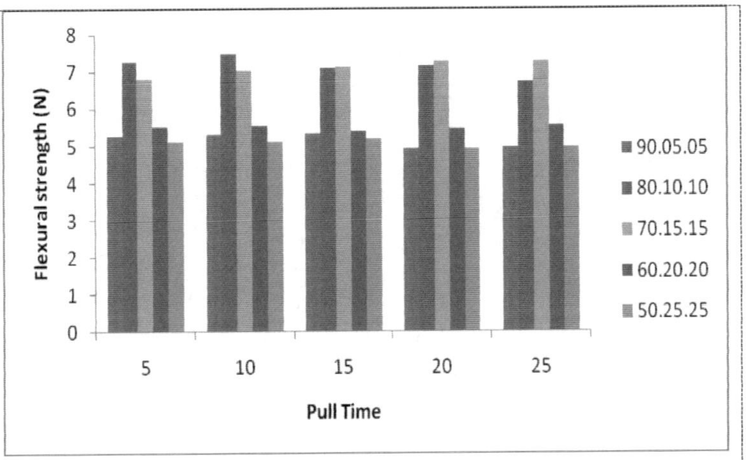

Figure 4. Flexural strength graph.

3.2 Thermogravimetric analysis (TGA)

Heat analysis of the material provides some basic information about the thermal stability of the material. All measurements were carried out under the flow of nitrogen (30 ml / min), maintaining a constant heating rate of 10 °C / min and using alumina containers. Thermogravimetric analysis (TGA) is used to characterize the decomposition and thermal stability of various material conditions. Particularly in the TGA analysis changes in thermal stability are examined in terms of percentage drop in sample weight as a function of temperature [7,8].

TGA is a technique used for the study of thermal stability/decomposition on a composite matrix. The thermal stability of the fibers is a very important parameter for the processing and use of materials. The composite preparation requires the mixing of fibers (filler) and matrix at high temperatures, so that the degradation effect on the cellulosic material properties can produce undesirable ones. A number of research reports have been investigated on the study of thermal stability of natural fiber composite fibers through TGA. The result of TGA in this research can be seen in figure 5.

In this study the results of the Thermogravimetric analysis (TGA) of composite matrix has been investigated and studied that the percentage of weight loss of samples with temperature increase are interrelated. The loss of mass as temperature increases, for the heating process to increase 10 °C / min, we can see in the graph the results of the analysis show the transition temperature ranging from 278.02 - 310.26 °C and Range for the final transition ranging from 600.86 – 601.58 °C, from room temperature to 600 °C.

From the graph above shows that the best temperature value in the heat-resistant composite matrix with mixing of 50: 25: 25 starting at the start temperature of 360.55 °C and the degradation start-up temperature starting from 446.63 - 492.20 °C and ending at temperature 521.18C so get the midpoint value 467.22 °C with get value weight loss -6.730 mg, here below can be seen in figure 6 graph of result of analysis of Thermogravimetri (TGA) mixing 50:25:25 below.

IOP Conf. Series: Materials Science and Engineering **334** (2018) 012083 doi:10.1088/1757-899X/334/1/012083

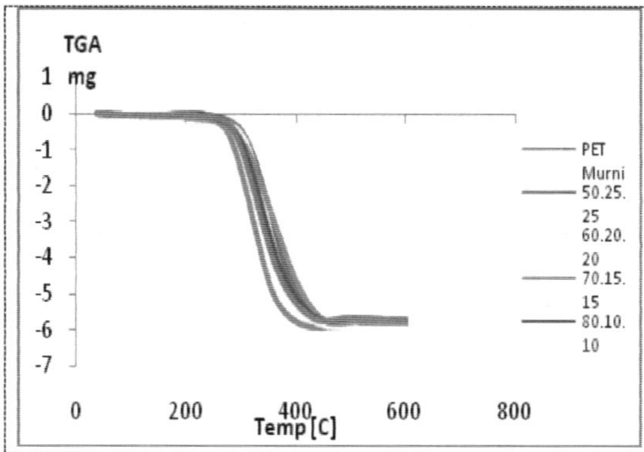

Figure.5 Graph The results of TGA analysis.

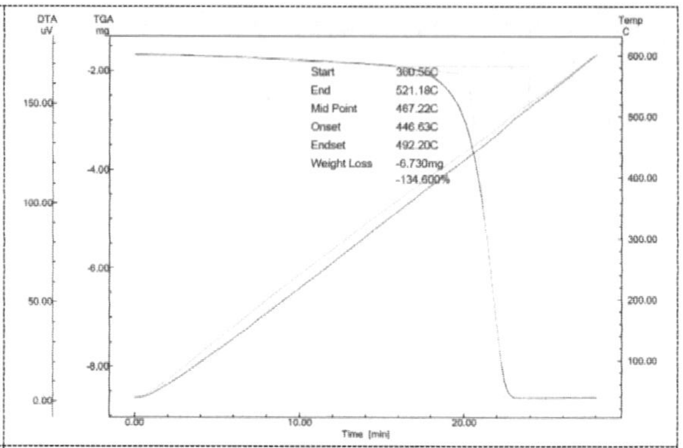

Figure.6 Graph The results of TGA mixing analysis 50:25:25 (20 minutes).

3.3 SEM (Scanning Electron microscope) Analysis

Structure morphology composites matrix was investigated by scanning electron microscopy LEO 1455 VPSEM. It is an excelent technique for examining morphology examination the surface morphology fiber. The best sample at the tensile test is used to investigated structure morphology can be seen in figure 7.

Figure. 7. Structure morphology for sample 50:25:25 wt% with hold time 20 minutes.

Figure 7 it's shown morphology composites for sample 50:25:25 wt% as the best sample at tensile test. Its observed that the fibers were broken and pull out from the shell while the matrix doesn't flow into porous fiber. It can be affect the mechanical properties composite not strongly because intercial adhesion between fiber and matrix were broken.

4. Conclusion

The result of tensile test analysis according to ASTM D 638 has get best result on mixing composition of 50:25:25 with 20 minutes holding time with tensile strength value 167 MPa. The result of Bending Strength test according to ASTM D 790 get best result on 80:10:10 mixing composition with 10 minutes hold time of its load value 93,1509 N and flexural strength value 5,147 MPa. The results of the Thermogravimetric analysis (TGA) analysis showed a heat-resistant matrix is a mixing composition of 50:25:25 with a starting temperature of 360.55 °C and the starting degradation temperature starting at 446.63 - 492.20 °C and ending at 521 , 18 °C so get the midpoint value 467,22 °C with get value weight loss -6,730.

References

[1] World growth. 2011. The economic benefit of palm oil to Indonsia. Pages 1-27 (www.worldgrowth.org)
[2] A.K.M. Moshiul Alam, M. B. (2012). Structures and performances of simultaneous ultrasound and alkali treated oil palm empty fruit bunch fiber reinforced poly (lactic acid) composites. *Composites: Part A* , **43** , 1921-1929.
[3] Ferdy Arif P, Sumarjr, Dedi Dwilaksana, (2014) The Effect Of Temperature And Holding Time On The Palm Fiber Composite On Mechanical Properties Of Polypropylene Matrix On Injection Molding Process, Jurusan Teknik Mesin, Fakultas Teknik, Universitas Jamber.
[4] Zaimahwati, Agusnar, H., Rihayat, T., Reflianto, D., Gea, S. (2015). The manufacture of palm oil-based polyurethane nanocomposite with organic montmorillonite nanoparticle as paint coatings, *International Journal of ChemTech Research*, **7, 5**, 2537-2544
[5] Ishaka, M.R. (2013). Sugar palm (Arengapinnata): Its fibres, polymers and composites. *Carbohydrate Polymers*, **91**, 699–710.
[6] Rosni Binti Yosoff, (2016) Tensile And Flexural Properties Of Polylactic Acid-Based Hybrid Green Composites Reinforced By Kenaf, Bamboo And Coir Fibers, *Industrial Crops And Products* , **94**, 562-573.
[7] Rihayat, T., Suryani, S., Zaimahwati, X. (2014). Preparation and properties and application of renewable source (palm oil polyol) based polyurethanes coatings and its thermal stability improvement by clay nanocomposites. *Advanced Materials Research*, **887-888**, 566-569
[8] Rihayat, T., Suryani, S., Zaimahwati, Z. (2014). Effects of heat treatment on the properties of polyurethane/clay nanocomposites paint. *Applied Mechanics and Materials*, **525**, 97-100

Morphology of one-time coated palladium-alumina composite membrane prepared by sol-gel process and electroless plating technique

R Sari[1,3], R Dewi[1], Pardi[1], L Hakim[2] and S Diana[1]

[1] Lhokseumawe State Polytechnic, 24301 Lhokseumawe, North Aceh, Indonesia
[2] Malikussaleh University, 141 Reuleut, North Aceh, Indonesia

E-mail: ratnasari@pnl.ac.id

Abstract. Palladium coated porous alumina ceramic membrane tube was obtained using a combination of sol-gel process and electroless plating technique. The thickness, structure and composition of palladium-alumina composite membrane were analyzed by transmission electron microscopy (TEM), scanning electron microscopy (SEM), energy-dispersive X-ray (EDX), and atomic force microscopy (AFM). Palladium particle size was 6.18 to 7.64 nm. Palladium membrane with thickness of approximately 301.5 to 815.1 nm was formed at the outer surface of the alumina layer. EDX data confirmed the formation of palladium-alumina membrane containing 45% of palladium. From this research it shows the combination of sol-gel process and electroless plating technique with one-time coating can produce a homogeneous and smoother palladium nano layer film on alumina substrate.

1. Introduction

Demand for hydrogen based fuels has increased due to the depletion of fossil fuel reserves and the large emission of greenhouse gas. The usage of hydrogen as a commodity feedstock for chemical, petrochemical and metallurgical process is well known. Hydrogen has gained interest as a clean alternative energy due to the recent and rapid advancement of fuel cells and internal combustion engines. Membrane technology has started to be applied for hydrogen separation in the recent years [1].

The membrane-based separation technique has received a considerable attention in petrochemical, pharmaceutical, food and biotechnology areas [2]. Energy saving and reduction in initial capital investment are the two main reasons for the preference for membrane separation rather than conventional separation processes, e.g. adsorption, absorption, distillation, etc.

Metal composite membranes are expected to play a key role in hydrogen separation. Palladium (Pd) based composite membranes have gained in importance in recent years due to their increased hydrogen permeation rates. Pd-based composite porous membranes consist of a thin dense layer of Pd or a Pd alloy on a porous support in the form of tubes or discs of porous glass, ceramics or stainless steel [3]. [1]

One of the deposition techniques frequently applied on ceramic materials is dip coating. Because of its simple procedure, sol-gel process to produce thin film has grown rapidly [4]. It consists of a

[3] To whom any correspondence should be addressed.

substrate removal, where solid film from condensation reaction is deposited from the fluid sol by drainage and evaporation. Dip coating process produced a good and flexible powder form dispersion (sol) which can improve the characteristic of the support [5].

Electroless plating is a method for metallic films plating on a support involving redox reaction. Grain growth in Pd forms a film, in which the microstructure has an important effect on the membrane morphology. The desired morphological aspects are compactness and equal growth of the deposits in perpendicular and lateral direction to the surface of substrate [6]. Deposition of Pd and Pd alloy membranes on porous substrates by electroless plating technique has been popularly used with its good coating ability and the simplicity of experimental. Sari [7] reported that a Pd membrane synthesized with plating time of 7 hours formed a smooth and thick layer of Pd 7.146 μm.

2. Methodology/ Experimental

2.1. Materials

α- Alumina membrane tube with an outer diameter, length, thickness and pore size of 1 cm, 8 cm, 0.2 cm and 100 nm respectively is used. The chemical substances used in the sol solutions were aluminum nitrate nonahydrate (98%, Fluka), nitric acid (65% Merck), poly vinyl alcohol (98%, Merck), butanol and hydrochloric acid (37%, Merck). The sensitization solution was tin (II) chloride (Sigma-Aldrich) and hydrochloric acid (37%, Merck). The activation solution was Pd chloride (59.9%, Merck) and hydrochloric acid (37%, Merck). The electroless plating solution consisted of Pd chloride (59.9%, Merck), sodium phosphinate monohydrate (Merck), ethylene dinitrilo tetraacetic acid (Merck), and ammonium hydroxide (28%, Sigma-Aldrich).

2.2. Preparation of Pd Coated Alumina Ceramic Composite Membrane

Prior to sol-gel process, the α- Alumina membrane tube was cleaned by sequential washing with natrium hydroxide, deionized water, isopropanol and followed with drying at 120^{0}C for 1 day to remove any traces of hydrocarbon existing in the membrane pores that might cause non-uniform coating. The preparation of Pd coated alumina ceramic composite membrane involves two steps. First, aluminium nitrate nonahydrate,nitric acid, and polyvinyl alcohol were added into deionized water and temperature was maintained at 80-90^{0}C. Then, they were refluxed with stirring at 80-90^{0}C for 24 hours to get a homogenous mixture for immersion process to the α-Alumina membrane for 2 hours for plating. After that it was dried at room temperature for 24 hours. Finally, it was calcined at 600^{0}C for 3 hours. In a separate beaker, Pd chloride was dissolved in hot deionized water and hydrochloric acid. The addition of hydrochloric acid is expected to increase Pd chloride solubility. The membrane from the step above was immersed into the sensitization solution, $SnCl_2$ (1 g/L) and HCl (1 mL/L) for 2 minutes, and washed with deionized water for 30 s; it was then immersed into the activation solution, $PdCl_2$ (0.09 g/L) and HCl (1 mL/L) for another 2 minutes. Then, it was washed with deionized water for 30 s. This cycle was repeated for five times. Second, the electroless plating solution consisted of $PdCl_2$, NH_4OH, EDTA, and $NaPO_2H_2$ was used for plating process, carried out at room temperature for 1 hour without repetition. The ceramic membrane with a layer of Pd coating was then heated overnight at 373 K and stored in the desiccator to protect it from ambient humidity.

2.3. Characterization

Transmission electron microscope (TEM) (Philips CM12) was used to observe the shape and size of the palladium particles while the atomic force microscope (AFM) image for membrane morphology was analysed with AFM Ntegra Prima NT-MDT. The cross-sectional image, thickness and morphology of Pd composite membrane were analysed using a scanning electron microscope (SEM, LEO 1450VP Zeiss) coupled with the energy-dispersive X-ray (EDX) spectrometer. Pd-alumina membrane was broken into pieces for its characterization.

3. Results and Discussion

The substrate material employed in this research is macroporous asymmetric alumina. Figure 1 (a) shows the SEM image of the porous α-alumina surface. It can be seen that the surface is rough and heterogeneous. The macro particles were irregular with large pores size. Figure 1 (b) shows a micrograph from AFM for alumina surface. The alumina has an average pore of size 0.023 to 0.513 μm, consists of small grains with mean diameter of 40.54 to 717.6 nm. It can be observed from Figure 1 (c) that the average roughness of the alumina support was 84.129 nm.

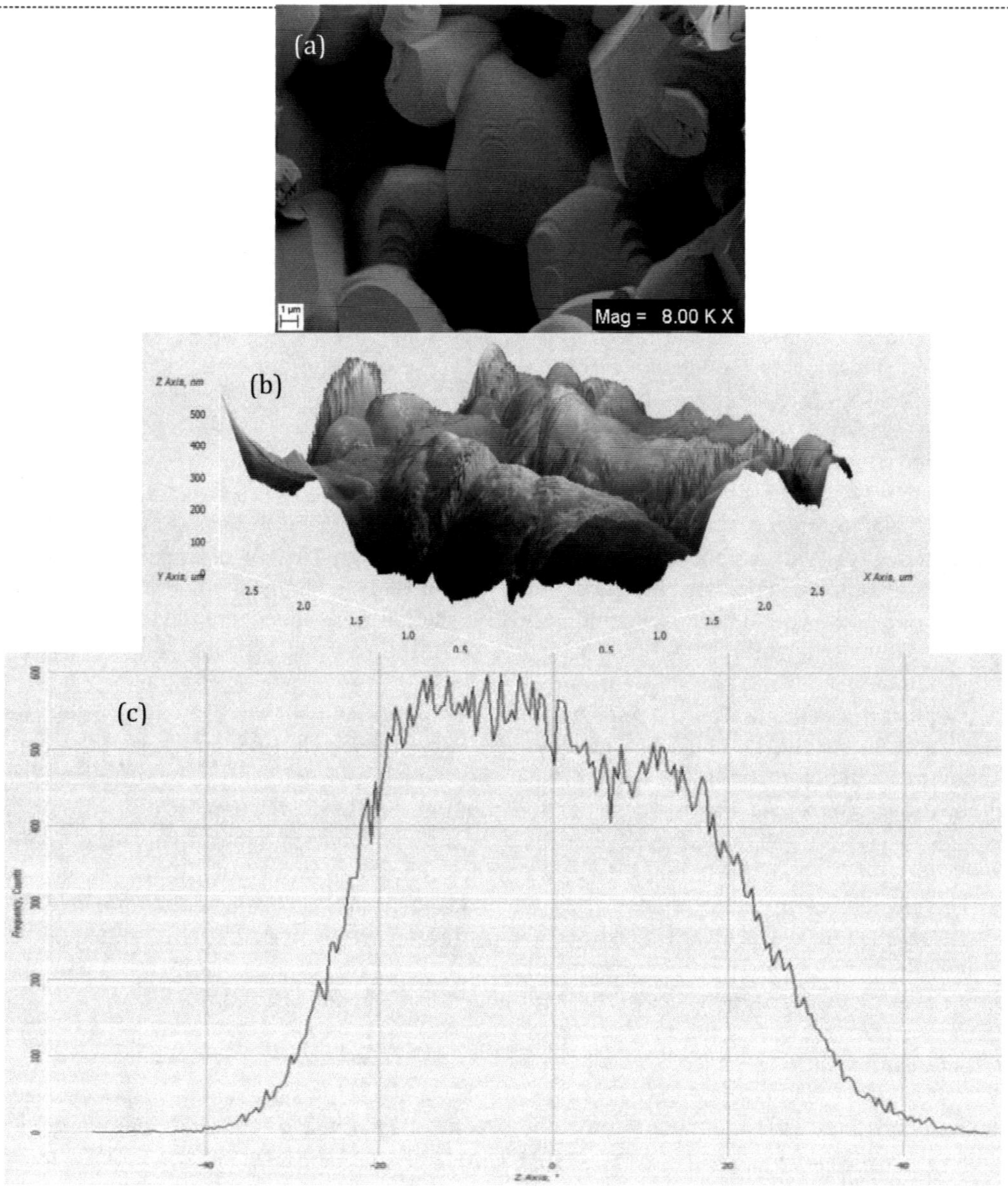

Figure 1. Surface micrograph of the α-alumina by (a) SEM and (b,c) AFM

IOP Conf. Series: Materials Science and Engineering **334** (2018) 012084 doi:10.1088/1757-899X/334/1/012084

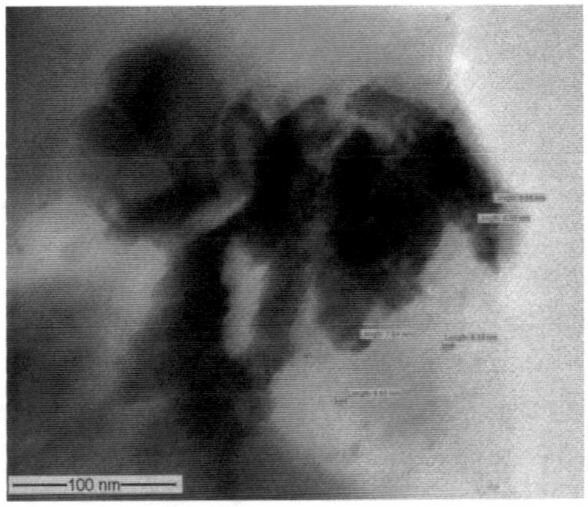

Figure 2. Transmission electron microscopy (TEM) image of Pd particles

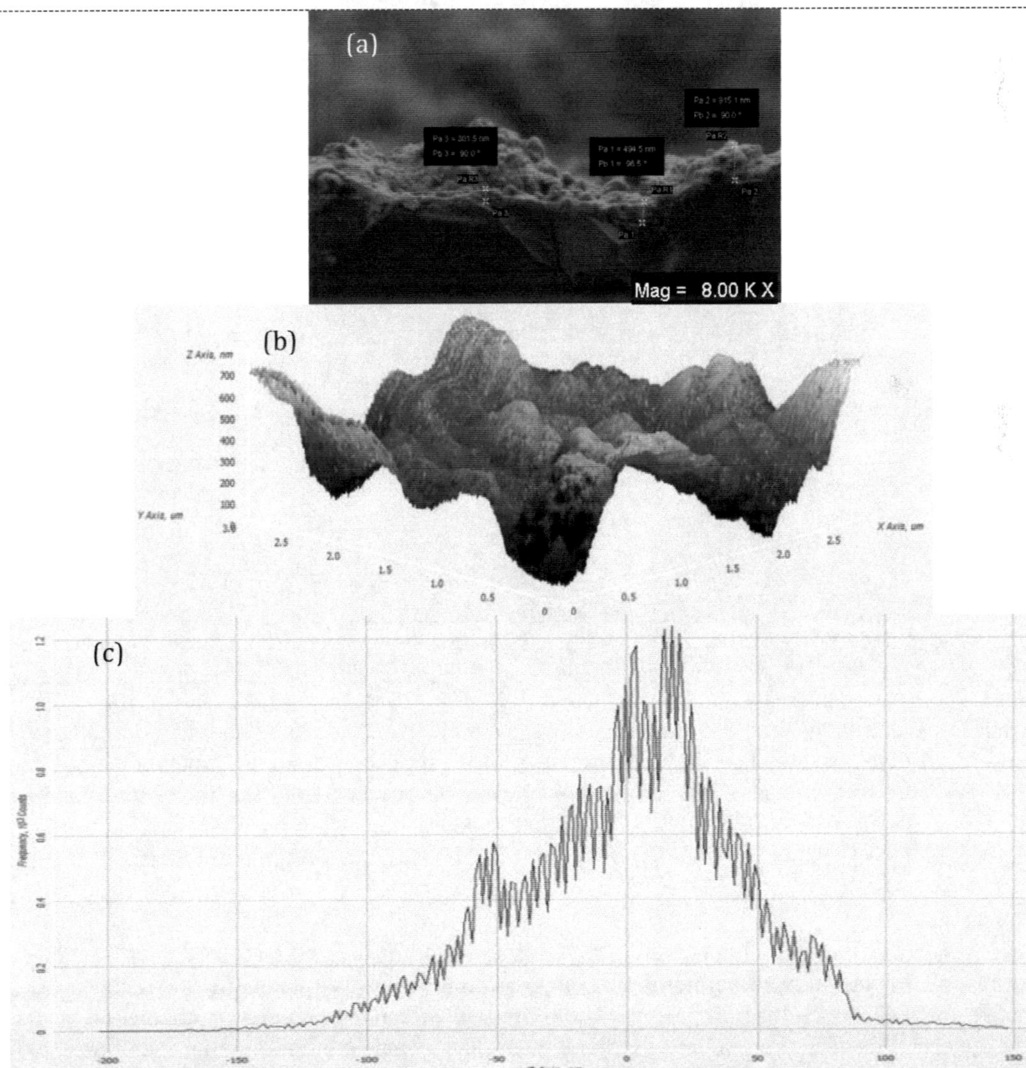

Figure 3. Surface micrograph of the Pd-alumina by (a) SEM and (b,c) AFM

Fine and homogeneous Pd particles from electroless plating is obtained as shown in Figure 2. From contacts between Pd particles with alumina depict structure of Pd-alumina composite membrane where Pd particles were observed as black small dots on the brighter alumina support. Pd particles were observed as black small dots on the brighter alumina support. These images reveal Pd particles to be well dispersed on the surface of alumina supports. The Pd particles are spherical in shape with diameters ranging from 6.18 to 7.64 nm.

After plating process, the top surface of membrane shows a uniform film (Figure 3.a). The substrate dipped into the plating bath had large activated surface area. When the surface area was occupied by samples, a big number of Pd particles was observed in the plating process and they continuously attached to each other on the surface and pores of alumina. α-alumina has been completely covered by a compact layer of Pd. The Pd-alumina is black. Pd adheres well onto the alumina substrate. Range of Pd film thickness were between 301.5 and 815.1 nm. The deposited Pd displays irregular surface as a result of rough alumina support. The thicker part of Pd film indicates the areas where they penetrated α-alumina pores. The adhesion strength between Pd layer and porous ceramic alumina surface is dependant on mechanical bonding strength [9]. This was consistent with the results reported by Zheng [8] and Li [3]. Pd-alumina has an average pore size of 0.017 to 0.119 μm, consists of small grains with mean diameter of 12.61 to 539.1 nm (Figure 3.b). The average roughness of the Pd-alumina support was 43.686 nm (Figure 3.c).

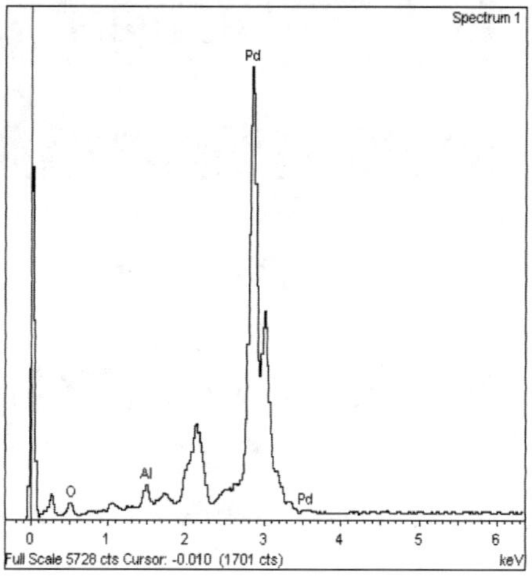

Figure 4. EDX measurement on the film of Pd-alumina membrane

The EDX spectrum of the Pd-alumina membrane surface detected Pd along with Al and O (Figure 4). The result obtained by EDX data confirmed the formation of Pd-alumina membrane containing 45% of Pd. The oxygen peak observed on the film membrane validated the formation of an oxide layer. A monolayer of oxygen is adsorbed rapidly at room temperature by dissociative adsorption of oxygen on Pd. Guo [10] also reported similar trend for Pd plating on α-alumina support.

4. Conclusion

A uniform thin Pd film on α-alumina substrate was fabricated by a combination of sol-gel process and electroless plating technique. The morphology of palladium film was a microstructure, smoother, finer and more uniform. The EDX profile of palladium-alumina membrane indicates that other elements (Al and O) except Pd were present in this membrane. The composite Pd-alumina membrane exhibit excellent adherence between Pd layer and the alumina support.

Acknowledgement
The authors gratefully acknowledge the support for this work by Ministry of Research Technology and Higher Education, Indonesia in Hibah Fundamental grant.

References

[1] H. Lu, L. Zhu, W. Wang, W. Yang and J. Tong 2015 Int. J. Hydrogen Energy **40**, 3548-3556
[2] D. Alique, M. Imperatore, R. Sanz, J. A. Calles and M. Giacinti Baschetti 2016 Int. J. Hydrogen Energy **41**, 18901-19498
[3] Y. Li, W. Ding, X. Jin, J. Yu, X. Hu and Y. Huang, Int. J. Hydrogen Energy **40**, 3528-3557 (2015).
[4] A. L. Ahmad and N. N. N. Mustafa 2007 Int. J. Hydrogen Energy 32, 2010-2021
[5] A. Bottino, M. Broglia, G. Capannelli, A. Comite, P. Pinacci, M. Scrignary and F. Azzurri 2014 Int. J. Hydrogen Energy **39**, 4717-4724
[6] S. K. Ryi, N. Xu, A. Li, C. J. Lim and J. R. Grace 2010 Int. J. Hydrogen Energy **35**, 2328-2335
[7] R. Sari, Z. Yaakob, M. Ismail, W. R. W. Daud and L. Hakim 2013 Ceram. Int. **39**, 3211-3219
[8] L. Zheng, H. Li and H. Xu 2016 Int. J. Hydrogen Energy **41**, 1002-1009
[9] S. Yun and S. T. Oyama 2011 J. Membr. Sci. 375, 28-45
[10] Y. Guo, Y. Jin, H. Wu, L. Zhou, Q. Chen, X. Zhang, X. Li 2014 Int. J. Hydrogen Energy **39**, 21044-21052

IOP Conf. Series: Materials Science and Engineering **334** (2018) 012085 doi:10.1088/1757-899X/334/1/012085

The effect of starch-garlic powder ratio on degradation rate of Gadung starch bioplastic

L Mairiza[1,2], Mariana[1], M Ramadhany[1], and C A Feviyussa[1]

[1] Department of Chemical Engineering, Faculty of Engineering, Syiah Kuala University, Darussalam, Banda Aceh 23111, Indonesia.

Email: liamairiza@yahoo.com

Abstract. Bioplastic is one of the solutions for environmental problems caused by plastics waste. Utilization of toxic gadung starch in the manufacturing of bioplastic would be as an alternative, due to gadung bulb has high starch content, and it is still not used optimally. This research aimed to learn about the using of gadung starch-mixed with garlic powder of making biodegradable plastic packaging. Also, to observe the duration of degradation, as a level of biodegradability of plastic film produced. The method used making this bioplastic was casting method. The variables used in this study were the ratios of starch and powdered garlic, were 10:0; 8:2; 6:4, and the concentration of garlic powder were 2%; 4%; 6%; and 8 %. The degradation test was done by soil burial test. The results of the soil burial test shown that the film was more rapidly degraded at ratio of 6: 4 compared to the ratio of 8: 2 and 10: 0. The results shown that bioplastic at the starch-garlic powder ratio of 10: 0 was decomposed in 21 days, at the the ratio of 8:2 was 15 days, while at the ratio of 6:4, the plastic film was degraded in the 11 days.

1. Introduction

Synthetic plastic waste nowadays still has been being one of the contributor of environmental pollution in Indonesia, even the world. Synthetic plastics are undegraded thus they would pollute the environment in hundreds even thousands years. Synthetic plastic is made by petroleum, which it is a non-renewable resource [1]. These conditions lead to the using of synthetic plastic packaging would not be used in a long time. One of the solution to this problem is, replacing the raw material of the conventional plastic into a material which easily degraded in nature, called bioplastic. In the recent years for the development of bioplastics from renewable resources to maintain sustainable development of economic and ecological attractive technology [2].

As the plant tuber, gadung has starch in a huge quantities. But some researches reported that the toxic compounds in gadung would cause serious health complications. Among types of bulbs, gadung is one of the unutilized due to alkaloids toxic called dioscorin [3]. Therefore, in this study, the use of gadung starch in the bioplastic manufacturing would be an added value of gadung starch.

A lot of types of bioplastic have different base of material, ranging from starch, cellulose and polyester. The manufacturing of bioplastic made from starch is the favorite choice due to easy and simple to do. Starch is used as a cheap biopolymer that is totally biodegradable, ultimately up to

[2] To whom any correspondence should be addressed.

IOP Conf. Series: Materials Science and Engineering **334** (2018) 012085 doi:10.1088/1757-899X/334/1/012085

carbon dioxide and water [4]. The previous researches which related to this research topic were conducted, based Ganyong Starch [5], rice starch [6] Research conducted by [7] which studied the effect of starch weight and volume of plasticizer on potato starch bioplastic film.

In this research, gadung starch was used due to has a fairly large content of starch compared to the starch from other types of plants bulbs. In addition, gadung bulbs are still rarely consumed by people caused by their toxicity, so this research would increase a value for gadung starch.

This research aimed to learn about the using of gadung starch-mixed with garlic powder, in the process of making biodegradable plastic packaging. Also, to observe the duration of degradation (time needed), as a level of biodegradability of plastic film produced.

2. Materials and Methods

2.1 Procedure

The main material used in this study were gadung starch and garlic powder. Other material used were liquid glycerol, glacial acetic acid, and aquadest. The apparatus used in this study were 300 ml and 100 ml baker glass, 100 ml and 10 ml measuring cup, blender, dropper, spatula, funnel, magnetic stirrer, hot plate (Cimarex), thermometer, glass mold and biodegradation testing container .

2.2 Methods

The gadung bulbs were peeled, then washed to remove toxins and sap from the tubers until clean. The washed bulb was blended to get the slurry. Then the slurry was filtered using a filter cloth and the powder of starch was filtered and the water was removed. The drying of starch was done using sunlight for 1-2 days. Starch was obtained in the form of fine powder.

The process of making plastic film from starch and garlic powder was made with two steps. First step, starch solution was made by mixing starch and aquadest, with 10% (w / v) starch. Garlic solution was made by mixing glacial acetic acid and garlic powder, respectively were 2%, 4%, 6%, and 8%. Both solutions were mixed and stirred using a magnetic stirrer until reach the temperature of 75^{0}C. Then glycerol was added as much as 70% (w/w) of the starch weight used. After the homogeneous was reached, the mixture was poured into a baker glass, then dried for 24 hours at room temperature.

Biodegradation test was done by using soil burial test method. 2 x 2 cm of sample was buried in the soil and the physical changes of plastic was observed every day until the plastic was degraded perfectly. The duration of the decomposition was recorded.

3. Results and Discussion

The soil burial test process or commonly called biodegradable test is needed to study the level of biodegradability of plastic film produced. Bioplastic biodegradation analysis was done through visual film observation. The results shown that the resulting bioplastic films were clearly biodegradable. These results in accordance with biodegradable plastic standards that should decomposed less than 180 days (ASTM D 6400 and EN 13432).

Ratio of 10:0					
Day 1	Day 3	Day 5	Day 7	Day 9	Day 11
			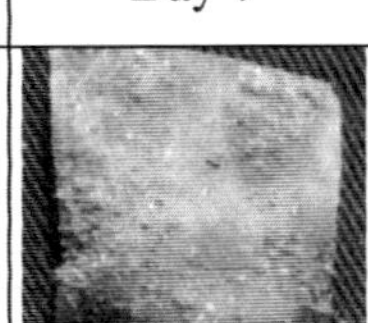	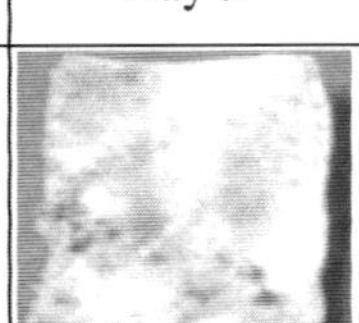	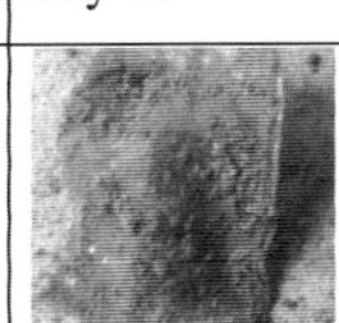
Day 13	Day 15	Day 17	Day 19	Day 21	

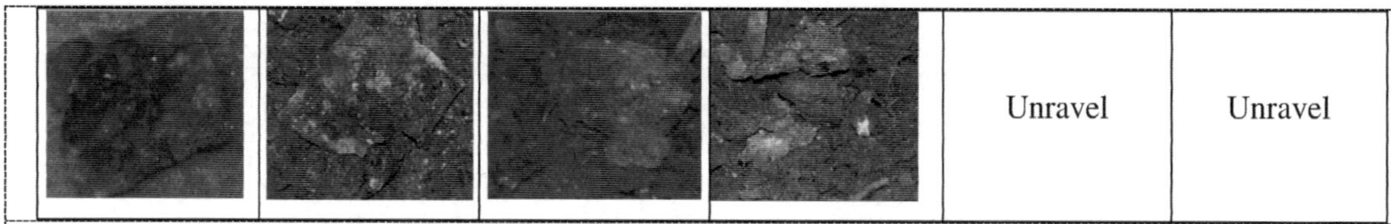

Figure 1. The duration of degradation of bioplastic film at gadung starch-garlic powder ratio of 10: 0.

Figure 1 shown that the bioplastic at the starch-garlic powder ratio of 10: 0 was decomposed in 21 days. The pictures below show the durations of bioplastic degradation at 8: 2 and 6: 4 ratios.

Figure 2. The duration of degradation of bioplastic film at gadung starch-garlic powder ratio of 8:2 (figure a) and 6:4 (figure b).

The effect of garlic powder shown that the increasing of the rate of degradation, which shown that on ratio of 6: 4, the plastic film was degraded in the 11 days. The addition of onion powder on the film would increase the number of OH-group. Increasing the number of OH-group, would cause more hydrophilicity and would easier to bind water molecules contained in the soil, which would cause increasing of free volume of the film so the contact surface area between plastic molecules and degradation enzymes would be increased.

The results of the soil burial test in Figure 2.a and 2.b show that at the ratio of starch-garlic powder film at 6: 4 ratio had faster degradation rather than the ratio of 8: 2 and 10: 0. The duration of degradation at the concentration ratio of 6: 4 i.e. 2%, 4%, 6%, and 8% respectively were 15 days, 13 days, 11 days, and also 11 days. While the duration of degradation at the ratio of 8: 2, with the same concentrations, respectively were 17 days, 17 days, 15 days and 15 days.

4. Conclusions

Bioplastics film at garlic powder ratio of 10: 0 was degraded perfectly at day 21. The duration of bioplastic degradation at starch-garlic powder ratio of 8:2 and garlic concentration of 2%, 4%, 6%, and 8% respectively were 17 days, 17 days, 15 days, and 15 days, while of bioplastic at starch-garlic powder ratio of 6:4 and garlic concentration of 2%, 4%, 6%, and 8% respectively were 15 days, 13 days, 11 days, and 11 days. All of duration of degradation resulted were indicate that gadung bioplastic was certainly safe for the environment.

Acknowledgements

The authors would like to thank the DIKTI which has facilitated this research by the grant.

References

[1] Akbar,F., Anita, Z., dan Harahap, H. 2013. Effect of time biodegradation of biodegradation of bioplastic film from singkong peel to mechanical properties. Journal of chemical engineering USU. 2(2):11-15

[2] Schlemmer, D., Angelica, R. S. and Sales, M. J. A., 2010, *Morphological and Thermo mechanical*
Characterization of Thermoplastic Starch/ Montmorillonite Nanaocomposites. Composite Structures, 92: 2066-2070

[3] Kumoro,C.A., Amalia,R., Retnowati,D.S., Budiyanti, C.S., dan Ratnawati, R. 2014. Preparation and Physocochemical characterization of modified (Acetylated) gadung (Dioscorea Hispida Dennst) Flours. Chemistry & Chemical Engineering Bitevhnology, Food insustry journal. 15(2):135-148.

[4] Mehta., Varda., Darshan., M., Nishith, D. 2014. Can a starch Based plastic be an option of enviromental friendly plastic?. Journal of global biosciences 3(3):681-685.

[5] Saputro, A.N.C., dan Ovita,A.L. 2017. Synthesis and Charactizations of bioplastic from kitosan-Ganyong starch. Journal of kimia dan pendidikan kimia. 2(1):13-21

[6] Wittaya,T. 2012. *Rice Starch-Based Biodegradable Films: Properties Enhancement.* Structure and Function of Food Engineering. Journal of intech)5):103-105.

[7] Radhiyatullah,A., Indriani,N., dan Ginting,M.H.S. 2015. Effect of starch's mass and volume of palsticizer glicerol on characteristics of bioplastics from potato starch. Journal of chemical engineering USU 4(3):35-39